Euclid's *Elements* in Hebrew Garb

Études sur le Judaïsme Médiéval

Fondées par

Georges Vajda

Rédacteur en chef

Paul B. Fenton

Dirigées par

Phillip I. Lieberman
Benjamin Hary
Katja Vehlow

TOME XCVIII

The titles published in this series are listed at *brill.com/ejm*

Euclid's *Elements* in Hebrew Garb

Critical Editions
of the Translation by Moses Ibn Tibbon
and the Translation Ascribed to Rabbi Jacob,
with an Introduction and Glossary
Books III–VI

By

Ofer Elior

BRILL

LEIDEN | BOSTON

הקרן הלאומית למדע
المؤسّسة الإسرائيلية للعلوم
Israel Science Foundation

Funding: Israel Science Foundation 2020

The Library of Congress Cataloging-in-Publication Data is available online at https://catalog.loc.gov
LC record available at https://lccn.loc.gov/2024046018

Typeface for the Latin, Greek, and Cyrillic scripts: "Brill". See and download: brill.com/brill-typeface.

ISSN 0169-815X
ISBN 978-90-04-70779-5 (hardback)
ISBN 978-90-04-70780-1 (e-book)
DOI 10.1163/9789004707801

This book is printed on acid-free paper and produced in a sustainable manner.

Contents

Preface

Several years ago I began studying the medieval Hebrew transmission of Euclid's *Elements*. This research was undertaken as part of "Studies in the History of Medieval Mathematics in Hebrew and Judeo-Arabic," a project headed by Ruth Glasner of the Hebrew University of Jerusalem and supported by the Israel Science Foundation (Grant No. 1139/20). A primary aim of the research is producing a modern critical edition of two main medieval Hebrew translations of the *Elements*, that ascribed to Rabbi Jacob and that by Moses Ibn Tibbon. Brill published the first volume of this edition, comprising Books I and II, in 2021.[1] The edition of Rabbi Jacob's translation was displayed alongside the first Latin translation of the *Elements*, ascribed to Adelard of Bath; Ibn Tibbon's translation was accompanied by an approximation of the Arabic text that was its source.

This second volume spans Books III–VI. Like the first volume, it accompanies the edited texts with a study of their sources and language and a glossary of technical terms. Here too I offer an approximation of the Arabic text underlying Ibn Tibbon's translation. Alongside Rabbi Jacob's translation, however, I present the text of an Arabic manuscript of the *Elements*, Paris, BULAC, MS Ara. 606, instead of Adelard's version. As far as I know this manuscript, only recently brought to the attention of scholars in the field, is the only Arabic *Elements* that includes a version that is either identical or very similar to the (presumably Arabic) version on which Rabbi Jacob's translation is based.

1 Elior, *Euclid's* Elements. The approximation of the Arabic source of **MIT**'s Book II there mistakenly provides Definition II.1 instead of Definition II.2. The correct approximation of Definition II.2 appears in the present volume in the Corrigendum.

Acknowledgements

I warmly thank Prof. Glasner for her most insightful guidance and her kind and continuous support throughout this intellectual endeavor. I express my boundless gratitude to Dr. Tariq Rajab for his invaluable copyediting of both the approximation of the Arabic source used by Moses Ibn Tibbon and the text of Ara. 606 Books III–VI. I extend my further thanks to Dr. Rajab for making me aware of the existence of this Arabic manuscript. I would also like to extend a debt of gratitude to Lenn Schramm, who edited the English sections of this edition.

Abbreviations

For abbreviations used in referring to Hebrew and Arabic manuscripts of the *Elements*, see Part II, Sections I.2–3.

A	the Latin translation of the *Elements* by Adelard of Bath (commonly known as "Adelard I")
ARAB	the Arabic text that was the basis for **MIT** and whose approximate reconstruction is included in the present edition (see Volume I, Part II, I.2.4)
B	the Arabic version of the *Elements* in Paris, BULAC, MS Ara. 606
E	the numbering in Engroff's critical edition of **I/T** Book V[1]
H	Heiberg's numbering
IMHM	The Institute of Microfilmed Hebrew Manuscripts, the National Library of Israel, Jerusalem
I/T	the version of the *Elements* transmitted in most extant Arabic MSS of the *Elements*
JBM	the Hebrew translation of the *Elements* by Jacob ben Makhir
MIT	the Hebrew translation of the *Elements* by Moses Ibn Tibbon
N	my numbering (see Part II, Section I.1)
RJ	the Hebrew translation of the *Elements* ascribed to Rabbi Jacob.

1 See Engroff, "The Arabic Tradition."

PART I

Introduction

CHAPTER I

Introduction

The translations printed here are two of the several Hebrew translations of Euclid's *Elements* produced in the Late Middle Ages. One translation, referred here by the siglum **RJ**, survives in four copies, all of them are partial, the earliest is dated to the fourteenth century. In two copies it is ascribed to a certain "Rabbi Jacob"; the ascription to "Jacob" appears in a poem attached at the end of third copy. The reference, as conjectured by Tony Lévy, is to the scholar and translator Rabbi Jacob Anatoli (d. ca. 1247). If Anatoli was indeed the translator, **RJ** was probably produced during the first half of the thirteenth century in southern France or in Italy. **RJ**'s circulation seems to have been limited, perhaps confined to the Italian peninsula. This was not a result of a disinterest in the *Elements* among Hebrew readers, but apparently of their preference for a different translation.

That translation, referred to here as **MIT**, was produced in southern France by Moses Ibn Tibbon (d. after 1274 or 1283), one of the most significant and prolific Jewish translators of the Middle Ages. Ibn Tibbon completed **MIT** in 1270. We do not know whether it was commissioned by some individual or community. Clearly, though, it was a fruit of the growing interest in science and philosophy among Jewish scholars in Southern Europe of that period. We also cannot know whether **MIT** was the first Hebrew translation of the *Elements*; **RJ** may have preceded it. If so, we may speculate that Ibn Tibbon was familiar with **RJ**. His acquaintance with **RJ** could provide him not only with valuable assistance for translating the *Elements*—especially technical terminology—but also a motive for undertaking this translation project. Because **RJ** is a hybrid version of the *Elements*, which seems to omit some parts of it, a full and reliable translation was a *desideratum*; **MIT**, as a whole, fills the need. Be that as it may, the numerous extant copies of **MIT** (at least nineteen) reveal that it was more popular than **RJ** and in fact the most widespread medieval Hebrew translation of the *Elements*. **MIT** had its own problems, however. It is extremely literal translation that defies smooth reading; nor is it free of mistranslations and inconsistencies. These shortcomings may have been why Ibn Tibbon's nephew, Jacob ben Makhir (1236–1308), a renowned translator in his own right, produced his own translation of the *Elements* (completed in 1289, also in southern France). **JBM** generally avoids the textual problems of **MIT** and is more readable, but draws heavily on it. This made it hard for Hebrew readers to distinguish the two versions and led to the circula-

 | DOI:10.1163/9789004707801_002

tion of "hybrid" exemplars of the Hebrew *Elements*, based on both **MIT** and **JBM**.[1]

In what follows I discuss the conclusions of my analysis of the parts of **RJ** and **MIT** presented in the present volume and following the "discovery" of the Arabic version preserved in Ara. 606 (**B**). I precede this discussion with introductory remarks about the main source of both translations, namely the primary Arabic transmission of the *Elements*.

1 For a fuller discussion of the historical context in which **RJ** and **MIT** were produced, their translators, and dissemination, see the introduction to Elior, *Euclid's* Elements.

CHAPTER II

The Primary Arabic Transmission

The primary Arabic transmission is preserved in some two dozen manuscript copies.[1] In all of these copies, except for **B**, the two first books are as a whole very similar, with respect to both text and diagrams—though with a fair number of micro-variants and several notable macro-variants. Some variants are unique to copies that belong to particular branches or families of transmission. One of these families comprises four copies: **M**, **P**, **R1**, and **R2**.[2] Another family comprises copies **C**, **T1**, and **T3**.[3] Nonetheless, as noted by Sonja Brentjes, all in all "Books I and II," as preserved in the Arabic copies of the *Elements*, "show a coherent, homogenous, and regular pattern within their text."[4]

These two families are also discernable in Books III–IV, with many micro- as well as macro-variants.[5] In addition to these families, there are notable variants in certain other copies; the most important are found in several propositions in **O2**.[6] Yet in these books, too, the primary Arabic transmission tends to a single version of Books III–IV; three additions by Thābit ibn Qurra (d. 901) are attached to this version in almost all copies.[7] There are, however, notable differences between the terminology, phrasing and diagrams in Books I–II and those in subsequent books—not only Books III–IV but also Books V–VI.[8]

1 Some of these copies include only a part or parts of the work. For a list see Elior, *Euclid's Elements*, 76, n. 33. No critical edition of the Arabic *Elements* based on its extant copies has been published. Book V was edited by John Engroff and Books VII–IX by Gregg De Young in their doctoral dissertations. Neither is based on all extant copies. See Engroff, "The Arabic Tradition"; De Young, "The Arithmetic Books."

2 On this branch of transmission, see Elior, *Euclid's* Elements, 47, and the studies listed ibid. n. 37. In Propositions I.3–14, **R2** has phrasings that distinguish it from all other copies. See De Young, "The Latin Translation of *Euclid's* Elements attributed to Gerard of Cremona," 366.

3 See De Young, "Two Hitherto Unknown Arabic Euclid Manuscripts."

4 See Brentjes, "Observations," 49.

5 Notably, **M**, **P**, **R1**, and **R2** all provide additional cases in no fewer than six propositions (as compared to only one proposition in Books I–II). In all four copies, the additions begin by identifying their origin in al-Ḥajjāj's version (on these version see below). See De Young, "New Traces," 650–657.

6 Book III: Propositions 7–8, 12–21, 23–29, 32–34 (with additional cases in 24, 32, 34); Book IV: Proposition 8.

7 They are attached to Propositions III.9–10, 30. See Engroff, "The Arabic Tradition," 22–24.

8 See Elior, "The Hebrew Translation," Appendix B.

 | DOI:10.1163/9789004707801_003

At the start of Book v, **Cam**, **O1**,[9] **O2**, U, **T4**, and S (at least in part) begin to share variants that distinguish them from all other copies. This divergence, which continues at least until Proposition VIII.18, delineates a new grouping of the Arabic copies: these five or six belong to one family, and the other copies belong to another family. Following De Young, I refer to the latter as Family A and to the former as Family B.[10] peculiar variants in **M**, **P**, **R1**, and **R2** and others in **C**, **T1**, and **T3** classify them as sub-families of Family A, here designated A-1 and A-2. According to Engroff, Families A and B are not all that different in Book v: "Much of the time [...] [they] seem all but congruent."[11] Engroff further noted that they "show few vocabulary differences in Book v."[12] Nevertheless they are clearly distinguishable, as manifested by many micro-variants. More significantly, there are additions or omissions of entire passages; differences in the order of propositions; alternative phrasings of some definitions as well as of propositions or (especially) parts of propositions, such as the enunciations and proofs; inclusion and exclusion of additions by Thābit; and differences in the diagrams.[13]

Until recently, it was widely assumed that **I/T**, namely the version or versions of the *Elements* found in the Arabic manuscripts excluding **B**, all go back to a revision by Thābit of a translation by Isḥāq ibn Ḥunayn (d. 910/1). It was also thought that certain variants in Books V–VIII of the two families, or in parts of them, transmit texts or diagrams that originate in other Arabic version or versions of the *Elements*, by al-Ḥajjāj ibn Yūsuf ibn Maṭar (fl. 786–833), which differ from the **I/T**. These assumptions were difficult to prove, because al-Ḥajjāj's versions were commonly believed not to be extant in their full form. Several years ago, Brentjes called these assumptions and beliefs into question with regard to Books III–IX.[14] Today the identity of those responsible for the extant Arabic versions of the *Elements* is a matter of dispute. However, two important facts were and remain certain, even after Brentjes' doubts. First, many medieval readers regarded the text transmitted in most of the Arabic manuscripts other than **B**, or at least most of this text, as the work of Isḥāq and Thābit. This is why I refer

9 At least some of the divergent copies may constitute a separate branch of transmission even before Book v. This is a matter for a separate investigation.

10 De Young, "The Latin Translation of *Euclid's* Elements attributed to Gerard of Cremona," suggests a more detailed division of the extant Arabic copies into families.

11 See Engroff, "The Arabic Tradition," 272.

12 See ibid., 274.

13 See Engroff's and De Young's dissertations. Differences of all these kinds are found in Book VI.

14 See Brentjes, "Who."

to this text as **I/T**. Secondly, many medieval readers were familiar with a version of the *Elements* which they ascribed to al-Ḥajjāj; according to them as well as to other sources, that version differed from **I/T** in its scope, arrangement, phrasing, technical terminology, and diagrams.[15]

The recent scholarly attention to **B** marked an important and perhaps dramatic development in the study of the Arabic transmission of the *Elements*. Historians working in the field seem to have been unaware of this manuscript, catalogued in Paris since the nineteenth century, or not to have considered it a member of the primary Arabic transmission of the *Elements*. That its version deserves special attention is already apparent from the title given to book headings in **B**: instead of كتاب أوقليدس في الأصول, which appears on **B**'s title page and is common in copies of **I/T**, it is كتابُ الأَرْكان.[16] Further inspection reveals numerous other differences. More importantly, anyone familiar with the scholarship on the differences among the Arabic versions realizes that many variants that medieval sources ascribed to the "Hajjaj version" are found in **B**. De Young reported that dozens of medieval reports about the text and diagrams in al-Hajjaj's version are found in **B**.[17] Thus it is clear that **B** is the closest we have to the "Hajjaj version." Moreover, and more germane in the present context, **B** is the closest extant version to the source of **RJ**'s proofs, and diagrams. Nevertheless, it cannot be assumed that **B** is identical to any medieval version of the *Elements* based on al-Ḥajjāj, not only because **B** is prone to the shortcomings of every manuscript copy—first and foremost, omissions, some of them substantial—but also because the version it transmits had apparently been revised from al-Ḥajjāj's text.[18]

15 See De Young, "The Arabic Version"; De Young, "MS Paris, BULAC."

16 See De Young, "MS Paris, BULAC," 7–8. This rare translation of the Greek Στοιχεῖα and its appearance in the manuscript were noticed by Rashed in his *Lexique historique de la langue scientifique arabe* (Rashed 2017, 12).

17 See De Young, "MS Paris, BULAC." I am grateful to Professor De Young for sharing this paper with me prior to its publication.

18 According to De Young, **B** is "either an edition of or a version derived from" al-Ḥajjāj's work (ibid., 84).

CHAPTER III

RJ

In order to produce the current volume of the edition, I systematically compared Books I–VI in **RJ** with the parallel sections in both **I/T** and **B**. The conclusions enhanced my understanding of the sources of **RJ**. In a separate study I have discussed these conclusions in detail.[1] Here I provide a brief summary.

The version of the *Elements* used for **RJ** was produced in two phases. The first phase was the merger of two different and complementary parts of Euclid's *magnum opus*, each taken from a different Arabic version, into a single hybrid version. The definitions and enunciations originated in **I/T** (those of Books V–VI specifically from Family A); they may derive from a copy that omitted some passages ascribed to Thābit, as well as other sections found in most extant copies of **I/T** but absent from **B**. The proofs and diagrams were taken from a version very close to **B** and probably to the version known to medieval readers as "Ḥajjāj's version."

In the second phase each section underwent changes, consciously based on versions that their author refers to as "Thābit's version" and "Ḥajjāj's version," as well as on Avicenna's *Uṣūl al-handasah* and other currently unidentified sources. Among the more important changes were revisions of **I/T**'s enunciations—*inter alia* their completions and also their replacement by alternative enunciations found in **B**—to bring them closer to the phrasing in the leading source for the proofs. In addition, **RJ** interpolates entire sections that were apparently absent from the leading source for the proofs. The most notable such additions in Books I–VI are the four propositions I.45, III.11, III.36, and VI.11. The "corrections" to the source included modifications of phrasing, most of them minor, as well as the addition of explanatory statements. Finally, a person who refers to himself as *ha-maʿatiq* added glosses to the proofs of Propositions II.5, II.6, and IV.5.

It cannot be determined whether the Hebrew translator was responsible for the merger of the two sections and/or the changes to them, or whether the hybrid version, before or after the changes, came from the Arabic tradition of the *Elements* and was translated as found into Hebrew. Although there are reasons to believe that if the hybrid version was created specifically for the Hebrew transmission, then one person, namely Rabbi Jacob, translated both

1 See Elior, "The Hebrew Translation."

 | DOI:10.1163/9789004707801_004

sections, it is not impossible that he translated only one of them and relied on an existing translation of the other. At any rate, if Rabbi Jacob created the hybrid Hebrew translation he was probably also responsible for the revisions made in the second phase. Terminological considerations, too, suggest that it was he who added glosses to the main text.

A comparison of the phrasing and technical terminology used in the definitions and enunciations with those found in the proofs highlights the differences between **RJ**'s two main sources. In addition, the partial reliance on an Arabic version that was substantially different from **I/T** resulted in a Hebrew translation whose proofs and diagrams diverge from those in **MIT** and **JBM**, which (at least in Books I–VI) were based wholly on **I/T**. Nonetheless, even in the proofs there is notable terminological similarity between **RJ** and **JBM** and even more so between **RJ** and **MIT**.

CHAPTER IV

MIT

Books I–II of **MIT** disclose that this translation is based on an Arabic exemplar of Family A-1 and provide no firm evidence for the use of additional source texts.[1] Books III–VI tell a similar story.[2] These Books also show that **MIT**'s source included all of Thābit's additions as found in the majority of Arabic copies.[3] Again like Books I–II, Books III–VI display variants that cannot be found in the direct Arabic transmission on which **MIT** was based; a few of these are found in Family B.[4] Other mostly micro-variants have no source in the Arabic MSS I have inspected. We may assume that some of these were found in **MIT**'s Vorlage; others are *contresens* or due to the translator's carelessness (quite a few such cases were also found in Books I–II).[5] There are not enough significant variants of this kind to suggest that Ibn Tibbon was familiar with a copy of the *Elements* that originated in Family B.

MIT's diagrams also generally follow those of the Arabic on Books III–IV, and specifically those of Family A on Books V–VI. All of the exceptions involve different labeling.[6] Many of them occur in propositions that do not specify the labels or structure of the accompanying diagram.[7] In such cases there is much variation among the Arabic MSS.

As for language, the present study revealed no substantive novelties in **MIT**: all my general observations about **MIT**'s language in Books I–II are valid for

1 See Elior, *Euclid's* Elements, 47–49.

2 Macro-variants that **MIT** shares with Family A-1 include: an addition to Definition V.5 (absent from **R1**; See Lévy, "Une version," 219); an order of propositions VI.23–26 that differs from the order in other copies of Family A (De Young, "MS Paris, BULAC," 56–57); and an alternative phrasing of the enunciation of Proposition VI.25 (Lévy, "Une version," 225).

3 These are found in Propositions III.9–10, 30. The additions to Definitions V.19 (H 17), VI.3 (H4), VI.5, and Porism VI.8, although not explicitly attributed to Thābit, may nonetheless be by him. See Engroff, "The Arabic Tradition," 24–29.

4 Salient examples of such variants include: second part of Definition III.11 (all copies of Family B save for **T4**, as well as **B**, **D** and **T2**); an addition to the enunciation of Proposition III.36 (not found in any of the Arabic MSS consulted); Definition V. 17 (all copies of Family B save for **T4**); and Definition VI.3 (all copies of Family B except for **Cam**).

5 See Elior, *Euclid's* Elements, 51.

6 These are found in Propositions III.6, 8, 9 (Thābit's addition), 10 (Thābit's addition), 11 (additional case), 12, 14, 23–24, 34, 35; IV.13; and VI.22, 29. See the apparatus for further details.

7 These include Propositions III.6. 9 (Thabit's addition), 11 (additional case), 12, 14, 23–24; IV.13.

 | DOI:10.1163/9789004707801_005

Books III–VI. Here I suffice with two of these. First, in those four books Ibn Tibbon continued to employ his primary methodology of extremely literal and word-by-word, translation.[8] The second observation pertains to the relation between the language of **MIT** and **RJ**. As I demonstrated in the first volume, despite the substantial terminological similarity between the two translations, there are also notable differences.[9] As I have suggested elsewhere, if one translation drew directly on the other, it is more likely that it was **RJ** that was used as a source for **MIT**: it makes more sense that Ibn Tibbon would decide to produce a homogeneous and full translation of the *Elements* if he was already familiar with a hybrid and perhaps incomplete translation such as **RJ**, rather than the opposite.[10]

8 See Elior, *Euclid's* Elements, 49–51.

9 See ibid., 62–64; Elior, "The Hebrew Translation." In **RJ** Books I–II I found several unique technical terms. Not only are they different from the terms employed in **MIT**, they are also rare or completely absent from the texts studied by Sarfatti or cannot be reasonably derived from those texts. See Elior, *Euclid's* Elements, 37–39. I did not trace such terms in Books III–VI.

10 See Elior, *Euclid's* Elements. 64–65; Elior, "The Hebrew Translation."

PART II

The Edition

CHAPTER I

Introduction

The general layout of the present edition, the manuscripts consulted, and the methodologies employed for editing **RJ**, **MIT**, and **ARAB** and producing the diagrams are in principle the same as those for the first volume.[1] The discussion that follows presents some exceptions, makes additional remarks about the editorial procedures, and describes the editing and presentation of **B** in this volume.

1.1 Numbering and Order

In addition to differences in the number of definitions and propositions, the order of some propositions in Books III–VI is not the same in the translations presented here as in Heiberg. There are similar but fewer differences between **RJ** and **MIT**. In the present edition, the numbering and order of the definitions and propositions follow **MIT**. I have indicated where other texts provide a different numbering, as well as Engroff's numbering of the definitions in Book V where my numbering of the Hebrew differs.

1.2 MIT

I have collated Books III–VI against all extant manuscripts of **MIT**. The results were similar to those of collating the copies for the edition of Books I–II, with two exceptions. First, three MSS, all previously considered to be "orphans," were found to be related to certain families of **MIT**'s transmission: Paris 1010 and St. Petersburg B 16 to Family A, and Madrid 5474 to Family B. Second, I found no evidence of a link between Oxford, Add. 20746 and Montecassino 510. To establish the edition I used four of the six MSS I employed for the edition of Books I–II, as follows:

ו Vienna, Österreichische Nationalbibliothek, Cod. hebr. 194 (IMHM F 1456)
ט Paris, private collection (IMHM F 39116)

1 See the detailed discussion in Elior, *Euclid's* Elements, 69–80.

 | DOI:10.1163/9789004707801_006

פ Paris, BnF, héb. 1010 (IMHM F 15712)
ק Montecassino, Archivio di Stato 510 (F 34894)

I also used

ב Oxford, Bodleian Library, Add. 20746 (Margoliouth 1001; IMHM F 5053; Lévy 22), copied in the fifteenth century in a Sephardi script. An owner's annotation (fol. 1a) attests to its presence in Germany in the first half of the seventeenth century.[2]

ט St. Petersburg, Rossiyarkaya Akademiya Nauk, Institut Vostokovedeniya, B 16 (IMHM F 52919). The beginning of this copy (Definition I.1 to the middle of Proposition I.34) is written in an eighteenth-century Karaite script. The rest was written in the fifteenth century in a Sephardi-Byzantine script by פרוטו קדוטי for Abraham Ibn Migas ha-Levi.[3]

I.3 ARAB

For the primary Arabic transmission of Books III–VI as well as to establish **ARAB** I consulted the same group of copies of the Arabic *Elements* used to produce the edition of Books I–II, as follows:

C Dublin, Chester Beatty Library, MS 3035 (begins at Prop. I.39)
Cam Cambridge, University Library, MS Addit. 1075 (beginning of introduction is missing)
D Dunedin, New Zealand, Otago Museum, MS De Beer 8 (beginning of introduction is missing)
M Madrid, Escorial Library, MS arabe 907
O1 Oxford, Bodleian Library, MS Huntington 435
O2 Oxford, Bodleian Library, MS Thurston 11
P Paris, BnF, héb. 1381
Q Kastamonu, Genel Kütüphane 73
R1 Rabat, al-Maktaba al-Malakiyya, al-Khizāna al-Ḥasaniyya, MS 53
R2 Rabat, al-Maktaba al-Malakiyya, al-Khizāna al-Ḥasaniyya, MS 1101
S St. Petersburg, Rossiyarkaya Akademiya Nauk, Institut Vostokovedeniya, MS C 2145
T1 Tehran, Majlis Shūrā, MS 200
T2 Tehran, Majlis Shūrā, MS 2060
T3 Tehran, Majlis Shūrā, MS 15480 (beginning of introduction is missing)

---

2 See; Margoliouth, *Catalogue*, III (Sections II–VII): 319; Lévy, Lévy, "Les *Éléments*," 85.

3 See the online catalogue of the IMHM.

T4 Tehran, Malik, MS 3586
U Uppsala, Universitetsbibliotek, MS O. Vet. 20
I also looked at
Cop Copenhagen, Kongelige Biblioteket, Mehren LXXXI (begins in Book V).
In Books V–VI, **ARAB** is based on **I/T** Family A. I used Engroff's edition for Book V and to compare **I/T** Family A of this Book to Family B. His edition of Family A is based *inter alia* on two copies I have not inspected: Istanbul Fātiḥ 3439 and Rampur, Riḍā Library, Arshī MS 3656; In the critical apparatus of the present edition these manuscripts are referred to as F and **Ram** respectively.[4] The MSS consulted by Engroff did not include though six copies I examined: **D**, **P**, **Q**, **T2**, **T3** (Family A), and T4 (Family B). For the approximation of **ARAB** I followed the same methodologies I employed for Books I–II, with respect to my choices of variants to be inserted into **ARAB** and to the content of the apparatus and its sigla. In a few cases I have disagreed with Engroff's editorial choices or corrected his readings. In the apparatus these are marked with a number sign (#).

1.4 B

B, with partial punctuation, is presented in parallel to **RJ**.[5] I have intervened in the text only where I found clear grammar or spelling errors, or mistakes in indicating diagram labels: in those cases the correct variant replaces the erroneous variant and given in parenthesis; the apparatus indicates the mistaken variant. The apparatus also reports differences between the diagrams in **B** and **RJ**. That in **B** is presented along with **RJ**'s only in Proposition III.36, where **B**, unlike **RJ**, discusses cases.

1.5 The Diagrams

As in the first volume, the diagrams in the present edition ignore differences in metrics that are found in the manuscripts. This is particularly important for the diagrams in Books V–VI. As a rule, when a proposition in these Books addresses a ratio between geometrical figures, the accompanying diagram in the Arabic

4 According to both Engroff and De Young, this copy belongs to Family A.
5 Though the punctuation is fuller than that in **ARAB**. Dr. Tariq Rajab, the copyeditor of the Arabic texts in this edition, deemed a fuller punctuation necessary because the text is based on a single manuscript and is wordier than **I/T**.

and Hebrew manuscripts do not reflect this ratio. By contrast, I endeavored to produce diagrams that do reflect the ratios mentioned in the text. Again as in the first Volume, I generally did not document cases where a diagram in the manuscripts or in some of them present particularizations of polygons which are included in the diagram, e.g., a right-angled instead of an acute-angled triangle, or a rectangle instead of a parallelogram. There are many such cases in Book VI.

The following table presents, for each label used in the Hebrew diagrams, the Arabic label it usually translates.

א	ב	ג	ד	ה	ז	ח	ט	כ	ל	מ	נ	ס	ע	פ	ק	ר	ש	ת	ת'
ا	ب	ج	د	ه	ز	ح	ط	ك	ل	م	ن	س	ع	ف	ق	ر	ش	ت	ث

1.6 Sigla

For convenience's sake I list here the sigla employed in the critical apparatus:

[א] uncertain reading
[..] Impossible reading
⟨א⟩ marginalia in the same hand
⟨⟨א⟩⟩ marginalia in a different hand
+א addition before the lemma
א+ addition after the lemma
א̶ deletion
* the lemma
ליתא the lemma is missing
mistake in Engroff's edition

In another set of notes I recorded differences in the versions of the diagrams transmitted in the Hebrew manuscripts. Here the following symbols and abbreviations are employed:

↔ swapping of two letters
$ mirror image
⊥ perpendicular(s)
⊥̸ not perpendicular

The Edition

[Book III]

המאמר השלישי מספר אקלידס

[1] העגולים השוים הם שאלכסוניהם שוים זה לזה, ואשר יהיו הקוים שיצאו ממרכזיהם אל הקוים המקיפים בהם שוה [!] קצתם לקצת.

[2] והקו הישר שיקרא ממשש העיגול הוא הפוגש העיגול, וכאשר הוצא לשני הצדדין לא יחתוך אותו.

[3] והעגולים שיאמר שקצתם ממשש לקצת הם אשר יפגוש קצתם קצת ולא יחתכו.

المَقالَةُ الثّالثةُ مِنْ كِتابِ الأَرْكان

[1] الدَّوائِرُ المُتَساوِيَةُ هي الَّتي أَقطُرُها مُتَساوِيَةٌ، وَالخُطوطُ الَّتي تَخْرُجُ مِنْ مَراكِزِها إِلى الخُطوطِ المُحيطَةِ بِها هِيَ أَيضًا مُتَساوِيَةٌ.

[2] وَإِذا جازَ خطٌّ عَلى دائِرَةٍ، وَيَماسُّها مِنْ خارِجِها وَلا يَقْطَعُ مِنْها شيئًا، فإِنَّهُ يُقالُ لَهُ المُماسُّ لِلدَّائِرَةِ.

[3] وَإِذا كانَتِ الدَّوائِرُ يَماسُّ بَعْضُها بَعْضًا، وَلَمْ تَقْطَعْ واحِدَةٌ مِنْها الأُخْرى، فَإِنَّهُ يُقالُ لِتِلْكَ الدَّوائِرِ المُتَماسَّةُ.

 | DOI:10.1163/9789004707801_007

المَقَالَةُ الثَّالِثَةُ مِنْ كِتَابِ أُقْلِيدِس

[1] الدَّوائِرُ المُتَساوِيَةُ هي الَّتِي أقطارُها مُساوٍ بَعْضُها لِبَعْض، أو الَّتِي تكونُ الخُطوطُ الَّتِي تَخْرُجُ مِنْ مَرَاكِزِها إلى الخُطوطِ المُحِيطَةِ بِهَا مُساوٍ بَعْضُها لِبَعْض.

[2] والخطُّ المُسْتَقِيمُ الَّذي يُقالُ له مُماسّ الدَّائِرَة هو الَّذي يَلْقَى الدَّائِرَةَ، وإذا أُخْرِجَ إلى كِلْتَيِ الجِهَتَيْنِ لم يَقْطَعْها.

[3] والدَّوائِرُ الَّتِي يُقالُ إنَّ بَعْضَها مُماسٌّ لِبعضٍ هِيَ الَّتِي يَلْقَى بَعْضُها بَعْضًا ولا تَتَقاطَعُ.

המאמר השלישי מספר אקלידס[1]

[1] העגולים השוים הם אשר קוטריהם שוים קצתם אל קצת[2], או אשר יהיו הקוים אשר[3] יצאו ממרכזיהם[4] אל הקוים[5] המקיפים בהם[6] שוים קצתם לקצת[7].

[2] והקו הישר אשר יקרא לו ממשש לעגולה הוא[8] אשר יפגש[9] לעגולה[10] וכאשר הוצא לכל אחד משני הצדדים לא יחתכה[11].

[3] והעגולות אשר יקרא קצתם ממששת לקצת הם אשר יפגשו קצתם לקצת ולא יחתכו.

---

1 אקלידס] +החכם **בק** 2 קצת] קצתם **ז** 3 יהיו ... אשר] +* **ב** 4 ממרכזיהם] ממ⟨ר⟩כזהם **ב** 5 הקוים] קוים **ט** 6 בהם] בם **ק** 7 שוים ... לקצת] קצתם שוים לקצת **ס** 8 הוא] ~~הוא~~+ **ב** 9 יפגש] יפגוש **ט** 10 יפגש לעגולה] יפג⟨ו⟩ש ל⟨ה⟩עגלה **ק** 11 לא יחתכה] ⟨לא⟩ יחתכה **ב** ⟨לא⟩ יחתכנה **ק**

[4] ויאמר שמרחק הקוים הישרים מהמרכז בעגול—שוה, כשיהיו העמודים היוצאים אליהם מהמרכז שוים.

[4] وَإذا كانَ في الدَّائِرَةِ خُطوطٌ، وَكانَتِ الأَعْمِدَةُ الَّتي تَخْرُجُ مِنَ المَرْكَزِ إلى تِلْكَ الخُطوطِ مُتَساوِيَةً، فَإِنَّ أَبْعادَ الخُطوطِ مِنَ المَرْكَزِ سَواءٌ.

[5] והקו שיאמר שמרחקו מהמרכז רחוק הוא אשר יהיה העמוד הנופל עליו גדול.

[5] وَأَبْعَدُها مِنَ المَرْكَزِ هُوَ الَّذي عَمودُهُ أَطْوَلُ.

[6] וחתכת העגול הוא אשר יקיף בה קו ישר, כשיקרא[1]—שיאמר[2] לו—מיתר, וחלק מהקו המקיף, כשתקרא—יאמר[3] לו—קשת.

[6] وَالقِطْعَةُ مِنَ الدَّائِرَةِ يُحيطُ بِها خَطٌّ مُسْتَقيمٌ وَطائِفَةٌ مِنَ الخَطِّ المُحيطِ يُقالُ لَهُ القَوْسُ.

[7] וזוית חתכת העגול היא אשר יקיף בה קו ישר וקשת מהקו המקיף בעיגול.

[7] وَزاوِيَةُ القِطْعَةِ يُحيطُ بِها الوَتَرُ وَخَطُّ القَوْسِ.

[8] והזוית שהיא תוך חתיכת העגול היא אשר יקיפו בה שני קוים ישרים שידבקו בנקודה שהיא רשומה על קשת החתכה כמו שנזדמן ובקצות הקו הישר שהוא תושבת החתכה.

[8] وَإِذا تَعَلَّمَتْ نُقْطَةٌ عَلى خَطِّ القَوْسِ وَأُخْرِجَ مِنْها خَطّانِ إِلى طَرَفَي الوَتَرِ وَصارَ الوَتَرُ قاعِدَةً لَهُما.

[9] וכאשר החזיקו הקוים הישרים המקיפים בזוית קשת מה, יאמר על הזוית שהיא על הקשת ההיא.

[9] فَإِنَّ الزَّاوِيَةَ الَّتي عِنْدَ النُّقْطَةِ وَالخَطّانِ يُحيطانِ بِها مُرَكَّبَةٌ عَلى خَطِّ القَوْسِ.

---

[1] כשיקרא] בשיקרא [2] שיאמר] ⟨⟨נס׳⟩⟩+
[3] יאמר] ⟨⟨נס׳⟩⟩+

[4] ويُقالُ إنّ بُعْدَ الخُطوطِ المُسْتقيمةِ مِنَ المَرْكَزِ في الدّائِرَةِ مُساوٍ، إذا كانَتْ الأَعْمِدَةُ المُخْرَجَةُ إلَيْها مِنَ المَرْكَزِ مُتَساوِيَةً.

[4] ויאמר[1] כי[2] רוחק הקוים הישרים מן המרכז בעגולה שוה, כאשר היו העמודים המוצאים[3] אליה מן המרכז שוים.

[5] والخَطّ الّذي يُقالُ إنّ بُعْدَهُ مِنَ المَرْكَزِ أعظم[1]، هو الّذي يكونُ العَمودُ الواقِعُ عَلَيْهِ أعظم[2].

[5] והקו אשר יאמר כי מרחקו מן המרכז יותר גדול, הוא אשר יהיה העמוד הנופל עליו יותר גדול[4].

[6] وقِطْعَةُ الدّائِرَةِ هي شَكْلٌ يُحيطُ بِهِ خَطٌّ مُسْتَقيمٌ^، وَقَوْسٌ مِنَ الخطِّ المُحيطِ بالدّائِرَةِ^.

[6] וחתיכת העגולה היא[5] אשר יקיף בה קו ישר, יקרא המיתר[6], והחלק מן הקו המקיף יקרא הקשת.

[7] وزاويةُ قِطْعَةِ الدّائِرَةِ هي الّتي يُحيطُ بها خطٌّ مُسْتَقيمٌ وَقَوْسٌ مِنَ الخَطِّ المُحيطِ بالدّائِرَةِ.

[7] וזוית חתיכת העגולה היא אשר יקיפו[7] בה קו[8] ישר וקשת מן הקו המקיף בעגולה.

[8] والزّاوِيَةُ الّتي في قِطْعَةِ الدّائِرَةِ هي الّتي يُحيطُ بها خطّانِ مُسْتَقيمانِ يُوصِلانِ مِنْ[3] نُقْطَةٍ—تُعَلَّمُ كَيْفَما اتَّفَقَتْ—على قَوْسِ القِطْعَةِ، وبَيْنَ طَرَفَيِ الخَطِّ المُسْتَقيمِ الّذي هو قاعِدَةُ القِطْعَةِ.

[8] והזוית אשר בחתיכת העגולה היא[9] אשר יקיפו בה שני קוים ישרים יגיעו מן נקודה—תרשם איך שתפול—על קשת החתיכה[10], ובין שני[11] קצות הקו הישר אשר הוא תושבת החתיכה.

[9] وَإذا جازَ الخَطّانِ المُسْتَقيمانِ المُحيطانِ بِالزّاوِيَةِ قَوْسًا فَإنَّ الزّاوِيَةَ يُقالُ لَها الّتي على تِلْكَ القَوْسِ.

[9] וכאשר יסבב השני קוים[12] הישרים המקיפים בזוית קשת[13], הנה הזוית יקרא לה אשר על הקשת ההיא.

---

1 ויאמר] ונאמר **פ** 2 כי] ׳כי **ב** 3 המוצאים] המוציאים **ב** 4 גדול] ⟨⟨*⟩⟩ **ב** 5 היא] הוא **ז** 6 המיתר] ~~המישר~~+ **ב** 7 יקיפו] יקיף **ס** 8 קו] ~~שני~~+ **ב** 9 היא] הוא **ס** 10 החתיכה] ~~העגולה~~+ **ק** 11 שני] שתי **בוטפק** 12 השני קוים] ⟨⟨ה⟩⟩שני קוים **ב** שני הקוים **וק** 13 קשת] +⟨⟨מה⟩⟩ **בק** +מה **ס**

---

1 اعظم] اكثر **O2** 2 اعظم] اطول **MO2R2**
3 من] **MR1R2ST4** بين

[10] וחותך העגול הוא התמונה שיקיפו בה שני קוים ישרים המקיפים בזוית על מרכז העגול וקשת שיחזיקו בה אלו[1] הקוים מן העגול.

[10] وَالشَّكلُ الَّذي يُقالُ لَهُ القِطاعُ هُوَ الَّذي يُحيطُ بِهِ خَطّانِ يَخْرُجانِ مِنَ المَرْكَزِ ويَنْتَهِيانِ إلى الخَطِّ المُحيطِ. وَتَكونُ الزّاوِيَةُ الَّتي يُحيطانِ بِها مُرَكَّبةً عَلى مَرْكَزِ الدّائِرَةِ.

[11] וחתכות העגול המתדמות הן המחזיקות זויות שוות.

[11] وقِطَعُ الدَّوائِرِ إذا كانَتِ الزَّوايا المُرَكَّبةُ عَلى قِسِيِّها مُتَساوِيَةً والقِطَعُ مُتَشابِهةً. وَإذا كانَتِ القِطَعُ مُتَشابِهةً فَالزَّوايا المُرَكَّبَةُ عَلى قِسِيِّها مُتَساوِيَةً.

[1] אלו] א[ל]לו

[10] وقِطاعُ الدَّائِرَةِ هو الشَّكلُ الَّذي يُحيطُ بِهِ خَطّان مُسْتَقيمانِ يُحيطانِ بِزاوِيَةٍ على مَرْكَزِ الدَّائِرَةِ وَقَوْسٍ يحوزُها ذانِكَ الخطّانِ مِنَ الدَّائِرَة.

[10] וחותך העגולה הוא[1] התמונה[2] אשר יקיפו[3] בה שני קוים ישרים יקיפו בזוית על מרכז העגולה וקשת יקיפוה אותם[4] שני הקוים מן העגולה.

[11] وَقِطَعُ الدَّوائِرِ المُتَشابِهَة هي الَّتي تَقْبَلُ زوايا مُتَساوِيَةً[1]^.

[11] וחתיכות העגולות המתדמות הם אשר יקבילו זויות שוות[5]. וכאשר[6] יהיו זויותיהן אשר יקבילו שוות הנה הם מתדמות.

---

[1] متساوية] +وقيل هي التي زواياها مساو بعضها لبعض Cam +وبعضها هي التي زواياها مساو بعضها لبعض DO2ST2 +ويقال هي التي زواياها مساو بعضها لبعض O1

---

[1] הוא] היא בס [2] התמונה] ⟨ה⟩תמונה ט [3] יקיפו] ⟨⟨*⟩⟩ ב [4] יקיפוה אותם] יקיפו⟨⟨ה⟩⟩ אותם ב יקיפו אותה ו יקיפו ס [5] שוות] ~~נצבות~~ ⟨⟨*⟩⟩ פ [6] וכאשר] +* ט

[Proposition III.1]

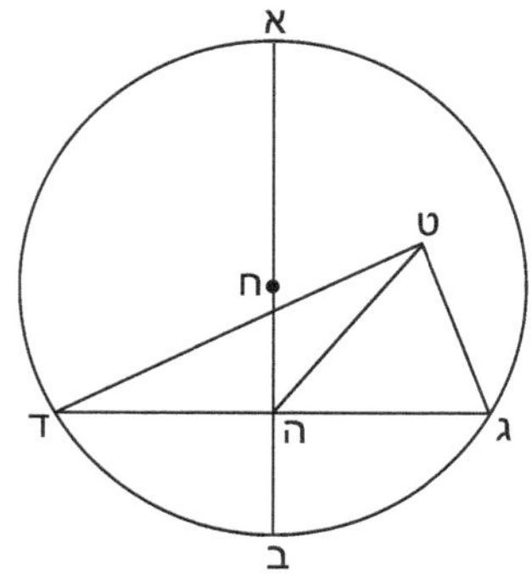

נרצה לבאר איך נמצא מרכז עגול מונח.

ונשים העגול הידוע עגול א״ב. וכאשר רצינו למצוא מרכזו, נוציא בו קו כמו שיפול, ויכלו קצותיו אל הקו המקיף, והוא קו ג״ד. ונחלוק אותו לחציים על נקודת ה׳. ונוציא מה׳ קו על זוית נצבה, והוא קו ה״א. (ונוציאהו)[1] אל ב׳. ונחלק קו א״ב לחציים על נקודת ח׳. ואומר שח׳ הוא מרכז העגול.

نُرِيدُ أَنْ نُبَيِّنَ كَيْفَ نَجِدُ مَرْكَزَ دائِرَةٍ مَعْلومَةٍ.

مثالُهُ: أَنَّا نَجْعَلُ الدَّائِرَةَ المَعْلومَةَ دائِرَةَ ا ب. وَنُرِيدُ أَنْ نَجِدَ مَرْكَزَها. فَنَخُطُّ فيها خَطًّا كَيْفَما وَقَعَ. يَنْتَهي طَرَفاهُ إِلى الخَطِّ المُحيطِ، وَهُوَ خَطُّ ج د. وَنَقْسِمُهُ بِنِصْفَيْنِ عَلى نُقْطَةِ ه. وَنُخْرِجُ مِنْ ه خَطًّا عَلى زاوِيَةٍ قائِمَةٍ، وَهُوَ خَطُّ ه ا. وَنُخْرِجُهُ إِلى ب. وَنَقْسِمُ خَطَّ ا ب بِنِصْفَيْنِ عَلى نُقْطَةِ ح. فَأَقولُ إِنَّ ح مَرْكَزُ الدَّائِرَةِ.

[1] ונוציאהו] ונוציא⟨⟨הו⟩⟩

نُرِيدُ أن نُبَيِّنَ كَيْفَ[1] نَجِدُ مَرْكَزَ دائِرَةٍ مَفْروضَةٍ.

فَلْتَكُن الدَّائِرَةُ المَفْروضَةُ[2] اب. ونُرِيدُ أَنْ نَجِدَ مَرْكَزَها. فَنَخُطُّ فيها[3] وَتَرًا كَيْفَما وَقَعَ، وهو ج د. وَنَقْسِمُهُ بِنِصْفَيْنِ على نُقْطَةِ ال ه. ونُخْرِجُ مِنْ ال ه خَطَّ ه ا على زوايا قائِمَةٍ مِنْ خطّ ج د، ونُخْرِجُهُ إلى ب. وَنَقْسِمُ ا ب بِنِصْفَيْنِ^ على ح. فَأَقولُ إنَّ ح مَرْكَزُ الدَّائِرَةِ، لا يُمْكِنُ غيره.

נרצה לבאר איך נמצא מרכז עגולה מונחת.

תהיה[1] העגולה המונחת א״ב[2]. ונרצה למצוא מרכזה. הנה נקוה בה מיתר איך שיפול, והוא ג״ד. ונחלקהו בשני חציים על ה׳. ונוציא מה׳[3] קו ה״א על זויות[4] נצבות מקו ג״ד, ונוציאהו[5] אל ב׳[6]. ונחלק א״ב[7] על ח׳[8]. הנה אומר כי ח׳ מרכז העגולה[9], ואי אפשר זולתו.

[1] نبين كيف] **MPR1R2** [2] المفروضة] **DT4** [3] فيها] بها **R2** +دايرة

שרטוט—**בוטק**: $ [1] תהיה] ותהיה **ט** [2] המונחת א״ב] ~~מונחת תהיה העגולה~~+ **ב** +⟨ובנ״א ונחלק א״ב בשני חצאין על ח׳ הנה אומר כי ח׳ מרכז העגולה ואי אפשר וכו׳⟩ **ס** +~~ותהיה מרכזה~~ **פ** [3] מה׳] ⟨*⟩ **ק** [4] זויות] זוית **ב** [5] ונוציאהו] ונוציאה **טפ** [6] ב׳] +וד׳ **פ** [7] א״ב] +⟨⟨בחצאים⟩⟩ **פ** [8] ח׳] ⟨⟨*⟩⟩ **ב** [9] העגולה] ~~נקודת ט׳ ונוציא מג׳~~+ **ב** +בשני חציים **ס**

وَبُرْهانُه: أَنَّهُ لا يُمْكِنُ غَيْرُه، فَإِنْ كانَ يُمْكِنُ فَلْتَكُنْ نُقْطَةُ ط [!]. وَنَصِلُ (ط ب ج ه د)[1]. ج ه مِثْلُ ه د. ه ط مُشْتَرَكٌ. فَجَميعُ ج ه ه ط مِثْلُ د ه ه ط. فَقاعِدَةُ ج ط مِثْلُ قاعِدَةِ ط د. فَالزّاوِيَةُ الَّتي مِنْ ج ه ط مِثْلُ الَّتي مِنْ د ه ط. وَإِذا قامَ خَطٌّ عَلى خَطٍّ فَصارَتِ الزّاوِيَتانِ اللَّتانِ عَنْ جَنَبَتَيْهِ مُتَساوِيَتَيْنِ، فَهُما قائِمَتانِ. وَكُلُّ الزَّوايا القائِمَةِ مُساوِيَةٌ بَعْضُها لِبَعْضٍ. فَالزّاوِيَةُ الَّتي مِنْ ج ه ط وَالَّتي مِنْ ج ه ح مُتَساوِيَتانِ. العُظْمى مِثْلُ الصُّغْرى—هذا خُلْفٌ لا يُمْكِنُ. فَقَدِ اسْتَبانَ أَنَّ ط لَيْسَ بِمَرْكَزٍ لِدائِرَةٍ وَلا غَيْرُهُ مِنَ المَواضِعِ إِلّا نُقْطَةَ ح. ح مَرْكَزُ دائِرَةِ ا ب.

ואי אפשר שיהיה זולתו. ואם היה אפשר, תהיה נקודת ט׳ המרכז. ונדביק ט׳ בג׳ ובה׳ ובד׳. ולפי שג״ה שוה לה״ד, וה״ט משותף, יהיה יחד ג״ה וה״ט שוה לד״ה וה״ט יחד. ותושבת ג״ט שוה לתושבת ט״ד. ואם כן זויות [!] גה״ט שוה לזוית דה״ט. וכאשר עמד קו ישר על קו ישר והיו שתי זויות אשר לשני צדיו שוות—הן נצבות. ואם כן תהיין זוית גה״ט וזוית דה״ט נצבות. ואם כן זוית דה״ט ודה״א נצבות שוות—הגדולה שוה לקטנה, זה חלוף לא יתכן. והתבאר אם כן שאין ט׳ מרכז העגול א״ב ולא זולתי מהמקומות, רק נקודת ח׳. וח׳ הוא מרכז עגול א״ב.

فَقَدْ تَبَيَّنَ أَنَّهُ لا يُمْكِنُ أَنْ يَكونَ خَطّانِ في دائِرَةٍ يَقْطَعُ أَحَدُهُما الآخَرَ بِنِصْفَيْنِ عَلى زَوايا قائِمَةٍ إِلّا، وَالخَطُّ الَّذي يَقْطَعُ الآخَرَ بِنِصْفَيْنِ عَلى زاوِيَةٍ قائِمَةٍ يَجوزُ عَلى مَرْكَزِ الدّائِرَةِ.

וכבר התבאר שלא יהיו שני קוים בעגול יחתוך אחד מהם האחר לחציים על זוית נצבה שלא יעבר אחד מהם על מרכז העגול.

وَذٰلِكَ ما أَرَدْنا بَيانَهُ.

וזה מה שרצינו לבאר.

[1] ط ب ج ه د] ط ج ه د

فَإِنْ أَمْكَنَ^، فَلْيَكُنْ الْمَرْكَزُ نُقْطَةَ ط. ونُخْرِج ط ج وط ه وط د. نخَطّ ج ه مثل خَطّ ه د. ونَجْعَلُ ه ط مُشْتَرَكًا. نخَطَّا ج ه و ه ط مثل خَطَّيْ د ه و ه ط. وقاعِدَة ج ط مثل قاعِدَة ط د. فزاوية ج ه ط مثل زاوية د ه ط. فهُمَا إِذَنْ قائِمَتانِ. وزاوية د ه ا أيضًا قائِمَة^. فَزاوِيَتا د ه ط د ه ا مُتَساوِيَتانِ—العُظْمَى مُساوِيَة للصُّغرى[1]. هذا خُلْفٌ. فَلَيْسَ ط مَرْكَز الدَّائِرَة ولا غيرها مِنَ النُّقَطِ سِوَى نُقْطَة ح.

כי אלו היה אפשר זה, הנה יהיה המרכז נקודת ט׳. ונוציא ט״ג וט״ה וט״ד. הנה קו ג״ה כמו קו ה״ד. ונשים ה״ט משותף. הנה קו ג״ה וה״ט כמו שני קוי ד״ה וה״ט. ותושבת ג״ט כמו תושבת ט״ד. הנה זוית גה״ט כמו זוית דה״ט. הנה שתיהן[1] אם כן נצבות. וזוית דה״א גם כן נצבת. וכבר התאמת[2] כי כל הזויות הנצבות[3] שוות. הנה אם כן שתי זויות דה״ט דה״א שוות—הגדולה לקטנה. וזה שקר. אם כן אין ט׳ מרכז העגולה ולא זולתה מן הנקודות בלתי נקודת ח׳.

وَذَلِكَ ما أَرَدْنا أَنْ نُبَيِّنَ.

וזה מה שרצינו ביאורו[4].

وقَدْ تَبَيَّنَ أنَّهُ لا يكونُ وَتَرانِ في دائِرَة يَقْطَعُ أَحَدُهُما الآخَرَ بِنِصْفَيْنِ على زاوِيَةٍ قائِمَةٍ، إلّا وَهُوَ يَجوزُ على مَرْكَزِ الدَّائِرَة[2].

וכבר התבאר שלא יהיו שני מיתרים[5] בעגולה יחתוך אחד מהם האחר בשני חציים על זוית נצבת, אלא שהוא[6] יעבור על מרכז העגולה.

---

[1]مساوية للصغرى] MPR1R2 [2]الدايرة] CamDO2ST4

[1]שתיהן] שניהם **ק** [2]התאמת] התאומת **ב** התבאר **ס** [3]הנצבות] ~~השוות~~ ⟨⟨ה⟩⟩נצבות **ס** [4]ביאורו] לבארו **ב** לבאר **טס** [5]שני מיתרים] מיתרים שנים **ק** [6]שהוא] שיהוא **ב**

[Proposition III.2]

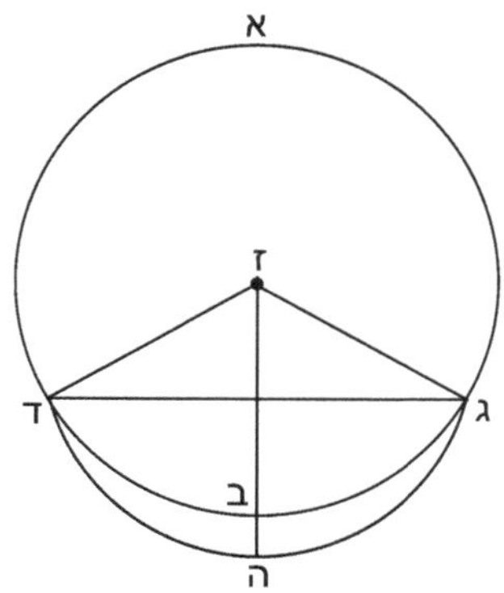

כשנרשמו על מקיף העגול שתי נקודות והוצא מאחת מהן קו ישר אל האחרת, יפול הקו בתוך העגול.

إِذا عُلِّمَ عَلى الخَطِّ المُحِيطِ بالدَّائِرَةِ نُقْطَتانِ فَالخَطُّ المُستَقِيمُ الَّذي يَصِلُ إِحْداهُما بِالأُخْرى، يَقَعُ في داخِلِ الدَّائِرَةِ.

המשל בזה: שנרשום על עגול א״ב שתי נקודות, והן ג׳ ד׳. ואומר שהקו הישר שידביק ג׳ בד׳ יפל בתוך העגול.

مِثالُهُ: أَنَّا نُعَلِّمُ عَلى دائِرَةٍ اب نُقْطَتَيْنِ، وَهُما ج د. فَأَقولُ إِنَّ الخَطَّ الَّذي يَصِلُ ج د يَقَعُ في داخِلِ الدَّائِرَةِ.

إذا تَعَلَّمَ على مُحِيطِ الدَّائِرَةِ[1] نُقْطَتانِ، وأُخْرِجَ مِنْ إحْداهُما خَطٌّ مُسْتَقِيمٌ إلى الأُخْرَى، فَإِنَّهُ يَقَعُ في داخِلِ الدَّائِرَةِ.

مِثالُهُ: إِنَّا نُعَلِّمُ على دائِرَةِ ا ب نُقْطَتَيْ ج د. ونُخْرِجُ خَطَّ ج د مُسْتَقِيمًا. فَأَقُولُ إِنَّهُ قَدْ وَقَعَ في داخِلِ الدَّائِرَةِ لا يُمْكِنُ غيره.

כאשר תרשום על מקיף העגולה שתי נקודות, והוצא[1] מאחת משתיהן[2] קו ישר אל[3] האחרת, הנה הוא יפול תוך העגולה.

המשל[4]: אנחנו נרשום על עגולת א״ב שתי נקודות ג׳ ד׳. ונוציא קו ג״ד ישר. הנה אומר כי הוא כבר[5] נפל[6] תוך העגולה אי אפשר זולתו.

---

[1] محيط الدايرة] P

שרטוט—ס: ג ← ב ונקודה ב׳ היא נקודת חיתוך ב״ד וא״ה [1] והוצא] והוציא ס [2] משתיהן] משתיהם ב [3] אל] ~~על~~+ ב [4] המשל] +בזה ז [5] כבר] ליתא ס [6] נפל] יפול ז

ולא יתכן זולת זה. ואם היה אפשר שיפול חוץ לעגול, כמו קו גה״ד, יפל. ויהיה מרכז העגול ז׳. ונדביק ז׳ בג׳ ובד׳. ונוציא מז׳ קו אל קשת ג״ד כמו שהוא, והוא ז״ב. ונוציאנו אל ה׳. ולפי שג״ז שוה לז״ד, תהיה זוית זג״ה שוה לזוית זד״ה. וזוית בה״ד גדולה מזוית זג״ה, לפי שהיא חיצונה למשלש זג״ה. ואמנם זוית זג״ה שוה לזוית זד״ה. וצלע ז״ג שוה לצלע ז״ד. וזוית בה״ד גדולה מזוית זד״ה. והגדולה מזויות כל משלש היא שיהיה מיתרה הצלע הארוך. ואם כן צלע ז״ד ארוך מצלע ז״ה. ואמנם ז״ד שוה לז״ב. ואם כן ז״ב ארוך מז״ה—הקצר ארוך מן הארוך, זה חלוף לא יתכן. התבאר כבר שהקו שידביק ג׳ בד׳ לא יפל חוץ לעגול ולא על קו המקיף בו. ואמנם יפל בתוכו, כמו ג״ד.

وَبُرْهانُهُ: أَنَّهُ لا يُمْكِنُ غَيْرُهُ إِلّا ذَلِكَ. فَإِنْ يُمكِنُ أَن يَقَعَ خارِجًا مِنْها مِثْل ج ه د، فَلْيَقَعْ. وَمَرْكَزُ الدَّائِرَةِ (ز)[1]. فَنَصِلُ ز ج. وَنُخْرِجُ مِنْ ز خطًّا إِلى قَوْسِ ج د كَيْفَما وَقَعَ، وَهُوَ ز ب. وَنُخْرِجُ إِلى ه. ج ز مِثْلُ ز د. [..] وَالزَّاوِيةُ الَّتي من ب ه د [..] لأَنَّ ضِلْعَ ج ز مِثْل ضِلْعِ ز د. فَالزَّاوِيةُ الَّتي مِنْ ب ه د أَعْظَمُ مِنَ الَّتي مِنْ ز د ه. وَأَعْظَمُ مِنْ زوايا المُثَلَّثِ يُوتِرُها الضِّلعُ الأَطْوَل. فَصارَ ز د أَطْوَلَ مِنْ ضِلْعِ ز ه. وَلَكِنَّ ز د مِثْلُ ز ب، ز ب أَطْوَلُ مِنْ ز ه. الأَقْصَرُ أَطْوَلُ—هذا خُلْفٌ لا يُمْكِنُ. فَقَدِ اسْتَبانَ أَنَّ الخطَّ الَّذي يَصِلُ ج د لا يَقَعُ خارِجًا مِنَ الدَّائِرَةِ وَلا عَلَى الخطِّ المُحيطِ بِها، وَلَكِنْ يَقَعُ في داخِلِها، مِثْلُ خَطِّ ج د.

וזה מה שרצינו לבאר.

وَذَلِكَ ما أَرَدْنا أَنْ نُبيِّنَ.

[1] [ز] ליתא

فَإِنْ يُمْكِنْ، فَلْيَقَعْ خارجًا مِنْها مِثْل خطّ ج ه د. وَلْيَكُنْ مَرْكَزُ الدَّائِرَةِ ز. وَنُخرج ز ج و ز د. ونُخرج مِنْ ز إلى قَوْس ج د[1] خطّ ز ب كَيْفَما وَقَعَ. ونُخرجهُ إلى ه. فزاوية ب ه د[2] أعظم مِنْ زاوية ز ج ه، لأنّها خارِجَة مِنْ مثلَّثِ ز ج ه. ولكنَّ زاوية ز ج ه مثل زاوية ز د ه، لأنَّ ضلع ز ج مثل ضلع ز د. فزاوية ز ه د أعظم مِنْ زاوية ز د ه. والزَّاوِيَةَ العُظْمَى مِنْ كُلِّ مُثَلَّثٍ يوتِرُها الضِّلْعُ الأَطْوَلُ. فَضِلْعَ ز د أطول مِنْ ضلع ز ه. ولكنَّ ز د مثل ز ب. ف ز ب أطول مِنْ ز ه—الأقصر أعظم مِنَ الأطوَلِ^، هذا خُلْفٌ غير مُمْكِن[3]. فَقَدِ اسْتَبَانَ أَنَّ الخطّ المُسْتَقِيمَ الَّذي يَخْرُجُ مِنْ ج إلى د لا يَقَعُ خارِجًا مِنَ الدَّائِرَةَ. وكَذَلِكَ أَيْضًا تَبَيَّنَ أَنَّهُ لا يَقَعُ على الخطِّ المُحِيطِ بِها. فَهُوَ يَقَعُ في داخِلِها مثل ج د.

وذَلِكَ ما أَرَدْنا أَنْ نُبَيِّنَ.

[1] ج د] DMO2PSU
[2] ب ه د] CamDMR2T4
[3] غير ممكن] Cam

שאם יהיה אפשר, הנה יפול[1] חוץ ממנה כמו קו גה״ד[2]. ויהיה מרכז העגולה ז׳. ונוציא ז״ג וז״ד[3]. ונוציא מן ז׳ אל קשת ג״ד קו ז״ב איך[4] שיפול. ונוציאהו אל ה׳. הנה זוית בה״ד יותר גדולה מזוית זג״ה מפני שהיא חוץ ממשולש זג״ה. אבל[5] זוית זג״ה כמו זוית זד״ה, מפני כי צלע ז״ג כמו צלע ז״ד. הנה זוית זה״ד יותר גדולה מזוית זד״ה. והזוית היותר גדולה מכל משולש יהיה מיתרה הצלע היותר ארוך. הנה צלע ז״ד יותר ארוך מצלע ז״ה. אבל ז״ד כמו ז״ב. אם כן ז״ב יותר ארוך מן ז״ה—היותר קצר יותר ארוך מן היותר[6] גדול, זה שקר אי אפשר. הנה כבר התבאר כי הקו הישר אשר יצא מן ג׳ אל ד׳ לא יפול חוץ מן העגולה. וכן גם כן התבאר[7] שהוא לא יפול על הקו המקיף בה. הנה הוא אם כן יפול בתוכה כמו ג״ד.

וזה מה[8] שרצינו לבאר[9].

[1] הנה יפול] שיפול **ס** [2] גה״ד] ~~ג״ד~~+ **ב**
[3] ונוציא ... וז״ד] ⟨⟨*⟩⟩ **ב** [4] איד] ⟨⟨*⟩⟩
[5] אבל] א⟨ב⟩ל **ט** [6] היותר] הצלע+ **ק** **פ**
[7] וכן ... התבאר] וכן גם התבאר **ב** וכן יתבאר גם כן **ק** [8] מה] ⟨*⟩ **ט** [9] לבאר] ביאורו **בפק**

[Proposition III.3]

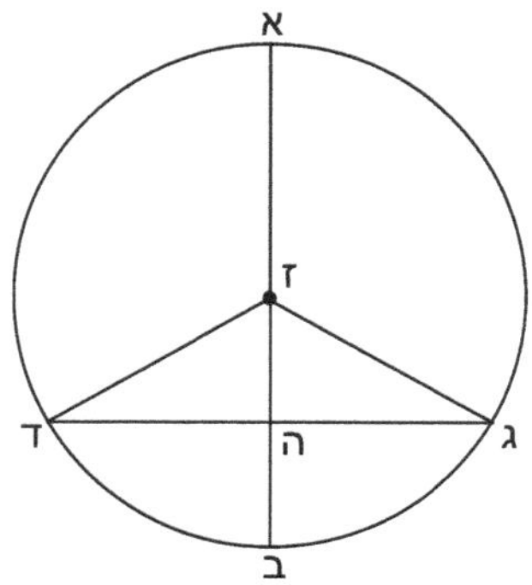

כשנפל מיתר בעגול בזולת המרכז, ויצא מן המרכז קו יחתוך המיתר לחציים, יחתכהו על זוית נצבה. ואם חתכו על זוית נצבה, יחתכהו לחציים.

إذا وَقَعَ خَطٌّ في دائِرَةٍ على غَيْرِ المَرْكَزِ، وَأُخْرِجَ مِنَ (المَرْكَزِ)[1] خَطٌّ يَقْطَعُ الخَطَّ الآخَرَ بِنِصْفَيْنِ، فَإِنَّهُ يَقْطَعُهُ على زاوِيَةٍ قائِمَةٍ. وَإِنْ قَطَعَهُ على زاوِيَةٍ قائِمَةٍ، فَإِنَّهُ يَقْطَعُهُ بِنِصْفَيْنِ.

המשל בזה: שנוציא בעגול א״ב קו עליו ג״ד. ולא יעבר על המרכז. ואלכסון העגול א״ב. ואומר שאם קו א״ב חתך את קו ג״ד לחציים, בהכרח יחתכהו על זוית נצבה. ואם חתכו על זוית נצבה, יחתכהו לחציים. ויחתכהו תחלה לחציים. ואומר שהוא יחתכהו על זוית נצבה.

مِثالُهُ: أَنَّا نَخُطُّ في دائِرَةِ اب خَطًّا عَلَيْهِ ج د، لا يَجُوزُ عَلَى المَرْكَزِ. (وَقُطْرُ الدّائِرَةِ)[2] اب. فَأَقُولُ إِنَّ خَطّ ز ه إِنْ قَطَعَ ج د بِنِصْفَيْنِ، فَإِنَّهُ يَقْطَعُهُ على زاوِيَةٍ قائِمَةٍ. فَإِنْ قَطَعَهُ على زاوِيَةٍ (قائِمَةٍ)[3]، فَإِنَّهُ يَقْطَعُهُ بِنِصْفَيْنِ. فَلْيَقْطَعْهُ أَوَّلًا بِنِصْفَيْنِ. فَأَقُولُ إِنَّهُ يَقْطَعُهُ على زاوِيَةٍ قائِمَةٍ.

[1] المركز] المخرج [2] وقطر الدائرة] ويقطع دائرة [3] قائمة] ليתא

إذا وَقَعَ وَتَرٌ في دائِرَةٍ على غَيْرِ الْمَرْكَزِ، وَخَرَجَ مِنَ الْمَرْكَزِ خطٌّ فقَطَعَ الوَتَرَ بنِصْفَيْنِ، فإِنَّهُ يَقْطَعُهُ على زاوِيَةٍ قائِمَةٍ. وإنْ قَطَعَهُ على زاوِيةٍ قائِمَةٍ، فإِنَّهُ يَقْطَعُهُ بنِصْفَيْنِ.

כאשר נפל מיתר בעגולה[1] על זולת המרכז, ויצא מן המרכז[2] קו וחתך המיתר בשני חציים, הנה הוא יחתכהו על זוית נצבת. ואם חתך אותו על זוית נצבת, הנה[3] יחתכהו בשני חציים.

مِثالُهُ: أنَّهُ وَقَعَ في دائِرَةِ اب وَتَرُ جد على غَيْرِ الْمَرْكَزِ. والقُطْرُ اب. فَأَقولُ إنَّ اب إنْ قَطَعَ جد بِنِصْفَيْنِ، فَإِنَّهُ يَقْطَعُهُ على زاوِيَةٍ قائِمَةٍ. وَإنْ قَطَعَهُ على زاوِيَةٍ قائِمَةٍ، فَإِنَّهُ قَطَعَهُ بِنِصْفَيْنِ. فليَقْطَعْهُ أوَّلًا بِنِصْفَيْنِ على نُقْطَةِ ه. فَأَقولُ إنّهُ يَقْطَعُهُ على زاوِيَةٍ قائِمَةٍ.

המשל בו: כי נפל בעגולת א״ב מיתר ג״ד על זולת המרכז. והקוטר א״ב. הנה אומר כי א״ב אם חתך ג״ד בשני חציים, הנה הוא יחתכהו על זוית נצבת. ואם חתכו על זוית נצבת, הנה הוא חתכו בשני חציים. ויחתכהו תחלה בשני[4] חציים על נקודת ה׳. הנה אומר כי יחתכהו על זוית נצבת.

שרטוט—**ס**: ה״ב חסר [1]בעגולה] העגולה **ב** [2]ויצא ... המרכז] ⟨⟨*⟩⟩ **ב** [3]הנה] +~~הוא~~ **ב** [4]בשני] ~~על~~+ **ט**

מופת זה: שמרכז העגול ז׳. ונדביק ז׳ בג׳ ובד׳. ויהיה ג״ה שוה לה״ד. וה״ז משותף. וקו [!] ג״ה וה״ז שוים לד״ה וה״ז, כל אחד לדומה לו. ותושבת ג״ז שוה לתושבת ז״ד. ואם כן זוית גה״ז שוה לזוית דה״ז. וכאשר עמד קו על קו והיו שתי הזויות אשר בשני צדיו שוות, הן[1] נצבות. והתבאר ששתי זויות גה״ז ודה״ז נצבות. וכבר חתך קו א״ב לקו ג״ד על זוית נצבה.

بُرهانُه: أَنَّ مَرْكَزَ الدَّائِرَةِ ز. وَنَصِلُ ز ج ز د. ج ه مِثْل ه د. ه ز مُشْتَرَك. فخَطَّا ج ه ه ز كُلُّ واحِدٍ مِنْهُما مِثْل نَظيرِه. وقاعِدَةُ ج ز مِثْل (القاعِدَة)[1] ز د. فَالزَّاوِيَةُ الَّتي مِنْ (ج ه ز)[2] مِثْل الَّتي مِنْ د ه ز. وَإِذا قامَ خَطٌّ عَلى خَطٍّ، فَصارَتِ الزَّاوِيَتانِ اللَّتانِ عَنْ جَنَبَتَيْهِ مُتَساوِيَتَيْنِ، فَهُما قائِمَتانِ. وَالزَّاوِيَتانِ اللَّتانِ مِنْ ج ه ز وَمِنْ د ه ز قائِمَتانِ. فَقَدْ قَطَعَ خَطُّ ز ه خَطَّ ج د عَلى زاوِيَةٍ قائِمَةٍ.

ויחתוך א״ב לג״ד על זוית נצבה. ואומר שכבר חתכו לחציים.

فَلْيَقْطَعْ خَطُّ ز ه أَيْضًا خَطَّ ج د عَلى زاوِيَةٍ قائِمَةٍ. فَأَقولُ إِنَّهُ قَطَعَهُ بِنِصْفَيْنِ.

מופת זה: לפי שג״ז שוה לז״ד, תהיה זוית זג״ד[2] שוה לזוית זד״ג[3]. ואמנם זוית גה״ז וזוית דה״ז נצבות. יהיו גז״ה ודה״ז שני משולשים. ושתי זויות זג״ה וגה״ז שוות לשתי זויות זד״ה ודה״ז, כל אחת לדומה לה. ואם כן יהיה כל צלע שוה לדומה לו—צלע ג״ז שוה לצלע ז״ד. והצלע שהוא מיתר שתי הזויות השוות—משותף, והוא ז״ה. ויהיו הצלעים הנשארים שוים—ג״ה שוה לה״ד. והתבאר שקו א״ב חתך ג״ד לחציים.

بُرْهانُهُ: أَنَّ ج ز مِثْل د ز. فالزَّاوِيَةُ الَّتي مِنْ ز ج د مِثْلُ الَّتي مِنْ ز د ج. وَلكِنَّ الَّتي مِنْ ج ه ز مِثْل الَّتي مِنْ د ه ز، مُساوِيَتانِ لِقائِمَتَيْنِ. ج ز ه د ه ز مُثَلَّثانِ. فَالزَّاوِيَتانِ اللَّتانِ من ز ج ه وَمِنْ ج ه ز مِثْل اللَّتَيْنِ مِنْ ز د ه وَمِنْ د ه ز، كُلُّ واحِدَةٍ مِثْلُ نَظيرَتِها. وَكُلُّ ضِلْعٍ مِثْلُ نَظيرِهِ—فَضِلْعُ ج ز مِثْل ضِلْعِ ز د. وَالضِّلْعُ الَّذي يوتِرُ الزَّاوِيَتَيْنِ اللَّتَيْنِ هُما مُتَساوِيَتانِ—مُشْتَرَكٌ، وَهُوَ ه ز. وَالضِّلْعانِ الباقِيانِ مُتَساوِيانِ—ج ه مِثْل ه د. فَخَطُّ ا ب قَدْ قَطَعَ ج د بِنِصْفَيْنِ.

---

[1] הן] נהן [2] זג״ד] זגד‹‹ה›› [3] זד״ג] * ‹‹זד״ה››

[1] القاعدة] نظيرة [2] ج ه ز] ج ه

بُرْهانُه: أَنَّا نَجْعَلُ المَرْكَزَ ز. ونُخْرِجُ ز ج و ز د. نَخُطّ ج ه مثل ه د. ونَجْعَلُ ه ز مُشْتَرَكًا. فَخَطَّا ج ه و ه ز مثل خَطَّي د ه و ه ز. وقاعِدَةُ ج ز مثل قاعِدَةِ ز د. فزاوية ج ه ز مثل زاوية د ه ز. فَهُما إذَنْ[1] قائِمَتانِ. فقَدْ قَطَعَ ا ب ج د بِنِصْفَيْنِ على زوايا قائِمَةٍ[2].

המופת: כאשר נשים המרכז ז׳, ונוציא ז״ג וז״ד[1], הנה קו ג״ה כמו ה״ד[2]. ונשים ה״ז משותף. הנה קו ג״ה וה״ז כמו קוי ד״ה וה״ז. ותושבת ג״ז כמו תושבת ז״ד. הנה זוית גה״ז כמו זוית דה״ז. הנה[3] שניהם אם כן נצבות. הנה כבר חתך א״ב ג״ד בשני חציים על זויות[4] נצבות.

وأيضًا فليَقْطَعْ ا ب ج د على زوايا قائِمَةٍ، فَأَقولُ إنَّه يَقْطَعُه بِنِصْفَيْنِ.

וגם כן הנה[5] יחתוך א״ב ג״ד על זויות נצבות[6]. הנה אומר כי הוא יחתכהו[7] בשני חציים.

بُرْهانُه: أنّ ج ز مثل ز د. فزاوية ز ج د مثل زاوية ز د ج. ولَكِنَّ زاوية ج ه ز وزاوية د ه ز قائِمَتانِ. فَزاوِيَتا ز ج ه و ج ه ز مِنْ مُثَلَّث ز ج ه مثل زاوِيَتَيّ ز د ه و د ه ز مِنْ مُثَلَّث ز د ه. وضلع ج ز مثل ضلع ز د وخطّ ه ز مُشْتَرَكٌ^. فالضِّلْعانِ الباقِيانِ، وَهُما ج ه و ه د، مُتَساوِيانِ. فَقَدْ قَطَعَ ا ب ج د بِنِصْفَيْنِ.

המופת: כי ג״ז כמו ז״ד. אם כן זוית זג״ד כמו זוית[8] זד״ג. אבל זוית[9] גה״ז וזוית דה״ז נצבות. אם כן שתי זויות זג״ה וגה״ז ממשולש זג״ה כמו שתי זויות זד״ה ודה״ז ממשולש זד״ה[10]. וצלע ג״ז כמו צלע ז״ד[11]. אם כן שתי הצלעות[12] הנשארות, והם ג״ה וה״ד, שוות. הנה כבר חתך א״ב ג״ד בשני חציים.

[1]اذن] MT2T4 اذا [2]بنصفحين ... قايمة] MPR1R2U

[1]וז״ד] ז״ד **וס** [2]ה״ד] קו+ **ו** [3]הנה] ואם כן **ס** [4]זויות] זוית **ב** [5]כן הנה] הנה ⟨⟨אם⟩⟩ **ב** [6]וגם ... נצבות] ⟨*⟩ **ט** [7]יחתכהו] יח⟨⟨ת⟩⟩כהו **ק** [8]זוית] ליתא **ק** [9]זוית] ~~זוית~~+ **ס** [10]ממשולש זד״ה] ⟨⟨*⟩⟩ **ב** ליתא **טפק** [11]ג״ז ... ז״ד] ז״ד ⟨⟨כמו ג״ז⟩⟩ **ב** [12]הצלעות] צלעות **ו**

וזה מה שרצינו לבאר. وَذَلِكَ ما أَرَدْنا بيَانَهُ.

ونשלם ביאורו[1].

وَذَلِكَ ما أَرَدْنا أَنْ نُبَيِّنَ.

[1] ונשלם ביאורו] וזה מה שרצינו לבאר **ט**

[Proposition III.4]

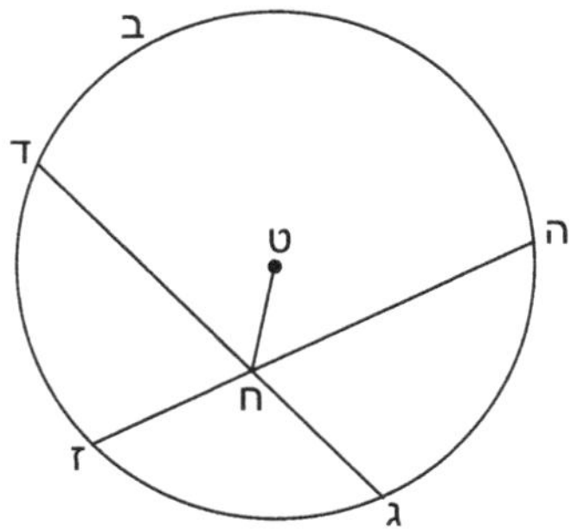

כל שני מיתרים בעיגול שלא יעברו על המרכז, לא יחלוק כל אחד מהם האחר לחציים.

המשל בזה: שנעשה בעגול (א״ב)[1] שני קוים, ולא יעברו על המרכז, והם ג״ד וה״ז. ויחתוך כל אחד מהם האחר על נקודת ח׳. ואומר שלא יחתוך כל אחד מהם האחר לחציים.

كُلُّ خَطَّيْنِ في دائِرَةٍ يَقْطَعُ أَحَدُهُما الآخَرَ وَلا يَجُوزانِ عَلى المَرْكَزِ، فَإِنَّهُ لَيْسَ يُمْكِنُ أَنْ يَقْطَعَ كُلُّ واحِدٍ مِنْهُما الآخَرَ بِنِصْفَيْنِ.

فَمِثالُهُ: أَنَّا نَخُطُّ في[1] دائِرَةِ اب خَطَّيْنِ لا يَجُوزانِ عَلَى المَرْكَزِ، وَهُما ج د ه ز، يَقْطَعُ كُلُّ واحِدٍ مِنْهُما الآخَرَ عَلَى نُقْطَةِ ح. فَأَقُولُ إِنَّهُ لا يَقْطَعُ كُلُّ واحِدٍ مِنْهُما الآخَرَ بِنِصْفَيْنِ.

שרטוט—B: סומנה | על היקף המעגל, ונראה שבתרגום הרב יעקב נשמטה א׳ [1]א״ב] ב׳

[1]في] ⟨*⟩ على

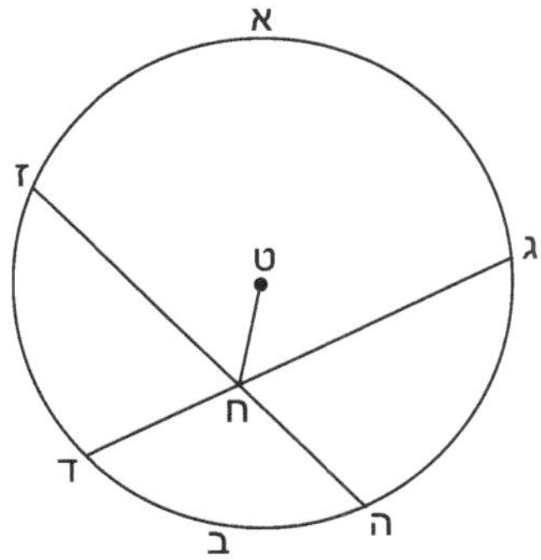

كُلُّ وَتَرَيْنِ في دائِرَةٍ يَقْطَعُ أَحَدُهُما الآخَرَ وَلا يَجوزانِ على الْمَرْكَزِ، فَلَيْسَ يَقْطَعُ كُلُّ واحِدٍ مِنْهُما الآخَرَ بِنِصْفَيْنِ.

כל[1] שני מיתרים בעגולה יחתוך אחד משניהם[2] האחר ולא[3] יעברו על המרכז, הנה לא יחתוך כל[4] אחד משניהם האחר בשני חציים.

مِثالُهُ: أَنَّ وَتَرَيْ جـ د ه ز في دائِرَةِ ا ب. وَقَدْ قَطَعَ أَحَدُهما[1] الآخَرَ على ح، ولَيْسَ يَجُوزانِ على الْمَرْكَزِ. فَأَقولُ إنَّهُ لا[2] يَقْطَعُ كُلُّ واحِدٍ مِنْهُما الآخَرَ بِنِصْفَيْنِ.

דמיון זה: כי שני מיתרי ג״ד וה״ז בעגולת[5] א״ב. וכבר חתך אחד משניהם האחר על ח׳, ולא יעברו על המרכז. הנה אומר כי הוא לא יחתוך כל אחד משניהם האחר בשני חציים.

---

[1] احدهما] CamDO2T1T2T4 واحد منهما
[2] لا] DO2T4 لم

שרטוט—**ט**: ב׳ חסרה [1] כל] כאשר היו+ **ק** [2] משניהם] מהם **ט** [3] האחר ולא] ליתא **פ** [4] כל] ליתא **בוס** [5] בעגולת] ⟨*⟩ **ט** בעגולות **פ**

ואם היה אפשר, יחתוך כל אחד מהם האחר לחציים על ח׳. ומרכז העגול ט׳. ונדביק ט׳ בח׳[1]. וכבר יצא ממרכז העגול קו אל ח׳ וחלק אותו לחציים. ואם כן חלקו על זוית נצבה[2]. ואם כן זוית דח״ט נצבה. ועוד, קו ט״ח שיצא מן המרכז כבר חלק קו ה״ז שאינו עובר על המרכז לחציים. ואם כן חלקו על זוית נצבה. ואם כן זוית זח״ט נצבה. ואם כן זוית דח״ט שוה לזוית זח״ט—הגדולה שוה לקטנה, זה חלוף[3]. ואם כן התבאר שקוי ג״ד וה״ז לא יחלוק אחד מהם האחר לחציים.

וזה מה שרצינו לבאר.

بُرهانُهُ: أَنَّهُ لا يُمْكِنُ ذَلكَ. فَإِنْ أَمْكَنَ، فَلْيَقْطَعْ كُلُّ واحِدٍ مِنْهُما الآخَرَ بِنِصْفَيْنِ عَلى نُقْطَةِ ح. ومَرْكَزُ الدَّائِرَةِ ط. وَنَصِلُ ط ح. فَقَدْ خَرَجَ مِنْ مَرْكَزِ الدَّائِرَةِ خَطٌّ إِلى ح، فَقَطَعَ ج د بِنِصْفَيْنِ، فهو يَقْطَعُهُ عَلى زاوِيَةٍ قائِمَةٍ. فَالزَّاوِيَةُ الَّتي من د ح ط إِذا قائِمَةٌ. وَأَيْضًا فَإِنَّ خَطَّ ط ح الَّذي خَرَجَ مِنَ المَرْكَزِ قد قَطَعَ خَطَّ ه ز الَّذي لَمْ يَجُزْ عَلى المَرْكَزِ بِنِصْفَيْنِ عَلى زاوِيَةٍ قائِمَةٍ. فَالزَّاوِيَةُ الَّتي مِنْ ز ح ط قائِمَةٌ. فَالَّتي مِنْ د ح ط مِثْلُ الَّتي مِنْ ز ح ط[1]، الصُّغْرى مِثْلُ الكُبْرى—هذا خُلْفٌ لا يُمْكِنُ. فَقَد اسْتَبانَ أَنَّ خَطَّيْ ج د ه ز لَم يَقْطَعْ كُلُّ واحِدٍ مِنْهُما الآخَرَ بِنِصْفَيْنِ.

وَذَلِكَ ما أَرَدْنا بَيانَهُ.

[1]בח׳] בז׳ [2]נצבה] ⟨⟨ר״ל⟩⟩ ~~מצורת ג׳ מזה השער~~
[3]חלוף] +⟨⟨לא יתכן⟩⟩

[1]زح ط] +اد ز

ولا يُمْكِنُ ذَلِكَ. فَإِنْ أَمْكَنَ ذَلِكَ، فَلْيَتَقَاطَعَا على حَ بِنِصْفَيْنِ. وَلْيَكُنْ الْمَرْكَزُ طَ. فَنُخْرِجُ ط ح. فقد خَرَجَ مِنَ الْمَرْكَزِ خَطٌّ إلى حَ فَقَطَعَ ج د بِنِصْفَيْنِ. فَهُوَ يَقْطَعُهُ على زاويةٍ قائِمَةٍ. فزاوية د ح ط قائِمَةٌ. وأيضًا فَإِنَّ ط ح قَطَعَ ه ز بِنِصْفَيْنِ فَهُوَ يَقْطَعُهُ على زاوِيَةٍ قائِمَةٍ. فزاوية ز ح ط قائِمَةٌ، وأيضًا فَإِنَّ ط ح قَطَعَ ه ز بِنِصْفَيْنِ، فَهُوَ يَقْطَعُهُ على زوايا قائِمَةٍ. فزاوية ز ح ط قائِمَةٌ^. فزاوية د ح ط مثل زاوية ز ح ط—الصُّغْرى مثل الكُبْرى. هذا خُلْفٌ. فَقَدِ اسْتَبَانَ أَنَّ وَتَرَيْ ج د و ه ز لم يَقْطَعْ كُلُّ واحدٍ مِنْهُما الآخَرَ بِنِصْفَيْنِ.

ואי אפשר זה. שאם היה אפשר, הנה יחתכו על ח׳[1] בשני חציים. ויהיה המרכז ט׳. ונוציא ט״ח. הנה כבר יצא מן המרכז קו אל ח׳ וחתך ג״ד[2] בשני חציים. הנה הוא יחתכהו על זוית נצבת. אם כן זוית דח״ט נצבת. וזוית זח״ט נצבת גם כן, מפני כי ט״ח כבר חתכו[3] ז״ה בשני חציים. הנה זוית דח״ט כמו זוית זח״ט—הקטן[4] כמו הגדול, זה שקר. הנה כבר התבאר כי שני מיתרי ג״ד וה״ז לא יחתוך כל אחד משניהם האחר בשני חציים.

وذَلِكَ ما أَرَدْنا أَنْ نُبَيِّنَ.

וזה מה שרצינו לבאר.

[1]ח׳] [א׳] ב׳ ח׳ **ב** [2]ג״ד] ג״ח **ס** [3]חתכו] חתך **זק** [4]הקטן] הקטון **בפק**

[Proposition III.5]

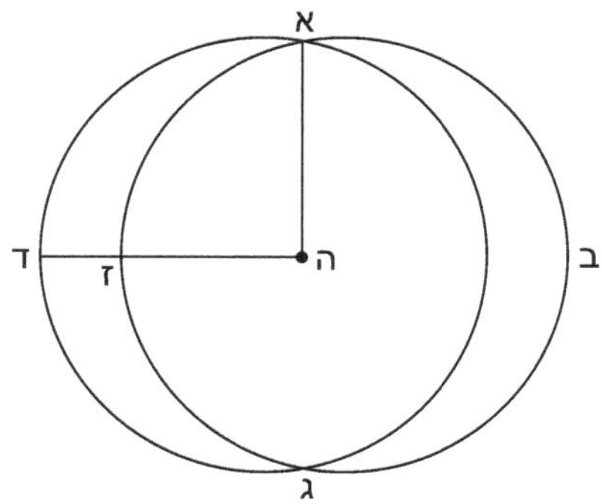

כל שני עגולים שיחתוך זה את זה, אין מרכזם אחד.

كُلُّ دائِرَتَيْنِ تَقْطَعُ إِحْداهُما الأُخْرى فَلَيْسَ مَرْكَزُهُما واحِدًا.

המשל בו: שעגולי א״ב וג״ד כל אחד מהם חלק האחר על נקודת א׳ וג׳. ואומר שאי איפשר שיהיה להם מרכז אחד.

مثالُهُ: أَنَّ دائِرَةَ ا ب ج وَدائِرَةَ ا د ج قَدْ قَطَعَتْ كُلُّ واحِدَةٍ مِنْهُما الأُخْرى عَلَى نُقْطَةِ ا ج. فَأَقولُ إِنَّ مَرْكَزَهُما لَيْسَ بِواحِدٍ.

كُلّ دائِرَتَيْنِ تَتَقاطَعانِ، فَلَيْسَ مَرْكَزاهُما بِواحِدٍ.

مِثالُهُ: أَنَّ دائِرَتَيّ اب جد تَتَقاطَعانِ على نُقْطَتَيْ اج، فَأَقولُ إنَّ مَرْكَزَيْهِما ليسا بِواحِدٍ.

כל שתי[1] עגולות תחתכנה[2], הנה אין שני מרכזיהם אחד.

דמיונו: כי שתי עגולות א״ב ג״ד תחתכנה[3] על שתי נקודות א׳ ג׳, הנה אומר כי מרכזיהם אינם אחד.

---

[1] כל שתי] *̲+ **ט** [2] תחתכנה] תתחתכנה **ס** תתחת⟨כ⟩נה **ק** [3] תחתכנה] תח⟨ת⟩כנה **ט** תתחתכנה **סק**

ואם היה איפשר, יהיה מרכזם ה׳. ונדביק א׳ בה׳. ונוציא מה׳ קו אל קשת אד״ג כמו שיפל, והוא קו ה״ד. ונקודת ה׳ היא מרכז עיגול אב״ג. ואם כן א״ה שוה לה״ז. וה״ד ארוך מה״ז. ואם כן ה״ד ארוך מא״ה. ועוד נקודת ה׳ מרכז עגול אג״ד. ואם כן א״ה שוה לה״ד. וכבר התבאר שה״ד ארוך מא״ה—זה חלוף. ואם כן אין מרכז שני העגולים אחד.

וזה מה שרצינו לבאר.

بُرهانُهُ: أَنَّهُ لا يُمْكِنُ ذَلِكَ. فَإِنْ أَمْكَنَ، فَلْيَكُنْ مَرْكَزُهُما ه. ونَصِلُ ا ه. ونُخْرِجُ مِنْ ه خَطًّا إِلى قَوْسِ ا د ج كَيْفَما وَقَعَ، وَهُوَ خَطُّ د ه. فَنُقْطَةُ ه هِيَ مَرْكَزُ دائِرَةِ ا ب ج. (ا ه)[1] مِثْل ه ز. ه د أَطْوَلُ مِنْ ا ه. وَأَيضًا نُقْطَةُ ه مَرْكَزُ دائِرَةِ ا ج د. ا ه مِثْل ه د. وَقَدْ كانَ تَبَيَّنَ أَنَّهُ أَطْوَلُ مِنْهُ—هذا خُلْفٌ لا يُمْكِنُ. فَلَيْسَ مَرْكَزُ الدَّائِرَتَيْنِ بِواحِدٍ.

وَذلِكَ ما أَرَدْنا بَيانَهُ.

[1] ا ه] اه د ا د ن

فَإِنْ أَمْكَنَ فَلْيَكُنْ مَرْكَزَاهُما ه. ونُخْرِجُ ا ه.^ ونُخْرِجُ مِنْ نُقْطَةِ ه^ خطَّ ه د إلى قَوْسِ ا د ج، كَيْفَما وَقَعَ. فنُقْطَة ه مَرْكَز دائِرَة ا ب. فخَطّ ا ه مثل خطّ ه ز. وخطّ ه د أطول مِنْ ا ه^. وأيضًا فإِنَّ نُقْطَةَ ه مَرْكَز دائِرَة ج د^، فخَطّ ا ه مثل خطِّ ه د. وَقَدْ كانَ تبيَّنَ أنّهُ أَطْوَلَ مِنْهُ^. فهَذا خُلْفٌ^. فَلَيْسَ مَرْكَز الدائِرَتَيْنِ بِواحِدٍ.

وذَلِكَ ما أَرَدْنا أن نُبيِّنَ.

שאם היה אפשר[1], הנה יהיו[2] מרכזיהם ה׳. ונוציא מן ה׳ קו א״ה אל נקודת א׳. הנה מן המבואר כי הוא כבר הגיע תכליתו אל מקיף שתי העגולות יחד. ונוציא קו ה״ד אל קשת אד״ג[3], איך שיפול. הנה נקודת ה׳ מרכז עגולת א״ב. אם כן קו א״ה כמו קו ה״ז. וגם כן הנה מפני כי נקודת ה׳ מרכז עגולת[4] אד״ג, יהיה קו א״ה שוה לקו ה״ד. וכבר[5] התבאר כי קו א״ה שוה לקו ה״ז. והשוים לדבר אחד[6] הנה הם שוים. הנה[7] קו ה״ד אם כן שוה לקו ה״ז[8]—הגדול[9] כמו הקטן[10], זה שקר בלתי אפשר. אם כן אין מרכז שתי העגולות אחד.

וזה מה שרצינו ביאורו[11].

---

1 אפשר] אחד+ **ס** 2 יהיו] יהיה **ט** 3 אד״ג] ⟨א⟩ד״ג **ט** 4 א״ב ... עגולת] ~~א״ג~~ ⟨⟨*⟩⟩ **ב** 5 וכבר] +* **ס** 6 אחד] ⟨*⟩ **ב** 7 הנה] ⟨⟨*⟩⟩ **פ** 8 לקו ה״ז] לה״ז **ס** 9 והשוים ... הגדול] ⟨*⟩ **ט** 10 כמו הקטן] לקטן **ט** 11 ביאורו] לבאר **ט** לבארנו **ס**

[Proposition III.6]

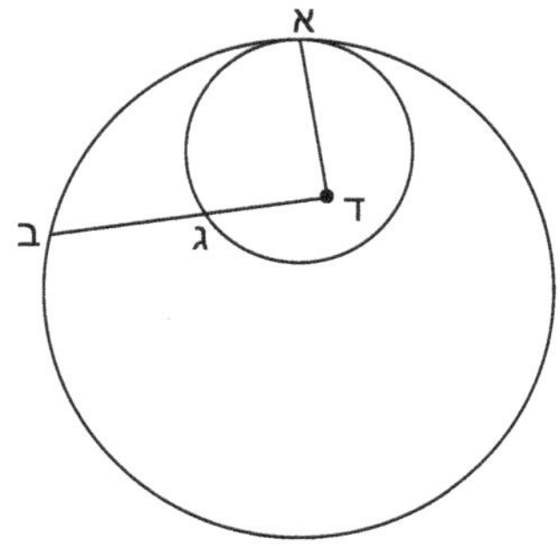

כל שני עגולים שיהיה אחד מהם ממשש האחר אין מרכזם אחד.

كُلُّ دائِرَتَيْنِ تَمَاسُّ إِحْداهُما الأُخرى فَلَيْسَ مَرْكَزُهُما بِواحِدٍ.

המשל בו: ששני עגולי א״ב וא״ג ממשש זה את זה על נקודת א׳. ואומר שאי אפשר שיהיה להם מרכז אחד.

مِثالُهُ: أَنَّ دائِرَتَيْ اب اج تَمَاسُّ إِحْداهُما الأُخْرى عَلَى نُقْطَةِ ا. فَأَقولُ إِنَّ مَرْكَزَيْهِما لَيْسَا بِواحِدٍ.

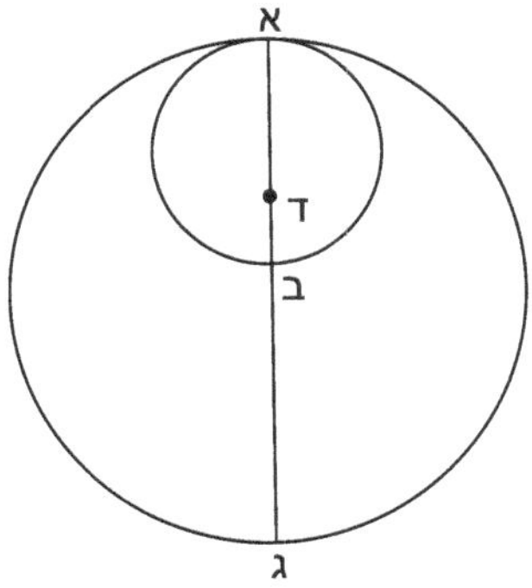

كُلّ دائِرَتَيْنِ تَتَماسّانِ، فَلَيْسَ مَرْكَزاهُمَا بِواحِدٍ.

כל שתי[1] עגולות תתמששנה[2], הנה אין מרכזי שניהם[3] אחד.

مِثالُهُ: أَنَّ دائِرَتَيّ اب و اج تَماسّانِ على ا. فَأَقولُ إنّ مَرْكَزَيْهما ليسا بِواحِدٍ.

דמיונו: כי שתי עגולות א"ב וא"ג תתמששנה[4] על א'. הנה אומר כי מרכזי שניהם[5] אינו אחד.

שרטוט—בכל כתבי היד הערביים ب ↔ ج

[1] שתי] שני ו [2] תתמששנה] תמששנה ס תממשנה פ [3] מרכזי שניהם] מרכזיהם ט [4] תתמששנה] תמששנה ס תממשנה פ [5] מרכזי שניהם] מרכזיהם שני ⟨⟨שניהם⟩⟩ ב

ואם היה אפשר, יהיה מרכזם אחד—ד׳. ונדביק א׳ בד׳. ונוציא מד׳ קו אל עגול א״ב כמו שיפל, והוא ד״ב. ונקודת ד׳ היא מרכז לעיגול א״ב. ואם כן א״ד שוה לד״ב. ועוד, נקודת ד׳ מרכז לעגול א״ג. ואם כן א״ד שוה לד״ג. וא״ד שוה לד״ב. ואם כן ד״ג שוה לד״ב—הקטן שוה לגדול, זה חלוף לא יתכן.

بُرْهانُهُ: أَنَّهُ لا يُمْكِنُ ذَلِكَ. فَإِنْ كانَ يُمْكِنُ، فَلْيَكُن مَرْكَزاهُما دَ. وَنَصِلُ ا د. وَنُخْرِجُ مِنْ دَ خَطًّا إِلى دائِرَةِ ا ب، كَيْفَما وَقَعَ، وَهُوَ د ب. فَنُقْطَةُ دَ هِيَ مَرْكَزُ دائِرَةِ ا ب. ا د إِذًا مِثْل د ب. وَأَيضًا نُقْطَةُ د هي مَرْكَزُ دائِرَةِ ا ب ج. ا د إِذًا مِثْل د ج. ا د مِثْل د ب. ب د مِثْل د ج، الأَطْوَل مِثْل الأَقْصَر—هذا خُلْفٌ لا يُمْكِنُ.

ואם כן התבאר שכל שני עגולים שימשש זה את זה אי אפשר שיהיה להם מרכז אחד.

فَقَدِ اسْتَبانَ أَنَّ كُلَّ دائِرَتَيْن مُتَماسَّتَيْن فلَيْسَ مَرْكَزاهُما بِواحِدٍ.

וזה מה שרצינו לבאר.

وَذَلِكَ ما أَرَدْنا بَيانَهُ.

ولا يُمْكِنُ ذَلِكَ. فَإِنْ أَمْكَنَ، فَلْيَكُنْ مَرْكَزاهُما^ نُقْطَةَ دَ. ونُخرج ا دَ. ونُخرج مِنْ دَ خطًّا إلى دائِرَةِ ا بَ كَيْفَما وَقَعَ، وهو د ب^. فنُقْطَةَ دَ مَرْكَزُ دائِرَةِ ا بَ. نَخَطّ ا دَ مثل خطّ د ب. ونُقْطَةَ دَ مَرْكَزُ دائِرَةِ ا جَ. نَخَطّ ا دَ مثل د ج. و ا دَ قَدْ كانَ مثل د ب. ف ب دَ مثل د ج—الأطول مثل الأَقْصَرِ، هذا خُلْفٌ.

فَقَدِ اسْتَبَانَ أَنَّ كُلّ دائِرَتَيْنِ تَماسّان فَلَيْسَ مَرْكَزاهُما بِواحِدٍ.

وذَلِكَ ما أَرَدْنا أَنْ نُبَيِّنَ.

ואי אפשר זה[1]. שאם היה אפשר, הנה יהיה[2] מרכז שניהם אחד, והוא[3] נקודת ד׳. ונוציא א״ד. ונוציא מן ד׳ קו אל עגולת א״ב איך שיפול, והוא ד״ב. ונעבירהו אל עגולת א״ג איך שיפול, והוא ד״ג. הנה נקודת ד׳[4] מרכז עגולת א״ב. הנה קו[5] א״ד כמו קו ד״ב. ונקודת ד׳ מרכז עגולת א״ג. הנה קו א״ד[6] כמו קו ד״ג. וא״ד כבר היה כמו ד״ב. הנה ב״ד כמו ד״ג—הארוך כמו הקטן, זה שקר.

הנה כבר התבאר כי כל שתי עגולות תתמששנה[7] הנה אין מרכזיהם אחד.

וזה מה שרצינו ביאורו.

---

[1] זה] ליתא ו [2] הנה יהיה] נאמר שיהיה ו
[3] והוא] והיא ק [4] ד׳] ד״ג ⟨⟨נ׳ נקודת ד׳⟩⟩
ו [5] קו] נקודת * ~~א״ב~~ ס [6] א״ד] +* ט
[7] תתמששנה] יתמששו ק

[Proposition III.7]

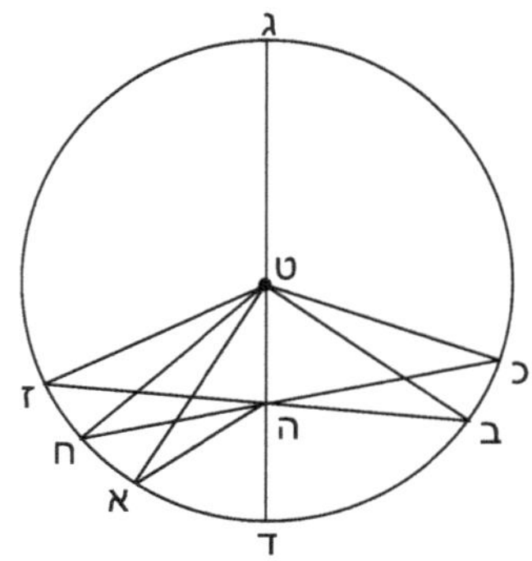

כל נקודה אשר בעגול שאינה על המרכז, ויצאו ממנה קוים אל הקו המקיף, יהיה הארוך שבהם אשר יעבר על המרכז, והקצר שבהם שלמות הקוטר מה שבין הנקודה ההיא וקצה הקוטר, ואשר קרב מן הקו אשר יעבר על המרכז ארוך מאשר רחק, ושני קוים בלבד אצל צדי הקו הקצר—שוים.

إِذا عُلِّمَ على قُطْرِ الدَّائِرَةِ نُقْطَةٌ على غَيْرِ المَرْكَزِ، وأُخْرِجَتْ مِنْها خُطوطٌ إِلى الخَطِّ المُحيطِ، فَإِنَّ أَطْوَلَ تِلْكَ الخُطوطِ الَّذي يَجوزُ عَلَى المَرْكَزِ، وأَقْصَرَها هُوَ فَضْلُ القُطْرِ عَلَى هذا الخَطِّ الأَطْوَلِ الَّذي يَجوزُ على المَرْكَزِ، وما قَرُبَ مِنْ خَطِّ المَرْكَزِ فهُوَ أَطْوَلُ مِمَّا هُوَ أَبْعَدُ مِنْهُ مِنَ الخُطوطِ الأَوَّل فَالأَوَّل، وَخَطَّانِ عن جَنَبَتَيِ الخَطِّ الأَقْصَرِ فَقَط مُتَساوِيانِ.

كُلُّ نُقْطَةٍ في دائِرَةٍ على غَيْرِ الْمَرْكَزِ، تَخْرُجُ مِنْها خُطوطٌ إِلى الخَطِّ المُحِيطِ، فَإِنَّ أَطْوَلَها الَّذي يَجوزُ على الْمَرْكَزِ، وأَقْصَرَها تَمامُ القُطْرِ مِنْ بَيْنِ النُقْطَةِ وَطَرَفِ القُطْرِ[1]، وَما قَرُبَ مِنَ الخَطِّ الَّذي يَجوزُ على الْمَرْكَزِ فَهُوَ أَطْوَلُ مِمّا بَعُدَ مِنْهُ، وَخَطّانِ فَقَط عَنْ جَنْبَتَي الخَطِّ الأَقْصَرِ—مُتَساوِيانِ.

כל נקודה בעגולה[1] על זולת המרכז, יצאו ממנה קוים אל הקו המקיף, הנה היותר ארוך מהם אשר יעבור על המרכז, והיותר קצר מהם שלמות[2] הקוטר מבין הנקודה וקצה[3] [4] הקוטר, ומה שקרב מן הקו אשר יעבור על המרכז, הנה הוא יותר ארוך ממה שרחק[5] ממנו, ושני קוים לבד משני[6] צדדי הקו היותר קצר[7]—שוים.

[1] وطرف القطر] MPR1R2

שרטוט—ס: א ↔ ז ה ↔ ט [1] נקודה בעגולה] נקוד~~ת~~⟨ה⟩ בעגול~~ת~~⟨ה⟩ ב [2] שלמות] +⟨⟨נ׳ המשלים⟩⟩ ק [3] הקוטר ... וקצה] ליתא ז [4] וקצה] והנה קצה ס [5] שרחק] ⟨⟨*⟩⟩ פ [6] משני] אשר+ ק [7] קצר] +~~מהם~~ ק

مِثالُهُ: أَنَّا نُعَلِّمُ عَلى قُطْرِ دائِرَةِ ا ب الَّذي عَلَيْهِ ج د نُقْطَةَ ه. وَلَيْسَتْ عَلى مَرْكَزِ الدَّائِرَةِ. وَنُخْرِجُ مِنْها خُطوطًا إِلى الخَطِّ المُحيطِ، وَهِيَ ه ز ه ح ا ه. فَأَقولُ إِنَّ أَطْوَلَها ج ه الَّذي يَجوزُ عَلى المَرْكَزِ، وَأَقْصَرَها الخَطّ الباقي مِنَ القُطْرِ، وَهُوَ د ه. وَهُوَ تَمامُ القُطْرِ. وَالخُطوط الباقِيَةَ أَقْرَبُها مِنْ خَطّ المَرْكَزِ هُوَ أَطْوَلُها، وَما بَعُدَ مِنْهُ فَهُوَ أَقْصَرُ. نَخَطُّ ه ز أَطْوَلُ مِن ه ح. ه ح أَطْوَلُ مِنْ ه ا. ه ا أَطْوَلُ مِنْ ه د. وَخَطّانِ عَنْ جَنَبَتَيْ ه د مُتَساوِيانِ فَقَط.

המשל בזה: שנרשום על קוטר עגול א״ב, שהוא קו ג״ד, נקודת ה׳, ואינה על המרכז. ונוציא ממנה קוים אל הקו המקיף, והם ה״ז, וה״ח, וה״א. ואומר: שהארוך שבהם קו ג״ה שעובר על המרכז; והקצר שבהם הקו הנשאר מן הקוטר, והוא ה״ד; והקוים הנשארים כפי שעור מרחקם מהקו העובר על המרכז; ויהיה ה״ז ארוך מה״ח, וה״ח ארוך מה״א; ושני קוים אצל צדי ה״ד בלבד—שוים.

مِثَالُهُ: أَنَّهُ خَرَجَ مِنْ نُقْطَةِ ه، مِنْ دائِرَةِ^ ا ب، وَهِيَ على غَيْرِ الْمَرْكَزِ، خُطوطٌ إلى الخطِّ المُحيط، وهي ه ز ه ح ه ا ه ج ه د. و ه ج هو الخطُّ الَّذي يَمُرُّ بالْمَرْكَزِ. و ه د تمام القُطْرِ. فَأَقولُ إنّ ه ج أطولها^، وَأَقصرها ه د، وهو تمام القُطْرِ، وأمّا الخُطوط الباقية—فَإِنَّ ه ز[1] أطول مِنْ ه ح، و ه ح أطول مِنْ ا ه، وخَطّانِ فَقَطْ عَنْ جَنْبَتَيْ هذا^ الخطِّ[2] الأَقْصَر—مُتَساوِيانِ.

דמיונו: כי[1] יצאו מנקודת ה׳, מן קוטר[2] עגולת א״ב, והיא על זולת[3] המרכז, קוים אל הקו המקיף, והם ה״ז, ה״ג, וה״ח, וא״ה[4], וה״ד. וה״ג[5] הוא הקו אשר ילך במרכז[6]. וה״ד שלמות הקוטר. הנה אומר: כי ה״ג היותר[7] ארוך; והיותר קצר מהם ה״ד[8], והוא שלמות[9] הקוטר; ואולם הקוים הנשארים—הנה ה״ז[10] יותר ארוך מן ה״ח, וה״ח יותר ארוך מן א״ה; ושני קוים לבד משני צדדי הקו היותר קצר—שוים[11].

[1] ه ز] +منها MPR1R2
[2] هذا الخط] MPR1R2

[1] כי] +כאשר ס [2] מן קוטר] ⟨ה⟩קוטר ט
[3] על זולת] זולת על פ [4] וה״ח וא״ה] ה״ח א״ה ק
[5] וה״ג] ⟨⟨*⟩⟩ פ [6] במרכז] ⟨⟨*⟩⟩ פ
[7] היותר] הוא+ ק [8] ה״ד] ~~ח״ד~~+ ס
[9] שלמות] שני+ ס [10] ואולם ... שוים] ליתא ז
[11] ה״ז] ~~ד״ג~~+ ק

מופת זה: שמרכז העגול הוא ט׳. ונדביק ט׳ בז׳ ובח׳ ובא׳. ולפי שכל משולש המחובר משני צלעיו יחד—ארוך מהצלע הנשאר, יהי ז״ט וט״ה יחד ארוך מה״ז. וז״ט שוה לט״ג. ואם כן ג״ה ארוך מה״ז. וז״ט שוה לט״ח. וט״ה משותף להם. ואם כן ז״ט[1] וט״ה שוים לח״ט וט״ה, כל אחד לדומה לו. וזוית זט״ה גדולה מזוית חט״ה. ואם כן תושבת ז״ה גדולה מתושבת ח״ה. ולפי זה גם כן יהיה ח״ה ארוך מא״ה. ועוד א״ה וה״ט יחד ארוך מא״ט. וא״ט שוה לט״ד. ואם כן א״ה וה״ט יחד ארוך מט״ד. ונשליך ט״ה המשותף. וישאר א״ה ארוך מה״ד. ואם כן הארוך שבקוים קו ה״ג שעבר על המרכז[2]. והקצר שבהם קו ה״ד הנשאר מן הקוטר. אמנם הקוים הנשארים—הקרוב מהם אל ה״ג הוא הארוך יותר, מאשר הוא הרחוק ממנו, שז״ה ארוך מה״ח, וה״ח ארוך מה״א.

بُرهانُه: أَنَّ مَرْكَزَ الدَّائِرَةِ ط. وَنَصِلُ (ط ب ز ح ا)[1]. وكُلّ مُثَلَّثٍ فَإِنَّ كُلّ ضِلْعَيْنِ مِنْ أَضْلاعِهِ أَطْوَلُ مِنَ الضِّلْعِ الباقي—ز ط ط ه أَطْوَلُ من ه ز. ز ط مثل ط ج. ج ه أَطْوَلُ من ه ز. وأَيْضًا ز ط مثل ط ح. ط ه مُشْتَرَكٌ. فكِلا خَطَّيْ ز ط ط ه مثل (كِلا)[2] خَطَّيْ ح ط ط ه، كُلٌّ واحِدٍ مثل نَظيرِهِ والزَّاوِيَةَ الَّتي مِنْ ز ط ه أَعْظَمُ مِنَ الَّتي مِنْ ح ط ه. قاعِدَةُ ز ه أَطْوَلُ من قاعِدَةِ ح ه. وَلِذلِكَ صارَ ح ه أَطْوَلُ من ا ه. وأَيْضًا ا ه ه ط مَجْموعانِ أَطْوَلُ من ا ط. ا ط مثل ط د. ا ه ه ط مَجْموعانِ أَطْوَلُ من د ط. فَنُلْقي ط ه المُشْتَرَكَ، فَيَبْقى ا ه أَطْوَلُ ه د. فأَطْوَلُ الخُطوطِ ه ج الَّذي جازَ عَلى المَرْكَزِ. وَأَقصَرُها خَطُّ ه د الباقي مِنَ القُطْرِ. فَأَمّا الخُطوطُ الباقِيَة فَإِنَّ أَقْرَبَها إِلى خَطِّ ه ج هو أَطْوَلُ مِمَّا هو أَبْعَد مِنْهُ، ز ه أَطْوَلُ من ه ح، ه ح أَطْوَلُ مِنْ (ه ا)[3].

---

[1]ז״ט] א״ט ⟨⟨*⟩⟩ [2]המרכז] ~~הקוים~~+

---

[1]ط ب ز ح ا] ط ز ح ا [2]كلا] كلي [3]ه ا] ا ح

بُرهانُهُ: أنّا نَجْعَلُ المَرْكَزَ ط. ونُخْرِجُ ط ز، و ط ح، و ط ا. وكُلّ ضِلْعَيْنِ مِنْ مُثَلَّث فهُما أَطْوَل مِنَ الضِّلع الباقي. ف ز ط و ط ه أطول مِنْ ه ز. وخطّ ز ط مثل ط ج. ف ج ه أطول مِنْ ه ز. و ز ط مثل ط ح. و ط ه مُشْتَرَك. فخطّا ز ط و ط ه مثل خطَّي ح ط و ط ه. وزاوية ز ط ه أعظم مِنْ زاوية ح ط ه. فقاعدَة ز ه أَطْوَلُ مِنْ قاعِدَةِ ح ه. وكذلكَ أَيْضًا تَبَيَّنَ أنَّ ح ه أطول مِنْ ا ه. وأيضًا فإنَّ ا ه و ه ط مجموعَيْنِ أَطْوَل مِنْ ا ط. و ا ط مثل ط د. ف ا ه و ه ط مجموعَيْنِ أطول مِنْ ط ه ه د. فيُلْقى^ ط ه المُشْتَرَك. فيَبْقى ا ه أطول مِنْ ه د. فأَطْوَل الخُطوط^ الّتي تَخْرُجُ مِنْ نُقْطَة ه—ه ج، وهُوَ الّذي جازَ على المَرْكَز. وأقْصَرُها—ه د، الّذي هو تَمامُ القُطرِ. والباقية—ما قرُبَ مِنْها مِنْ ه ج، أَطْوَل ممّا بَعُدَ. فخطّ ه ز أطْوَل مِنْ ه ح. و ه ح أطْوَل مِنْ ا ه.

ומופתו: שנשים המרכז ט׳. ונוציא ט״ז, וט״ח, וא״ט[1]. וכל שתי צלעות ממשולש שניהם[2] יותר ארוכים מן הצלע הנשאר. הנה ז״ט וט״ה יותר ארוכים מן ה״ז. וקו ז״ט כמו ט״ג[3]. הנה ג״ה יותר ארוך מן ה״ז[4]. וז״ט כמו ט״ח. וט״ה משותף[5]. הנה שני קווי[6] ז״ט וט״ה כמו שני קוי ח״ט וט״ה. וזוית זט״ה יותר גדולה מזוית חט״ה. הנה תושבת ז״ה יותר ארוכה מתושבת ח״ה. וכן גם[7] כן התבאר כי ח״ה יותר ארוך מן א״ה. וגם כן הנה[8] א״ה[9] וה״ט מקובצים יותר ארוך מן א״ט. וא״ט כמו ט״ד. הנה א״ה וה״ט מחוברים יותר ארוך מן ט״ה וה״ד[10]. ויפול מהם ט״ה המשותף. וישאר א״ה יותר ארוך מן ה״ד[11]. הנה היותר ארוך מן הקוים אשר יצאו מנקודת ה׳—ה״ג, והוא אשר עבר על המרכז. והיותר קצר מהם—ה״ד, אשר הוא שלמות הקוטר. והנשארים—מה שקרב מהם[12] מקו ה״ג, יותר ארוך ממה שרחק. הנה קו ה״ז יותר ארוך מן ה״ח, וה״ח יותר ארוך מן א״ה.

---

[1] וא״ט] א״ט **בו** [2] שניהם] הנה+ **ז** [3] ט״ג] +⟨⟨וה״ט משותף⟩⟩ **ק** [4] וקו ... ה״ז] ליתא **ז** ⟨⟨*⟩⟩ **פ** [5] הנה ז״ט ... משותף] ⟨⟨הנה ז״ט וט״ה יותר ארוכים מן ה״ז וז״ט כמו ט״ה וט״ה משותף הנה⟩⟩ **ב** [6] קוי] ליתא **פ** קוים **ק** [7] וכן גם] וגם **פ** [8] הנה] ⟨*⟩ **ט** [9] א״ה] ~~א״ט~~ ⟨⟨א״ה⟩⟩ **ק** [10] וה״ד] +~~ויפול מהם ט״ה וה״ד~~ **ט** [11] ה״ד] +וקו גם כן ה״ט וט״ז יותר ארוכים מן ה״ז וט״ז כמו ט״ג אם כן ה״ט וט״ג יותר ארוכים מן ה״ז **ז** [12] מהם] ליתא **ק**

וכמו כן נבאר ששני קוים בלבד אצל צדי קו ה״ד הקצר—שוים.

המופת בו: שנעמיד על קו (ט״ד)[1], על נקודת ט׳, זוית שוה לזוית אט״ה, והיא זוית הט״ב. ונוציא מה׳ קו אל ב׳. וא״ט שוה לט״ב. וט״ה משותף. ואם כן א״ט וט״ה שוים לט״ב וט״ה, כל אחד לדומה לו. וזוית אט״ה שוה לזוית הט״ב. ואם כן תושבת א״ה שוה לתושבת ה״ב. ואומר שאי אפשר שיהיה קו במקום מן העגול שיצא מנקודת ה׳ אל הקו המקיף, שיהיה שוה לקו א״ה, כי אם קו ה״ב. ואם היה אפשר, יהיה קו ה״כ. ונדביק כ׳ בט׳. וא״ט שוה לט״כ. וט״ה משותף. ואם כן א״ט וט״ה שוים לכ״ט וט״ה, כל אחד לדומה לו. ותושבת א״ה שוה לתושבת ה״כ. ואם כן זוית אט״ה שוה לזוית כט״ה—הגדולה שוה לקטנה, זה חלוף לא יתכן. ואם כן התבאר שלא יתכן שיצא מנקדת ה׳ קו אחר שוה לקוי א״ה וה״ב זולתם. והתבאר ששני קוי א״ה וה״ב שהם אצל צדי קו ה״ד הקצר, הם שוים. והארוך שבקוים קו ג״ה שעובר על המרכז. והקצר שבהם קו ה״ד הנשאר מן הקוטר. והקוים הנשארים—הקרוב מהם על קו העובר על המרכז, ארוך מהרחוק ממנו, ז״ה ארוך מה״ח, וה״ח ארוך מא״ה.

[1] ט״ד] ~~א״ג~~ ד״ג

وَخَطّانِ فقط يَكُونانِ عَنْ جَنَبَتَيْ خَطِّ ه د الأَقْصَرِ مُتَساوِيَيْنِ.

بُرهانُهُ: أَنَّا نُقِيمُ على خَطِّ ط د نُقْطَةَ ط مثل الزَّاوِيَةِ الَّتِي مِنْ ا ط ه، وَهِيَ الَّتِي من ط ه ب. ونُخْرِجُ مِنْ ه خَطًّا إلى ب. ا ط مثل ط ب. ط ه مُشْتَرَكٌ. فَكِلَيْ خَطَّيْ ا ط ط ه مثل كِلَيْ خَطَّيْ ب ط ط ه، كُلُّ واحِدٍ مثل نَظِيرِهِ. فالزَّاوِيَةُ الَّتِي من ا ط ه مثل الَّتِي من ه ط ب. فَقاعِدَةُ ا ه مثل قاعِدَةِ ه ب. فَأَقولُ إِنَّهُ لا يُمْكِنُ أَنْ يَكونَ خَطٌّ في مَوْضِعٍ مِنَ الدَّائِرَةِ يَخْرُجُ مِنْ نُقْطَةِ ه إلى الخَطِّ المُحِيطِ مثل خَطِّ ا ه إِلّا خَطّ ه ب. ولا يُمْكِنُ ذلكَ. فَإِنْ أَمْكَنَ، فَلْيَكُنْ خَطّ ه ك. وَنَصِلُ ط ك. ا ط مثل ط ك. ط ه مُشْتَرَكٌ. فَكِلَيْ خَطَّيْ ا ط ط ه مثل كِلَيْ خَطَّيْ ك ط ط ه، كُلُّ واحِدٍ مثل نَظِيرِهِ. وقاعِدَةُ ا ه مثل قاعِدَةِ ه ك. فالزَّاوِيَةُ الَّتِي من ا ط ه إِذًا مثل الَّتي من ك ط ه. وَلٰكِنَّ الَّتِي من ا ط ه مثل الَّتي من ب ط ه. فَالَّتِي من ك ط ه مثل الَّتِي من ب ط ه. الكُبرى مثل الصُّغرى—هذا خُلْفٌ لا يُمْكِنُ. فَلَيْسَ يُمْكِنُ أَنْ يَخْرُجَ مِنْ نُقْطَةِ ه خَطٌّ آخَرُ مثل خَطِّ ا ه ومثل خَطِّ ه ب غَيْرهما. فكلا خَطَّيْ ا ه ه ب اللَّذَيْنِ عَنْ جَنَبَتَيْ خَطِّ ه د الأَقْصَرِ مُتَساوِيانِ. فَأَطْوَلُ الخُطوطِ خَطُّ ج ه الَّذي جازَ عَلَى المَرْكَزِ وأَقْصَرُها خَطّ ه د الَّذي بَقِيَ مِنَ القُطْرِ. وَالخُطوط الباقِيَة (فَأَقْرَبها)[1] مِنْ

[1] فاقربها] قاقربها

وأَقولُ إنَّ خطَّيْنِ فقَطْ يكُونانِ عَنْ جَنْبَتَي هذا[1] الأَقْصَر—مُتَساوِيانِ.

بُرهانُهُ: أنّا نُقيمُ على نُقْطَةِ ط مِنْ خطّ ط ه زاويةً^ مثل زاوِيَةِ ا ط ه، وهي زاوية ه ط ب. ونُخْرِج ه ب. نخطّ ا ط مثل ط ب. ونَجْعَلُ ط ه مُشْتَرَكًا. نخطّا ا ط و ط ه مثل خطَّي ب ط و ط ه. وزاوية ط ا ه مثل زاوية ه ط ب. فقاعِدَة ا ه مثل قاعدَة ه ب. وأَقولُ أيضًا، إنّهُ لَيْسَ يمْكِنُ أَنْ يَخْرُجَ مِنْ ه إلى الخطّ المُحِيطِ مثل ا ه، إلّا ه ب. فإنْ أَمْكَنَ، فلْيَكُنْ ه ك. ويخرج ط ك. نخطّ ا ط مثل ط ك. ونَجْعَلُ ط ه مُشْتَرَكًا. نخطّا ا ط و ط ه مثل خطَّي ك ط و ط ه. وقاعِدَة ا ه مثل قاعِدَة ه ك. فزاوية ا ط ه مثل زاوية ك ط ه. ولَكِنَّ زاوية ا ط ه مثل زاوية ب ط ه—فزاوِيَة ك ط ه مثل زاوية ب ط ه—العُظمى مثل الصُّغرى، هذا خُلْفٌ^. فلَيْسَ يمْكِنُ أَنْ يَخْرُجَ مِنْ ه خطّ آخر مثل أَحَدِ خطَّي ا ه و ه ب، غيرهما، نخطّا ا ه و ه ب اللَّذانِ هُما عَنْ جَنْبَيّ الأقصر—فقَطْ مُتَساوِيانِ.

ואומר כי שני קוים לבד יהיו משני צדדי זה הקצר[1]—שוים.

המופת: אנחנו נעמיד על נקודת ט׳ מקו ט״ה[2] כמו זוית אט״ה, והיא[3] זוית הט״ב. ונוציא ה״ב. הנה קו א״ט כמו ט״ב. ונשים ט״ה[4] משותף. הנה קו א״ט וט״ה כמו שני קוי ב״ט וט״ה. וזוית אט״ה כמו הט״ב[5]. אם כן תושבת א״ה[6] כמו תושבת ה״ב[7]. ואומר גם כן, כי[8] אי אפשר שיצא מן ה׳ אל הקו המקיף כמו א״ה, כי אם ה״ב. ואם יהיה אפשר, יהיה ה״כ. ויצא ט״כ. הנה קו א״ט כמו קו[9] ט״כ. ונשים ה״ט משותף. הנה קו א״ט וט״ה כמו שני קוי כ״ט וט״ה. ותושבת א״ה כמו תושבת ה״כ. הנה זוית אט״ה כמו זוית כט״ה. אבל[10] זוית אט״ה כמו זוית בט״ה—הנה זוית כט״ה כמו זוית[11] [12] בט״ה[13]—הגדולה כמו הקטנה, זה שקר אי[14] אפשר. הנה אי אפשר[15] שיצא מן ה׳ קו[16] אחר כמו אחד[17] משני קוי א״ה וה״ב, זולתם. הנה שני קוי א״ה וה״ב אשר הם משני צדדי[18] היותר קצר—לבד[19] שוים.

---

[1] הקצר] הקוטר ס [2] ט״ה] +זוית ט ליתא ס +⟨⟨זוית⟩⟩ פ [3] והיא] והוא ס [4] ונשים ט״ה] *+ ב [5] הט״ב] זוית+ ט [6] א״ה] ⟨⟨*⟩⟩ ק [7] כמו ... ה״ב] ⟨⟨*⟩⟩ ב [8] כי] ליתא ס [9] קו] ליתא בוס [10] אבל] א⟨ב⟩ל ט [11] זוית ... זוית] ⟨⟨*⟩⟩ ק [12] כמו זוית] ⟨⟨שוה לזוית⟩⟩ ב [13] בט״ה] +אבל זוית אט״ה כמו זוית בט״ה הנה זוית בט״ה כמו זוית בט״ה ז [14] אי] אם כן+ ז [15] הנה ... אפשר] ליתא זס [16] קו] הקו ב ⟨⟨*⟩⟩ פ [17] אחד] ליתא ז [18] צדדי] +⟨⟨הקוטר⟩⟩ פ ⟨⟨הקו⟩⟩ ק [19] לבד] ליתא ז

---

[1] هذا] D ه د

خَطِّ المَرْكَزِ أَطْوَلُ مِمَّا هو أَبْعَد مِنْهُ—زه أَطْوَل من ه ح، ه ح أَطْوَلُ من ا ه. وَخَطَّانِ فَقَط مُتَساوِيانِ عَنْ جَنَبَتَيْ خَطِّ ه د الأَقْصَرِ وَهُما ا ه ه ب.

וזה מה שרצינו.

وَذَلِكَ ما أَرَدْنا بَيانَهُ.

وذَلِكَ ما أَرَدْنا أَنْ نُبَيِّنَ.

וזה מה שרצינו לבאר[1].

[1] וזה ... לבאר] ונשלם ביאורו **ט** וזה מה שרצינו ביאורו **בק**

[Proposition III.8]

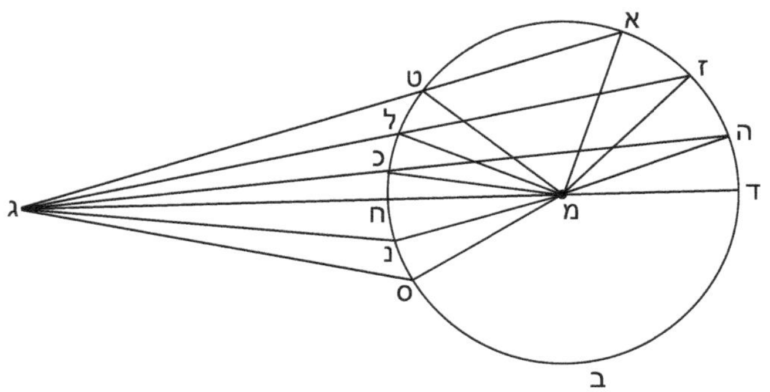

כל נקודה שהיא חוץ לעיגול, שיצאו ממנה קוים אל העיגול, יהיה הארוך מאשר באו בתוך העגול וחתך אותו—העובר מהם על המרכז. ואשר קרב מהקוים הנשארים אל הקו העובר על המרכז, ארוך מאשר רחק ממנו. והקצר שבקוים אשר יכלו אל מקיף העיגול ולא יבאו בתוכו, הוא הקו אשר בין הנקודה ובין קצה הקוטר. והקרוב אליו מן הקוים הנשארים קצר מאשר רחק ממנו. ושני קוים בלבד אצל צדי הקצר—שוים.

إذا وُضِعَتْ نُقْطَةٌ خارِجَةٌ مِنَ الدَّائِرَةِ[1]، ثُمَّ أُخْرِجَ مِنْها خُطوطٌ إِلى الدَّائِرَةِ، فَجازَ واحِدٌ مِنَ الخُطوطِ عَلى مَرْكَزِ الدَّائِرَةِ فَسائِرِ الخُطوطِ كَيْفَما وقعت فَأَمَّا ما وَقَعَ في داخِلِ الدَّائِرَةِ مِنَ الخُطوطِ. فَإِنَّ أَطْوَلَها الخَطُّ الَّذي يَجوزُ عَلى المَرْكَزِ، وَأَطْوَلَ الباقِيَةِ أَقْرَبُها إِلى خَطِّ المَرْكَزِ الأَوَّلِ، فَالأَوَّل كُلّ واحد أَطْوَل مِنَ الآخَرِ وأَمّا الخُطوط الَّتي تَقَعُ عَلى الخَطِّ المُحيطِ مِنْ خارِجٍ. فَإِنَّ أَقصَرَها الخَطُّ الَّذي بَيْنَ يَدَي النُّقطَةِ وَبَيْنَ قُطْرِ الدَّائِرَةِ. وَأَقْصَرُ الباقِيَةِ أَقْرَبُها إِلَيْهِ وما قَرُبَ إِلَيْه، فَهُوَ أَقْصَرُ مِمَّا هُوَ أَبْعَدُ مِنْهُ. وَخَطَّانِ فَقَط عَنْ جَنبَتَي الخَطِّ الأَقْصَرِ مُتَساوِيانِ.

[1] الدائرة] الخط+

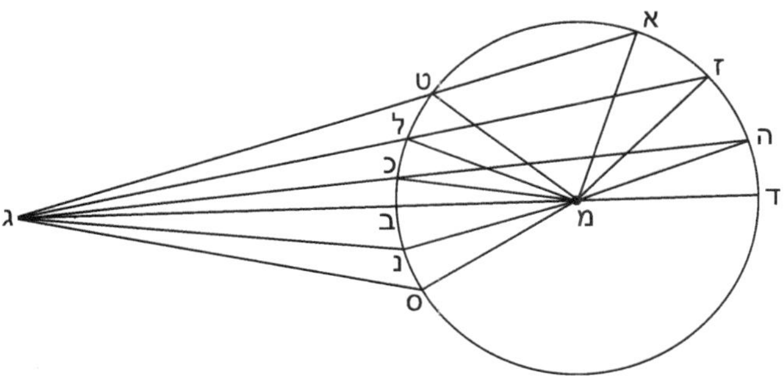

كُلُّ نُقْطَةٍ خارِجَةٍ مِنْ دائِرَةٍ، تَخْرُجُ مِنْها خُطوطٌ إلى الدَّائِرَةِ، فَإِنَّ أَطْوَلَ ما دَخَلَ الدَّائِرَةَ وَقَطَعَها مِنَ الخُطوطِ، هُوَ الَّذي يَجوزُ على المَرْكَزِ. وَما قَرُبَ مِنَ الخُطوطِ الباقِيَةِ مِنَ الخَطِّ الَّذي يَمُرُّ بالمَرْكَزِ، أطول مِمّا بَعُدَ. وأقصر الخُطوط الَّذي ينتهي إلى الدَّائِرَةِ ولا يَدْخُلُها، هُوَ الخَطُّ الَّذي بَيْنَ النُّقْطَةِ وَبَيْنَ طَرَفِ القُطْرِ. وَما قَرُبَ مِنْهُ مِنَ الخُطوطِ الباقِيَةِ، أقصر مِمّا بَعُدَ. وَخَطّانِ فَقَطْ عَنْ جَنْبَتَي الخَطِّ الأَقْصَرِ^—مُتَساوِيانِ.

כל נקודה יוצאת מעגולה[1] יצאו ממנה קוים אל העגולה, הנה היותר ארוך ממה שיכנס בעגולה ויחתכה מן הקוים, הוא אשר יעבור על המרכז. ומה שקרב מן הקוים הנשארים מן הקו אשר ילך על המרכז, יותר ארוך ממה שרחק. והיותר קצר שבקוים אשר יכלה אל העגולה ולא יכנס בה, הוא הקו אשר בין הנקודה ובין קצה[2] הקוטר. ומה שקרב ממנו מן[3] הקוים הנשארים, יותר קצר ממה שרחק. ושני קוים לבד משני צדדי הקוטר שוים[4].

שרטוט—**ס**: ב ← ה, ה ← ח; **בוטסק**: $; בכתבי היד הערביים ממשפחה A-1 ن ← ب, ובכל שאר כתבי היד ب ← ح, ובחלקם ن נכתבה בתור ب [1]מעגולה] חוץ+ **ס** [2]קצה] קצת **ב** [3]מן] אל **ט** [4]שוים] השוים **ט**

המשל בו: שנרשום נקודת ג׳ חוץ לעגול א״ב. ונוציא ממנה קוים אל העיגול, והם ג״ד וג״ה וג״ז וג״א. ואחד מן הקוים, והוא ג״ד, יעבר על המרכז. ואומר שהארוך שבהם הוא ג״ד. והקצר מהקוים שחוץ לעגול—ג״ח, שהוא בין הנקודה ובין קצה הקוטר. ואמנם הקוים הנשארים מאשר באו בתוך העגול—הקו שהוא קרוב אל הקוטר הוא ארוך[1] מהרחוק, קו ג״ה ארוך מג״ז, וג״ז ארוך מג״א. ואמנם אשר יכלו אל הקו המקיף שהם חוץ לעיגול—הקצר שבהם ג״ח. והנשארים—הקרוב אל ג״ח הוא הקצר מן הרחוק, קו ג״ב קצר מג״ל, וג״ל קצר מג״ט[2]. ושני קוים בלבד אצל צדי הקצר הם שוים.

مِثالُهُ: أَنَّا نَضَعُ خارِجًا مِنْ دائِرَةِ ا ب نُقْطَةً عَلَيْها ج. وَنُخْرِجُ مِنْها خُطوطًا إلى دائِرَةِ ا ب، وَهِيَ ج د ج ه ج ز ج (ج ا)[1]. فَأَحَدُ الخُطوطِ، وَهُوَ ج د، يَجوزُ عَلى المَرْكَزِ. فَأَقولُ إِنَّ خَطَّ (ج د)[2] هُوَ أَطْوَلُها. وَأَمَّا الخُطوطُ الباقِيَةُ فَإِنَّهُ ما وَقَعَ مِنْها في داخِلِ الدَّائِرَةِ. فَالخَطُّ الَّذي هُوَ أَقْرَبُ إِلى القُطْرِ هو أَطْوَلُ مِنْ غَيْرِهِ، نَخَطُّ ج ه أَطْوَلُ مِنْ خَطِّ ج ز، ج ز أَطْوَلُ مِنْ ج ا. وَأَمَّا الخُطوطُ الَّتي تَقَعُ عَلى الخَطِّ المُحيطِ مِنْ خارِجٍ، فَأَقصَرُها ج ح الَّذي يَقَعُ بَيْنَ نُقْطَةِ ج وبَيْنَ قُطْرِ الدَّائِرَةِ. وَالباقِيَةُ مِنْها فَأَقْرَبُها مِنَ الخَطِّ الأَقْصَرِ هُوَ أَقْصَرُ مِمَّا هُوَ أَبْعَدُ مِنْهُ. نَخَطُّ ج ك أَقْصَرُ مِنْ ج ل، ج ل أَقْصَرُ مِنْ ج ط. وَخَطَّانِ فَقَط عَنْ جَنْبَتَيِ الخَطِّ الأَقْصَرِ مُتَساوِيانِ.

[1]ארוך] הארוך [2]מג״ט] [...] ⟨*⟩

[1]ج ا] ا [2]ج د] ج ه

مِثَالُهُ: أَنَّهُ خَرَجَ مِنْ نُقْطَةِ ج إلى دائِرَةِ ا ب خُطوطٌ، وهِيَ ج د وج ه وج ز وج ا. والخطّ الَّذي يَجوزُ على الْمَرْكَزِ—ج د. فَأَقولُ إِنَّهُ أَطول الخُطوطِ الدّاخِلَةِ؛ وأَنَّ ج ه أطول مِنْ ج ز؛ وأَنّ ج ز أطول مِنْ ج ا؛ وأَنّ أقصر الخُطوط الخارجة هُوَ خطّ ج ح الَّذي بَيْنَ النُّقْطَةِ وبَيْنَ طَرَفِ القُطْرِ^؛ وأَنّ ج ك أقصر مِنْ ج ل، و ج ل أقصر مِنْ ج ط؛ وَخَطّانِ فَقَطْ عن جَنْبَيْ ج ح^ الأَقْصَر مُتَساوِيانِ.

דמיונו: כי הוצא מנקודת ג׳ אל[1] עגולת א״ב קוים, והם ג״ד וג״ה וג״ז וא״ג[2]. והקו אשר יעבור על המרכז—ג״ד. הנה אומר שהוא היותר ארוך שבקוים[3] הנכנסים; וכי ג״ה יותר ארוך מן ג״ז; וכי ג״ז יותר ארוך מן א״ג; וכי היותר קצר שבקוים היוצאים אשר בין הנקודה ובין קצה[4] הקוטר הוא ג״ב; וכי ג״כ יותר קצר מן[5] [6] ג״ל, וג״ל[7] יותר קצר מן ג״ט[8]; ושני קוים לבד משני צדדי ג״ב היותר קצר—שוים.

---

[1] אל] +תוך **ס** [2] וג״ז וא״ג] ⟨⟨*⟩⟩ **פ**
[3] שבקוים] מן **ס** [4] קצה] קצת **ב** [5] וכי ... מן] ⟨⟨וכי ג״ב יותר קצר מג״ד [..]⟩⟩ **ק** [6] מן] ⟨⟨מן ג״כ וכי ג״ב⟩⟩ יותר מן **ב** +ג״כ וכי ג״ב יותר קצר מן **ז** +ג״כ וכי ג״ב יותר קצר מן **ס** +וכי ג״ב יותר קצר מן **ק** [7] וג״ל] וכי ג״ל **ט**
[8] ג״ט] ~~ב״ט~~+ **ס**

ויהיה מרכז העגול מ׳. ונוציא ממ׳ קוים אל נקודות[1] ה׳ וז׳ וא׳ וט׳ ול׳ וכ׳, והם קוי מ״ה ומ״ז ומ״א ומ״ט ומ״ל ומ״כ. ושני קוי מ״ה ומ״ג יחד ארוך מקו ג״ה. ומ״ה שוה למ״ד. ואם כן ג״ד ארוך מג״ה. ומ״ה שוה למ״ז. ומ״ג משתף [!]. ואם כן קוי ה״מ[2] ומ״ג שוים לקוי[3] ז״מ ומ״ג, כל אחד לדומה לו. וזוית המ״ג גדולה מזוית זמ״ג. ואם כן תושבת ה״ג גדולה מתושבת ז״ג. וכמו כן נבאר שז״ג גדול מא״ג. ועוד מ״כ וכ״ג יחד גדול ממ״ג. ומ״כ שוה למ״ח. וישאר כ״ג[4] ארוך מח״ג. וח״ג קצר מכ״ג. ואם כן במשלש גל״מ עמדו על צלע מצלעיו, והוא מ״ג, שני קוים מ״כ וכ״ג. ואם כן מ״ל ול״ג ארוכים ממ״כ וכ״ג. אמנם מ״כ שוה למ״ל. ואם כן נשאר ל״ג ארוך מכ״ג. ולפי זה יהיה ט״ג ארוך מל״ג. והתבאר שהארוך שבקוים שיבואו בתוך העגול הוא ג״ד העובר על המרכז, והקטן שבחצונים ג״ח, שהוא בין הנקודה ובין קצה הקוטר. ואמנם האחרים מן הפנימיים—הקרוב מהם אל הקו העובר על המרכז הוא הארוך: ג״ה ארוך מג״ז, וג״ז ארוך מג״א. והקוים החיצונים—הקרוב מהם אל קו ג״ח [!], הקצר הוא הקצר מהרחוק אליו: כ״ג קצר מל״ג, ול״ג קצר מג״ט.

بُرْهانُهُ: أَنَّ مَرْكَزَ الدَّائِرَةِ م. وَقَد أُخْرِجَ مِنْ م خُطوطٌ إِلى نُقْطَةِ ه ز ا ط ل ك، وَهِيَ خُطوطُ م ه ز م م ا م ط م ل م ك. نَخَطّ م ه م ج مَجْموعانِ أَطْوَلُ مِنْ ج ه. م ه مِثْلُ (م د)[1]. د ج أَطْوَلُ مِنْ ج ه. م ه مِثْلُ ز م. م ج مُشْتَرَكٌ. فَكِلا خَطَّيْ (ز م)[2] م ج مِثْلُ كِلَيْ خَطَّيْ ز ه م ج، كُلٌّ واحِدٍ مثل نَظيرِهِ. وَالزَّاوِيَةُ الَّتي مِنْ ه م ج أَعْظَمُ مِنَ الَّتي مِنْ ز م ج. فَقاعِدَةُ ه ج أَطْوَلُ مِنْ قاعِدَةِ ز ج. و كَذَلِكَ ز ج أَطْوَلُ مِنْ ا ج. وَأَيْضًا م ك ك ج مَجْموعانِ أَطْوَلُ مِنْ م ج. م ك مثل م ح. وبَقِيَ (ك ج)[3] أَطْوَلُ مِنْ ح ج. ج ح أَقْصَرُ مِنْ ج ك. وَمُثَلَّثُ (ل ج م)[4] خَرَجَ مِنْ (ضِلْعٍ)[5] مِنْ أَضْلاعِهِ، وَهُوَ م ج، خَطّانِ م ك ك ج. م ل ل ج أَطْوَلُ مِنْ (م ك)[6] ك ج. فَأَمّا م ك فَمِثْلُ م ل. وَبَقِيَ ل ج أَطْوَلُ مِنْ ج ك. وَكَذَلِكَ ج ط أَطْوَلُ من ج ك. فَأَطْوَلُ الخُطوطِ الَّتي إِلى الدَّائِرَةِ هُوَ خَطُّ ج د الَّذي يَجوزُ عَلى المَرْكَزِ، وَأَقْصَرُ الخارِجَةِ خَطُّ ج ح الَّذي بَيْنَ النُّقْطَةِ وَبَيْنَ[7] القُطْرِ. وَأَمَّا الخُطوطُ الأُخَرُ (الَّتي)[8] مِنْ داخِلِ الدَّائِرَةِ فَإِنَّ الخَطّ الَّذي يَكونُ أَقْرَبُ إِلى خَطِّ المَرْكَزِ هو أَطْوَلُ: ج ه أَطْوَلُ مِنْ ج ز، ج ز أَطْوَلُ من ج ا. وَالخُطوطُ الخارِجَةُ مِنَ[9] الدَّائِرَةِ

[1] נקודות] נקוד⟨⟨ו⟩⟩ת [2] ה״מ] ~~ה[ם]~~ ⟨⟨ה⟩⟩״מ
[3] לקוי] ⟨⟨ל⟩⟩קוי [4] כ״ג] ~~ה״ג כד~~ ⟨⟨*⟩⟩

[1] م د] ز ج [2] ز م] ه م [3] ك ج] ك ح [4] ل ج م] ل د م [5] ضلع] ضلعي [6] م ك] ج ك [7] وبين] و⟨بين⟩
[8] التي] الذي [9] من] ⟨*⟩

بُرْهانُهُ: أَنَّا نَجْعَلُ الْمَرْكَزَ م. ونُخْرِجُ مِنْهُ خُطوطَ م ه و م ز و م ا و م ط و م ل و م ك. فَخَطَّا م ه و م ج مَجْموعَيْنِ، أطول مِنْ ج ه. و م ه مثل م د. ف د ج أطول مِنْ ج ه. و م ه مثل م ز. ونَجْعَلُ م ج مُشْتَرَكًا. فَخَطَّا ه م و م ج مثل خَطَّيْ ز م و م ج. وزاوية ه م ج أعظم مِنْ زاوية ز م ج. فَقاعِدَةُ ه ج أَطول مِنْ قاعِدَةِ ز ج. وكَذَلِكَ أيضًا تَبَيَّنَ أَنَّ ز ج أطول مِنْ ا ج. وَأَيْضًا فَإِنَّ م ك و ك ج أطول مِنْ م ج. و م ك مثل م ح^. فَيَبْقَى ك ج أطول مِنْ ج ح. ف ج ح أقصر مِنْ ك ج. وفي مُثَلَّثِ ل م ج ضِلْعانِ قائِمانِ على قاعِدَةِ م ج، وَهُما م ك و ك ج. وَقَدِ الْتَقَيا في داخِلِهِ. ف م ل و ل ج أطول مِنْ م ك و ك ج. فَأَمَّا م ك مثل^ م ل. ويَبْقَى ل ج أطول مِنْ ج ك. وكذَلِكَ أَيْضًا تَبَيَّنَ أَنَّ ج ط أطول مِنْ ج ل. فأَطْوَلُ الخُطوطِ الدَّاخِلَةِ—ج د. ثُمَّ ج ه أطول مِنْ ج ز. ثُمَّ ج ك أطول مِنْ ج ا. وأقصر الخارِجَةِ—ج ح^. ثم ج ك أقصر مِنْ ج ل. ثُمَّ ج ل أقصر مِنْ ج ط.

המופת: אנחנו נשים המרכז נקודת מ׳. ונוציא ממנה קוים מ״ה ומ״ז ומ״א ומ״ט ומ״ל ומ״כ. הנה שני קוי[1] מ״ה ומ״ג מקובצים[2], יותר ארוך מן ג״ה. ומ״ה כמו מ״ד. הנה[3] ד״ג יותר ארוך מן ג״ה. ומ״ה כמו מ״ז. ונשים מ״ג משותף[4]. הנה שני קוי ה״מ ומ״ג[5] כמו שני קוי[6] ז״מ ומ״ג[7]. וזוית המ״ג יותר גדולה מזוית זמ״ג. הנה תושבת ה״ג יותר ארוכה מתושבת ז״ג. וכן גם[8] כן התבאר כי ז״ג יותר ארוך מן א״ג. וגם[9] כן הנה מ״כ וכ״ג יותר ארוך מן מ״ג. ומ״כ כמו מ״ב. ונשאר ג״ב[10] יותר ארוך מן ג״ב. הנה ג״ב[11] יותר קצר[12] מן כ״ג. ובמשולש למ״ג שתי צלעות נצבות על תושבת מ״ג, והם מ״כ וכ״ג. והנה נפגשו בתוכו. הנה מ״ל ול״ג יותר ארוך מן מ״כ וכ״ג. ואולם מ״כ הוא כמו מ״ל. ונשאר ל״ג יותר ארוך מן ג״כ[13]. וכן גם כן התבאר[14] כי ג״ט יותר ארוך מן ג״ל. הנה היותר ארוך שבקוים הנכנסים—ג״ד. ואחר כן ג״ה יותר ארוך[15] מן ג״ז. ואחר[16] כן ג״ז יותר[17] ארוך מן א״ג. והיותר קצר שביוצאים—ג״ב. אחר[18] כן ג״כ יותר[19] קצר מן ג״ל. אחר כן[20] ג״ל יותר קצר מן ג״ט.

---

1 קוי] קוים **ב** +~~מ״ח~~ **ק** 2 מקובצים] +ומ״ג משותף **ס** 3 הנה] ⟨⟨*⟩⟩ **ק** 4 משותף] משותפים **טפ** 5 ה״מ ומ״ג] מ״ג וה״מ **ק** 6 קוי] ⟨⟨*⟩⟩ **ט** 7 ז״מ ומ״ג] ה״מ ומ״נ כמו שני קוי+ **ס** 8 וכן גם] וגם גם **פ** 9 וגם] וכן גם **ו** 10 ג״כ] [ז״כ] כן גם **ט** גם כן+ **ס** ⟨⟨*⟩⟩ **פ** 11 הנה ג״ב] ⟨*⟩ **ו** 12 קצר] ~~ארוך~~+ **ב** 13 ג״כ] מ״כ וכ״ג **ס** 14 וכן ... התבאר] ~~וכן גם כן התבאר~~ ואולם מ״כ כמו מ״ל ונשאר ל״ג יותר ארוך מג״ב+ **ס** 15 ג״ה ... ארוך] ג״ה ארוך **ס** היותר ארוך הוא ג״ה **ק** 16 ואחר] ואחרי **ט** 17 יותר] ליתא **ס** 18 אחר] ואחר **ס** 19 ג״כ יותר] ⟨*⟩ **ב** 20 כן] ⟨*⟩ **ט**

فَأَقْرَبُها مِنْ خَطِّ ج ح الأَقْصَرُ هُوَ أَقْصَرُ مِمّا هُوَ أَبْعَدُ مِنْهُ: ك ج أَقْصَرُ مِنْ ل ج، ل ج أَقْصَرُ مِنْ ج ط.

וכמו כן נבאר ששני קוים בלבד אצל צדי ג״ח הקצר שוים.

وَأَقولُ إِنَّهُ قَدْ وَقَعَ عَنْ جَنبَتَيْ خَطِّ ج ح الأَقْصَر خَطّانِ مُتَساوِيانِ فَقَط.

המופת בו: שנעמיד על קו מ״ג, על נקודת מ׳, זוית שוה לזוית כמ״ג, והוא [!] זוית נמ״ג[1]. ונוציא מנקודת ג׳ קו אל נ׳, והוא קו ג״נ. וכ״מ שוה למ״נ. ומ״ג משתף [!]. ואם כן שני קוי כ״מ ומ״ג שוים לשני קוי מ״נ ומ״ג, כל אחד לדומה לו. וזוית כמ״ג שוה לזוית נמ״ג. ואם כן תושבת כ״ג שוה לתושבת נ״ג. ואומר שאי איפשר שיצא מנקודת ג׳ קו אחר [!] שוה לכ״ג ונ״ג זולתם.

بُرهانُهُ: أَنّا نُقيمُ عَلى م ج عَلى نُقْطَةِ م مثل الزّاوِيَةِ الَّتي من ك م ج، وَهِيَ الَّتي من ن م ج. وَنُخْرِجُ مِنْ نُقْطَةِ ج خَطًّا إِلى ن، وَهُوَ خَطُّ ج ن. ك م مثل م ن. م ج مُشْتَرَكٌ. فَكِلا خَطَّيْ ك م م ج مثل كِلَيْ خَطَّيْ م ن م ج، كُلّ واحِدٍ مثل نَظيرِهِ. والزّاوِيَةُ الَّتي من ك م ج مثل الَّتي من (ن م ج)[1]. فَقاعِدَةُ ك ج مثل قاعِدَةِ ج ن. وَأَقولُ إِنَّهُ لا يَكونُ خَطٌّ آخَر مِثْلَهُما.

[1] כמ״ג] ~~אב״ג~~

[1] ن م ج] ن ج م

وَأَقولُ إِنَّ خَطَّيْنِ فَقَطْ عن جَنْبَتَيْ ج ح الأقصر مُتَساوِيانِ.

بُرْهانُهُ: أنّا نُقيمُ^ على نُقْطَةِ م مِنْ خَطِّ م ج زاوية ن م ج مثل زاوية ك م ج. ونُخْرِجُ ج ن. نَخُطّ ك م مثل خطّ م ن. ونَجْعَلُ م ج مُشْتَرَكًا. نَخُطّا ك م و م ج مثل خَطَّيْ م ن و م ج. وزاوية ك م ج مثل زاوية ن م ج. فَقاعِدَة ك ج مثل قاعِدَة ج ن. فَأَقولُ إنَّهُ لا يَخْرُجُ مِنْ ج خطٌّ آخَر مِثْل كُلِّ واحِدٍ مِنْ ك ج ن ج.

ואומר כי שני קוים לבד משני צדדי ג״ב היותר קצר שוים[1] [2].

המופת: שתעמיד[3] על נקודת מ׳ מקו מ״ג זוית נמ״ג כמו זוית כמ״ג. ונוציא ג״נ[4]. הנה קו כ״מ כמו קו מ״נ. ונשים ג״מ משותף. הנה שני קוי[5] כ״מ ומ״ג כמו שני קוי מ״נ ומ״ג. וזוית כמ״ג כמו זוית נמ״ג. אם כן תושבת כ״ג כמו תושבת ג״נ. הנה אומר[6] כי לא יצא מן ג׳ קו אחר כמו כל אחד מן כ״ג ונ״ג[7].

---

[1]ואומר ... שוים] ⟨⟨*⟩⟩ ק [2]שוים] הקוים+ ס [3]שתעמיד] נעמיד ב שנעמיד זק [4]ג״נ] ⟨⟨*⟩⟩ ב [5]קוי] ⟨⟨*⟩⟩ ב [6]אומר] +~~כי~~ ~~אומר~~ ז [7]ונ״ג] ⟨⟨*⟩⟩ ב

ואם היה אפשר, יהיה שוה להם ג״ס. ונוציא מנקודת מ׳ קו אל ס׳. וכ״מ שוה למ״ס. ומ״ג משתף. ואם כן שני קוי כ״מ ומ״ג שוים לקוי ס״מ ומ״ג, כל אחד לדומה לו. ותושבת ג״כ שוה לתושבת ג״ס. ואם כן זוית גמ״כ שוה לזוית גמ״ס. וזוית גמ״ס שוה לזוית גמ״נ—הגדולה שוה לקטנה, זה חלוף לא יתכן. ואם כן התבאר שאי אפשר שיצא מנקודת ג׳ קו שוה לכ״ג וקו ג״נ זולתם.

וזה מה שרצינו לבאר.

فَإِنْ كانَ يُمْكِنُ، فَلْيَكُنْ، وَهِيَ ج س. ونُخْرِجُ مِنْ نُقْطَةِ م خَطًّا إلى س. ك م مثل م س. م ج مُشْتَرَكٌ. فَكِلا خَطَّيْ ك م م ج مثل كِلَيْ خَطَّي س م م ج، كُلُّ واحِدٍ مثل نَظِيرِهِ. فَقاعِدَةُ ك ج مثل قاعِدَةِ ج س. وَالزَّاوِيَةُ الَّتِي من ج م ك مثل الزَّاوِيَةِ الَّتِي من ج م س. وَلَكِنَّ الَّتِي من ج م ك مثل الَّتِي من ج م ز[1]. فَالَّتِي من ج م ن مثل الَّتِي من ج م س—الكُبْرى مثل الصُّغْرى، هَذا خُلْفٌ لا يُمْكِنُ. فَقَدِ اسْتَبانَ أَنَّهُ لا يُمْكِنُ أَنْ نُخْرِجَ مِنْ نُقْطَةِ ج خَطًّ آخَر مثل خَطِّ ك ج وَخَطّ ج ن غَيْرُهُما.

وَذَلِكَ ما أَرَدْنا بَيانَهُ.

---

[1] ولكن التي من ج م ك مثل التي من ج م ز] ⟨*⟩

فَإِنْ أَمْكَنَ—فَلْيَخْرُجْ، وهو ج س. ونُخْرِجُ مِنْ م خطّ إلى س. نَخَطّ ك م م ج مثل خَطَّيْ م س و م ج^. ونَجْعَلُ م ج مُشْتَرَكًا. نَخَطّا ك م و م ج مثل خَطَّيْ س م و م ج. وقاعِدَة ك ج مثل قاعِدَة ج س. فزاوية ج م ك مثل زاوية ج م س. وَقَدْ كانَتْ زاوية ج م ك مثل زاوية ج م ن. فزاوية ج م س مثل زاوية ج م ن—الكُبرى مِثْل الصُّغرى، هذا خُلْفٌ. فَلَيْسَ يُمْكِنُ أَنْ يَخْرُجَ مِنْ ج^ خطٌّ آخَر مِثْل ك ج غير ج ن^.

שאם היה אפשר—יוצא, והוא ג״ס[1]. ונוציא מן מ׳ קו[2] אל ס׳. הנה קו כ״מ כמו[3] קו מ״ס. ונשים מ״ג[4] משותף. הנה שני קוי כ״מ ומ״ג כמו שני קוי ס״מ ומ״ג. ותושבת כ״ג כמו תושבת ג״ס. הנה זוית גמ״כ כמו זוית גמ״ס. וכבר היתה זוית גמ״כ כמו זוית גמ״נ. הנה זוית[5] גמ״ס כמו זוית גמ״נ—הגדולה כמו[6] הקטנה, זה שקר. הנה אי[7] אפשר שיצא קו אחר כמו כ״ג זולת[8] נ״ג. הנה שני קוי[9] כ״ג נ״ג אשר הם משני צדדי הקו היותר קצר לבד—שוים.

وذَلِكَ ما أَرَدْنا أَنْ نُبيِّنَ^.

וזה מה שרצינו[10].

[1] ג״ס] ג״כ **ט** [2] קו] ליתא **ו** [3] כמו] ⟨⟨*⟩⟩ **ב** [4] מ״ג] +~~מ״כ~~ **ו** [5] זוית] +~~גמ״נ כמו זוית~~ **ק** [6] כמו] ⟨*⟩ א״ד **ט** [7] הנה אי] אי **ו** הנה ⟨אי⟩ **ט** [8] זולת] ליתא **ב** [9] קוי] ליתא **ס** [10] שרצינו] +*̲ **ב**

[Proposition III.9]

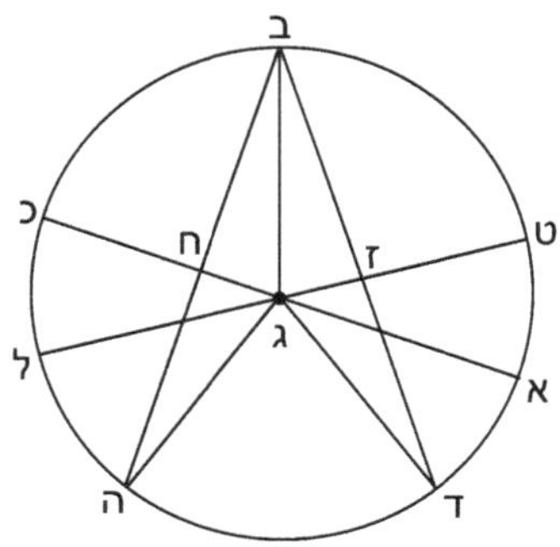

כשיצאו מנקודה שבתוך העגול קוים אל הקו המקיף, יותר משנים, והיו הקוים שוים, הנקודה היא מרכז העיגול.

إِذا وُضِعَتْ نُقْطَةٌ في داخِلِ الدَّائِرَةِ، وَأُخْرِجَ مِنْها خُطوطٌ أَكْثَرَ مِنْ خَطَّيْنِ إِلى الخَطِّ المُحِيطِ، فكانَتِ الخُطوطُ مُتَساوِيَةً، فالنُّقْطَةُ هِيَ مَرْكَزُ الدَّائِرَةِ.

המשל בו: שנרשום בעיגול א״ב נקודת ג׳. ונוציא ממנה קוים יותר משנים אל הקו המקיף, והם ג״ד וב״ג וג״ה. והם שוים. ואומר שג׳ הוא מרכז העיגול.

مِثالُهُ: أَنَّا نَضَعُ في دائِرَةِ اب النُّقْطَة عَلَيْها ج. وَنُخْرِجُ مِنْها خُطوطًا أَكْثَرَ مِنْ خَطَّيْنِ إِلى الخَطِّ المُحِيطِ، وَنَجْعَلُها مُتَساوِيَةً، وَهِيَ خُطوطُ ج د ج ب ج ه. فَأَقولُ إِنَّ ج مَرْكَزُ دائِرَةِ اب.

שרטוט—B: ו ← מ, ל ← ו, ז מונחת בנקודת חיתוך ג ט עם האנך הנצב עליו מנקודה ב, ה ו-ז (או ב?) הושלמו בדיו אחרת

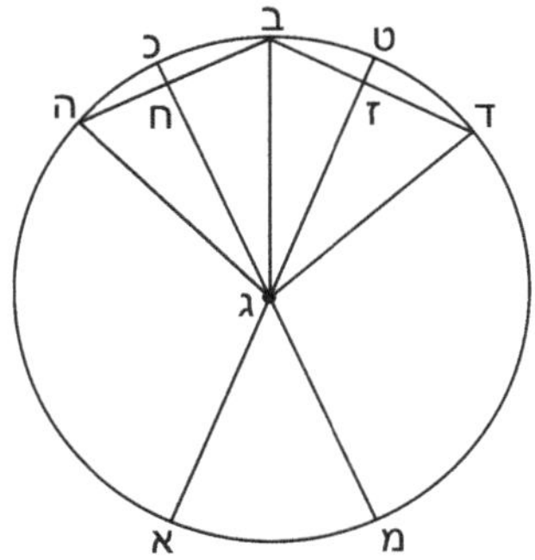

إذا أُخْرِجَ مِنْ نُقْطَةٍ في دائِرَةٍ أَكْثَرُ مِنْ خَطَّيْنِ إلى الخَطِّ المُحِيطِ، وكانَتِ الخُطُوطُ مُتَساوِيَةً، فَإِنَّ النُّقْطَةَ مَرْكَزُ الدَّائِرَةِ.

כאשר הוצא מנקודה[1] בעגולה יותר משני קוים אל הקו המקיף, והיו הקוים שוים, הנה הנקודה מרכז העגולה.

مِثالُهُ: أَنَّهُ خَرَجَ مِنْ نُقْطَةِ ج مِنْ دائِرَةِ ا ب خُطُوطٌ مُتَساوِيَةٌ، أَكْثر مِنْ خَطَّيْنِ، وَهِيَ ج د وج ب وج ه. فَأَقولُ إِنَّ نُقْطَةَ ج مَرْكَزُ دائِرَةِ ا ب.

המשל בו: שיוצא מנקודת[2] ג׳ מעגולת[3] א״ב קוים שוים[4], יותר משני קוים, והם ג״ד וג״ב וג״ה. הנה אומר כי נקודת ג׳ מרכז עגולת א״ב.

שרטוט ראשון—**ס**: ג׳ חסרה
שרטוט שני—**ס**: ה״ב וז״ב חסרים: **בוסק**: מובא שרטוט נוסף, שהרדיוסים בו מונחים באופן שונה ממה שהם מונחים בשרטוט כאן, וחסרים בו קווים ג״א וג״מ; השרטוט (הראשי) זהה לשרטוט בכתבי היד הערביים **CamDO2UST4**, ובכתבי היד האחרים ز ↔ ا   [1] מנקודה] ליתא **ו**   [2] מנקודת] מנקודה **ב**   [3] מעגולת] מעגולה **ב**   [4] שוים] +ישרים **ס**

מופת זה: שנדביק ד׳ בב׳, וב׳ בה׳. ונחלק קוי ד״ב וב״ה כל אחד מהם לחציים על ח׳ וז׳. ונדביק ז׳ בג׳, וח׳ בג׳[1]. ונוציא ג״ז וג״ח בשני הצדדים אל הקו המקיף. וד״ז שוה לז״ב. וז״ג משותף. ואם כן שני קוי ד״ז ז״ג שוים לשני קוי ז״ב וז״ג, כל אחד לדומה לו. ותושבת ג״ד שוה לתושבת ג״ב. ואם כן זוית דז״ג שוה לזוית בז״ג. וכאשר עמד קו על קו, והיו שתי הזויות אשר אצל צדי הקו שוות, הן נצבות. ואם כן זויות דז״ג ובז״ג נצבות. וכבר נפל בעגול א״ב קו ט״ל, וחלק מיתר[2] ד״ב לחציים על זויות נצבות. ואם כן מרכז העגול הוא בקו ל״ט. ולפי זה היה גם כן מרכז העגול על קו כ״א. ואם כן על החלוק המשותף לקוי ל״ט וכ״א נפל מרכז העגול. והחלוק המשותף היא הנקודה אשר עליה חתך כל אחד מקוי ט״ל וכ״א האחר, והיא נקודת ג׳. ואם כן ג׳ הוא מרכז העיגול.

וזה מה שרצינו[3].

بُرهانُه: أنّا نَصِلُ ب د ب ه. ونَقسِمُ خَطَّيْ د ب ب ه كُلّ واحِدٍ بِنِصفَيْنِ عَلَى ح ز. ونَصِلُ ز ج ح ج. ونُخْرِجُ ج ز ج ح في كِلْتَي الجِهَتَيْنِ إِلى الخَطِّ المُحيطِ بالدَّائِرَةِ. د ز مثل ز ب. ز ج مُشْتَرَكٌ. فكِلَيْ خَطَّيْ ب ز ز ج مثل كِلَيْ خَطَّيْ د ز ز ج، كُلّ واحِدٍ مثل نَظيرِه. وقاعِدَة ج د مثل قاعِدَة ج ب. فالزَّاوِيَة الَّتي من د ز ج مثل الَّتي من ب ز ج. وَإِذا قامَ خَطٌّ عَلَى خَطَّيْنِ فكانَت الزَّاوِيَتَيْنِ اللَّتانِ عن جَنَبَتَي الخَطّ مُتَساوِيَتَيْنِ، فَهُما قائِمَتانِ. فالزَّاوِيَتانِ اللَّتانِ من د ز ج ومن ب ز ج قائِمَتان. فَقَدْ وَقَعَ في دائِرَةِ ا ب خَطُّ ط ا وَقَطَعَ د ب بِنِصْفَيْنِ عَلَى زاوِيَتَيْنِ قائِمَتَيْنِ. فَمَرْكَزُ الدَّائِرَة عَلَى خَطِّ ا ط. وَكَذَلِكَ صارَ مَرْكَزُ الدَّائِرَةِ أَيضًا عَلَى خَطِّ ك م. فَعَلَى القَطْعِ المُشْتَرَكِ لِخَطَّيْ ط ا ك م وَقَعَ مَرْكَزُ الدَّائِرَة. وَالقَطْعُ المُشْتَرَكُ هِيَ النُّقطَةُ الَّتي عَلَيْها قَطَعَ خَطُّ ط ا ك م كُلّ واحِدٍ مِنْهُما الآخَرَ، وَهِيَ نُقْطَةُ ج. فَنُقْطَةُ ج مَرْكَزُ دائِرَةِ ا ب.

وَذلِكَ ما أَرَدْنا بَيانَهُ.

[1]וח׳ בג׳] ⟨⟨*⟩⟩ [2]מיתר] קו ⟨⟨*⟩⟩ [3]שרצינו] +⟨⟨תאבת מביא מופת עליה מצורת [ז׳] ובחלוף ודבר בטל והדבר פשוט כאשר תדמה נקודת ג׳ הנקודה שאינה על המרכז ויצאו ממנה קוים וכו׳ הכל כאשר תעיין מצורה ההיא צורת [ז׳]⟩⟩

بُرْهانُهُ: أنّا نُخرِجُ د ب و ب ه. وَنَقْسِمُ خَطَّيْ د ب وب ه بِنِصْفَيْنِ على ح ز. ونُخْرِجُ ج ز ج ح. ونُنفِذُهما في الجِهَتَيْنِ إلى الخطِّ المُحِيطِ. فخطّ د ز مثل خطّ ز ب. ونَجْعَلُ ز ج مُشْتَرَكًا. فخطّا د ز و ز ج مثل خَطَّيْ ب ز و ز ج. وقاعِدَة ج د مثل قاعِدَة ج ب. فزاوية د ز ج مثل زاوية ب ز ج. فَهُما إذن قائِمَتانِ. فَقَدْ قَطَعَ خَطُّ ا ط وَتَرَ ب د بِنِصْفَيْنِ على زاوِيَتَيْنِ قائِمَتَيْنِ. فَمَرْكَزُ الدّائِرَةِ على ا ط. وكذلك أيضًا^ تَبَيَّنَ أنَّ المَرْكَزَ على ك م. فالمَرْكَزُ يَقَعُ على القَطْعِ المُشْتَرَك لِخَطَّيْ ا ط وك م، فهو نُقْطَةُ ج.

וְذَلِكَ ما أَرَدْنا أَنْ نُبَيِّنَ.

המופת: אנחנו נוציא ד״ב וב״ה. ונחלק שני קוי[1] ד״ב וב״ה בשני חציים על ח׳ ז׳. ונוציא ג״ז וג״ח. ונעבירם בשני הצדדים אל הקו המקיף. הנה קו ד״ז כמו קו[2] ז״ב. ונשים ז״ג משותף. הנה שני קוי[3] ד״ז וז״ג כמו שני קוי ב״ז וז״ג. ותושבת ג״ד כמו תושבת ג״ב. אם כן זוית דז״ג כמו זוית בז״ג. הנה שניהם[4] אם כן שתי נצבות[5]. הנה כבר חתך קו א״ט[6] מיתר ב״ד בשני חציים על זויות נצבות. אם כן מרכז העגולה[7] על א״ט. וכן התבאר כי המרכז על כ״מ[8]. והנה המרכז יפול על החתך המשותף לשני הקוים א״ט וכ״מ, והוא נקודת ג׳.

וזה מה שרצינו לבאר.

[1]קוי] קוים **ב** [2]קו] ליתא **ק** [3]קוי] ⟨⟨*⟩⟩ **ט** [4]שניהם] שתיהם **ס** [5]נצבות] זויות+ **ו** [6]א״ט] +⟨⟨ד״ת קו ה״ט⟩⟩ **ק** [7]העגולה] +⟨⟨ד״ת על ה״ט⟩⟩ **ק** [8]כ״מ] כ״ד **ס** +⟨⟨ד״ת על כ״ד⟩⟩ **ק**

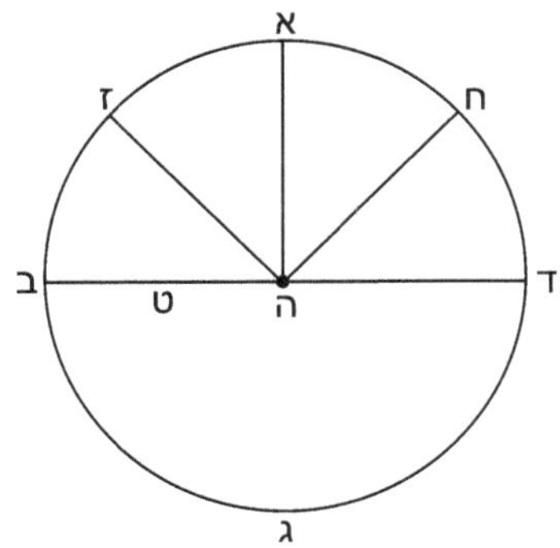

قَالَ ثَابِتٌ: وَجَدْنا في بَعْضِ النُّسَخِ اليونانِيَّة لِهَذا الشَّكلِ بُرْهانًا آخَر.

אמר תאבת: מצאנו בקצת הנוסחאות[1] מן היוניים לזאת התמונה מופת אחר.

[1] הנוסחאות] נסחאות ז

وهو: أنّا نجعل الدّائرة ا ب ج د. والنّقطة الّتي في داخلها نقطة ه. ونخرج من نقطة ه إلى الدّائرة خطوطًا متساويةً، وهي ه ز و ه ا و ه ح. فأقول إنّ نقطة ه مركز دائرة ا ب ج د. فإن لم يكن كذلك، فليكن مركزها نقطة ط، إن أمكن ذلك. ونصل خطّ ط ه. وننفذه في الجهتين إلى نقطتي د ب. فلأنّه قد تعلّم^ في داخل دائرة ا ب ج د نقطة كيفما وقعت، وهي ه، وأخرجت منها خطوط ه ب و ه ز و ه ا و ه ح إلى الدّائرة، كيفما وقعت، وخطّ ب ه منها يمرّ بالمركز— يكون خطّ ب ه أطولها، وخطّ ه د أقصرها، وخطّ ز ه أطول من خطّ ا ه، وخطّ ا ه أطول من خطّ ه ح. ولكنّ هذه الخطوط الثّلاثة قد كانت^ متساويةً. وهذا غير ممكن. فليست نقطة ط مركز دائرة ا ب ج د. وكذلك أيضًا يتبيّن أنّه لا يكون مركزها نقطة^ أخرى سوى نقطة ه. فنقطة ه مركز دائرة ا ب ج د.

وذلك ما أردنا أن نبيّن.

והוא: שאנחנו נשים העגולה אבג״ד. והנקודה אשר בתוכה נקודת ה׳. ונוציא מנקודת ה׳[1] אל העגולה קוים שוים[2], והם ה״ז וא״ה[3] וה״ח. הנה אומר כי נקודת ה׳ מרכז עגולת[4] אבג״ד. ואם לא יהיה כן, הנה יהיה מרכזה ט׳, אם[5] יהיה אפשר זה. ונגיע קו ט״ה. ונעבירהו בשני הצדדים אל שתי נקודות[6] ד׳ ב׳. הנה מפני שכבר הרשמנו[7] בתוך עגולת אבג״ד נקודה[8] איך מה שתיפול, והיא ה׳, והוצאו[9] ממנה קוים[10] ה״ב וה״ז[11] וא״ה וה״ח אל העגולה[12], איך מה שנפלו, וקו ב״ה מהם ילך במרכז— יהיה קו ב״ה היותר ארוך מהם, וקו היותר קצר מהם[13], וקו ז״ה יותר ארוך מקו א״ה, וקו א״ה יותר ארוך מקו ה״ח. אבל אלו הקוים השלשה שוים. וזה בלתי אפשר. אם כן אין[14] נקודת ט׳ מרכז לעגולת[15] אבג״ד. וכן גם כן התבאר שאי אפשר שיהיה מרכזה מנקודה[16] אחרת זולת נקודת ה׳. אם כן נקודת ה׳ מרכז עגולת אבג״ד.

וזה מה שרצינו לבאר[17].

---

[1] ה׳] ⟨*⟩ ס [2] שוים] +ה״ב ב [3] וא״ה] ⟨⟨*⟩⟩ ב [4] עגולת] העגולת ב עגולה פ [5] אם] +* ט [6] נקודות] הנקודות ו [7] הרשמנו] הרשמו בספ הרשם ו [8] נקודה] נקודת ב [9] והוצאו] והוציאו ב [10] קוים] קוי ק [11] וה״ז] ה״ז ק [12] העגולה] העגולת ב [13] היותר ... מהם] ⟨⟨*⟩⟩ ק [14] אין] ⟨⟨*⟩⟩ ב [15] לעגולת] עגולה ב [16] מנקודה] מנקודת ב [17] לבאר] לבארו ט

[Proposition III.10]

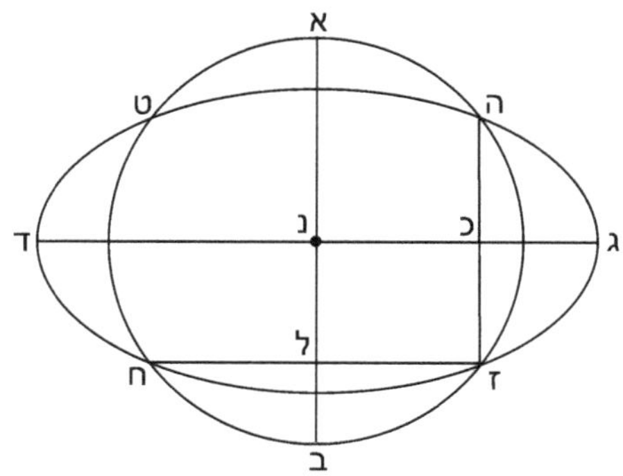

לא יתכן שיחלוק עגול לעגול אחר רק בשני מקומות.

لا تَقْطَعُ دائِرَةٌ دائِرَةً في أَكْثَرَ مِنْ مَوْضِعَيْنِ.

ואם היה אפשר, יחלוק עגול א״ב לעגול ג״ד ביותר משני מקומות על נקודת ה׳ וז׳ ח׳ וט׳. ונדביק ה׳ בז׳, וז׳ בח׳. ונחלוק קוי ה״ז וז״ח כל אחד לחציים על נקודת כ׳[1] ול׳. ונוציא מקוי ה״ז וז״ח, מנקודת כ׳ ול׳, שני קוים על זוית נצבה, והם כ״ד וא״ל. ונוציאם עד ב׳ וג׳.

مِثالُهُ: دائِرَتَيْنِ ا ب ج د المُتقاطِعَتانِ. أَقولُ إِنَّهُما لا يَتَقاطَعانِ في أَكْثَرَ مِنْ مَوْضِعَيْنِ.

بُرهانُهُ: أَنَّهُ لا يُمْكِنُ ذَلِكَ. فَإِنْ كانَ ذَلِكَ يُمْكِنُ، (فَلْتَقْطَعْ)[1] دائِرَةُ ا ب دائِرَةَ (ج د)[2] في أَكْثَرَ مِنْ مَوْضِعَيْنِ عَلى نُقْطَةِ ح ه ز (ط)[3]. وَنَصِلُ ه ز ز ح. وَنَقْسِمُ خَطَّيْ ه ز ز ح كُلَّ واحِدٍ بِنِصْفَيْنِ عَلى نُقْطَتَيْ ك ل. وَنُخْرِجُ مِنْ خَطَّيْ ه ز ز ح مِنْ نُقْطَتَيْ ك ل خَطَّيْنِ عَلى زَوايا [!] قائِمَة، وَهُما ك د ا ل. وَنُخْرِجُهُما إِلى ب ج.

שרטוט—B: ا ↔ ب, ו—ج د בכתיבה אחרת
[1] כ׳] ה׳+

[1] فلتقطع] فليقطع [2] ج د] ا ج د [3] ط] ליתא

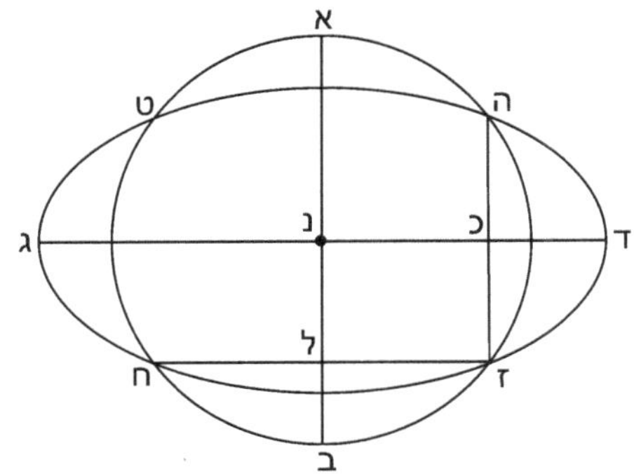

لا يُمْكِنُ أَنْ تقْطَعَ دائِرَةٌ دائِرَةً في أَكْثَر مِنْ مَوْضِعَيْنِ.

אי אפשר שתחתוך עגולה עגולה[1] ביותר משני[2] מקומות.

فَإِنْ أمْكَنَ، فلْتَقْطَعْ دائِرَةُ ا ب دائِرَةَ ج د في أكثر مِنْ موضِعَيْنِ على نُقْطَة ه ز ح ط. ونُخْرِجُ خَطَّيْ ه ز و ز ح. وَنَقْسِمهُما بِنِصْفَيْنِ على كَ لَ. ونُخْرِج ك د^ و ل ا على زاوِيتَيْنِ قائِمَتَيْنِ. ونُنْفِذهُما إلى ب ج.

שאם היה אפשר זה, הנה[3] תחתוך עגולת א״ב עגולת ג״ד[4] ביותר משני[5] מקומות על נקודת ה׳ ז׳ ח׳ ט׳. ונוציא שני קוי ה״ז וז״ח. ונחלקם בשני חציים על כ׳ ל׳. ונוציא כ״ג ול״א על זויות[6] נצבות. ונעביר שניהם אל ב׳ ד׳.

שרטוט ראשון—**פ**: נראה שנעשו שני נסיונות שרטוט בלתי מוצלחים שרטוט שני—**וט**: ה׳ חסרה; **פ**: ט ← ז, ד ← א, ה ← ג, ג ← ד, א ← ה; **ס**: ח׳ על רדיוס נוסף במעגל דה״ז, בין כ׳ לג׳, וה׳ חסרה; בכתבי היד הערביים האותיות מונחות במספר אופנים שונים זה מזה והשונים כולם מאופן הנחתן בתרגום אבן תיבון

[1] עגולה] ליתא **ב** [2] משני] משתי **בוטספ** [3] זה הנה] זההוה **ב** הנה **ט** [4] עגולת ג״ד] ליתא **ב** [5] משני] משתי **בוטספ** [6] זויות] זוית **ב**

וקו א״ב, שהוא (אלכסון)[1] לעגול א״ב, כבר חלק קו ז״ח לחציים על זויות נצבות. ואם כן מרכז עגול א״ב הוא על קו א״ב[2]. ועוד קו ג״ד, שהוא לעגול ג״ד, כבר חלק קו ז״ה לחציים על זויות נצבות. ואם כן מרכז עגול ג״ד הוא על קו ד״ג. וכבר התבאר שמרכז עגול א״ב הוא על קו א״ב. ומצד זה היה מרכז שני העגולים על קו ג״ד על הנקודה שעליה יחלוק כל אחד משני הקוים האחר, והיא נקודת נ׳. ואם כן נקודת נ׳ היא מרכז שני עגולי א״ב וג״ד. ונמצאו שני עגולים יחתוך אחד מהם האחר, ומרכזם על נקודה אחת—זה חלוף לא יתכן.

ואם כן לא יחלוק עגול לעגול אחר ביותר משני מקומות.

וזה מה שרצינו לבאר.

فَخَطُّ اب قَدْ قَطَعَ زح بِنِصْفَيْنِ عَلى زاوِيَتَيْنِ قائِمَتَيْنِ. فَمَرْكَزُ دائِرَةِ اب عَلى خَطِّ اب. وَلِمِثْلِ ذَلِكَ مَرْكَزُ دائِرَةِ ج د عَلى خَطِّ اب. وَأَيضًا خَطُّ ج د قَدْ قَطَعَ خَطَّ ه ز بِنِصْفَيْنِ عَلى زاوِيَتَيْنِ قائِمَتَيْنِ. فَمَرْكَزُ دائِرَةِ اب أَيضًا عَلى خَطِّ د ج. وَبِمِثْلِ ذَلِكَ مَرْكَزُ دائِرَةِ ج د عَلى خَطِّ ج د. فَقَدِ اسْتَبانَ أَنَّ مَرْكَزَيْ دائِرَتَيْ اب ج د عَلى خَطِّ اب. فَعَلى النُّقطَةِ الَّتي عَلَيْها (يَقْطَعُ)[1] أَحَدُ الخَطَّيْنِ الآخَرَ، وَهِيَ نُقْطَةُ نَ، مَرْكَزُ دائِرَةِ اب (ج د)[2]. فَدائِرَتانِ تَقْطَعُ إِحْداهُما الأُخرى يَكونُ مَرْكَزاهُما عَلى نُقْطَةٍ واحِدَةٍ—هذا خُلْفٌ لا يُمْكِنُ.

فَلَيْسَ تَقْطَعُ دائِرَةٌ دائِرَةً إِلّا في مَوْضِعَيْنِ فَقَط.

وَذلِكَ ما أَرَدْنا بَيانَهُ.

[1]אלכסון] ליתא [2]א״ב] +~~ועוד, קו ג״ד שהוא לעגול ג״ד, כבר חלק קו ז״ה לחציים על זויות נצבות. ואם כן מרכז עגול א״ד הוא על קו א״ב~~

[1]يقطع] نقطع [2]ج د] ج ز ا

فَخَطّ ا ب في دائِرَة ا ب قد قَطَعَ زح بِنِصْفَيْنِ على زاوِيةٍ قائِمَة. فمَرْكَزُ الدَّائِرَة على ا ب. وخطّ ج د أيضًا في دائِرَة ا ب قَطَعَ زه بِنِصْفَيْنِ على زاوِيةٍ قائِمَة. فمَرْكَزُ دائِرَة ا ب على ج د. وَقَدْ تَبَيَّنَ أنَّ مَرْكَزَ دائِرَة ا ب على خطّ ا ب. فمَرْكَزُها على النُّقْطَةِ المُشْتَرَكَة لِخَطَّيْ ا ب و ج د. ولَيْسَ يَشْتَرِكانِ في نُقْطَةٍ سِوى نُقْطَة ن. فنُقْطَةُ ن مَرْكَزُ دائِرَة ا ب. وبِمِثْلِ ذَلِكَ تَبَيَّنَ أنَّ مَرْكَزَ دائِرَة ج د نُقْطَة ن. فنُقْطَةُ ن هي مَرْكَزُ دائِرَتَيْ ا ب ج د. وَهُما يَتَقاطَعانِ، يكونُ مَرْكَزاهُما واحِدًا—هذا خُلْفٌ[1].

הנה קו א״ב בעגולת א״ב כבר חתך ז״ח בשני חציים על זוית נצבת. אם כן מרכז העגולה על א״ב. וג״ד גם כן בעגולת[1] א״ב חתך ז״ה בשני חציים על זוית נצבת. אם כן מרכז עגולת א״ב על ג״ד. וכבר התבאר כי מרכז עגולת א״ב על קו א״ב[2]. אם כן מרכזה על הנקודה המשותפת לשני קוי א״ב וג״ד. ואינם משותפים בנקודה זולת נקודת נ׳. אם כן נקודת[3] נ׳ היא מרכז עגולת א״ב. וכמו כן התבאר שמרכז עגולת ג״ד היא נקודת[4] נ׳. הנה נקודת נ׳ מרכז שתי עגולות א״ב וג״ד. ושתיהן[5] תחתכנה[6]. אם כן אין מרכז שתיהן[7] אחד.

فَلَيْسَ يُمْكِنُ ^ أَنْ تَقْطَعَ دائِرَةً دائِرَةً على أكثرِ مِنْ نُقْطَتَيْنِ.

אם כן לא תחתוך עגולה עגולה[8] ביותר משתי מקומות[9] [10].

[1] يكون ... خلف] MPR1R2

[1] בעגולת] על⟨ב⟩עגולת **פ** [2] אם ... א״ב] ⟨⟨*⟩⟩ **ק** [3] נקודת] מרכז+ **ו** [4] נקודת] +מרכז **ז** [5] ושתיהן] ושתיהם **בוטס** שתיהן+ **ק** [6] תחתכנה] תתחכנה **ב** תתחתכנה **ז** [7] שתיהן] שניהם **ב** שתיהם **וטפ** [8] עגולה] ליתא **ב** [9] מקומות] +וזה מה שרצינו לבאר **וט** [10] התבאר ... מקומות] ⟨⟨*⟩⟩ **ס**

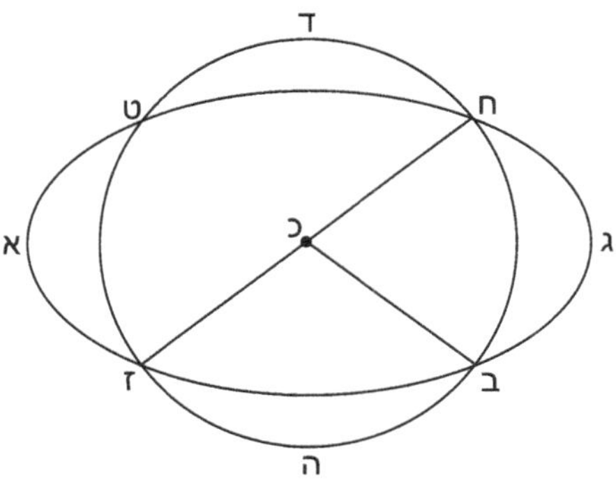

قالَ ثابتٌ: وَجَدْنا في بَعْضِ النُّسَخِ اليونانيّةِ لِهذا الشَّكْلِ بُرهانًا آخَرَ.

وهو أنّا نَجْعَلُ دائِرَةَ ا ب ج تَقْطَعُ دائِرَةَ د ه ز على أكثرَ مِنْ نُقْطَتَيْنِ، وهي نُقَطُ ح ب ز ط. وَلْيَكُنْ مَرْكَزُ دائِرَةِ ا ب ج نُقْطَةَ كَ. وَنَصِلُ خُطوطَ ب كَ وح كَ و ز كَ. فَلِأنَّهُ قَدْ تَعَلَّمَتْ نُقْطَةٌ على دائِرَةِ د ه ز، وهي نُقْطَةُ كَ^، وأُخْرِجَتْ مِنْها خُطوطٌ^ إلى دائِرَةِ د ه ز أكثرَ مِنْ خَطَّيْنِ، وَكانَتْ مُتَساوِيَةً، وَهِيَ خُطوطُ كَ بَ و كَ زَ و كَ حَ، فنُقْطَةُ كَ مَرْكَزُ دائِرَةِ د ه ز. وَقَدْ كانَتْ أيضًا مَرْكَزَ دائِرَةِ ا ب ج. فقد تَقاطَعَتْ دائِرَتانِ وَصارَ مَرْكَزاهُما واحِدًا، وَهُوَ نُقْطَةُ كَ. وذلِكَ غيرُ مُمْكِنٍ. فَلَيْسَ تَقْطَعُ دائِرَةٌ دائِرَةً على أكثرَ مِنْ نُقْطَتَيْنِ.

وذلِكَ ما أَرَدْنا أَنْ نُبَيِّنَ.

אמר תאבת[1]: מצאנו בקצת הנסחאות מהיונים לתמונה הזאת מופת אחר.

והוא, שאנחנו נשים עגולת אב״ג[2] תחתוך עגולת דה״ז על יותר משתי נקודות, והם נקודות ח׳ ב׳ ז׳ ט׳. ויהיה מרכז עגולת אב״ג נקודת כ׳. ונגיע קוי ב״כ וח״כ וז״כ. הנה מפני כי כבר רשמת[3] נקודה[4] בעגולת דה״ז, והוצאת ממנה אל עגולת דה״ז יותר משני קוים, והיו שוים, והם קוי[5] כ״ב וכ״ז[6] וכ״ח, הנה נקודת כ׳ מרכז עגולת דה״ז. וכבר היתה גם כן מרכז עגולת אב״ג. הנה כבר נחתכו שתי עגולות והיה[7] מרכזי[8] שתיהן אחד, והוא[9] נקודת כ׳[10]. וזה בלתי אפשר. הנה לא תחתוך[11] עגולה עגולה ביותר משני[12] מקומות.

~~וזה מה שרצינו ביאורו~~[13].

[1]תאבת] תבאת **ז** [2]אב״ג] אבג״ד **בוס** אבג״ד **פ** [3]רשמת] הרשמת **בוס** [4]נקודה] נקודת **בטפ** נקודות **ז** [5]קוי] קוים **ב** [6]וכ״ז] ליתא **ב** [7]והיה] והיו **ט** [8]מרכזי] מרכזם **ק** [9]והוא] והיא **ס** [10]כ׳] ⟨*⟩ **פ** [11]תחתוך] יחתוך **ז** [12]משני] משתי **בוטספ** [13]ביאורו] לבאר **טס**

[Proposition III.11]

[H11–12]

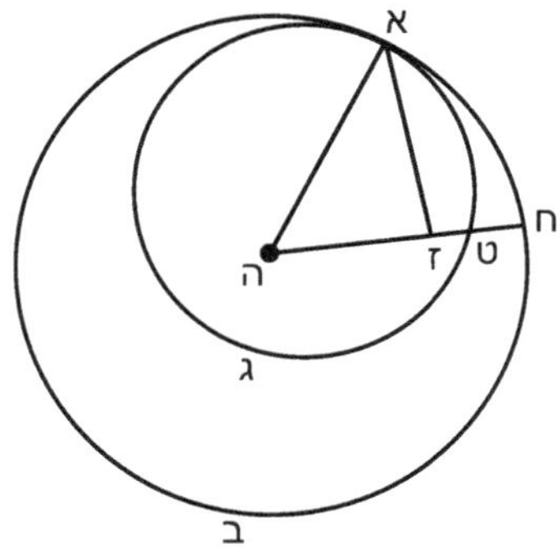

כל שני עגולים שימשש זה את זה, יהיה הקו הישר[1] שיעבר על מרכזיהם הולך ביושר[2] על מקום המשוש, יהיה המשוש מבפנים או מחוץ.

كُلُّ دائِرَتَيْنِ مُتَماسَّتَيْنِ، فَإِنَّ الخَطَّ الَّذي يَجوزُ عَلَى مَرْكَزَيهِما يَمُرُّ عَلى النُّقطَةِ الَّتي عَلَيْها تَماسُّ إِحدى الدَّائِرَتَيْنِ الأُخْرى.

[1] הישר] ⟨⟨*⟩⟩ [2] ביושר] ⟨⟨*⟩⟩

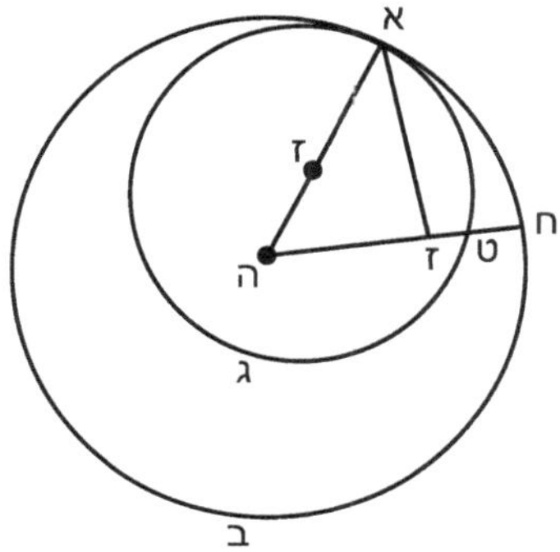

كُلُّ دائِرَتَيْنِ مُتَماسَّتَيْنِ، فَإِنَّ الخطّ الَّذي يَجوزُ على مَرْكَزَيهِما—يَمُرُّ بِمَوْضِعِ التَّماسِّ، مِنْ داخِلٍ تَماسَّا أو مِنْ خارِجٍ[1].

כל שתי עגולות תתמששנה[1], הנה הקו אשר יעבור על מרכזי[2] שתיהן[3]—יעבור במקום המשוש, מבפנים היה משושם או מחוץ[4].

שרטוט ראשון—**בק**: $; **פ**: במקום ז״א הוצא קו מהנקודה ז׳ שעל א״ה אל ט׳, וח׳ נמצאת על המשך ז״ט   שרטוט שני—**ט**: ח׳ ← ד׳; **ס**: ה׳ חסרה, המקום של ח׳ אינו ברור, והמעגלים אינם משיקים; מהשרטוטים בכתבי היד הערביים רק זה בכתב יד T3 זהה לזה בתרגום אבן תיבון, ובשאר כתבי היד אין אחידות במבנה השרטוט ובאותיות המונחות בו   [1]תתמששנה] תמששנה **ס**   [2]מרכזי] מרכז **וס**   [3]שתיהן] שתיהם **בטפ** שניהם **ו** שתיהם **ס**   [4]מחוץ] מבחוץ **וט**

---

[1]من ... خارج] **CamMPR$_1$R$_2$U**

המשל בזה: שעגול א״ב וא״ג[1] מושש זה את זה על נקודת א׳. ומרכז עגול א״ג הוא ז׳. ומרכז עגול א״ב הוא ה׳. ואומר שהקו שיעבר על ה׳ וז׳ יפל על א׳ שהוא מקום המשוש.

مثالُهُ: أَنَّ دائِرَتَيْ ا ب ا ج (تَماسّانِ)[1] عَلى نُقْطَةِ ا الَّتي عَلَيْها تَماسُّ إِحْدى الدائِرَتَيْنِ الأُخْرى.

ואי אפשר אלא זה. ואם היה אפשר, יפול כמו [..][2]. ונדביק א׳ בז׳ וא׳ בה׳. ויהיו א״ז וז״ה יחד ארוך מה״א. וא״ז שוה לז״ט, לפי שז׳ מרכז עגול א״ג. ואם כן ה״ט ארוך מה״א. וה״א שוה לה״ח, לפי שה׳ מרכז עגול א״ב. ואם כן ה״ט ארוך מה״ח—זה חלוף לא יתכן. ואם כן התבאר שהקו שידביק ה׳ בז׳ ויעבר עליהם, אין מוצאו כמו ה״ח, ולא יפול במקום אחר, כי אם על נקודת א׳ אשר עליה משוש שני העגולים.

بُرهانُهُ: أَنَّهُ لا يُمْكِنُ إِلّا ذلِكَ. فَإِنْ كانَ يُمْكِنُ أَنْ يَقَعَ عَلى غَيْرِ ذَلِكَ، فَلْيَقَعْ مثل ه ز ط ح. وَنَصِلُ ا ز ا ه. ا ز ز ه أَطْوَلُ من ه ا. ا ز مثل ز ط، لأَنَّ ز هو مَرْكَزُ دائِرَةِ ا ج. ه ط أَطْوَلُ من ه ا. ه ا مثل ه ح، لأَنَّ ه مَرْكَزُ دائِرَةِ ا ب. ه ط أَطْوَلُ من ه ح، الأَقْصَرُ أَطْوَلُ—هذا خُلْفٌ لا يُمْكِنُ. فَقَدِ اسْتَبانَ أَنَّ الخَطَّ الَّذي يَصِلُ بَيْنَ المَرْكَزَيْنِ وَيَجوزُ عَلَيْهِما لَيْسَ خُروجُهُ مثل ه ح، ولا (يَقَعُ)[2] في مَوْضِعٍ آخَرَ إِلّا عَلى نُقْطَةِ ا الَّتي عَلَيْها تَماسُّ الدائِرَتَيْنِ.

وَذلِكَ ما أَرَدْنا بَيانَهُ.

---

[1]וא״ג] ~~וג״ד~~ [2]...] ~~[..ט״ח]~~ ⟨⟨[..ט״..ח]⟩⟩

[1]تماسّان] يماسان [2]يقع] يقطع

مِثَالُهُ: أَنَّ دائِرَةَ اب تماسُّها دائِرَةَ اج^ على نُقْطَةِ ا. وَلْيَكُنْ مَرْكَزُ دائِرَةِ اب—ه، ومَرْكَزُ دائِرَةِ اج—ز. فَأَقولُ إِنَّ الخطّ الَّذي يَجوزُ على ه ز يَنْتَهي إلى نُقْطَةِ ا، لا يُمْكِنُ غيره.

فَإِنْ أَمْكَنَ، فَلْيَقَعْ مِثْل ه ز ط ح. ونُخْرِج خَطَّيْ ا ز ا ه. نَخَطّا ا ز و ز ه مَجْمُوعَيْنِ أطول مِنْ ا ه. وا ز مثل ز ط. ف ه ط أطول مِنْ ا ه. و ا ه مثل ا ح. ف ه ط أطول مِنْ ه ح—هذا خُلْفٌ. فَقَدِ اسْتَبَانَ أَنَّ الخطّ الَّذي يَجوزُ على نُقْطَتَيْ ه ز لَيْسَ خُروجه مِثْل خُروجِ ه ح. ولا يَقَعُ في مَوْضِعٍ آخر إلّا في نُقْطَةِ ا حَيْثُ تَتَمَاسُّ الدَّائِرَتانِ.

המשל בו: כי עגולת א״ב תמששה עגולת א״ג[1] על נקודת א׳. ויהיה מרכז עגולת א״ב—ה׳. ומרכז עגולת א״ג—ז׳. הנה אומר כי הקו אשר יעבור על ה׳ ז׳ יכלה אל נקודת א׳[2], אי אפשר זולתו.

שאם היה אפשר, הנה יפול כמו ה״ז ט״ח. ונוציא שני קוי א״ז וא״ה. הנה שני קוי א״ז וז״ה מקובצים יותר ארוך מן א״ה. וא״ז כמו ז״ט. אם כן ה״ט יותר ארוך מן א״ה. וא״ה כמו ה״ח. אם כן[3] ה״ט יותר ארוך[4] מן ה״ח—זה שקר. הנה כבר התבאר כי הקו אשר יעבור על שתי נקודות ה׳ ז׳ אין יציאתו כמו יציאת ה״ח. ולא יפול במקום אחר כי אם בנקודת[5] א׳ כאשר תתמששנה[6] שתי העגולות.

---

[1] א״ג] א״ב **ס** [2] א׳] +ויהיה מרכז עגול [אב״ה] **ס** [3] כן] +* **פ** [4] מן ... ארוך] ⟨⟨*⟩⟩ **ק** [5] בנקודת] בנקודה **ס** [6] תתמששנה] תמששנה **וס**

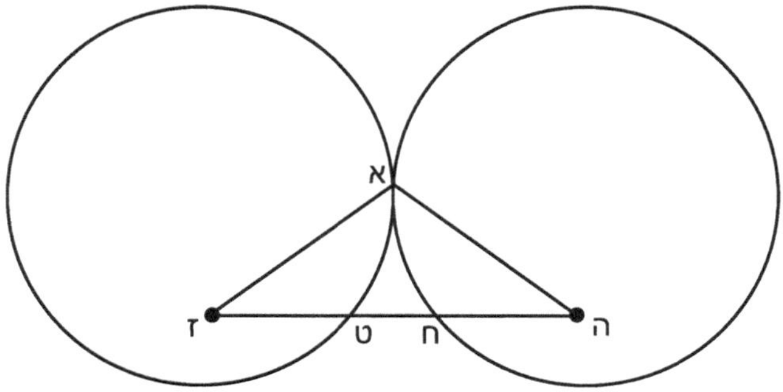

לא נזכר בהעתקת זה האיש אל חג׳ג׳ משוש העיגולים מחוץ. ודרך המופת בהם אחד. ולשון בן סינא: ״ויהיה המשוש מחוץ והמדבק שני המרכזים החט״ז. ויהיה ה״א א״ז יחד ארוך מה״ז. ואולם א״ה כמו ה״ת. וא״ז הוא כמו ז״ט. אם כן כל קו ז״ה ארוך מא״ז א״ה. וכבר היה קצר—זה שקר[1], רצו׳ לומ׳ ה״ח ז״ט יחד ארוך מה״ח ט״ז—זה חלוף לא יתכן.״

וזה מה שרצינו לבאר.

[1] ארוך מה״ז ... שקר] ⟨⟨ד״ת *⟩⟩

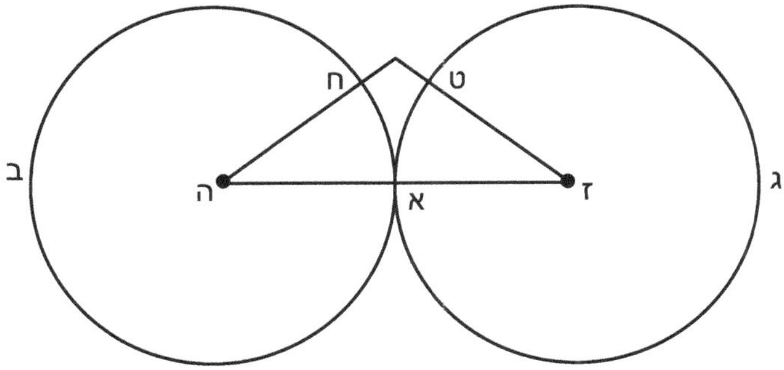

وَأَيْضًا فَلْنَجْعَلْ دائِرَةَ اَجَ مُماسَّةً لِدائِرَةِ اَبَ مِنْ خارِجٍ على نُقْطَةِ اَ. وَلْيَكُنْ مَرْكَزُ دائِرَةِ اَبَ—ﻩَ، ومَرْكَزُ دائِرَةِ اَجَ—زَ. فَأَقولُ إِنَّ الخطّ الَّذي يَجوزُ على ﻩَ زَ يمرُّ [في] نُقْطَةِ اَ.

فَإِنْ لم يَكُنْ كَذَلِكَ، فَلْيَقَعْ مثل خطّ ﻩ ح ط ز. ونُخْرِجُ خَطَّيْ اَزَ و اَﻩَ. نَخَطّا اَزَ و اَﻩَ مَجْموعَيْنِ أطول مِنْ خطّ زَﻩَ^. فَأَمَّا اَﻩَ فَهُوَ مثل ﻩ ح. وَأَمَّا اَزَ فَهُوَ مثل زَطَ. نَخَطّا ﻩ ح و زَطَ مَجْموعَيْنِ أطول مِنْ خطّ ﻩَزَ^. وَهُما أصغر مِنْهُ—هذا خُلْفٌ^. فَالخطّ الَّذي يَصِلُ بَيْنَ نُقْطَتَيْ ﻩَ زَ يَمُرُّ بِنُقْطَةِ اَ.

وذَلِكَ ما أَرَدْنا أَنْ نُبَيِّنَ.

וגם כן הנה אנחנו נשים עגולת א״ג תמשש עגולת א״ב מחוץ על נקודת א׳. ויהיה מרכז עגולת א״ב—ה׳, ומרכז עגולת א״ג—ז׳. הנה אומר כי הקו אשר יעבור על ה׳ ז׳ יעבור בנקודת א׳.

ואם לא יהיה כן[1], הנה יפול כמו קו החט״ז[2]. ונוציא שני קוי א״ז וא״ה. הנה שני[3] קוי א״ז וא״ה מקובצים, יותר ארוך מקו זח״ה הישר[4]. ואולם א״ה[5] הנה הוא כמו ה״ח. ואולם א״ז הנה הוא כמו ז״ט. אם כן שני קוי ה״ח וז״ט מחוברים יותר ארוך מקו הח״ז. ושניהם יותר קטנים ממנו—זה שקר אי אפשר. הנה הקו אשר יגיע[6] בין שתי נקודות ה׳ ז׳ יעבור בנקודת א׳.

וזה מה שרצינו לבאר.

1 כן] ליתא **ב** 2 החט״ז] הט״ז **בופק** 3 שני] שניה~~ם~~ **ט** 4 הישר] ליתא **ו** 5 ואולם א״ה] ⟨⟨*⟩⟩ **ק** 6 יגיע] יגיעו **ס**

[Proposition III.12]

[H13]

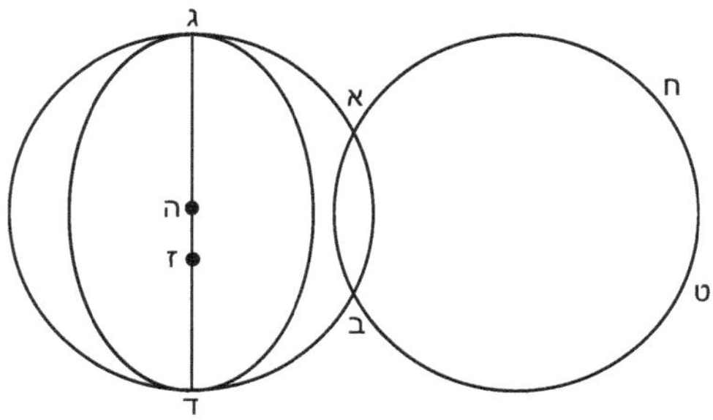

לא ימשש עגול לעגול אחר כי אם במקום אחד—יהיה אחד מהם בתוך האחר, או חוץ ממנו.

لا تَماسُّ دائِرَةٌ دائِرَةً في مَوْضِعَيْنِ وَلا في أَكْثَرَ، وَلٰكِنْ تَماسُّهُما في مَوْضِعٍ واحِدٍ كانَتْ داخِلَ الدّائِرَةِ أَو خارِجَةً مِنْها.

שרטוט—B: הנחת האותיות וקו ג״ד שונות מאלו בתרגום ר״י, כדלקמן:

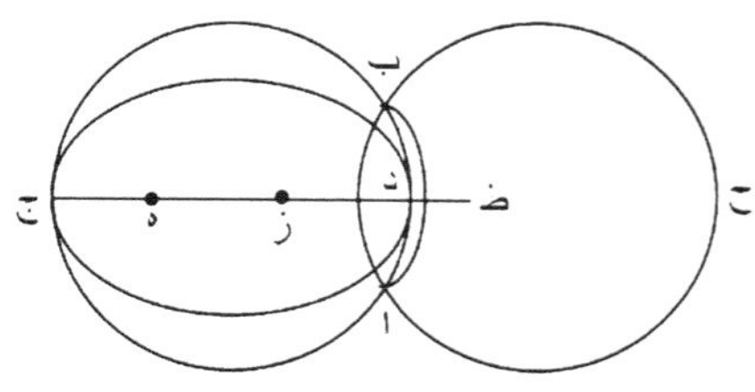

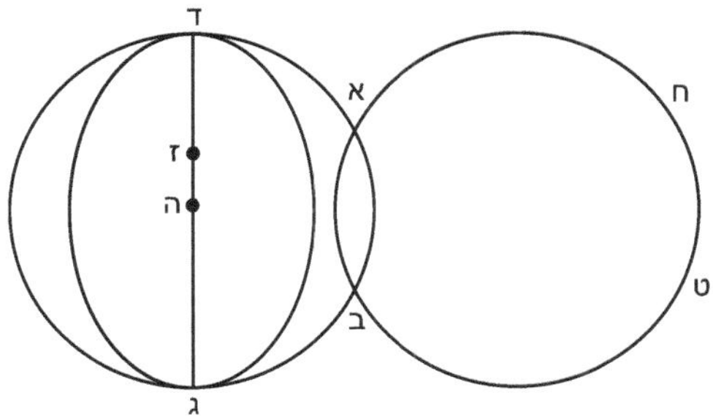

لا تَماسُّ دائِرَةٌ دائِرَةً إلّا في مَوْضِعٍ واحِدٍ، إنْ كانَتْ إحْداهُما داخِلَةً في الأُخْرَى أَو خارِجَةً مِنْها.

לא תמשש עגולה עגולה[1] כי אם במקום אחד, כאשר תהיה אחת משתיהן[2] תוך האחרת או חוץ ממנה.

שרטוט—**בוטספק**: $; ז: ה׳ על קשת ג״ד הימנית, ד׳ על קשת ג״ד השמאלית; יש שונות רבה בשרטוטים בכתבי היד הערביים מצד המבנה ומצד הנחת האותיות, והשרטוט בכתב יד **O2** קרוב מכולם לשרטוט בתרגום אבן תיבון ובו | ← ב [1]עגולה] ליתא **ב** [2]משתיהן] משתיהם **ט**

ואם היה אפשר, ימשש עגול ג״ד לעגול א״ב בשני מקומות מבפנים על נקודת ג׳ וד׳. ומרכז עגול א״ב—ה׳, ומרכז עגול ג״ד—ז׳. והקו שידביק ה׳ בז׳ ויצא עד הקו המקיף יפול על מקומות המשושים. ונדביק ה׳ בז׳. ונוציא הקו על מקומות המשושים[1].

مِثالُهُ: دائِرَتا ا ب ج د المُتَماسَّتانِ. أَقولُ إِنَّهُما لا تَتَماسّانِ في أَكْثَرَ مِنْ مَوْضِعٍ واحِدٍ.

بُرهانُهُ: أَنَّهُ لا يُمْكِنُ إِلّا ذلِكَ. فَإِنْ كانَ يُمْكِنُ، فَلْتَتَماسُّ دائِرَةُ ج د دائِرَةَ ا ب في مَوْضِعٍ مِنْ داخِلٍ عَلَى نُقْطَتَيْ ج د. وَمَرْكَزُ دائِرَةِ ا ب—ه. وَمَرْكَزُ دائِرَةِ ج د—ز. وَالخَطّ الَّذي يَصِلُ ه ز يَخْرُجُ إِلى الخَطِّ المُحيطِ وَيَقَعُ عَلَى المُرَبَّعِ الَّذي فيهِ تَتَماسُّ الدّائِرَتَيْنِ. فَنَصِلُ ه ز، وَنُخْرِجُ الخَطَّ إِلى النُّقْطَتَيْنِ اللَّتَيْنِ عَلَيْهِما تَتَماسُّ الدّائِرَتانِ.

[1] ונדביק ... המשושים] ⟨⟨*⟩⟩

فَإِنْ أَمْكَنَ أَنْ تَمَاسَّها في مَوْضِعَيْنِ أو أكثر، فلتماسّ دائِرَةَ ج د دائِرَةَ ا ب في مَوْضِعَيْنِ مِنْ داخِلٍ على نُقْطَتَيْ ج د. وَلْيَكُنْ مَرْكَزُ دائِرَةِ ا ب نُقْطَةَ ه، ومَرْكَزُ دائِرَةِ ج د نُقْطَةَ ز. والخطّ الَّذي يَجُوزُ على ه ز يَقَعُ حَيْثُ تَتَمَاسُّ الدَّائِرَتانِ. فَنُخْرِجُ خَطَّ ه ز. ونُخْرِجُهُ إلى نُقْطَتَيْ ج د، ومَرْكَزُ دائِرَةِ ا ب—ه.

שאם היה אפשר שתמששה[1] בשתי מקומות או יותר, הנה תמשש עגולת ג״ד עגולת[2] א״ב[3] בשני[4] מקומות מבפנים על שתי[5] נקודות ג׳ ד׳. ויהיה מרכז עגולת א״ב נקודת ה׳, ומרכז עגולת ג״ד נקודת[6] ז׳. הנה הקו אשר יעבור על ה׳ ז׳ יפול במקום אשר תתמששנה[7] שתי העגולות. ונוציא קו ה״ז. ונוציאהו אל שתי נקודות ג׳ ד׳, ומרכז עגולת[8] א״ב—ה׳.

[1]שתמששה] שמתמששנה **ב** [2]ג״ד עגולת] ליתא **ב** [3]בשתי ... א״ב] +~~הנה קו ד׳ ה״ג כמו~~ **פ** ⟨*⟩ **ק** [4]בשני] בשתי **בוטספ** [5]שתי] ליתא **ו** [6]נקודת] ליתא **ט** [7]תתמששנה] תמששנה **וס** [8]ומרכז עגולת] הנה מפני שמרכז **ס**

ולפי שמרכז עגול א״ב הוא ה׳, יהיה ג״ה שוה לה״ד. וה״ד ארוך מז״ד. הנה ג״ה ארוך מז״ד[1]. וג״ז ארוך מג״ה, וכמה הוא ארוך ג״ז מז״ד. ועוד, לפי שמרכז עגול ג״ד הוא ז׳, יהיה ג״ז שוה לז״ד. וכבר התבאר שג״ז ארוך הרבה מז״ד—זה חלוף לא יתכן. ואם כן התבאר שלא ימשש עגול ג״ד לעגול א״ב כי אם במקום אחד. וימשש מחוץ, אם היה אפשר, עגול ט״ח לעגול א״ב, בב׳ מקומות על שתי נקדות א׳ ב׳[2]. והקו שידביק ב׳ בא׳ יפול בתוך עגול א״ב וחוץ מעגול ח״ט—זה חלוף לא יתכן, לפי שכל שתי נקודות שעל קשת עגול הקו שידביק אחת מהן באחרת לא יפל חוץ לעגול. [!]

وזה מה שרצינו לבאר.

فَمَرْكَزُ ا ب هٍ. (ج ه)[1] مثل د ه. ج ه أَطْوَلُ مِنْ ز د بِفَصْلِ ه ز. فَمَا أَكْثَر فَصْل ج ز عَلَى ز د. وأَيضًا مَرْكَزُ دائِرَةِ ج د هو ز. ج د مثل ز د. وَقَدِ اسْتَبانَ أَنَّ ج ز أَطْوَلُ كثيرًا مِنْ ز د—هذا خُلْفٌ لا يُمْكِنُ. فلَيْسَ تَماسُّ دائِرَةٌ دائِرَةً إِلّا في مَوْضِعٍ واحِدٍ. فَلَتَماسّ مِنْ خارِجٍ، إِنْ كانَ يُمْكِنُ، في مَوْضِعَيْنِ، مثل دائِرَةِ طً ح لِدائِرَةِ ا ب. فَالخَطّ الَّذي يَصِلُ بِهِ (ا ب)[2] يَقَعُ في دائِرَةِ ا ب وَخارِجًا مِنْ دائِرَةِ ح ط—هذا خُلْفٌ لا يُمْكِنُ، لأَنَّ كُلّ نُقْطَتَيْنِ تَقَعانِ عَلى قَوْسِ دائِرَةٍ فالخَطّ الَّذي يَصِلُ إِحْداهُما بالأُخرى لا يَكونُ إِلّا في داخِلِ الدَّائِرَةِ. فلَيْسَتْ تَماسُّ دائِرَةٌ دائِرَةً إِلّا في مَوْضِعٍ واحِدٍ كانَتْ داخِلَ الدَّائِرَةِ أَو خارِجَها.

وَذلِكَ ما أَرَدْنا بَيانَهُ.

---

[1] וה״ד ... מז״ד] ⟨⟨*⟩⟩ [2] בב׳ ... ב׳] ⟨⟨*⟩⟩

[1] ج ه] ליתא [2] ا ب] اً

فخطّ ه ج مثل خطّ ه د. و ه د أطول مِنْ ز د. ف ج ز أطول كثيرًا مِنْ خطّ ز د. وأيضًا فإِنَّ مَرْكَزَ دائِرَة ج د نُقْطَة ز. فخطّ ز د مثل خطّ ز ج. وَقَدْ كانَ اسْتَبَانَ أَنّ ج ز أطول كثيرًا[1] مِنْ ز د—هذا خُلْفٌ. فَلَيْسَ تَماسُّ دائِرَةُ ج د دائِرَةَ ا ب إلّا في مَوْضِعٍ واحِدٍ فقط. فَلْتَماسّ خارِجًا، إنْ أَمْكَنَ ذَلِكَ في مَوْضِعَيْنِ^، مثل دائِرَة ح ط لدائِرَة ا ب في مَوْضِعَيْنِ على نُقْطَة ا ب[2]. فالخطّ الَّذي يَخْرُجُ مِنْ ا إلى ب يَقَعُ في[3] دائِرَة ا ب وخارِجًا مِنْ دائِرَة ح ط—هذا خُلْفٌ، لأنّ كُلّ^ نُقْطَتَيْنِ تَقَعانِ على قَوْسِ دائِرَةٍ، فإِنَّ الخطّ الَّذي يَخْرُجُ مِنْ إِحْداهُما إلى الأُخرى يَقَعُ في داخِلِ الدّائِرَة.

הנה קו ה״ג[1] כמו קו ה״ד. וד״ה[2] יותר ארוך מן ז״ד. אם כן ג״ז יותר ארוך הרבה[3] מן ז״ד. וגם כן הנה מרכז עגולת ג״ד נקודת ז׳[4]. אם כן קו ז״ד כמו קו[5] ז״ג. וכבר היה שהתבאר[6] כי ג״ז יותר ארוך הרבה מן ז״ד—וזה שקר. הנה[7] אם כן לא תמשש עגולת ג״ד עגולת א״ב כי אם במקום אחד לבד. ותתמששנה מחוץ, אם יהיה אפשר זה, כמו עגולת ח״ט לעגולת א״ב בשתי מקומות על שתי נקודות א׳ ב׳. הנה הקו[8] אשר יצא מן א׳ אל ב׳ יפול בעגולת א״ב וחוץ מעגולת ח״ט—זה[9] שקר, מפני כי שתי נקודות תפולנה[10] על[11] קשת עגולה[12], הנה הקו אשר יצא מאחת משתיהן אל האחרת יפול בתוך העגולה.

فَلَيْسَ تَماسُّ دائِرَةٌ دائِرَةً إلّا في مَوْضِعٍ واحِدٍ، مِنْ داخِلٍ أو مِنْ خارِجٍ.

وذَلِكَ ما أَرَدْنا أَنْ نُبيِّنَ.

הנה לא תמשש עגולה[13] בעגולה[14] כי אם במקום אחד, מבפנים או מבחוץ.

וזה מה שרצינו ביאורו[15].

1 أطول كثيرا] **MPR2** 2 في ... ب] **D** 3 في] **O2S** داخلا في **MPR1R2** في داخل **CCamDT1T2T3U**

1 ה״ג] ה״ב **ב** 2 וד״ה] ליתא **ב** 3 הרבה] ⟨*⟩ **ז** ⟨⟨*⟩⟩ **פ** 4 נקודת ז׳] ⟨ז׳⟩ **ט** 5 קו] ליתא **ט** 6 היה שהתבאר] היה התבאר **ב** התבאר **ס** 7 הנה] ליתא **פ** 8 הקו] קו+ **ב** +הישר **ספ** 9 זה] וזה **ק** 10 תפולנה] ⟨⟨כאשר⟩⟩+ **ק** 11 על] +* **ס** 12 עגולה] עגולת **ב** 13 הנה ... עגולה] ⟨*⟩ **ט** 14 בעגולה] לעגולה **ז** 15 ביאורו] לבאר **טפ**

[Proposition III.13]

[H14]

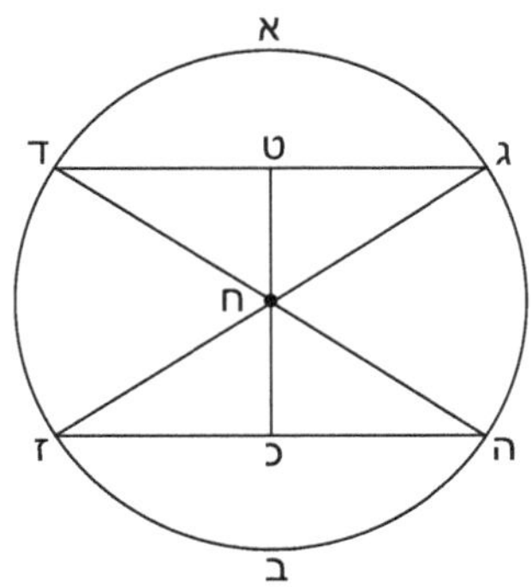

الخُطوطُ الَّتي تكونُ في الدَّائِرَةِ وَتَنْتَهي إِلى الخَطِّ المُحيطِ بِها، إِذا كانَتْ مُتَساوِيَةً فَإِنَّ بُعْدَها مِنَ المَرْكَزِ سَواءٌ، وَإِنْ كانَ بَعْضُها[1] مِنَ المَرْكَزِ سواءٌ، وَتَنْتَهي إِلى الخَطِّ المُحيطِ بِها فَهِيَ مُتَساوِيَةٌ.

כשיפלו בעיגול מיתרים שוים, יהיו מרחקיהם מן המרכז שוים. ואם היו מרחקיהם מן המרכז שוים, יהיו הם שוים.

مِثالُهُ: أَنَّ في دائِرَةِ اب خَطَّيْنِ مُتَساوِيَيْنِ، وَيَنْتَهِيانِ إِلى الخَطِّ المُحيطِ، وَهُوَ ج د ه ز. فَأَقولُ إِنَّ بُعْدَهُما مِنَ المَرْكَزِ سَواءٌ.

המשל בזה: שבעגול א״ב שני קוים שוים[1], והם ג״ד וה״ז. ואומר שמרחקיהם מהמרכז שוים.

[1] كان بعضها] ⟨⟨*⟩⟩

[1] שוים] ⟨⟨*⟩⟩

إذا وَقَعَتْ في دائِرَةٍ أوتارٌ مُتَساوِيَةٌ، فَإِنَّ أَبْعادَها مِنَ الْمَرْكَزِ مُتَساوِيَةٌ. وإنْ كانَتْ أَبْعادُها مِنَ الْمَرْكَزِ مُتَساوِيَةً، فهي مُتَساوِيَةٌ.

כאשר נפלו בעגולה מיתרים[1] שוים, הנה[2] מרחקיהם[3] מן המרכז שוים. ואם היו מרחקיהם מן המרכז שוים, הנה הם שוים.

مِثالُهُ: أنّهُ وَقَعَ في دائِرَةِ اب وَتَرانِ مُتَساوِيانِ، وَهُما ج د ه ز. فَأَقولُ إنَّ بُعْدَهُما مِنَ الْمَرْكَزِ سَواءٌ.

המשל: שהוא נפל[4] בעגולת א״ב שני[5] מיתרים שוים, והם ג״ד וה״ז[6]. הנה אומר כי מרחק שניהם מן המרכז שוה.

שרטוט—**פ**: משולשים גח״ד הח״ז $, ב׳ חסרה [1]מיתרים] ⟨⟨שני⟩⟩+ **ק** [2]הנה] ליתא **בו** [3]מרחקיהם] מרחקם **ט** [4]שהוא נפל] **בו** שנפל **ט** [5]שני] ליתא **בס** ליתא **ופק** [6]וה״ז] ~~וח״ז~~+ **ק**

המופת בו: שמרכז העגול ח׳. ונוציא מח׳ אל קוי ג״ד ה״ז שני עמודים, והם ח״ט וח״כ. ונדביק ח׳ בג׳ וה׳ ובד׳ ובז׳. וג״ד שוה לה״ז. וח״ג שוה לח״ה. ואם כן שני קוי ד״ג וג״ח שוים לשני קוי ז״ה וה״ח, כל אחד לדומה לו. ותושבת ד״ח שוה לתושבת ז״ח. ואם כן זוית טג״ח שוה לזוית כה״ח. וזוית גט״ח נצבה, והיא שוה לזוית חכ״ה. ואם כן שתי זויות כל אחת [!] ממשולשי גט״ח וחכ״ה שוות לשתי זויות האחר, כל אחת לדומה לה. וצלעי ג״ח וה״ח, שהן מיתרי הזויות השוות—שוות. ואם כן שני הצלעים הנשארים שוים לשני הצלעים הנשארים, כל אחד לדומה לו, והזוית הנשארת שוה לזוית הנשארת. ואם כן ח״ט שוה לח״כ. ואם כן מרחק ג״ד וה״ז מן המרכז שוה, לפי שהקוים שמרחקם מן המרכז שוה הם כשיהיו העמודים שיצאו עליהם [!] מן המרכז שוים.

بُرهانُهُ: أَنَّ مَرْكَزَ الدَّائِرَةِ ح. فَنُخْرِجُ مِنْ ح إلى خَطَّي ج د ه ز عَمودَيْنِ، وَهُما ح ط ك ح. وَنَصِلُ (ج ح ح ه ح د ح ز)[1]. و ج د مثل ه ز. ج ح مثل ح ه. فَكِلَا خَطَّي ج د ج ح مثل كِلَيْ خَطَّي ز ه ح ه، كُلُّ واحد مثل نَظيرِه. وَقاعِدَةُ د ح مثل قاعِدَةِ ز ح. فَالزَّاوِيَةُ الَّتي مِنْ ط ج ح مثل الزَّاوِيَةِ الَّتي مِنْ ك ه ح. وَالَّتي مِنْ ج ط ح قائِمَةٌ، وَهِيَ مثل القائِمَةِ الَّتي من ح ك ه. فَزاوِيَتا كُلِّ واحِدٍ مِنْ مُثَلَّثَيْ ح ط ج ح ه ك مثل زاوِيَتَيِ الآخَرِ، كُلُّ زاوِيَةٍ مثل نَظيرَتِها. وَضِلْعا ج ح (ح ه)[2] اللَّذانِ يُوتِرانِ الزَّاوِيَتَيْنِ المُتَساوِيَتَيْنِ—مُتَساوِيَتانِ. فَالضِّلْعانِ الباقِيانِ مثل الضِّلْعَيْنِ الباقِيَيْنِ، كُلّ واحِدٍ مثل نَظيرِه. والزَّاوِيَةُ الباقِيَةُ مثل الزَّاوِيَةِ الباقِيَةِ. ح ط مثل ح ك. فَبُعْدُ ج د ه ز مِنَ المَرْكَزِ سَواءٌ، لِأَنَّ الخُطوطَ الَّتي بُعْدُها مِنَ المَرْكَزِ سَواءٌ، وَهِيَ الَّتي تَكونُ الأَعْمِدَةُ الَّتي تَخْرُجُ إِلَيْها مِنَ المَرْكَزِ مُتَساوِيَةً.

---

[1] ج ح ح ه ح د ح ز] ج ح ه د [2] ح ه] ج ه

بُرْهانُهُ: أنّا نَجْعَلُ المَرْكَزَ نُقْطَةَ ح. ونُخْرِجُ مِنْهُ إلى وَتَرَيْ ج د و ه ز عَمودَيْ ح ط و ح ك. ونُخرج ح ه ح د ح ز ح ج. فخطّ ج د مثل ه ز. وج ح مثل ه ح. فخطّا د ج و ج ح مثل خطَّيْ ز ه و ه ح. وقاعِدَة د ح مثل قاعِدَة ز ح. فزاوية ط ج ح مثل زاوية ك ه ح. وزاوية ج ط ح مثل زاوية ه ك ح، لأنّهما قائمتان. فزاويتا مُثلّث ج ط ح مثل زاويتَي مُثلّث ه ك ح. وضِلْعا ج ح ه ح مُتَساوِيان. فالضِّلْعانِ الباقِيانِ مثل الضِّلْعَيْنِ الباقِيَيْنِ، كُلُّ واحدٍ مثل نَظيرِهِ. والزَّاوِيَةُ الباقِيَةُ مثل الزَّاوِيَةِ الباقِيَةِ. ف ط ح مثل ك ح. وَهُما العَمودانِ. فَبُعْدُ ج د و ه ز مِنَ المَرْكَزِ سَواء.

המופת: כי נשים[1] המרכז נקודת ח׳. ונוציא ממנו אל שני מיתרי ג״ד וה״ז[2] שני עמודי ח״ט וכ״ח. ונוציא ח״ה וח״ד וח״ז וח״ג. הנה ג״ד כמו ה״ז. וג״ח כמו ה״ח[3]. הנה שני קוי ד״ג וג״ח[4] כמו שני קוי[5] ז״ה וה״ח. ותושבת ד״ח כמו תושבת ז״ח. אם כן זוית טג״ח כמו זוית כה״ח[6]. וזוית גט״ח כמו זוית הכ״ח, מפני כי שתיהן נצבות. אם כן שתי זויות[7] משולש גט״ח כמו שתי זויות משולש הכ״ח. ושתי צלעות ג״ח ה״ח שוות. הנה שתי הצלעות[8] הנשארות כמו שתי הצלעות[9] הנשארות, כל אחת[10] כמו הנכחית[11] לה. והזוית[12] הנשארת כמו הזוית הנשארת[13]. אם כן ט״ח כמו כ״ח. ושניהם שני העמודים[14]. הנה מרחק[15] ג״ד וה״ז מן המרכז שוה.

[1]כי נשים] שנשים **ז**
[2]וה״ז] +~~שנים~~ **פ**
[3]שני ... ה״ח] ליתא **ב**
[4]וג״ח] ~~וה״ח~~+ **ק**
[5]קוי] ליתא **ב** ⟨⟨*⟩⟩ **ק**
[6]כה״ח] דה״ח **ס**
[7]זויות] ~~זוית~~+ **ט**
[8]הצלעות] צלעות **בוס**
[9]הצלעות] צלעות **וטס**
[10]אחת] +~~לגילה~~ **ס**
[11]הנכחית] הנכחי **ט** הנכחי⟨י⟩ת **ק**
[12]והזוית] וזוית **ז**
[13]כל ... הנשארת] ליתא **ב**
[14]העמודים] עמודים **ט**
[15]מרחק] +~~שניהם~~ **פ**

ועוד: קו [!] ג״ד וה״ז מרחקם מן המרכז שוה. ואומר שהם גם כן שוים.

وأَيضًا: خَطَّا ج د ه ز بُعْدُهُما مِنَ المَرْكَزِ سَواءٌ. وَيَنْتَهِيانِ إِلى الخَطِّ المُحيطِ. فَأَقولُ إِنَّهُما مُتَساوِيانِ.

המופת בו: לפי שכבר יצא ממרכז עגול א״ב קו ח״ט אל קו ג״ד שאינו הולך על המרכז, וחלק לחציים על זוית נצבה, אם כן ג״ט[1] שוה לט״ד. וד״ג כפל ג״ט. וכמו כן ז״ה כפל ה״כ. וג״ח שוה לח״ה. ואשר יהיה מהכאת ג״ח בעצמו שוה לאשר יהיה מהכאת ה״ח בעצמו. ואמנם אשר יהיה מהכאת ג״ח בעצמו שוה לאשר יהיה מג״ט וט״ח, כל אחד בעצמו, לפי שזוית גט״ח נצבה. ואשר יהיה מהכאת ה״ח בעצמו שוה לאשר יהיה מהכאת ה״כ וכ״ח, כל אחד בעצמו. ואם כן אשר יהיה מהכאת ג״ט וט״ח, כל אחד בעצמו, שוה לאשר יהיה [מ]ה״כ וכ״ח, כל אחד בעצמו. ואשר יהיה מהכאת ט״ח [!] בעצמו שוה לאשר יהיה מהכאת ח״כ בעצמו. ונשאר אשר יהיה מהכאת ג״ט בעצמו שוה לאשר יהיה מהכאת ה״כ בעצמו. ואם כן[2] ג״ט שוה לה״כ. וג״ד כפל ג״ט. וה״ז כפל כ״ה. וכל שוים שנכפלו הם שוים.

بُرْهانُهُ: أَنَّهُ قَدْ خَرَجَ مِنْ مَرْكَزِ دائِرَةِ ا ب خَطُّ ح ط إِلى خَطِّ ج د الَّذي لَمْ يَجُزْ على المَرْكَزِ، وَقَطَعَهُ بِنِصْفَيْنِ على زاوِيَةٍ قائِمَةٍ. ج ط مثل ط د. د ج مِثْلا ج ط. وَكَذَلِكَ صارَ ز ه مِثْلَيْ ه ك. ج ح مثل ه ح. فالَّذي يكونُ مِنْ ضَرْبِ ج ح في نَفْسِهِ مثل الَّذي يكونُ مِنْ ضَرْبِ ح ه في نَفْسِهِ. ولكن الَّذي يَكونُ مِنْ ضَرْبِ ج ح في نَفْسِهِ مثل الَّذي يكونُ مِنْ ضَرْبِ ج ط ط ح، كُلُّ واحِدٍ في نَفْسِهِ. والَّذي يكونُ مِنْ ضَرْبِ ه ح في نَفْسِهِ مثل الَّذي يكونُ مِنْ ضَرْبِ ه ك ك ح، كُلّ واحِدٍ في نَفْسِهِ. فالَّذي يَكونُ مِنْ ضَرْبِ ج ط ط ح، كُلّ واحِدٍ في نَفْسِهِ، مثل الَّذي يَكونُ مِنْ ضَرْبِ ه ك ك ح، كُلّ واحِدٍ في نَفْسِهِ. وَالَّذي يَكونُ مِنْ ضَرْبِ ط ح في نَفْسِهِ مثل الَّذي يَكونُ مِنْ ضَرْبِ ك ح في نَفْسِهِ. فَيَبْقَى الَّذي يَكونُ مِنْ ضَرْبِ ج ط في نَفْسِهِ مثل الَّذي يَكونُ مِنْ ضَرْبِ ه ك في نَفْسِهِ. ج ط مثل ه ك. ج د مثل ج ط. ه ز مِثْلا ك ه. وكُلّ مُتَساوِيَيْنِ أَضْعافًا كانَا مُتَساوِيَيْنِ.

וזה מה שרצינו באורו.

وَذلِكَ ما أَرَدْنا بَيانَهُ.

---

[1] ג״ט] ~~[ג]״ט~~+ [2] ואם כן] ~~{ואם}~~ ⟨⟨*⟩⟩

وَأَيْضًا فَإِنَّا نَجْعَلُ بُعْدَ وَتَرَيْ ج د و ه ز مِنَ الْمَرْكَزِ سَواء. فَأَقولُ إِنَّهما مُتَساوِيانِ.

וגם כן הנה נשים מרחק שני מיתרי ג״ד וה״ז מן המרכז שוה. ואומר כי שניהם שוים.

بُرْهانُه: أَنَّهُ قَدْ خَرَجَ مِنَ الْمَرْكَزِ خَطُّ ط ح إلى ج د، فَقَطَعَهُ على زاويةٍ قائِمَةٍ. فَهُوَ إِذَنْ[1] يَقْطَعُهُ بِنِصْفَيْنِ. نَخَطّ ج ط مثل خطّ ط د. و د ج مِثْلا ج ط. وَلِذَلِكَ يكونُ ز ه مِثْلا ه ط. وج ح مثل ه ح. فالْمُرَبَّع الكائِن مِنْ ج ح[2] مثل الْمُرَبَّعَيْن الكائِنَيْن مِنْ ج ط ط ح، لأَنَّ زاويةَ ج ط ح قائِمَةٌ. والْمُرَبَّع الكائِن مِنْ ه ح مثل الْمُرَبَّعَيْن الكائِنَيْن مِنْ ه ك ك ح، لأَنَّ زاوية ه ك ح قائِمَة. فالْمُرَبَّعان الكائِنان مِنْ ط ح ط ج مثل الْمُرَبَّعَيْن الكائِنَيْن مِنْ ه ك و ك ح. والْمُرَبَّع الكائِن مِنْ ط ح مثل الْمُرَبَّع الكائِن مِنْ ك ح. ويَبْقَى الْمُرَبَّعُ الكائِنُ مِنْ ج ط مثل الْمُرَبَّع الكائِن مِنْ ه ك. نخطّ ح ط مثل خطّ ه ك. وج د مثل ج ط. و ه ز مثل ك ه. وكُلّ مُتَساوِيَيْن إذا أُضْعِفا كانا مُتَساوِيَيْن. فوَتَرَ ج د مثل وَتَرِ ه ز.

המופת: כי כבר הוצא[1] מן המרכז קו ט״ח אל ג״ד, וחתכו על זוית נצבת. והוא אם כן יחתכהו בשני חציים. אם כן קו ג״ט[2] כמו קו ט״ד. וד״ג כפל ג״ט[3]. ולכן יהיה ז״ה כפל ה״כ. וג״ח כמו ח״ה. אם כן המרובע ההווה מן ג״ח כמו שני המרובעים ההווים מג״ט[4] ט״ח, מפני כי[5] זוית גט״ח נצבת. והמרובע ההווה מן ה״ח כמו שני המרובעים ההווים[6] מן ה״כ[7] וכ״ח, מפני כי זוית הכ״ח נצבת. הנה שני המרובעים ההווים מן ט״ח ט״ג כמו שני המרובעים ההווים מן ה״כ וכ״ח[8]. והמרובע ההווה מן ט״ח כמו המרובע ההווה מן כ״ח. ונשאר המרובע ההווה מן ג״ט כמו המרובע ההווה מן ה״כ. הנה קו ח״ט כמו קו ה״כ. וג״ד[9] כפל ג״ט. וה״ז כפל כ״ה. וכל שני שוים[10] כאשר נכפלו היו שוים. אם כן מיתר ג״ד כמו מיתר ה״ז[11].

وذَلِكَ ما أَرَدْنا أَنْ نُبَيِّنَ.

וזה מה שרצינו לבאר[12].

---

[1]اذن] **MR1T2T4**   [2]ح ج] **SU** +مثل المربع الكائن من خط ه ح ولكن المربع الكائن من خط ج ح

---

[1]הוצא] הובא **ס**   [2]ג״ט] ~~ג״ד~~ ⟨⟨ג״ט⟩⟩ **ק**   [3]ג״ט] +~~אם כן~~ **ק**   [4]מג״ט] ~~מג״ח~~ **ס**   [5]כי] ליתא **ב**   [6]ההווים] ליתא **בפ**   [7]ה״כ] ח״כ **ב**   [8]מפני … וכ״ח] ליתא **ז**   [9]ח״ט … וג״ד] ד״ט כמו קו ה״כ וכן ג״ט כמו ה״ב וג״ד **ס** +~~כמו~~ **ק**   [10]שוים] קוים+ **ס** ⟨⟨קוים⟩⟩ **פ**   [11]ג״ד … ה״ז] ח״ז וג״ד שוים **ב**   [12]לבאר] באורו **בפק**

[Proposition III.14]

[H15]

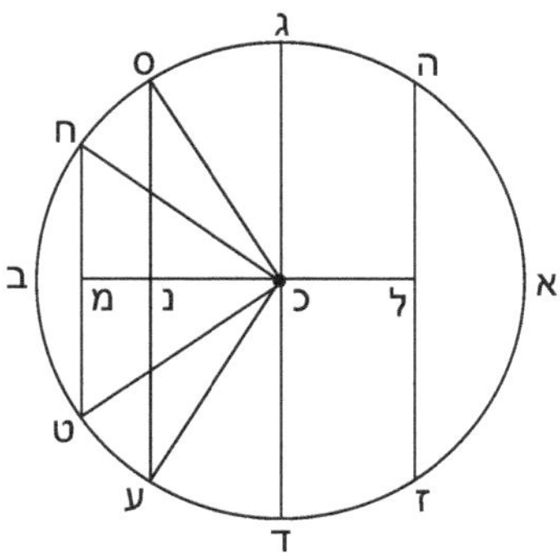

כשנפלו בעגול מיתרים, הארוך שבהם הוא אלכסון העגול, והנשארים—הקרוב מהם אל האלכסון הוא ארוך מן הרחוק.

الخُطُوطُ المُستَقيمَةُ الَّتي تكونُ في الدَّائِرَةِ فأَطْوَلُها قُطْرُ الدَّائِرَةِ. وَكُلُّ خَطٍّ هُوَ أَقْرَبُ إِلى القُطْرِ هُوَ أَطْوَلُ مِمَّا هُوَ أَبْعَدُ مِنْهُ.

המשל בו: שעגול א״ב אלכסונו ג״ד, ובו שני קוים אחרים, והם ה״ז וח״ט. וה״ז קרוב לאלכסון יותר מח״ט. ואומר שהאלכסון שהוא ג״ד הוא הארוך שבהם, וה״ז ארוך מח״ט.

مِثالُهُ: أَنَّ دائِرَةَ ا ب قُطْرُها ج د. وَفيها خَطَّانِ آخَرانِ ه ز ح ط، ه ز أَقْرَبُ إِلى القُطْرِ مِنْ ح ط. فَأَقولُ إِنَّ القُطْرَ الَّذي هو ج د أَطْوَلُها، وَأَنَّ ه ز أَطْوَلُ مِنْ ح ط.

שרטוט—לא סומנו א׳ וב׳

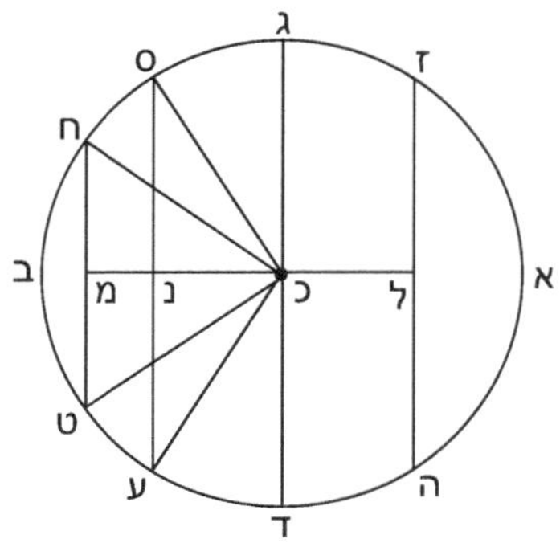

إذا وَقَعَتْ في دائِرَةٍ أَوْتارٌ فَإِنَّ أَطْوَلَها قُطْرُ الدَّائِرَةِ والباقية فالأقرب مِنْها إلى الْمَرْكَزِ أَطْوَلُ مِنَ الأَبْعَدِ.

כאשר נפלו בעגולה[1] מיתרים, הנה היותר ארוך מהם קוטר העגולה, והנשארים הנה היותר קרוב מהם אל המרכז יותר ארוך מן היותר רחוק[2].

مِثالُهُ: أَنَّهُ وَقَعَ في دائِرَةَ اب وَتَرانِ، وَهُما ه ز ح ط، والقُطْرُ ج د. و ه ز أقربُ إلى الْمَرْكَزِ مِنْ ح ط. فَأَقولُ إنّ قُطْرَ ج د—أطولها، وإنَّ ه ز أطول مِنْ ط ح.

המשל בו[3]: כי נפלו בעגולת א״ב שני מיתרים, והם ז״ה וח״ט[4], והקוטר ג״ד[5]. וז״ה יותר קרוב אל המרכז מן ח״ט[6]. הנה אומר כי קוטר ג״ד—היותר ארוך מהם, ושה״ז יותר ארוך מן ט״ח.

שרטוט—**ב**: א ↔ ב; שונות בהנחת האותיות בכתבי היד הערביים, ואין אחד שבו היא זהה לזו שבתרגום אבן תיבון; עיקר ההבדלים נוגעים להחלפת מקום באותיות א׳ וב׳ ובאותיות ה׳ וז׳

[1] בעגולה] בעגולת **ב** [2] מהם ... רחוק] ליתא **ב** [3] בו] ליתא **ס** [4] ז״ה וח״ט] ~~ה״זז״ח~~ וח״ט **ס** [5] והקוטר ג״ד] ליתא **ק** [6] ח״ט] והקוטר ג״ד **ק**

מופת זה: שמרכז העגול—כ׳. ונוציא מן המרכז שני עמודים על ה״ז וח״ט, והם כ״ל וכ״מ. וז״ה קרוב אל המרכז מח״ט. ואם כן מ״כ ארוך מל״כ. ונחלק ממ״כ קו שוה לכ״ל והוא כ״נ. ונוציא מנקודת נ׳ קו נכוחי לקו ח״ט, והוא ס״ע. ויהיו ה״ז וס״ע מרחקם מן המרכז שוה. ואם כן הם שוים[1], וה״ז שוה לס״ע. וכ״ס שוה לג״כ. וכ״ע שוה לכ״ד. ואם כן ג״ד ארוך מס״ע. וס״ע שוה לה״ז. ואם כן ג״ד ארוך מה״ז. וס״כ שוה לכ״ח. וכ״ע שוה לכ״ט. ואם כן שני קוי כ״ע וס״כ שוים לשני קוי ח״כ וכ״ט, כל אחד לאחר. וזוית סכ״ע גדולה מזוית חכ״ט. ואם כן תושבת ס״ע גדולה מתושבת ח״ט. ואמנם ס״ע שוה לה״ז. ואם כן ה״ז ארוך מח״ט. ואם כן התבאר שהקוים אשר בעגול א״ב—הארוך שבהם הוא האלכסון, והוא ג״ד; וה״ז שהוא הקרוב אליו ארוך מח״ט שהוא רחוק ממנו.

וזה מה שרצינו לבאר.

[1]שוים] +⟨⟨ונדביק נ׳ בס׳ ובע׳ ובח׳ ובט׳⟩⟩

بُرهانُه: أَنَّ مَرْكَزَ الدَّائِرَةِ—كَ. فَنُخْرِجُ مِنَ المَرْكَزِ عَمودَيْنِ إِلى خَطَّيْ ه ز ح ط، وَهُما ك ل م ك، م ك أَطْوَلُ مِنْ ل ك. ح ط أَبْعَدُ مِنَ القُطْرِ مِنْ ز ه. فَيُفْصَلُ مِنْ ك م مثل ك ل، وَهُوَ ك ن. وَنُخُطُّ عَلى نُقْطَةِ نَ خَطًّا مُوازِيًا لِخَطِّ ه ز[1]، وَهُوَ خَطُّ س ع. س ع ه ز بُعْدُهُما مِنَ المَرْكَزِ سَواءٌ، فَهُما مُتَساوِيانِ، ه ز مثل س ع. وَنَصِلُ ك س ك ع ك ط ك ح. س ك ك ع مَجْموعانِ أَطْوَلُ مِنْ س ع. س ك مثل ك ج. ك ع مثل ك د. ج د أَطْوَلُ من س ع. س ع مثل ه ز. ج د أَطْوَلُ من ه ز. س ك مثل ك ح. ع ك مثل ك ط. فَكِلَا خَطَّيْ س ك ك ع مثل كِلَيْ خَطَّيْ ح ك ك ط، كُلٌّ واحِدٍ مِثْلُ الآخَرِ. والزَّاوِيَةُ الَّتي من س ك ح أَعْظَمُ مِنَ الَّتي من ح ك ط. فَقاعِدَةُ س ع أَطْوَلُ مِنْ قاعِدَةِ ح ط. وَلَكِنَّ س ع مثل ه ز. ه ز أَطْوَلُ من ح ط. فَالخُطوطُ الَّتي في دائِرَةِ ا ب—أَطْوَلُها القُطْرُ، وَهِيَ ج د ه ز، وهو أَقْرَبُ إِلَيْهِ أَطْوَلُ مِنْ ح ط الَّذي هو أَبْعَدُ مِنْهُ.

وَذَلِكَ ما أَرَدْنا بَيانَهُ.

[1]ه ز] اولי צ״ל ح ط

بُرْهانُهُ: أنّا نَجْعَلُ الْمَرْكَزَ نُقْطَةَ كَ. ونُخْرِجُ مِنْهُ عَمُودَيْنِ إلى وَتَرَيْ ه ز وح ط، وَهُما كَ ل وكَ م. فوتَر ز ه أقرب إلى الْمَرْكَزِ مِنْ ح ط. ف م كَ أطول مِنْ ل كَ. ونَفْصِلُ مِنْ كَ م مثل ل كَ، وهو كَ ن. ونَخُطُّ على نَ خطًّا موازيًا لِخَطّ ط ح، وهو س ع. فَخَطّا ه ز و س ع بُعْدُهما مِنَ الْمَرْكَزِ سَواءٌ. فَهُما مُتَساوِيان. فخطّ ه ز مثل س ع. ونُخْرِجُ خُطوطَ كَ س و كَ ح و كَ ع و كَ ط. فخطّا س كَ و كَ ع مجْموعَيْنِ أطْوَلُ مِنْ س ع. و س كَ مثل كَ ج. و كَ ع مثل كَ د. فخطّ ج د أطول مِنْ س ع. و س ع مثل ه ز. ف ج د أطول مِنْ ه ز. و س كَ مثل كَ ح. و ع كَ مثل كَ ط. فَخَطّا س كَ و كَ ع مثل خَطَّيْ كَ ح و كَ ط. وزاوية س كَ ع أعظم مِنْ زاوية ط كَ ح. فقاعِدَة س ع أطول مِنْ قاعِدَة ط ح. ولكن س ع مثل ه ز. ف ه ز أطول مِنْ ط ح. فالأوتار التي وَقَعَت في دائِرَة ا ب— أطولها القُطْر، وهو ج د؛ و ه ز، الأقرب مِنَ الْمَرْكَزِ، أطول مِنْ ط ح الأبعد.

وذَلِكَ ما أَرَدْنا أَنْ نُبيِّنَ.

המופת: כי נשים המרכז נקודת כ׳. ונוציא ממנו[1] שני עמודים אל שני מיתרי ה״ז וח״ט, והם כ״ל וכ״מ. הנה מיתר ז״ה יותר קרוב אל המרכז מן ח״ט. אם כן מ״כ[2] יותר ארוך מן ל״כ. ונבדיל מן כ״מ[3] כמו ל״כ, והוא כ״נ. ונקוה על נ׳ קו נכחי לקו ט״ח, והוא ס״ע. הנה שני קוי ה״ז וס״ע[4] מרחק שניהם מן המרכז שוה. אם כן שניהם שוים. אם כן קו ה״ז כמו ס״ע. ונוציא קוים כ״ס וכ״ח וכ״ע[5] וכ״ט. הנה קו [!] ס״כ וכ״ע מחוברים יותר ארוך מן ס״ע. וס״כ[6] כמו כ״ג[7]. וכ״ע כמו כ״ד[8] [9]. הנה קו ג״ד יותר ארוך מן ס״ע. וס״ע כמו ה״ז. אם כן[10] ג״ד יותר ארוך מן[11] ה״ז. וס״כ כמו כ״ח. וע״כ כמו כ״ט. הנה שני קוי ס״כ וכ״ע כמו שני קוי כ״ח וכ״ט. וזוית עכ״ס יותר גדולה מזוית טכ״ח. אם כן תושבת ס״ע יותר ארוכה מתושבת ט״ח. אבל ס״ע כמו ה״ז. אם כן ה״ז יותר ארוך מן ט״ח. הנה[12] המיתרים אשר[13] נפלו בעגולת[14] א״ב—היותר ארוך מהם הקוטר[15], והוא ג״ד; וה״ז, היותר קרוב מן המרכז, יותר ארוך מן ט״ח היותר רחוק.

ונשלם ביאורו.

---

[1] ממנו] ממנה **ס** [2] מ״כ] ~~מ״זמ״ג~~+ **ב** [3] כ״מ] ל״מ **בוק** ⟨כ⟩~~ל~~״מ **פ** [4] וס״ע] ס״ע **ס** [5] וכ״ח וכ״ע] וכ״ע וכ״ח **ק** [6] וס״כ] וס״ע **ב** [7] כ״ג] ה״ז **ב** כ״ע **ו** [8] וכמו ... כ״ד] ליתא **ב** [9] כ״ד] ~~ג״ד~~ * **ס** [10] ג״ד ... כן] ⟨⟨*⟩⟩ **ק** [11] וס״ע ... מן] ליתא **ב** [12] הנה] +שני **ס** אם כן **פ** [13] המיתרים אשר] המיתרים **ב** ⟨ה⟩מיתרים **ו** המיתרים כאשר **ס** [14] בעגולת] בעגולה **ס** [15] הקוטר] הוא+ **ס**

[Proposition III.15]

[H16]

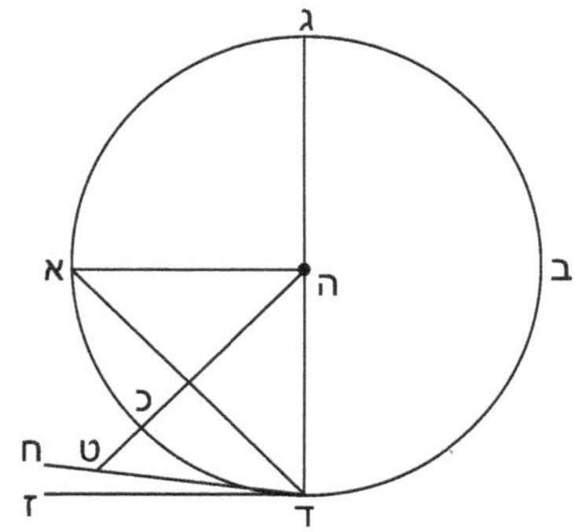

כשהוצא מקצה קוטר עגול קו ישר על זוית נצבה, הוא יוצא חוץ לעיגול, ולא יפל בינו ובין קו המקיף קו אחר ישר, ותהיה זוית חצי העגול גדולה מכל זוית חדה ישרת הקוים, ותהיה הזוית שיקיפו בה זה הקו והקו המקיף קטנה מכל זוית חדה ישרת הקוים.

إِذا خَرَجَ مِنْ طَرَفِ قُطْرِ الدَّائِرَةِ خَطٌّ عَلى زاوِيَةٍ قائِمَةٍ، فَإِنَّ الخَطَّ الَّذي يَكونُ خارِجًا مِنَ الدَّائِرَةِ، وَلا يَقَعُ بَيْنَهُ وبَيْنَ الخَطِّ المُحيطِ خَطٌّ آخَرَ، وَتَكونُ الزَّاوِيَةُ الَّتي (يُحيطُ)[1] بِها ذَلِكَ الخَطُّ والخَطُّ المُحيطُ أَصْغَرَ مِنْ كُلِّ زاوِيَةٍ حادَّةٍ، وَالزَّاوِيَةُ الَّتي تَلِيها مِنْ داخِلِ الدَّائِرَةِ الَّتي يُحيطُ بِها خَطُّ القُطْرِ وَالخَطُّ المُحيطُ أَعْظَمُ مِنْ كُلِّ زاوِيَةٍ حادَّةٍ.

המשל בזה: שנצייר עגול עליו א״ב וקוטרה [!] ג״ד. ואומר שהקו שיצא מקצה ג״ד מנקודת ד׳ על זוית נצבה יפל חוץ לעגול.

مِثالُهُ: أَنَّا نَخُطُّ دائِرَةً عَلَيْها ا ب وقُطْرُها ج د. فَأَقولُ إِنَّ الخَطَّ الَّذي يَخْرُجُ مِنْ طَرَفِ خَطِّ ج د مِنْ نُقْطَةِ د عَلى زاوِيَةٍ قائِمَةٍ تَقَعُ خارِجًا مِنَ الدَّائِرَةِ.

שרטוט—B: ك ← ط ط ← ح, ك מונחת בחיתוך א ד ה ط, ו-ح היא קצה ד ح

[1] يحيط] تحيط

إذا أُخْرِجَ مِنْ طَرَفِ قُطْرِ الدَّائِرَةِ[1] خطٌّ مُسْتَقيمٌ على زاوِيَةٍ قائِمَةٍ، فَإِنَّهُ يكونُ خارِجًا مِنَ الدَّائِرَةِ، ولا يَقَعُ بَيْنَهُ وبَيْنَ المُحِيطِ^ خَطٌّ آخَرُ مُسْتَقيمٌ. وَتَكونُ زاوِيَةُ نِصْفِ الدَّائِرَةِ أعظم مِنْ كُلِّ زاوِيَةٍ حادَّةٍ مُسْتَقيمة الخَطَّيْنِ. وتَكونُ الزَّاوِيَةُ الَّتِي يُحِيطُ بِها ذَلِكَ الخطُّ والخطُّ المُحِيط—أصغر مِنْ كُلِّ زاوِيَةٍ حادّةٍ مُسْتَقيمة الخَطَّيْنِ.

كאשר הוצא מקצה קוטר העגולה קו ישר[1] על זוית נצבת[2], הנה יהיה יוצא מן העגולה. ולא יפול בינו ובין הקו מן[3] המקיף קו אחר ישר. ותהיה זוית חצי העגולה יותר גדולה מכל זוית חדה ישרת שני הקוים. ותהיה הזוית אשר יקיף בה הקו[4] ההוא והקו המקיף— יותר קטנה מכל זוית חדה ישרת שני הקוים.

مِثالُهُ: أنّ دائِرَةَ $\overline{\text{ا ب}}$ خَرَجَ مِنْ طَرَفِ قُطْرِها، وهو خطّ $\overline{\text{ج د}}$، خطٌّ مُسْتَقيمٌ^ على زاوِيَةٍ قائِمَةٍ مِنْ نُقْطَةِ $\overline{\text{د}}$. فَأَقولُ إنّه يَقَعُ خارِجًا مِنَ الدَّائِرَةِ، لا يُمْكِنُ غير ذَلِكَ.

המשל[5]: כי עגולת א״ב יצא מקצה קוטרה, והוא[6] קו ג״ד, קו ד״ז על זוית נצבת מנקודת ד׳. הנה אומר כי[7] הוא יפול חוץ מן העגולה, אי אפשר זולת זה.

[1] الدايرة] MO2PQR1R2S

שרטוט—ס: ז״ד מוארך מצד ד׳ עד כ׳, אין מופיעה כ׳ על ה״ט, והוסף כ״ה [1]קו ישר] ⟨⟨*⟩⟩ פ [2]נצבת] נצבה ק [3]מן] ליתא ק [4]בה הקו] ליתא ב [5]ותהיה … המשל] בו ז ליתא ט [6]והוא] והיא פ [7]אומר כי] ⟨*⟩ ט

ואי אפשר אלא זה. ואם היה אפשר, יפול בתוכה כמו ד"א. ומרכז העגול הוא ה׳. ונדביק א׳ בה׳. וא"ה שוה לה"ד. וזוית הא"ד שוה לזוית הד"א. וזוית הד"א נצבה. וא"כ זוית הא"ד נצבה. ואם כן משלש הא"ד נצב שתי זויות—זה חלוף, לא יתכן. וכבר התבאר שהקו שיצא מקצה קו ג"ד מנקודת ד׳ על זוית נצבה אמנם יפל חוץ לעגול. ואם כן יפול כמו ד"ז.

بُرهانُهُ: أَنَّهُ لا يُمْكِنُ إِلّا ذلِكَ. فَإِنْ كانَ يُمْكِنُ، فَلْيَقَعْ داخِلَها مِثْلُ خَطِّ د ا. وَمَرْكَزُ الدّائِرَةِ ه. وَنَصِلُ ا ه. ا ه مثل ه د. وَالزّاوِيَةُ الَّتي من ه ا د مثل الَّتي من ه د ا. وَالَّتي من ه د ا قائِمَةٌ. فَالَّتي مِنْ (ه ا د)[1] قائِمَةٌ. فَمُثَلَّثُ ا ه د قائِمُ[2] الزّاوِيَتَيْنِ—هذا خُلْفٌ لا يُمْكِنُ. فَقَدِ اسْتَبانَ أَنَّ الخَطَّ الَّذي يَخْرُجُ مِنْ طَرَفِ خَطِّ ج د مِنْ نُقْطَةِ د عَلى زاوِيَةٍ قائِمَةٍ إِنَّما (تَقَعُ)[3] خارِجًا مِنَ الدّائِرَةِ. فَلْتَقَعْ مِثْلَ د ز.

ואומר שלא יפול קו אחר ישר[1] בין קשת (גא"ד)[2] ובין קו ד"ז[3]. ואם היה אפשר שיפול ביניהם קו אחר, יהיה ד"ח. ונוציא[4] מה׳ עמוד על קו ד"ח [!] והוא קו ה"ט. ותהיה זוית הט"ד נצבה. וזוית הד"ט היא קטנה מנצבה. אם כן[5] זוית הט"ד גדולה מזוית הד"ט. והזוית הגדולה במשולש יהיה לה מיתר הצלע הגדול ממנו. ואם כן ה"ד גדול מה"ט. וה"ד שוה לה"כ. ואם כן ה"כ גדול מה"ט—זה חלוף, לא יתכן. כבר התבאר שלא יפול קו אחר ישר בין קו קשת ד"א ובין קו ד"ז.

فَأَقولُ إِنَّه لا يَقَعُ بَيْنَ خَطِّ قَوْسِ ج ا د وبَيْنَ خَطِّ د ز خَطٌّ آخَرُ.

بُرهانُهُ: أَنَّهُ لا يُمْكِنُ ذَلِكَ. فَإِنْ أَمْكَنَ، فَلْيَقَعْ بَيْنَهُما مثل خَطِّ د ح. وَنُخْرِجُ مِنْ ه عمودًا إِلى خَطِّ د ح، وَهُوَ ه ط. وَالزّاوِيَةُ الَّتي مِنْ ه ط د قائِمَةٌ. فَالَّتي من ه د ط أَصْغَرُ مِنْ قائِمَةٍ. والَّتي من ه ط د أَعْظَمُ مِنَ الَّتي مِنْ ه د ط. وأَكْبَرُ زوايا المُثَلَّثِ يُوتِرُها الضِّلْعُ الأَطْوَلُ. ه د إِذًا أَطْوَلُ مِنْ ه ط. ه د مثل ك ه. ه ك أَطْوَلُ مِنْ ه ط—هذا خُلْفٌ لا يُمْكِنُ. فَقَدِ اسْتَبانَ أَنَّهُ لا يَقَعُ بَيْنَ خَطِّ قَوْسِ ج ا د وبَيْنَ خَطِّ د ز خَطٌّ آخَرُ.

[1]אחר ישר] ‹‹*›› [2]גא"ד] [ג]"ד‹‹ד"א››
[3]ד"ז] +~~קו אחר~~ [4]ונוציא] +~~קו~~ [5]אם כן] ‹‹*››

[1]ه ا د] ه د ا [2]قائم] +* [3]تقع] يقع

فَإِنْ أَمْكَنَ، فَلْيَقَعْ فِي داخِلِها مثل خَطّ د ا. وَلْيَكُنِ الْمَرْكَزُ ه. وَنُخْرِجُ ه ا، وهو مثل ه د. فزاوية ه ا د مثل زاوية ه د ا. وزاوية ه د ا قائِمَة. فزاوية ه ا د قائِمَة. فَمُثَلَّث ه ا د قائِم الزَّاوِيَتَيْنِ—هذا خُلْفٌ. فَقَدِ اسْتَبَانَ أَنَّ الخطّ الَّذي يَخْرُجُ مِنْ دَ—وهو طَرَفُ ج د—على زاويةٍ قائِمَةٍ، يَقَعُ خارِجًا مِنَ الدَّائِرَةِ. فَلْيَقَعْ مثل د ز.

שאם[1] היה אפשר, הנה יפול בתוכה כמו קו א״ד. ויהיה המרכז ה׳. ונוציא א״ה, והוא כמו ה״ד. אם כן זוית הא״ד[2] כמו זוית הד״א[3]. וזוית הד״א[4] נצבת. אם כן זוית הא״ד נצבת. אם כן משולש הא״ד נצב[5] שתי[6] הזויות—זה שקר. הנה כבר התבאר כי הקו אשר יצא מן ד׳—והוא קצה ג״ד—על זוית נצבת, יפול חוץ מן העגולה. ויפול[7] כמו ד״ז.

فَأَقولُ إِنَّهُ لا يَدْخُلُ بَيْنَهُ وَبَيْنَ قَوْسِ ج ا د خطٌّ آخر مُسْتَقيم. فإن أمكَنَ، فَلْيَقَعْ بَيْنَهُما مثل خطّ د ح. فَنُخْرِجُ مِنْ ه عمودًا إلى خطّ د ح، وهو ه ط. فزاوية ه ط د قائِمَة. وزاوية ه د ط أصغر مِنْ قائِمَة. فزاوية ه ط د أعظم مِنْ زاوية ه د ط. والزَّاوِيَةَ العُظمى مِنْ مُثَلَّث يوتِرُها الضِّلعُ الأَطْوَلُ. فَضِلْع ه د أطول مِنْ ه ط. و ه د مثل ه ك. ف ه ك أطول مِنْ ه ط—هذا خُلْفٌ. فَقَدِ اسْتَبَانَ أَنَّهُ لا يَقَعُ بَيْنَ قَوْسِ ج ا د وَبَيْنَ خطّ د ز خطٌّ آخر مُسْتَقيم.

הנה אומר כי לא יכנס בינו ובין קשת גא״ד קו אחר ישר. שאם היה אפשר[8], הנה יפול[9] בין שניהם כמו קו ד״ח. ונוציא מן ה׳ עמוד אל קו ד״ח, והוא ה״ט. הנה[10] זוית הט״ד נצבת. וזוית הד״ט קטנה מנצבת. אם כן זוית הט״ד יותר גדולה מזוית הד״ט. והזוית היותר גדולה מכל משולש יהיה מיתרה הצלע היותר ארוך. אם כן צלע ה״ד יותר ארוך[11] מן ה״ט[12]. וה״ד כמו ה״כ. אם כן ה״כ יותר ארוך מן ה״ט—זה שקר. הנה[13] כבר התבאר כי לא יפול בין קשת גא״ד ובין קו[14] ד״ז קו אחר ישר.

---

[1]ההוא ... שאם] +* פ [2]הא״ד] *⟨הד״א⟩ ב [3]כמו ... הד״א] ליתא בס [4]וזוית הד״א] ⟨וזוית הא״ד⟩ ט [5]נצב] נצבת ק [6]שתי] ליתא ט [7]ויפול] +חוץ מן העגולה ט +⟨⟨חוץ מן העגולה⟩⟩ פ [8]אפשר] +זה ק [9]יפול] +* פ [10]ה״ט הנה] ⟨⟨ה״ט⟩⟩ ב [11]אם ... ארוך] ליתא ב [12]ה״ט] +⟨⟨והוא ה״ד⟩⟩ ב ה״ס ⟨⟨*⟩⟩ ק [13]הנה] ⟨⟨*⟩⟩ ט [14]קו] קשת ז

ואומר שהזוית הפנימית שיקיפו בה קו הקוטר והקו המקיף[1], והיא זוית כד״ה [!][2], גדולה מכל זוית חדה, והזוית החיצונה שעליה ד׳ קטנה מכל זוית חדה.

فَأَقولُ إِنَّ الزَّاوِيَةَ الدَّاخلةَ الَّتي يُحيطُ بِها خَطُّ القُطْرِ وَالخَطُّ وَهِيَ الَّتي مِنْ ك ه د، أَكْبَرُ مِنْ كُلِّ زاوِيَةٍ حادَّةٍ. وَالزَّاوِيَةُ الَّتي مِنْ ك د ز أَصْغَرُ مِنْ كُلِّ زاوِيَةٍ حادَّةٍ.

לפי שאם היתה זוית חדה גדולה מהזוית הפנימית או קטנה מן החיצונה אשר תארנום, (לא)[3] היה אפשר שיפול בין קו ז״ד ובין קשת גא״ד קו אחר. ואמנם התבאר שלא יפול ביניהם קו אחר ישר. ואם כן זוית חצי העגול, והיא זוית כד״ה [!] הפנימית, גדולה מכל זוית חדה. והנשארת החיצונה שעליה (ד׳)[4] קטנה מכל זוית חדה.

بُرهانُهُ: أَنَّهُ لَوْ كانَتْ، تَكونُ زاوية حادَّة هِيَ أَكْبَرُ مِنَ الزَّاوِيَةِ الدَّاخِلَةِ وَأَصْغَرُ مِنَ الخارِجَةِ اللَّتَيْنِ وَصَفْناهُما، لا يُمْكِنُ أَنْ يَقَعَ بَيْنَ خَطِّ ز د وَبَيْنَ قَوْسِ ج ا د خَطٌّ آخَر. لا يَقَعُ. فَزاوِيَةُ نِصْفِ الدَّائِرَةِ الَّتي عَلَيْها ك د ج أَكْبَرُ مِنْ كُلِّ زاوِيَةٍ حادَّةٍ. والباقِيَةُ الخارِجَةُ الَّتي عَلَيْها ك د ز أَصْغَرُ مِنْ كُلِّ زاوِيَةٍ حادَّةٍ.

והנה התבאר לך הנה שהקו היוצא מקצה קוטר העיגול על זוית נצבה הוא ממשש העיגול.

وَهُنالِكَ اسْتَبانَ أَنَّ الخَطَّ الَّذي يَخْرُجُ مِنْ طَرَفِ قُطْرِ الدَّائِرَةِ عَلَى زَوايا قائِمَةٍ (يَماسُّ)[1] الدَّائِرَةَ.

וזה מה שרצינו לבאר.

وَذَلِكَ ما أَرَدْنا بَيانَهُ.

[1] המקיף] +~~גדולד~~ [2] והיא כד״ה] טעות תרגום: הכוונה היא לקשת ולא לזווית [3] לא] ליתא [4] ד׳] [נ׳]

[1] يماس] تماس

وأقولُ إنَّ الزَّاوِيَةَ الدّاخِلَةَ الَّتي يُحيطُ بها القُطْرُ والخطُّ المُحيطُ الَّتي عَلَيّها كـ د ه. أعظم مِنْ كُلِّ زاوِيَةٍ حادَّةٍ مُسْتَقيمةِ الخَطَّيْنِ. وَأَنَّ زاوية كـ د ز الخارِجَة أصغر مِنْ كُلِّ زاويةٍ حادَّةٍ مُسْتَقيمةِ الخَطَّيْنِ^.

ואומר כי הזוית הנכנסת[1] אשר יקיפו בה הקוטר והקו המקיף, אשר עליה[2] כד״ה, יותר גדולה מכל זוית חדה ישרת שני הקוים[3].

لأَنَّهُ لَوْ كانَتْ زاوية حادَّة مُسْتَقيمة الخَطَّيْنِ أعظم مِنَ الدّاخلة أو^ أصغر مِنَ الخارِجَة اللَّتَيْنِ سَمَّيْنا، لأَمْكَنَ أنْ يَقَعَ بَيْنَ خطّ د ز وبَيْنَ قَوْس ج ا د خطٌّ مُسْتَقيمٌ، وَلكنَّهُ لا يَقَعُ. فزاوية نِصْف الدَّائِرَة الَّتي عليها ج د كـ، أعظم مِنْ كُلِّ زاوِيَةٍ حادَّةٍ مُسْتَقيمةِ الخَطَّيْنِ. والَّتي عَلَيّها كـ د ز[1] أصغر مِنْ كُلِّ زاوِيَةٍ حادَّةٍ مُسْتَقيمةِ الخَطَّيْنِ.

כי אלו היתה זוית חדה ישרת שני הקוים יותר גדולה מן הפנימית ויותר קטנה מן החיצונה אשר זכרנום בשם, היה אפשר שיפול בין קו ד״ז ובין קשת גא״ד קו ישר. אבל לא יפול. אם כן זוית חצי העגולה אשר עליה גד״כ, יותר גדולה מכל זוית חדה ישרת שני הקוים. ואשר עליה כד״ז יותר קטנה מכל זוית חדה ישרת שני הקוים.

وَهُنالِكَ اسْتَبانَ أَنَّ الخطَّ الَّذي يَخْرُجُ مِنْ طَرَفِ قُطْرِ الدَّائِرَةِ عَلى زاوِيَةٍ قائِمَةٍ يَماسُّ الدَّائِرَةَ.

ובכאן נתבאר כי כאשר הוצא מקצה הקוטר קו[4] חוץ מן העגולה אל זוית נצבת ימשש העגולה[5].

وذَلِكَ ما أَرَدْنا أَنْ نُبَيِّنَ.

וזה מה שרצינו לבאר[6].

[1] كـ د ز] CamMPR1R2 +الخارجة

[1] הזוית הנכנסת] זויות הנכנסות הפנימיות **ס** [2] עליה] צ״ל: עליו [3] הקוים] +וזוית כד״ז החיצונה יותר קטנה מכל זוית חדה ישרת שני הקוים **ס** [4] קו] ⟨⟨(*)⟩⟩ **פ** ליתא **בוטסק** [5] ובכאן ... העגולה] ליתא **בוטק** ⟨⟨הנה בכאן התבאר כי כאשר הוצא מקצה הקטר קו חוץ מעגולה על זוית נצבת הנה הוא ימשש העגולה⟩⟩ **פ** [6] וזה ... לבאר] וזה מה שרצינו באורו **בק**

[Proposition III.16]

[H17]

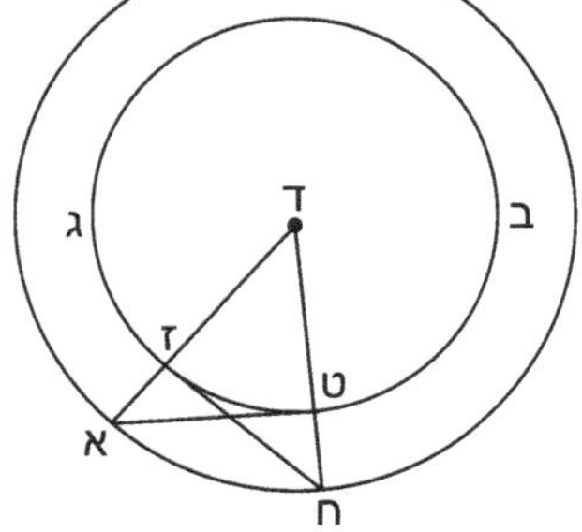

נרצה להוציא מנקודה ידועה אל עגול ידוע קו ימשש העגול.

ונשים הנקודה הידועה נקודת א׳. והעגול הידוע עגול ב״ג. וכאשר רצינו להוציא מנקודת א׳ קו שימשש עגול ב״ג[1], נוציא ממרכז עגול ב״ג, והוא ד׳, קו אל א׳. ונקיף על מרכז ד׳ עגול בשעור המרחק שבין ד׳ ובין א׳, והוא עגול א״ח. ונוציא מקו ד״א מנקודת ז׳ קו על זוית נצבה, והוא קו ז״ח. ונדביק ד׳ בח׳, וט׳ בא׳. ואומר שכבר הוצאנו מנקודת א׳ קו ממשש עגול ב״ג והוא קו א״ט.

نُرِيدُ أَنْ نُخْرِجَ مِنْ نُقْطَةٍ مَعْلومَةٍ خَطًّا يَمَاسُّ دائِرَةً مَعْلومَةً.

مِثالُهُ: أَنَّا نَجْعَلُ النُّقْطَةَ المَعْلومَةَ نُقْطَةَ ا. وَالدَّائِرَةَ المَعْلومَةَ دائِرَةَ ا ب ج د. وَنُرِيدُ أَنْ نُخْرِجَ مِنْ نُقْطَةِ ا خَطًّا يَمَاسُّ دائِرَةَ ب ج. فَنُخْرِجُ مِنْ مَرْكَزِ دائِرَةِ ب ج، وَهُوَ د، خَطًّا إِلى نُقْطَةِ ا. وَنَخُطُّ عَلى مَرْكَزِ د وَنُقَدِّرُ بُعْدَها بَيْنَ د ا دائِرَةَ ح ا. وَنُخْرِجُ مِنْ خَطِّ د ا مِنْ نُقْطَةِ ز مِنْهُ خَطًّا عَلى زاوِيَةٍ قائِمَةٍ، وَهُوَ خَطُّ ز ح. وَنَصِلُ (د ح)[1] ط ا. فَنَقولُ إِنَّا قَدْ أَخْرَجْنا مِنْ نُقْطَةِ ا خَطًّا يَمَاسُّ دائِرَةَ ا ب ج، وَهُوَ خَطُّ ا ط.

[1]וכאשר ... ב״ג] ⟨⟨*⟩⟩

[1]د ح] ز ح

نُرِيدُ أَنْ نُخْرِجَ مِنْ نُقْطَةٍ مَعْلومَةٍ[1] خَطًّا يُمَاسُّ الدَّائِرَةَ.

נרצה שנוציא מנקודה ידועה קו ימשש העגולה.

فَنَجْعَلُ النُقْطَةَ المَعْلومَةَ نُقْطَة ا. والدَّائِرَة دائِرَة ب ج. ونُرِيدُ أن نُخْرِجَ مِنْ نُقْطَةِ ا خَطًّا يُمَاسُّ دائِرَةَ ب ج. فَنُخْرِجُ مِنَ المَرْكَزِ^، وهو د، خطّ د ا. ونَخُطّ على مَرْكَزِ د وبِبُعْدِ د ا دائِرَةَ ا ح[2]. ونُخْرِج، مِنْ خطّ ا د مِنْ نُقْطَةِ ز مِنْهُ، خطًّا على زاوِيةٍ قائِمَةٍ، وهو ز ح. ونُخْرِج د ح و ط ا. نَخُطّ د ح مثل د ا. و د ز مثل د ط.

הנה נשים הנקודה הידועה נקודת א׳. והעגולה עגולת ב״ג. ונרצה שנוציא[1] מן א׳ קו ימשש עגולת ב״ג. הנה נוציא ממרכז העגולה, והוא ד׳, קו ד״א. ונקוה על מרכז ד׳ ובמרחק א״ד עגולת א״ח. ונוציא, מקו א״ד מנקודת ז׳ ממנו, קו[2] על זוית נצבת, והוא[3] ז״ח. ונוציא ד״ח[4] וט״א. הנה[5] קו ד״ח כמו קו[6] א״ד. וד״ז כמו ד״ט.

---

[1] معلومة] CT1T3 +الى دايرة معلومة CamMO2PR1R2UT2 [2] ا ح]

שרטוט—**בוטספק**: $; **ס**: ג׳ חסרה [1] ונרצה שנוציא] הנה נוציא ממרכז **ס** [2] קו] ליתא **ס** [3] והוא] הוא **ס** [4] ד״ח] מקו+ **ב** [5] הנה] +~~כבב~~ **ט** [6] קו] ליתא **טס**

מופת זה: שד״ח שוה לד״א, וד״ז שוה (לד״ט)[1]. ואם כן שני קוי ד״ח וד״ז שוים לשני קוי א״ד וד״ט, כל אחד לדומה לו. והזוית שיקיפו בה קוי ד״ח וד״ז וקוי א״ד וד״ט—אחת, והיא זוית ד׳. ואם כן תושבת ח״ז שוה לתושבת א״ט. ומשלש זד״ח שוה למשולש טד״א. והזויות הנשארות שוות לזויות הנשארות שהצלעים השוים הם להם מיתרים. ואם כן זוית דז״ח שוה לזוית דט״א[2]. וזוית דז״ח נצבת. ואם כן זוית דט״א נצבת, והן יחד נצבות. וד״ט[3] הוא קוטר העגול. והקו שיצא מקצה הקוטר על זוית נצבה הוא ממשש לעגול. וכבר התבאר שקו ט״א ימשש עגול ב״ג.

ונשלם מופתו.

[1]לד״ט] לח״ט [2]דט״א] [ד]ט״א [3]וד״ט] ⟨⟨*⟩⟩

بُرْهانُهُ: أَنَّ د ح مثل د ا. د ز مثل د ط. (فَكِلا)[1] خَطَّيْ د ح د ز مثل كِلَيْ خَطَّيْ ا د د ط، كُلّ واحدٍ مثل نَظيرِهِ. وَالزَّاوِيَةُ الَّتي يُحيطُ بِها خَطّا د ح د ز وخَطّا ا د د ط—واحِدَةٌ، وَهِيَ زاوِيَةُ د. فَقاعِدَةُ ح ز مثل قاعِدَةِ ا ط. ومُثَلَّثُ ز ح د مثل الَّتي مِنْ د ط ا [..]. وَهُما جَميعًا قائِمَتانِ. د ط خَطُّ القُطْرِ. وَالخَطُّ الَّذي يَخْرُجُ مِنْ طَرَفِ القُطْرِ عَلى زاوِيَةٍ قائِمَةٍ هو مُماسُّ الدَّائِرَةِ. فَقَدِ اسْتَبانَ أَنَّ خَطَّ ط ا يَماسُّ دائِرَةَ ب ج.

وَذَلِكَ ما أَرَدْنا بَيانَهُ.

[1] فكلا] فكلي

نَخَطّا د ح و د ز مِثْلُ خَطَّي ا د و د ط. وَالزَّاوِيَةُ الَّتي يُحِيطُ بِها خَطّا د ح و د ز وَخَطّا ا د و د ط—واحِدَةٌ. فَقاعِدَةُ ز ح مِثْلُ قاعِدَةِ ا ط. وَمُثَلَّثُ ز د ح مِثْلُ مُثَلَّثِ ط د ا. وَالزَّوايا الباقِيَةُ مِثْلُ الزَّوايا الباقِيَةِ الَّتي يُوتِرُها الأَضْلاعُ المتساوِيَةُ. فَزاوِيَةُ د ط ا مِثْلُ زاوِيَةِ د ز ح. فَهُما جَمِيعًا قائِمَتانِ. و د ط خَطُّ القُطْرِ. وَالخَطُّ الَّذي يَخْرُجُ مِنْ طَرَفِ القُطْرِ عَلى زاوِيَةٍ قائِمَةٍ مُماسٌّ لِلدَّائِرَةِ. فَقَدِ اسْتَبانَ أَنَّ خَطَّ ا ط مُماسٌّ لِدائِرَةِ ب ج.

وَذَلِكَ ما أَرَدْنا أَنْ نُبَيِّنَ.

הנה שני קוי[1] ד״ח וד״ז כמו שני קוי[2] א״ד וד״ט. והזוית[3] אשר יקיפו בה שני קוי ח״ד וד״ז ושני קוי א״ד וד״ט—אחת[4]. הנה תושבת ז״ח כמו תושבת א״ט. ומשולש זד״ח[5] כמו משולש טד״א. והזויות הנשארות כמו הזויות הנשארות אשר יהיו מיתריהם הצלעות השוות. אם כן[6] זוית דט״א כמו זוית דז״ח. הנה שתיהם[7] יחד נצבות. וד״ט קו הקוטר. והקו אשר יצא מקצה הקוטר על זוית נצבת[8] ימשש העגולה. הנה כבר התבאר כי קו[9] א״ט ימשש עגולת ב״ג[10].

וזה מה שרצינו לבאר[11].

---

[1]קוי] קוים **ק** [2]ד״ח ... קוי] ליתא **ז** [3]והזוית] והזויות **ס** [4]והזוית ... אחת] ⟨⟨וזויות ד׳ משותף⟩⟩ **ב** כמו שני קוי ז״ד וח״ד ויקיפו בזוית אחת **ז** [5]זד״ח] ~~זד״ט~~+ **ק** [6]כן] ליתא **ב** [7]שתיהם] שניהם **ט** [8]על ... נצבת] ליתא **ז** [9]קו] ליתא **ס** [10]עגולת ב״ג] עגולה ב״ט **ס** [11]לבאר] לבארו **בפ** ביאורו **ז**

[Proposition III.17]

[H18]

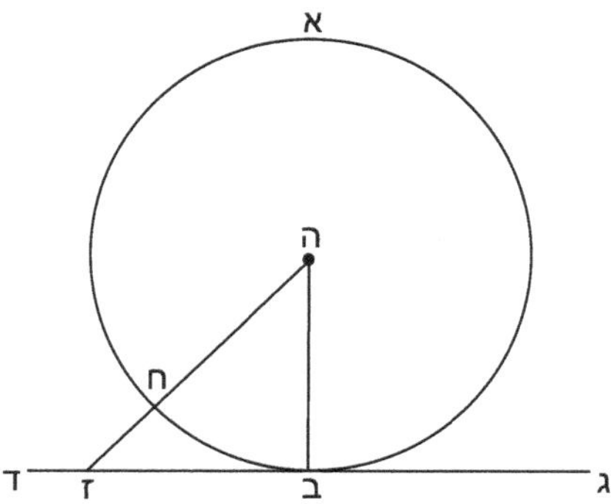

כל קו שימשש עגול ויצא ממקום משושו קו אל המרכז, יהיה הקו ההוא עמוד על הקו הממשש העגול.

إِذا كانَ خَطٌّ يَماسُّ الدَّائِرَةَ، وأُخْرِجَ مِنَ المَوْضِعِ الَّذي يَماسُّها فيهِ خَطٌّ إِلى المَرْكَزِ. فَإِنَّ الخَطَّ المُخْرَجَ يَكونُ عَمُودًا عَلى الخَطِّ المُماسِّ لِلدَّائِرَةِ.

המשל בו: שעגול א״ב ימששהו קו[1] גב״ד על נקודת ב׳. ומרכז העגול ה׳. ונדביק ה׳ בב׳. ואומר שקו ב״ה עמוד על קו ג״ד[2].

مِثالُهُ: أَنَّ (دائِرَةَ)[1] ا ب يَماسُّها خَطُّ (ج ب د)[2] عَلى نُقْطَةِ ب. وَمَرْكَزُ الدَّائِرَةِ ه. فَنَصِلُ ب ه. فَأَقولُ إِنَّ ه ب عَمودٌ وَقَعَ عَلى خَطِّ ج د.

שרטוט—ד׳ חסרה [1]קו] ~~עגול~~+ [2]ג״ד] ⟨⟨ג״⟩⟩ד

[1]دائرة] دائرتي [2]ج ب د] ب ج د

كُلُّ خَطّ يَمَاسُّ الدَّائِرَةَ[1]، ويَخْرُجُ مِنَ المَوْضِعِ الَّذي يَمَاسُّهَا فيهِ خَطٌّ إلى المرْكَزِ، فَإنَّ الخَطَّ يكونُ عَمُودًا على الخَطِّ المُمَاسِّ للدائِرَةِ.

כל קו ימשש[1] העגולה, ויצא מן המקום אשר ימששה בו קו אל המרכז, הנה הקו יהיה עמוד[2] על הקו הממשש[3] לעגולה.

مِثَالُهُ: أَنَّ خَطّ ج د يَمَاسُّ دائِرَةَ ا ب على نُقْطَةِ ب. والمَرْكَزَ ه. ونُخْرِجُ ب ه. فَأَقولُ إنَّهُ عَمودٌ على ج د، لا يُمْكِنُ غيره.

המשל בו: כי קו ג״ד ימשש עגולת[4] א״ב על נקודת ב׳. והמרכז ה׳. ונוציא ב״ה. הנה אומר שהוא[5] עמוד על ג״ד, אי אפשר זולתו.

[1] الدايرة] MPR1R2

שרטוט—**ס**: ח׳ חסרה [1] ימשש] +* **ט**
[2] עמוד] העמוד **ו** [3] הממשש] ימשש **ס**
[4] עגולת] לעגולת **ב** [5] שהוא] כי ב״ה **ט**

ואי איפשר אלא זה. ואם היה אפשר, נוציא מה׳ עמוד על קו ג״ד בלעדי ה״ב, והוא ה״ז. ואם כן זוית הז״ב נצבה. וזוית[1] הב״ז קטנה מנצבה. ואם כן זוית הז״ב גדולה מזוית הב״ז. והזוית הגדולה יהיה לה מיתר הצלע הגדולה, ואם כן ה״ב ארוך מה״ז. וה״ב שוה לה״ח. ואם כן ה״ח ארוך מה״ז—זה חלוף. ואם כן לא יתכן שיהיה ה״ז עמוד על קו ג״ד, ולא זולתו מן הקוים שיצאו מה׳ כי אם קו ה״ב.

וזה מה שרצינו.

بُرهانُهُ: أنَّهُ لا يُمْكِنُ غَيْرُهُ. فَإِنْ كانَ يُمْكِنُ، فَلْيُخْرَجْ مِنْ ه عَموداً إلى خَطِّ ج د غَيْرِ ه ب. وَهُوَ ه ز. فَالزَّاوِيَةُ الَّتِي مِنْ ه ز ب إذاً قائِمَةٌ. وَالَّتِي مِنْ ه ب ز حادَّةٌ. وَالَّتِي مِنْ ه ز ب إذاً أَعْظَمُ مِنَ الَّتِي مِنْ ه ب ز. فَالزَّاوِيَةُ العُظْمى يوتِرُها الضِّلْعُ الأَطْوَلُ، ه ب أَطْوَلُ من ه ز. ه ب مثل ه ح. ه ح أَطْوَلُ من ه ز—هذا خُلْفٌ لا يُمْكِنُ. فَلَيْسَ يُمْكِنُ أَنْ يكونَ ه ز عموداً على خَطِّ ج د، وَلا غَيْرُهُ مِنَ الخُطوطِ الَّتِي تَخْرُجُ مِنْ ه إلّا خَطّ ه ب.

وَذَلِكَ ما أَرَدْنا بَيانَهُ.

---

[1] וזוית] ו[...]זוית

فَإِنْ أَمْكَنَ، فَلْيَخْرُجْ مِنْ ه عَمُودٌ إلى خطّ ج د غير ب ه، وهو ه ز. فزاوية ه ز ب إِذَنْ[1] قائِمَةٌ. وزاوية ه ب ز حادّة. فزاوية ه ز ب أعظم مِنْ زاوية ه ب ز. والزَّاوِيَةُ العُظمى مِنْ كُلِّ مُثَلَّث—يوتِرُها الضِّلعُ الأَطْوَلُ. فَخَطّ ه ب أطول مِنْ ه ز. و ه ب مثل ه ح. ف ه ح أطول مِنْ ه ز. وهذا خُلْفٌ. فَلَيْسَ يُمْكِنُ أَنْ يكونَ ه ز عَمُودًا على ج د ولا غيره مِنَ الخطوط الَّتي تَخْرُجُ مِنْ ه.^

وذَلِكَ ما أَرَدْنا أَنْ نُبيِّنَ.

שאם היה אפשר[1], הנה יצא[2] מן ה׳ עמוד אל[3] קו ג״ד זולת ב״ה, והוא ה״ז. הנה זוית הז״ב אם כן נצבת. וזוית הב״ז חדה. אם כן זוית הז״ב יותר גדולה מזוית הב״ז. והזוית היותר גדולה מכל משולש—מיתרה הצלע היותר ארוך. הנה[4] קו ב״ה יותר ארוך מן ה״ז. וה״ב כמו ה״ח. אם כן[5] ה״ח יותר ארוך מן ה״ז. וזה שקר. אם כן[6] [7] אי אפשר[8] שיהיה ה״ז עמוד על ג״ד ולא זולתו מן הקוים אשר יצאו ממנו.

וזה מה שרצינו לבאר.

1 اذن] DT4

1 זולתו ... אפשר] ליתא **ב** +זולתו **ס** 2 יצא] יוצא **ט** 3 אל] על **ט** 4 הנה] +אין **ז** +⟨⟨אם כן⟩⟩ **ק** 5 כן] ליתא **ז** 6 ה״ח ... כן] ליתא **ס** 7 אם כן] ליתא **ז** 8 אם ... אפשר] +* **ז**

[Proposition III.18]

[H19]

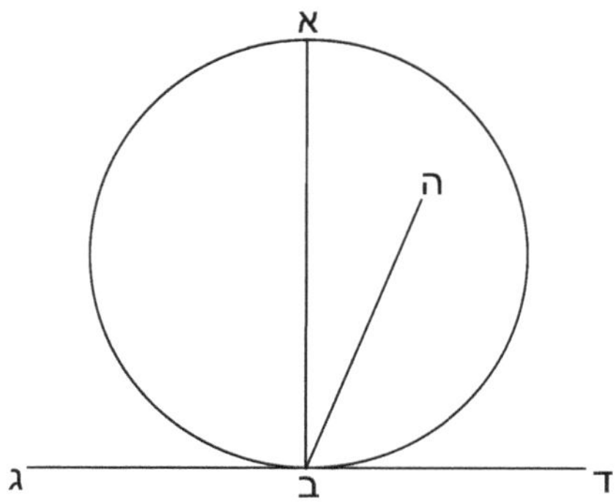

כשימשש קו לעגול ויצא ממקום משושו קו על זוית נצבה אל תוך העגול—עליו יהיה המרכז.

إِذا كانَ خَطٌّ يَماسُّ الدَّائِرَةَ وأُخْرِجَ مِنَ المَوْضِعِ الَّذي يَماسُّها فيهِ خَطٌّ عَلى زاوِيَةٍ قائِمَةٍ إِلى داخِلِ الدَّائِرَةِ، فَعَلى ذلكَ الخَطِّ يَكونُ مَرْكَزُ الدَّائِرَةِ.

המשל בו: שקו ג״ד ימשש עגול א״ב על נקודת ב׳. ויצא מב׳ קו על זוית נצבה, והוא א״ב. ואומר שהמרכז יפל על קו א״ב.

مِثالُهُ: أَنَّ خَطَّ ج د يَماسُّ الدَّائِرَةَ الَّتي عَلَيْها ا ب عَلى نُقْطَةِ ب. ونُخْرِجُ مِنْ ب خَطًّا عَلى زاوِيَةٍ قائِمَةٍ، وَهُوَ ا ب. فَأَقولُ إِنَّ المَرْكَزَ يَقَعُ عَلى خَطِّ ا ب.

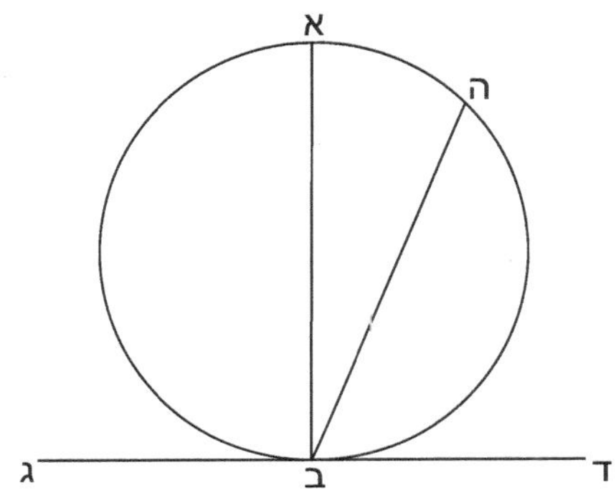

إذا ماسَّ^ خطٌّ دائِرَةً، وَخَرَجَ مِنْ حَيْثُ يماسُّها خَطٌّ على زاوِيَةٍ قائِمَةٍ إلى داخِلِ الدَّائِرَةِ، فَعَلَيْهِ يَكونُ الْمَرْكَزُ.

כאשר ימשש קו עגולה[1], ויצא מהמקום שימששה קו על זוית נצבת לפנים מן[2] העגולה[3], הנה עליו יהיה המרכז.

مِثالُهُ: أَنَّ خطّ ج د يَماسُّ دائِرَةَ ا ب على ب. وَيَخْرُجُ مِنْ نُقْطَةِ ب خطّ ا ب على زاوِيةٍ قائِمَةٍ. فَأَقولُ إنَّ الْمَرْكَزَ يَقَعُ على خَطِّ ا ب، لا يُمْكِنُ غيره.

המשל בו: כי קו[4] ג״ד ימשש עגולת א״ב על ב׳. ויצא מנקודת ב׳ קו א״ב על זוית נצבת. הנה אומר כי המרכז[5] יפול על קו א״ב, אי אפשר זולתו.

שרטוט—**בוטספק**: $: **בפ**: ח׳ במרכז המעגל; **פ**: ה׳ בתוך המעגל; בכתבי היד הערביים יש שונות במקום ە—על היקף המעגל או בתוכו; במקצתם ج ↔ د
[1]עגולה] העגולה **ו** [2]מן] ליתא **ס** [3]העגולה] העגולות **ב** העגולת **פ** [4]קו] ליתא **ב**
[5]המרכז] +שהוא ז׳ **ב**

بُرْهانُهُ: أَنَّهُ لا يُمْكِنُ إِلّا ذلِكَ. فَإِنْ كانَ يُمْكِنُ، فَلْيَقَعْ عَلى خَطِّ ه ب. فَالزّاوِيَةُ الَّتي من (ه ب د)[1] إِذًا قائِمَةٌ. وَالَّتي مِنْ (د ب ا)[2] قائِمَةٌ. فَالَّتي مِنْ ه ب د مثل الَّتي مِنْ (د ب ا)[3]، الكُبْرى مثل الصُّغْرى—هَذا خُلْفٌ لا يُمْكِنُ. فَلَيْسَ مَرْكَزُ دائِرَةِ[4] ا ب عَلى خَطِّ ه ب، وَلا عَلى غَيْرِهِ مِنَ الخُطوطِ إِلّا عَلى خَطِّ ا ب يكونُ مَرْكَزُ الدّائِرَةِ.

وَذَلِكَ ما أَرَدْنا بَيانَهُ.

ואי אפשר שיפול על זולתו. ואם היה אפשר, יפל על קו ה״ב. ואם כן זוית הב״ד נצבה. וזוית דב״א נצבה. ואם כן זוית הב״ד[1] שוה לזוית דב״א. ואם כן הקצת שוה לכל—זה חלוף מן[2] המוסכמות, שכל זויות נצבות שוות קצתם לקצת. ואם כן אין ה׳ מרכז[3] עגול א״ב. ולא יהיה בזולת קו א״ב. והתבאר שעל [!] קו א״ב הוא מרכז העגול.

וזה מה שרצינו.

[1] ه ب د] ه ب [2] د ب ا] ه ب ا [3] د ب ا] ب د ا
[4] دائرة] ⟨*⟩

[1] הב״ד] הג״ד
[2] מז] +⟨⟨[כל]⟩⟩ ⟨⟨[ע״כ]⟩⟩
[3] מרכז] ⟨*⟩

فَإِنْ أَمْكَنَ، فَلْيَقَعْ^ على خَطِّ ب ه. فزاوية ه ب د قائِمَة. وزاوية د ب ا قائِمَة. فَهُما مُتَساوِيَتانِ^—هذا خُلْفٌ. فَلَيْسَ يُمكِنُ الْمَرْكَزُ على ب ه ولا غير خطّ ا ب. فَمَرْكَزُ الدَّائِرَةِ على ا ب.

שאם יהיה אפשר, נאמר שיפול על קו ב״ה. אם כן זוית הב״ד נצבת. וזוית דב״א נצבת. ושתיהן[1] נצבות—זה שקר[2]. אם כן אין המרכז על ב״ה, ולא בזולת[3] קו א״ב. אם כן מרכז העגולה על א״ב.

فذَلِكَ ما أَرَدْنا أَنْ نُبَيِّنَ.

וזה מה שרצינו לבאר[4].

---

[1] ושתיהן] ושתיהם **וטס** [2] שקר] +אי אפשר **ק** [3] בזולת] על זולת **ז** זולת **ט** [4] לבאר] ולבאר **ב**

[Proposition III.19]

[H20]

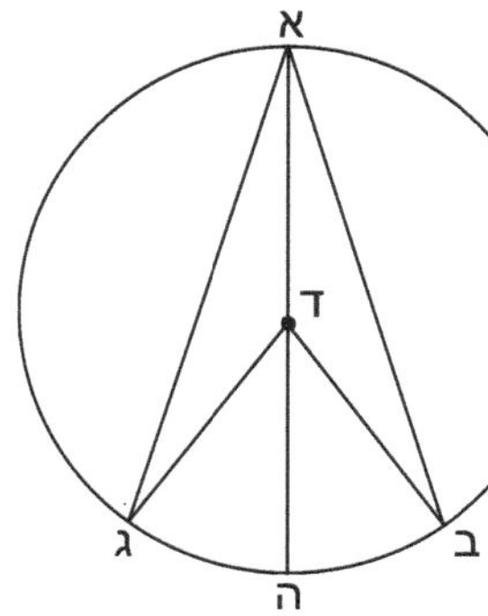

הזוית שיהיה[1] על מרכז העגול, כפל הזוית שיהיה על הקו המקיף, כשתהיה תושבתם קשת אחת.

الزَّاوِيَةُ الَّتِي تكونُ عَلى مَرْكَزِ الدَّائِرَةِ مِثْلا الزَّاوِيَةِ الَّتِي تَكونُ عَلى مُحِيطِها، إِذا كانَتْ قاعِدَتُهُما قَوْسًا واحِدًا.

המשל בו: שעל מרכז עגול אב״ג זוית בד״ג, ועל הקו המקיף זוית בא״ג. ותושבתם אחת, והיא קשת ב״ג. ואומר שזוית בד״ג כפל זוית בא״ג.

مِثالُهُ: أَنَّ عَلى مَرْكَزِ دائِرَةِ ا ب ج زاوِيَةً، وَهِيَ الَّتِي مِنْ ب د ج. وَعَلى الخَطِّ المُحِيطِ زاوِيَةً، وَهِيَ الَّتِي مِنْ ب ا ج. وَقاعِدَتُهُما جَمِيعًا واحِدَةٌ، وَهِيَ قَوْسُ ب ج. فَأَقولُ إِنَّ الزَّاوِيَةَ الَّتِي مِنْ ب د ج مِثْلا الَّتِي مِنْ ب ا ج.

שרטוט—B: ז̇ על היקף המעגל

1 שיהיה] ש⟨י⟩היה

الزَّاوِيَةُ الَّتِي تكونُ على مَرْكَزِ الدَّائِرَةِ، مِثْلا الَّتِي تَكونُ على الخطِّ المُحِيطِ، إذا كانَتْ قاعِدَتُهما قَوْسًا واحِدَةً.

הזוית אשר תהיה על מרכז העגולה, כמו שני פעמים מאשר תהיה על הקו המקיף, כאשר תהיינה תושבות שתיהן[1] קשת אחת.

مِثالُهُ: أنَّ على مَرْكَزِ دائِرَةِ ا ب ج زاوية ب د ج. وعلى الخطّ المُحِيط بها زاوية ب ا ج. وقاعدتهما جَمِيعًا—قَوْس ب ج. فَأَقولُ إنَّ زاوية ب د ج مِثْلا زاوية ب ا ج.

המשל בו: כי על מרכז עגולת בא״ג זוית בד״ג. ועל הקו המקיף בה זוית[2] בא״ג. ותושבת שתיהן[3] יחד—קשת[4] ב״ג. הנה אומר כי זוית בד״ג כפל זוית בא״ג.

[1] שתיהן] שתיהם **בופ** [2] בה זוית] **ב**זוית **ס**
[3] שתיהן] שתיהם **בוטפ** [4] קשת] קש⟨ת⟩ **ט**

המופת בו: שנדביק א׳ בד׳. ונוציא הקו עד ה׳. ותהיה זוית גד״ה כפל זוית דא״ג. וזוית בד״ה כפל זוית בא״ד. ואם כן הזוית כלה שהיא זוית בד״ג כפל בא״ג כלה[1].

بُرْهانُهُ: أَنَّا نَصِلُ ا د. وَنُخْرِجُ الخَطَّ إِلى ه. فَالزَّاوِيَةُ الَّتِي مِنْ ج د ه مِثْلا الَّتِي مِنْ ج ا د. وَالَّتِي مِنْ ب د ه مِثْلا الَّتِي مِنْ ب ا د. فَجَمِيعُ الَّتِي مِنْ ب د ج مِثْلا الَّتِي مِنْ ب ا ج.

וזה מה שרצינו.

وَذَلِكَ ما أَرَدْنا بَيانَهُ.

[1]כלה] כל⟨⟨ה⟩⟩

بُرْهانُهُ: أَنّا نُخْرِجُ خَطَّ ا د. ونُخْرِجُهُ إلى ه. فَخَطّا ا د و د ج مُتَساوِيانِ. فَزاوِيَتا د ا ج و د ج ا مُتَساوِيَتانِ[1]. وزاوية ه د ج مِثْلا زاوية د ا ج. وَلِذَلِكَ[2] زاوية ب د ه مِثْلا زاوية ب ا د. فَجَميع زاوية ب د ج مِثْلا جميع زاوية ب ا ج.

وذَلِكَ ما أَرَدْنا أَنْ نُبَيِّنَ.

המופת: כי נוציא קו א״ד. ונוציאהו אל ה׳. הנה שני קוי א״ד וד״ג שוים. אם כן שתי זויות דא״ג ודג״א שוות. וזוית הד״ג כפל זוית[1] דא״ג. ולכן[2] זוית בד״ה כפל זוית בא״ד. הנה כל[3] זוית בד״ג כפל כל[4] זוית בא״ג.

וזה מה שרצינו לבאר.

---

[1] فخطا ... متساويتان] **MPR1R2U** [2] ولذلك] **MPR1R2U**

---

[1] זוית] זויות **ב** ליתא **ו** [2] ולכן] הנה כל **ב** +⟨⟨נ״א וזוית הד״ג כמו שתי זויות דא״ג דג״א אם כן זוית הד״ב כפל זוית דא״ג וכו׳⟩⟩ **ק** [3] בד״ה ... כל] ליתא **ב** ⟨⟨*⟩⟩ **ק** [4] כל] ליתא **טס**

[Proposition III.20]

[H21]

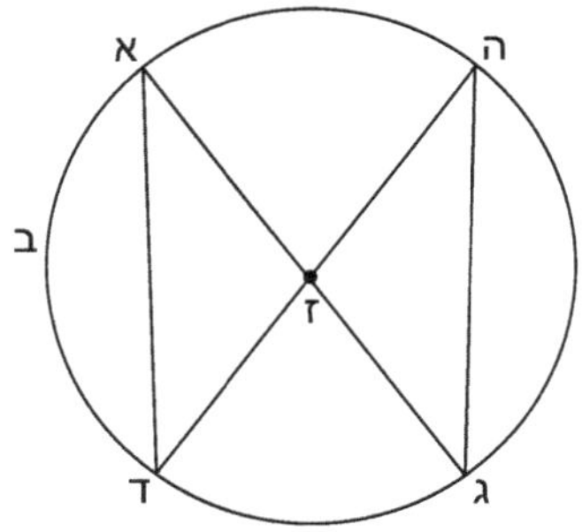

כשתהיינה בחתיכה אחת מעיגול שתי זויות—הן שוות.

המשל בזה: שעגול א״ב—בחתיכה[1] ממנו, והוא חתיכת גא״ד, שתי זויות, והן זוית גה״ד וזוית גא״ד. ואומר שהן שוות.

إذا كانَ في قِطْعَةٍ واحِدَةٍ مِنَ الدَّائِرَةِ زاوِيَتانِ عَلى قاعِدَةٍ واحِدَةٍ، فَهُما مُتَساوِيَتانِ.

مِثالُهُ: أَنَّ دائِرَةَ ا ب، في قِطْعَةٍ مِنْها، وَهِيَ قِطْعَةُ ج ا د. زاوِيَتانِ، وَهُما اللَّتانِ مِنْ (ج ه د)[1] وَمِنْ ج ا د. فَأَقولُ إِنَّهُما مُتَساوِيَتانِ.

שרטוט: B: ا ↔ ه ונקודת החיתוך של ج ز ز د מתחת נקודת החיתוך של ا د ج ه   [1] בחתיכה] בחת⟨י⟩כה

[1] ج ه د] ج ه

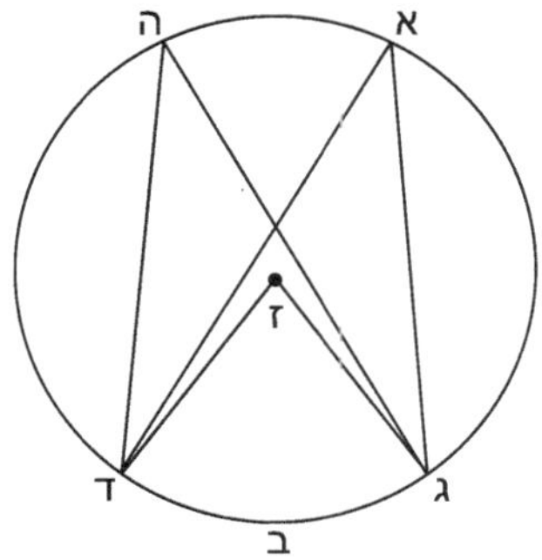

إِذا كانَ في قِطْعَةٍ واحِدَةٍ مِنْ دائِرَةٍ زاوِيَتانِ فَهُما مُتَساوِيَتانِ.

כאשר תהיינה[1] בחתיכה[2] אחת מעגולה[3] שתי זויות, הנה שתיהן שוות.

مِثَالُهُ: أَنَّ في دائِرَةِ ا ب، في قِطْعَةِ ج ا ه د مِنْها[٨]، زاوِيَتا ج ه د و ج ا د. فَأَقولُ إِنَّهُما مُتَساوِيَتانِ.

המשל בו[4]: כי בעגולת אב״ג חתיכת גאה״ד[5]—בה שתי זויות גה״ד וגא״ד. הנה אומר שהן שוות.

[1]תהיינה] תהיה **ב** [2]בחתיכה] בחתיכת **ב** [3]מעגולה] מעגולת **ב** [4]בו] ליתא **פ** [5]גאה״ד] ג״ד **ס**

המופת בו: שמרכז העגול נקודת ז׳. ונדביק ג׳ בז׳, וז׳ בד׳. ותהיה זוית גז״ד על מרכז העגול. וזוית גה״ד על הקו המקיף. ותושבתם אחת, והיא קשת ג״ד. ואם כן זוית גז״ד כפל זוית גה״ד. ותהיה גם כן כפל זוית גא״ד. ואם כן זוית גה״ד שוה לזוית גא״ד.

بُرْهانُهُ: أَنَّ مَرْكَزَ الدَّائِرَةِ نُقْطَةُ زَ. فَنَصِلُ ج ز ز د. فَالزَّاوِيَةُ الَّتي مِنْ ج ز د على مَرْكَزِ الدَّائِرَةِ، وَالَّتي مِنْ ج ه د على[1] الخَطِّ المُحيطِ، وَقاعِدَتُهُما واحِدَةٌ، وَهِيَ قَوْسُ ج د. فَالزَّاوِيَةُ الَّتي مِنْ ج ز د مِثْلا الَّتي مِنْ (ج ه د)[2]. وَكَذَلِكَ تَكونُ أَيْضًا مِثْلَ الَّتي مِنْ ج ا د. وَالَّتي مِنْ ج ه د مُساوِيَةٌ لِلَّتي مِنْ ج ا د.

וזה מה שרצינו לבאר.

وَذَلِكَ ما أَرَدْنا بَيانَهُ.

[1] مركز ... على] ⟨*⟩ [2] ج ه د] ه ج د

بُرْهانُهُ: أنَّا نَجْعَلُ الْمَرْكَزَ ز. ونُخْرِجُ ج ز و ز د. فزاوية ج ز د على الْمَرْكَزِ. وزاوية ج ه د على الخَطِّ المُحِيطِ. وقاعِدَتُهما قَوْس ج د^. فزاوية ج ز د مِثْلا زاوية ج ه د. ولِذَلِكَ[1] أيضًا مِثْلا زاوية ج ا د. فزاوية ج ه د مثل زاوية ج ا د.

وذَلِكَ ما أَرَدْنا أَنْ نُبَيِّنَ.

[1] ولِذَلِكَ] R1R2U

המופת בו: כי נשים המרכז ז׳[1]. ונוציא ג״ז וז״ד. הנה זוית גז״ד על המרכז[2]. וזוית גז״ד כפל זוית גה״ד. ולכן גם כן יהיה כפל זוית גא״ד. אם כן זוית גה״ד כמו זוית גא״ד.

וזה מה שרצינו לבאר[3].

[1] המרכז ז׳] ז׳ מרכז **ס** [2] המרכז] +⟨⟨וזוית גה״ד על הקו המקיף והגביל אותה קשת ג״ד אם כן זוית דז״ג כפל זוית [..]⟩⟩ **פ** +⟨⟨נ״א הנה זוית גז״ד על המרכז וזוית גה״ד על הקו המקיף והגביל אותה [..] ג״ד אם כן זוית גז״ד כפל זוית גה״ד וכו׳⟩⟩ **ק** [3] לבאר] לבארו **בפק**

[Proposition III.21]

[H22]

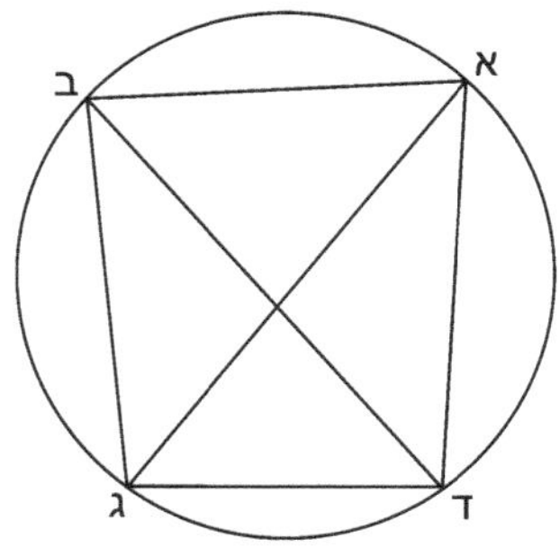

כל עגול שבו תמונת [!] בעלת ארבע צלעות, תהיין כל שתי זויות מזויותיו המקבילות זו את זו שוות לשתי זויות נצבות.

كُلُّ دائِرَةٍ يَقَعُ فيها سَطْحٌ ذو أَرْبَعَةِ أَضْلاعٍ، فَكُلّ زاوِيتَيْنِ مُتَقابِلَتانِ فيهِ مُساوِيَتانِ[1] لِزاوِيَتَيْنِ قائِمَتَيْنِ.

המשל בזה: שבתוך עגול אב״ג תמונה בעלת ארבע צלעות, והם א״ב וב״ג וג״ד וד״א. ואומר שכל שתי זויות המקבילות זו את זו מזוייות אבג״ד שוות לשתי זויות נצבות.

مثالُهُ: أَنَّا نُصَيِّرُ في دائِرَةِ ا ب ج د سَطْحًا ذا أَرْبَعَةِ أَضْلاعٍ، وَهِيَ ا ب ب ج ج د[2] د ا. فَأَقولُ إِنَّ كُلَّ زاوِيَتَيْنِ تَتَقابَلانِ[3] مِنْ زَوايا ا ب ج د مُساوِيَتانِ لِزاوِيَتَيْنِ قائِمَتَيْنِ.

---

[1] مساويتان] متساويتان [2] ج د] ⟨*⟩ [3] تتقابلان] يتقابلان

كُلُّ دائِرَةٍ يَقَعُ فيها شَكْلٌ ذو أَرْبَعَةِ أَضْلاعٍ، فكُلُّ زاوِيَتَيْنِ تَتَقابَلانِ مِنْ زَواياهُ، فَهُما مُساوِيَتانِ لِزاوِيَتَيْنِ قائِمَتَيْنِ.

כל[1] עגולה תפול בה תמונה[2] בעלת ארבע צלעות, הנה כל שתי זויות מתנגדות מזויותיה, הנה שתיהן שוות[3] לשתי זויות[4] נצבות.

مِثالُهُ: أَنَّ في دائِرَةِ ا ب ج د^ شَكْلٌ ذو أَرْبَعَةِ أَضْلاعٍ، وهو شَكْلُ ا ب ج د. فَأَقولُ إِنَّ كُلَّ زاوِيَتَيْنِ تَتَقابَلانِ مِنْ زوايا شَكْلِ ا ب ج د^ مُساوِيَتانِ لِزاوِيَتَيْنِ قائِمَتَيْنِ.

המשל בו: כי בעגולת אב״ג תמונה[5] בעלת ארבע צלעות, והיא תמונת אבג״ד. הנה אומר כי כל שתי זויות מתנגדות מזויותיה שוות לשתי נצבות.

שרטוט—בוסק: $; **פ**: שני שרטוטים, באחד ג ↔ ד [1]כל] ליתא **ב** [2]תמונה] תמונת **ז** [3]שוות] ⟨*⟩ **ז** [4]זויות] זוית **ב** ליתא **ז** [5]תמונה] תמונת **בפ**

המופת בו: שנדביק א׳ בג׳ וד׳ בב׳. וכל שתי זויות בחתיכת עגול הן שוות. ואם כן זוית בא״ג שוה לזוית בד״ג. ועוד זוית בד״א [!] שוה[1] לזוית אג״ב. ואם כן הזוית כלה שוה לזוית בא״ג וזוית בג״א. ונשים זוית גב״א משותפת. תהיין זוית בא״ג וזוית אב״ג וזוית בג״א יחד שוות לזוית אד״ג וזוית אב״ג. וזוית גב״א ובא״ג[2] ובג״א יחד שוות לשתי זויות נצבות. ואם כן זויות אד״ג ואב״ג שוות לשתי זויות נצבות. וכפי זה המופת יתבאר שזויות בא״ד ודג״ב שוות לשתי זויות נצבות.

וזה מה שרצינו לבאר.

بُرْهانُهُ: أَنَّا نَصِلُ ا ج ب د. فَكُلّ زاوِيَتَيْنِ في قِطْعَة واحِدَة مِنَ الدَّائِرَة فَهُما مُتَساوِيَتانِ. فَالزَّاوِيَةُ الَّتي مِنْ ب ا ج مثل الزَّاوِيَةِ الَّتي مِنْ ب ج د. وَأَيْضًا الَّتي مِنْ ا د ب مثل الَّتي مِنْ ا ج ب. فَجَميعُ الَّتي مِنْ ا د ج[1] مثل الَّتي مِنْ ب ا ج وَمِنْ ب ج ا. وَنَجْعَلُ الَّتي مِنْ ج ب ا مُشْتَرَكَةً. فَالَّتي مِنْ ب ا ج وَالَّتي مِنْ ا ج ب[2] وَالَّتي مِنْ ج ب ا جَميعًا مثل اللَّتَيْنِ مِنْ ا د ج وَمِنْ ا ب ج. فَالَّتي مِنْ ج ب ا وَمِنْ ب ا ج وَمِنْ ب ج ا مثل زَاوِيَتَيْنِ قائِمَتَيْنِ. وَاللَّتانِ مِنْ ا د ج وَمِنْ ا ب ج مثل زَاوِيَتَيْنِ قائِمَتَيْنِ.

وَذَلِكَ ما أَرَدْنا بَيانَهُ.

---

[1]שוה] ~~שמאב״ג~~+ [2]ובא״ג] *+

[1]ا د ج] ج ~~ا د~~ ‹‹*›› [2]ا ج ب] ‹ا›ج ~~ا~~ ب

بُرْهانُه: أنّا نُخْرِجُ ا ج و ب د. وكُلّ زاوِيَتَيْنِ في قِطْعَةٍ[1] فَهُما مُتَساوِيَتانِ. فزاوية ب ا ج مثل زاوية ب د ج. وزاوية ا د ب أيْضًا مثل زاوية ا ج ب. فجميع زاوية ا د ج مثل زاوِيَتَيْ ب ا ج و ب ج ا. ونَجْعَلُ زاوِيةَ ج ب ا مُشْتَرَكَةً. فزوايا ب ا ج و ا ج ب و ج ب ا[2] مثل زاوِيَتَيْ ا د ج ا ب ج. وزوايا ج ب ا ب ا ج ب ج ا جَمِيعًا مثل زاوِيَتَيْنِ قائِمَتَيْنِ. فَزاوِيَتا ا د ج ا ب ج المُتَقابِلَتانِ[3] مثل زاوِيَتَيْنِ قائِمَتَيْنِ. وبِمِثْلِ هَذِهِ الصِّفةِ تَعلَّمْ أنَّ زاوِيَتَيْ ب ا د و د ج ب مُساوِيَتانِ لزاوِيَتَيْنِ قائِمَتَيْنِ.

وذلكَ ما أَرَدْنا أنْ نُبيِّنَ.

المופת: אנחנו נוציא א״ג וב״ד. וכל שתי זויות בחתיכה[1] הנה שתיהן[2] שוות. הנה זוית בא״ג כמו זוית[3] בד״ג. וזוית אד״ב גם כן כמו זוית אג״ב. הנה כל זוית אד״ג כמו שתי זויות[4] אג״ב וגא״ב. ונשים זוית גב״א משותפת. אם כן[5] זויות[6] בא״ג ואג״ב וגב״א כמו שתי זויות אד״ג ואב״ג[7]. וזויות גב״א ובא״ג ובג״א יחד כמו שתי זויות נצבות[8]. הנה שתי זויות אד״ג ואב״ג המתנגדות [!] כמו שתי נצבות. ובכמו[9] הצורה הזאת תדע כי שתי זויות בא״ד ודג״ב שוות לשתי נצבות.

וזה מה שרצינו לבאר[10].

[1]قطعة] U +واحدة [2]وج ب ا] M +جميعا
[3]المتقابلتان] MPR1R2

[1]בחתיכה] +אחת **ס** [2]שתיהן] שתיהם **ט**
[3]בא״ג ... זוית] ⟨⟨*⟩⟩ **ב** [4]זויות] זוית **ב**
[5]כן] ⟨⟨*⟩⟩ **ב** [6]זויות] זוית **וס** [7]ואב״ג] ואג״ב **ו** [8]נצבות] +~~אם כן~~ **ס** +~~ובכמו הצורה הזאת תדע כי~~ **ק** [9]ובכמו] וכמו **ט** [10]וזה ... לבאר] ונשלם באורו **ט**

[Proposition III.22]

[H23]

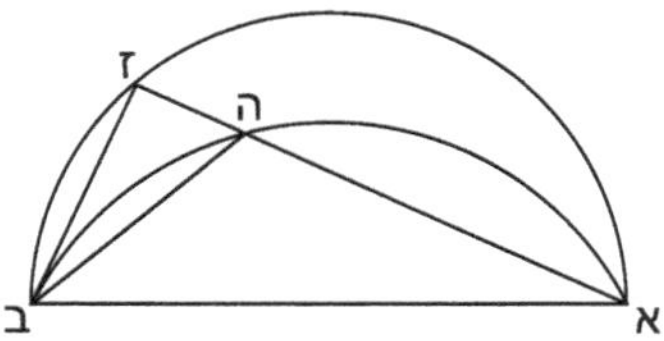

לא יתכן שתעמדנה על קו אחד ישר מונח שתי חתיכות מתדמות מחתיכות העגולים בצד אחד ותהיה אחת מהן גדולה מן האחרת.

لا يُمْكِنُ أَنْ يَقومَ[1] عَلى خَطٍّ واحِدٍ قِطْعَتانِ مِنْ دائِرَةٍ مُتَشابِهَتانِ في جِهَةٍ واحِدَةٍ، إِحْداهُما أَعْظَمُ مِنَ الأُخْرى.

مِثالُهُ: خَطُّ ا ب. فَأَقولُ إِنَّهُ لا يُمْكِنُ أَنْ يَقومَ عَلى خَطِّ ا ب قِطْعَتانِ مِنْ دائِرَةٍ مُتَشابِهَتانِ في جِهَةٍ واحِدَةٍ إِحْداهُما أَعْظَمُ مِنَ الأُخْرى.

ואם היה אפשר, נעמיד על קו א״ב שתי חתיכות עגול והן אה״ב ואז״ב, מתדמות בצד אחד. ואחת מהן, והיא חתיכת אז״ב, גדולה מן האחרת. ונרשום על חתיכת אה״ב נקודת ה׳ כמו שנפלה. ונדביק א׳ בה׳. ונוציא קו א״ה אל ז׳ מקשת חתיכת אז״ב. ונדביק ה׳ בב׳ וז׳ בב׳.

بُرْهانُهُ: أَنَّهُ لا يُمْكِنُ. فَإِنْ كانَ يُمْكِنُ، فَلْيَقُمْ عَلى خَطِّ ا ب قِطْعَتانِ مِنْ دائِرَةٍ، وَهُما ا ه ب (ا ز ب)[2] مُتَشابِهَتانِ في جِهَةٍ واحِدَةٍ، إِحْداهُما أَعْظَمُ مِنَ الأُخْرى، وَهِيَ قِطْعَةُ (ا ز ب)[3]. فَنُعَلِّمُ عَلى قِطْعَةِ ا ه ب نُقْطَةَ ه كَيْفَما وَقَعَتْ. وَنَصِلُ ا ه. وَنُخْرِجُ خَطَّ ا ه إِلى ز مِنْ قَوْسِ قِطْعَةِ ا ز ب. وَنَصِلُ ه ب ز ب.

[1]ان يقوم] ⟨⟨*⟩⟩ [2]ا ز ب] ا ز [3]ا ز ب] ا ف

لا يُمْكِنُ أَنْ يقومَ على خَطٍّ واحدٍ مُسْتَقيمٍ قِطْعَتان مُتَشابِهَتانِ مِن قِطَعِ الدَّوائِرِ في جِهَةٍ واحِدَةٍ، إحْداهُما أعظم مِنَ الأُخرى.

אי אפשר שיעמוד על קו[1] אחד ישר[2] שתי[3] חתיכות מתדמות מחתיכות העגולות בצד אחד[4], ואחת משתיהן יותר גדולה מן האחרת.

فَإِنْ كانَ يُمْكِنُ، فَلْيَقُمْ^ على خَطِّ ا ب المُسْتَقيمِ قِطْعَتانِ مُتَشابِهَتانِ مِنْ قِطَعِ الدَّوائِرِ في جِهَةٍ واحِدَةٍ، إحْداهُما أعظم مِنَ الأُخرى. وَهُما قِطْعَتا ا ه ب و ا ز ب. والعُظْمى مِنْهُما قِطْعَةُ ا ز ب. وتَتَعَلَّمُ على قَوْسِ ا ه ب نُقْطَةَ ه. ونُخْرِجُ خَطَّ ا ه. ونُخْرِجُهُ إلى ز. ونُخْرِجُ خَطَّيْ ه ب ز ب.

שאם היה אפשר, נאמר שיקום על קו א״ב הישר שתי חתיכות דומות מחתיכות העגולות ובצד אחד, אחת מהן יותר גדולה[5] מן האחרת. והם שתי חתיכות אה״ב ואז״ב[6]. והיותר[7] גדולה משתיהן[8] חתיכת אז״ב. ונרשום על קשת אה״ב נקודת ה׳. ונוציא[9] קו א״ה. ונוציאהו עד ז׳. ונוציא שני קוי ה״ב וז״ב.

שרטוט—ס: ה׳ אינה על א״ז; **פ**: $ [1]קו] +ישר **פ** [2]ישר] +* **ב** [3]שתי] שני **ס** [4]אחד] אחת **ס** [5]גדולה] גדול **ב** [6]ואז״ב] ו⟨א⟩זה״ב **ט** [7]והיותר] +* **ט** [8]משתיהן] משתיהם **טס** [9]ונוציא] +⟨על⟩ ~~קשת אה״ב נקדת ה׳~~ **ק**

וחתיכת אה״ב דומה לחתיכת אז״ב. ואם כן זוית אה״ב שוה לזוית אז״ב—החיצונה שוה לפנימית, זה לא יתכן.

وَقِطْعَةُ ا ه ب تُشْبِهُ قِطْعَةَ ا ز ب. فَالزَّاوِيَةُ الَّتي مِنْ ا ه ب مثل الَّتي مِنْ ا ز ب، الخارِجَةُ مثل الدَّاخِلَةِ—هذا خُلْفٌ.

ואם כן התבאר שלא תעמדנה על קו אחת [!] שתי חתיכות עגול בצד אחד מתדמות, אחת מהן גדולה מן האחרת.

فَلَيْسَ يُمْكِنُ أَنْ يَقومَ عَلى خَطٍّ واحِدٍ قِطْعَتانِ مِنْ دائِرَةٍ في جِهَةٍ واحِدَةٍ مُتَشابِهَتانِ، إِحْداهُما أَعْظَمُ مِنَ الأُخْرى.

וזה מה שרצינו באורו.

وَذلِكَ ما أَرَدْنا بَيانَهُ.

فقِطْعَةُ ا ه ب تُشْبِهُ قِطْعَةَ ا ز ب. فزاويةُ ا ه ب مثل زاويةِ ا ز ب—الخارِجَة مِنَ المُثَلَّثِ مثل الدّاخِلَةِ، هذا خُلْفٌ.

הנה[1] חתיכת[2] אה״ב דומה לחתיכת אז״ב. אם כן זוית אה״ב כמו זוית אז״ב—היוצאת[3] מן המשולש כמו הפנימית, זה שקר.

فَلَيْسَ يقومُ على خَطٍّ واحِدٍ مُسْتَقِيمٍ قِطْعَتانِ مُتَشابِهَتانِ مِنْ قِطَعِ الدَّوائِرِ في جِهَةٍ واحِدَةٍ، إحْداهُما أعظم مِنَ الأُخرى.

אם כן לא יעמדו על קו אחד ישר שתי[4] חתיכות דומות מחתיכות העגולים בצד אחד, ואחת[5] מהן[6] יותר גדולה מן האחרת.

وذَلِكَ ما أَرَدْنا أَنْ نُبَيِّنَ.

וזה מה שרצינו לבאר[7].

---

[1]הנה] +~~חתכה אחת והיא~~ ק [2]חתיכת] חתיכות ס [3]היוצאת] היוצאה ב [4]שתי] שני ס [5]ואחת] אחת ט [6]מהן] מהם ז משתיהן ס [7]לבאר] ליתא ס

[Proposition III.23]

[H24]

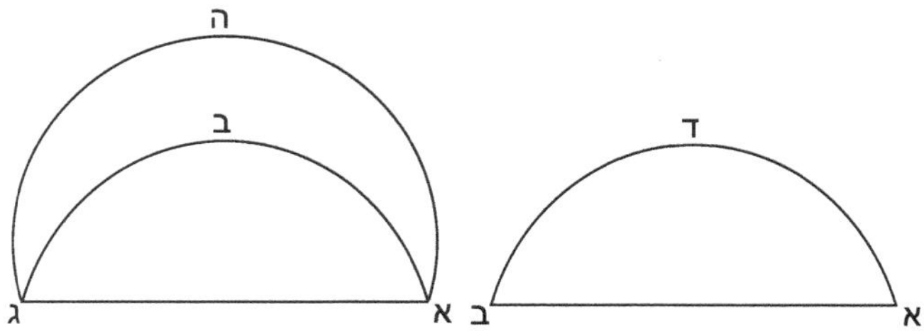

כשתהיין חתיכות עגולים מתדמות, והיו על קוים ישרים שוים, תהיין החתיכות שוות.

إذا كانَتْ قِطَعُ دَوائِرِ مُتَشابِهَةً، وَكانَتْ عَلى خُطوطٍ مُتَساوِيَةٍ، فَالقِطَعُ أَيضًا مُتَساوِيَةٌ.

המשל בו: ששתי חתיכות עגולים, והם אב״ג ואד״ב, מתדמות, והם על שני קוים שוים, והם א״ב א״ג. ואומר ששתי החתיכות[1] שוות.

مِثالُهُ: أَنَّا نَخُطُّ قِطْعَتَينِ مِنْ دائِرَةَ عَلى خَطَّيْنِ مُتَساوِيَيْنِ—خَطَّيْ ا ب ا ج—مُتَشابِهَيْنِ، عَلى إِحْداهُما ا ب ج وَعَلى الأُخْرى ا د ب. فَأَقولُ إِنَّ القِطْعَتَينِ مُتَساوِيَتانِ.

[1] החתיכות] ⟨ה⟩חתיכות

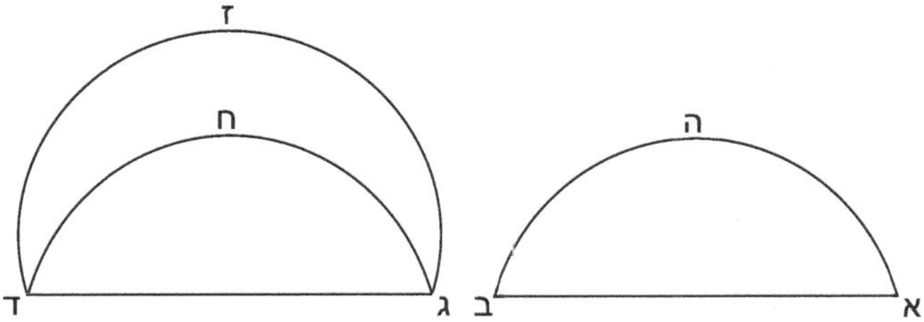

إذا كانَتْ قِطَعُ الدَّوائِرِ[1] مُتشابِهَةً، وَكانَتْ على خُطوطٍ مُسْتَقيمَةٍ مُتَساوِيَةٍ، فَإِنَّ القِطَعَ أيضًا مُتَساوِيَةٌ.

مِثالُهُ: أَنَّ قِطْعَتَيْ ا ه ب وج ز د مُتَشابِهَتانِ على خَطَّيْنِ مُسْتَقيمَيْنِ مُتَساوِيَيْنِ، وَهُما ا ب وج د. فَأَقولُ إِنَّ القِطْعَتَيْنِ مُتَساوِيَتَيْنِ.

כאשר יהיו חתיכות העגולות דומות, והיו על קוים ישרים שוים, הנה החתיכות[1] גם כן שוות.

המשל בו: כי שתי חתיכות אה״ב וגז״ד דומות על שני קוים ישרים שוים[2], והם א״ב וג״ד. הנה אומר כי שתי החתיכות[3] שוות.

[1]الدوائر] دوائر MPQR_1R_2SU

שרטוט—מקום האותיות ז׳ וח׳ זהה למקומן בעדי הנוסח MR_2T_2Q; בשאר עדי הנוסח הערביים ז ↔ ח [1]החתיכות] חתיכות ב [2]הנה ... שוים] ⟨⟨*⟩⟩ פ [3]החתיכות] חתיכות ב

המופת בו: כי[1] כשנרכיב חתיכת אד״ב על חתיכת אב״ג, ותושבת א״ב על תושבת א״ג, ושמנו נקודת א׳ על נקודת א׳, נפלה נקודת ב׳ על נקודת ג׳, לפי שא״ב שוה לא״ג. ויפל קו קשת אד״ב על קו קשת אב״ג, לפי שהם[2] שתי חתיכות מתדמות. שאם לא היה נופל עליו, יהיה נופל מחוץ, כמו אה״ג, או מבפנים. והיו[3] עומדות על קו א״ג שתי חתיכות מתדמות בצד אחד, ואחת מהן גדולה מן האחרת— וזה לא יתכן. ואם כן חתיכת אד״ב שוה לחתיכת אב״ג.

وזה מה שרצינו באורו.

بُرْهانُهُ: أَنَّا إذا رَكَّبْنا قِطْعَةَ ا د ب عَلى قِطْعَةِ ا ب ج، وَقاعِدَةَ ا ب عَلى قاعِدَةِ ا ج، وَوَضَعْنا نُقْطَةَ ا عَلى نُقْطَةِ ا، وَقَعَتْ نُقْطَةُ ب عَلى نُقْطَةِ ج، لأَنَّ ا ب مثل ا ج. ويَقَعُ قَوْسُ ا د ب عَلى قَوْسِ ا ب ج، لِأَنَّ القِطْعَتَينِ مُتَشابِهَتانِ. فَإِنْ لَمْ يَقَعْ عَلَيْها فَإِنَّها سَتَقَعُ مِنْ خارِجٍ، مثل قَوْسِ ا ه ج. فَتَقومُ عَلى ا ج قِطْعَتانِ مِنْ دائِرَةٍ— ا ب ج وَالأُخْرَى ا ه ج، مُتَشابِهَتانِ، في جِهَةٍ واحِدَةٍ، إِحْداهُما أَعْظَمُ مِنَ الأُخْرَى. وَهَذا ما لا يُمْكِنُ. فَقِطْعَةُ ا د ب مثل قِطْعَةِ ا ب ج.

وَذَلِكَ ما أَرَدْنا بَيانَهُ.

[1]כי] ~~שנדביק~~+ [2]שהם] ⟨⟨*⟩⟩ [3]והיו] ~~ואין~~⟨⟨*⟩⟩

بُرْهانُهُ: أَنَّا إِذا رَكَّبْنا قِطْعَةَ ا ه ب على قِطْعَةِ ج ز د، وقاعِدَةَ ا ب على قاعِدَةِ ج د، وَقَعَتْ قَوْسُ ا ه ب على قَوْسِ ج ز د. فَإِنْ لم تَقَعْ عَلَيْها، وَوَقَعَتْ مثل ج ح د، وكانَتْ قاعِدَةُ ا ب قَدْ وَقَعَتْ على قاعِدَةِ ج د، فَقَدْ قامَ على خَطِّ ج د المُسْتَقِيمِ قِطْعَتانِ مُتَشابِهَتانِ مِنْ قِطَعِ الدَّوائِرِ في جِهَةٍ واحِدَةٍ، إِحْداهُما أعظم مِنَ الأُخرى، وَهُما ج ز د ج ح د—هذا خُلْفٌ، غير مُمْكِنٍ[1]. فقِطْعَةُ ا ه ب تَقَعُ على قِطْعَةِ ج ز د. فَهِيَ إِذَنْ مُساوِيَةٌ لَها.

وذَلِكَ ما أَرَدْنا أَنْ نُبَيِّنَ.

המופת: כי אנחנו כאשר הרכבנו חתיכת אה״ב על חתיכת גז״ד, ותושבת א״ב על תושבת ג״ד, נפלה קשת אה״ב על קשת גז״ד. שאם לא תפול עליו, ונפל כמו גח״ד, והיה[1] תושבת א״ב כבר[2] נפלה[3] על תושבת ג״ד, הנה כבר עמד על קו ג״ד הישר שתי חתיכות דומות מחתיכות העגולות[4] בצד אחד, ואחת משתיהן[5] יותר גדולה[6] מן האחרת, והם גז״ד וגח״ד—זה[7] שקר, אי אפשר. אם כן חתיכת אה״ב תפול על חתיכת גז״ד. הנה היא אם כן שוה אליה[8].

וזה מה שרצינו לבאר[9].

---

[1] غير ممكن] MPR1R2

---

[1] והיה] והיא **ס** [2] כבר] ליתא **טפ** [3] נפלה] ליתא **בוטפ** נפל **ס** [4] העגולות] העגולה **ז** [5] משתיהן] משתיהם **ט** [6] יותר גדולה] גדולה יותר **ק** [7] זה] וזה **ק** [8] אליה] לה **ק** [9] לבאר] לבארו **בק** ליתא **ס**

[Proposition III.24]

[H25]

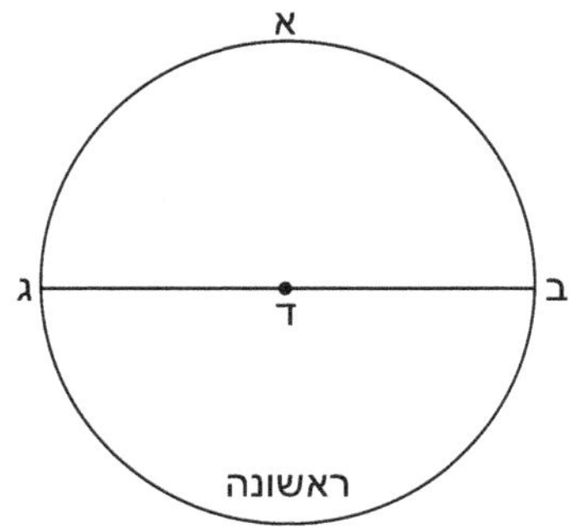

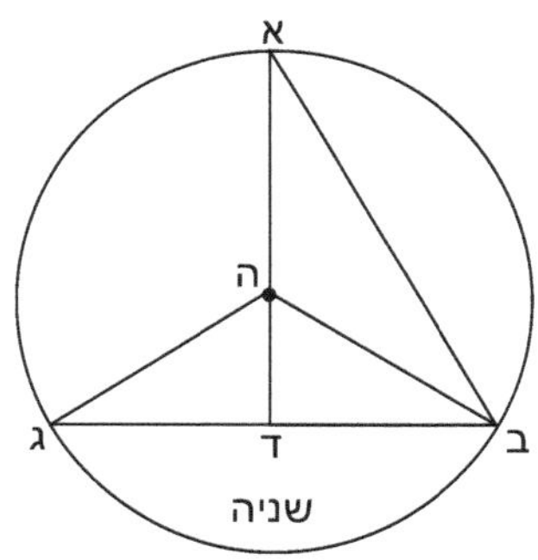

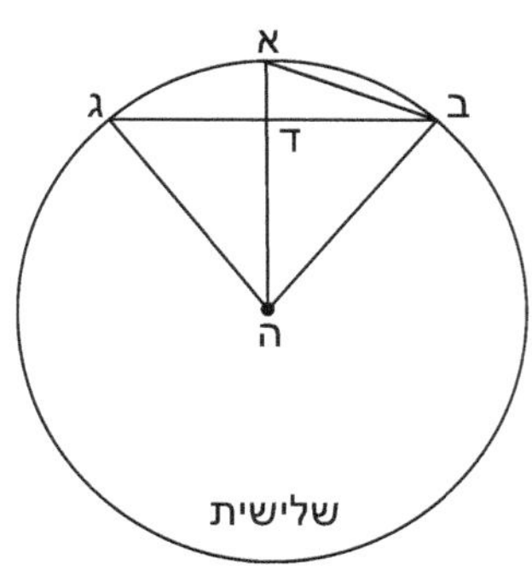

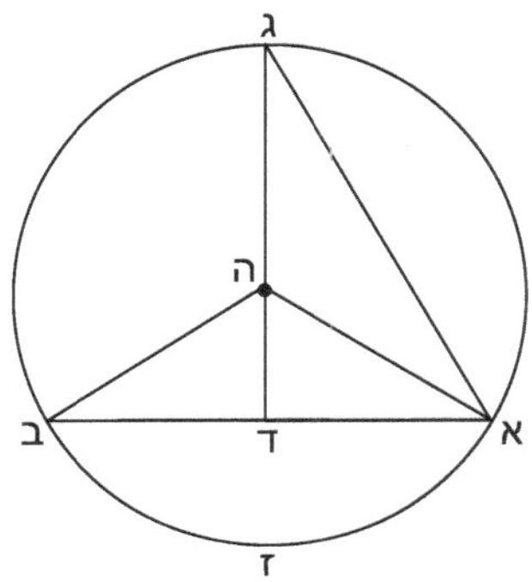

نُرِيدُ أن نُتَمِّمَ قِطْعَةً مَعْلُومَةً مِنْ دائِرَةٍ[1].

فَلْتَكُنِ القِطْعَةُ المَعْلُومَةُ قِطْعَةَ^ اج ب. فإذا أَرَدْنا أن نُتَمِّمَ دائِرَتَها، فإنَّا نَصِلُ اب. وَنَقْسِمُهُ بِنِصْفَيْنِ على د. وَنُخْرِجُ مِنْ نُقْطَةِ د عَمُودًا على اب، وهو دج. وَنَصِلُ اج. وَنُقِيمُ على خَطِّ اج المُسْتَقِيمِ، على نُقْطَةِ ا مِنْهُ، زاوِيَةً مُساوِيَةً لِزاوِيَةِ اج د، المُسْتَقِيمَةِ الخَطَّيْنِ^، وهي زاوِيَةُ ج ا ه.

נרצה שנשלים חתיכה ידועה מעגולה.

ותהיה החתיכה[1] הידועה אג״ב. הנה כאשר רצינו להשלים עגולתה, הנה נגיע א״ב. ונחלקהו בשני חציים על ד׳. ונוציא מנקודת ד׳ עמוד על א״ב, והוא ד״ג. ונגיע א״ג. ונעמיד על קו א״ג הישר, על נקודת א׳ ממנו, זוית[2] שוה לזוית אג״ד, ישרת שתי הצלעות, והיא זוית גא״ה.

---

[1] نريد ... دائرة] MPR1R2U

שרטוט—**ב**: נוסף ב״ג; **ס**: נוסף ג״ז; השרטוט אינו זה לשרטוטים בכתבי היד הערביים, שברובם اب ه ج נחתכים

[1] החתיכה] ~~העגולה~~ ⟨⟨החתיכה⟩⟩ **בפ** העגולה **טק** חתיכת העגולה **ס** [2] זוית] זו״ה ~~לנ׳~~ **פ**

إِذا كَانَتْ قِطْعَةٌ مِنْ دائِرَةٍ وَأَرَدْنا أَنْ نُتِمَّ دائِرَتَها، إِنْ كانَتِ القِطْعَةُ نِصْفَ دائِرَةٍ أو أَكْثَرَ أَو أَصْغَرَ.

נרצה להשלים חתיכה ידועה מעגול.

مِثالُهُ: أَنَّا نَجْعَلُ القِطْعَةَ قِطْعَةَ ا ب ج. وَنَجْعَلُها أَوَّلًا نِصْفَ دائِرَةٍ. وَنُرِيدُ أَنْ نُتِمَّ دائِرَتَها. فَنَقْطَعُ خَطَّ ب ج بِنِصْفَيْنِ عَلَى نُقْطَةِ (د)[1]. (د)[2] مَرْكَزُ الدَّائِرَةِ. فَعَلَى مَرْكَزِ (د)[3] نُقَدِّرُ بُعْدَ د ب د ج. وَنُتِمُّ قِطْعَةَ دائِرَةِ ا ب ج الأُولى، الَّتِي هِيَ نِصْفُ دائِرَةٍ.

ל״א: כשתהיה חתכת עיגול, ונרצה להשלים העגול, בין שתהיה החתיכה חצי העיגול או יותר מחצי עגול או פחות.

ונשים החתיכה חתיכת אב״ג. ונשים אותה תחלה חצי עגול. וכאשר נרצה להשלים העגול, נחלוק קו ב״ג לחציים על נקודת ד׳. וד׳ הוא מרכז העגול. ונקיף על מרכז ד׳ ובשיעור מרחק ד״ב וד״ג, ונשלים העגול הראשון.

---

[1] د] ج [2] د] ج [3] د] ج

שרטוט—B: בבל השרטוטים אין מושלמת קשת בא״ג למעגל, ובשרטוט השלישי בה״ג היא קשת ואילו ב״א א״ג קווים

وزاوية د ج ا أصغر مِنْ قائِمَة. فَزاوِيَتا د ج ا وج ا ه أصغر مِنْ قائِمَتَيْنِ. وَالخُطوطُ الَّتي تَخْرُجُ مِنْ أَقَلَّ مِنْ زاوِيَتَيْنِ قائِمَتَيْنِ وتَبْعُدُ إلى غَيْرِ نِهايَةَ—لا بُدَّ أَنْ^ تَلْتَقي[1]. وَلْيَلْتَقِي^ خَطّا ا ه ج د على نُقْطَة ه. وَنَصِلُ خطّ ب ه. فَلِأَنَّ زاوِيَةَ ج ا ه مُساوِيَةٌ لِزاوِيَةِ ا ج ه، يكونُ ضِلْعُ ا ه مُساوِيًا لِضِلْعِ ه ج. وذَلِكَ أَنَّ زاوِيَتَيْ مُثَلَّث ه ا ج، اللَّتَيْنِ على القاعِدَةِ، مُتَساوِيَتانِ. وَأَيْضًا فإِنَّ ضِلْعَ ا د مُساوٍ لِضِلْعِ د ب، وخَطّ د ه مُشْتَرَك، فَكِلا خَطَّيْ^ ا د د ه مُساوِيانِ لِكِلَيْ خَطَّيْ ب د د ه، كُلّ واحدٍ لِنَظيرِهِ، وزاوية ا د ه مُساوِيَة لِزاوِيَةِ ب د ه—فقاعِدَة ا ه مُساوِيَة لقاعِدَة ب ه. وَقَدْ كانَ تبيَّنَ أَنَّ خطّ ا ه مُساوٍ لخطّ ه ج. فخُطوط ا ه ه ج ه ب الثَّلاثة مُتَساوِيَة. فَإِذا جَعَلْنا نُقْطَةَ ه مَرْكَزًا، وَخَطَطْنا دائِرَةً بِأَبْعاد ه ا ه ج ه ب، مَرَّتْ بِالنُّقَطِ الثَّلاث[2]. فَنَخُطُّ هذه الدَّائِرَةَ، وَهِيَ دائِرَةُ ا ز ب. فَقَدْ خَطَطْنا الدَّائِرَةَ إلى[3] ا ج ب[4] قِطْعَةً مِنْها، وَهِيَ دائِرَةُ ا ز ب.

وذَلِكَ ما أَرَدْنا أَنْ نُبيِّنَ.

[1] اذا كانت قطعة معلومة من دائرة واردنا ان نتمم دائرتها وزاوية ... تلتقي] MPR1R2U
[2] الثلاث] MPR1R2U
[3] الى] U التي
[4] ا ج ب] قطعة+ MPR1R2

וזוית דג״א יותר קטנה מנצבת. אם כן שתי זויות דג״א וגא״ה יותר קטנות משתי נצבות. והקוים אשר יצאו מפחות משתי זויות[1] נצבות וירחקו אל לא תכלית—יפגשו. ונאמר שיפגשו שני קוי א״ה[2] ג״ד על נקודת ה׳. ונגיע קו ב״ה. הנה מפני כי זוית גא״ה שוה לזוית אג״ה, יהיה צלע א״ה שוה לצלע ה״ג. וזה כי[3] שתי זויות[4] משולש הא״ג, אשר הן[5] על התושבת, שוות. וגם כן הנה[6] מפני כי צלע א״ד שוה[7] לצלע ד״ב, וקו ד״ה משותף, וכל אחד משני קוי א״ד ד״ה[8] שוים לכל אחד משני קוי ב״ד ד״ה, כל אחד לנכחי אליו[9], וזוית אד״ה שוה לזוית בד״ה—הנה תושבת א״ה שוה לתושבת ב״ה. והיה כבר התבאר כי קו א״ה שוה לקו ה״ג. הנה קוי[10] א״ה ה״ג ה״ב השלשה[11] שוים. וכאשר שמנו נקודת ה׳ מרכז, וסבבנו עגולה במרחק ה״א ה״ג ה״ב[12], הלכה בנקודות[13] השלש[14]. ונקוה העגולה הזאת, והיא עגולת אז״ב. הנה קוינו אל עגולת אג״ב חתיכה ממנה, והיא עגולת[15] אז״ב[16].

וזה מה שרצינו לבאר[17].

[1] זויות] זוית פ [2] א״ה] +~~וא״ד~~ פ [3] כי] +הם ס [4] זויות] ⟨⟨*⟩⟩ ב [5] הן] ליתא ו [6] הנה] ליתא ט [7] שוה] ושוה ב [8] ד״ה] ד״ז ס [9] אליו] לו ט [10] קוי] קו ס [11] השלשה] ~~שוה~~+ ב [12] ה״ג ה״ב] ה״ב ה״ג ק [13] בנקודות] בנקודת בס [14] השלש] ה׳ ס [15] עגולת] ~~עגולת~~ ⟨⟨חתכת⟩⟩ ב חתיכת ו [16] הנה ... אז״ב] ⟨*⟩ ט [17] לבאר] לבארו בפ ליתא ס

ועוד חתיכת אב״ג השנייה, והיא יותר מחצי עגול. וכאשר נרצה להשלים העגול נחלוק קו ב״ג לחציים על ד׳. ונוציא מד׳ קו על זוית נצבה אל קשת בא״ג, והוא קו א״ד. ונדביק ב׳ בא׳. ובא״ג גדול מחצי עגול. ואם כן א״ד ארוך מב״ד. ואם כן זוית דב״א גדולה מזוית בא״ד. ונעמיד מקו א״ב על נקודת ב׳ זוית שוה לזוית בא״ד, והיא זוית אב״ה. הנה זוית אד״ב נצבה. אם כן דא״ב ודב״ה קטנו׳ מב׳ נצבו׳. והקוי׳ שיצאו מפחו׳ מב׳ נצבו׳ וירחקו ללא תכלית—יפגשו. הנה יפגשו קווי א״ד ב״ה על נקודת ה׳[1]. ואם כן צלע א״ה שוה לצלע ה״ב. ונדביק ה׳ בג׳. וב״ד שוה לד״ג. וד״ה משותף. ואם כן שני צלעי ב״ד וד״ה שוים לשני צלעי ג״ד וד״ה, כל אחד לדומה לו. וזוית בד״ה שוה לזוית גד״ה. ואם כן תושבת ב״ה שוה לתושבת ה״ג. וא״ה שוה לה״ב. ואם כן הקוים השלשה א״ה וה״ב וה״ג שוים. ואם כן ה׳ הוא מרכז העיגול. ועל ה׳ נשלים עגול החתיכה בשיעור מרחק א׳ וב׳ וג׳, והיא החתיכה השנייה שהיא גדולה מחצי עגול. והוא עגול אבז״ג[2].

[1] הנה ... ה׳] ⟨⟨*⟩⟩ [2] אבז״ג] ⟨⟨*⟩⟩

ثُمَّ نَجْعَلُ القِطْعَةَ الثَّانِيَةَ، أَكْبَرَ مِنْ نِصْفِ دائِرَةٍ، وَهِيَ قِطْعَةُ ا ب ج. وَنُرِيدُ أَنْ نُتِمَّ دائِرَتَها. فَيَقْسِمُ خَطَّ ب ج بِنِصْفَيْنِ عَلَى د. وَنُخْرِجُ مِنْ د خَطًّا عَلَى زاوِيَةٍ قائِمَةٍ إِلَى قَوْسِ ب ا ج، وَهُوَ خَطُّ ا د. وَنَصِلُ ب ا. فَقِطْعَةُ ب ا ج أَعْظَمُ مِنْ نِصْفِ دائِرَةٍ، ا د أَطْوَلُ مِنْ ب د. فَالزَّاوِيَةُ الَّتِي مِنْ د ب ا أَعْظَمُ مِنَ الَّتِي مِنْ ب ا د. وَنُقِيمُ عَلَى خَطِّ ا ب عَلَى نُقْطَةِ ب زاوِيَةً مثل الزَّاوِيَةِ الَّتِي مِنْ ب ا د، وَهِيَ الَّتِي مِنْ ا ب ه. وَنُخْرِجُ مِنْها خَطًّا مَتَى يَلْتَقِي خَطُّ ا د عَلَى نُقْطَةِ ه، وَنَصِلُ ه ج. فَأَقُولُ إِنَّ عَلَى مَرْكَزِ ه وَبِبُعْدِ ه د ه ا ه ج نُتِمُّ الدَّائِرَةَ.

بُرْهانُهُ: أَنَّ الزَّاوِيَةَ الَّتِي ب ا ه مثل الَّتِي مِنْ ا ب ه. فَضِلْعُ ا ه مثل ضِلْعِ ه ب. ب د مثل د ج. د ه مُشْتَرَكٌ. (فكلا)[1] ضِلْعَيْ ب د د ه مثل كِلَيْ ضِلْعَيْ ج د د ه، كُلُّ واحد مثل نَظِيرِهِ. وَالزَّاوِيَةُ الَّتِي مِنْ ب د ه مثل الَّتِي مِنْ ج د ه. فَقاعِدَةُ ب ه مثل قاعِدَةِ ه ج. ا ه مثل ه ب. فَالأَضْلاعُ (الثَّلاثَةُ)[2] ا ه ه ب ه ج مُتَساوِيَةٌ. فَعَلَى مَرْكَزِ ه (وَبِبُعْدِ)[3] ا ب ج نُتِمُّ قِطْعَةَ دائِرَةِ ا ب ج الثَّانِيَةِ الَّتِي هِيَ أَكْبَرُ مِنْ نِصْفِ دائِرَةٍ.

[1] فكلا] فكلي [2] الثلاثة] الثلثة [3] وببعد] ونبعد

עוד נשים החתיכה קטנה מחצי עגול, והיא חתיכת אב״ג השלישית. וכאשר רצינו להשלים העגול, נחלוק קו ב״ג לחציים על ד׳. ונוציא מד׳ קו על זוית נצבה אל קשת בא״ג, והוא קו ד״א. ונדביק א׳ בב׳. ולפי שחתיכת אב״ג [!] קטנה מחצי עגול, יהיה ד״א קצר מד״ב. ואם כן זוית אב״ד קטנה מזוית בא״ד. ונעמיד על קו א״ב, על נקודת ב׳, זוית שוה לזוית בא״ד, והיא זוית אב״ה. ונוציא קו א״ד עד שיפגוש בקו ב״ה על נקודת ה׳. ונדביק ה׳ בג׳. ותהיה זוית אב״ה שוה לזוית בא״ה. ואם כן צלע א״ה שוה לצלע ה״ב. וב״ד שוה לד״ג. וד״ה משותף. ושני קוי ב״ד וד״ה שוים לשני קוי ג״ד וד״ה, כל אחד לדומה לו. וזוית בד״ה שוה לזוית גד״ה. ואם כן תושבת ה״ב שוה לתושבת ה״ג. וא״ה שוה לה״ב. ואם כן הקוים השלשה שוים—א״ה וה״ב וה״ג. ואם כן ה׳ מרכז העיגול. ועל מרכז ה׳ ובמרחק א׳ וב׳ וג׳ נשלים עגול חתכת אב״ג השלישית, שהיא קטנה מחצי עגול, והוא עגול בא״ג[1].

ונשלם המופת.

[1] בא״ג] ⟨⟨[בה]ג״א⟩⟩

ثُمَّ نَجْعَلُ القِطْعَةَ الثَّالِثَةَ أَصْغَرَ مِنْ نِصْفِ دائِرَةٍ، وَهِيَ قِطْعَةُ ا ب ج. ونُرِيدُ أَنْ نُتِمَّ دائِرَتَها. فَنَقْسِمُ خَطَّ ب ج بِنِصْفَيْنِ عَلى د. وَنُخْرِجُ مِنْ د خَطًّا عَلى زاوِيَةٍ قائِمَةٍ إِلى قَوْسِ ب ا ج. ونَصِلُ ا ب. فَقِطْعَةُ ب ا ج أَصْغَرُ مِنْ نِصْفِ دائِرَةٍ، د ا أَقْصَرُ مِنْ د ب. وَالزَّاوِيَةُ الَّتِي مِنْ ا ب د أَصْغَرُ مِنَ الَّتِي مِنْ ب ا د. فنُقِيمُ عَلى خَطِّ ا ب على نُقْطَةِ ب زاوِيَةً مثل الزَّاوِيَةِ الَّتِي مِنْ ب ا د، وَهِيَ الَّتِي مِنْ ا ب ه. وَنُخْرِجُ خَطَّ ا د حَتَّى يَلْتَقِيَ بِخَطِّ ه ب عَلى نُقْطَةِ ه. وَنَصِلُ ه ج. فَأَقُولُ إِنَّ عَلى مَرْكَزِ ه وَبِبُعْدِ ب ا ج (تَتِمُّ)[1] الدَّائِرَةُ.

بُرْهانُهُ: أَنَّ الزَّاوِيَةَ الَّتِي مِنْ ا ب ه مثل الَّتِي مِنْ ب ا ج. فَضِلْعُ ا ه مثل ضِلْعِ ه ب. وَضِلْعُ ب د مثل ضِلْعِ د ج. د ه مُشْتَرَكٌ. (وَكِلا)[2] ب د د ه مثل كِلَيْ ج د د ه، كُلُّ واحِدٍ مثل نَظِيرِهِ. وَالزَّاوِيَةُ الَّتِي مِنْ ب د ه مثل الزَّاوِيَةِ الَّتِي مِنْ ج د ه. وَقاعِدَةُ ه ب مثل قاعِدَةِ ه ج. ا ه مثل ه ب. فَالأَضْلاعُ (الثَّلاثَةُ)[3] ا ه (ه ج)[4] ه ب مُتَساوِيَةٌ. فَعَلى مَرْكَزِ ه وَبِبُعْدِ ا ب ج تَتِمُّ دائِرَةُ قِطْعَةِ ا ب ج الثَّالِثَةِ الَّتِي هِيَ أَصْغَرُ مِنْ نِصْفِ دائِرَةٍ.

وَذَلِكَ ما أَرَدْنا بَيانَهُ.

[1] تتم] يتم [2] وكلا] وكلي [3] الثلاثة] الثلثة [4] ه ج] ب ج

[Proposition III.25]

[H26]

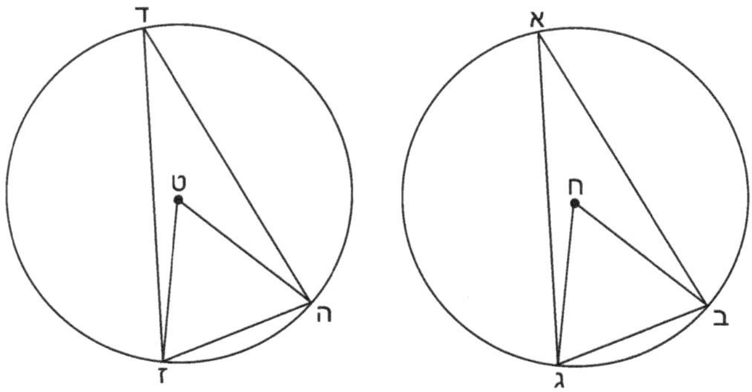

כשתהיין זויות שוות בעגולים שוים[1], הן על קשתות שוות, בין שתהיין על המרכז, בין שתהיין על הקו המקיף.

המשל בו: שנצייר שני עיגולים שוים, עליהם אב״ג ודה״ז. ומרכזיהם ח׳ וט׳. ונעמיד על שני המרכזים שתי זויות שוות, והן זוית בח״ג וזוית הט״ז. ואומר שקשת ב״ג שוה לקשת ה״ז.

إِذا كانَ في الدَّوائِرِ المُتَساوِيَةِ زَوايا مُتَساوِيَةٌ، فَهِيَ عَلى قُسِيٍّ مُتَساوِيَةٍ، إِنْ كانَتِ الزَّوايا عَلى مَراكِزِ الدَّوائِرِ، (أَو كانَتْ)[1] عَلى الخُطوطِ المُحيطَةِ بِها.

مِثالُهُ: أَنَّا نَخُطُّ دائِرَتَيْنِ مُتَساوِيَتَيْنِ عَلَيْهِما ا ب ج د ه ز. وَمَرْكَزاهُما ح ط. وَنُقيمُ عَلى المَرْكَزَيْنِ زاوِيَتَيْنِ مُتَساوِيَتَيْنِ، وَهُما الَّتي مِنْ ب ح ج والَّتي مِنْ ه ط ز. فَأَقولُ: قَوْسُ ب ج مثل قَوْسِ ه ز.

[1]שוים] ⟨*⟩

[1]أو كانت] وكانت

إِذا كانَتْ في دَوائِرَ مُتَساوِيَةٍ زَوايا مُتَساوِيَةٌ، فَهِيَ على قُسِيٍّ مُتَساوِيَةٍ—على المَرْكَزِ كانَتْ أَوْ على الخَطِّ المُحِيطِ[1].

مِثالُهُ: أَنَّ دائِرَتَيْ ا ب ج د ه ز مُتَساوِيَتانِ. ومَرْكَزاهُما ح ط. وَفِيهِما زاوِيَتانِ مُتَساوِيَتانِ على المَرْكَزِ[2]، وَهُما ب ح ج ه ط ز. فَأَقولُ إِنَّ قَوْسَ ب ج مثل قَوْسِ ه ز.

כאשר היו[1] בעגולות[2] שוות שתי זויות[3] שוות[4], הנה הן[5] על שתי קשתות שוות—על המרכז היו[6] או על הקו[7] המקיף.

דמיונו: כי שתי עגולות אב״ג דה״ז שוות. ומרכזי[8] שתיהן ח׳ ט׳. ובשתיהן שתי זויות שוות על המרכז, והם בח״ג הט״ז[9]. הנה אומר כי קשת ב״ג כמו קשת ה״ז.

---

[1]الخط المحيط] Q الخطوط المحيطة [2]المركز] CT1T2T3 المركزين

שרטוט—**ט**: מא׳ מורד אנך לב״ג, ומד׳ מורד אנך לה״ז [1]היו] היה **ס** [2]בעגולות] בעגולת **ב** [3]זויות] זוית **ב** [4]שוות] ⟨⟨*⟩⟩ **ב** שו⟨ו⟩ת **פ** [5]הן] ליתא **ט** [6]היו] היה **ס** [7]הקו] קו **ס** [8]ומרכזי] ומרכז **וס** [9]הט״ז] חט״ז **ס** +~~מפני~~ **פ**

מופת זה: שנרשום על קשתות בא״ג והד״ז שתי נקודות, נקודת א׳ ונקודת ד׳. ונדביק א׳ בב׳ ובג׳, וד׳ בה׳ ובז׳. (ושני העיגולים שווים)[1]. והקוים היוצאים ממרכזיהם אל הקוים המקיפים בהם שוים. ואם כן ח״ב שוה לט״ה, וח״ג שוה לט״ז. ושני קוי ח״ב וח״ג שוים לשני קוי ט״ה וט״ז. וזוית בח״ג שוה לזוית הט״ז. ואם כן תושבת ב״ג שוה לתושבת ה״ז. והזוית אשר על המרכז כפל הזוית אשר על הקו המקיף. ואם[2] כן זוית בח״ג כפל זוית בא״ג. וכמו כן זוית הט״ז כפל זוית הד״ז[3]. ואמנם זוית בח״ג שוה לזוית הט״ז. וכמו כן זוית בא״ג שוה לזוית הד״ז. ואם כן חתיכת העגול שעליה בא״ג שוה לאשר עליה הד״ז. וקשת בא״ג שוה לקשת הד״ז, לפי שהקשתות מתדמות. וכאשר התדמו הקשתות והיו על קוים שוים, הן שוות, כאשר[4] קדם בתמונה השלישית[5] לזו, ושתי[6] הזויות שוות. ואם כן ב״ג שוה לה״ז.

بُرْهانُه: أَنَّا نُعَلِّمُ عَلى قَوْسِ ب ا ج وقَوْسِ ه د ز نُقْطَتَيْنِ وَهُما ا د. وَنَصِلُ (ا ب ا ج)[1]. فَالدَّائِرَتانِ مُتَساوِيَتانِ. وَالخُطوطُ الَّتي تَخْرُجُ مِنْ مَرْكَزَيْهِما إِلى الخَطَّيْنِ المُحيطَيْنِ بِهِما مُتَساوِيَةٌ. ح ب مثل ط ه. ح ج مثل ط ز. (فَكِلا)[2] خَطَّيِ ب ح ح ج مثل كِلَيْ خَطَّيْ ه ط ط ز، كُلُّ واحِدٍ مثل الآخَرِ. فَالزَّاوِيَةُ الَّتي مِنْ ب ح ج مثل الزَّاوِيَةِ الَّتي مِنْ ه ط ز. وَقاعِدَةُ ب ج مثل قاعِدَةِ ه ز. وَالزَّاوِيَةُ الَّتي عَلى المَرْكَزِ مِثْلا الزَّاوِيَةِ الَّتي عَلى الخَطِّ المُحيطِ، إِذا كانَتْ قاعِدَتُهُما واحِدَةً. فَالزَّاوِيَةُ الَّتي مِنْ ب ح ج مِثْلا الزَّاوِيَةِ الَّتي مِنْ ب ا ج. وَكَذٰلِكَ تَكونُ الَّتي مِنْ ه ط ز مِثْلا الَّتي مِن ه د ز. وَلٰكِنَّ الَّتي مِنْ ب ح ج مثل الَّتي مِنْ ه ط ز. فَالَّتي من ب ا ج مِثْلا الَّتي من ه د ز. وَالقِطْعَةُ الَّتي عَلَيْها ب ا ج مثل الَّتي عَلَيْها ه د ز. وقَوْسُ ب ا ج مثل قَوْسِ ه د ز. (وَالزَّاوِيَتانِ)[3] مُتَساوِيَتانِ. فقَوْسُ ب ج مثل قَوْسِ ه ز.

[1]ושני ... שווים] ושתי הזויות שוות [2]ואם] ~~כשת[..] זאת~~+ [3]הד״ז] ~~הד״ח~~+ [4]כאשר] ⟨⟨ר״ל⟩⟩+ [5]השלישית] אפשר שמילה זו מחוקה [6]ושתי] ו⟨⟨[ע״כ]⟩⟩+

[1]ا ب ا ج] ا ب ج [2]فكلا] فكلي [3]والزاويتان] والدائرتان

بُرْهانُهُ: أنّا نُخْرِجُ مِنْ نُقْطَتَيْ ا د مِنْ قَوْسَيْ ب ا ج ه د ز، خُطوطَ ا ب ا ج د ه د ز. نَخُطّ ح ب مثل ط ه، و ح ج مثل ط ز، لِأَنَّهُما في دَوائِرَ مُتَساوِيَةٍ[1]. وَكِلَا ب ح^ ح ج مثل كِلَا ه ط ط ز. وزاوية ب ح ج مثل زاوية ه ط ز. فَقاعِدَة ب ج مثل قاعِدَةِ ه ز. وزاوية ب ح ج مِثْلا زاوية ب ا ج. وزاوية ه ط ز مِثْلا زاوية ه د ز. وَقَدْ كانَتْ زاوية ب ح ج مثل زاوية ه ط ز. فزاوية ب ا ج مثل ه د ز. فقِطْعَة ب ا ج تُشْبِهُ قِطْعَةَ ه د ز. وَهُما على قاعِدَتَيْنِ^ مُتَساوِيَتَيْنِ. فَقَوْسُ ب ا ج مثل قَوْسِ ه د ز. والدَّائِرَتانِ مُتَساوِيَتانِ. فَقَوْسُ ب ج مثل قَوْسِ ه ز الباقِيَة[2].

[1] لانهما ... متساوية] MPR1R2U [2] الباقية] MPR1R2SU

המופת: אנחנו נוציא משתי נקודות א׳ ד׳ משתי קשתות בא״ג הד״ז, קוי[1] א״ב א״ג ד״ה ד״ז. הנה קו ח״ב כמו ט״ה, וח״ג כמו ט״ז, מפני כי שניהם בעגולות שוות. הנה כל אחד[2] מב״ח ח״ג כמו כל אחד מן ה״ט ט״ז. וזוית בח״ג כמו זוית הט״ז. אם כן תושבת ב״ג כמו תושבת ה״ז. וזוית בח״ג כפל זוית[3] בא״ג. וזוית הט״ז כפל זוית הד״ז. וכבר היתה זוית בח״ג כמו זוית הט״ז. אם כן זוית בא״ג כמו זוית הד״ז. אם כן[4] חתיכת בא״ג דומה לחתיכת הד״ז. ושתיהן על שתי חתיכות[5] שוות. אם כן[6] קשת בא״ג כמו קשת הד״ז. ושתי העגולות שוות. אם כן קשת[7] ב״ג הנשארת כמו קשת ה״ז הנשארת.

[1] קוי] ליתא **ו** [2] אחד] אחת **בופ** [3] זוית] זו⟨י⟩ת **פ** [4] זוית בא״ג ... כן] ⟨*⟩ **ט** [5] חתיכות] *⟨⟨תושבות⟩⟩ **ק** [6] כן] ⟨⟨*⟩⟩ **ט** [7] קשת] בקשת **ב**

וכבר התבאר שכשתהיין זויות שוות בעיגולים שוים, הן על קשתות שוות, אם שתהיין על המרכז, או על קו המקיף.

وזה מה שרצינו.

فَقَدِ اسْتَبانَ أَنَّ الزَّوايا المُتَساوِيَةَ الَّتي في الدَّوائِرِ المُتَساوِيَةِ تكونُ عَلى قُسِيٍّ مُتَساوِيَةٍ، إِنْ كانَتِ الزَّوايا عَلى المَراكِزِ، أَو كانَتْ عَلى القُسِيِّ.

وَذَلِكَ ما أَرَدْنا بَيانَهُ.

فَقَدِ اسْتَبَانَ أَنَّ الزَّوايا المُتَساوِيَةَ الَّتي في دَوائِرَ مُتَساوِيَةٍ تَكونُ على قِسِيٍّ مُتَساوِيَةٍ—إِنْ كانَتْ على المَرْكَزِ أو كانَتْ على الخُطوطِ المُحِيطَةِ.

وذَلِكَ ما أَرَدْنا أَنْ نُبَيِّنَ.

הנה כבר התבאר כי השתי זויות[1] השוות אשר[2] בעגולות שוות[3] תהיינה על שתי קשתות שוות—אם היו על המרכז או היו על הקו המקיף.

וזה מה שרצינו לבארו[4].

---

[1]השתי זויות] השתי הזויות בז שתי הזויות פ [2]אשר] ליתא ק [3]שוות] השוות ק [4]וזה ... לבארו] וזה מה שרצינו ביאורו ז ונשלם באורו ט זה מה שרצינו ס

[Proposition III.26]

[H27]

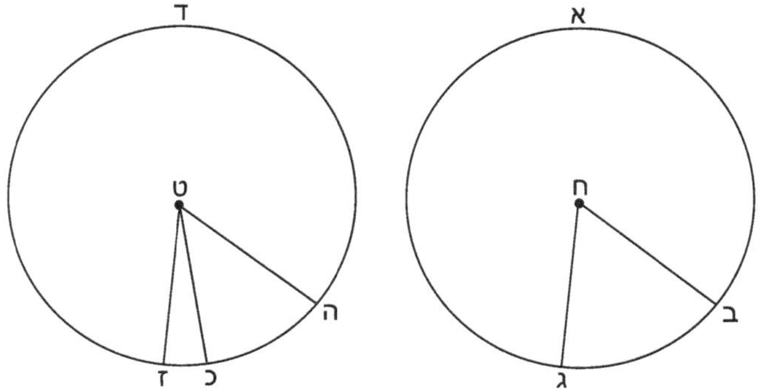

כשתהיין בעגולים שוים זויות על קשתות שוות[1], הזויות הן שוות, בין שתהיין על המרכזים בין שתהיין על הקוים המקיפים.

המשל בו: שנצייר שני עיגולים שוים אב״ג ודה״ז. ונעמיד על קשתות ב״ג וה״ז השוות שתי זויות. ונשים אותם תחלה על שני המרכזים, והם זוית בח״ג וזוית הט״ז. ואומר שהן שוות.

إِذا كانَتْ في الدَّوائِرِ المُتَساوِيَةِ قُسِيٌّ ما مُتَساوِيَةٌ، فَإِنَّ الزَّوايا الَّتي تَكونُ عَلى تِلْكَ القُسِيِّ مُتَساوِيَةٌ، إِنْ كانَتِ الزَّوايا عَلى المَراكِزِ أَو كانَتْ على القُسِيِّ.

مِثالُهُ: أَنَّا نَخُطُّ دائِرَتَيْنِ مُتَساوِيَتَيْنِ، وَهُما دائِرَتا ا ب ج د ه ز. ونُقِيمُ عَلى قَوْسِ ب ج ه ز المُتَساوِيَتَيْنِ زاوِيَتَيْنِ. وَنَجْعَلُهُما أَوَّلًا عَلى المَرْكَزَيْنِ، وَهُما اللَّتانِ مِنْ ب ط ج ومِنْ (ه ح ز)[1]. فَأَقولُ إِنَّهُما مُتَساوِيَتانِ.

שרטוט—B: ح ↔ ط [1]שוות] ⟨*⟩

[1] ه ح ز] ح ه ز

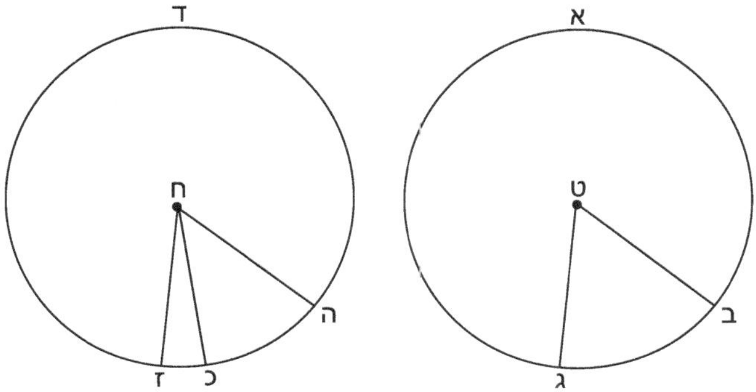

إذا كانَتْ في دَوائِرَ مُتَساوِيَةٍ زَوايا على قُسِيٍّ مُتَساوِيَةٍ، فإِنَّ الزَّوايا مُتَساوِيَةٌ— كانَتْ على الْمَرْكَزِ أو على الخُطوطِ المُحِيطَةِ[1]^.

مِثالُهُ: أَنَّ دائِرَتَيْ ا ب ج د ه ز مُتَساوِيَتانِ. وقَوْسا ب ج ه ز مُتَساوِيَتانِ، وعليهما زاويتا ب ط ج ه ح ز. وَهُما على الْمَرْكَزَيْنِ أو لا. فَأَقولُ إِنَّهُما مُتَساوِيَتانِ، لا يُمْكِنُ إِلّا ذَلِكَ.

כאשר תהיינה בעגולות שוות שתי זויות על שתי[1] קשתות שוות, הנה הזויות שוות—היו על המרכז או היו[2] על הקו המקיף.

המשל בו: כי[3] שתי עגולות אב״ג דה״ז שוות. ושתי קשתות ב״ג ה״ז שוות, עליהן[4] שתי זויות בט״ג הח״ז. והם על המרכז או לא[5]. הנה אומר כי שתיהן[6] שוות, אי אפשר כי אם זה.

[1] الخطوط المحيطة] **CamMPR1R2U** القسي

שרטוט—**בס**: הוספו א״ב א״ג ד״ה ד״ז; **פ**: הוספו ב״ג ה״ז [1] שתי] ליתא **ב** [2] היו] ליתא **ספ** [3] כי] ליתא **פ** [4] עליהן] עליהם **וט** [5] לא] על המקיף **ס** ~~לא~~ ⟨⟨על הקו מקיף⟩⟩ **ק** [6] שתיהן] שתיהם **ב**

ולא יתכן זולת זה. ואם היה אפשר, תהיה זוית בח״ג קטנה מזוית הט״ז. ונעמיד על קו ה״ט, על נקודת ט׳[1], זוית שוה לזוית בח״ג, והיא זוית הט״כ. והזויות השוות אשר בעגולים השוים הן על קשתות שוות, בין שתהיין על המרכזים, בין שתהיין על הקוים המקיפים. ואם כן קשת ה״כ שוה לקשת (ב״ג)[2]. ואמנם קשת ב״ג שוה לקשת ה״ז. ואם כן ה״כ שוה לה״ז—הקטנה שוה לגדולה, זה חלוף, לא יתכן. ואם כן אין זוית בח״ג בלתי שוה לזוית הט״ז. ואם כן היה [!] שוה לה. וכמו כן יתבאר אם היו הזויות על הקוים המקיפים שהיו שוות.

וזה מה שרצינו לבאר.

[1] ט׳] א׳ ⟨⟨*⟩⟩ [2] ב״ג] ה״ג

بُرهانُهُ: أَنَّهُ لا يُمْكِنُ غَيْرُهُ. فَإِنْ كانَ يُمْكِنُ، فَلْتَكُنِ الَّتي مِنْ ب ط ج أَصْغَرَ مِنَ الَّتي مِنْ ه ح ز. وَنُقيمُ عَلى خَطِّ ه ح عَلى نُقْطَةِ ح زاوِيَةً قائِمَةً مِثْلَ الزّاوِيَةِ الَّتي مِنْ ب ط ج، وَهِيَ الَّتي مِنْ ه ح ك. وَالزَّوايا الْمُتَساوِيَةُ الَّتي في الدَّوائِرِ الْمُتَساوِيَةِ هِيَ عَلى قِسِيٍّ مُتَساوِيَةٍ، إِذا كانَتْ عَلى الْمَراكِزِ أَو كانَتْ عَلى الْقِسِيِّ. فَقَوْسُ ب ج إِذًا مِثْلُ قَوْسِ ه ك. وَلَكِنَّ ب ج مِثْلُ ه ز. ه ز إِذًا مِثْلُ ه ك، الْعُظْمَى مِثْلُ الصُّغْرَى—هذا خُلْفٌ لا يُمْكِنُ. فَلَيْسَتِ الزّاوِيَةُ الَّتي مِنْ ب ط ج بِغَيْرِ مُساوِيَةٍ الَّتي مِنْ (ه ح ز)[1]. وَلَكِنَّها مُساوِيَةٌ لَها. فَقَدْ تَبَيَّنَ أَيْضًا أَنَّهُما لَوْ كانَتا عَلى الْقَوْسَيْنِ لَكانَتا مُتَساوِيَتَيْنِ.

وَذَلِكَ ما أَرَدْنا بَيانَهُ.

[1] ه ح ز] ح ه ز

فَإِنْ أَمْكَنَ، فَلْتَكُنْ^ زاوِيَةُ ب ط ج أصغر مِنْ زاوِيَةِ ه ح ز. ونُقيمُ على نُقْطَةِ ح، مِنْ خطّ ه ح، زاوية ه ح ك مثل زاوِيَةِ ب ط ج. فقَوْسُ ب ج مِثْلِ قَوْسِ ه ك. وَلكِنَّ قَوْس ب ج قَدْ كانَتْ مِثْل قَوْسِ ه ز. فقَوْسُ ه ز مثل قَوْسِ^ ه ك—العُظمى مثل الصُّغرى، هذا خُلْفٌ. فَلَيْسَت زاوية ب ط ج بِغَيْرِ مُسَاوِيَةٍ لِزَاوِيَةِ ه ح ز. فَهِيَ مُسَاوِيَةٌ لَها. والزَّاوِيتانِ اللَّتانِ تكُونانِ على قَوْسِ[1] ب ا ج ه د ز نِصْفُ زاوِيَتَيْ^ ب ط ج ه ح ز المُتَساوِيَتَيْنِ. فَهُما إِذَنْ مُتَساوِيَتانِ.

وذلِكَ ما أَرَدْنا أَنْ نُبيِّنَ.

שאם היה אפשר, נאמר שתהיה זוית[1] בט״ג יותר קטנה מזוית הח״ז. ונעמיד על נקודת[2] ח׳, מקו ה״ח, זוית הח״כ[3] כמו זוית בט״ג[4]. הנה קשת ב״ג כמו קשת ה״כ. אבל קשת ב״ג כבר היה כמו קשת ה״ז. הנה קשת ה״ז[5] כמו ה״כ—הגדולה כמו הקטנה, זה שקר. הנה[6] אם כן אין זוית[7] בט״ג בלתי שוה לזוית הח״ז. אם כן היא שוה לה. ושתי הזויות אשר תהיינה על קשת בא״ג הד״ז חצי זוית בט״ג הח״ז השוות. הנה הן[8] אם כן שוות.

וזה מה שרצינו לבאר[9].

[1] قوس] CMR1T1T3

[1] זוית] ליתא **ז** [2] על נקודת] **מ**⟨על⟩ נקודת **פ** [3] הח״כ] ליתא **ב** הח״ג **ט** [4] בט״ג] +והיא זוית כח״ה **ב** [5] ה״ז] ⟨*⟩ **פ** [6] הנה] ליתא **טס** [7] אין זוית] אין **בס** ⟨אין⟩ **ז** [8] השוות ... הן] הנה **ט** [9] לבאר] לבארו **פ**

[Proposition III.27]

[H28]

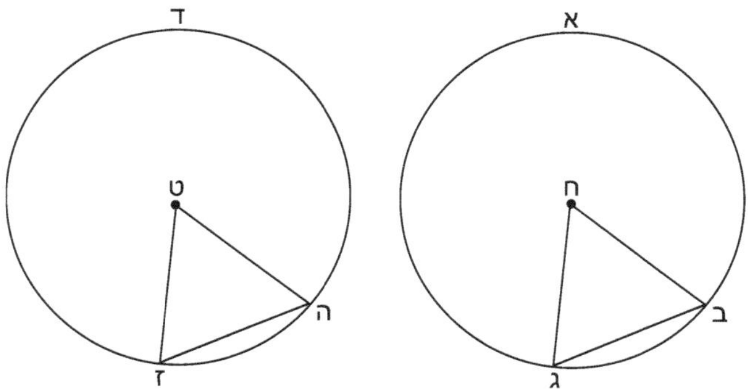

המיתרים השוים אשר בעיגולים שוים יחלקו קשתות שוות—הגדולה לגדולה, והקטנה לקטנה. ואם היה קצת המיתרים ארוך מקצת, תהיה הקשת אשר יחתוך אותה המיתר הארוך גדולה מהקשת שיחתוך אותה המיתר הקטן.

الخُطوطُ المُتَساوِيَةُ الَّتي في الدَّوائِرِ المُتَساوِيَةِ تَقْطَعُ قُسِيًّا مُتَساوِيَةً. وَإِنْ كانَ بَعْضُ الخُطوطِ أَطْوَلُ مِنْ بَعْضٍ، فَالقَوْسُ الَّذي يَقْطَعُها الخَطُّ الأَطْوَلُ أَعْظَمُ مِنَ القَوْسِ الَّتي يَقْطَعُها الخَطُّ الأَقْصَرُ.

המשל בו: שנצייר שני עגולים שוים, והם עגול בא״ג והד״ז. ונעביר בתוכם שני קוים ב״ג וה״ז. ואומר שקשת בא״ג שוה לקשת הד״ז.

مِثالُهُ: أَنَّا نَخُطُّ دائِرَتَيْنِ مُتَساوِيَتَيْنِ عَلَيْهِما ا ب ج د ه ز. وَنَخُطُّ فيهِما خَطَّيْنِ مُتَساوِيَيْنِ، وَهُما ب ج ه ز. فَأَقولُ إِنَّ قَوْسَ ب ج مُساوِيَةٌ لِقَوْسِ ه ز.

שרטוט—B: ח ↔ ט

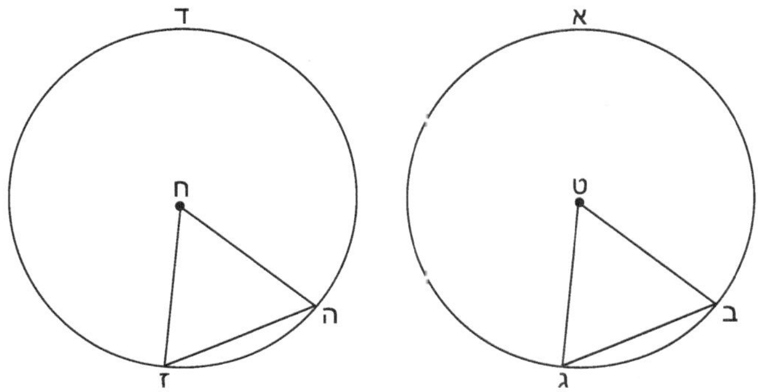

الأَوتارُ المُتَساوِيَةُ الَّتي في الدَّوائِرِ المُتَساوِيَةِ تَفْصِلُ قُسِيًّا مُتَساوِيَةً—العُظمى لِلْعُظمى وَالصُّغرى لِلصُّغرى.

המיתרים השוים אשר בעגולות השוות יבדילו קשתות שוות—הגדולה לגדולה, והקטנה לקטנה.

مِثالُهُ: أَنَّ دائِرَتَيْ ا ب ج د ه ز مُتَساوِيَتانِ. وَفيهِما وَتَرا ب ج ه ز مُتَساوِيانِ. فَأَقولُ إِنَّهُما يَفْصِلانِ قُسِيًّا مُتَساوِيَةً—أَمّا قَوْسُ ب ج فَمِثْلُ قَوْسِ ه ز، وأَمّا قَوْسُ ب ا ج فَمِثْلُ قَوْسِ ه د ز.

המשל בו: כי שתי עגולות אב״ג דה״ז שוות. ושני מיתרי ב״ג ה״ז הם שוים. הנה אומר כי שניהם יבדילו[1] קשתות שוות—אולם קשת ב״ג כמו קשת ה״ז, ואולם[2] קשת בא״ג כמו קשת הד״ז.

שרטוט—**פ**: מעגל הד״ז מימין למעגל בא״ג [1]יבדילו] יבדיל **בטפ** [2]ואולם] + * **ק**

ויהיו מרכזי העגולים ח׳ וט׳. ונדביק ח׳ בב׳ ובג׳, וט׳ בה׳ ובז׳. ותהיין שתי הזויות שוות, לפי שהקוים היוצאים מן המרכזים אל הקוים המקיפים הם שוים. ואם כן ח״ב שוה לט״ה. וח״ג שוה לט״ז. ותושבת ב״ג שוה לתושבת ה״ז. ואם כן זוית בח״ג שוה לזוית הט״ז. והזויות השוות אשר בעגולים שוים הן על קשתות שוות, אם שתהיין על המרכז או על קו המקיף. ואם כן קשת ב״ג שוה לקשת ה״ז. והעגול שוה לעגול. ואם כן קשת בא״ג הנשארת שוה לקשת הד״ז הנשארת.

וזה מה שרצינו לבאר.

بُرْهانُهُ: أَنَّ دائِرَةَ ب ا ج مُساوِيَةٌ ل ه د ز. وَمَرْكَزُ الدَّائِرَتَيْنِ نُقْطَةُ ح وَنُقْطَةُ ط. فَنَصِلُ ط بِنُقْطَةِ ب ط ج. وَنَصِلُ أَيْضًا ح ه ح ز. (فَالزَّاوِيَتانِ)[1] مُتَساوِيَتانِ. وَالخُطُوطُ الَّتي تَخْرُجُ مِنْ مَرْكَزَيْهِما إِلى الخَطَّيْنِ المُحيطَيْنِ بِهِما مُتَساوِيَةٌ. فَخَطُّ ب ط مِثلُ ح ه. وَخَطُّ ط ج مِثلُ ح ز. (فَكِلا)[2] خَطَّيْ ب ط ط ج مِثلُ كِلَيْ خَطَّيْ ه ح ح ز، كُلُّ واحِدٍ مِثلُ الآخَرِ. وَقاعِدَةُ ب ج مِثلُ قاعِدَةِ ه ز. فَالزَّاوِيَةُ الَّتي مِنْ ب ط ج مِثلُ الزَّاوِيَةِ الَّتي مِنْ ه ح ز. وَالزَّوايا المُتَساوِيَةُ في الدَّوائِرِ المُتَساوِيَةِ تَكونُ عَلى قُسِيٍّ مُتَساوِيَةٍ، إِذا كانَتْ عَلى المَراكِزِ أَو عَلى القُسِيِّ. فَقَوْسُ ب ج مِثلُ قَوْسِ ه ز. وَالدَّائِرَةُ مِثلُ الدَّائِرَةِ. فَقَوْسُ ب ا ج الباقِيَةُ مِثلُ قَوْسِ ه د ز الباقِيَةِ.

وَذَلِكَ ما أَرَدْنا بَيانَهُ.

[1] فالزاويتان] فالدائرتان [2] فكلا] فكلي

وَلْيَكُنِ المَرْكَزَانِ ط ح. ونُخْرِجُ ط ب ط ج و ح ه و ح ز. فَخَطّ ط ب مثل خطّ ه ح. وخطّ ط ج مثل خَطّ ح ز. فكِلا خَطَّيْ ب ط ط ج مثل كِلا خَطَّيْ ه ح ح ز. وقاعِدَة ب ج مثل قاعِدَة ه ز. فزاوية ب ط ج مثل زاوية ه ح ز. وقَوْس ب ج مثل قَوْسِ ه ز. والدَّائِرَة مثل الدَّائِرَة. فقوس ب ا ج مثل قَوْسِ ه د ز.

ויהיו שני המרכזים ט׳ ח׳. ונוציא ט״ב ט״ג וח״ה[1] וח״ז. הנה קו ט״ב כמו קו ה״ח. וקו ט״ג כמו קו ח״ז. הנה כל שני קוי ב״ט ט״ג כמו כל[2] שני קוי ה״ח ח״ז. ותושבת ב״ג כמו תושבת ה״ז. אם כן זוית בט״ג כמו זוית הח״ז. וקשת[3] ב״ג כמו קשת ה״ז. והעגולה כמו העגולה[4]. אם כן קשת בא״ג כמו קשת הד״ז[5].

وذلِكَ ما أَرَدْنا أَنْ نُبَيِّنَ.

ונשלם ביאורו[6].

[1] וח״ה] וז״ה+ **ס** [2] כל] ליתא **בוס** [3] וקשת] נקשת **ב** [4] כמו העגולה] ליתא **ב** [5] הד״ז] הד״ג **ס** [6] ונשלם ביאורו] וזה מה שרצינו לבארו **ט**

[Proposition III.28]

[H29]

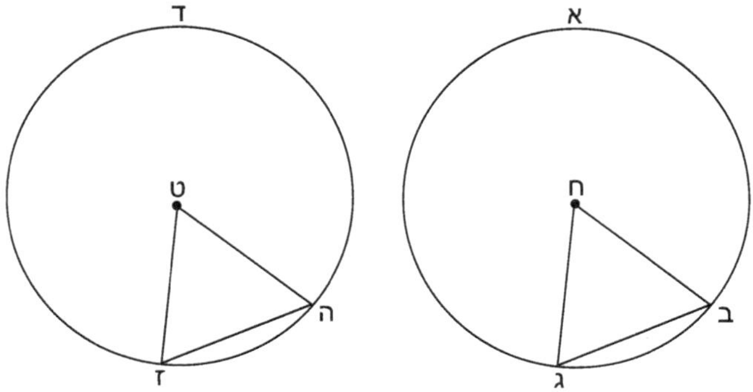

הקשתות השוות מן העגולים השוים יחלקו אתם [!] מיתרים שוים.

המשל בו: *שנצייר שני עגולים שוים*, והם עגולי בא״ג והז״ד. ונחלוק מהם שתי קשתות שוות, קשת ב״ג וקשת ה״ז. ואומר שהקוים שהם מיתרים לקשתות ההן—שוים, והם קוי ב״ג וה״ז.

إِذا كَانَتْ فِي الدَّوائِرِ المُتَساوِيَةِ قُسِيٌّ مُتَساوِيَةٌ، فَإِنَّ الأَوْتارَ الَّتِي تَقَعُ عَلى تِلْكَ القُسِيِّ مُتَساوِيَةٌ.

مِثالُهُ: أَنَّا نَخُطُّ دائِرَتَيْنِ مُتَساوِيَتَيْنِ عَلَيْهِما ا ب ج (د ه ز)[1]. وَنَقْطَعُ مِنْهُما قَوْسَيْنِ مُتَساوِيَيْنِ، قَوْس ب ج وَقَوْس ه ز. فَأَقُولُ إِنَّ الخَطَّيْنِ اللَّذَيْنِ يُوتِرانِ القَوْسَيْنِ مُتَساوِيانِ، وَهُما خَطّا ب ج ه ز.

שרטוט—B: ח ↔ ט

[1] د ه ز] د و ز

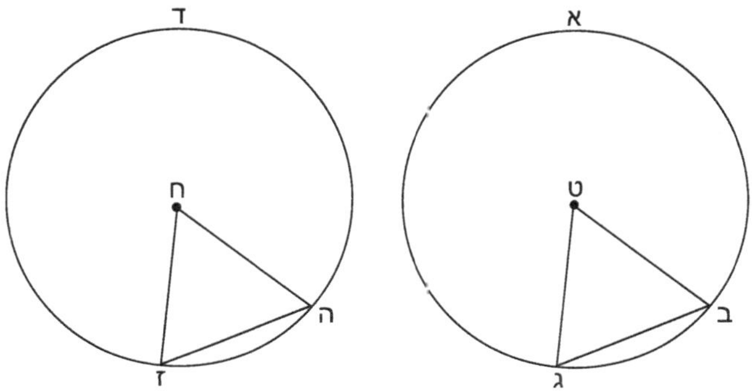

القُسِيُّ المُتَساوِيَةُ مِنَ الدَّوائِرِ المُتَساوِيَةِ، تَفْصِلُها أَوْتارٌ مُتَساوِيَةٌ.

השתי קשתות השוות מן העגולות השוות, יבדילום[1] מיתרים שוים.

مِثالُهُ: أَنَّ دائِرَتَيْ ا ب ج د ه ز مُتَساوِيَتانِ. وَمِنْها قَوْسا ب ج و ه ز مُتَساوِيَتانِ. فَأَقولُ إِنَّ وَتَرَيْهِما، وَهُما ب ج ه ز، مُتَساوِيانِ.

ויהיו שתי עגולות[2] אב״ג דה״ז שוות. ומשתיהן שתי קשתות ב״ג ה״ז שוות. הנה אומר כי שני מיתריהם, והם ב״ג ה״ז, שוים.

[1]יבדילום] יבדילו ס [2]עגולות] העגולות ס

ויהיו מרכזי העגולים ח׳ וט׳. ונדביק ח׳ בב׳ ובג׳, וט׳ בה׳ ובז׳. והקשתות השוות אשר הם מעיגולים שוים תפלנה עליהן זויות שוות, בין שתהיין על המרכז או על הקו המקיף. ואם כן זוית בח״ג שוה לזוית הט״ז. ושני העגולים שוים. והקוים היוצאים ממרכזיהם אל הקוים המקיפים—שוים. ואם כן ב״ח שוה לה״ט. וח״ג שוה לט״ז. וזוית בח״ג שוה לזוית הט״ז. ואם כן תושבת ב״ג שוה לתושבת ה״ז.

وזה מה שרצינו לבאר.

بُرْهانُه: أَنَّ مَرْكَزَيْ الدَّائِرَتَيْن ط ح. فَنَصِلُ (ط ب ب ج، ح ب ه ز)[1]. وَالقُسِيُّ المُتَساوِيَةُ الَّتي في الدَّوائِرِ المُتَساوِيَةِ تَقَعُ عَلى (زوايا)[2] مُتَساوِيَة، إِنْ كانَتِ الزَّوايا عَلى المَراكِزِ وَعَلى[3] القُسِيِّ. فَالزَّاوِيَةُ الَّتي مِنْ ب ط ج مِثْل الَّتي مِنْ (ه ح ز)[4]. وَالدَّائِرَتان مُتَساوِيَتانِ. فَالخُطوطُ الَّتي تَخْرُجُ مِنْ مَرْكَزَيْهِما إِلى الخَطَّيْنِ المُحيطَيْنِ بِهِما هِيَ مُتَساوِيَةٌ. ب ط مثل ه ح. ط ج مثل ح ز. (وكِلا)[5] خَطَّي ب ط ط ج مثل كِلَيْ خَطَّي ح ه ح ز، كُلُّ واحِد مثل الآخَرِ. وَالزَّاوِيَةُ الَّتي مِنْ ب ط ج مثل الَّتي مِنْ ه ح ز. فَقاعِدَةُ ب ج مثل قاعِدَةِ ه ز.

وَذَلِكَ ما أَرَدْنا بَيانَهُ.

[1] ط ب ب ج ح ب ه ز] ط ب ج ح ه ز [2] زوايا] زاويا [3] وعلى] נראה שצ״ל أو على [4] ه ح ز] ح ه ز [5] وكلا] وكلي

بُرْهانُهُ: أنّا نَجْعَلُ المَرْكَزَيْنِ ط ح. ونُخْرِجُ ط ب ط ج ح ه ح ز. نَخَطّ ط ب مثل خَطِّ ح ه. وخطّ ط ج مثل خطّ ح ز. نَخَطّا ب ط ط ج مثل خَطَّيْ ه ح ح ز. والزَّوايا الَّتي تَقَعُ في القِطَعِ المُتَساوِيَةِ مُتَساوِيَةٌ. فزاوية ب ط ج مثل زاوية ه ح ز. فقاعِدَةُ ب ج مثل قاعِدَةِ ه ز. فَقَدِ اسْتَبانَ أنَّ وَتَرَيْ ب ج ه ز مُتَساوِيانِ.

המופת: כי נשים המרכזים ט׳ ח׳. ונוציא ט״ב ט״ג ה״ח ח״ז. הנה קו ט״ב כמו קו ח״ה. וקו ט״ג כמו קו ח״ז. הנה שני קוי ב״ט ט״ג[1] כמו שני קוי ה״ח ח״ז. והזויות אשר יפלו בחתיכות השוות—שוות. אם כן זוית בט״ג כמו זוית הח״ז. אם כן תושבת ב״ג כמו תושבת ה״ז. הנה כבר התבאר כי שני מיתרי[2] ב״ג ה״ז שוים.

وذَلِكَ ما أَرَدْنا أَنْ نُبَيِّنَ.

וזה מה שרצינו לבאר[3].

[1] ט״ג] ⟨*⟩ פ [2] מיתרי] מיתרים ס [3] וזה ... לבאר] ונשלם ביאורו ס

[Proposition III.29]

[H30]

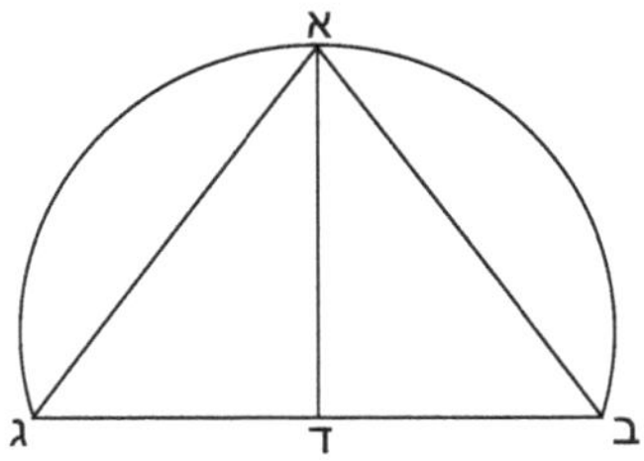

נרצה לחלוק קשת ידועה לחציים.

ונשים הקשת הידועה קשת בא״ג. וכאשר נרצה לחלק אותה, נדביק ב׳ בג׳. ונחלק קו ב״ג לחציים על נקודת ד׳. ונוציא מד׳ קו על זוית נצבה אל קשת בא״ג, והוא קו ד״א. ואומר שכבר חלקנו קשת בא״ג לחציים על נקודת א׳.[1]

نُرِيدُ أَنْ نَقْطَعَ قَوْسًا مَعْلومًا بِنِصْفَيْنِ.

مِثالُهُ: أَنَّا نَجْعَلُ قَوْسَ ا ب ج مَعْلومَةً. وَنُرِيدُ أَنْ نَقْطَعَها بِنِصْفَيْنِ. فَنَصِلُ ب ج. وَنَقْسِمُ خَطَّ ب ج بِنِصْفَيْنِ عَلَى نُقْطَةِ د. وَنُخْرِجُ مِنْ د خَطًّا عَلَى زاوِيَةٍ قائِمَةٍ عَلَى قَوْسِ ب ا ج، وَهُوَ خَطُّ د ا. فَأَقولُ إِنَّا قَدْ قَطَعْنا القَوْسَ المَعْلومَةَ بِنِصْفَيْنِ عَلَى نُقْطَةِ ا.

---

[1] א׳] +ה׳

نُرِيدُ أَنْ نَقْطَعَ قَوْسًا مَعْلومَةً بِنِصْفَيْنِ.

فَنَجْعَلَ القَوْسَ المَعْلومَةَ ب ا ج. ونُرِيدُ أَنْ نَقْطَعَهَا بِنِصْفَيْنِ. فَنُخْرِجَ وَتَرَ^ ب ج. ونَقْطَعَهُ بِنِصْفَيْنِ على د. ونُخْرِجَ مِنْ د إلى قَوْسِ ب ا ج خَطَّ د ا على زاوِيَةٍ قائِمَةٍ، على خَطِّ ب ج.

נרצה לחתוך קשת[1] ידועה בשני[2] חציים.

הנה נשים הקשת הידועה בא״ג. ונרצה לחתוך אותה בשני חציים. הנה נגיע ב״ג. ונחלקהו בשני חציים על ד׳. ונוציא מן ד׳ אל קשת בא״ג קו א״ד[3] על זוית נצבת[4], על קו ב״ג.

[1] קשת] קשתי **ב** [2] בשני] בשתי **ב** [3] א״ד] אל ד׳ **ט** [4] נצבת] נצבה **טספ**

המופת בו: שנדביק א׳ בב׳ ובג׳. וב״ד שוה לד״ג. וד״א משותף. ושני קוי ב״ד ד״א שוים לד״ג וד״א. וזוית בד״א שוה לזוית גד״א. ואם כן תושבת א״ב שוה לתושבת א״ג. והקוים השוים אשר בעגולים שוים יחלקו קשתות שוות. ואם כן קשת א״ב שוה לקשת א״ג. וקשת בא״ג כבר נחלקה לחציים.

וזה מה שרצינו לבאר.

بُرهانُه: أنَّا نَصِلُ (ا ب ب ج)[1]. ب د مثل د ج. د ا مُشْتَرَكٌ. (فكِلا)[2] خَطَّيْ ب د د ا مثل كِلَيْ خَطَّيْ ج د د ا، كُلُّ واحِدٍ مثل نَظيرِه. وَالزَّاوِيَةُ الَّتي من ب د ا مثل الزَّاوِيَةِ الَّتي مِنْ ج د ا. فَقاعِدَةُ ا ب مثل قاعِدَةِ ا ج. وَالخُطوطُ المُتَساوِيَةُ تَقْطَعُ قُسِيًّا مُتَساوِيَةً. وَالخَطُّ (الأَطْوَلُ)[3] يَقْطَعُ القَوْسَ العُظْمَى، وَالأَقْصَرُ يَقْطَعُ القَوْسَ الصُّغْرَى. فَقَوْسُ ا ب مثل قَوْسِ ا ج. فَقَوْسُ ب ا ج المَعْلومَةُ قَدْ قُطِعَتْ بِنِصْفَيْنِ عَلَى نُقْطَةِ ا.

وَذَلِكَ ما أَرَدْنا بَيَانَهُ.

[1] ا ب ب ج] ا ب ج [2] فكلا] فكلي [3] الاطول] الأُول

وَنُخْرِجَ خَطَّيْ اب اج. فَخَطّ ب د مثل خَطِّ ج د. وَنَجْعَلَ اد مُشْتَرَكًا. فَخَطّا ب د د ا مثل خَطَّيْ ج د د ا. وزاوية ب د ا مثل زاوية ج د ا. فقاعِدَة ا ب، الَّتي هِيَ وَتَرٌ، مثل قاعِدَةِ ا ج، الَّتي هِيَ وَتَرٌ، فقَوْس ا ب إذَنْ[1] مثل قَوْسِ ا ج. فَقَدْ قَطَعْنا قَوْسَ ب ا ج بِنِصْفَيْنِ على نُقْطَةِ ا.

ונוציא שני קוי א״ב א״ג[1] [2]. הנה קו ב״ד כמו קו[3] ג״ד. ונשים[4] א״ד משותף. הנה שני קוי ב״ד[5] ד״א כמו שני קוי ג״ד ד״א[6] [7]. וזוית בד״א כמו זוית גד״א. אם כן תושבת א״ב, אשר הוא מיתר[8], כמו תושבת א״ג[9], אשר היא[10] מיתר[11]. הנה קשת א״ב אם כן כמו קשת א״ג. הנה כבר חתכנו קשת בא״ג בשני חציים על נקודת א׳.

وَذَلِكَ ما أَرَدْنَا أَنْ نُبَيِّنَ.

וזה מה שרצינו לבאר.

[1] إذن] DPR1 إذا

[1] ונוציא ... א״ג] ליתא ס [2] א״ג] ~~ג״ד~~+ ק
[3] קו] ליתא ס [4] ונשים] הנה נשים בו ליתא פ [5] ב״ד] ג״ד ב [6] א״ד ... ד״א] ⟨⟨*⟩⟩ פ [7] כמו ... ד״א] ⟨⟨*⟩⟩ ב [8] מיתר] +קשת א״ב ס [9] אשר ... א״ג] ⟨⟨*⟩⟩ ק [10] היא] הוא וק [11] מיתר] +קשת א״ג ס

[Proposition III.30]

[H31]

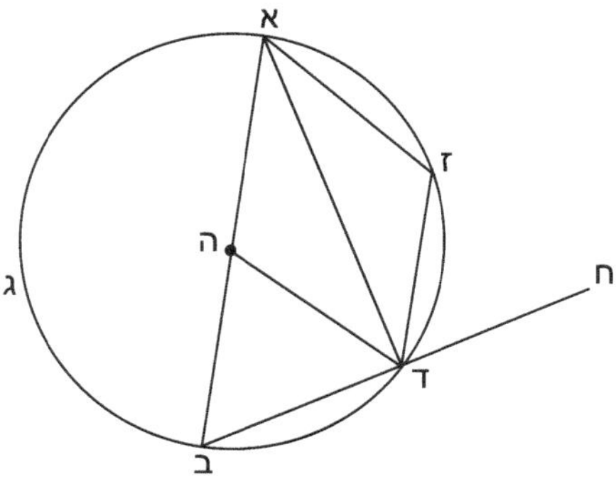

نُرِيدُ أَنْ نُبَيِّنَ [..] أَنَّ الزَّاوِيَةَ[1] المُسْتَقِيمَةَ الخَطَّيْنِ إِذا كانَتْ مُرَكَّبَةً عَلَى القَوْسِ، وَكانَتْ فِي نِصْفِ دائِرَةٍ، فَهِيَ قائِمَةٌ. وَإِذا كانَتْ فِي قِطْعَةٍ أَصْغَرَ مِنْ نِصْفِ دائِرَةٍ، فَإِنَّها أَكْبَرُ مِنْ قائِمَةٍ. وَإِذا كانَتْ فِي قِطْعَةٍ أَعْظَمَ مِنْ نِصْفِ دائِرَةٍ، فَهِيَ أَصْغَرُ مِنْ قائِمَةٍ. وَزاوِيَةُ القِطْعَةِ مِنَ الدّائِرَةِ الَّتِي هِيَ أَعْظَمُ مِنْ نِصْفِها هِيَ أَعْظَمُ مِنْ قائِمَةٍ. وَزاوِيَةُ القِطْعَةِ مِنَ الدّائِرَةِ الَّتِي هِيَ أَصْغَرُ مِنْ نِصْفِها هِيَ أَصْغَرُ مِنْ قائِمَةٍ.

כשתהיה בחתיכת עגול זוית ישרת הקוים מורכבת על קשת, והיתה החתיכה חצי עגול, תהיה הזוית ההיא נצבה. ואם היתה גדולה מחצי עגול, תהיה חדה. ואם היתה קטנה מחצי עגול, תהיה נרוחת. אמנם זוית החתכה שהיא גדולה מחצי עגול, תהיה נרוחת; אבל זוית החתכה שהיא קטנה מחצי עגול, היא חדה.

[1] ان ... الزاوية] ⟨⟨*⟩⟩

إذا كانَتْ في قِطْعَةِ دائِرَةٍ[1] زاوِيَةٌ مُسْتَقيمَةُ الخَطَّيْنِ مُرَكَّبَةٌ على القَوْسِ، وكانَتِ القِطْعَةُ نِصْفَ دائِرَةٍ، فَإِنَّ الزّاوِيَةَ قائِمَةٌ. وإِنْ كانَتْ أَعْظَمَ مِنْ نِصْفِ دائِرَةٍ، فَهِيَ حادَّةٌ. وإِنْ كانَتْ أصغر مِنْ نِصْفِ دائِرَةٍ، فَهِيَ مُنْفَرِجَةٌ. وأَمّا زاوِيَةُ القِطْعَةِ الَّتي هِيَ أعظم^ مِنْ نصف دائِرَةٍ. فإِنَّها مُنْفَرِجَةٌ^. وأمّا زاويةُ القِطْعَةِ الَّتي هِيَ أَصْغَرُ^ مِنْ نِصْفِ دائِرَةٍ، فَإِنَّها حادَّةٌ^.

[1] قطعة دائرة] MPR1R2

כאשר היתה בחתיכת[1] העגולה זוית ישרת שני הקוים מורכבת על הקשת, והיתה החתיכה חצי עגולה, הנה הזוית[2] נצבת[3]. ואם היתה יותר גדולה מחצי עגולה, הנה היא חדה. ואם היתה יותר קטנה מחצי עגולה, הנה היא נרחבת. ואולם זוית החתיכה אשר היא יותר[4] קטנה מחצי עגולה, הנה היא חדה; וזוית החתיכה אשר היא[5] יותר גדולה מחצי עגולה[6], הנה היא נרחבת.

שרטוט—**בוטסק**: $ [1] בחתיכת] בחתיכה **בופ** בחתיכת העגולה **ס** בחתכת ⟨עגולה⟩ **ק** [2] הזוית] ⟨⟨ה⟩⟩זוית **ב** [3] נצבת] הנצבת **ס** [4] יותר] ⟨⟨*⟩⟩ **פ** [5] היא] ליתא **ט** [6] הנה ... עגולה] ⟨⟨*⟩⟩ **ב**

המשל בזה: שנצייר עגול עליו אב״ג. וקטרו א״ב. ונוציא מא׳ קו אל ד׳ כמו שיפל. ואומר שהזוית שתפל בחצי עגול היא נצבה, והיא זוית אד״ב. וכאשר תפל בפחות מחצי עגול היא גדולה מנצבה. ואשר תפל ביותר מחצי עגול היא קטנה מנצבה.

مِثالُهُ: أَنَّا نَخُطُّ دائِرَةً عَلَيْها ا ب ج. وَقُطْرُها ا ب. وَنُخْرِجُ مِنْ ا خَطًّا إِلى د، كَيْفَما وَقَعَ. فَأَقولُ إِنَّ الزَّاوِيَةَ الَّتي تَقَعُ في نِصْفِ دائِرَةٍ قائِمَةٌ، وَهِيَ الَّتي تَقَعُ في ا د ب. وَالَّتي تَقَعُ في أَصْغَرَ مِنْ نِصْفِ دائِرَةٍ، هِيَ أَكْبَرُ مِنْ قائِمَةٍ. وَالَّتي تَقَعُ في أَعْظَمَ مِنْ نِصْفِ دائِرَةٍ، هِيَ أَصْغَرُ مِنْ قائِمَةٍ.

مِثَالُهُ: أَنَّ دائِرَةَ ا ب ج قُطْرُها خَطُّ ا ب. ونَتَعَلَّمُ على القَوْسِ نُقْطَةً د كَيْفَما وَقَعَتْ. ونُخْرِجُ مِنْها وَتَرَيْ د ا د ب. فَأَقولُ إِنَّ زاوِيَةَ ا د ب الَّتي في نِصْفِ الدَّائِرَةِ—قائِمَةٌ. وإِنْ كانَتِ الزَّاوِيَةُ في أَعْظَمَ مِنْ نِصْفِ دائِرَةٍ، فَهِيَ حادَّةٌ. وإِنْ كانَتْ في أَصْغَرَ مِنْ نِصْفِ دائِرَةٍ—فإِنَّها مُنْفَرِجَةٌ.

דמיון זה: כי עגולת אב״ג קוטרה א״ב. ונרשום[1] על הקשת נקודת ד׳ איך שתפול[2]. ונוציא ממנה מיתרי א״ד ד״ב. הנה אומר כי זוית אד״ב אשר[3] בחצי העגולה[4]—נצבת. וכאשר היתה הזוית ביותר גדולה מחצי עגולה[5], הנה היא חדה. וכאשר היתה ביותר קטנה מחצי עגולה[6] הנה היא נרחבת.

[1] קוטרה ... ונרשום] ליתא **ב** [2] נקודת ... שתפול] נקודה איך שיפול והיא ד׳ **ז** [3] אשר] +היא **פ** [4] העגולה] +~~נרחבת~~ **פ** [5] העגולה] ליתא **ז** עגולה **טס** [6] עגולה] העגולה **בפק**

המופת בו: שנרשום על קשת א״ד נקודה כמו שתפול, והיא נקודת ז׳. ומרכז העגול ה׳. ונדביק א׳ בז׳, וז׳ בד׳, וב׳ בד׳, וה׳ בד׳. וה׳ מרכז עגול אב״ג. ואם כן ה״ד שוה לה״ב. וזוית הב״ד שוה לזוית הד״ב. ושתי זויות הב״ד והד״ב כפל זוית הד״ב. ואמנם הזוית החיצונה, והיא זוית אה״ד, שוה לשתי הזויות יחד, רצו׳ לומ׳ זוית הב״ד והד״ב. ואם כן זוית אה״ד כפל[1] זוית הד״ב. ולפי זה היתה זוית דה״ב כפל זוית הד״א. וזויות אה״ד ודה״ב יחד כפל זוית אד״ב. ואמנם זוית דה״א ודה״ב יחד שוות לשתי זויות נצבות. ואם כן זוית אד״ב נצבה. והיא בחצי עגול, והוא שעליו אזד״ב. וזוית דב״א קטנה מנצבה. והיא ביותר מחצי עגול שעליו דבג״א. וכל עגול שיפול בתוכו שטח בעל ארבע צלעות, כל שתי זויות מזויותיו שהן מקבילות זו את זו הן שוות לשתי זויות נצבות. וזוית דב״ה קטנה מנצבה. תשאר[2] זוית דז״א גדולה מנצבה. והיא בפחות מחצי עגול שעליו אז״ד. ואומר שזוית[3] החתכה שהיא גדולה מחצי עגול, והיא מקשת א״ד ומקו ד״א, והיא זוית בד״א, היא גדולה מנצבה, והוא אגב״ד [!], שהוא גדול מחצי עגול. וזוית החתיכה שהיא קטנה מחצי עגול, והיא מקשת ז״ד וקו ד״א, והיא זוית זד״א, קטנה מנצבה.

[1] כפל] ~~כפל זוית אה״ד~~+ [2] תשאר] ~~והיא ביותר~~ ~~מחצי עגול~~+ [3] שזוית] שזו⟨י⟩ת

بُرهانُهُ: أَنَّا نعلِّمُ على قَوْسِ ا د نُقْطَةً كَيْفَما وَقَعَتْ، وَهِيَ ز. ومَرْكَزُ الدَّائِرَةِ ه. وَنَصِلُ ا ز ز د[1] د ب د ه. (ه)[2] مَرْكَزُ دائِرَةِ ا ب ج. ه د مثل ه ب. فالزَّاوِيَةُ الَّتِي من ه ب د مثل الَّتِي من ه د ب. فَاللَّتانِ مِنْ ه ب د ومِنْ ه د ب مِثْلا الَّتِي مِنْ ه ب د. وَلَكِنَّ الزَّاوِيَةَ الخارِجَةَ مِنَ المُثَلَّثِ، وَهِيَ الَّتِي من ا ه د، مثل[3] الزَّاوِيَتَيْنِ جَمِيعًا اللَّتَيْنِ مِنْ ه ب د ومِنْ ه د ب. فَالَّتِي مِنْ ا ه د مِثْلا الَّتِي مِنْ ه د ب. وَلِذَلِكَ صارَتِ الَّتِي مِنْ (د ه ب)[4] مثل الَّتِي مِنْ ه د ا. فَالَّتِي مِنْ ه د ا وَالَّتِي مِنْ ه د ب[5] جَمِيعًا مِثْلا الَّتِي مِنْ ا د ب. وَلَكِنَّ اللَّتانِ مِنْ (د ه ا)[6] د ه ب مثل زاوِيَتَيْنِ قائِمَتَيْنِ. وَالَّتِي مِنْ ا د ب قائِمَةٌ. وَهِيَ في نِصْفِ دائِرَةِ الَّتِي عَلَيْها ا ز د ب. فَالزَّاوِيَةُ الَّتِي مِنْ د ب ا أَصْغَرُ مِنْ قائِمَةٍ. وَكُلُّ دائِرَةٍ يَقَعُ فيها سَطْحٌ ذو أَرْبَعَةِ أَضْلاعٍ، فَكُلّ زاوِيَتَيْنِ تَقابَلانِ فيهِ مُساوِيَتَيْنِ لِزاوِيَتَيْنِ قائِمَتَيْنِ. فَالَّتِي مِنْ د ب ا وَمِنْ د ز ا مُساوِيَتانِ لِزاوِيَتَيْنِ قائِمَتَيْنِ. لَكِنَّ الَّتِي مِنْ (د ب ا)[7] أَصْغَرُ مِنْ قائِمَةٍ. وَبَقِيَتِ الَّتِي مِنْ د ز ا أَكْبَرَ مِنْ قائِمَةٍ. وَهِيَ في قِطْعَةٍ أَصْغَرَ مِنْ نِصْفِ دائِرَةِ الَّتِي عَلَيْها ا ز د. وَأَقولُ إِنَّ الزَّاوِيَةَ القِطْعَةِ الَّتِي هِيَ أَكْبَرُ مِنْ قائِمَةٍ نصف دائِرَةٍ، وَهِيَ الَّتِي يُحَطُّ بِها قَوْسُ ا ب د ووَتَرُ د ا، أَعْظَمُ

[1] ز د] +ه ب [2] ه] ליתא [3] مثل] +~~التي من~~ ~~د ه ب~~ [4] د ه ب] ه د ب [5] فالتي … ه د ب] נראה שצ״ל: فالتي من ا ه د والتي من د ه ب [6] د ه ا] ه د ا [7] د ب ا] ب د ا

بُرْهانُهُ: أنَّا نَتَعَلَّمُ على قَوْسِ ا د نُقْطَةً ز كَيْفَما وَقَعَتْ. ونَجْعَلُ الْمَرْكَزَ ه. ونُخْرِجُ خُطوطَ ا ز ز د ه د. نَخَطّ ه د مثل خطّ ه ب، لأنَّ الْمَرْكَزَ نُقْطَةُ ه. فزاوية ه ب د مثل زاوية ه د ب. فَهُما مِثْلا زاوية ه د ب. لكنَّ زاوية ا ه د الخارِجَةَ مِنَ الْمُثَلَّثِ، مثل زاوِيَتَيْ ه د ب ه ب د جَميعًا. فَهِيَ مِثْلا زاوية ه د ب. ولِذَلِكَ تكونُ زاوِيَةُ د ه ب مِثْلَيْ زاوية ه د ا. وزاوِيَتا د ه ا د ه ب^ قائِمَتانِ. فزاوية ا د ب الَّتي في نِصْفِ دائِرَةِ ا ز د ب قائِمَةٌ. وزاوية د ب ا أَصْغَرُ مِنْ قائِمَةٍ، وَهِيَ في قَوْسٍ أعظم مِنْ نِصْفِ دائِرَةٍ[1]، وَهِيَ قَوْسُ د ب ج ا. وَكُلّ شَكْلٍ ذي أَرْبَعَةِ أَضْلاعٍ يَقَعُ في دائِرَةٍ، فَإِنَّ كُلّ زاوِيَتَيْنِ مُتَقابِلَتانِ مِنْ زواياهُ مُتَقابِلَتانِ مُعادِلَتانِ لِزاوِيَتَيْنِ قائِمَتَيْنِ. فَزاوِيَتا ا ب د د ز ا مُعادِلَتانِ لِقائِمَتَيْنِ. وزاوية د ب ه أَصْغَرُ مِنْ قائِمَةٍ. تَبْقَى زاوِيَةُ د ز ا أعظم مِنْ قائِمَةٍ، وَهِيَ في قَوْسِ ا ز د ب، الَّتي هِيَ أَصْغَرُ مِنْ نِصْفِ الدَّائِرَةِ. وَأَقولُ إِنَّ الزَّاوِيَةَ الَّتي يُحيطُ بها قَوْسُ ب د ووَتَرُ ا د مُنْفَرِجَةٌ، وَهِيَ زاوِيَةُ قِطْعَةِ اج ب د، الَّتي هِيَ أَعْظَمُ مِنْ نِصْفِ دائِرَةٍ، وَإِنَّ الزَّاوِيَةَ الَّتي يُحيطُ بِها قَوْسُ ا ز د ووَتَرُ د ا حادَّةٌ، وَهِيَ زاوِيَةُ قِطْعَةِ ا ز د، الَّتي هِيَ أَصْغَرُ مِنْ نِصْفِ دائِرَةٍ.

[1] دائرة] O2

המופת: אנחנו נרשום על קשת א״ד נקודת ז׳ איך שתפול. ונשים המרכז ה׳. ונוציא קוים[1] א״ז ד״ז ד״ה. הנה[2] קו ה״ד כמו קו ה״ב, מפני כי המרכז נקודת ה׳. הנה[3] זוית הב״ד כמו זוית הד״ב. הנה שתיהן כפל זוית[4] הד״ב. אבל זוית אה״ד החיצונה מן המשולש, כמו שתי זויות הד״ב הב״ד[5] יחד[6]. הנה היא[7] כפל זוית הד״ב. ולכן תהיה זוית[8] דה״ב כפל זוית הד״א. ושתי זויות דה״א דה״ב כמו שתי זויות נצבות. אם כן זוית אד״ב אשר בחצי עגולת אזד״ב נצבת. וזוית דב״א יותר קטנה מנצבת, והיא בקשת יותר גדולה מחצי עגולה, והיא קשת דבג״א. וכל תמונה בעלת ארבע צלעות תפול בעגולה, הנה כל שתי זויות מקבילות מזויותיה שוות לשתי נצבות. אם כן שתי זויות אב״ד דז״א שוות לשתי נצבות. וזוית דב״ה[9] יותר קטנה מנצבת. ונשארה[10] זוית[11] דז״א יותר גדולה מנצבת, והיא בקשת אז״ד, אשר היא יותר קטנה מחצי העגולה. ואומר כי הזוית אשר תקיף בה קשת ב״ד ומיתר א״ד נרחבת[12], והיא זוית[13] חתיכת אגב״ד, אשר היא יותר גדולה מחצי העגולה, ושהזוית אשר[14] יקיפו בה קשת אז״ד[15] ומיתר א״ד חדה, והיא זוית חתיכת אז״ד, אשר היא יותר קטנה מחצי עגולה[16].

[1] קוים] הקוים ו [2] הנה] +~~שני~~ ק [3] הנה] ליתא בוס ⟨⟨*⟩⟩ פ [4] זוית] זויות ס [5] הב״ד] +הפנימיות ס [6] יחד] ליתא ב [7] היא] ~~הוא~~ ⟨⟨שתיהן⟩⟩ ב הוא וטפק [8] זוית] ~~כפל~~+ פ [9] דב״ה] דב״א ס [10] ונשארה] ונשאר בספ [11] זוית] +*̲ ק [12] נרחבת] +~~לו נראה אד״ה~~ ב [13] זוית] ⟨*⟩ פ [14] אשר] +היא יותר גדולה אשר ס [15] אז״ד] ז״ד ס [16] מחצי עגולה] העגולה בו עגולת ט ~~מנצבת~~+ ק

مِنْ قائِمَةٍ. وَأَنَّ زاوِيَةَ القِطْعَةِ الَّتي هِيَ أَصْغَرُ مِنْ نِصْفِ دائِرَةٍ، وَهِيَ الَّتي تُحيطُ بِها قَوْسُ ا ز د وَوَتَرُ ا د هِيَ أَصْغَرُ مِنْ قائِمَةٍ.

بُرْهانُهُ: أَنّا نُخْرِجُ خَطَّ ب د عَلى اسْتِقامَةٍ إِلى نُقْطَةِ ح. فَمِنْ أَجْلِ أَنَّ الزّاوِيَةَ المُسْتَقيمَةَ الخَطَّيْنِ الَّتي مِنْ ا د ب قائِمَةٌ، صارَتِ الَّتي يُحيطُ بِها خَطُّ ا د الوَتَرَ (د ب)[1] قائِمَةً. فَإِذا كانَت الَّتي مِنْ ح د ا قائِمَةَ، كانَتِ الَّتي يُحيطُ بِها خَطُّ ز د القَوْس وَخَطّ د ا الوَتَرَ أَصْغَر مِنْ قائِمَةٍ، وَهِيَ زاوِيَةُ القِطْعَةِ الَّتي هِيَ أَصْغَرُ مِنْ نِصْفِ دائِرَةٍ. وتَكونُ الزّاوِيَةُ الَّتي يُحيطُ بِها خَطُّ ا ب د القَوْسَ وَخَطّ د ا الوَتَرَ أَعْظَم مِنْ قائِمَةٍ، وَهِيَ زاوِيَةُ القِطْعَةِ الَّتي هِيَ أَكْبَرُ مِنْ نِصْفِ دائِرَةٍ.

وَذَلِكَ ما أَرَدْنا بَيانَهُ.

המופת בו: שנוציא ב״ד אל ח׳. ולפי שזוית אד״ב נצבה, תהיה זוית זד״ב[1] גדולה מנצבה. והיא זוית החתכה שהיא גדולה מחצי עגול. ולפי ששתי הזויות, רצו׳ לומ׳ זוית אד״ב וזוית אד״ח, שוות לשתי זויות נצבות, וזוית אד״ב נצבה, ותשאר זוית אד״ח נצבה, היתה אם כן זוית זד״א, שהיא קשת [!] ז״ד[2] וקו ד״א, קטנה מזוית אד״ח. ואם כן הוא קטן [!] מנצבה. והיא זוית החתיכה שהיא קטנה מחצי עגול.

וזה מה שרצינו לבאר.

---

[1]زד״ב] ~~א״ד~~ ⟨⟨*⟩⟩ ~~וקשת ד״ב~~ ⟨⟨[*]⟩⟩

[2]ז״ד] ~~ח״ד~~

---

[1]د ب] د ح

بُرْهانُهُ: أَنّا نُخْرِجُ وَتَرَ ب د إلى ح. فَلِأَنَّ زاوِيَةَ ا د ب قائِمَةٌ، صارَتِ الزَّاوِيَةُ[1] الَّتِي يُحِيطُ بِها وَتَرُ ا د وقَوْسُ د ب مُنْفَرِجَةً. وَلأَنَّ زاوِيَتَيّ ا د ب ا د ح[2] مثل^ قائِمَتَيْنِ، وزاوية ا د ب قائِمَة. تبْقى زاويةُ ا د ح قائِمَةً^. فَصارَتِ الزّاوِيَةُ الَّتِي يُحِيطُ بِها قَوْسُ ز د ووَتَرُ ا د حادَّةً^.

המופת: אנחנו נוציא[1] מיתר ב״ד אל ח׳. הנה מפני כי זוית אד״ב נצבת, היתה הזוית[2] אשר יקיפו בה מיתר א״ד וקשת ד״ב נרחבת. ומפני כי שתי זויות אד״ב אד״ח נצבות, היתה הזוית אשר יקיפו בה קשת ז״ד ומיתר א״ד יותר קטנה מזוית אד״ח.

وذَلِكَ ما أَرَدْنا أَنْ نُبَيِّنَ.

וזה מה שרצינו לבאר.

[1] الزاوية] CamMPR1R2S [2] ا د ب ا د ح] ز د ب ز د ح MPR1R2ST2U

[1] נוציא] מוציא ב [2] הזוית] זוית ו

قالَ ثابتٌ: وَجَدْنا في بَعْضِ النُّسَخِ اليونانيَّةِ لِهَذا الشَّكْلِ بُرْهانًا آخَرَ على أنَّ زاويةَ ادب قائِمَةٌ. وَهُوَ: أنَّ خطَّ ا ه مثل خطِّ ه د. فزاوية ه ا د مثل زاوية ه د ا. وخطّ ه د مُساوٍ لخطّ ب ه. فزاوية ه ب د مثل زاوية ه د ب. فجميع زاوية ا د ب مُسَاوِيَة لزاوِيَتَيْنِ ب ا د ا ب د وزاوية ا د ح أَيْضًا مُسَاوِيَة لِزَاوِيَتَيّ ب ا د ا ب د. فزاوية ا د ب مُسَاوِيَة لِزَاوِيَةِ ا د ح. فَهِيَ إذَنْ[1] قائِمَةٌ.

وذَلِكَ ما أَرَدْنا أَنْ نُبَيِّنَ.

אמר תאבת: מצאנו[1] בקצת הנוסחאות מהיונים[2] לתמונה הזאת מופת אחר על היות זוית אד״ב נצבת.

והוא: כי[3] קו[4] א״ה כמו קו ה״ד. אם כן זוית הא״ד כמו[5] זוית הד״א. וקו ה״ד שוה לקו ב״ה. אם כן זוית הב״ד כמו זוית[6] הד״ב. הנה כל זוית אד״ב שוה לשתי זויות בא״ד אב״ד[7]. וזוית אד״ח גם כן שוה לשתי זויות בא״ד אב״ד. הנה זוית אד״ב[8] שוה לזוית אד״ח. הנה היא[9] אם כן נצבת.

וזה מה שרצינו לבאר.

---

[1] إذن] CT1T2T3

---

[1] מצאנו] ⟨*⟩ ו [2] מהיונים] מן+ פ [3] אד״ב ... כי] הד״א ב [4] קו] ליתא ט [5] כמו] +* ב
[6] זוית] ליתא ספ [7] כמו ... אב״ד] ליתא ב
[8] אד״ב] ~~אב״ד~~+ ק [9] היא] ליתא ט

[Proposition III.31]

[H32]

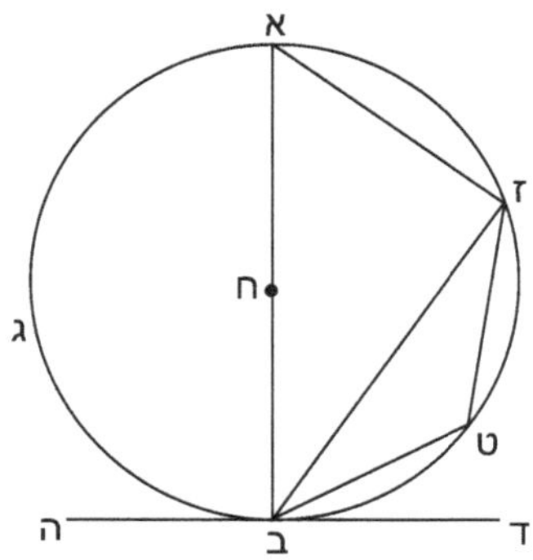

כל קו ממשש עגול שיצא ממקום משושו קו ישר שיחתוך העגול ולא יעבר על המרכז, תהיין שתי הזויות אשר אצל צדי הקו שוות לשתי זויות שתפלנה בשתי חתיכות העגול המתחלפות להן.

إذا كانَ خَطٌّ يَماسُّ الدَّائِرَةَ، وأُخْرِجَ مِنَ المَوْضِعِ الَّذي يَماسُّها فيهِ خَطٌّ آخَرُ يَقْطَعُ الدَّائِرَةَ، وَلَمْ يَجُزْ عَلى المَرْكَزِ. فَإِنَّ الزَّاوِيَتَيْنِ اللَّتَيْنِ تَقَعانِ[1] في قِطْعَتَيِ الدَّائِرَةِ عن جَنْبَتَيِ الخَطِّ المُخْرَجِ، وَهُما عَلى الخَطِّ المُماسِّ، هُما مِثْلُ الزَّاوِيَتَيْنِ اللَّتَيْنِ (تَقَعانِ)[2] في قِطْعَتَيِ الدَّائِرَةِ المُتَبادَلَتَيْنِ.

---

[1] تقعان] يقعان [2] تقعان] يقعان

كُلُّ خَطٍّ يَمَاسُّ دائِرَةً^ وَيَخْرُجُ مِنْ حَيْثُ يَمَاسُّهَا خَطٌّ مُسْتَقِيمٌ، يَقْطَعُ الدَّائِرَةَ، وَلا يَمُرُّ بِالْمَرْكَزِ، فَإِنَّ الزَّاوِيَتَيْنِ اللَّتَيْنِ عَنْ جَنْبَيْهِ مثل الزَّاوِيَتَيْنِ اللَّتَيْنِ تَقَعانِ فِي قِطْعَتَي الدَّائِرَةِ الْمُبادِلَتَيْنِ لَهُما.

כל קו ימשש העגולה ויצא מן המקום אשר ימששה קו ישר, יחתוך העגולה, ולא יעבור במרכז, הנה השתי זויות[1] משני[2] צדדיו כמו השתי זויות אשר תפולנה בשתי[3] חתיכות העגולה המומרות[4] להם.

שרטוט—**טס**: חסרה ג׳; **ס**: ד״ה צויר פעמיים [1]השתי זויות] שתי הזויות **טפ** [2]משני] אשר+ **טק** [3]בשתי] בשני **ב** [4]המומרות] המומר⟨ו⟩ת **ו**

המשל בזה: שעגול אב״ג ימששהו קו ד״ה על נקודת ב׳. ויצא מב׳ קו אל ז׳ שלא יעבר על המרכז ויחתך העגול. ואומר ששתי הזויות[1] שאצל צדי קו ב״ז, שהן זוית זב״ד וזוית זב״ה, שוות לשתי זויות אשר תפלנה בשתי[2] חתיכות העגול המתחלפות להן, זוית זב״ד שוה לזוית אשר תפל בחתיכת זאג״ב, וזוית זב״ה שוה[3] לזוית אשר תפל בחתכת זט״ב.

مِثالُهُ: أَنَّ دائِرَةَ ا ب ج يَمَاسُّها خَطُّ د ه على نُقْطَةِ ب. وَنُخْرِجُ مِن ب خَطًّا إِلى ز لا يَجوزُ عَلى المَرْكَزِ وَيَقْطَعُ الدَّائِرَةَ. فَأَقولُ إِنَّ الزَّاوِيَتَيْنِ اللَّتَيْنِ عَنْ جَنَبَتَي خَطِّ ب ز اللَّتَيْنِ مِنْ ز ب د وَمِنْ ز ب ه هُما مِثْلُ الزَّاوِيَتَيْنِ اللَّتَيْنِ (تَقَعانِ)[1] في قِطْعَتَي الدَّائِرَتَيْنِ المُتَبادَلَتَيْنِ، الزَّاوِيَةُ الَّتي من ز ب د مِثْلُ الزَّاوِيَةِ الَّتي تَقَعُ في قِطْعَةِ ز ا ب. والزَّاوِيَةَ الَّتي مِنْ ز ب ه مِثْلُ الزَّاوِيَةِ الَّتي تَقَعُ في قِطْعَةِ[2] ز ط ب.

[1]הזויות] ⟨ה⟩זויות [2]בשתי] ~~בשתי החתיכות~~+ [3]לזוית ... שוה] ⟨*⟩

[1]تقعان] يقعان [2]قطعة] +ز ا ب والزاوية التي من ز ب ط مثل الزاوية التي تقع في قطعة

مِثَالُهُ: أَنَّ خطّ د ه يَماسُّ دائِرَةً ا ب ج على ال ب. وخَرَجَ مِنْ ب خطُّ ب ز. فَقَطَعَ الدَّائِرَةَ على غَيْرِ الْمَرْكَزِ. فَأَقولُ إِنَّ زاوِيَتَيْ ز ب د ز ب ه مثل الزَّاوِيَتَيْنِ اللَّتَيْنِ تَقَعانِ في قِطْعَتَيْ ز ط ب ز ا ب.

המשל בו: כי קו ד״ה ימשש עגולת אב״ג[1] על ב׳. ויצא מב׳ קו ב״ז, וחתך העגולה על זולת המרכז. הנה אומר כי שתי זויות זב״ד זב״ה כמו השתי זויות[2] אשר יפלו בשתי[3] חתיכות זט״ב זא״ב[4].

[1] אב״ג] ה+ פ [2] השתי זויות] השתי הזויות ט שתי זויות ס [3] בשתי] בשני ס [4] זא״ב] וזא״ב ק

המופת בו: שנרשום על קשת ז״ב נקדת ט׳ כמו שתפל. ומרכז העגול ח׳. ונדביק ח׳ בב׳ בקו. ונמשכו אל נקדת א׳ שבקשת זאג״ב. ונדביק ב׳ בט׳, וט׳ בז׳, וז׳ בא׳. ועיגול אב״ג ימששהו קו דב״ה. וכבר יצא מן המרכז קו אח״ב והוא עמוד על קו דב״ה. וזוית חב״ה נצבה. והזוית אשר תפל בחצי עגול נצבה. ואם כן זוית אז״ב נצבה, והוא[1] בחצי עגול. וזוית חב״ה נצבה. והיא שוה לזוית אז״ב. ונשים זוית זב״א משותפת. ותהיה זוית זב״ה שוה לזוית [!] אז״ב וזב״א. וזויות דב״ז וזב״ה הן שוות לשתי זויות נצבות. ואמנם השלש זויות—זוית בא״ז ואז״ב וזב״א—הן שוות לשתי זויות נצבות, והן יחד שוות לזוית דב״ז וזב״ה. וזוית זב״ה שוה לזוית אז״ב וזוית זב״א. ואם כן הזוית הנשארת, והיא זוית זב״ד, שוה לזוית זא״ב. והיא בחתיכת זאג״ב המתחלפת. וכל עיגול שיפל בו שטח בעל ארבע צלעות, כל שתי זויות מזויותיו המקבילות זו את זו שוות לשתי זויות נצבות. ואם כן זוית זט״ב וזוית זא״ב שוות לשתי זויות נצבות. וזוית זב״ד שוה לזוית זא״ב—תהיה הזוית הנשארת, שהיא זוית זט״ב, שוה לזוית זב״ה, והיא בחתכת זט״ב המתחלפת.

بُرهانُهُ: أنَّا نُعَلِّمُ على قَوْسِ ز ب نُقْطَةَ ط كَيْفَما وَقَعَتْ. ومَرْكَزُ الدَّائِرَةِ (ح)[1]. وَنَصِلُ ح ب بخط ح ب، ونُخْرِجُهُ إلى نُقْطَةِ ا مِنْ قَوْسِ ز ا ج. وَنَصِلُ ب ط ط ز (ز ا)[2]. ودائِرَةُ ا ب ج يماسُّها خَطُّ ب ه. وَقَدْ أُخْرِجَ مِنَ المَرْكَزِ خَطُّ ح ب، فهو عَمودٌ عَلى خَطِّ د ب ه. فَالزَّاوِيَةُ الَّتي من ح ب ه قائِمَةٌ، وَهِيَ مِثْلُ الَّتي مِنْ ا ز ب لِأَنَّها في نِصْفِ دائِرَةٍ[3]. وَنَجْعَلُ الَّتي مِنْ ز ب ا مُشْتَرَكَةً. فَجَميعُ الَّتي مِنْ ز ب ه مِثْل اللَّتَيْنِ مِنْ ا ز ب ومِنْ ز ب ا. واللَّتانِ مِنْ د ب ز وَمِنْ ز ب ه مِثْلُ زاوِيَتَيْنِ قائِمَتَيْنِ. وَلَكِنَّ الثَّلاث[4] اللَّواتي مِنْ ب ا ز ومِنْ ا ز ب وَمِنْ ز ب ا هُنَّ مِثْل زَاوِيَتَيْنِ قائِمَتَيْنِ، وَهِيَ جَميعًا مِثْل اللَّتَيْنِ من د ب ز وَمِنْ ز ب ه. وَالَّتي مِنْ ز ب ه مِثْل اللَّتَيْنِ مِنْ ا ز ب ومِنْ ز ب ا. فَالزَّاوِيَةُ الباقِيَةُ الَّتي مِنْ ز ب د مِثْل المُبادِلَةِ لَها اللَّتي مِنْ ز ا ب، وَهِيَ في قِطْعَةِ ز ا ج ب. وَكُلّ دائِرَةٍ[5] يَقَعُ فيها سَطْحٌ ذو أَرْبَعَةِ أَضْلاعٍ فَكُلّ زاوِيَتَيْنِ تتقابلانِ[6] فيهِ هُما مُساوِيَتانِ لِزَاوِيَتَيْنِ قائِمَتَيْنِ. وَالَّتي مِنْ ز ط ب والَّتي مِنْ ز ا ب مِثْل زَاوِيَتَيْنِ قائِمَتَيْنِ. والَّتي مِنْ ز ب د مثل الَّتي مِنْ ز ا ب. فَالزَّاوِيَةُ الباقِيَةُ الَّتي مِنْ ز ط ب مِثْل الباقِيَةِ المُبادِلَةِ لَها الَّتي مِنْ ز ب ه، وَهِيَ في قِطْعَةِ ز ط ب.

---

[1] והוא] אפשר שמילה זו מתוקנת ל״והיא״

[1] ح] ه ح [2] ز ا] ליתא [3] لانها في نصف دائرة] 〈〈*〉〉 [4] الثلاث] الثلث [5] دائرة] زاوية [6] تتقابلان] يتقابلان

بُرْهانُهُ: أَنَّا نَتَعَلَّمُ على قَوْسِ ز ب نُقْطَةَ ط كَيْفَما وَقَعَتْ. وَلْيَكُنِ الْمَرْكَزُ ح. ونُخْرِجُ ب ح. ويُبْعِدُهُ إلى ا. ونُخْرِجُ خُطوطَ ب ط ط ز ز ا. نخطّ د ب ه يَماسُّ دائِرَةَ ا ب ج. وَقَدْ خَرَجَ مِنَ الْمَرْكَزِ خَطُّ ح ب. فَهُوَ عَمُودٌ على خَطِّ د ب ه. فزاوية ح ب ه قائِمَةٌ. وزاوية ا ز ب قائِمَة. لأَنَّها في نِصْفِ الدَّائِرَةِ. وَنَجْعَلُ زاوية ز ب ا مُشْتَرَكَةً. فجميع زاوِيَةِ ز ب ه مثل زاوِيَتَيْ ا ز ب ز ب ا. وزاوِيَتا د ب ز ز ب ه مثل قائِمَتَيْنِ. وَلَكِنَّ زوايا ب ا ز ا ز ب ز ب ا الثَّلاث^، مثل قائِمَتَيْنِ. فَهِيَ جَمِيعًا مثل زاوِيَتَيْ د ب ز ز ب ه. وزاوية ز ب ه مثل زاوِيَتَيْ ا ز ب ز ب ا. فزاوية ز ب د الباقِيَة مثل زاوية ز ا ب، وهي في قِطْعَةِ ز ا ب. وَكُلّ شَكْلٍ ذي أَرْبَعَةِ أَضْلاعٍ في دائِرَةٍ فَإِنَّ كُلّ زاوِيَتَيْنِ مُتَقابِلَتَيْنِ مِنْ زواياهُ مثل قائِمَتَيْنِ. فَزاوِيَتا ز ط ب ز ا ب مثل قائِمَتَيْنِ. فَهُما مثل زاوِيَتَيْ ز ب ه ز ب د[1]. وزاوية ز ب د مثل زاوية ز ا ب. فزاوية ز ط ب الباقية مثل زاوية ز ب ه الباقية[2]. وهي في قِطْعَةِ ز ط ب.

המופת: אנחנו נרשום על קשת ז״ב נקודת ט׳ איך שתפול. ויהיה המרכז ח׳. ונוציא ב״ח. ויעבור אל א׳. ונוציא קוים ב״ט ט״ז א״ז[1]. הנה[2] קו דב״ה ימשש עגולת אב״ג. וכבר הוצא מן המרכז קו ח״ב. הנה[3] הוא עמוד על קו דב״ה[4]. אם כן זוית חב״ה נצבת. וזוית אז״ב נצבת, כי היא בחצי העגולה. ונשים זוית זב״א משותפת[5]. הנה כל זוית זב״ה[6] כמו שתי זויות אז״ב זב״א. ושתי זויות דב״ז זב״ה[7] כמו שתי[8] נצבות. אבל זויות[9] בא״ז אז״ב אב״ז[10] המשולש [!], כמו שתי נצבות. אם כן הם יחד כמו שתי זויות דב״ז זב״ה. וזוית[11] זב״ה כמו שתי זויות[12] [13] אז״ב זב״א. אם כן זוית זב״ד הנשארת כמו זוית זא״ב, והיא בחתיכת[14] זא״ב. וכל תמונה בעלת ארבע צלעות בעגולה הנה כל שתי זויות מקבילות מזויותיה כמו[15] שתי נצבות. אם כן שתי[16] זויות זט״ב זא״ב כמו שתי[17] נצבות. ושתיהם כמו שתי זויות זב״ה זב״ד. וזוית זב״ד כמו זוית זא״ב. אם כן זוית זט״ב הנשארת כמו זוית זב״ה[18] הנשארת, והיא בחתיכת זט״ב.

---

1 فهما ... ز ب د] MPR1R2U 2 الباقية] ⟨*⟩ S

---

1 א״ז] ליתא ו 2 הנה] +אומר ב 3 הנה] +אומר כי ס 4 ימשש ... דב״ה] ליתא ב 5 משותפת] +~~אם כן~~ פ 6 זב״ה] ~~זב״א~~+ ק 7 זב״ה] ~~הב״א~~+ ק 8 זויות אז״ב ... שתי] ליתא ב 9 זויות] זוית בס 10 אב״ז] מן ב +⟨⟨מן⟩⟩ ק 11 וזוית] ~~וזבית~~ פ 12 דב״ז ... זויות] +*̲ ט 13 זויות] זוית ב 14 בחתיכת] ⟨ב⟩חתכת ק 15 כמו] +~~כן~~ ט 16 שתי] +נצבות ב 17 שתי] +~~זב״א אם כן זוית זב״ד הנשארת~~ פ 18 זב״ה] +~~הג״ד~~ ב

وَذَٰلِكَ ما أَرَدْنا بَيانَهُ.	וזה מה שרצינו לבאר.

وذَلِكَ ما أَرَدْنا أَنْ نُبيِّنَ.

וזה מה שרצינו לבאר[1].

[1]לבאר] ביאורו פ

[Proposition III.32]

[H33]

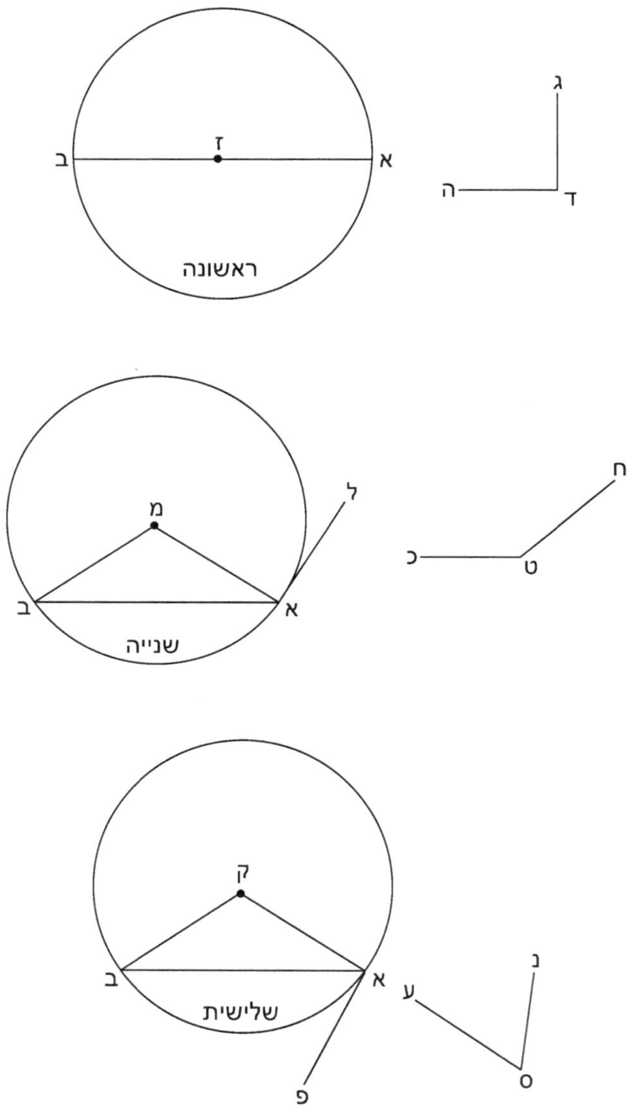

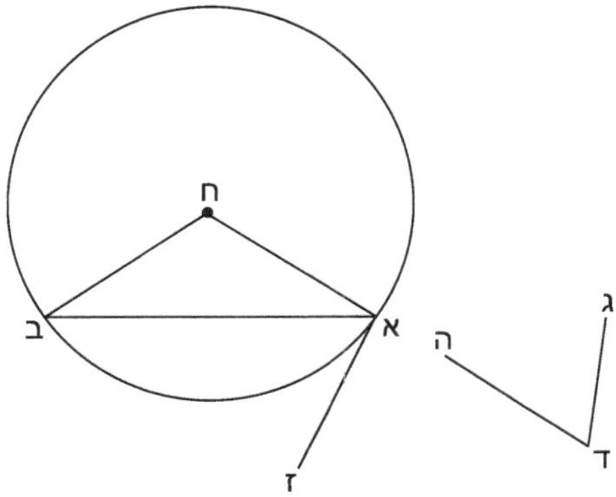

نُرِيدُ أَنْ نَعْمَلَ على خَطٍّ مَعْلومٍ قِطْعَةَ دائِرَةٍ تَقْبَلُ زاوِيَةً مثل زاوِيَةٍ مَفْروضَةٍ مُسْتَقيمَةِ الخَطَّيْنِ

فَلْنَجْعَلِ الخَطَّ المَعْلومَ خَطَّ ا ب. وَالزَّاوِيَةَ المَعْلومَةَ ج د ه. ونُرِيدُ أَنْ نَعْمَلَ على خَطِّ ا ب قِطْعَةَ دائِرَةٍ تَقْبَلُ زاوِيَةً مثل زاوِيَةِ ج د ه.

נרצה לעשות על קו ידוע חתיכת עגולה[1] תקביל[2] זוית כמו זוית[3] מונחת ישרת שני[4] הקוים.

הנה[5] נשים הקו הידוע א״ב. והזוית הידועה גד״ה. ונרצה לעשות על קו א״ב חתיכת עגולה[6] תקביל[7] זוית כמו זוית גד״ה[8].

[1]עגולה] עגולת ב [2]תקביל] תקבל ז [3]כמו זוית] *+ פ [4]שני] +חתיכת עגולה תקביל זוית כמו זוית מונחת ישרת שני ב ⟨⟨ה⟩⟩שני ט [5]הנה] +* פ [6]עגולה] עגולת ב עגולת ה ס [7]תקביל] תקבל ז [8]כמו ... גד״ה] גד״א ב

נרצה לעשות על קו ישר ידוע חתכת עגול תקביל זוית שוה לזוית ידועה ישרת הקוים; ל״א: אם שהיתה הזוית נצבה או קטנה מנצבה או גדולה.

نُرِيدُ أَنْ نُبَيِّنَ كَيْفَ نَعْمَلُ قِطْعَةَ دائِرَةٍ عَلى خَطٍّ مَعْلوم (تَقْبَلُ)[1] مِثْلَ زاوِيَةٍ مَعْلومَةٍ، إِنْ كانَتِ الزّاوِيَةُ قائِمَةً، أَو أَكْبَرَ، أَو أَصْغَرَ مِنْ قائِمَةٍ.

ונשים הקו הידוע קו א״ב. ונשים הזוית הידועה תחילה נצבה, והיא זוית גד״ה הראשונה. וכאשר רצינו שנעשה על קו א״ב הידוע חתכת עגול שתקביל זוית שוה לזוית גד״ה הראשונה, נחלק קו א״ב לחציים על נקודת ז׳. ונקיף על מרכז ז׳ ובשעור מרחק ז״ב חצי עגול. וכבר עשינו על קו א״ב חתיכת עגול שתקביל זוית שוה לזוית גד״ה[1] הראשונה הידועה.

مِثالُهُ: أَنّا نَجْعَلُ الخطَّ المَعْلومَ خَطَّ ا ب. وَنَجْعَلُ الزّاوِيَةَ المَعْلومَةَ أَوَّلًا قائِمَةً، وَهِيَ الَّتي مِنْ ج د ه. وَنُرِيدُ أَنْ نَعْمَلَ عَلى خَطِّ ا ب المَعْلوم قِطْعَةَ دائِرَةٍ (تَقْبَلُ)[2] مِثْلَ زاوِيَةِ ج د ه الأُولى. فَنَقْسِمُ خَطَّ ا ب بِنِصْفَيْنِ عَلى نُقْطَةِ ز. وَنَخُطُّ عَلى مَرْكَزِ ز. وَنُقَدِّرُ بُعْدَ ب ز نصفَ دائِرَةٍ عَلَيْها[3] ا ب. فَقَد عَمِلْنا عَلى ا ب قِطْعَةَ دائِرَةٍ تَقْبَلُ مِثْلَ زاوِيَةِ ج د ه الأُولى المَعْلومَة.

עוד נשים הזוית הידועה גדולה מנצבה, והיא זוית חט״כ השנייה הידועה. ונעמיד על קו א״ב השני, על נקדת א׳, זוית שוה לזוית חט״כ, והוא זוית בא״ל. ונוציא מקו א״ל, מנקדת א׳, קו על זוית נצבה, והוא א״מ. ונעמיד על קו א״ב, על נקודת ב׳, זוית שוה לזוית בא״מ, והיא זוית אב״מ. ולפי שזוית בא״מ שוה לזוית אב״מ, יהיה[2] צלע א״מ שוה לצלע מ״ב. ונקיף על מרכז מ׳ ובשעור מרחק א׳ וב׳, עגול שיעבר על שאר הנקודות, והוא עגול א״ב השני. ואומר שכבר עשינו על קו א״ב הידוע חתכת עגול שתקביל זוית שוה לזוית חט״כ הידועה, והיא חתיכת א״ב השנייה.

ثُمَّ نَجْعَلُ الزّاوِيَةَ المَعْلومَةَ أَكْبَرَ مِنْ قائِمَةٍ، وَهِيَ الَّتي مِنْ ح ط ك الثّانِيَة. وَنُقيمُ عَلى خَطِّ ا ب الثّاني المَعْلوم عَلى نُقْطَةِ ا مِثْلَ زاوِيَةِ ح ط ك الثّانِيَةِ المَعْلومَة، وَهِيَ الَّتي مِنْ ب ا ل. وَنُخْرِجُ مِنْ خَطِّ ا ل مِنْ نُقْطَةِ ا خَطًّا عَلى زاوِيَةٍ قائِمَةٍ، وَهُوَ ا م. وَنُقيمُ عَلى خَطِّ ا ب عَلى نُقْطَةِ ب مِثْلَ الزّاوِيَةِ الَّتي مِنْ ب ا م. فَضِلْعُ ا م مِثْلُ ضِلْعِ م ب. فَعَلى مَرْكَزِ م وَنُقَدِّرُ بُعْدَ ا ب بِخَطِّ دائِرَةٍ تَجوزُ عَلى سائِرِ النُّقَطِ، وَهِيَ دائِرَةُ ا ب الثّانِيَةُ. فَأَقولُ إِنّا قَدْ عَمِلْنا عَلى خَطِّ ا ب الثّاني المَعْلوم قِطْعَةَ دائِرَةٍ تقبل مِثْلَ زاوِيَةِ ح ط ك المَعْلومَةِ، وَهِيَ قِطْعَةُ ا ب الثّانِيَة.

שרטוט—B: בשרטוט השלישי ק ← ן, ף ← ק
[1]גד״ה] ~~גד״א~~ ⟨⟨*⟩⟩ [2]יהיה] יהי⟨ה⟩

[1]تقبل] يقبل [2]تقبل] يقبل [3]عليها] +*

فَنُقِيمُ على خطّ اب، على نُقْطَة ا مِنْهُ، زاويةَ ب ا ز مثل زاوِيَةِ ج د ه. ونُخْرِجُ مِنْ خطّ ا ز، مِنْ نُقْطَة ال ا مِنْهُ، خَطَّ ا ح على زاويةٍ قائمَة. وَنُقِيمُ على خطّ ا ب، على نُقْطَةِ ب مِنْهُ، زاوِيَةً ا ب ح مثل زاوية ب ا ح. وَلْيَلْتَقِ ضِلْعا ا ح ح ب على نُقْطَةِ ح. لأَنَّهُما خَرَجا مِنْ أَقَلّ مِنْ زاوِيَتَيْنِ قائِمَتَيْنِ. فَضِلْع ا ح مثل ضِلْعِ ح ب. وَنَجْعَلُ نُقْطَةَ ح^ مَرْكَزًا. وَنَخُطُّ دائِرَةً بِبُعْدِ ا ح. والْمَرْكَزَ ح. فزاوية ز ا ح قائِمَة. نَخَطّ ا ز يَماسُّ دائِرَةَ ا ب. وَقَدْ خَرَجَ مِنْ حَيْثُ يَماسُّها خَطُّ ا ب فقَطَعَ الدَّائِرَةَ على غَيْرِ الْمَرْكَزِ. فَفِي جَنْبَيْهِ زاوِيَتانِ مثل الزَّاوِيَتَيْنِ اللَّتَيْنِ تَقَعانِ في قِطْعَتَيِ الدَّائِرَةِ الْمُتَبادِلَتَيْنِ لَهُما. فزاوية ز ا ب مثل الَّتي تَقَعُ في قِطْعَةِ ا ب. وَلْتَكُنْ ز ا ب مثل زاوية ج د ه. فزاوية ج د ه مثل الَّتي تَقَعُ في قِطْعَةِ ا ب.

وذَلِكَ ما أَرَدْنا أَنْ نُبَيِّنَ.

הנה נעמיד על קו א״ב, על נקודת א׳ ממנו, זוית[1] בא״ז כמו זוית גד״ה. ונוציא מקו א״ז, מנקודת א׳ ממנו, קו א״ח על זוית נצבת. ונעמיד על קו א״ב, על נקודת ב׳ ממנו[2], זוית אב״ח כמו זוית בא״ח. ויפגשו שתי צלעות[3] א״ח ח״ב על נקודת ח׳, מפני כי שתיהן יצאו מפחות משתי זויות נצבות. אם כן צלע א״ח כמו צלע ח״ב. ונשים מרכז[4]. ונקיף עגולה במרחק א״ב. והמרכז ח׳. הנה זוית זא״ח נצבת. הנה קו א״ז ימשש[5] עגולת א״ב. וכבר יצא ממקום משושם קו א״ב וחתך העגולה על זולת[6] המרכז. הנה בשני[7] צדדיו שתי זויות כמו השתי[8] זויות אשר יפלו בשתי חתיכות העגולה המומרות לשתיהן. אם כן זוית זא״ב כמו אשר תפול בחתיכת א״ב. אבל זוית זא״ב כמו זוית גד״ה. אם כן זוית גד״ה[9] כמו אשר תפול בחתיכת א״ב[10].

וזה מה שרצינו לבאר[11].

---

[1] זוית] הזוית **ב** [2] ב׳ ממנו] ממנו ב׳ **פ** [3] שתי צלעות] שני הצלעות **ו** [4] מרכז] +ח׳ **ס** +⟨ח׳⟩ **ק** [5] ימשש] ימש⟨⟨ש⟩⟩ **ט** [6] זולת] זוי⟨ו⟩ת **ט** [7] בשני] בשתי **בוספק** [8] השתי] שתי **ט** [9] אם ... גד״ה] ליתא **ב** [10] זוית זא״ב ... א״ב] ליתא **ו** [11] וזה ... לבאר] ונשלם ביאורו **ו**

بُرهانُهُ: أَنَّ المَرْكَزَ م. وَالزَّاوِيَةُ الَّتِي مِنْ ل ا م [..][1] تَمَاسُّ دائِرَةَ ا ب. وَقَد أُخْرِجَ مِنَ المَوْضِعِ الَّذي يَماسُّها فيهِ خَطُّ ا ب فَقَطَعَ الدَّائِرَةَ، وَلَمْ يَجُزْ عَلى المَرْكَزِ. فَالزَّاوِيَتانِ اللَّتانِ عَنْ جَنَبَتَيْهِ مِثْلِ اللَّتَيْنِ تَقَعانِ في قِطْعَتَي الدَّائِرَةِ المُتَبادَلَتَيْنِ. فَالزَّاوِيَةُ الَّتي مِنْ ل ا ب مِثْلِ الَّتي تَقَعُ في قِطْعَةِ الدَّائِرَةِ الَّتي مِنْ ا ب الصُّغْرَى، وَهِيَ الزَّاوِيَةُ المُبادِلَةُ لِزاوِيَةِ ل ا ب. وَلَكِنَّ الَّتي مِنْ ل ا ب مِثْلِ الَّتي مِنْ ح ط ك. فَالَّتي مِنْ ح ط ك مِثْلِ الَّتي تَقَعُ في قِطْعَةِ ا ب الصُّغْرَى.

מופת זה: שהמרכז הוא מ׳. וזוית לא״מ נצבה. ואם כן קו א״ל ממשש עגול א״ב. וכבר יצא ממקום משושו קו א״ב וחתך העיגול. והזויות אשר אצל צדיו שוות לזויות אשר תפלנה בחתיכות העגול המתחלפות להן. ואם כן זוית לא״ב שוה לזוית אשר תפול בחתיכת א״ב. ואמנם זוית לא״ב שוה לזוית חט״כ. ואם כן זוית חט״כ שוה לזוית שתפול בחתכת א״ב הקטנה.

ثُمَّ نَجْعَلُ الزَّاوِيَةَ المَعْلومَةَ أَصْغَرَ مِنْ قائِمَةٍ، وَهِيَ الَّتي مِنْ ن س ع الثَّالِثَةَ. وَنُقيمُ عَلى خَطِّ ا ب الثَّالِثِ المَعْلومِ عَلى نُقْطَةِ ا مِثْلِ ن س ع الثَّالِثَةِ المَعْلومَةِ، وَهِيَ الَّتي مِنْ (ب ا ف)[2]. وَنُخْرِجُ مِنْ خَطِّ ا ف مِنْ نُقْطَةِ ا خَطًّا عَلى زاوِيَةٍ قائِمَةٍ، وَهُوَ خَطُّ ا ف. وَنُقيمُ عَلى خَطِّ (ا ب)[3] عَلى نُقْطَةِ ب مِثْلَ الزَّاوِيَةِ الَّتي مِنْ ب ا ف، وَهِيَ الَّتي مِنْ ق ب ا. فَضِلْعُ ق ا مِثْلِ ضِلْعِ ق ب. فَعَلى مَرْكَزِ ق وَبِبُعْدِ ا ب نَخُطُّ الدَّائِرَةَ الَّتي تَجُوزُ عَلَيْها سائِرُ النُّقَطِ، وَهِيَ دائِرَةُ ا ب الثَّالِثَةَ. فَأَقولُ إِنَّا قَدْ عَمِلْنا عَلى خَطِّ ا ب المَعْلومِ الثَّالِثِ قِطْعَةَ دائِرَةٍ تَقبلُ مِثْلَ زاوِيَةِ ن س ع المَعْلومِ، وَهِيَ قِطْعَةُ ا ب الكُبْرَى.

עוד נשים הזוית הידועה קטנה מנצבה, והיא זוית נס״ע. ונעמיד על קו א״ב השלישי, על נקודת א׳, זוית שוה לזוית נס״ע, והיא זוית בא״פ. ונוציא מקו א״ב [!] מנקודת א׳, קו על זוית נצבה, והיא קו א״ק. ונעמיד על קו א״ב, על נקודת ב׳, זוית שוה לזוית קא״ב, והיא זוית קב״א[1]. ולפי שזוית קב״א[2] שוה לזוית קא״ב, יהיה צלע ק״א שוה לצלע ק״ב. ונקיף עגול על מרכז ק׳ ובמרחק א׳ וב׳, ויעבר על שאר הנקודות, והוא עגול א״ב השלישי. ואומר שכבר עשינו על קו א״ב הידוע השלישי חתכת עגול תקביל זוית שוה לזוית נס״ע.

[1] [..] חסרות כאן כמה מילים [2] ب ا ف] ب ا ق
[3] ا ب] ا ف

[1] קב״א] קצ״א [2] קב״א] קצ״א

המופת בו: שהמרכז ק׳. וזוית קא״פ[1] נצבה. ואם כן קו א״פ ממשש עגול א״ב. וכבר יצא ממקום משושו קו א״ב וחתך העגול. ואם כן שתי הזויות אשר אצל צדיו שוות לשתי הזויות אשר תפלנה בחתכת העגול המתחלפות להן. ואם כן זוית בא״פ שוה לזוית אשר תפל בחתכת א״ב הגדולה. ואמנם זוית בא״פ שוה לזוית נס״ע. ואם כן זוית נס״ע שוה לאשר תפל בחתכת א״ב הגדולה. וכבר עשינו על קו א״ב השלישי חתכת עגול שתקביל זוית שוה לזוית נס״ע.

וזה מה שרצינו לבאר.

بُرْهانُهُ: أَنَّ المَرْكَزَ ق. وَالزَّاوِيَةُ الَّتِي مِنْ ق ا ف[1] قائِمَةٌ. نَخَطُّ ا ف يَماسُّ دائِرَةَ ا ب. وَقَدْ أُخْرِجَ مِنَ المَوْضِعِ الَّذي يَماسُّها فيهِ خَطُّ ا ب فَقَطَعَ الدَّائِرَةَ وَلَمْ يَجُزْ عَلَى المَرْكَزِ. وَالزَّاوِيتانِ اللَّتانِ عَنْ جَنَبَتَيْهِ مِثْلُ الزَّاوِيَتَيْنِ اللَّتَيْنِ (تَقَعانِ)[2] في قِطْعَتَي الدَّائِرَةِ المُتَبادَلَتَيْنِ. فَالزَّاوِيَةُ الَّتِي مِنْ ف ا ب مِثْل الَّتي تَقَعُ في قِطْعَةِ ا ب الكُبْرَى. وَلَكِنَّ الَّتي مِنْ ف ا ب مِثْل الَّتي مِنْ ن س ع. فَالَّتي مِن ن س ع مِثْل الَّتي تَقَعُ في قِطْعَةِ ا ب الكُبْرَى. فَقَدْ عَمِلْنا عَلَى خَطِّ ا ب الثَّالِثِ المَعْلوم قِطْعَةَ دائِرَةِ ا ب الكُبْرَى تقبل مِثْلَ ع س ن.

وَذَلِكَ ما أَرَدْنا بَيانَهُ.

[1] קא״פ] [..] ⟨⟨*⟩⟩

[1] ق ا ف] ب ا ف [2] تقعان] يقعان

[Proposition III.33]

[H34]

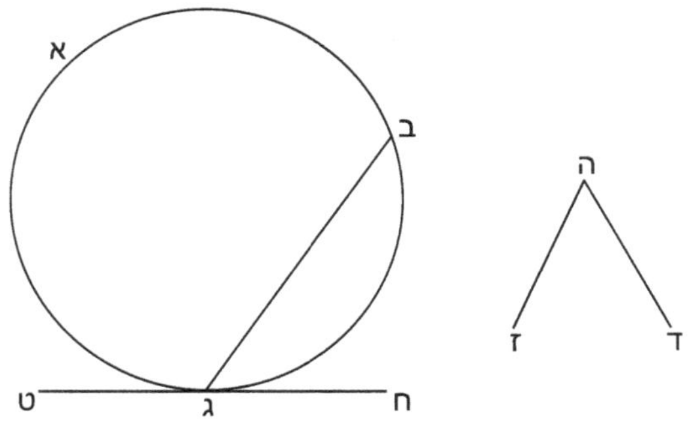

نُرِيدُ أَنْ نُبَيِّنَ كَيْفَ نَفْصِلُ مِنْ دائِرَةٍ مَعْلومَةٍ قِطْعَةً تقبل مِثْل زاوِيَةٍ مَعْلومَةٍ.

נרצה לחלק מעגול ידוע חתכה שתקביל זוית ישרת הקוים שוה לזוית ידועה ישרת הקוים.

مِثالُهُ: أَنَّا نَجْعَلُ الدَّائِرَةَ المَعْلومَةَ دائِرَةَ ا ب ج. وَالزَّاوِيَةَ المَعْلومَةَ الَّتي من د ه ز. وَنُرِيدُ أَنْ نَفْصِلَ مِنْ دائِرَةِ ا ب ج قِطْعَةً تقبل مِثْل زاوِيَةِ د ه ز. فَنُجِيزُ عَلى ج خَطًّا يَماسُّ دائِرَةَ ا ب ج، وَهُوَ ح ج ط. ونُقِيمُ عَلى خَطِّ ح ج ط عَلى نُقْطَةِ ج مِثْلَ الزَّاوِيَةِ الَّتي مِنْ د ه ز[1] المَعْلومَة، وَهِيَ الَّتي مِنْ ح ج ب. فَأَقولُ إِنَّا قَدْ فَصَلْنا مِنْ دائِرَةِ ا ب ج المَعْلومَةِ قِطْعَةً تقبل مِثْل زاوِيَةِ د ه ز المَعْلومَة، وَهِيَ قِطْعَةُ ج ا ب.

ונשים העגול הידוע עגול אב״ג. והזוית הידועה זוית דה״ז. וכאשר רצינו שנחלק מעגול אב״ג חתכה שתקביל זוית שוה לזוית דה״ז, נעביר על ג׳ קו ממשש העגול, והוא קו חג״ט. ונעמיד על קו ח״ג [!], על נקודת ג׳, זוית שוה לזוית דה״ז הידועה, והיא זוית חג״ב [!]. ואומר שכבר חלקנו מעגול אב״ג חתכה שתקביל זוית שוה לזוית דה״ז, והיא חתכת גא״ב.

1 د ه ز] ه د ز

نُرِيدُ أَنْ نَفْصِلَ مِنْ دائِرَةٍ مَعْلُومَةٍ قِطْعَةً تُقابِلُ زاوِيَةً مُسْتَقِيمَةَ الخَطَّيْنِ مثل زاوِيَةٍ^ مَعْلُومَةٍ مُسْتَقِيمَةِ الخَطَّيْنِ.

נרצה שנבדיל מעגולה ידועה חתיכה תקביל זוית[1] ידועה ישרת שני הקוים[2].

فَنَجْعَلَ الدَّائِرَةَ المَعْلُومَةَ دائِرَةَ ا ب ج. وَالزَّاوِيَةَ المَعْلُومَةَ المُسْتَقِيمَةَ الخَطَّيْنِ زاوية د ه ز. وَنُرِيدُ أَنْ نَفْصِلَ مِنْ دائِرَةِ ا ب ج قِطْعَةً تُقابِلُ زاوِيَةً مثل زاوِيَةِ د ه ز.

ונשים העגולה הידועה עגולת אב״ג. והזוית הידועה ישרת שני הקוים זוית דה״ז. ונרצה שנבדיל מעגולת אב״ג חתיכה תקביל זוית כמו זוית דה״ז[3].

[1] זוית] +⟨⟨ישרת הקוים כמו זוית⟩⟩ **ק**
[2] הקוים] קוים **ק** [3] דה״ז] ~~גה״ז~~+ **ק**

מופת זה: שעגול אב״ג ימששהו קו חג״ט. וכבר יצא ממקום משושו קו ג״ב, וחתך העגול. ואם כן שתי הזויות אשר אצל צדיו שוות לשתי הזויות עד שתפלנה בחתכת העגול המתחלפות. ואם כן זוית חג״ב שוה לזוית אשר תפל בחתכת גא״ב. [..] זוית [..] שוה לזוית דה״ז. ואם כן זוית דה״ז שוה ל[..] בחתכת גא״ב[1]. וכבר חתכנו מהעגול הידוע חתכת גא״ב מקבלת זוית שוה לזוית דה״ז.

וזה מה שרצינו לבאר.

[1] [..] ... גא״ב] 〈〈*〉〉

بُرْهانُهُ: أَنَّ دائِرَةَ ا ب ج يَماسُّها خَطُّ ح ج ط. وَقَد أُخْرِجَ مِنَ المَوْضِعِ الَّذي يَماسُّها فيهِ خَطُّ ج ب، فَقَطَعَ الدّائِرَةَ اللَّتانِ [..][1] عن جَنْبَتَيْهِ مِثْل الزّاوِيَتَيْنِ اللَّتَيْنِ تَقَعانِ في قِطْعَتَي الدّائِرَةِ المُتَبادِلَتَيْنِ. فَالزّاوِيَةُ الَّتي مِنْ ح ج ب مِثْل الَّتي تَقَعُ في قِطْعَةِ ج ا ب. وَلٰكِنَّ الَّتي مِنْ ح ج ب مِثْل الَّتي مِنْ د ه ز. فَالَّتي مِنْ د ه ز مِثْل الَّتي تَقَعُ في قِطْعَةِ ا ب. فَقَدْ فَصَلْنا مِنَ الدّائِرَةِ المَعْلومَةِ قِطْعَةَ ج ا ب وَهِيَ تَقْبَلُ مِثْل زاوِيَةِ د ه ز.

وَذٰلِكَ ما أَرَدْنا بَيانَهُ.

[1] [..] חסרות כאן כמה מילים

فَنُجيزُ على نُقْطَةِ ج خَطَّ ح ج ط مُماسًّا لِدائِرَةِ ا ب ج. ونُقيمُ على خَطِّ ج ح على نُقْطَةِ ج مِنْهُ زاوِيَةَ ب ج ح مثل زاوِيَةِ د ه ز. فَخَطُّ ح ج ط يُماسُّ دائِرَةَ ا ب ج. وقَدْ خَرَجَ مِنْ حَيْثُ يُماسُّها خَطُّ ب ج فَقَطَعَ الدّائِرَةَ على غَيْرِ الْمَرْكَزِ[1]. فَعَنْ جَنْبَتَيْهِ زاوِيَتانِ مثل الزّاوِيَتَيْنِ[2] اللَّتَيْنِ تَقَعانِ في قِطْعَتَي الدّائِرَةِ الْمُبادِلَتَيْنِ لَهُما. فزاوية ب ج ح مثل الَّتي تَقَعُ في قِطْعَةِ ج ا ب. وَلٰكِنَّ زاوِيَةَ ب ج ح مثل زاوِيَةِ د ه ز. فزاوية د ه ز مثل الَّتي تَقَعُ في قِطْعَةِ ج ا ب[3]. فَقَدْ فَصَلْنا مِنْ دائِرَةِ ا ب ج الْمَعْلومة قِطْعَةَ ج ا ب تُقابِلُ زاوِيَةً مثل زاوِيَةِ د ه ز[4].

وذٰلِكَ ما أَرَدْنا أَنْ نُبَيِّنَ.

ונעביר על נקודת ג׳ קו חג״ט ממשש לעגולת אב״ג. ונעמיד על קו ג״ח על נקודת ג׳ ממנו זוית[1] בג״ח כמו זוית דה״ז. הנה קו חג״ט ימשש עגולת אב״ג. וכבר יצא מהמקום שימששה קו ב״ג יחתוך העגולה על זולת המרכז. הנה[2] משני צדדיו שתי זויות כמו שתי[3] זויות אשר יפלו בשתי חתיכות העגולה המומרות לשתיהן[4]. אם כן זוית בג״ח כמו אשר[5] תפול בחתיכת[6] גא״ב. אבל זוית בג״ח כמו זוית דה״ז. אם כן זוית דה״ז כמו אשר תפול בחתיכת[7] גא״ב[8]. הנה כבר הבדלנו מעגולת אב״ג הידועה חתיכת גא״ב תקביל זוית כמו זוית[9] דה״ז.

וזה מה שרצינו[10] לבאר.

---

1 على ... المركز] Cam　2 الزاويتين] CamS
3 ولكن ... ج ا ب] CamDSU　4 د ه ز] S
+المعلومة

---

1 זוית] ליתא **ס**　2 הנה] +~~מכל~~ **ב**　3 שתי] השתי **ק**　4 לשתיהן] לשתיהם **בספ** ליתא **ט**　5 כמו אשר] כאשר **ס**　6 בחתיכת] בחתיכה **ו**　7 בחתיכת] בחתיכה **ו**　8 גא״ב] +תקביל **ב** +זוית בג״ח **ו**　9 זוית] ליתא **ו** ~~ז״ה~~+ **פ**　10 וזה ... שרצינו] +* **פ**

[Proposition III.34]

[H35]

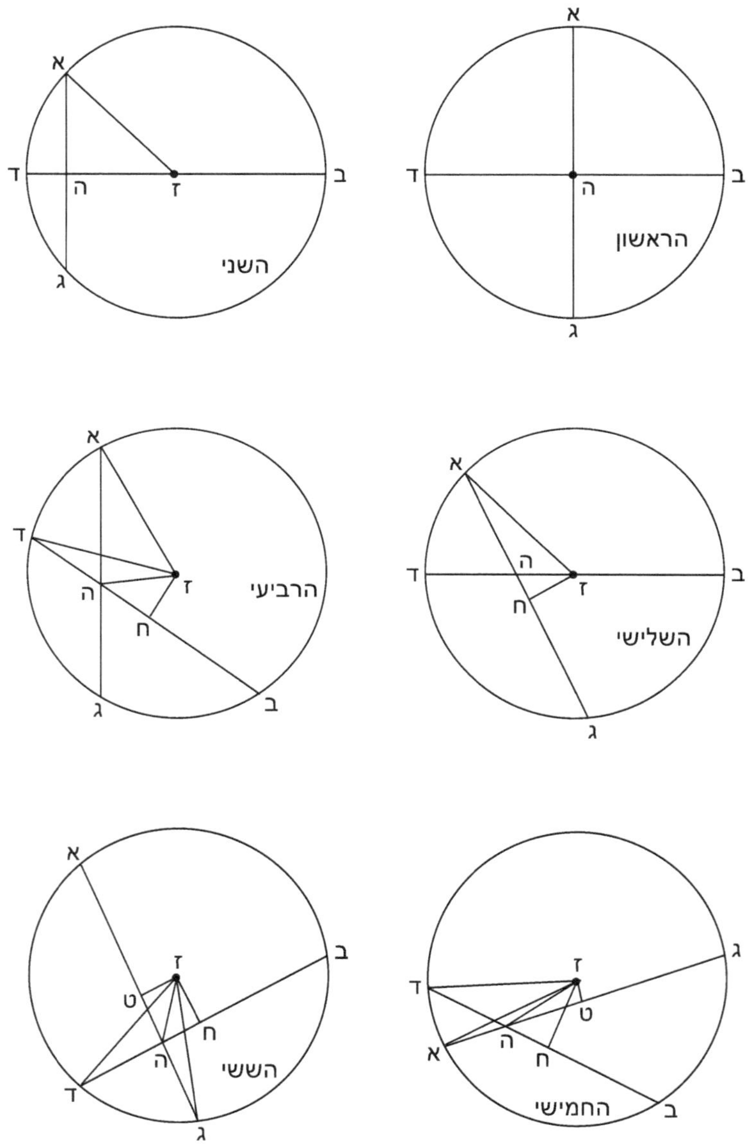

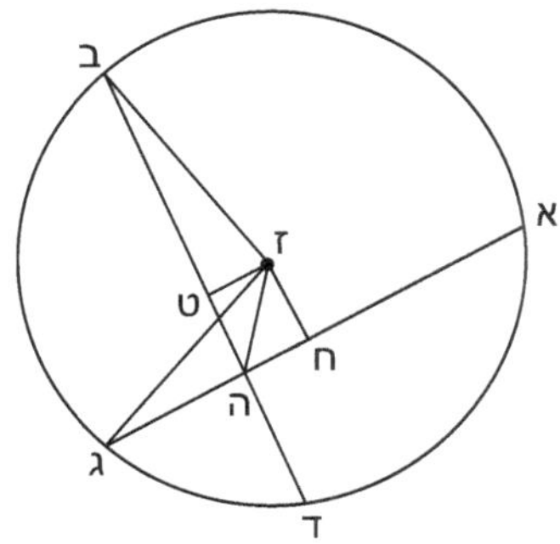

كُلُّ وَتَرَيْنِ يَتَقاطَعانِ في دائِرَةٍ، فَإِنَّ السَّطْحَ القائِمَ الزَّوايا الَّذي يُحِيطُ بِهِ قِسْما أَحَدِ الوَتَرَيْنِ، مثل السَّطْحِ القائِمِ الزَّوايا الَّذي يُحِيطُ بِهِ قِسْما الوَتَرِ الآخَرِ.

כל שני מיתרים יתחתכו[1] בעגולה, הנה[2] השטח הנצב הזויות אשר יקיפו בו שני חלקי אחד משני המיתרים[3], כמו השטח[4] הנצב הזויות אשר יקיפו בו שני חלקי המיתר האחר[5].

مِثالُهُ: أَنَّ وَتَرَيْ اج ب د يَتَقاطَعانِ في دائِرَةِ اب ج د على نُقْطَةِ ه. فَأَقولُ إِنَّ السَّطْحَ القائِمَ الزَّوايا الَّذي يُحِيطُ بِهِ خَطّا ا ه ه ج مُساوٍ لِلسَّطْحِ القائِمِ الزَّوايا الَّذي يُحِيطُ بِهِ خَطّا د ه ه ب.

המשל בו: שני מיתרי א״ג ב״ד יתחתכו בעגולת אבג״ד על נקודת ה׳. הנה אומר כי השטח הנצב הזויות אשר יקיפו בו שני קוי א״ה ה״ג שוה לשטח הנצב הזויות אשר יקיפו בו שני קוי ד״ה ה״ב.

---

[1] יתחתכו] יחתכו **ט** [2] הנה] +* **ב** [3] משני המיתרים] משני מיתרים **ב** מן המיתרים **ז** [4] אחד ... השטח] המיתר האחר שוה לשטח **ס** [5] האחר] ⟨⟨ה⟩⟩אחר **ט** +⟨⟨נ״א אחד מן המיתרים כמו השטח הנצב הזויות אשר יקיפו בו שני חלקי המיתר האחר וכו׳⟩⟩ **ס**

כל שני מיתרים שיחתך זה את זה בעגול, יהיה השטח נצב הזויות שיקיפו בו חלקי אחד מן המיתרים, שוה לשטח נצב הזויות שיקיפו בו חלקי המיתר האחר.

كُلُّ خَطَّيْنِ في دائِرَةٍ يَقْطَعُ أَحَدُهُما الآخَرَ، فَقَدِ انْقَسَمَ كُلُّ واحِدٍ مِنْهُما عَلى النُّقْطَةِ الَّتي عَلَيْها تَقاطَعا، وَالَّذي يَكونُ مِنْ ضَرْبِ أَحَدِ القِسْمَيْنِ في الآخَرِ في أَحَدِ الخَطَّيْنِ مِثْلَ الَّذي يَكونُ مِنْ ضَرْبِ أَحَدِ القِسْمَيْنِ في الآخَرِ مِنَ الخَطِّ الآخَرِ.

המשל בזה: שבעגול אבג״ד הראשון שני קוים א״ג וב״ד. וחתך כל אחד מהם האחר על נקדת ה׳. ואומר שאשר יהיה מהכאת ב״ה בה״ד שוה לאשר יהיה מהכאת א״ה בה״ג[1]. ואם היה מרכז העגול ה׳, כבר התבאר מה שאמרנו. לפי שה״א וה״ב וה״ג וה״ד הם שוים. ואם כן מה שיהיה ב״ה בה״ד שוה לאשר יהיה מהכאת א״ה בה״ג.

مِثالُهُ: أَنَّ في دائِرَةِ ا ب ج د الأُولى (خَطّا)[1] (ا ج)[2] ب د. فَقَدْ قَطَعَ كُلُّ واحِدٍ مِنْهُما الآخَرَ عَلى نُقْطَةِ ه. فَأَقولُ إِنَّ الَّذي يَكونُ مِنْ ضَرْبِ ب ه في ه د مِثْلَ الَّذي يَكونُ مِنْ ضَرْبِ ا ه في ه ج. وَإِنْ كانَ مَرْكَزُ الدَّائِرَةِ ه، فَقَدِ اسْتَبانَ الَّذي ذَكَرْنا، لِأَنَّ ا ه ه ب ه ج ه د تَكونُ مُتَساوِيَةً. وَالَّذي يَكونُ مِنْ ضَرْبِ ب ه في ه د مِثْلَ الَّذي يَكونُ مِنْ ضَرْبِ ا ه في ه ج.

שרטוט—B: בשרטוט הראשון חסרה ה׳ (ه) והציון ألأول מופיע בשרטוט השני [1]בה״ג] + ~~ועוד בעגול אבג״ד השני שני קוים~~

[1] خطا] خطي [2] ا ج] ه ا ج

بُرْهانُهُ: أنّا نَجْعَلُ مَرْكَزَ الدّائِرَةِ ا ب ج د نُقْطَةَ ز. وَنَصِلُ خَطَّ ز ه. وَنُخْرِجُ مِنْ نُقْطَةِ ز إلى خَطَّيْ ا ج ب د عَمُودَيْ ز ح ز ط. وَنَصِلُ خَطَّيْ ز ج ز ب. فَلِأَنَّهُ قَدْ مَرَّ بِمَرْكَزِ دائِرَةِ ا ب ج د خَطٌّ مُسْتَقِيمٌ، وَهُوَ خَطُّ ز ح، فَقَطَعَ ا ج على زوايا قائِمَةٍ، فَيَكُونُ قَدْ قَسَمَهُ بِنِصْفَيْنِ. فَخَطُّ ا ح مثل خَطِّ ج ح. وَلأَنَّ خَطَّ ا ج قَدْ قُسِمَ بِنِصْفَيْنِ على نُقْطَةِ ح، وَبِقِسْمَيْنِ غَيْرِ مُتَساوِيَيْنِ على نُقْطَةِ ه، يَكُونُ السَّطْحُ القائِمُ الزَّوايا الَّذي يُحِيطُ بِهِ خَطّا ا ه ه ج، مَعَ المُرَبَّعِ الكائِنِ مِنْ ه ح، مُساوِيًا لِلمُرَبَّعِ الكائِنِ مِنْ ج ح. وَنَجْعَلُ المُرَبَّعَ الكائِنَ مِنْ ز ح مُشْتَرَكًا. فَالسَّطْحُ القائِمُ الزَّوايا الَّذي يُحِيطُ بِهِ خَطّا ا ه ه ح، مَعَ المُرَبَّعَيْنِ الكائِنَيْنِ مِنْ خَطَّيْ ه ح ح ز، مُساوٍ لِلمُرَبَّعَيْنِ الكائِنَيْنِ مِنْ خَطَّيْ ز ح ح ج. لَكِنَّ المُرَبَّعَيْنِ الكائِنَيْنِ مِنْ خَطَّيْ ز ح ح ه مُساوِيانِ لِلمُرَبَّعِ الكائِنِ مِنْ ز ه، لأَنَّ زاوِيَةَ ز ح ه قائِمَةٌ. وَالمُرَبَّعانِ الكائِنانِ مِنْ خَطَّيْ ز ح ح ج مُساوِيانِ لِلمُرَبَّعِ الكائِنِ مِنْ ز ج. فَالسَّطْحُ القائِمُ الزَّوايا الَّذي يُحِيطُ بِهِ خَطّا ا ه ه ج. مَعَ المُرَبَّعِ الكائِنِ مِنْ ز ه مُساوٍ لِلمُرَبَّعِ الكائِنِ مِنْ ز ج. وَكَذَلِكَ أَيْضًا تَبَيَّنَ أَنَّ السَّطْحَ القائِمَ الزَّوايا الَّذي يُحِيطُ بِهِ خَطّا د ه ه ب، مَعَ المُرَبَّعِ الكائِنِ مِنْ ز ه، مُساوٍ لِلمُرَبَّعِ الكائِنِ مِنْ خَطِّ ز ب. وَالمُرَبَّعُ الكائِنُ مِنْ ز ب مُساوٍ لِلمُرَبَّعِ الكائِنِ مِنْ ز ج.

המופת: שנשים מרכז[1] עגולת אבג״ד נקודת ז׳. ונגיע קו ז״ה. ונוציא מן נקודת ז׳ אל שני קוי א״ג ב״ד שני עמודי ז״ח ז״ט. ונגיע שני קוי ז״ג ז״ב. הנה מפני שכבר[2] יצא ממרכז עגולת אבג״ד קו ישר, והוא קו ז״ח, וחתך א״ג על זויות נצבות[3], יהיה כבר[4] חלקו בשני[5] חציים. הנה קו א״ח כמו קו ג״ח. ומפני כי א״ג כבר נחלק בשני חציים על נקודת ח׳[6], ובשני חלקים בלתי שוים על נקודת ה׳, יהיה השטח הנצב הזויות אשר יקיפו בו שני קוי[7] א״ה ה״ג, עם המרובע ההוה מן ה״ח, שוה למרובע ההוה מן ג״ח. ונשים המרובע ההוה מן ז״ח משותף. הנה השטח הנצב הזויות אשר יקיפו בו שני קוי א״ה ה״ג, עם שני המרובעים ההוים משני קוי ה״ח ח״ז, שוה לשני המרובעים ההווים[8] משני קוי ז״ח ח״ג. אבל שני המרובעים ההוים משני קוי ז״ח ח״ה שוים למרובע ההווה מן ז״ה, מפני כי זוית זח״ה[9] נצבת. אם כן שני המרובעים ההווים משני קוי ז״ח ח״ג שוים למרובע ההווה מן ז״ג[10]. אם כן השטח הנצב הזויות אשר יקיפו בו שני קוי א״ה ה״ג, עם המרובע ההוה מן ז״ה, שוה למרובע ההווה מן ז״ג. וכן גם כן התבאר כי השטח[11] הנצב הזויות אשר יקיפו בו[12] שני קוי ד״ה ה״ב, עם המרובע ההווה מן ז״ה, שוה למרובע ההווה מן קו ז״ב. והמרובע ההווה מן ז״ב שוה למרובע ההווה מן ז״ג.

[1] מרכז] המרכז **ז** +~~שנשים~~ **פ** [2] שכבר] כי כבר **ט** [3] זויות נצבות] זוית נצבת **ב** [4] כבר] ⟨⟨*⟩⟩ **ק** [5] בשני] ⟨⟨*⟩⟩ **ז** [6] ח׳] ליתא **ס** [7] שני קוי] ליתא **ט** [8] ההווים] ההווה **בפ** [9] זח״ה] זח״א **בוט** [10] מפני ... ז״ג] ליתא **ס** [11] השטח] ליתא **ס** [12] בו] ליתא **ז**

ועוד בעגול אבג״ד השני, שני קוים, חתך כל[1] אחד מהם האחר על נקודת ה׳. ואומר שהעולה מהכאת א״ה בה״ג שוה לאשר יהיה לעולה מהכאת ב״ה בה״ד. ולא יהיה מרכז העגול ה׳. ואמנם יהיה על אחד משני הקוים. ויהיה על ז׳ מקו ב״ד.

وَأَيْضًا في دائِرَة ا ب ج الثَّانِيَةِ خَطَّا ب د ا ج. قد قَطَعَ كُلُّ واحِدٍ مِنْهُما الآخَرَ على نُقْطَةِ ه. فَأَقولُ إِنَّ الَّذي يَكونُ مِنْ ضَرْبِ ب ه في ه د مِثْلُ الَّذي يَكونُ مِنْ ضَرْبِ ا ه في ه ج. وَلا نَجْعَلُ مَرْكَزَ الدَّائِرَةِ ه. وَلَكِنْ نَجْعَلُهُ عَلى أَحَدِ الخَطَّيْنِ، وَهُوَ خَطُّ ب د على نُقْطَةِ (ز)[1].

והמופת על מה שאמרנו: שנדביק ז׳ בא׳. וב״ד יחתוך ג״א לחציים או לא לחציים. ונשימהו תחלה שיחלקהו לחציים. וא״ה שוה לה״ג. והעולה מהכאת א״ה בעצמו שוה לעולה מהכאת א״ה בה״ג. וקו ב״ד כבר נחלק לחציים על ז׳, ובשני חלקים שאינם שוים על ה׳. ואם כן העולה מהכאת ב״ה בה״ד וז״ה בעצמן [!] יחד, שוה לעולה מהכאת ז״ד בעצמו, כמו[2] שזכרנו בתמונה החמשית מן המאמר השני[3]. והעולה מהכאת ז״ד בעצמו, הוא שוה לעולה מהכאת ז״א בעצמו. אם כן העולה מהכאת ב״ה בה״ד וז״ה בעצמו יחד, שוה לעולה מהכאת ז״א בעצמו. והעולה מהכאת ז״א בעצמו, הוא שוה לעולה מהכאת א״ה וה״ז כל אחד בעצמו יחד. והעולה מא״ה וה״ז כל אחד בעצמו שוה לעולה מב״ה בה״ד וז״ה בעצמו יחד. ויושלך המשותף, שהוא ז״ה בעצמו, וישאר העולה מב״ה בה״ד שוה לעולה מא״ה בעצמו. והעולה מא״ה בעצמו, שוה לעולה מא״ה בה״ג, לפי שהם שוים. ואם כן התבאר שהעולה מא״ה בה״ג, שוה לעולה מב״ה בה״ד.

بُرْهانُهُ: أَنَّا نَصِلُ ز ا. ب د يَقْطَعُ[2] ا ج بِنِصْفَيْنِ أَو بِغَيْرِ نِصْفَيْنِ. فَنَجْعَلُهُ أَوَّلًا نَقْطَعُهُ [!] بِنِصْفَيْنِ. ا ه مِثْلُ ه ج. والَّذي يَكونُ مِنْ ضَرْبِ ا ه في نَفْسِهِ مِثْلُ الَّذي يَكونُ مِنْ ضَرْبِ ا ه في ه ج. فَخَطُّ ب د قد قُسِمَ بِنِصْفَيْنِ عَلى ز وَبِقِسْمَيْنِ مُخْتَلِفَيْنِ عَلى ه. فالَّذي يَكونُ مِنْ ضَرْبِ ب ه في ه د ز ه في نَفْسِهِ جَميعًا مِثْلُ الَّذي يَكونُ مِنْ ضَرْبِ ز د في نَفْسِهِ. والَّذي يَكونُ مِنْ ضَرْبِ ز د في نَفْسِهِ مِثْلُ الَّذي يَكونُ مِنْ ضَرْبِ ز ا في نَفْسِهِ. والَّذي يَكونُ مِنْ ضَرْبِ ب ه في ه د ز ه في نَفْسِهِ جَميعًا مِثْلُ الَّذي يَكونُ مِنْ ضَرْبِ ز ا في نَفْسِهِ. والَّذي يَكونُ مِنْ ضَرْبِ ز ا في نَفْسِهِ هو مِثْلُ ما يَكونُ في ضَرْبِ ا ه ه ز كُلّ واحِدٍ في نَفْسِهِ جَميعًا. فالَّذي يَكونُ مِنْ ضَرْبِ ا ه ز ه، كُلّ واحِدٍ في نَفْسِهِ جَميعًا، مِثْلُ ما يَكونُ مِنْ ضَرْبِ ب ه في ه د ز ه في نَفْسِهِ جَميعًا. فَتُلْقِي المُشْتَرَكَ الَّذي مِنْ ز ه في نَفْسِهِ. فَيَبْقَى الَّذي يَكونُ مِنْ ضَرْبِ ب ه في ه د مِثْلُ الَّذي يَكونُ مِنْ ضَرْبِ ا ه في نَفْسِهِ. والَّذي

[1] כל] ⟨*⟩ [2] כמו] ⟨⟨[ר״ל]⟩⟩+ [3] השני] +⟨⟨ע״כ⟩⟩

[1] ز] ب [2] يقطع] نقطع

فالسَّطْح القائِم الزَّوايا الَّذي يُحيطُ بِهِ خَطّا ا ه ه ج، مَعَ المُرَبَّع الكائِنِ مِنْ ز ه، مُساوٍ للسَّطْح القائِمِ الزَّوايا الَّذي يُحيطُ بِهِ خَطّا د ه ه ب، مَعَ المُرَبَّع الكائِنِ مِنْ ز ه. فَإِذا نَقَّصْنا المُشْتَرَكَ، وَهُوَ المُرَبَّع الكائِن مِنْ خَطّ ز ه، بَقِيَ السَّطْحُ القائِمُ الزَّوايا الَّذي يُحيطُ بِهِ خَطّا ا ه ه ج، مُساوِيًا لِلسَّطْح القائِمِ الزَّوايا الَّذي يُحيطُ بِهِ خَطّا د ه ه ب.

وذَلِكَ ما أَرَدْنا أَنْ نُبَيِّنَ.

אם כן השטח הנצב הזויות אשר יקיפו בו שני קוי[1] א״ה ה״ג, עם המרובע ההוה מן ז״ה, שוה לשטח[2] הנצב הזויות אשר יקיפו בו שני קוי ד״ה ה״ב, עם המרובע ההווה מן ז״ה. וכאשר חסרנו המשותף, והוא המרובע ההווה מן ז״ה[3], נשאר השטח הנצב הזויות אשר יקיפו בו[4] שני קוי א״ה ה״ג, שוה לשטח הנצב הזויות אשר יקיפו בו שני קוי ד״ה ה״ב.

וזה מה שרצינו לבאר[5].

[1]שני קוי] ליתא **ט** [2]לשטח] למרובע+ **ב**
[3]וכאשר ... ז״ה] ⟨*⟩ **ט** ליתא **ס** [4]בו] ליתא **ט** [5]לבאר] לבארנו **ס**

يَكونُ مِنْ ضَرْبِ ا ه في نَفْسِهِ مِثْلِ الَّذي يَكونُ مِنْ ضَرْبِ ا ه في ه ج، لِأَنَّهما مُتَساوِيانِ. فَقَدْ تَبَيَّنَ أَنَّ الَّذي يَكونُ [..][1] ا ه في ه ج مِثْل ما يَكونُ مِنْ ضَرْبِ ب ه في ه د.

وأَيضًا: في دائِرَةِ ا ب ج د الثَّالِثَة خَطّا (ب د)[2] ا ج. يَقْطَعُ كُلُّ واحد مِنْهُما الآخَرَ على نُقْطَةِ ه. فَأَقولُ إِنَّ الَّذي يَكونُ مِنْ ضَرْبِ ب ه في ه د مِثْل الَّذي يَكونُ مِنْ ضَرْبِ ا ه في ه ج. وَلا نَجْعَلُ خَطَّ ب د يَقْطَعُ خَطَّ ا ج بِنِصْفَيْنِ. وَلٰكِنْ يَقْطَعُهُ بِقِسْمَيْنِ مُخْتَلِفَيْنِ. وَنَجْعَلُ قِسْمَهُ الأَطْوَلَ (ه ج)[3]. ومَرْكَزَ الدَّائِرَةِ ز.

ועוד בעגול אבג״ד השלישי שני קוים ב״ד וא״ג, כל אחד מהם יחתוך האחר על נקודת ה׳. ואומר שהעולה ממערכת ב״ה בה״ד שוה לעולה מא״ה בה״ג. ולא נשים קו ב״ד שיחתך א״ג לשני חלקים שוים, אבל יחתכהו לשני חלקים שאינם שוים. ונשים חלקו הארוך ה״ג. ומרכז העיגול ז׳.

[1] ..] חסרות כאן כמה מילים

[2] ب د] ا ب د

[3] ه ج] ه ج د

והמופת על מה שאמרנו: שנוציא מז׳ עמוד שיפול על קו א״ג, והוא ז״ח. [!] וכבר יצא מן המרכז עמוד. ואם כן חתך קו א״ג לשני חלקים שוים. ואם כן א״ח שוה לח״ג. וקו א״ג כבר נחלק בשני חלקים שוים על ח׳, ובשני חלקים שאינם שוים על ה׳. והעולה מא״ה בה״ג וה״ח בעצמו יחד, שוה לעולה מח״א בעצמו. ונשים העולה מח״ז בעצמו משותף. ויהיה העולה מג״ה בה״א וז״ח וה״ח בעצמו יחד, שוה לעולה מז״ח וח״א, כל אחד בעצמו יחד. והעולה מז״ח וה״ח כל אחד בעצמו הוא שוה לעולה מז״ה בעצמו, לפי שזוית זח״ה נצבה. והעולה מז״ח וח״א כל אחד בעצמו יחד, שוה לעולה מז״א בעצמו. והעולה מג״ה בה״א וה״ז בעצמו יחד, שוה לעולה מז״א בעצמו. העולה [..] ב״ה וה״[ז] בעצמו יחד, שוה לעולה מז״א בעצמו. והעולה מג״ה בה״א וה״ז בעצמו יחד, שוה לעולה מז״א בעצמו. ואם כן העולה מב״ה בה״ד הוא שוה לעולה מג״ה בה״א.

بُرهانُهُ: أنّا نُخْرِجُ مِنْ ز عَمودًا يَقَعُ على خَطّ اج، وَهُوَ زح. وَنَصِلُ زا. فَقَدْ خَرَجَ مِنْ د المَرْكَزِ عَمودُ زح. فَقَطَعَ[1] خَطَّ اج بِنِصْفَيْنِ عَلى ح. اح مِثْلُ ح ج. نَخَطُّ (اج)[2] (قَدْ)[3] قُسِمَ بِنِصْفَيْنِ عَلى ح وَبِقِسْمَيْنِ مُخْتَلِفَيْنِ عَلى ه. فَالَّذي يَكونُ مِنْ ضَرْبِ اه في ه ج و ه ح في نَفْسِهِ جَميعًا مِثْلُ الَّذي يَكونُ مِنْ ضَرْبِ ح ا في نَفْسِهِ. وَنَجْعَلُ الَّذي يَكونُ مِنْ ضَرْبِ ح ز العَمود في نَفْسِهِ مُشْتَرَكًا. فَالَّذي يَكونُ مِنْ ضَرْبِ ج ه في ه ا و زح في نَفْسِهِ و ه ح في نَفْسِهِ جَميعًا، مِثْلُ الَّذي يَكونُ مِنْ ضَرْبِ زح و ح ا، كُلّ واحدٍ في نَفْسِهِ جَميعًا. والَّذي يَكونُ مِنْ ضَرْبِ زح و ح ه، كُلّ واحدٍ في نَفْسِهِ، مِثْلُ الَّذي يَكونُ مِنْ ضَرْبِ زه في نَفْسِهِ، لِأَنَّ الزَّاوِيَةَ الَّتي من زح ه قائِمَةٌ. فَالَّذي يَكونُ مِنْ ضَرْبِ ج ه في ه ا و ه ز في نَفْسِهِ جَميعًا مِثْلُ الَّذي يَكونُ مِنْ ضَرْبِ زا في نَفْسِهِ. وَكَذَلِكَ يَصيرُ الَّذي يَكونُ مِنْ ضَرْبِ ب ه في (ب د)[4] زه في نَفْسِهِ مِثْلُ الَّذي يَكونُ مِنْ ضَرْبِ زا في نَفْسِهِ. فَنُلْقي المُشْتَرَكَ الَّذي يَكونُ مِنْ ضَرْبِ د ه في نَفْسِهِ. فَيَبْقَى الَّذي يَكونُ مِنْ ضَرْبِ ب ه في ه د مِثْلُ الَّذي يَكونُ ه ج في ه ا.

[1] فقطع] فقد+ [2] اج] ح ج [3] قد] فقد [4] ب د] ه ج

ועוד בעגול אבג״ד הרביעי שני קוים א״ג וב״ד, כל אחד מהם יחתוך האחר על נקדת ה׳. ואומר שהעולה מב״ה בה״ד שוה לעולה מג״ה בה״א. ולא נשים המרכז על אחד מקוי א״ג וב״ד, אבל במה שבין ב׳ ה׳ א׳, על נקודת ז׳. וקו ב״ד יחתוך קו ג״א תחלה לשני חלקים שוים. וחלק ב״ד הארוך הוא ב״ה.

وَأَيْضًا في دائِرَةِ ا ب ج الرَّابِعَة خَطَّا ا ج ب د، يَقْطَعُ كُلُّ واحِدٍ مِنْهُما الآخَرَ عَلى نُقْطَةِ ه. فَأَقولُ إِنَّ الَّذي يَكونُ مِنْ ضَرْبِ ب ه في ه د مِثْل الَّذي يَكونُ مِنْ ضَرْبِ ج ه في ه ا. فَلا نَجْعَلُ المَرْكَزَ عَلى واحِدٍ مِنْ خَطَّي ا ج ب د، وَلَكِنْ فِيما بَيْنَ (ب ه ا)[1] عَلى نُقْطَةِ ز. وَخَطُّ ب د يَقْطَعُ خَطَّ ج ا أَوَّلًا بِنِصْفَيْنِ. وَقِسْمُ ب د الأَطْوَلُ[2] هو ب ه.

והמופת על מה שאמרנו: שנוציא מז׳ שהוא המרכז עמוד אל קו בה״ד, והוא ז״ח. ונדביק ז׳[1] בד׳, וז׳ בה׳. וקו ב״ד כבר נחלק לשני חלקים שוים על נקדת ח׳, ולשני חלקים בלתי שוים על נקודת ה׳. ואם כן העולה מב״ה בה״ד וה״ח בעצמו שוה לעולה מח״ד בעצמו. ונשים ז״ח העמוד משותף.

بُرْهانُهُ: أَنَّا نُخْرِجُ مِنْ ز المَرْكَزِ عَمودًا إِلى خَطِّ ب ه د. وَنَصِلُ (ز ب ا د ه)[3]. فَخَطّ ب د قَدْ قُسِمَ بِنِصْفَيْنِ عَلى نُقْطَةِ ح وَبِقِسْمَيْنِ مُخْتَلِفَيْنِ عَلى نُقْطَةِ ه. فالَّذي يَكونُ مِنْ ضَرْبِ (ب ه)[4] في ه د ه ح في نَفْسِهِ جَميعًا مِثْل الَّذي يَكونُ مِنْ ضَرْبِ (ح د)[5] في نَفْسِهِ. وَنَجْعَلُ ز ح العَمودَ في نَفْسِهِ مُشْتَرَكًا.

[1] ז׳] ז׳ ב וז׳

[1] ب ه ا] ب ه ه ا [2] الاطول] هو+ [3] ز ب ا د ه] ز ا د ه [4] ب ه] د ه [5] ح د] ح ه

ויהיה העולה מב״ה בה״ד וז״ח בעצמו וח״ה בעצמו יחד, שוה לעולה מז״ח וח״ד, כל אחד בעצמו יחד. ואמנם העולה מז״ח וח״ה, כל אחד בעצמו יחד, שוה לעולה מז״ה בעצמו. ואמנם[1] העולה מז״ח וח״ד, כל אחד בעצמו יחד, שוה[2] לעולה מז״ד בעצמו, לפי שזוית זח״ד נצבה. ואם כן העולה מב״ה בה״ד וז״ה בעצמו יחד, שוה לעולה מז״ד. וז״ד שוה לז״א. והעולה[3] מז״א בעצמו שוה לעולה מז״ה וה״א, כל אחד בעצמו יחד, לפי שזוית זה״א נצבה. ואם כן העולה מב״ה בה״ד וז״ה בעצמו יחד, שוה לעולה מז״ה וה״א, כל אחד בעצמו יחד. ויושלך המשותף, והוא העולה מז״ה בעצמו. וישאר העולה מב״ה בה״ד שוה לעולה מא״ה בעצמו. ואמנם העולה מא״ה בעצמו, שוה לעולה מא״ה בה״ג, לפי שהם שוים. ואם כן העולה מא״ה בה״ג שוה לעולה מב״ה בה״ד.

1 ואמנם] ~~והעולה מז״א וח״ד כל אחד בעצמו יחד~~+

2 שוה] +⟨⟨ואם כן העולה מ[..] והעולה מ[..] שוה לעולה מז״ח וח״ד והעולה מ[..] וח״[..] וכו׳⟩⟩

3 והעולה] ⟨⟨ואם כן העולה מב״ה בה״ד וז״ה בעצמו יחד, שוה לעולה מז״ה בעצמו והעולה מ[א״ה]⟩⟩+

فالَّذي يكونُ مِنْ ضَرْبِ ب ه في ه د و ز ح في نَفْسِهِ و ه ح في نَفْسِهِ جَميعًا هُوَ مِثْلُ[1] الَّذي يَكونُ مِنْ ضَرْبِ ز ح و ح د، كُلّ واحدٍ في نَفْسِهِ جَميعًا. وَلكِنَّ الَّذي يكونُ مِنْ ضَرْبِ ز ح و ح ه، كُلّ واحدٍ في نَفْسِهِ جَميعًا، مِثْل ما يَكونُ مِنْ ضَرْبِ ز ه في نَفْسِهِ. وَالَّذي يكونُ مِنْ ضَرْبِ ز ح و ح د، كُلّ واحدٍ في نَفْسِهِ جَميعًا، مِثْل الَّذي يَكونُ مِنْ ضَرْبِ ز د في نَفْسِهِ، لِأَنَّ الزَّاوِيَةَ الَّتي مِنْ ز ح د قائِمَةٌ. فالَّذي يكونُ مِنْ ضَرْبِ ب ه في ه د و ز ه في نَفْسِهِ جَميعًا، مِثْل الَّذي يكونُ مِنْ ضَرْبِ ز د في نَفْسِهِ. و ز د في نَفْسِهِ مِثْل ز ا. والَّذي يكونُ مِنْ ضَرْبِ ز ا في نَفْسِهِ مِثْل ما يكونُ مِنْ ضَرْبِ ز ه و ه ا، كُلّ واحدٍ مِنْهُما في نَفْسِهِ جَميعًا، لِأَنَّ الزَّاوِيَةَ الَّتي من ز ه ا قائمة. وَالَّذي يكونُ مِنْ ضَرْبِ ب ه في ه د و ز ه في نَفْسِهِ جَميعًا مِثْل الَّذي يكونُ مِنْ ضَرْبِ ز ه و ه ا، كُلّ واحدٍ مِنْهُما في نَفْسِهِ مَجْموعَيْنِ. فتلقي المشترك الَّذي يكونُ مِنْ ضَرْبِ ز ه في نَفْسِهِ. فَيَبْقَى الَّذي يكونُ مِنْ ضَرْبِ ب ه في ه د مِثْل الَّذي يكونُ مِنْ ضَرْبِ ا ه في نَفْسِهِ. لكِنَّ الَّذي يكونُ مِنْ ضَرْبِ ا ه في نَفْسِهِ مِثْل الَّذي يكونُ مِنْ ضَرْبِ ا ه في ه ج لِأَنَّهُما مُتَساوِيانِ فالَّذي يكونُ مِنْ ضَرْبِ ا ه في ه ج مِثْل الَّذي يكونُ مِنْ ضَرْبِ ب ه في ه د.

1 مثل] ⟨*⟩

ועוד בעגול אבג"ד החמשי שני קוים ב"ד וא"ג, יחתוך כל אחד מהם האחר על נקודת ה׳. ואומר שהעולה מב"ה בה"ד שוה לעולה מא"ה בה"ג. ולא נשים קו ב"ד שיחתוך א"ג לשני חלקים שוים. אבל יהיה חלק ב"ד הארוך—ב"ה, וחלק א"ג הארוך—ג"ה. ומרכז העגול ז׳.

והמופת על מה שאמרנו: שנוציא מז׳ שני עמודים אל קוי ב"ד וא"ג, והם ז"ח וז"ט[1]. ונדביק ז׳ בה׳ ובד׳ ובא׳. וכבר יצא מן המרכז קו וחתך קו א"ג שלא יעבר על המרכז על זוית נצבה. ואם כן ג"ט שוה לט"א. וכמו כן ב"ח שוה לח"ד. וקו א"ג כבר נחלק לשני חלקים שוים על ט׳, ולשני חלקים בלתי שוים על ה׳. ואם כן העולה מג"ה בה"א וט"ה בעצמו יחד, שוה לט"א[2] בעצמו. ונשים העולה מז"ט בעצמו משותף. ויהיה העולה מג"ה בה"א וז"ט בעצמו וט"ה בעצמו, יחד, שוה לעולה מז"ט וט"א, כל אחד בעצמו, יחד. ואמנם העולה מז"ט וט"ה כל אחד בעצמו יחד, שוה לעולה מז"א בעצמו [!]. ואם כן העולה מג"ה בה"א וז"ה בעצמו יחד, שוה לעולה מז"ט וט"א יחד ולעולה [..][3]. והעולה מז"ט וט"א כל אחד בעצמו יחד, שוה לעולה מז"א בעצמו, לפי שזוית זט"א[4] נצבה.

1וז"ט] וא⟨ז⟩"ט 2לט"א] ~~לעולה מז"ט וט"א כל אחד בעצמו יחד~~+ 3ואם ... [..]] ⟨⟨*⟩⟩ 4זט"א] א⟨ז⟩ט"א

وَأَيْضًا في دائِرَةِ ا ب ج د الخامِسَةِ خَطَّا ب د ا ج، يَقْطَعُ كُلُّ واحِدٍ مِنْهُما الآخَرَ عَلى نُقْطَةِ ه. فَأَقولُ إِنَّ الَّذي يَكونُ مِنْ ضَرْبِ ب ه في ه د مِثْلُ الَّذي يَكونُ مِنْ ضَرْبِ ا ه في ه ج. وَلا نَجْعَلُ خَطَّ ب د يَقْطَعُ خَطَّ ا ج بِنِصْفَيْنِ. وَلٰكِنْ يَكونُ قِسمِ ب د الأَطْوَلُ—ب ه، وَقِسْمُ ا ج الأَطْوَلُ—ج ه. وَمَرْكَزُ الدّائِرَةِ ز.

بُرْهانُهُ: أَنّا نُخْرِجُ مِنْ ز عَمودَيْنِ إِلى خَطَّي ب د ا ج، وَهُما ز ح و ز ط. وَنَصِلُ ز ه و د ا. فَقَدْ خَرَجَ مِنَ المَرْكَزِ[1] خَطُّ ز ط. فَقَدْ قَطَعَ خَطَّ ا ج الَّذي لَمْ تَجُزْ عَلى المَرْكَزِ بِنِصْفَيْنِ عَلى زاوِيَةٍ قائِمَةٍ. ج ط مِثْلُ ط ا. وَكَذٰلِكَ ب ح مِثْلُ ح د. فَخَطُّ ا ج قَدْ قُسِمَ بِنِصْفَيْنِ عَلى ط، وبِنِصْفَيْنِ مُخْتَلِفَيْنِ عَلى ه. فالَّذي يَكونُ مِنْ ضَرْبِ ج ط في ه ا و ط ه في نَفْسِهِ جَميعًا، مِثْلُ الَّذي يَكونُ مِنْ ضَرْبِ ج ه في ه ا و ط ه في نَفْسِهِ جَميعًا، مِثْلُ الَّذي يَكونُ مِنْ ضَرْبِ ط ا في نَفْسِهِ. وَنَجْعَلُ الَّذي يَكونُ مِنْ ضَرْبِ ز ط في نَفْسِهِ مُشْتَرَكًا. فالَّذي يَكونُ مِنْ ضَرْبِ ج ه في ه ا ط ز في نَفْسِهِ و ط ه في نَفْسِهِ[2] جَميعًا، هُوَ مِثْلُ ما يَكونُ مِنْ ضَرْبِ ز ط ط ا، كُلّ واحِدٍ في نَفْسِهِ جَميعًا. وَلٰكِنْ ما الَّذي يَكونُ مِنْ ضَرْبِ ز ط ط ه، كُلّ واحِدٍ في نَفْسِهِ جَميعًا، مِثْلُ الَّذي يَكونُ مِنْ ضَرْبِ [..] ز ه في نَفْسِهِ. والَّذي يَكونُ مِنْ ضَرْبِ ز ط ط ا،

1المركز] ⟨*⟩ 2و ط ه ... نفسه] ⟨*⟩

كُلّ واحِدٍ في نَفْسِهِ جَميعًا، مِثْلَ الَّذي يَكونُ مِنْ ضَرْبِ ز ا في نَفْسِهِ.

ואם כן העולה מג״ה בה״א וז״ה עצמו שוה לעולה מז״א בעצמו. ומצד זה יהיה העולה מב״ה בה״ד וז״ה בעצמו יחד, שוה לעולה מז״א בעצמו. ויושלך המשותף, והוא העולה מז״ה בעצמו. ישאר העולה מב״ה בה״ד שוה לעולה מג״ה בה״א.

وَمِنْ أَجْلِ ذٰلِكَ يَصيرُ الَّذي يَكونُ مِنْ ضَرْبِ ب ه في ه د و ز ه في نَفْسِهِ مَجْموعَيْنِ، مِثْلَ الَّذي يَكونُ مِنْ ضَرْبِ ز ا في نَفْسِهِ. فَتُلْقِي المُشْتَرَكَ الَّذي يَكونُ مِنْ ضَرْبِ ز ه في نَفْسِهِ. فَيَبْقَى الَّذي يَكونُ مِنْ ضَرْبِ ب ه في ه د مِثْلَ الَّذي يَكونُ مِنْ ضَرْبِ ج ه في ه ا.

ועוד בעגול אבג״ד השישי, שני קוים ב״ד וא״ג, שיחתוך כל אחד מהם האחר על נקודת ה׳. ואין אחד מהם עובר על המרכז. ואומר שהעולה מב״ה בה״ד שוה לעולה מא״ה בה״ג.

وَأَيْضًا في دائِرَةِ ا ب ج د السّادِسَةِ خَطّا ب د و ا ج يَقْطَعُ كُلُّ واحِدٍ مِنْهُما الآخَرَ عَلى نُقْطَةِ ه. وَلْيَكُنِ المَرْكَزُ فيما بَيْنَ القِسْمَيْنِ الأَطْوَلَيْنِ ب ه ه د. فَأَقولُ الَّذي يَكونُ مِنْ ضَرْبِ ب ه في ه د مِثْلَ الَّذي يَكونُ مِنْ ضَرْبِ ج ه في ه ا.

והמופת: שמרכז העגול הוא ז׳. ונוציא מז׳ שני עמודים אל קוי ב״ד וג״א, והם ז״ח וז״ט. ונדביק ז׳ בג׳ ובה׳ ובד׳. וקו ז״ח יצא מן המרכז וחתך קו ב״ד שאינו עובר על המרכז על זוית נצבה. ואם כן ב״ח שוה לח״ד. וקו ב״ד כבר נחלק לשני חלקים שוים על ח׳, ולשני חלקים בלתי שוים על ה׳. ואם כן העולה מב״ה בה״ד וח״ה בעצמו יחד, שוה לעולה מח״ד בעצמו. ונשים העולה מז״ח בעצמו משותף. ויהיה העולה מב״ה בה״ד (וה״ח)[1] בעצמו וד״ח בעצמו, יחד, שוה לעולה מז״ח וח״ד כל אחד בעצמו יחד. והעולה מז״ח וח״ה כל אחד בעצמו, יחד, שוה לעולה מז״ה בעצמו[2]. והעולה מז״ח וח״ד כל אחד בעצמו יחד, שוה לעולה מז״ד בעצמו, לפי שזוית זח״ד נצבה.

بُرْهانُهُ: أَنَّ مَرْكَزَ الدَّائِرَةِ زَ. فَنُخْرِجُ مِنْ زَ عَمودَيْنِ إِلى خَطِّ ب د ج ا، وَهُما ز ح ز ط. وَنَصِلُ ز ج ز ه ز د. فَنَخُطُّ ز ح أُخْرِجَ مِنَ المَرْكَزِ فَقَطَعَ خَطَّ ب د الَّذي لَمْ يَجُزْ عَلى المَرْكَزِ بِنِصْفَيْنِ عَلى زاوِيَةٍ قائِمَةٍ. فَخَطُّ ب ح مِثْلُ ح د. فَخَطُّ ب د قَدْ قُسِمَ بِنِصْفَيْنِ عَلى ح. وَبِقِسْمَيْنِ مُخْتَلِفَيْنِ عَلى ه. فَالَّذي يَكونُ مِنْ ضَرْبِ ب ه في ه د و ه ح في نَفْسِهِ مَجْموعَيْنِ، مِثْلُ الَّذي يَكونُ مِنْ ضَرْبِ ح د في نَفْسِهِ. وَنَجْعَلُ الَّذي يَكونُ مِنْ ضَرْبِ ز ح في نَفْسِهِ مُشْتَرَكًا. فَالَّذي يَكونُ مِنْ ضَرْبِ ب ه في[1] ه د و ز ح في نَفْسِهِ و ه ح في نَفْسِهِ جَميعًا، مِثْلُ الَّذي يَكونُ مِنْ ضَرْبِ ز ح ح د، كُلِّ واحِدٍ في نَفْسِهِ مَجْموعَيْنِ. وَلَكِنَّ الَّذي يَكونُ مِنْ ضَرْبِ ز ح ح ه، كُلِّ واحِدٍ في نَفْسِهِ مَجْموعَيْنِ، مِثْلُ الَّذي يَكونُ مِنْ ضَرْبِ ز ه في نَفْسِهِ. وَالَّذي يَكونُ مِنْ ضَرْبِ ز ح ح د، كُلِّ واحِدٍ في نَفْسِهِ مَجْموعَيْنِ، مِثْلُ الَّذي يَكونُ مِنْ ضَرْبِ ز د في نَفْسِهِ، لِأَنَّ الزَّاوِيَةَ الَّتي مِنْ ز ح ه قائِمَةٌ.

[1] וה״ח] וז״ח [2] בעצמו] +⟨⟨אם כן [..] שוה לעולה מז״ח בח״ד. והעולה מז״ח וכו׳⟩⟩

[1] في] في هذا+

ואם כן העולה מב״ה בה״ד וז״ה בעצמו, שוה לעולה מז״ד בעצמו. ומצד זה יהיה העולה מא״ה בה״ג וז״ה בעצמו, יחד, שוה לעולה מז״ג בעצמו. וז״ג שוה לז״ד. ויושלך המשותף, והוא העולה מז״ה בעצמו. ואם כן ישאר העולה מב״ה בה״ד, שוה לעולה מא״ה בה״ג.

וזה מה שרצינו.

فالَّذي يكونُ مِنْ ضَرْبِ ب ه[1] في ه د و ز ه في نَفْسِهِ جَميعِها، مِثْل الَّذي يكونُ مِنْ ضَرْبِ ز د في نَفْسِهِ. ومِنْ أَجْلِ ذَلكَ يَصيرُ الَّذي يكونُ مِنْ ضَرْبِ ا ه في ه ج و ز ه في نَفْسِهِ مَجْموعَيْنِ، مِثْل الَّذي يكونُ مِنْ ضَرْبِ ز د في نَفْسِهِ. فَتُلْقِي المُشْتَرَكَ الَّذي يكونُ مِنْ ضَرْبِ ز ه في نَفْسِهِ. فَيَبْقَى الَّذي يكونُ مِنْ ضَرْبِ ب ه في ه د مِثْل الَّذي يكونُ مِنْ ضَرْبِ ا ه في ه ج.

وَذَلكَ ما أَرَدْنا بَيانَهُ.

[1] ب ه] ز ه

[Proposition III.35]

[H36]

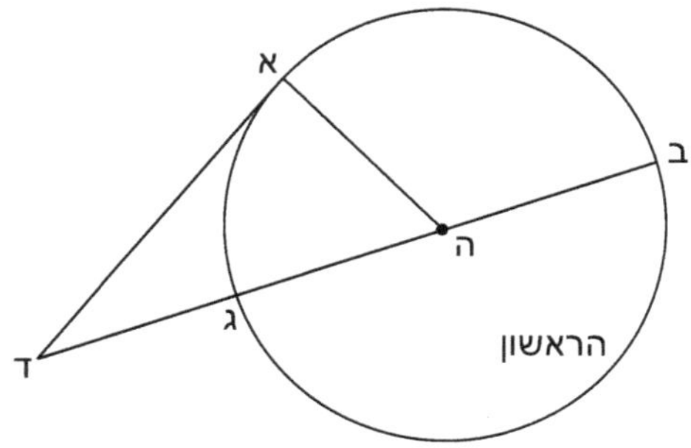

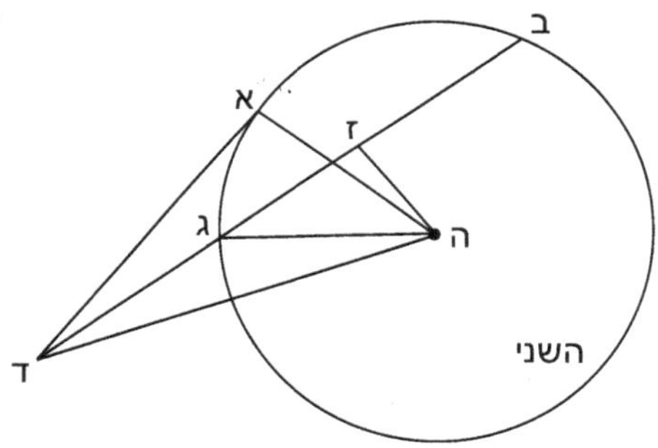

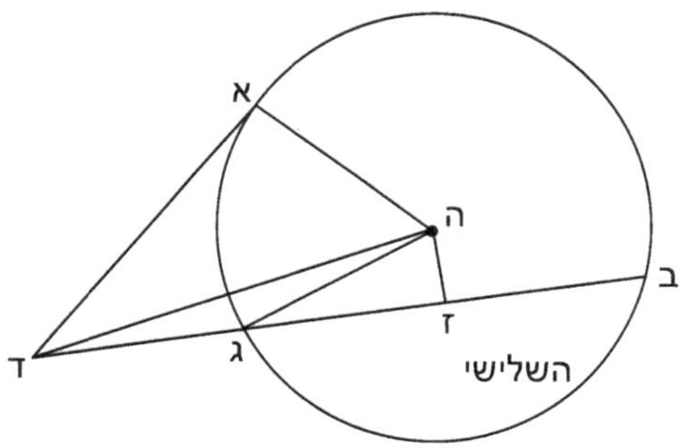

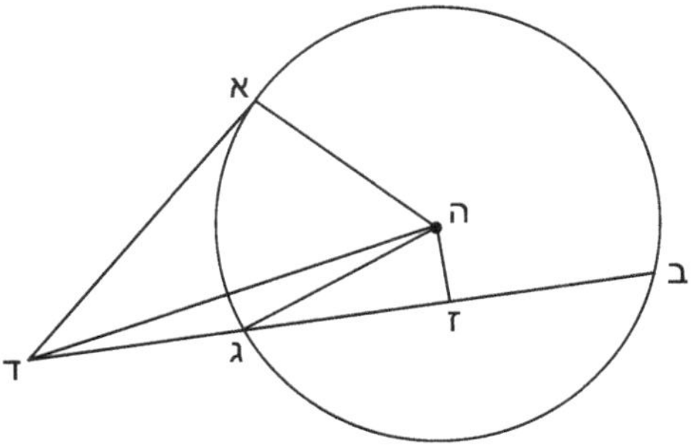

إذا تَعَلَّمَتْ نُقْطَةٌ خارِجَةٌ مِنْ دائِرَةٍ، وَأُخْرِجَ مِنْها خَطّانِ مُسْتَقيمانِ إلى الدَّائِرَةِ، أَحَدُهُما يَقْطَعُها، وَالآخَرُ يُماسُّها، فَإِنَّ السَّطْحَ القائِمَ الزَّوايا الَّذي يُحيطُ بِهِ الخَطُّ كُلُّهُ الَّذي يَقْطَعُ الدَّائِرَةَ والخطّ الَّذي يَقَعُ مِنْها خارِجَ الدَّائِرَةِ، مُساوٍ لِلمُرَبَّعِ الكائِنِ مِنَ الخَطِّ المُماسِّ.

כאשר רשמת נקודה חוץ מעגולה, והוצאו[1] ממנה שני קוים ישרים אל העגולה[2], אחד משניהם יחתכה והאחר ימששה, הנה השטח הנצב הזויות אשר יקיף[3] בו הקו כלו אשר[4] יחתוך העגולה עם הקו אשר יפול ממנו[5] חוץ לעגולה, שוה למרובע ההוה מן הקו הממשש.

فَلْتَكُنِ الدَّائِرَةُ ا ب ج. والنُقْطَة الَّتي تَعَلَّمَتْ خارِجًا مِنْها نُقْطَة د. ولنُخْرِجْ مِنْها إلى دائِرَةِ ا ب ج خَطَّيْ د ا د ب المُسْتَقيمَيْنِ. وَلْيَكُنْ خَطُّ د ب مِنْهُما قاطِعًا لِلدَّائِرَةِ. وخطّ د ا مُماسًّا لَها. فَأَقولُ إنَّ السَّطْحَ القائِمَ الزَّوايا الَّذي يُحيطُ بِهِ خَطّا ب د د ج، مُساوٍ لِلمُرَبَّعِ الكائِنِ مِنْ خَطِّ ا د.

ותהיה העגולה אב״ג. והנקודה אשר נרשמה חוץ ממנה נקודת ד׳. ונוציא ממנה אל עגולת אב״ג[6] שני קוי ד״א ד״ב הישרים. ויהיה קו ד״ב מהם חותך לעגולה. וקו א״ד ממשש[7] אותה. הנה אומר כי השטח הנצב הזויות אשר יקיפו בו שני קוי[8] ב״ד ד״ג, שוה למרובע ההוה מן קו[9] א״ד.

שרטוט—**בוטק**: $;**ס**: ז ↔ ג; בכתבי היד הערביים השרטוט זהה לשרטוט השני בכתב יד B [1]והוצאו] והוציאו **ב** [2]העגולה] +~~עם הקו אשר~~ **ב** [3]יקיף] יקיפו **ס** ~~יקיפו~~+ [4]אשר] כאשר **ב** [5]ממנו] ממנה⟨ו⟩ **פ** [6]אב״ג] אב״⟨ג⟩ **פ** [7]ממשש] ימשש **ז** [8]שני קוי] ליתא **ט** [9]קו] ליתא **ס**

כשנרשמה[1] נקדה חוץ לעגול והוצאו ממנה שני קוים אל העגול—אחד מהם יחתך אותו והאחר ימששהו—יהיה שטח [!] נצב הזויות שיקיפו בו הקו כלו שיחתך העגול והקו הנופל ממנו חוץ לעיגול, שוה למרובע הקו הממשש.

إذا كانَتْ نُقْطَةٌ خارِجَة مِنْ دائِرَة، وَأُخْرِجَ مِنْها خَطّانِ إلى الدّائِرَةِ—أَحَدُهُما يَقْطَعُها والآخَرُ يُماسُّها، فَإِنَّ الّذي يَكونُ مِنْ ضَرْبِ جَميعِ الخَطِّ الّذي يَقْطَعُ الدّائِرَةَ فيما كانَ خارِجًا مِنَ الدّائِرَةِ مِنْهُ، مِثْلُ الّذي يَكونُ مِنْ ضَرْبِ الخَطِّ المُماسِّ للدّائِرَةِ في نَفْسِه.

המשל בזה: שנרשום חוץ לעיגול אב״ג הראשון נקדת ד׳. ונוציא ממנה שני קוים אל העגול, אחד מהם יחתך העגול, והוא ד״ב, והאחר ימשש[2] אותו, והוא ד״א. ואומר שהעולה מב״ד בד״ג שוה לעולה מד״א בעצמו. ונשים מרכז העגול על קו ד״ב או למעלה ממנו או למטה ממנו. ונשימהו תחילה על קו ב״ד על נקודת ה׳ מן העגול הראשון.

مِثالُهُ: أَنّا نُعَلِّمُ خارِجًا مِنْ دائِرَةِ $\overline{\text{ا ب ج}}$ الأُولى نُقْطَةَ د. وَنُخْرِجُ مِنْها خَطَّيْنِ إلى الدّائِرَة، أَحَدُهُما يَقْطَعُها، وَهُوَ خَطُّ ($\overline{\text{د ج ب}}$)[1]، وَالآخَرُ يُماسُّها، وَهُوَ خَطُّ $\overline{\text{د ا}}$. فَأَقولُ إِنَّ الّذي يَكونُ مِنْ ضَرْبِ $\overline{\text{ب د}}$ في $\overline{\text{د ج}}$ مِثْلُ الّذي يَكونُ مِنْ ضَرْبِ $\overline{\text{د ا}}$ في نَفْسِه. وَنَجْعَلُ مَرْكَزَ الدّائِرَةِ عَلى خَطِّ $\overline{\text{ب د}}$ أَوْ فَوْقَهُ أَوْ تَحْتَهُ. فَنَجْعَلُهُ أَوَّلًا عَلى خَطِّ $\overline{\text{ب د}}$ عَلى نُقْطَةِ ه مِنَ الدّائِرَةِ الأُولى.

והמופת על מה שאמרנו: שנדביק א׳ בה׳. וקו א״ד ימשש עגול אב״ג. ויצא ממקום משושו קו אל המרכז, והוא א״ה. ואם כן א״ה עמוד על א״ד. וזוית הא״ד נצבה. והעולה מה״ד בעצמו שוה לעולה מה״א וא״ד, כל אחד בעצמו, יחד. וקו ג״ב כבר נחתך[3] לשני חלקים שוים על נקודת ה׳. ונוסף בארכו קו ג״ד.

وَبُرْهانُهُ: أَنّا نَصِلُ $\overline{\text{ا ه}}$. فَخَطُّ $\overline{\text{ا د}}$ يُماسُّ دائِرَةَ $\overline{\text{ا ب ج}}$. وَقَدْ أُخْرِجَ مِنَ المَوْضِعِ الّذي يُماسُّها فيهِ خَطٌّ إلى المَرْكَزِ، وَهُوَ $\overline{\text{ا ه}}$. وَ $\overline{\text{ا ه}}$ عَمودٌ عَلى خَطِّ $\overline{\text{ا د}}$. فَالزّاوِيَةُ الّتي من $\overline{\text{ه ا د}}$ قائِمَةٌ. وَالّذي يَكونُ مِنْ ضَرْبِ $\overline{\text{ه د}}$ في نَفْسِهِ هُوَ مِثْلُ الّذي يَكونُ مِنْ ضَرْبِ $\overline{\text{ه ا}}$ $\overline{\text{ا د}}$، كُلّ واحِدٍ مِنْهُما في نَفْسِهِ جَميعًا. وَخَطُّ $\overline{\text{ج ب}}$ قَدْ قُطِعَ بِنِصْفَيْنِ عَلى نُقْطَةِ ه. وَقَدْ يَزيدُ في طُولِهِ $\overline{\text{ج د}}$.

1 כשנרשמה] ⟨⟨*⟩⟩ 2 ימשש] ימששהו 3 נחתך] ⟨⟨נ⟩⟩חתך

1 $\overline{\text{د ج ب}}$] $\overline{\text{د ب ج}}$

بُرْهانُهُ: أنّا نَجْعَلُ المَرْكَزَ ه. ونُخْرِجُ مِنْ نُقْطَةِ ه إلى خَطِّ ب د عَمُودَ ه ز. ونَصِلُ خُطوطَ ه ا ه ج ه د. فَلأَنَّ في دائِرَةِ ا ب ج خَطًّا يَمُرُّ بالمَرْكَزِ وهو ه ز، وقَدْ قَطَعَ ب ج على زاوِيَةٍ قائِمَةٍ، فَهُوَ يَقْطَعُهُ بِنِصْفَيْنِ. فَخَطُّ ب ز مُساوٍ لِخَطِّ ز ج. وَلِأَنَّ خَطَّ ب ج قد قُسِمَ بِنِصْفَيْنِ على نُقْطَةِ ز، ووصل بِهِ خطّ ج د على اسْتِقامَةٍ، يَكُونُ السَّطْحُ القائِمُ الزَّوايا الَّذي يُحِيطُ بِهِ خَطّا ب د د ج، مَعَ المُرَبَّعِ الكائِنِ مِنْ ز ج، مُساوِيًا لِلمُرَبَّعِ الكائِنِ مِنْ ز د^. وَنَجْعَلُ المُرَبَّعَ الكائِنَ مِنْ ز ه مُشْتَرَكًا. فَالسَّطْحُ القائِمُ الزَّوايا الَّذي يُحِيطُ بِهِ خَطّا ب د د ج، مَعَ المُرَبَّعَيْنِ الكائِنَيْنِ مِنْ ه ز ز ج، مُساوٍ لِلمُرَبَّعَيْنِ الكائِنَيْنِ مِنْ خَطَّيْ ز ه ز د. لَكِنَّ^ المُرَبَّعَيْنِ الكائِنَيْنِ مِنْ خَطَّيْ ه ز ز ج، مُساوِيانِ لِلمُرَبَّعِ الكائِنِ مِنْ ه ج، لأَنَّ زاوِيَةَ ه ز ج قائِمَةٌ. وَالمُرَبَّعانِ الكائِنانِ مِنْ خَطَّيْ ه ز ز د، مُساوِيانِ لِلمُرَبَّعِ الكائِنِ مِنْ ه د، لأَنَّ زاوِيَةَ ه ز د قائِمَةٌ. فالسَّطْحُ القائِمُ الزَّوايا الَّذي يُحِيطُ بِهِ خَطّا ب د د ج، مَعَ المُرَبَّعِ الكائِنِ مِنْ ه ج، مُساوٍ لِلمُرَبَّعِ الكائِنِ مِنْ ه د. وَلَكِنَّ المُرَبَّعَ الكائِنَ مِنْ ه د مُساوٍ لِلمُرَبَّعَيْنِ الكائِنَيْنِ مِنْ خَطَّيْ ه ا ا د، لأَنَّ زاوِيَةَ ه ا د قائِمَةٌ.

המופת: שנשים[1] המרכז ה׳. ונוציא מנקודת ה׳ אל קו ב״ד עמוד ה״ז. ונגיע קוי א״ה ג״ה ה״ד[2]. הנה[3] מפני כי בעגולת אב״ג קו יצא[4] מן המרכז והוא ה״ז, וכבר חתך ב״ג על זוית נצבת, הנה הוא חתכו בשני חציים. הנה קו ב״ז שוה לקו ז״ג. ומפני כי קו ב״ג כבר נחלק בשני חלקים על נקודת ז׳, והגיע[5] בו קו ג״ד על יושר, יהיה השטח הנצב הזויות אשר יקיפו בו שני קוי ב״ד ד״ג עם המרובע ההוה מן ז״ג, כמו המרובע ההווה מן ד״ז בעצמו. ונשים המרובע[6] ההוה מן ז״ה משותף. הנה השטח הנצב הזויות[7] אשר יקיפו בו שני קוי ב״ד ד״ג, עם שני המרובעים ההוים מן ה״ז ז״ג, שוה[8] לשני המרובעים ההוים משני קוי ז״ה ז״ד[9]. ושני המרובעים ההוים משני קוי ז״ה ז״ג[10], שוים למרובע ההווה מן ה״ג[11], מפני כי זוית הז״ג[12] נצבת. ושני[13] המרובעים ההוים משני קוי ה״ז ז״ד, שוים למרובע ההווה מן ה״ד, מפני כי זוית הז״ד נצבת[14] [15]. הנה השטח הנצב הזויות אשר יקיפו בו שני קוי[16] ב״ד ד״ג, עם המרובע ההווה מן ה״ג, שוה למרובע ההוה מן ה״ד. אבל המרובע ההוה מן ה״ד שוה לשני המרובעים ההוים משני קוי ה״א א״ד, מפני כי זוית הא״ד נצבת.

---

[1] שנשים] כי נשים **ט** [2] ה״ד] ליתא **ב** [3] הנה] ⟨*⟩ **פ** [4] יצא] יוצא **ו** [5] והגיע] ונגיע **ב** וכבר הגיע **ו** [6] המרובע] ליתא **ב** [7] הנצב הזויות] הזוית **ב** [8] שוה] שוים **וספ** [9] מן ... ז״ד] משני קוים ז״ה ז״ג שוים למרובע ההווה מן ה״ג מפני כי זוית הז״ג נצבת **ב** [10] ז״ג] ז״ד **ב** [11] ה״ג] א״ד **ב** ה״ד **ס** [12] הז״ג] חז״ד **ב** הז״ד **ס** [13] ושני] וגם כן שני **ס** [14] המרובעים ... נצבת] ליתא **ב** [15] משני ... נצבת] מן ג״ז ז״ה שוים למרובע ההוה מן ה״ג **ס** [16] שני קוי] ליתא **ט**

ואם כן העולה ממערכת ב״ד בד״ג וג״ה בעצמו יחד, שוה לעולה מה״ד בעצמו, והוא שוה לעולה מה״א וא״ד, כל אחד בעצמו, יחד[1]. ואם כן העולה מב״ד בג״ד וה״ג בעצמו, שוה[2] לעולה מא״ה וא״ד כל אחד בעצמו. והעולה מג״ה בעצמו, שוה לעולה מא״ה בעצמו, לפי שהם שוים. ונשאר העולה מב״ד בד״ג שוה לעולה מא״ד בעצמו.

فالَّذي يكونُ مِنْ ضَرْبِ ب د في د ج ج ه في نَفْسِهِ، هُوَ مِثْل ما يكونُ [!] ه ا ا د، كُلّ واحِدٍ في نَفْسِهِ جَمِيعًا. فالَّذي يكونُ مِنْ ضَرْبِ ه ج في نَفْسِهِ مِثْل الَّذي يكونُ مِنْ ضَرْبِ ا ه في نَفْسِهِ، لأَنَّهُما مُتَساوِيانِ. فَيَبْقَى الَّذي يكونُ مِنْ ضَرْبِ ب د في د ج مِثْل الَّذي يكونُ مِنْ ضَرْبِ ا د في نَفْسِهِ.

ועוד חוץ לעיגול אב״ג השני נקדת ד׳. ויצאו מד׳ שני קוים אל העגול[3], אחד מהם יחתוך אותו, והוא קו ד״ב, והאחר ימשש אותו, והוא קו ד״א. ואומר שהעולה מב״ד בד״ג שוה לעולה מא״ד בעצמו. ולא נשים המרכז על קו ב״ג, אבל ממעל לו, על ה׳.

وأَيضًا خارِجًا مِنْ دائِرَةِ ا ب ج الثَّانِيَةِ نُقْطَةُ د. وَنُخْرِجُ مِنْ نُقْطَةِ د خَطَّيْنِ إلى الدَّائِرَةِ، أَحَدُهما يَقْطَعُها، وَهُوَ خَطُّ د ج ب، وَالآخَرُ يُماسُّها، وَهُوَ د ا. فَأَقولُ إِنَّ الَّذي يكونُ مِنْ ضَرْبِ ب د في د ج هُوَ مِثْل الَّذي يكونُ مِنْ ضَرْبِ ا د في نَفْسِهِ. وَلا نَجْعَلُ المَرْكَزَ على خَطِّ ب ج، وَلَكِنْ فَوْقَهُ.

---

[1] יחד] + ~~וקו ג״ב כבר חתך לשני חלקים~~ [2] שוה] ש⟨ו⟩ה [3] העגול] ה⟨עגול⟩

فالسَّطْحُ القائِمُ الزَّوايا الَّذي يُحيطُ بِهِ خَطّا ب د د ج، مَعَ المُرَبَّعِ الكائِنِ مِنْ ه ج، مُساوٍ لِلمُرَبَّعَيْنِ الكائِنَيْنِ مِنْ ه ا ا د. والمُرَبَّعُ الكائِنُ مِنْ ه ج مُساوٍ لِلْمُرَبَّعِ الكائِنِ مِنْ ا ه، وذَلِكَ أَنَّهُما أُخْرِجا مِنْ مَرْكَزِ الدَّائِرَةِ إلى الخَطِّ المُحيطِ بِها. فَيَبْقَى السَّطْحُ القائِمُ الزَّوايا الَّذي يُحيطُ بِهِ خَطّا ب د د ج، مُساوِيًا لِلْمُرَبَّعِ الكائِنِ مِنْ ا د.

وذَلِكَ ما أَرَدْنا أَنْ نُبَيِّنَ.

הנה השטח הנצב הזויות אשר יקיפו בו שני קוי[1] ב״ד ד״ג, עם המרובע ההוה מן ה״ג, שוה לשני המרובעים ההווים מן א״ה א״ד. והמרובע ההוה מן ה״ג שוה למרובע ההוה מן[2] א״ה, וזה שהם יצאו מן מרכז[3] העגולה אל הקו המקיף בה[4]. הנה נשאר השטח הנצב הזויות אשר יקיפו בו שני קוי ב״ד ד״ג, שוה למרובע ההווה[5] מן א״ד.

ונשלם ביאורו[6].

---

1 שני קוי] ליתא **ט** 2 מן] ⟨⟨*⟩⟩ **ט** 3 מרכז] המרכז **בו** 4 המקיף בה] ⟨*⟩ **ט** 5 ההווה] ליתא **ב** 6 ונשלם ביאורו] וזה מה שרצינו לבאר **ט**

והמופת על מה שאמרנו: שנוציא מה׳ עמוד אל קו בג״ד, והוא ה״ז. ונדביק ה׳[1] בא׳, וה׳ בג׳, וה׳ בד׳. וקו א״ד ימשש עגול אב״ג. וכבר יצא ממקום משושו קו אל המרכז, והוא קו א״ה. ואם כן א״ה עמוד על קו א״ד. ואם כן העולה מה״ד בעצמו שוה לעולה מה״א וא״ד כל אחד בעצמו, לפי שזוית דא״ה נצבה. וקו ה״ז יצא מן המרכז, וחתך קו ב״ג שלא יעבר על המרכז על זוית נצבה. ואם כן ב״ז שוה לז״ג[2]. וקו ב״ג כבר נחלק לשני חלקים שוים על ז׳. ונוסף בארכו קו ג״ד. ואם כן העולה מב״ד בג״ד וז״ג בעצמו, יחד, שוה לעולה מז״ד[3] בעצמו. ונשים ה״ז בעצמו משותף.

بُرْهانُه: أنَّا نُخْرِجُ مِنْ ه عَمودًا إلى خَطِّ ب ج د، وَهُوَ ه ز. ونَصِلُ ه ا ه ج ه د. نَخَطُّ ا د يَماسُّ دائِرَةَ ا ب ج. وَقَد أُخْرِجَ مِنَ المَوْضِعِ الَّذي يَماسُّها فيهِ خَطٌّ إلى المَرْكَزِ، وَهُوَ خَطُّ ا ه. و ه ا عَمودٌ وَقَعَ عَلى خَطِّ ا د. فالَّذي يَكونُ مِنْ ضَرْبِ ه د في نَفْسِهِ مِثْلُ الَّذي يَكونُ مِنْ ضَرْبِ ه ا ا د، كُلِّ واحِدٍ في نَفْسِه، لِأَنَّ الزّاوِيَةَ الَّتي مِنْ ه ا د قائِمَةٌ. وَخَطُّ ه ز أُخْرِجَ مِنَ المَرْكَزِ فَقَطَعَ خَطَّ ب ج الَّذي لَمْ يَجُزْ عَلى المَرْكَزِ بِنِصْفَيْنِ عَلى زاوِيَةٍ قائِمَةٍ. ب ز مِثْلُ ز ج. نَخَطُّ ب ج قَدْ قُطِعَ بِنِصْفَيْنِ عَلى ز. وَزيدَ في طُولِهِ خَطُّ ج د فالَّذي يَكونُ مِنْ ضَرْبِ ب د في ج د و ز ج في نَفْسِهِ جَميعًا، مِثْلُ الَّذي يَكونُ مِنْ ضَرْبِ[1] ز د في نَفْسِهِ. وَنَجْعَلُ ه ز في نَفْسِهِ مُشْتَرَكًا.

1ה׳] 〈*〉 2לז״ג] ~~לג׳+~~ 3מז״ד] ~~מה״ז+~~

1من ضرب] 〈*〉

ויהיה העולה מב״ד בד״ג וג״ז בעצמו וה״ז בעצמו, יחד, שוה לעולה מה״ז וז״ד כל אחד בעצמו יחד. והעולה מה״ז וז״ג[1] כל אחד בעצמו יחד, שוה לעולה מה״ג בעצמו, לפי שזוית הז״ג נצבה. ואם כן העולה מב״ד בג״ד וז״ג בעצמו יחד, שוה לעולה מה״ז וז״ד יחד. והעולה מה״ז וז״ד כל אחד בעצמו יחד שוה לעולה מה״ד בעצמו, לפי שזוית הז״ג נצבה. ואם כן העולה מב״ד בד״ג וה״ג [!] בעצמו יחד, שוה לעולה מה״ד בעצמו. והעולה מה״ד בעצמו שוה לעולה מה״א וא״ד כל אחד בעצמו יחד. ואמנם העולה מה״ג בעצמו שוה לעולה מה״א בעצמו, לפי שהם שוים. ונשאר העולה מב״ד בד״ג שוה לעולה מא״ד בעצמו. ועוד חוץ לעיגול אב״ג השלישי נקודת ד׳. ויצאו מד׳ שני קוים אל העגול, אחד מהם יחתך אותו, והוא קו דג״ב, והאחר ימששהו, והוא קו ד״א. ואומר שהעולה מב״ד בד״ג שוה לעולה מא״ד בעצמו. ונשים מרכז העיגול תחת קו דג״ב, על נקודת ה׳.

[1] וז״ג] ⟨⟨*⟩⟩

فالَّذي يكونُ مِنْ ضَرْبِ ب د في د ج زج[1] في نفْسِهِ ه ز في نفْسِهِ جَميعًا، مِثْل الَّذي يكونُ مِنْ ضَرْبِ ه ز ز د، كُلّ واحِدٍ في نفْسِهِ جَميعًا. فالَّذي يكونُ مِنْ ضَرْبِ ه ز في نفْسِهِ ز د في نفْسِهِ جَميعًا، مِثْل الَّذي يكونُ مِنْ ضَرْبِ ه ج في نفْسِهِ. فَصَارَ الَّذي يكونُ مِنْ ضَرْبِ ب د في د ج ه ج في نفْسِهِ مِثْل الَّذي يكونُ مِنْ ضَرْبِ ه د في نفْسِهِ. والَّذي يكونُ مِنْ ضَرْبِ ه ا ا د كُلّ واحِدٍ في نفْسِهِ. مِثْل الَّذي يكونُ مِنْ ضَرْبِ ه د في نفْسِهِ. ه ج في نفْسِهِ مِثْل ه ا في نفْسِهِ فيَبْقَى الَّذي يكونُ مِنْ ضَرْبِ ب د في د ج مِثْل الَّذي مِنْ ضَرْبِ د ا في نفْسِهِ.

وأَيضًا خارِجًا مِنْ دائِرَة ا ب ج الثَّالثَة نُقْطَةُ د. ونُخْرِجُ مِنْ د خَطَّيْنِ إِلى الدَّائِرَة أَحَدُهُما يَقْطَعُها، وهو خَطُّ د ج ب، والآخَرُ يَماسُّها وهو د ا فأَقولُ إِنَّ الَّذي يكونُ مِنْ ضَرْبِ ب د في د ج مِثْل الَّذي يكونُ مِنْ ضَرْبِ ا د في نفْسِهِ. ونَجْعَلُ المَرْكَزَ تَحْتَ خَطِّ د ج ب عَلى نُقْطَةِ ه.

[1] زج] ز د

והמופת על זה: שנוציא מה׳ עמוד על קו ב״ג, והוא ה״ז. ונדביק ה׳ בא׳ ובג׳ ובד׳. וקו ה״ז יצא מן המרכז וחתך קו ב״ג על זוית נצבה. ואם כן ב״ז שוה לז״ג. וקו ב״ג כבר נחלק לשני חלקים שוים על ז׳. ונוסף בארכו קו ג״ד. ואם העולה מב״ד בד״ג וז״ג בעצמו יחד שוה לעולה מז״ד בעצמו. ונשים העולה מז״ה בעצמו משותף. ויהיה העולה מב״ד בד״ג וה״ז בעצמו וז״ג בעצמו יחד, שוה לעולה מה״ז וז״ד כל אחד בעצמו יחד. ואמנם העולה מה״ז וז״ג כל אחד בעצמו יחד, שוה לעולה מה״ג בעצמו. ואם כן העולה מה״ז וז״ד כל אחד בעצמו יחד, שוה לעולה מה״ד בעצמו, לפי שזוית הז״ד נצבה. ואם כן העולה מב״ד בד״ג וג״ה בעצמו יחד, שוה לעולה מה״ד בעצמו. אמנם העולה מה״ד בעצמו שוה לעולה מה״א וא״ד כל אחד בעצמו יחד, לפי שזוית הא״ד נצבה.

بُرْهانُهُ: أَنَّا نُخْرِجُ مِنْ ه عَمودًا عَلى خَطِّ ب ج، وَهُوَ ه ز. وَنَصِلُ ه ا ه ج ه د. فَخَطُّ ه ز قَدْ أُخْرِجَ مِنَ المَرْكَزِ فَقَطَعَ خَطَّ (ب ج)[1] بِنِصْفَيْنِ عَلى زاوِيَةٍ قائِمَةٍ. ب ز مِثْلُ ز ج. وَخَطُّ ب ج فَقَدْ قُطِعَ بِنِصْفَيْنِ وَزِيدَ في طُولِهِ خَطُّ ج د. فالَّذي يَكونُ مِنْ ضَرْبِ ب د في د ج. و ز ج في نَفْسِهِ جَميعًا مِثْلُ الَّذي يَكونُ مِنْ ضَرْبِ ز د في نَفْسِهِ. وَنَجْعَلُ الَّذي يَكونُ مِنْ ضَرْبِ ز ه في نَفْسِهِ مُشْتَرَكًا. فالَّذي يَكونُ مِنْ ضَرْبِ (ب د)[2] في د ج و ه ز في نَفْسِهِ و ز ج في نَفْسِهِ جَميعًا، هُوَ مِثْلُ الَّذي يَكونُ مِنْ ضَرْبِ ه ز ز د، كُلُّ واحِدٍ في نَفْسِهِ جَميعًا. وَلٰكِنَّ الَّذي يَكونُ مِنْ ضَرْبِ ه ز ز ج، كُلُّ واحِدٍ في نَفْسِهِ جَميعًا، مِثْلُ الَّذي يَكونُ مِنْ ضَرْبِ ه ج في نَفْسِهِ. والَّذي يَكونُ مِنْ ضَرْبِ ه ز ز د، كُلُّ واحِدٍ في نَفْسِهِ جَميعًا، مِثْلُ الَّذي يَكونُ مِنْ ضَرْبِ ه د في نَفْسِهِ، لِأَنَّ الزّاوِيَةَ الَّتي مِنْ ه ز د قائِمَةٌ. فَالَّذي يَكونُ مِنْ ضَرْبِ ب د في د ج (وج ه)[3] في نَفْسِهِ جَميعًا، مِثْلُ الَّذي يَكونُ مِنْ ضَرْبِ ه د في نَفْسِه. وَالَّذي يَكونُ مِنْ ضَرْبِ ه د في نَفْسِه[4] هُوَ مِثْلُ الَّذي يَكونُ مِنْ ضَرْبِ ه ا ا د، كُلُّ واحِدٍ مِنْهُما في نَفْسِهِ جَميعًا، لِأَنَّ الزّاوِيَةَ الَّتي من ه ا د قائِمَةٌ.

1 ب ج] ب د   2 ب د] ز د   3 وج ه] وج
4 والذي ... نفسه] 〈*〉

ואם כן העולה מב״ד בד״ג וג״ה בעצמו יחד, שוה לעולה מה״א וא״ד כל אחד בעצמו יחד. ואמנם העולה מה״ג בעצמו שוה לעולה מה״א בעצמו. ישאר העולה מב״ד בד״ג שוה לעולה מא״ד בעצמו.

וזה מה שרצינו לבאר.

فالَّذي يكونُ مِنْ ضَرْبِ ب د في د ج ج ه في نَفْسِهِ جَميعًا، مِثْل الَّذي يكونُ مِنْ ضَرْبِ ه ا ا د، كُلّ واحِدٍ في نَفْسِهِ جَميعًا. وَلَكِنَّ الَّذي يَكونُ مِنْ ضَرْبِ ه ج في نَفْسِهِ، مِثْل الَّذي يَكونُ مِن ضَرْبِ ه ا في نَفْسِهِ. فَيَبْقَى الَّذي يَكونُ مِنْ ضَرْبِ ب د في د ج مِثْل الَّذي يَكونُ مِنْ ضَرْبِ ا د في نَفْسِهِ.

وَذَلِكَ ما أَرَدْنا بَيانَهُ.

[Proposition III.36]

[H37]

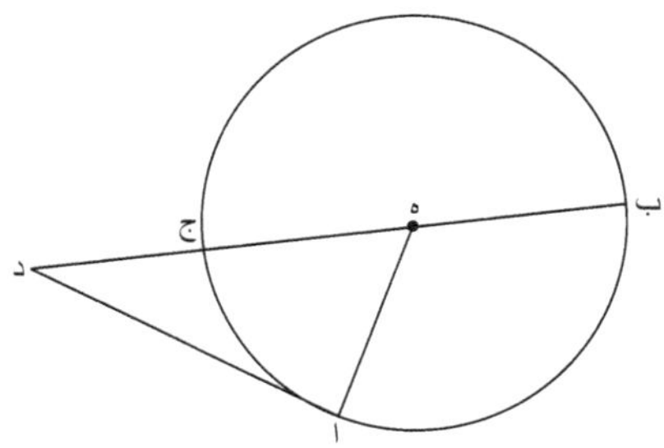

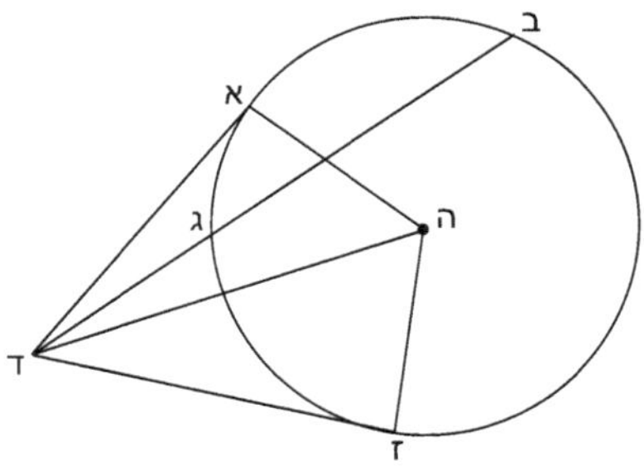

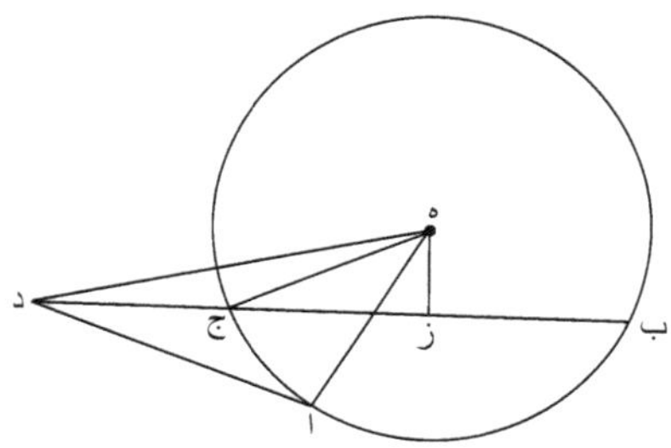

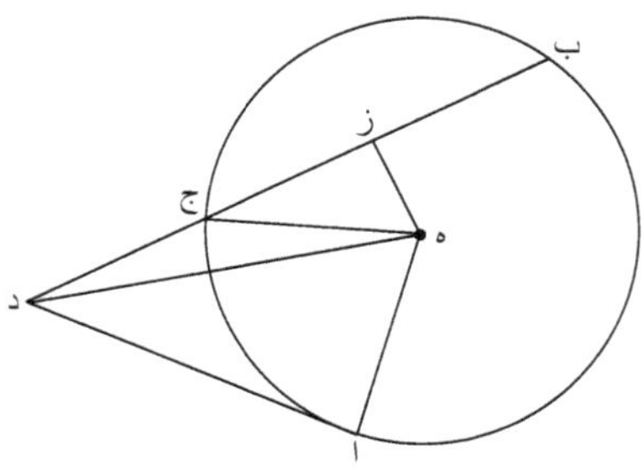

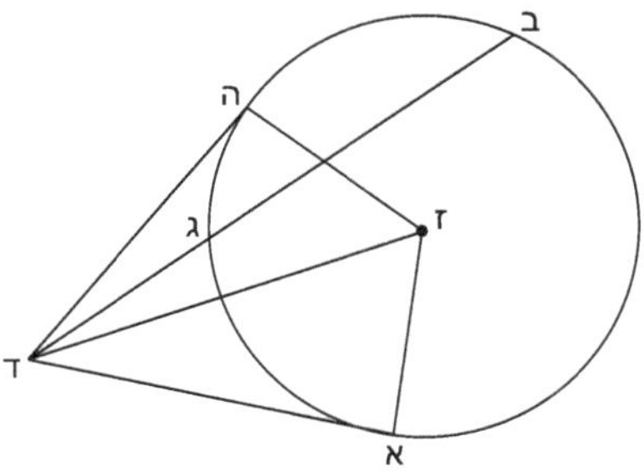

إذا كانَتْ دائِرَةٌ، وتُعَلَّمُ خارِجًا عَنْها نُقْطَةٌ، وأُخرِجَ مِنْها خَطّانِ مُسْتَقيمانِ إلى الدَّائِرَةِ، أَحَدُهُما يَقْطَعُها وَالآخَرِ يَنْتَهِي إِلَيْها، وكانَ السَّطْحُ القائِمُ الزَّوايا الَّذي يُحيطُ بِهِ الخَطُّ الَّذي يَقْطَعُها كُلّه والقِطْعَة الَّتي تَقَعُ مِنْهُ خارِجًا عَنِ الدَّائِرَةِ، مُساوِيًا لِلْمُرَبَّعِ الكائِنِ مِنَ الخَطِّ الآخَرِ الَّذي يَنْتَهي إلى الدَّائِرَةِ، فَإِنَّ الخَطَّ الَّذي يَنْتَهي إِلَيْها مُمَاسٌّ لِلدَّائِرَةِ^.

כאשר היתה עגולה, ונרשמה חוץ ממנה נקודה, והוציאו ממנה שני קוים ישרים אל העגולה, אחד משניהם יחתכה והאחר יכלה אליה, והיה השטח הנצב הזויות אשר יקיף בו הקו אשר יחתכה כלו והחתיכה אשר יפול ממנו[1] חוץ מן העגולה, שוה למרובע ההוה מן הקו האחר אשר יכלה אל העגולה, הנה הקו אשר יכלה אליה ממשש לעגולה.

וכאשר יצאו שני קוים מנקודה[2] אחת ושניהם[3] ימששו[4] העגולה, הנה הם[5] שוים[6] [7].

שרטוט—פ: ב״ד ז״א נחתכים   [1]ממנו] ממנה⟨ו⟩ ו   [2]קוים מנקודה] קוי הנקודה ב   [3]ושניהם] ושנים ב   [4]ימששו] ישׁשה ס   [5]הנה הם] שניהם ס   [6]וכאשר ... שוים] ליתא ו ⟨*⟩ ט   [7]שוים] +וכאשר יצאו הקוים מנקודה אחת יחתכו העגולה הנה השטחים אשר מכל אחד מהם מן הקו כלו עם החתיכה אשר יפול מעט חוץ מן ~~הקו~~ העגולה שוים קצתם אל קצתם ס

כשתהיה נקדה חוץ לעגול, ויצאו ממנה שני קוים, אחד מהם יחתך אותו והאחר יכלה אליו, ויהיה השטח נצב הזויות שיקיפו בו הקו החותך את העגול כלו, והקו הנופל ממנו חוץ לעגול שוה למרובע הקו הכלה אל העגול, יהיה הקו הכלה[1] ממשש העגול.

إذا وُضِعَتْ نُقْطَةٌ خارِجَةٌ مِنَ (مَرْكَزِ)[1] الدَّائِرَةِ، وأُخْرِجَ مِنْها خَطَّانِ إلى الدَّائِرَةِ، أَحَدُهُما يَقْطَعُها وَالآخَرُ يَنْتَهي إِلَيْها، وكانَ إِذا ضُرِبَ الخَطُّ الَّذي يَقْطَعُ الدَّائِرَةَ فيما كانَ خارِجًا مِنَ الدَّائِرَةِ مِنْهُ مِثْلَ ما يكونُ مِنْ ضَرْبِ الخَطِّ المُنْتَهِي إِلى الدَّائِرَةِ في نَفْسِهِ، فَإِنَّ الخَطَّ المُنْتَهِي إِلى الدَّائِرَةِ في نَفْسِهِ يَماسُّها.

זה לא נזכר בנסחת חאג׳ג׳. ובלשון בן סינא: ונאמר בהפך—כשהיה העניין בהכאה כמו שאמרנו, יהיה הקו שלא יונח חותך—ממשש העגול. אמנם בצורה הראשונה, לפי שהכאת ד״ב בד״ג שוה להכאת ד״א בעצמו. והכאת ה״ג בעצמו שוה לא״ה בעצמו. וכל הכאות זה כהכאות אותן. אמנם הכאת ב״ד בד״ג וה״ג בעצמו, שוה להכאת ד״ה בעצמו. וד״ה בעצמו שוה לד״א בעצמו וה״א בעצמו. ואם כן זוית א׳ היא נצבה. וכמו כן בדרך זה יתבאר התמונה האחרת. וזה מה שרצינו.

مِثالُهُ: أَنَّا نُعَلِّمُ خارِجًا مِنْ دائِرَةِ ا ب ج نُقْطَةً عَلَيْها د. وَنُخْرِجُ مِنْها خَطَّيْنِ إِلى دائِرَةِ ا ب ج، أَحَدُهُما يَقْطَعُها، وَهُوَ د ب، وَالآخَرُ يَنْتَهي إِلَيْها، وَهُوَ ا د. وَنَجْعَلُ المَرْكَزَ أَوَّلًا عَلى خَطِّ ب د عَلى نُقْطَةِ ه مِنْهُ. وَالَّذي يَكونُ مِنْ ضَرْبِ ب د في د ج مِثْلَ الَّذي يَكونُ مِنْ ضَرْبِ ا د في نَفْسِهِ. فَأَقولُ إِنَّ ا د يَماسُّ الدَّائِرَةَ.

שרטוט—על אף ההתייחסות ל״צורה הראשונה״ אין מופיע שרטוט נוסף; **B**: בשרטוט השני ז ← ב וחסרה אות על חיתוך המשך קו ד ב במעגל [1]הכלה] נראה שחסרה כאן המילה ״אליה״ או מילים אחרות במשמעות שווה

[1] مركز] المركز

فَلْتكُنِ الدَّائِرَةُ ا ب ج. وَلْتُعلَّمْ خارِجًا عَنْها نُقْطَةُ ال د. ويَخْرُجُ مِنْها إلى^ دائِرَةِ ا ب ج خَطّا د ا و د ب المُسْتقيمان. وَلْيَكُنْ د ب قاطِعًا لَها. و د ا مُنتهِيًا لَها. وَلْيكُنِ السَّطْحُ القائِمُ الزَّوايا الَّذي يُحيطُ بِهِ خَطّا ب د د ج، مُساوٍ لِلْمُرَبَّعِ الكائِنِ مِنْ د ا. فَأقولُ إنَّ خطّ د ا مُماسٌ لِدائِرَةِ ا ب ج.

ותהיה העגולה אב״ג[1]. ונרשמה חוץ ממנה נקודת ד׳. ויצאו ממנה אל מקיף עגולת אב״ג שני קוי א״ד ד״ב הישרים. ויהיה[2] ד״ב חותך אותה, וד״א כלה אליה. ויהיה השטח הנצב הזויות אשר יקיפו בו שני קוי ב״ד ד״ג, שוה למרובע ההוה מן א״ד. הנה אומר כי קו א״ד ממשש לעגולת אב״ג.

فَنُخْرِجُ مِنْ نُقْطَةِ د خَطًّا مُماسًّا لِدائِرَةِ ا ب ج، عَلَيْهِ د ه. ونَجْعَلُ مَرْكَزَ دائِرَةِ ا ب ج نُقْطَةَ ز. وَنَصِلُ خُطوطَ ز ا ز د ز ه. فَلأَنَّ السَّطْحَ القائِمَ الزَّوايا الَّذي يُحيطُ بِهِ خَطّا ب د د ج مُساوٍ لِلْمُرَبَّعِ الكائِنِ مِنْ د ا، والسَّطحُ القائِمُ الزَّوايا الَّذي يُحيطُ بِهِ خَطّا ب د د ج مُساوٍ لِلْمُرَبَّعِ الكائِنِ مِنْ خَطِّ د ه، أَيْضًا[1] يكونُ المُرَبَّعُ الكائِنُ مِنْ خَطِّ د ا مُساوِيًا لِلْمُرَبَّعِ الكائِنِ مِنْ د ه. فَخطُّ د ا مُساوٍ لخطِّ د ه. وَلأَنَّ خطَّ ز ا مُساوٍ لِخَطِّ ز ه—وذَلِكَ لأَنَّهُما خَرَجا مِنْ مَرْكَزِ الدَّائِرَةِ إلى الخَطِّ المُحيطِ—وخطّ ز د مُشْتَرَكًا، يَكونُ كِلا خَطَّيْ ا ز ز د مُساوِيَيْنِ لِكِلَيْ خَطَّيْ ه ز ز د، كُلّ واحد لِنَظيرِه. وَقاعِدَةُ د ا مُساوِيَةٌ لِقاعِدَةِ د ه. تكونُ زاوِيَةُ ا ز د مُساوِيَةً لِزاوِيَةِ ه ز د. ومُثَلَّثُ ا ز د مُساوٍ لِمُثَلَّثِ ز د ه. وَسائِرُ الزَّوايا مُساوِيَةٌ لِسائِرِ الزَّوايا.

ונוציא מנקודת ד׳ קו ממשש לעגולת אב״ג, והוא ד״ה. ונשים מרכז עגולת אב״ג נקודת ז׳. ונגיע קוי א״ז ז״ד ז״ה. הנה מפני כי השטח הנצב הזויות אשר יקיפו בו שני קוי ב״ד ד״ג שוה למרובע ההוה מקו א״ד, והשטח הנצב הזויות אשר יקיפו בו שני קוי ב״ד ד״ג שוה למרובע ההוה[3] מן ד״ה, גם כן יהיה המרובע ההוה[4] מן קו א״ד שוה למרובע ההוה מן ד״ה[5]. אם כן[6] קו א״ד שוה לקו ד״ה. ומפני[7] כי קו א״ז שוה לקו ז״ה—וזה כי הם יצאו מן המרכז אל הקו המקיף—וקו ד״ז משותף, יהיו כל שני קוי א״ז ז״ד שוים לכל שני קוי ה״ז ז״ד, כל אחד לנכחי לו. ותושבת א״ד שוה לתושבת ד״ה. תהיה[8] זוית אז״ד[9] שוה לזוית הז״ד. ומשולש אז״ד[10] שוה למשולש זד״ה[11]. ושאר הזויות שוות לשאר הזויות[12].

---

[1]ايضا] MPR1R2

---

[1]אב״ג] +~~שני הקוים הישרים~~ ב [2]ויהיה] ותהיה ס [3]מקו ... ההוה] ⟨⟨*⟩⟩ ב ליתא ס [4]ההוה] ליתא פ [5]ד״ה] ~~קו א״ד שוה למרובע ההוה מן ד״ה~~ ⟨⟨*⟩⟩ ב [6]כן] +~~יהיה המרובע ההוה~~ ב [7]ומפני] ⟨⟨ו⟩⟩מפני ב [8]תהיה] ותהיה ב [9]אז״ד] +~~אה״ד~~ ק [10]אז״ד] זד״ה ב [11]זד״ה] זד״א ב [12]הזויות] +נ״א כל אחד נכחי לו אשר יהיו מיתריהם הצלעות השוות אם כן זוית זא״ד שוה לזוית זה״ד וכו׳ ס

אמנם בנסחת תאבת נמצא המופת על דרך אחרת. והוא שהוציא [!] מנקדת ד׳ לצד האחר קו ממשש העגול, וממקום המשוש הוציא קו בתוך העיגול אל המרכז. והוציא קו אחר מן המרכז אל מקום שכלה הקו שלא יחתך העגול. והוציא קו אחד משותף מן המרכז אל הנקדה החצונה. והראה שהקו היוצא מן המרכז אל תכלית הקו שלא יחתוך העגול הוא עושה עמו זוית נצבת, במופת מהצורה השמינית מן השער הראשון. והצורה [זה] קרובה, וזאת צורת העניין הזה.

נשלם המאמר השלישי מספר אקלידס בכח השם ועזרו. ולו לבדו התהלה.

بُرْهانُهُ: أَنَّ مَرْكَزَ الدَّائِرَةِ ه. فَنَصِلُ ه ا. خَطُّ ب ج قَدْ قُسِمَ بِنِصْفَيْنِ عَلَى ه. وَزِيدَ في طولِهِ خَطُّ ج د. وَضَرْبُ ب د في د ج ه ج في نَفْسِهِ جَميعًا، مِثْل ضَرْبِ ه د في نَفْسِهِ. ه ج مِثْل ه ا. وَضَرْبُ ب د في د ج مِثْل ضَرْبِ ا د في نَفْسِهِ. فَضَرْبُ ب د في د ج ه ج في نَفْسِهِ جَميعًا، مِثْل ضَرْبِ ا د ا ه، كُلّ واحِدٍ في نَفْسِهِ جَميعًا. وَلٰكِنَّ ضَرْبَ ب د في د ج ه ج في نَفْسِهِ جَميعًا، مِثْل ه د في نَفْسِهِ. فَضَرْبُ ا د في نَفْسِهِ ا ه في نَفْسِهِ جَميعًا، مِثْل ه د في نَفْسِهِ. فَزاوِيَةُ ه ا د قائِمَةٌ. وَإِذا أُخْرِجَ مِنْ طَرَفِ قُطْرِ الدَّائِرَةِ خَطٌّ عَلَى زاوِيَةٍ قائِمَةٍ، فَالخَطُّ المُخْرَجُ يَماسُّ الدَّائِرَةَ. فَخَطُّ ا د مُماسٌّ لِدائِرَةِ ا ب ج الأُولى.

وَأَيضًا خارِجًا مِنْ دائِرَةِ ا ب ج الثَّانِيَةِ. نُقْطَةُ (د)[1]. وَنُخْرِجُ مِنْها خَطَّيْنِ إِلى الدَّائِرَةِ، أَحَدُهُما يَقْطَعُها، وَهُوَ د ب، وَالآخَرُ يَنْتَهِي إِلَيْها، وَهُوَ د ا. وَلا نَجْعَلُ المَرْكَزَ عَلى خَطِّ ب ج. وَلٰكِنْ نَجْعَلُهُ فَوْقَهُ، عَلى ه. فَالَّذي يَكونُ مِنْ ضَرْبِ ب د في د ج مِثْل الَّذي يَكونُ مِنْ ضَرْبِ ا د في نَفْسِهِ. فَأَقولُ إِنَّ ا د يَماسُّ الدَّائِرَةَ.

1 د] ج

فَزاوِيَتا زا د ا د ز مُساوِيَتانِ لزاوِيَتَيْ ز ه د ه د ز، كُلّ واحِدٍ لِنَظيرَتِها الَّتي توتِرُها الأَضْلاعُ المُتَساوِيَةُ. فزاوية زا د مُساوِيَة لِزاوِيَة ز ه د. وزاوية ز ه د قائِمَة[1]. فزاوية زا د قائِمَة. وخطّ ا ز إذا أُخْرِجَ فَهُوَ قُطْرٌ. وَقَدْ أُخرِجَ مِنْ طَرَفِ خَطِّ د ا على زاوِيَةٍ قائِمَةٍ. فخطّ د ا مُماسّ لِدائِرَة ا ب ج.

وذلِكَ ما أَرَدْنا أَنْ نُبَيِّنَ.

تَمَّتِ المَقالَةُ الثّالِثَةُ مِنْ كتاب أُقليدِس.

אם כן שתי זויות זא״ד אד״ז שוות לשתי זויות זה״ד הד״ז, כל אחד לנכחי לו אשר יהיו[1] מיתריהם הצלעות השוות. אם כן זוית זא״ד שוה לזוית[2] זה״ד. וזוית זה״ד נצבת. הנה[3] זוית זא״ד נצבת[4]. וקו א״ז כאשר הוצא הנה הוא קוטר. וכבר הוצא מקצה קו א״ד[5] על זוית נצבת[6]. הנה קו א״ד ממשש לעגולת אב״ג.

וזה מה שרצינו לבאר.

נשלם המאמר השלישי מספר אקלידס החכם[7] [8].

[1]قايمة] MR1R2 +لانه قد خرج من طرف القطر خط مماس للدايرة

[1]יהיו] יהיה ק [2]שוה לזוית] ⟨שוה⟩ לזוית ט [3]הנה] אם כן ז [4]הנה ... נצבת] ליתא ב [5]קו א״ד] ליתא ט [6]נצבת] נצבה ק [7]נשלם ... החכם] ליתא ט [8]החכם] +תהלה לנותן ליעף כח ועצמה בפ +תל״ח ז +ת״ל ס ליתא ק

بُرْهانُهُ: أَنَّ مَرْكَزَ الدَّائِرَةِ ه. وَنُخْرِجُ مِنْ ه عَمُودًا إِلى خَطِّ ب ج، وهُوَ ه ز. وَنَصِلُ ه ا ه د ه ج. فَخَطُّ ب ج قَدْ قُسِمَ بِنِصْفَيْنِ عَلى ز، وَزِيدَ في طولِهِ خَطُّ ج د. فَالَّذي يكونُ مِنْ ضَرْبِ ب د في د ج ج ز في نَفْسِهِ، هُوَ مِثْل الَّذي يكونُ مِنْ ضَرْبِ ز د في نَفْسِه. وَنَجْعَلُ الَّذي يكونُ مِنْ ضَرْبِ ز ه العَمودِ في نَفْسِهِ مُشْتَرَكًا. فالَّذي يكونُ مِنْ ضَرْبِ ب د في د ج ز ج في نَفْسِهِ و ز ه في نَفْسِهِ، مِثْل الَّذي يكونُ مِنْ ضَرْبِ ز د في نَفْسِهِ ز ه في نَفْسِهِ جَميعًا. والَّذي يكونُ مِنْ ضَرْبِ ز ج في نَفْسِهِ ز ه في نَفْسِهِ جَميعًا[1]، مِثْل الَّذي يكونُ مِنْ ضَرْبِ ج ه في نَفْسِهِ، لِأَنَّ زاوِيَةَ ه ز ج قائِمَةٌ. فَصارَ الَّذي يكونُ مِنْ ضَرْبِ ز د في د ج ه ج في نَفْسِهِ جَميعًا، مِثْل الَّذي يكونُ مِنْ ضَرْبِ ه د في نَفْسِهِ. ه ج مِثْل ا ه. فَالَّذي يكونُ مِنْ ضَرْبِ ا ه في نَفْسِهِ ا د في نَفْسِهِ جَميعًا، مِثْل الَّذي يكونُ مِنْ ضَرْبِ ب د في د ج د ه ج في نَفْسِهِ جَميعًا[2].

[1] جميعا] +ولكن الذي يكون من ضرب ز د في نفسه و ز ه في نفسه جميعا, مثل الذي يكون من ضرب ز د في نفسه [2] جميعا] 〈*〉

فَالَّذي يكونُ مِنْ ضَرْبِ ب د في ج د ه ج في نَفْسِهِ مِثْل الَّذي يكونُ مِنْ ضَرْبِ ه د في نَفْسه. فالَّذي يكونُ مِنْ ضَرْبِ ا ه في نَفْسِهِ ا د في نَفْسهِ مِثْل الَّذي يكونُ مِنْ ضَرْبِ ه د في نَفْسه. فَالزَّاوِيَةُ الَّتي مِنْ ه ا د قائِمَةٌ إِذًا. وإذا أُخْرِجَ مِنْ طرَ[ف] [قُطْرِ] الدَّائِرَةِ خَطٌّ عَلى زاوِيَةٍ قائِمَةٍ، فَالخَطُّ المُخْرَجُ مُمَاسٌّ لِلدَّائِرَةِ. فَخَطُّ ا د مُمَاسٌّ لِدائِرَةِ ا ب ج الثَّانِيَةِ.

وَأَيضًا خارِجًا مِنْ دائِرَةِ ا ب ج الثَّالِثَةِ، نُقْطَةُ د. وَنُخْرِجُ مِنْها خَطَّيْنِ إِلى الدَّائِرَةِ، أَحَدُهُما يَقْطَعُها، وَهُوَ د ب، وَالآخَرُ يَنْتَهي إِلَيْها، وَهُوَ ا د. والَّذي يَكونُ مِنْ ضَرْبِ ب د في د ج مِثْل الَّذي يَكونُ مِنْ ضَرْبِ ا د في نَفْسهِ. فَأَقولُ إِنَّ ا د مُمَاسٌّ لِدائِرَةِ ا ب ج.

وَنَجْعَلُ مَرْكَزَ الدَّائِرَةِ تَحْتَ ب ج د عَلى نُقْطَةِ ه.

بُرْهانُهُ: أَنَّا نُخْرِجُ مِنْ ه عَمُودًا عَلى خَطِّ ب ج، وَهُوَ ه ز. وَنَصِلُ ه ا ه د ه ج. فَخَطُّ ب ج قَدْ قُسِمَ بِنِصْفَيْنِ عَلى ز، وَزِيدَ في طُولِهِ خَطُّ ج د. فَالَّذي يَكونُ مِنْ ضَرْبِ ب د في ج د و ج ز في نَفْسِهِ جَميعًا، مِثْل الَّذي يَكونُ مِنْ ضَرْبِ ز د في نَفْسِهِ. وَنَجْعَلُ الَّذي يَكونُ مِنْ ضَرْبِ ه ز في نَفْسِهِ مُشْتَرَكًا. فَالَّذي يَكونُ مِنْ ضَرْبِ ب د في د ج ج ز في نَفْسِهِ و ز ه في نَفْسِهِ جَميعًا، مِثْل الَّذي يَكونُ مِنْ ضَرْبِ ز د في نَفْسِهِ ز ه

في نَفْسِهِ جَميعًا. وَلَكِنَّ الَّذي يَكونُ مِنْ ضَرْبِ ز ه في نَفْسِهِ ز ج في نَفْسِهِ جَميعًا، هو مِثْل الَّذي يَكونُ مِنْ ضَرْبِ د ج في نَفْسِهِ. والَّذي يَكونُ مِنْ ضَرْبِ ز ه في نَفْسِهِ ز د في نَفْسِهِ جَميعًا، مِثْل الَّذي يَكونُ مِنْ ضَرْبِ ه ج[1] في نَفْسِهِ، لأَنَّ زاوِيَةَ ه ز ج قائِمَةٌ. فَصَارَ الَّذي يَكونُ مِنْ ضَرْبِ ب د في د ج ه ج في نَفْسِهِ جَميعًا، مِثْل الَّذي يَكونُ مِنْ ضَرْبِ ه د في نَفْسِهِ. ا ه مِثْل ه ج. فالَّذي يَكونُ مِنْ ضَرْبِ ا ه في نَفْسِهِ و ا د في نَفْسِهِ جَميعًا، مِثْل الَّذي يَكونُ مِنْ ضَرْبِ ب د في ج د ه ج في نَفْسِهِ جَميعًا. والَّذي يَكونُ مِن ضَرْبِ ب د في ج د ه ج في نَفْسِهِ، مِثْل الَّذي يَكونُ مِنْ ضَرْبِ ه د في نَفْسِهِ. فالَّذي يَكونُ مِن ضَرْبِ ا د في نَفْسِهِ ا ه في نَفْسِهِ، مِثْل ه د في نَفْسِهِ. فَالزّاوِيَةُ الَّتي مِنْ ه ا د إِذًا قائِمَةٌ. وإذا خَرَجَ مِنْ طَرَفِ قُطْرِ الدَّائِرَةِ خَطٌّ عَلى زاوِيَةٍ قائِمَةٍ، فَالخَطُّ المُخْرَجُ مُماسٌّ للدَّائِرَةِ.

وَذَلِكَ ما أَرَدْنا بَيانَهُ.

---

[1] ه ج] د ه

[Book IV]

המאמר הרביעי מאקלידס

[1] יאמר שהתמונה מצויירת בתוך תמונה כשתהיה כל זוית מזויותיה ממששת לכל צלע מצלעות התמונה שהיא מצויירת בתוכה.

[2] ויאמר שהתמונה מצויירת סביב תמונה שכשיהיה כל אחד מצלעיה ממשש לכל זוית מזויות התמונה שהיא מצויירת סביבה.

المَقالةُ الرَّابِعَةُ مِنْ كِتابِ الأَرْكان

[1] إِذا كانَ شَكْلٌ في شَكْلٍ، وَكانَتْ زَوايا الشَّكْلِ الدَّاخِلِ يَماسُّها أَضْلاعُ الشَّكْلِ الخَارِجِ، فَإِنَّ الخَارِجَ يُقالُ لَهُ المُحيطُ بِالدَّاخِلِ.

المَقالةُ الرَّابِعَةُ مِنْ كِتابِ أُقْليدِس

[1] يُقالُ إِنَّ الشَّكلَ مَرْسومٌ فِي الشَّكْلِ إِذا كانَتْ كُلُّ واحِدَةٍ مِنْ زَواياهُ مُماسَّةً لِكُلِّ واحِدٍ مِنْ أَضْلاعِ الشَّكْلِ الَّذي هُوَ مَرْسومٌ فيهِ.

[2] وَيُقالُ إِنَّ الشَّكلَ مَرْسومٌ حَوْلَ الشَّكْلِ إِذا كانَ كُلُّ واحِدٍ مِنْ أَضْلاعِهِ مُماسًّا لِكُلِّ واحِدَةٍ مِنْ زَوايا الشَّكْلِ الَّذي هُوَ مَرْسومٌ حَوْلَهُ.

המאמר הרביעי מספר אקלידס[1]

[1] יאמר כי התמונה מורשמת בתמונה כאשר תהיה כל אחת מזויותיה ממששת[2] לכל אחת מצלעות התמונה אשר היא נרשמת בה.

[2] ויאמר כי התמונה נרשמת סביב התמונה כשתהיה כל אחת מצלעותיה ממששת[3] לכל אחת[4] מזויות[5] התמונה[6] אשר היא נרשמת סביבה.

---

[1] מספר אקלידס] +החכם **בפ** ליתא **טס** [2] ממששת] ממשת **ב** [3] מצלעותיה ממששת] מצלע תיה ממשש **ב** [4] אחת] ⟨⟨*⟩⟩ **ב** [5] מזויות] מזוי⟨ו⟩ת **ט** [6] התמונה] ליתא **פ**

[Proposition IV.1]

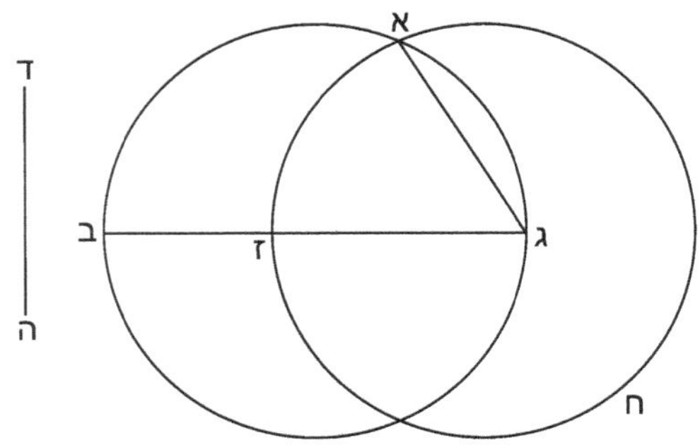

נרצה לצייר בעגול ידוע מיתר[1] שוה לקו ישר ידוע שאיננו גדול מקוטר העגול הידוע.

نُرِيدُ أَنْ نُبَيِّنَ[1] كَيْفَ نَخُطُّ فِي دائِرَةٍ مَعْلومَةٍ خَطًّا مُساوِيًا لِخَطٍّ مَعْلومٍ لَيْسَ بِأَطْوَلَ مِنْ قُطْرِ الدّائِرَةِ يَقْطَعُ قَوْسًا مِنْها.

ונשים העגול הידוע עגול[2] אב״ג. והקו הידוע קו ד״ה. וכאשר רצינו שנצייר בעגול אב״ג מיתר שוה לד״ה ובלתי גדול מקטרו, נוציא קוטר העגול והוא ב״ג. ואם היה קו ד״ה שוה לקו[3] ב״ג, כבר יצא מה שרצינו. ואם היה קצר ממנו, יהיה ד״ה שוה לז״ג. ונקיף על מרכז ג׳ ובמרחק ז׳ עגול ז״ח. ונדביק א׳ בג׳. ואומר שכבר עשינו מיתר בעגול אב״ג שוה לד״ה, והוא א״ג.

مِثالُهُ: أَنّا نَجْعَلُ الدّائِرَةَ المَعْلُومَةَ دائِرَةَ ا ب ج. وَالخَطَّ المَعْلومَ خَطَّ د ه. وَنُرِيدُ أَنْ نَخُطَّ فِي دائِرَةِ ا ب ج خَطًّا مِثْلَ د ه، يَقْطَعُ قَوْسًا مِنَ الدّائِرَةِ، لَيْسَ بِأَطْوَلَ مِنْ قُطْرِها. فَنُخْرِجُ قُطْرَ الدّائِرَةِ، وَهُوَ ب ج. فَإِنْ كانَ خَطُّ د ه مِثْلَ خَطِّ ب ج، فَقَدْ تَبَيَّنَ ما أَرَدْنا. وَإِنْ كانَ أَقْصَرَ مِنْهُ، فَلْيَكُنْ د ه مِثْلَ[2] ز ج، فَعَلَى مَرْكَزِ ج وَبِبُعْدِ ج ز نَخُطُّ دائِرَةَ ا ز ح. وَنَصِلُ ا ج. فَأَقُولُ إِنّا قَدْ خَطَطْنا دائِرَةَ ا ب ج مِثْلَ خَطِّ د ه يَقْطَعُ قَوْسًا مِنَ الدّائِرَةِ لَيْسَ بِأَطْوَلَ مِنَ القُطْرِ، وَهُوَ ا ج.

שרטוט—B: ح היא נקודת החיתוך של העיגולים שאינה ו [1]מיתר] +~~זוית~~ [2]עגול] ~~קו~~+ [3]לקו] ~~ל[ד]״ה~~+

[1]أن نبين] +* [2]مثل] مثل د ه+

نُريدُ أنْ نَخُطَّ في دائِرَةٍ مَعْلومَةٍ وَتَرًا مُساوِيًا لِخطٍّ مُستقيمٍ مَعْلومٍ لَيْسَ بِأَعْظَمَ مِنْ قُطْرِ الدَّائِرَةِ.

נרצה שנקוה בעגולה ידועה מיתר שוה לקו ישר ידוע אינו יותר גדול מקוטר העגולה.

فَنَجْعَلَ الدَّائِرَةَ ا ب ج. والخَطَّ المُسْتَقيمَ المَعْلومَ الَّذي ليس بِأَعْظَمَ مِنْ قُطرِ الدَّائِرَةِ—خطّ د ه. ونُريدُ أَنْ نخُطَّ في دائِرَةِ ا ب ج وَتَرًا مُساوِيًا لِخطّ د ه.

הנה נשים העגולה הידועה[1] עגולת אב״ג. והקו הישר הידוע[2] אשר לא יהיה יותר גדול מקוטר העגולה—קו ד״ה. ונרצה שנקוה בעגולת[3] אב״ג מיתר שוה לקו ד״ה.

שרטוט—**ט**: ד״ה בימין השרטוט; **ס**: ג׳ על היקף מעגל אז״ח [1]הידועה] ⟨*⟩ **ו** [2]הידוע] הידועה **ב** [3]בעגולת] ⟨⟨ב⟩⟩עגולת **ט**

بُرْهانُهُ: أَنَّ ج مَرْكَزُ دائِرَةِ ا ز ح. ز ج مِثْل ج ا. ز ج مِثْل د ه. د ه مِثْل ا ج. فَقَدْ وَقَعَ في دائِرَةِ ا ب ج المَعْلُومَةِ خَطٌّ مِثْل خَطِّ د ه المَعْلُومِ يَقْطَعُ قَوْسًا مِنَ الدَّائِرَةِ لَيْسَ بِأَطْوَلَ مِنَ القُطْرِ، وَهُوَ ا ج.

מופת זה: לפי שג׳ מרכז עגול ז״ח. אם כן[1] ז״ג שוה לג״א. וז״ג שוה לד״ה. ואם כן (ג״א)[2] שוה לד״ה. וכבר נפל בעגול אב״ג הידוע מיתר שוה לד״ה, בלתי גדול מקטרו, אבל הוא קטן ממנו, והוא א״ג.

وَذَلِكَ ما أَرَدْنا بَيانَهُ.

וזה מה שרצינו.

[1]אם כן] ⟨⟨*⟩⟩ [2]ג״א] ג״ח

فَنُخْرِجَ قُطرَ الدَّائِرَةِ، وهو ب ج. فَإِنْ كانَ د ه مثل ب ج، فَقَدْ كانَ ما أَرَدْنا. وإِنْ كانَ أقصر[1]، فَلْيَكُنْ ز ج مثل د ه. ونَجْعَلُ ج مَرْكَزًا. ونَخُطّ بِبُعْدِ ج ز دائِرَة ا ز ح. ونُخرِج خطّ ا ج. فَمَرْكَزُ دائِرَة ا ز ح نُقْطَةَ ج. فخطّ ج ز مثل خطّ ج ا. وج ز^ مثل د ه. ف ا ج مثل د ه. فقد خَطَطْنا في دائِرَةِ ا ب ج وَتَرًا مثل خطّ د ه الَّذي ليس بِأَعْظَم مِنْ قُطْرِها^.

وذَلِكَ ما أَرَدْنا أَنْ نُبَيِّنَ.

הנה נוציא קוטר העגולה, והוא ב״ג. ואם היה ד״ה כמו ב״ג, כבר היה מה שרצינו. ואם היה יותר קצר, יהיה ז״ג כמו ד״ה. ונשים ג׳[1] מרכז, ובמרחק[2] ג״ז עגולת זא״ח[3]. ונוציא קו ג״א. הנה ג״ז כמו ד״ה[4]. אם[5] כן א״ג כמו ד״ה. הנה כבר קוינו בעגולת אב״ג מיתר כמו קו ד״ה שאינו יותר גדול מן הקוטר.

וזה מה שרצינו לבאר[6].

[1]اقصر] MPR1R2 +منه

[1]ג׳] ⟨*⟩ פ [2]ובמרחק] ונקוה על המרכז ג׳+ ב ונקוה ס ⟨ונקוה⟩ ק [3]זא״ח] א״ח בופ [4]ד״ה] ~~ד״א~~+ פ [5]אם] ⟨⟨וא״ג כמו ג״ז מפני כי נקודת ג׳ מרכז עגלת זא״ח אם כן⟩⟩+ ק [6]לבאר] ליתא ס

[Proposition IV.2]

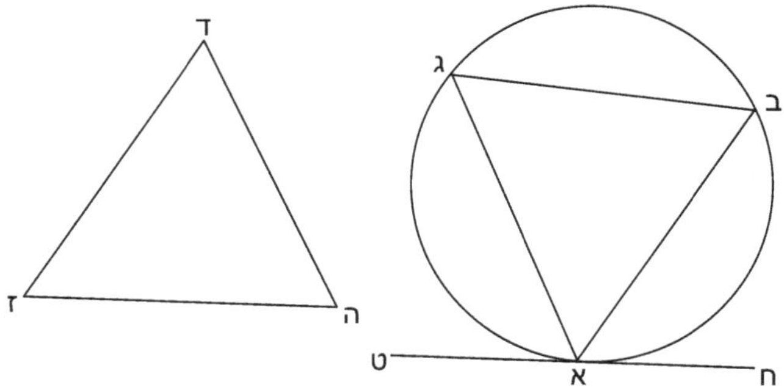

נרצה לעשות בתוך עגול ידוע משלש שזויותיו שוות לזויות משלש ידוע.

ונשים העגול הידוע עגול אב״ג, והמשלש הידוע משלש דה״ז. וכאשר רצינו לעשות בעגול אב״ג משלש שתהיינה זויותיו שוות לזויות משולש דה״ז, נעביר על נקודת א׳ קו ממשש העגול, והוא קו חא״ט. ונעמיד על הקו הממשש, על נקודת א׳, זוית שוה לזוית דה״ז מן המשלש, והיא זוית חא״ב. ועוד נעמיד על נקדת א׳ זוית שוה לזוית דז״ה האחרת מן המשולש, והיא זוית טא״ג. ונדביק ב׳ בג׳. ואומר שכבר עשינו בעגול אב״ג הידוע משלש שזויותיו שוות לזויות משלש דה״ז, והוא משלש אב״ג.

نُرِيدُ أَنْ نُبَيِّنَ كَيْفَ نَعْمَلُ فِي دائِرَةٍ مَعْلومَةٍ مُثَلَّثًا تَكونُ زواياهُ مُتَساوِيَةً لِزَوايا مُثَلَّثٍ مَعْلومٍ.

مِثالُه: أَنَّا نَجْعَلُ الدَّائِرَةَ المَعْلومَةَ دائِرَةَ ا ب ج، والمُثَلَّثَ المَعْلومَ مُثَلَّثَ د ه ز. ونُرِيدُ أَنْ نَعْمَلَ فِي دائِرَةِ ا ب ج مُثَلَّثًا تَكونُ (زواياهُ)[1] مُساوِيَةً لِزَوايا مُثَلَّثٍ مَعْلومٍ د ه ز. فَنُجِيزُ عَلَى نُقْطَةِ ا خَطًّا يَماسُّ الدَّائِرَةَ، وَهوَ خَطُّ ح ا ط. ونُقِيمُ عَلَى الخَطِّ المُماسِّ عَلَى نُقْطَةِ ا زاوِيَةً مِثْلَ زاوِيَةِ المُثَلَّثِ الَّتِي مِنْ د ه ز، وَهِيَ الَّتِي مِنْ ح ا ب. ونُقِيمُ أَيضًا عَلَى خَطِّ ح ا ط عَلَى نُقْطَةِ ا زاوِيَةً مِثْلَ زاوِيَةِ المُثَلَّثِ الأُخْرَى الَّتِي مِنْ (د ه ز)[2]، وَهِيَ الَّتِي مِنْ ط ا ج. وَنَصِلُ ب ج. فَأَقولُ إِنَّا قَد عَمِلْنا فِي دائِرَةِ ا ب ج المَعْلومَةِ مُثَلَّثًا مُساوِيَةً زَواياهُ لِزَوايا مُثَلَّثِ ه د ز، وَهوَ مُثَلَّثُ ا ب ج.

[1] زواياه] زوايا [2] د ه ز] ه د ز

نُرِيدُ أَنْ نَعْمَلَ فِي دائِرَةٍ مَعْلومَةٍ مُثَلَّثًا مُسَاوِيَةً زَاوِياهُ لِزَوايا مُثَلَّثٍ مَعْلومٍ.

נרצה לעשות בעגולה ידועה משולש שוות זוויתיו לזויות[1] משולש ידוע.

فَنَجْعَلَ الدَّائِرَةَ المَعْلومَةَ دائِرَةَ ا ب ج، والمُثَلَّثَ المَعْلومَ مُثَلَّثَ د ه ز. ونُرِيدُ أَنْ نَعْمَلَ فِي دائِرَةِ ا ب ج مُثَلَّثًا مُسَاوِيَةً زَوَاياهُ لِزَوايا مُثَلَّثِ د ه ز.

הנה נשים העגולה[2] הידועה עגולת אב״ג, והמשולש הידוע משולש דה״ז. ונרצה שנעשה בעגולת אב״ג משולש שוות זויותיו לזויות[3] משולש[4] דה״ז.

שרטוט—**ט**: דה״ז מופיע בצד הימני של השרטוט; **פ**: ב״ג וח״ט $ [1]זוויתיו לזויות] זויותיו לזוית **ב** זויות לזויות **ס** [2]העגולה] ⟨⟨*⟩⟩ **ק** [3]לזויות] לזוית **ב** [4]משולש] +ידוע **ק**

מופת זה: לפי שקו חא״ט ממשש עגול אב״ג, וכבר יצא [!] ממקום משושו שני קוים יחתכו העגול, והם א״ב וא״ג. וידוע ששתי הזויות אשר אצל כל קו שיצא ממקום העגול שוות לשתי הזויות אשר תפלנה בשתי חתכות העגול המתחלפות להן. אם כן זוית בא״ח שוה לזוית אג״ב, וזוית גא״ט שוה לזוית אב״ג[1]. ואמנם זוית טא״ג שוה לזוית דז״ה. וזוית חא״ב שוה לזוית דה״ז[2]. אם כן שתי זויות דז״ה ודה״ז שוות לשתי זויות אג״ב ואב״ג. ולפי שכל משלש שלש זויותיו שוות לשתי זויות נצבות, תשאר זוית הד״ז שוה לזוית בא״ג. ואם כן זויותיו[3] שוות לזויות משלש דה״ז הידוע.

וזה מה שרצינו.

بُرْهانُهُ: أَنَّ خَطَّ ح ا ط يَماسُّ دائِرَةَ ا ب ج. وَقَدْ أُخْرِجَ مِنَ المَوْضِعِ الَّذي يَماسُّها فيهِ خَطّانِ يَقْطَعانِ الدَّائِرَةَ، ا ب ا ج. فالزَّاوِيَتانِ اللَّتانِ عَنْ جَنْبَتَيْ كُلِّ خَطٍّ يَخْرُجُ مِنَ المَوْضِعِ الَّذي يَماسّ الخطَّ الدَّائِرَةَ، مِثْل الزَّاوِيَتَيْنِ اللَّتَيْنِ تَقَعانِ في قِطْعَتَي الدَّائِرَةِ المُتَبادَلَتَيْنِ. فَالزَّاوِيَةُ الَّتي مِنْ ب ا ج مِثْل الَّتي مِنْ (ا ج ب)[1]. والَّتي مِنْ ج ا ط مِثْل الَّتي مِنْ ا ب ج. والَّتي مِنْ ط ا ج مِثْل الَّتي مِنْ (د ز ه)[2]. والَّتي مِنْ ح ا ب مِثْل الَّتي مِنْ (د ه ز)[3]. فاللَّتانِ مِنْ (ه ز د)[4] د ه ز مِثْل اللَّتَيْنِ مِنْ (ا ج ب)[5] وَمِنْ ا ب ج. فَكُلّ مُثَلَّثٍ زَواياهُ الثَّلاثُ مِثْلُ قائِمَتَيْنِ فَبَقِيَتِ الَّتي مِنْ (ه د ز)[6] مِثْلُ الَّتي مِنْ ب ا ج. فَزَاويا [!] مُثَلَّث د ه ز مُساوِيَةٌ لِزَوايا مُثَلَّثِ ا ب ج المَعْمُولِ في دائِرَةِ ا ب ج.

وَذَلِكَ ما أَرَدْنا بَيانَهُ.

[1]שוה לזוית אג״ב ... אב״ג] ⟨⟨*⟩⟩ שוה לזוית א[ג״ב] [2]דה״ז] ~~בא״ב~~+ [3]זויותיו] לפני מילה זו שתיים או שלוש מילים שאי אפשר לקראן

[1]ا ج ب] ج ا ب [2]د ز ه] د ه ز [3]د ه ز] ه د ز [4]ه ز د] ه د ز [5]ا ج ب] ج ا ب [6]ه د ز] ه ز د

فَنُجيزَ على نُقْطَةِ ا خطّ ط ا ح مُماسًّا للدائِرَةِ. ونُقيمَ على نُقْطَةِ ا مِنْ خطّ ا ح زاوية ب ا ح مثل زاوية د ه ز. ونُقيمَ أيضًا على نُقْطَة ا مِنْ خطّ ا ط زاوية ط ا ج مثل زاوية د ز ه.^ ونُخْرِجَ ب ج. فخطّ ط ا ح يَماسُّ دائِرَةَ ا ب ج. وَقَدْ خَرَجَ مِنْ حيثُ يماسُّها خَطّا ا ب ا ج يَقْطَعانِ الدَّائِرَةَ. فَعَنْ جَنْبَتَيْ كُلِّ واحدٍ مِنْهُما زاوِيَتانِ مثل اللَّتَيْنِ تَقَعانِ في قِطْعَتَيِ الدَّائِرَةِ المُبادِلَتَيْنِ لَهُما^. فزاوية ب ا ح مثل زاوية ب ج ا. وزاوية ج ا ط مثل زاوية ا ب ج. وزاوية ج ا ط مثل زاوية د ز ه. وزاوية ب ا ح مثل زاوية د ه ز. فَزاوِيَتا د ه ز د ز ه مثل زاوِيَتَيّ ا ب ج ا ج ب. وبقِيَتْ زاويةُ ه د ز مثل زاويةِ ب ا ج. فَزَوايا مُثَلَّث د ه ز مُساوِيَةَ لزوايا مُثَلَّثِ ا ب ج المَعْمُول في دائِرَةِ ا ب ج.

وذَلِكَ ما أَرَدْنا أَنْ نُبَيِّنَ.

הנה נעביר על נקודת א׳ קו טא״ח ממשש לעגולה. ונעמיד על נקודת א׳ מקו א״ח זוית בא״ח כמו זוית דה״ז. ונעמיד גם כן על א״ט[1] על נקודת א׳ ממנו כמו זוית דז״ה, והיא זוית טא״ג. ונוציא ב״ג. הנה קו טא״ח ממשש לעגולת אב״ג[2]. וכבר יצא ממקום ימששה[3] קוי[4] א״ב א״ג יחתכו העגולה. הנה משני צדדי כל אחת מהן שתי זויות כמו אשר יפלו בשתי[5] חתיכות העגולה המומרות. אם כן זוית בא״ח כמו זוית בג״א. וזוית גא״ט כמו זוית אב״ג. וזוית גא״ט כמו זוית דז״ה. וזוית בא״ח כמו זוית דה״ז. אם כן שתי זויות דה״ז דז״ה כמו[6] שתי זויות אב״ג אג״ב. ונשארה זוית הד״ז כמו זוית בא״ג. אם כן זויות משולש דה״ז שוות לזויות משולש אב״ג העשוי בעגולת אב״ג.

וזה מה שרצינו לבאר.

[1]גם ... א״ט] ליתא **ב** [2]אב״ג] ~~אד״ח~~+ **ס** [3]ימששה] ימשש **ב** מישושו **ו** ימששו **ספ** מששו **ק** [4]קוי] קו **בוטפ** [5]בשתי] משתי **ב** בשני **ס** [6]כמו] ~~כמו שתי זויות דה״ז דז״ה~~+ **ק**

[Proposition IV.3]

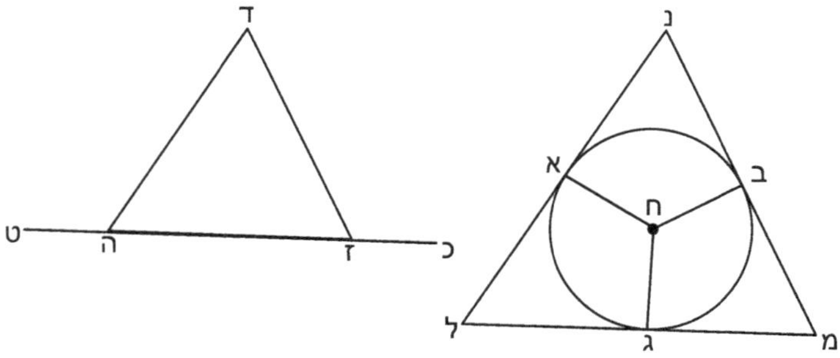

נרצה לעשות על עגול ידוע משולש שיקיף בו, וזויותיו שוות לזויות משולש ידוע.

نُرِيدُ أَنْ نُبَيِّنَ كَيْفَ نَعْمَلُ عَلى دائِرَةٍ مَعْلومَةٍ مُثَلَّثًا تُحيطُ بِهِ، تكونُ زَواياهُ (الثَّلاثُ)[1] مُساوِيَةً لِزَوايا مُثَلَّثِ الدَّائِرَةِ المَعْلومَةِ.

שרטוט—B: ב ↔ ג

[1] الثلاث] الثلث

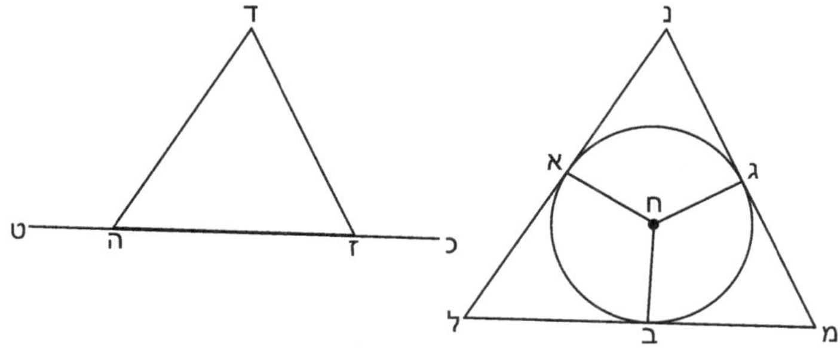

نُرِيدُ أَنْ نَعْمَلَ على دائِرَةٍ مَعْلومةٍ مُثَلَّثًا يُحِيطُ بِها، مُسَاوِيَةً زَوايَاهُ لِزَوَايا مُثَلَّثٍ مَعْلومٍ.

נרצה לעשות על עגולה ידועה[1] משולש יקיף בה, תהיינה זויותיו שוות לזויות משולש ידוע.

שרטוט—**פ**: $; **ק**: למ״נ $ [1]עגולה ידועה] עגולה הידועה **ב** העגולה הידועה **ו** עגולה הידועה **ס**

ונשים העגול הידוע עגול אב״ג, והמשולש הידוע משולש דה״ז. וכאשר רצינו לעשות על עגול אב״ג משולש מקיף בו, וזויותיו שוות למשולש דה״ז, נוציא קו ה״ז[1] לשני הצדדים אל ט׳ ואל כ׳. ויהיה מרכז העגול ח׳. ונוציא מח׳ קו אל הקו המקיף כמו שנפל, והוא ח״ב. ונעמיד על קו ח״ב, על נקודת ח׳, זוית שוה לזוית אחת חצונה ממשולש דה״ז, והיא זוית דה״ט. והזוית שנעמיד שוה לה תהיה זוית בח״ג. ונעמיד בצד האחר זוית בח״א שוה לזוית החיצונה האחרת מהמשלש, והיא זוית דז״כ. ונעביר על נקדת [!] א׳ וב׳ וג׳ קוים ממששים עגול [!] אב״ג, והם קוי ל״מ, מ״נ, נ״ל. ואומר שכבר עשינו על עגול אב״ג משלש שהוא מקיף בו, וזויותיו שוות[2] לזויות משלש דה״ז, והוא משולש למ״נ.

[1] ה״ז] ⟨⟨*⟩⟩ [2] שוות] שו⟨ו⟩ת

[..] دائرةَ ا ب ج، والمُثَلَّثُ المَعْلومُ مُثَلَّثُ د ه ز. وَنُريدُ أَنْ نَعْمَلَ عَلى دائِرَةِ ا ب ج مُثَلَّثًا يُحيطُ بِها، تَكونُ زَواياهُ مُساوِيَةً لِزَوايا مُثَلَّثِ ه د ز. فَنُخْرِجُ خَطَّ ه ز في كِلْتَي الجِهَتَيْنِ إِلى ط (وك)[1]. ومَرْكَزُ الدّائِرَةِ ح. وَنُخْرِجُ مِنْ ح خَطًّا إِلى الخَطِّ المُحيطِ كَيْفَما وَقَعَ، وَهُوَ خَطُّ ح ب. وَنُقيمُ عَلى خَطِّ ح ب عَلى نُقْطَةِ ح زاوِيَةً مِثْلَ زاوِيَةِ المُثَلَّثِ الخارِجَةِ الَّتي مِنْ د ه ط، وَهِيَ الَّتي مِنْ ب ح ج. وَنُقيمُ أَيْضًا عَلى نُقْطَةِ ح مِثْلَ زاوِيَةِ المُثَلَّثِ الخارِجَةِ الأُخْرى الَّتي مِنْ د ز ك، وَهِيَ الَّتي مِنْ ب ح ا. وَنُجيزُ عَلى نُقْطَةِ ا و ب و ج خُطوطًا مُماسَّةً لِدائِرَةِ ا ب ج، وَهِيَ خَطُّ ل م وَخَطُّ ن ل[2]. فَأَقولُ إِنّا قَدْ عَمِلْنا عَلى دائِرَةِ ا ب ج مُثَلَّثًا يُحيطُ بِها، مُساوِيَةً زَواياهُ لِزَوايا مُثَلَّثِ ه د ز، وَهُوَ مُثَلَّثُ ل م ن.

[1] وك] زك [2] ن ل] נראה שחסר כאן م ن

فَنَجْعَلَ الدَّائِرَةَ دائِرَة ا ب ج، والمُثَلَّثَ المَعْلومَ مُثَلَّث د ه ز. ونُريدُ أَنْ نَعْمَلَ على دائِرَةِ ا ب ج مُثَلَّثًا يُحيطُ بِها، مُسَاوِيَةً زواياهُ لزوايا مُثَلَّثِ د ه ز. فَنُخْرِجَ ه ز في كِلْتَيْ الجِهَتَيْنِ إلى ط ك. وَلْيَكُنِ المَرْكَزُ ح. ونُخْرِجُ مِنْهُ خطّ ب ح إلى الخطّ المُحيطِ كَيْفَما وَقَعَ. وَنُقيمُ على ح مِنْ خَطِّ ب ح زاويةً مثل زاويةِ د ه ط، وهي زاويةُ ب ح ج، ومثل زاويةِ د ز ك، وهي زاويةُ ب ح ا. ونُجيزُ على نُقَطِ ا ب ج خُطوطَ ل م م ن ن ل مُماسَّةً لدائِرَةِ ا ب ج.

הנה נשים העגולה[1] עגולת אב״ג, והמשולש הידוע משולש דה״ז. ונרצה לעשות על עגולת אב״ג משולש יקיף בה, שוות זויותיו לזויות משולש דה״ז. הנה נוציא ה״ז בכל אחד משני הצדדים אל ט׳ כ׳. ויהיה המרכז ח׳. ונוציא ממנו קו ח״ב אל המקיף איך שיפול. ונעמיד על ח׳[2] מקו ב״ח זוית כמו זוית דה״ט, והיא זוית בח״ג, וכמו זוית דז״כ, והיא זוית בח״א. ונעביר על נקודות ב׳ ג׳ א׳ קוים ל״מ מ״נ נ״ל ממששים[3] לעגולת[4] אב״ג.

---

[1] העגולה] +הידועה **ק** [2] ח׳] 〈〈*〉〉 **ק**

[3] ממששים] ממשים **טפ** [4] לעגולת] *+ **ב**

מופת זה: לפי שעגול אב״ג ימששהו קו נ״מ[1], ויצא ממקום משושו קו אל המרכז, והוא ח״ב, יהיה קו ח״ב עמוד על קו נב״מ[2], וזוית נב״ח[3] נצבה. ויהיה מב״ח נצבה גם כן. ומצד זה שתי הזויות שעל ג׳ הן נצבות, ושתי הזויות שעל א׳ גם כן נצבות. וכל שטח בעל ארבע צלעות—ארבע זויותיו שוות לארבע זויות נצבות. ואם כן זויות שטח אחב״נ שוות לארבע זויות נצבות. אמנם שתי הזויות שאצל א׳ וב׳ הן נצבות—תשארנה שתי הזויות שאצל ח׳ ונ׳, שהן זו לעמת זו, שוות לשתי זויות נצבות. ואמנם זוית דז״כ וזוית דז״ה שוות לשתי זויות נצבות. ואם כן שתי זויות ח׳ ונ׳ שזו לעמת זו, שוות לזויות דז״כ ודז״ה. וזוית דז״כ שוה לזוית אח״ב—תשאר זוית דז״ה שוה לזוית נ׳. ומצד זה זוית דה״ט וזוית דה״ז שוות לזוית ח׳ וזוית מ׳, שזו לעמת זו. וזוית דה״ט היא שוה לזוית ח׳—תשאר זוית מ׳ שוה לזוית דה״ז, לפי שכל משלש שלש זויותיו שוות לשתי זויות נצבות. וכבר התבאר שזוית דה״ז וזוית דז״ה שוות לזוית נ׳ וזוית מ׳—תשאר זוית ד׳ שוה לזוית ל׳. ואם כן משולש דה״ז זויותיו שוות לזויות משולש נל״מ המקיף בעגול אב״ג.

[1]נ״מ] ⟨⟨*⟩⟩ [2]נב״מ] ז׳ מתוקנת לנ׳ [3]נב״ח] ז׳ מתוקנת לנ׳

بُرْهانُهُ: أَنَّ دائِرَةَ ا ب ج يَماسُّها خَطُّ ل م. وَمِنَ المَوْضِعِ الَّذي يَماسُّها فيهِ خَرَجَ خَطٌّ إِلى المَرْكَزِ، وَهُوَ ح ب. (ح ب)[1] عَمودٌ عَلى (ل ب ب م)[2]. فَالزَّاوِيَةُ الَّتي مِنْ (ل ب ح)[3] قائِمَةٌ. والَّتي مِنْ م ب ح قائِمَةٌ أَيضًا. وَمِنْ أَجْلِ ذلِكَ اللَّتانِ عَلى ج قائِمَتانِ، وَاللَّتانِ عَلى ا قائِمَتانِ. وَكُلُّ ذي أَرْبَعَةِ أَضْلاعٍ فَإِنَّ زَواياهُ الأَرْبَعَ مُساوِيَةٌ لأَرْبَعِ زَوايا قائِمَةٍ. فَزَوايا سَطْحِ (ا ح ل ب)[4] الأَرْبَعُ مِثْلُ أَرْبَعِ زَوايا قائِمَةٍ. وَأَمَّا اللَّتانِ عِنْدَ ا ب فَقائِمَتانِ. وَبَقِيَتِ الَّتي عِنْدَ ح و ل المُتَقابِلَتانِ مُساوِيَتَيْنِ لِزاوِيَتَيْنِ قائِمَتَيْنِ. وَلكِنَّ الَّتي مِنْ د ز ك وَالَّتي مِنْ د ز ه مِثْلُ زاوِيَتَيْنِ قائِمَتَيْنِ. فَزاوِيَتا ح و ل المُتَقابِلَتانِ مُساوِيَتانِ الَّتي مِنْ د ز ك وَمِنْ د ز ه. وَالَّتي مِنْ د ز ك مِثْلُ الَّتي مِنْ (ا ح ب)[5]. وبَقِيَتِ الَّتي مِنْ د ز ه مِثْلُ الَّتي عِنْدَ ل. وَمِنْ أَجْلِ[6] ذلِكَ صارَتِ اللَّتانِ مِنْ د ه ط وَمِنْ د ه ز مِثْلَ اللَّتَيْنِ عِنْدَ ح و م. والَّتي مِنْ (د ه ط)[7] مِثْلُ الَّتي عِنْدَ ح. وبَقِيَتِ الَّتي عِنْدَ م مِثْلُ الَّتي من د ه ز. وكُلُّ مُثَلَّثٍ فَإِنَّ زَواياهُ الثَّلاثَ مِثْلُ زاوِيَتَيْنِ قائِمَتَيْنِ. وَاللَّتانِ من د ه ز ومن د ز ه مِثْلُ اللَّتَيْنِ عِنْدَ ل و م. وبَقِيَتِ الَّتي عِنْدَ د مِثْلُ الَّتي عِنْدَ ن. فَمُثَلَّثُ د ه ز مُساوٍ زَواياهُ لِزَوايا مُثَلَّثِ ن ل م المَعْمُولِ عَلى دائِرَةِ ا ب ج المُحيطِ بِها.

[1]ح ب] ج ب [2]ل ب ب م] ا ب م ز [3]ل ب ح] ب ا ح [4]ا ح ل ب] ا ج ل ب [5]ا ح ب] ا ح [6]اجل] ⟨*⟩ [7]د ه ط] ه د ط

فخطّ ل م يماسُّ دائرةَ ا ب ج. وقدْ خرجَ منْ حيْثُ يماسُّها خطُّ ح ب إلى المركزِ، فهو عمودٌ على خطِّ ل ب م. فزاوية ل ب ح قائمَة. وزاوية م ب ح أيضًا قائمَة. وكذلكَ تكونُ[1] الزّاوِيتانِ اللّتيْنِ عِنْدَ ج قائمَتانِ والزّاويتانِ اللّتانِ عِنْدَ ا قائمَتانِ^. وكُلّ شَكْلٍ ذي أرْبعَةِ أضْلاعٍ فإنَّ زَواياهُ الأربع مُعادِلَةَ^ لِأرْبَعِ زوايا قائمَة. فزوايا سَطْحِ ا ح ب ل مثل أربع زوايا قائمَة. ولكِنَّ اللّتيْنِ عِنْدَ ا ب قائمَتانِ. فبَقِيَ اللّتان عِنْدَ ح ل المُتَقابِلَتانِ مثل قائمَتيْنِ. ولكِنَّ زاوِيتَيْ د ز ك د ز ه مثل قائمَتيْنِ. فزاوِيتا ح ل مثل زاوِيتا د ز ك د ز ه. فزاوية د ز ك مثل زاوية ا ح ب. وبقِيَتْ زاويةُ د ز ه مثل زاوِيةِ ل. وكذلكَ أيضًا[2] تكونُ زاوِيتا د ه ط د ه ز مثل زاوِيتَيْ ح م. وزاوية د ه ط مثل زاوية ح. وبقِيَتْ زاوِيةُ م مثل زاوِيةِ د ه ز. وكُلّ مُثلَّثٍ فإنَّ زواياهُ الثّلاث مثل زاوِيتيْنِ قائمَتيْنِ. فزاوِيتا د ه ز د ز ه مثل زاوِيتَيْ ل م. وبقِيَتْ زاويةُ د مثل زاوِيةِ ن. فزوايا مُثلَّثِ د ه ز مُساوِيَة لِزوايا مُثلَّثِ ن ل م المَعْمُول على دائرَةِ ا ب ج المُحِيط بِها.

הנה קו ל״מ[1] ימשש עגולת אב״ג. וכבר הוצא ממקום ימששה[2] קו ח״ב אל המרכז, והוא עמוד על קו לב״מ. אם כן זוית לב״ח נצבת. וזוית מב״ח גם כן נצבת. וכן יהיו שתי הזויות[3] אשר אצל ג׳[4], וכן יהיו השתי זויות אשר אצל א׳[5] [6], כל אחת מהן[7] נצבת. וכל תמונה בעלת ארבע צלעות הנה זויותיה הארבע שוות לארבע זויות נצבות[8]. אם כן זויות[9] שטח אחב״ל כמו ארבע זויות נצבות[10] [11]. אבל[12] אשר אצל א׳ ב׳ נצבות. הנה נשארו אשר אצל ח׳ ל׳ המתנגדות[13] כמו שתי נצבות. אבל שתי זויות דז״כ דז״ה כמו[14] שתי נצבות. אם כן שתי זויות ח׳ ל׳ כמו שתי[15] זויות דז״כ דז״ה[16]. וזוית דז״כ כמו זוית ח׳. ונשארה[17] זוית דז״ה כמו זוית ל׳. וכן גם כן יהיו שתי זויות דה״ט דה״ז כמו שתי זויות ח׳ מ׳. וזוית דה״ט כמו זוית ח׳. ונשארה זוית מ׳ כמו זוית דה״ז. וכל[18] משולש הנה זויותיו[19] השלש כמו שתי נצבות[20]. ושתי זויות דה״ז דז״ה[21] כמו שתי זויות ל׳ מ׳. ונשארה זוית ד׳ כמו זוית נ׳. אם כן זויות[22] משולש דה״ז שוות לזויות נל״מ[23] העשוי על עגולת אב״ג המקיף בו.

[1] تكون] MPR$_1$R$_2$U [2] أيضا] MPR$_1$R$_2$

[1] ל״מ] +~~ל״א~~ **ט** [2] ימששה] משׁשה **פ** [3] הזויות] זויות **בו** [4] ג׳] +נצבת **ס** +נצבות **פק** [5] אצל א׳] ⟨⟨*⟩⟩ **פ** [6] א׳] ~~ג׳~~+ **ט** [7] מהן] מהם **בוטפ** [8] לארבע ... נצבות] לארבע שוות נצבות **ב** ⟨⟨*⟩⟩ **ק** [9] זויות] זוית **בס** זויות[יו] **ט** [10] אם ... נצבות] ליתא **ו** [11] שטח ... נצבות] הארבעה שוות לארבע זויות נצבות ~~אם כן זויות~~ **ט** [12] אבל] +⟨⟨שתי זויות⟩⟩ **ק** [13] המתנגדות] המנגדות **ט** [14] כמו] ליתא **ס** [15] נצבות אבל ... שתי] ליתא **ו** [16] דז״ה] +~~כמו שתי נצבות אם כן שתי זויות~~ **ק** [17] ונשארה] ונשאר **ספ** [18] וכל] +~~שתי זויות~~ **ק** [19] זויותיו] זוית **ס** [20] נצבות] זויות+ **ו** [21] דז״ה] ליתא **ב** [22] זויות] זוית **בס** [23] נל״מ] משולש+ **ט** ⟨⟨משולש⟩⟩+ **פק**

וזה מה שרצינו. وَذَلِكَ ما أَرَدْنا بَيانَهُ.

وذَلِكَ ما أَرَدْنا أَنْ نُبَيِّنَ.

וזה מה שרצינו לבאר.

[Proposition IV.4]

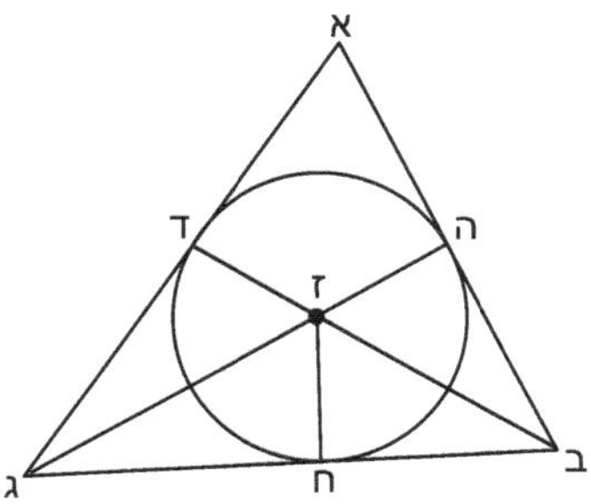

נרצה לעשות בתוך משלש ידוע עגול שיהיה המשולש סביבו.

نُرِيدُ أَنْ نُبَيِّنَ كَيْفَ نَعْمَلُ فِي مُثَلَّثٍ مَعْلُومٍ دائِرَةً يُحِيطُ بِها.

ונשים המשולש הידוע משולש אב״ג. וכאשר רצינו שנעשה בתוכו עגול שיקיף עליו המשולש, נחלק זוית ב׳ לחציים בקו ב״ד[1], וזוית ג׳ לחציים בקו ג״ה. ויפגשו זה את זה על נקדת ז׳. ונוציא מז׳ עמודים אל א״ב וב״ג וג״א, והם ז״ח ז״ד וז״ה. ונקיף עגול על מרכז ז׳ ובמרחק ד׳ וה׳ וח׳. והעגול הד״ח ימשש צלעות המשולש על נקודת[2] ד׳ וח׳ וה׳. ואומר כי כבר עשינו בתוך משולש אב״ג עגול שהוא מקיף עליו.

مِثالُهُ: أَنّا نَجْعَلُ المُثَلَّثَ المَعْلُومَ مُثَلَّثَ ا ب ج. وَنُرِيدُ أَنْ نَعْمَلَ فِي مُثَلَّثِ ا ب ج المَعْلُومِ دائِرَةً تُحِيطُ بِها. فَنَقْطَعُ زاوِيَةَ ب بِنِصْفَيْنِ بِخَطِّ ب د. وَزاوِيَةَ ج أَيْضًا بِنِصْفَيْنِ خَطِّ ج ه. وَيَلْتَقِيانِ عَلَى نُقْطَةِ د. وَنُخْرِجُ مِنْ ز أَعْمِدَةً إِلَى خَطِّ ا ب وَخَطِّ ب ج وَخَطِّ ج ا، وَهِيَ ز ح و ز د و ز ه، فَعَلَى مَرْكَزِ ز وَبِبُعْدِ د و ه و ح. وَنَعْمَلُ الدّائِرَةَ الَّتِي فِي مُثَلَّثِ ا ب ج الَّتِي تَماسُّ أَضْلاعَهُ عَلَى نُقْطَةِ د وح و ه. فَأَقُولُ إِنّا قَدْ خَطَطْنا فِي مُثَلَّثِ ا ب ج دائِرَةً يُحِيطُ بِها.

[1]ב״ד] +⟨⟨ד״ת נחלק זוית [..] וזוית [ג׳] לח[..] ובקוים [..] [וקוי] על [..] קוי [..]⟩⟩ [2]נקודת] דל״ת מנוקדת בחולם

نُرِيدُ أَنْ نَعْمَلَ فِي مُثَلَّثٍ مَعْلومٍ دائِرَةً يُحيطُ بِها.

נרצה שנעשה במשולש ידוע עגולה יקיף בה[1].

فَنَجْعَلَ المُثَلَّثَ المَعْلومَ مُثَلَّث ا ب ج. ونُرِيدُ أَنْ نَعْمَلَ فيهِ دائِرَةً يُحيطُ بِها. فَنَقْسِمَ زاوِيَةَ ا ب ج بِنِصْفَيْنِ بِخَطِّ ب د، وزاوية ا ج ب بِنِصْفَيْنِ بِخَطِّ ج ز. وَلْيَلْتَقِ^ هذان الخَطّانِ على نُقْطَةِ ز. ونُخْرِجَ مِنْ زَ إلى خُطوطِ ا ب ب ج ج ا أَعْمِدَةَ ز ه ز ح ز د. ونَجْعَل نُقْطَةَ زَ مَرْكَزًا. ونُديرِ بِبُعْدِ د ز دائِرَةً في مُثَلَّثِ ا ب ج. فَأَقولُ إِنَّها تَماسُّ أَضْلاعَهُ على نُقَطِ د ح ه.

הנה נשים המשולש הידוע משולש אב״ג. ונרצה שנעשה בו עגולה[2] יקיף בה[3]. הנה נחלק זוית אב״ג בשני[4] חציים בקו ב״ד, וזוית בג״א בשני חציים בקו ג״ז[5]. ונדביק שני הקוים האלו על ז׳[6]. ונוציא מן ז׳ אל קוי א״ב ב״ג א״ג עמודים ז״ה ז״ח ז״ד. ונשים נקודת ז׳ מרכז. ונקיף[7] במרחק ז״ד עגולה במשולש אב״ג. הנה אומר שהוא ימשש[8] צלעותיו על נקודות[9] ד׳ ח׳ ה׳.

---

[1]בה] בו **ב** [2]עגולה] עגולת **ב** [3]בה] +והנה נוציא מזוית אב״ג קו יחלק אותה לחציים ומזוית אג״ב קו יחלק אותה לחציים ויפגשו אלו הקוים על נקודת ז׳ ונוציא מן ז׳ **ס** +⟨והנה נוציא מזוית אב״ג קו יחלק אותה לחציים ומזוית אג״ב קו יחלק אותה לחציים ויפגשו אלו הקוים על נקודת ז׳ ונוציא מן ז׳⟩ **ק** [4]בשני] בשתי **ס** [5]ג״ז] ג״ה **וסק** [6]נחלק ... ז׳] +והנה נוציא מזוית אב״ג קו יחלק אותה לחציין ומן זוית אג״ב קו יחלק אותה לחצאים ומן זוית בא״ג קו יחלק אותה לחצאים ויפגשו אלו הקוים אל נקודת ז׳ **ב** [7]ונקיף] ~~והרחיק~~+ **פ** [8]שהוא ימשש] כי תמשש **ב** [9]נקודות] נקודת **וספ**

מופת זה: לפי שזוית דג"ז שוה לזוית זג"ח, וזוית גד"ז נצבה, והיא שוה לזוית גח"ז, ואם כן שתי זויות ממשלש דז"ג—והן זוית זג"ד וזוית זד"ג—שוות לשתי זויות זג"ח וגח"ז, כל אחת לדומה לה. ובמשלש צלע אחד שהוא מיתר שתי זויות שוות, והוא ג"ז, ואם כן צלעי הזויות הנשארות מן המשולש האחד שוות לשני צלעי המשלש האחר, כל אחד לדומה לו. ואם כן ז"ד שוה לז"ח[1]. ומצד זה ז"ח שוה לז"ה. ושלשה הקוים ז"ד וז"ח וז"ה שוים. והזויות אשר על נקודת ד' וח' וה' נצבות. ואם כן התבאר שכבר עשינו בתוך משלש אב"ג הידוע עגול שהמשלש סביבו, והוא עגול הד"ח.

וזה מה שרצינו לבאר.

بُرْهانُهُ: أَنَّ الزَّاوِيَةَ الَّتِي من د ج ز مِثْلُ الَّتِي مِنْ ز ج ح. وَالَّتِي مِنْ ج د ز قائِمَةٌ. وَهِيَ مِثْلُ الَّتِي مِن ج ح ز. فَزاوِيَتا مُثَلَّثِ (د ج ز)[1] اللَّتانِ من د ج ز وَمِنْ ج د ز مُساوِيَتانِ لِزاوِيَتَي مُثَلَّثِ ز ح ج اللَّتَيْنِ مِنْ ز ج ح وَمِنْ ج ح ز، كُلُّ واحِدَةٍ مِثْلُ نَظيرَتِها. وَفي المُثَلَّثَيْنِ ضِلْعٌ واحِدٌ مُشْتَرَكٌ، وَهُوَ ج ز يوتِرُ زاوِيَتَيْنِ مُتَساوِيَتَيْنِ. فَضِلْعا المُثَلَّثَيْنِ الباقِيانِ مِثْلُ ضِلْعَي المُثَلَّثِ الآخَرِ الباقِيَيْنِ، كُلُّ واحِدٍ مِثْلُ نَظيرِهِ. ز د مِثْلُ ز ح[2]. وَكَذَلِكَ ز ح مِثْلُ ز ه. فَالأَضْلاعُ (الثَّلاثُ)[3] د ز و ز ح و ز ه مُتَساوِيَةٌ. فَالزَّوايا الَّتي عِنْدَ نُقْطَةِ د و ح و ه قائِمَةٌ. فَقَدْ تَبَيَّنَ أَنّا قَدْ خَطَطْنا في مُثَلَّثِ ا ب ج المَعْلومِ دائِرَةً يُحيطُ بِها، وَهِيَ دائِرَةُ ه د ح.

وَذَلِكَ ما أَرَدْنا بَيانَهُ.

[1] לז"ח] ~~לח"ז~~+

[1] د ج ز] ه ج ز [2] ز ح] +ح ج مثل ح د [3] الثلاث] الثلث

بُرْهانُهُ: أَنَّ زاوِيَةَ د ج ز مثل زاوِيَةِ ز ج ح. وزاوية ج د ز قائِمَة، وهي مثل زاوية ج ح ز. فَزاوِيَتا د ج ز ج د ز مِنْ مُثَلَّث د ز ج مثل زاوِيَتَيْ ز ح ج ح ج ز مِنْ مُثَلَّث ز ح ج. وضلع ج ز مُشْتَرَك لَهُما، يوتِرُ زاوِيَتَيْنِ مُتَساوِيَتَيْنِ مِنْ زواياهُما. فَضِلْعا المُثَلَّثِ الباقِيَانِ مثل ضِلْعي المُثَلَّثِ الآخَرِ الباقِيَيْنِ، كُلّ واحدٍ مثل نَظِيرِهِ. فَضِلْع ز د مثل ضِلْعِ ز ح. وكَذَلِكَ أيضًا نُبَيِّنُ أَنَّ ز ح مثل ز ه. فَخُطوط ز د ز ه ز ح الثَّلاثة مُتَساوِيَةٌ. والزَّوايا الَّتي عِنْدَ نُقَطِ د ه ح قائِمَة. فالدَّائِرَة الَّتي تُدارُ على مَرْكَزِ ز وبِبُعْدِ ز د تَمُرُّ بِنُقْطَتَيْ ه ح، وَتَماسُّ أَضْلاعَ المُثَلَّثِ. فَقَدْ عَمِلْنا في مُثَلَّثِ ا ب ج المَعْلومِ دائِرَةَ ه د ح يُحيطُ بِها.

وذَلِكَ ما أَرَدْنا أَنْ نُبَيِّنَ.

מופתו: כי זוית דג״ז כמו זוית זג״ח. וזוית גד״ז נצבת, והיא כמו זוית גח״ז. אם כן שתי זויות דג״ז גד״ז[1] ממשולש דז״ג, כמו שתי זויות זג״ח גח״ז[2] ממשולש זח״ג. וצלע ג״ז משותף לשניהם, יהיה מיתר שתי זויות שוות מזויות שניהם[3]. אם כן שתי צלעות המשולש הנשארות כמו שתי צלעות המשולש האחר הנשארות, כל אחת לנכחית[4] לה. הנה צלע ד״ז כמו צלע ז״ח. וכן גם כן[5] התבאר כי ז״ח כמו ז״ה. אם כן קוי ז״ד ז״ח[6] ז״ה השלשה שוים. והזויות אשר אצל נקודות[7] ד׳ ה׳ ח׳ נצבות. אם כן העגולה אשר תסבוב על מרכז ז׳[8] ובמרחק ז״ד תלך[9] בשתי נקודות ה׳ ח׳, ותמשש צלעי[10] המשולש. הנה כבר עשינו במשולש אב״ג הידוע עגולת הד״ח יקיף בה[11].

וזה מה שרצינו לבאר.

---

[1] דג״ז גד״ז] ~~גד״ז גח״ז~~+ **פ** [2] גח״ז] ⟨⟨*⟩⟩ **ב** [3] מזויות שניהם] מזו משניהם **ב** [4] לנכחית] אצל הנכחית **טפק** [5] כן] ליתא **ב** [6] ז״ח] ליתא **ס** [7] נקודות] נקודת **ספ** [8] ז׳] ⟨*⟩ **ט** [9] תלך] הלך **ס** [10] צלעי] צלעו **ס** [11] יקיף בה] ליתא **ז**

[Proposition IV.5]

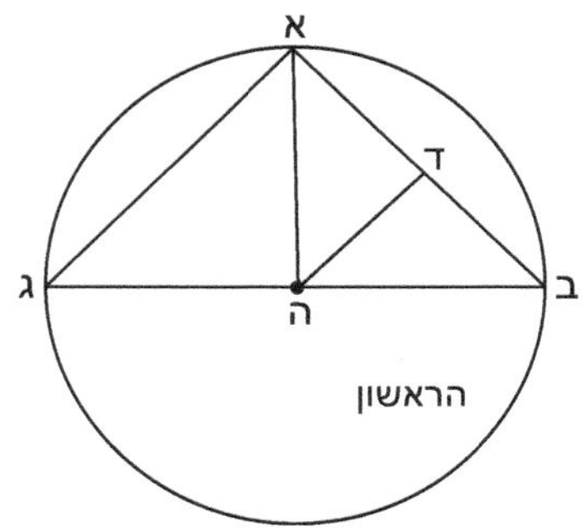

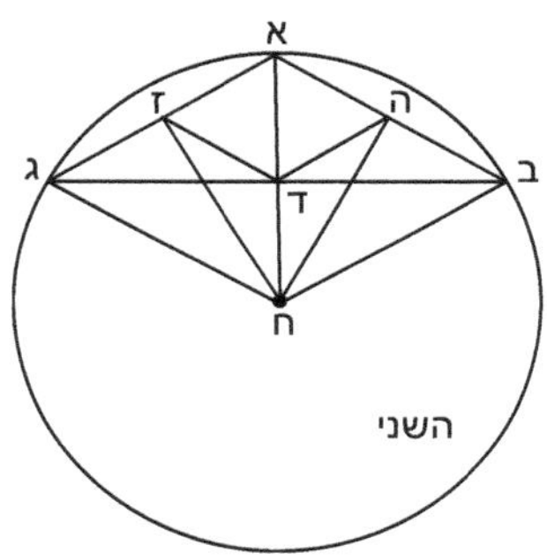

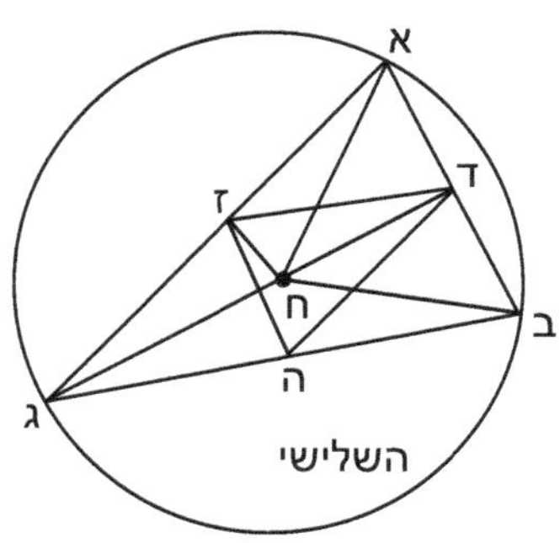

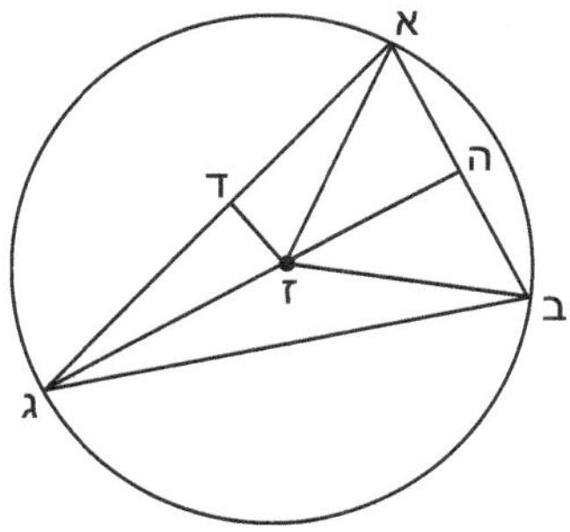

نُرِيدُ أَنْ نَعْمَلَ على مُثَلَّثٍ مَعْلومٍ دائِرَةً تُحِيطُ بِهِ.

فَلْيَكُنِ المُثَلَّثُ المَعْلومُ مُثَلَّثَ ا ب ج. ونُرِيدُ أَنْ نَعْمَلَ عَلَيْهِ دائِرَةً تُحِيطُ بِهِ. فَنَقْسِمَ ضِلْعَيْ^ ا ب ا ج بِنِصْفَيْنِ نِصْفَيْنِ^ على نُقْطَتَيْ د ه. ونُخْرِجَ مِنْهُما خَطَّيْنِ قائِمَيْنِ^ على خَطَّيْ ا ب ا ج على زَوايا قائِمَةٍ، وَهُما د ز ه ز. وَنَصِلُ[1] خُطوطَ ز ب ز ج ز ا.

[1] ونصل] MPR1R2

נרצה לעשות על משולש ידוע עגולה תקיף בו.

הנה יהיה המשולש הידוע משולש אב״ג. ונרצה לעשות עליו עגולה תקיף בו. הנה נחלק כל אחת משתי צלעות א״ב א״ג בשני חציים על שתי נקודות ד׳ ה׳. ונעמיד על שני קוי א״ב א״ג שני קוים על זויות נצבות[1], והם ד״ז ה״ז. ונדביק קוי ז״א ז״ג ז״ב.

[1] נצבות] +על שתי נקודות ד׳ ה׳ **בטס** +⟨על שתי נקודות ד׳ ה׳⟩ **וק**

נרצה לעשות על[1] משולש ידוע עגול שיקיף בו, בין שיהיה המשולש נצב הזויות או רחב הזויות או חד הזויות.

نُرِيدُ أَنْ نُبَيِّنَ كَيْفَ نَعْمَلُ عَلَى مُثَلَّثٍ مَعْلُومٍ دائِرَةً تُحِيطُ (بِهِ)[1]، كانَ المُثَلَّثُ قائِمَ الزّاوِيَةِ أَوْ مُنْفَرِجَ الزّاوِيَةِ أَوْ حادَّ الزّاوِيَةِ.

ונשים המשלש אב״ג. ונשיאהו נצב הזויות, והוא המשלש הראשון. וכאשר רצינו לעשות על משולש אב״ג הראשון נצב הזוית עגול מקיף בו, נחלוק צלע א״ב לחציים על ד׳, וצלע ב״ג לחציים על ה׳. ונדביק ה׳ בד׳, וא׳ בה׳. ונקיף עגול על מרכז ה׳ ובמרחק א׳ וב׳ וג׳. ואומר שכבר עשינו עגול[2] על משלש אב״ג הידוע עגול מקיף בו, והוא עגול אב״ג.

مِثالُهُ: أَنّا نَجْعَلُ المُثَلَّثَ المَعْلُومَ مُثَلَّثَ ا ب ج. وَنَجْعَلُهُ أَوَّلًا قائِمَ الزّاوِيَةِ، وَهُوَ المُثَلَّثُ الأَوَّلُ. وَنُرِيدُ أَنْ نَعْمَلَ عَلَى المُثَلَّثِ الأَوَّلِ القائِمِ الزّاوِيَةِ دائِرَةً تُحِيطُ بِهِ. فَنَقْطَعُ ضِلْعَ ا ب بِنِصْفَيْنِ عَلَى نُقْطَةِ د، وَضِلْعَ ب ج بِنِصْفَيْنِ عَلَى نُقْطَةِ ه. وَنَصِلُ ه د ا ه. فَعَلَى مَرْكَزِ ه وَبِبُعْدِ ا و ب و ج، بِخَطِّ [!] الدّائِرَةِ الَّتِي تُحِيطُ بِمُثَلَّثِ ا ب ج القائِمِ الزّاوِيَةِ.

מופת זה: לפי שקו א״ב כבר נחלק לחציים על ד׳, וב״ג על ה׳, ונדביק ד׳ בה׳, יהיה ד״ה נכוחי לא״ג[3]. וכבר נפל עליהם קו א״ב. ואם כן זוית החיצונה, והיא זוית בד״ה, שוה לפנימית לעומתה[4], והיא זוית בא״ג. וזוית בא״ג[5] נצבה. ואם כן זוית בד״ה נצבה. וזוית הד״א גם כן נצבה. וא״ד שוה לב״ד. וד״ה משותף. ואם כן שני קוי א״ד וד״ה שוים לשני קוי ב״ד וד״ה, כל אחד לדומה לו. וזוית (בד״ה)[6] שוה לזוית אד״ה. ואם כן תושבת א״ה שוה לתושבת ב״ה. וג״ה שוה (לה״ב)[7], ואם כן שוה [לה״א][8]. ואם כן שלשה קוי א״ה ה״ג וה״ב שוים. והתבאר אם כן שכבר עשינו על משולש אב״ג הראשון נצב הזויות עגול מקיף בו, והוא עגול אב״ג. וזה מה שרצינו.

بُرْهانُهُ: أَنَّ خَطَّ ا ب قَدْ قُطِعَ بِنِصْفَيْنِ عَلَى د، و ب ج بِنِصْفَيْنِ عَلَى ه. ووصلت د ه. ف د ه مُوازٍ لِخَطِّ ا ج. وَقَدْ وَقَعَ عَلَيْهِما خَطُّ ا ب فَالزّاوِيَةُ الخارِجَةُ الَّتِي مِنْ ب د ه مِثْلُ الدّاخِلَةِ الَّتِي تُقابِلُها الَّتِي من ب ا ج. وَالَّتِي من ب ا ج قائِمَةٌ. فَالَّتِي من ب د ه أَيْضًا قائِمَةٌ. وَالَّتِي من ه د ا قائِمَةٌ. وا د مِثْلُ ب د. و د ه مُشْتَرَكٌ. فَكِلَيْ خَطَّيِ ا د د ه [..]، كُلُّ واحِدٍ مِثْلُ نَظِيرِهِ. وَالزّاوِيَةُ الَّتِي من ب د ه مِثْلُ الَّتِي من ا د ه. فَقاعِدَةُ ا ج مِثْلُ قاعِدَةِ ه ب. ه ج مِثْلُ ه ب. فَالأَضْلاعُ ا ه ه ب ه ج مُتَساوِيَةٌ. فَقَدِ اسْتَبانَ أَنّا قَدْ عَمِلْنا عَلَى مُثَلَّثِ ا ب ج الأَوَّلِ القائِمِ الزّاوِيَةِ دائِرَةً تُحِيطُ بِهِ، وَهِيَ دائِرَةُ ا ب ج.

[1]על] ‹*›   [2]עגול] נראה שמילה זו מיותרת   [3]לא״ג] +ד״ת לפי שכשיצא מנקודת ד׳ אל נקודת ג׳ קו אחר יצא בהכרח קו ד״ה נכוחי לקו א״ג מצורת ל״ו מן הראשון וכן לכל שאר המשולשים והמעתיק לא שם ליבו לזה   [4]לעומתה] ~~שהיא נצמתת~~   [5]וזוית בא״ג] ‹‹*››   [6]בד״ה] בד״א   [7]לה״ב] לה״ג   [8]לה״א] לה״ב

[1]به] ‹‹*›› بها; קו מחיקה על המילה بها, ונראה שהוא נמתח ביד השונה מהיד שכתבה את המילה

فَلأَنَّ خطّ ا ه مُساوٍ لخطّ ه ب، وخطّ ز ه مُشْتَرَك، يكونُ كِلا خَطَّيْ ا ه ه ز مثل كِلَيْ خَطَّيْ ب ه ز ه، كُلّ واحدٍ مثل^ نظيره. وزاوية ا ه ز القائِمَة مُسَاوِيَة لِزَاوِيَةِ ب ه ز القائِمَة. فَقاعِدَة ا ز مُسَاوِيَة لِقاعِدَةِ ب ز. وَكَذَلِكَ أَيْضًا تَبَيَّنَ أَنَّ خطّ ا ز مُساوٍ لِخَطِّ ز ج. فيكونُ خَطُّ ب ز مُساوِيًا لِخَطِّ ج ز. فخُطوط ا ز ز ب ز ج مُتَساوِيَةٌ. فَإذا جَعَلْنا نُقْطَةَ ز مَرْكَزًا، وَأَدَرْنا دائِرَةً بِبُعْدِ ا ز مَرَّتْ بنُقْطَتَيْ ب ج^. فَنَخُطُّ هَذِهِ الدَّائِرَةَ وَلْيَكُنْ عَلَيْهِما ا ب ج. فَقَدْ عَمِلْنا على مُثَلَّثِ ب ج دائِرَةً تُحيطُ بِهِ، وهي دائِرَةُ ا ب ج^.

وذَلِكَ ما أَرَدْنا أَنْ نُبَيِّنَ.

הנה מפני כי קו א״ה שוה לקו ה״ב, וקו ז״ה משותף, יהיו[1] כל שני קוי א״ה ה״ז כמו כל[2] שני[3] קוי ב״ה ז״ה, כל אחד אצל הנכחי לו. וזוית אה״ז הנצבת שוה לזוית בה״ז הנצבת[4]. אם כן תושבת א״ז שוה לתושבת ב״ז. וכן גם כן התבאר כי קו א״ז שוה לקו ז״ג. אם כן יהיה קו ב״ז שוה לקו ז״ג. אם כן[5] קוי א״ז ז״ב ז״ג שוים. הנה כאשר שמנו נקודת ז׳ מרכז[6], וסבבנו במרחק א״ז עגולה הלכה בנקודות[7] א׳ ב׳ ג׳, הנה נקוה העגולה הזאת, ויהיו עליה אב״ג. הנה כבר עשינו על משולש אב״ג עגולה תקיף בו.

וזה מה שרצינו לבאר[8].

[1]יהיו] הנה+ ו [2]כל] ליתא ס [3]קוי ... שני] ליתא ק [4]הנצבת] +* ב [5]יהיה ... כן] ליתא ו [6]מרכז] ⟨*⟩ ו [7]בנקודות] בנקודת טס [8]וזה ... לבאר] ונשלם ביאורו ט

ועוד משולש אב״ג השני רחב הזויות. וכאשר רצינו לעשות עליו עגול מקיף בו, נחלוק כל צלע מצלעיו לחציים—צלע א״ב על ה׳, וצלע ב״ג על ד׳, וצלע א״ג על ז׳. ונדביק ה׳ בד׳ וז׳ בד׳.

וכבר התבאר שד״ה נכוחי לא״ג. וז״ד נכוחי לא״ב. ולפי שה״ד נכוחי לא״ג, וכבר נפל עליהם קו א״ב, תהיה הזוית החיצונה, והיא זוית בה״ד, גם כן נרוחת. וכמו כן זוית דז״ג נרוחת. ונוציא מקוי א״ב וא״ג, מנקודות ה׳ וז׳, שני קוים, כל אחד מהם על זויות נצבות, ויחתכו קו ב״ג. ויפגשו על נקודת ח׳, והם קו ה״ח וקו ז״ח. ונדביק ח׳ בב׳ ובג׳. וא״ה שוה לה״ב. וה״ח משותף גם כן. ושני קוי א״ה וה״ח שוים לשני קוי ב״ה וה״ח, כל אחד לדומה לו. וזוית בה״ח שוה לזוית אה״ח. ואם כן תושבת א״ח שוה לתושבת ב״ח. ומצד זה א״ח שוה לח״ג. ואם כן שלשה הקוים, והם א״ח וב״ח וג״ח, שוים. ונקיף עגול על מרכז ח׳ ובמרחק א׳ וב׳ וג׳ שיקיף במשלש אב״ג רחב הזויות. וזה מה שרצינו.

وأَيضًا مُثَلَّثُ ا ب ج الباقي مُنْفَرِجُ الزَّاوِيَةِ. ونُرِيدُ أَنْ نَعْمَلَ عَلَيْهِ دائِرَةً تُحيطُ بِهِ. فَنَقْطَعُ كُلَّ ضِلْعٍ مِنْ أَضْلاعِهِ بِنِصْفَيْنِ، ضِلْعَ ا ب وَضِلْعَ ب ج وَضِلْعَ ا ج، عَلَى ه و د و ز. وَنَصِلُ (ه د)[1] ز د.

فَقَدِ اسْتَبانَ أَنَّ (ه د)[2] مُوازٍ لِخَطِّ ا ج. (و د ز)[3] مُوازٍ لِخَطِّ ا ب. (ه د)[4] ا ج مُتَوازِيانِ، فَوَقَعَ عَلَيْهِما خَطُّ ا ب، فَالزَّاوِيَةُ الخارِجَةُ الَّتِي مِنْ (ب ه د)[5] مِثْلُ الدَّاخِلَةِ الَّتِي تُقابِلُها الَّتِي من ب ا ج المُنْفَرِجَةِ. ز د مُوازٍ لِخَطِّ ا ب. وَوَقَعَ عَلَيْها ا ج. فَالزَّاوِيَةُ الخارِجَةُ الَّتِي مِنْ ج ز د مِثْلُ الدَّاخِلَةِ الَّتِي مِنْ ج ا ب. المُنْفَرِجَةِ. فَيُخْرِجُ مِنْ خَطَّي ا ب ا ج مِنْ نُقْطَتَي (ز)[6] و ه خَطَّيْنِ، كُلُّ واحِدٍ مِنْهُما عَلى زاوِيَتَيْنِ قائِمَتَيْنِ. وَيَقْطَعانِ خط ب ج. وَيَلْتَقِيانِ عَلى نُقْطَةِ ح. وَهُما خَطَّا ه ح (و ز ح)[7] وَنَصِلُ (ح ب ب ج ا)[8]. فَعَلى مَرْكَزِ ح وَبِبُعْدِ ب و ا و ج نَخُطُّ الدَّائِرَةَ الَّتِي تُحيطُ مُثَلَّثَ ا ب ج الثَّاني المُنْفَرِجَ الزَّاوِيَةِ.

---

[1] ه د] ه ز [2] ه د] ز ه [3] و د ز] و ز [4] ه د] ه ز
[5] ب ه د] ب ه ز [6] ز] د [7] و ز ح] و د ح [8] ح
ب ب ج ا] ح ب و ج و ا

بُرْهانُهُ: أَنَّ خَطَّ ا ه مِثْلُ ه ب. و ه ج مُشْتَرَكٌ. فَعَلى خَطَّيْ ا ه و ه ح مِثْلُ كُلِّ خَطَّيْ ب ه ه ح، كُلّ واحِدٍ مِثْل نَظيرِهِ. وَالزّاوِيَةُ الَّتي مِنْ ب ه ح مِثْل الَّتي مِنْ ا ه ح. فَقاعِدَةُ ا ح مِثْل قاعِدَةِ (ح ب)[1]. وَكَذَلِكَ ا ح مِثْلُ ح ج. فَالأَضْلاعُ الثَّلاثُ ا ح و ب ح و ج ح مُتَساوِيَةٌ. فَعَلى مَرْكَزِ ح وَبِبُعْدِ ا ب و ج نَخُطُّ الدّائِرَةَ الَّتي تُحيطُ مُثَلَّثَ ا ب ج الثّاني المُنْفَرِجَ الزّاوِيَةِ.

وَأَيْضًا مُثَلَّثُ ا ب ج الثّالِثُ حادُّ الزّاوِيَةِ. وَنُريدُ أَنْ نَعمَلَ عَلَيْهِ دائِرَةً تُحيطُ بِهِ. فَنَقْطَعُ كُلَّ ضِلْعٍ مِنْ أَضْلاعِهِ بِنِصْفَيْنِ عَلى نُقْطَةِ د و ه و ز. وَنَصِلُ ه د ز ه (و د ز)[2]. فَقَدِ اسْتَبانَ أَنَّ خَطَّ د ه مُوازٍ لِخَطِّ ا ج. فَوَقَعَ عَلَيْهِما خَطُّ ا ب. فَالزّاوِيَةُ الخارِجَةُ الَّتي مِنْ ب د ه مِثْلُ الدّاخِلَةِ الَّتي تُقابِلُها الَّتي مِنْ ب ا ج. وَالَّتي مِنْ ب ا ج حادَّةٌ. فَالَّتي مِنْ ب د ه حادَّةٌ. وَكَذَلِكَ الَّتي مِنْ ه ز ج حادَّةٌ. وَنُخْرِجُ مِنْ خَطِّ ا ب مِنْ نُقْطَةِ د خَطًّا عَلى زاوِيَةٍ قائِمَةٍ تَجوزُ فيما بَيْنَ الزّاوِيَتَيْنِ اللَّتَيْنِ مِنْ ا د ز وَمِنْ ب د ه. وَنُخْرِجُ أَيْضًا من خَطِّ ا ج مِنْ نُقْطَةِ ز خَطًّا عَلى زاوِيَةٍ قائِمَةٍ تَجوزُ فيما بَيْنَ الزّاوِيَتَيْنِ من ا ز د ومن ج ز ه. وَيَلْتَقي هَذانِ الخَطّانِ المُخْرَجانِ عَلى نُقْطَةِ ح. وَنَصِلُ (ح ب ا ب ج)[3]. فَعَلى مَرْكَزِ ح وَبِبُعْدِ ب و ا و ج نَخُطُّ الدّائِرَةَ الَّتي تُحيطُ بِمُثَلَّثِ ا ب ج الثّالِثِ الحادِّ الزّاوِيَةِ.

ועוד משולש אב״ג חד הזויות השלישי— וכאשר רצינו לעשות עגול מקיף בו, נחלוק כל צלע מצלעיו לחציים על נקודת [!] ד׳ וה׳ וז׳. ונדביק ד׳ בה׳, וז׳ בה׳, וד׳ בז׳. וכבר התבאר שקו ד״ה נכוחי לקו א״ג. ונפל עליהם קו א״ב. ואם כן הזוית החיצונה, והיא זוית בד״ה, שוה לזוית הפנימית שכנגדה, והיא זוית בא״ג. וזוית בא״ג חדה. ואם כן זוית בד״ה חדה. ומזה הצד זוית הז״ג גם כן חדה. ונוציא מקו א״ב, מנקדת ד׳, קו על זוית נצבה, ויעבר על מה שבין שתי זויות אד״ז ובד״ה. ונוציא עוד מקו א״ג, מנקודת ז׳, קו על זוית נצבה, ויעבר על מה שבין שתי זויות אז״ד וגז״ה. ויפגשו שני קוים אלו המוצאים על נקודת ח׳. ונדביק ח׳ בא׳ ובב׳ ובג׳. וא״ד שוה לב״ד. וד״ח משותף. ושני קוי א״ד וד״ח שוים לשני קוי ב״ד וד״ח. וזוית בד״ח שוה לזוית אד״ח. ואם כן תושבת ב״ח שוה לתושבת א״ח. וכמו כן א״ח שוה לח״ג. ואם כן שלשה קוי ח״א וח״ב וח״ג שוים. ונקיף על מרכז ח׳ ובמרחק א׳ וב׳ וג׳ עגול מקיף במשולש אב״ג השלישי חד הזויות.

---

[1] ح ب] وج ح [2] ود ز] و ز [3] ح ب ا ب ج] ح ا وب وج

بُرْهانُهُ: أَنَّ خَطَّ ا د مِثْل ب د. و د ح مُشْتَرَكٌ. فَكِلَيْ خَطَّيْ ا د و د ح مِثْل كِلَيْ خَطَّيْ ب د و د ج، كُلّ واحِدٍ مِثْل نَظيرِه. والزَّاوِيَةُ الَّتي من ب د ح مِثْل الَّتي من ا د ح. فَقاعِدَة ب ح مِثْل قاعِدَةِ ا ح. وَكَذلِكَ ا ح مِثْل ح ج. فالأَضْلاعُ (الثَّلاثُ)[1] ح ا و ح ب[2] و ح ج مُتَساوِيَةٌ. فَعَلى مَرْكَزِ ح وَبِبُعْدِ ا و ب و ج نَخُطُّ الدَّائِرَةَ الَّتي تُحيطُ بِمُثَلَّثِ ا ب ج الثَّالِثِ الحادِّ الزَّاوِيَةِ المَعْلوم.

وَذلِكَ ما أَرَدْنا بَيانَهُ.

וזה מה שרצינו באורו.

[1] الثلاث] الثلث [2] وح ب] נראה שחסר وح ج

אמר המעתיק: נרא׳ לי שזאת התמונה אחד מהתלמידים שגה במופתה, לפי שלא ידע הדרך שנזכר עליה בנסחת תאבת, והיא להוציא הקוים על זויות נצבות על מקום חלוקי שתי הצלעות, אחר שנחלקו הצלעות לחציים, ולהוציא ממקום פגישת שתי הקוים ההם שלשה קוים אל זויות המשלש. ואמת שתאבת לא צייר רק תמונת משלש חד הזויות, והיא [..][1] לשתי התמונות הנשארות. לפי שידוע שאם המשלש הוא חד הזויות, המרכז ימצא בהכרח בתוך המשולש. ואם הוא תחת רחב[2] הזוית, ימצא חוץ למשולש. ואם לא, הוא[3] נצב הזוית, ימצא במיתר הזוית הנצבה. ופגישת שני הקוים המוצאים על זויות נצבות תהיה בהכרח במקום המרכז בכל תמונה ותמונה על הדרך הנזכר. וזה כלו ידוע מתמונת [ל׳] מהמאמר השלישי. וממנה יודע כי במשולש נצב הזויות אין צריך רק לחלוק מיתר הזוית הנצבה. ובהכרח תהיה נקדת מקום החלוק מרכז העגול המבוקש. והתלמיד ששגה בזה שגגתו היתה שלקח מופת מהתמונה השנית מהמאמר הששי. ואם היה צריך לה אקלידס בתמונה זאת לא היתה מקום התמונה הזאת במאמר זה, לפי שלא ילקח מופת רק מאשר קדמה ידיעתו. וכבר מצאתי שגיאות אחרות בנסחת הספר שהעתקתי ממנו[4]. וכפי שהבנתי אותם תקנתי הלשון בכל מקום. והמתקן אחרי תבא עליו ברכה.

---

[1] ..] ~~הנותרת~~ ⟨⟨[..]⟩⟩ [2] רחב] ⟨*⟩ [3] הוא] ⟨*⟩
[4] ממנו] ~~מתוכו~~+

[Proposition IV.6]

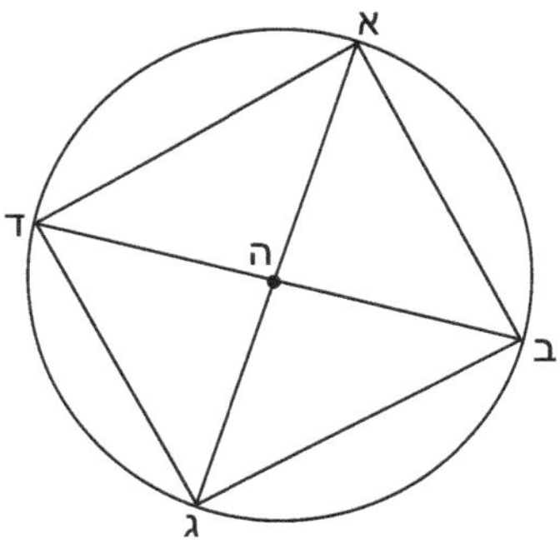

نُرِيدُ أَنْ نُبيِّنَ كَيْفَ نَعْمَلُ فِي دائِرَةٍ مَعْلومَةٍ مُسَطَّحًا مُرَبَّعًا يُحِيطُ بِهِ.

مِثالُهُ: أَنَّا نَجْعَلُ الدَّائِرَةَ المَعْلومَةَ دائِرَةَ ا ب ج. ونُرِيدُ أَنْ نَعْمَلَ فِيها سَطْحًا مُرَبَّعًا يُحِيطُ بِهِ. فَنُخْرِجُ قُطْرَيْها يَتقاطَعانِ عَلى زَوايا قائِمَةٍ، وَهُما ا ج ب د. وَنَصِلُ (ا ب ب)[1] و د و ج ب ب و د. فَأَقولُ إِنَّا قَدْ عَمِلْنَا في دائِرَةِ ا ب ج مُرَبَّعًا عَلَيْهِ ا ب ج.

נרצה לעשות בתוך עגול ידוע שטח מרובע יקיף בו העגול.

ונשים העיגול הידוע עגול אבג״ד. וכאשר רצינו שנעשה בתוכו מרובע שיקיף בו העגול, נוציא שני קטריו בדרך שיחלק כל אחד מהם האחר על זוית נצבה בנקדת ה׳, והם א״ג וב״ד. ונדביק א׳ בב׳, וא׳ בד׳, וג׳ בב׳ ובד׳. ואומר שכבר עשינו בתוך עגול אב״ג מרובע עליו אבג״ד.

[1] ا ب ب] ا ب

نُرِيدُ أَنْ نَعْمَلَ في دائِرَةٍ مَعْلومَةٍ سَطْحًا مُرَبَّعًا تُحِيطُ بِهِ.

נרצה לעשות בעגולה ידועה שטח מרובע תקיף בו.

فَنَجْعَلُ الدَّائِرَةَ المَعْلومَةَ دائِرَةَ ا ب ج. ونُرِيدُ أَنْ نَعْمَلَ فيها سَطْحًا مُرَبَّعًا تُحِيطُ بِهِ.

הנה נשים העגולה הידועה עגולת אב״ג. ונרצה לעשות בה שטח מרובע תקיף בו.

ומופת זה: לפי שב״ה שוה לה״ד, וה״א משותף, ושני קוי ב״ה וה״א שוים לשני קוי ד״ה וה״א, וזוית בה״א שוה לזוית דה״א, ואם כן תושבת א״ד שוה לתושבת א״ב. ומזה הצד ב״ג וג״ד שוים לא״ב א״ד. ואם כן שטח אבג״ד שוה הצלעות. והוא גם כן נצב הזויות, לפי שכל זוית שהיא על חצי עגול היא נצבה. ואם כן כל הזויות שעליהן א׳ וב׳ וג׳ וד׳[1] נצבות.

بُرْهانُهُ: أَنَّ ب ه مِثْلُ ه د. و ه ا مُشْتَرَكٌ. فَكِلَيْ خَطَّيْ ب ه ه ا مِثْل كِلَيْ خَطَّيْ د ه ه ا، كُلُّ واحِدٍ مِثْل الآخَرِ. وَالزَّاوِيَةُ الَّتِي مِنْ ب ه ا مِثْل الَّتِي من د ه ا. وقاعِدَةُ ا د مِثْلُ قاعِدَةِ ا ب. وِمِنْ أَجْلِ ذَلِكَ ب ج و ج د مِثْل ا ب و ا د. فَمُرَبَّعُ ا ب ج د مُتساوي الأَضْلاعِ قائِمُ الزَّوايا. وَالزَّاوِيَةُ الَّتِي عَلَى نِصْفِ الدَّائِرَةِ قائِمَةٌ. وَجَمِيعُ الزَّوايا الَّتِي عَلَيْها ا ب وج و د قائِمَةٌ.

והתבאר אם כן שכבר עשינו בתוך עגול אבג״ד מרבע שוה הצלעות נצב הזויות.

فَقَدِ اسْتَبانَ أَنَّا قَدْ عَمِلْنا في دَائِرَةِ ا ب ج د مُرَبَّعًا مُساوي الأَضْلاعِ قائِمَ الزَّوايا تُحِيطُ بِهِ الدَّائِرَةُ.

וזה מה שרצינו

وَذَلِكَ ما أَرَدْنا بَيانَهُ.

[1] וד׳] ⟨ו⟩ד׳

فَنُخْرِجَ فيها قُطْرَيْنِ يَتَقاطَعانِ على زوايا قائِمَةٍ، وَهُما ا ج ب د. ونُخْرِجَ خُطوطَ ا ب ا د ج ب ج د. فَخَطّ ب ه مثل خطّ ه د. وخطّ ه ا مُشْتَرَك. فَكِلا خَطَّيْ ب ه ه ا مثل كِلَيّ خَطَّيْ د ه ه ا. وزاوية ب ه ا مثل زاوية د ه ا. فَقاعِدَة ا ب مثل قاعِدَة ا د. وكَذَلِكَ أيضًا تَبَيَّنَ أنَّ ب ج مثل ج د، وإن ج د مثل ا د. فَمُرَبَّع ا ب ج د مُتساوي الأضلاع. والزَّوايا الَّتي في أَنْصافِ الدَّوائِرِ قائِمَةٌ. فَجميع الزَّوايا الَّتي عِنْدَ نُقَطِ ا ب ج د كُلّ واحِدَةٍ مِنْها[1] قائِمَة.

הנה נוציא בה שני קוטרים[1] יחתכו על זויות נצבות[2], והם א״ג ב״ד. ונוציא קוים[3] א״ב א״ד ב״ג ג״ד. הנה קו ב״ה כמו קו ה״ד. וקו א״ה משותף. הנה כל שני קוי[4] ב״ה א״ה כמו כל שני קוי ד״ה א״ה. וזוית[5] בה״א[6] כמו זוית דה״א. אם כן תושבת א״ב כמו תושבת א״ד. וכן גם כן התבאר כי ב״ג כמו ג״ד[7], וכן[8] ג״ד כמו א״ד. אם כן מרובע אבג״ד שוה הצלעות. והזויות אשר בחציי העגולות נצבות. הנה כל הזויות אשר אצל נקודות א׳ ב׳ ג׳ ד׳ כל אחת מהן נצבת[9].

فَقَدْ تَبَيَّنَ أنّا قَدْ عَمِلْنا في دائِرَةِ ا ب ج د^ المَعْلومَةِ مُرَبَّعًا.

הנה כבר התבאר שאנחנו כבר עשינו בעגולת אב״ג הידועה מרובע[10].

وذَلِكَ ما أَرَدْنا أَنْ نَعْمَلَ.

וזה מה שרצינו לבאר.

---

[1] منها] M

---

[1] קוטרים] קוים+ **ב** קוי+ **ס** ~~קוים~~+ **ק** [2] זויות נצבות] זוית ⟨⟨נצבה⟩⟩ ויהיה מקום חתוכה נקודת והיא מרכז ⟨⟨ה⟩⟩ **ב** [3] קוים] שני+ **ס** [4] קוי] ליתא **ט** [5] וזוית] וזויות **בפ** [6] בה״א] ~~בא״ה~~+ **ס** [7] כמו ג״ד] ליתא **ב** [8] וכן] ~~אם~~ כן **ק** [9] נצבת] +~~הנה כל הזויות~~ ⟨⟨אשר [אצל] נקודת א״ב ג״ד כל אחת מהן נצבת⟩⟩ **ס** [10] מרובע] +⟨⟨תקיף בו⟩⟩ **ק**

[Proposition IV.7]

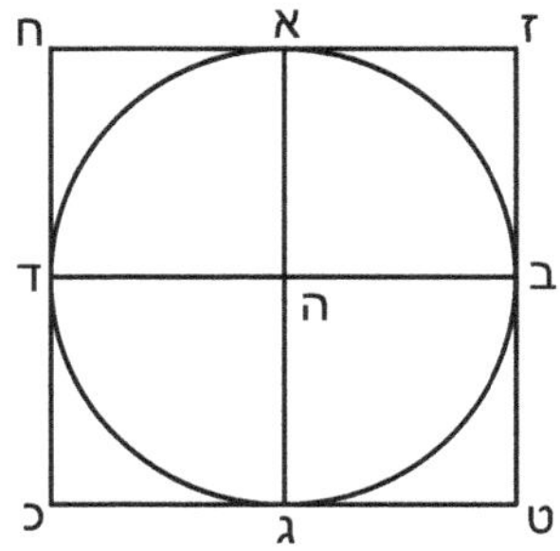

נרצה לעשות על עגול ידוע מרובע מקיף בעגול.

ונשים העגול הידוע עגול אבג״ד. וכאשר רצינו לעשות סביבו מרובע נוציא שני קטריו, ויחתך כל אחד מהם האחר על זוית נצבה— על ה׳, והם א״ג ב״ד. ונוציא מנקדת א׳ וב׳ וג׳ וד׳ קוים ממששים לעיגול, והם ז״ח וז״ט וט״כ וכ״ח. ואומר שכבר עשינו על עגול אבג״ד מרובע מקיף בעגול, והוא מרבע (זטכ״ח)[1].

نُرِيدُ أَنْ نُبَيِّنَ كَيْفَ نَعْمَلُ عَلَى دائِرَةٍ مَعْلومَةٍ مُرَبَّعًا يُحيطُ بِها.

مثالُهُ: أَنْ نَجْعَلَ الدَّائِرَةَ المَعْلومَةَ دائِرَةَ ا ب ج د. وَنُرِيدُ أَنْ نَعْمَلَ عَلَيْها مُرَبَّعًا يُحيطُ بِها. فَنُخْرِجُ قُطْرَيِ الدَّائِرَةِ يَتَقاطَعانِ عَلَى زَوايا قائِمَةٍ، وَهُما اج ب د. وَنُخْرِجُ مِنْ نُقْطَةِ ا و ب و ج و د خُطوطًا تُماسُّ الدَّائِرَةَ، وَهِيَ زح و زط و طك و كح.

---

[1] זטכ״ח] הטכ״ח

نُرِيدُ أَنْ نَعْمَلَ على دائِرَةٍ مَعْلومَةٍ مُرَبَّعًا يُحيطُ بِها.

نרצה לעשות על עגולה ידועה מרובע יקיף בה.

فَنَجْعَلَ الدَّائِرَةَ المَعْلومَةَ دائِرَةَ ا ب ج د. وَنُرِيدُ أَنْ نَعْمَلَ عَلَيْها مُرَبَّعًا يُحيطُ بِها. فَنُخْرِجَ فيها قُطْرَيْنِ يَتَقاطَعانِ على زوايا قائِمَةٍ، وَهُما ا ج ب د. وَنُخْرِجَ مِنْ نُقَطِ ا ب ج د خُطوطَ زح زط طك كح مُماسَّةً لِدائِرَةٍ. وَهُوَ أَنْ نُخْرِجَها على زوايا قائِمَةٍ على أَطْرافِ الأَقْطارِ[1].

הנה נשים[1] העגולה הידועה[2] עגולת אב״ג. ונרצה לעשות עליה מרובע יקיף בה. ונוציא בה שני קוטרים יחתכו על זויות נצבות, והם א״ג ב״ד. ונוציא מנקודות א׳ ב׳ ג׳ ד׳ קוי ז״ח ז״ט ט״כ כ״ח ממששים לעגולה, והוא שנוציאם[3] על זויות נצבות על קצות הקוטרים.

---

[1]وهو ... الأقطار] U وهو ان نخرجها على زوايا قائمة من أطراف الأقطار MPR2 ⟨⟨وهو ان نخرجها على زوايا قائمة من أطراف الأقطار⟩⟩ Cam

[1]נשים] נשלים ס [2]הידועה] ליתא ב [3]והוא שנוציאם] והיא שנוצאים ב

מופת זה: לפי שקו ז״ח ממשש העגול, וכבר יצא ממקום משושו קו אל המרכז, והוא ה״א, ואם כן א״ה עמוד על קו ז״ח. ושתי זויות זא״ה וחא״ה נצבות. ומצד זה תהיין הזויות אשר על ב׳ ועל ג׳ ועל ד׳ נצבות. ותהיין זוית זא״ה וזוית בה״א שוות לשתי זויות נצבות. ואם כן ה״ב וז״א נכוחיים. ולזה ב״ז וה״א נכוחיים. ואם כן שטח זבה״א נכוחי הצלעות. וצלעיו וזויותיו שזו לעמת זו שוות[1]. ואם כן ז״א שוה לב״ה. ומצד זה ב״ה שוה לג״ט. וב״ט שוה לה״ג. וב״ה שוה לז״א. ואם כן ז״א שוה לג״ט. וכמו כן א״ג שוה לז״ט. וא״ג שוה לח״כ[2]. ואם כן שטח זטכ״ח שוה הצלעות. והוא נצב הזויות, לפי שב״ה נכוחי לז״א, ונפל עליהם קו ב״ז. ואם כן שתי הזויות הפנימיות, והן זוית הב״ז וזוית אז״ב, שוות לשתי זויות נצבות. וזוית הב״ז נצבה. ותשאר זוית אז״ב נצבה. ומצד זה תהיין הזויות שעליהן ט׳ וכ׳ וח׳ [ז׳] נצבות. ואם כן שטח זטכ״ח מרובע שוה הצלעות נצב הזויות. וכבר עשינו על עגול אבג״ד מרובע מקיף בו.

וזה מה שרצינו.

برهانه: أن زح تماس الدائرة. وقد خرج من الموضع الذي يماسها فيه خط إلى المركز، وهو خط ه ا. اه عمود على خط زح. فالزاويتان اللتان من زاه ومن حاه قائمتان. ومن أجل ذلك صارت الزوايا التي على ب و ج و د قوائم[1]. فالتي من زاه والتي من ب ه ا مثل زاويتين قائمتين. فخط ه ب زا متوازيان. وكذلك ب ز ه ا متوازيان. فسطح ز ب ه ا متوازي الأضلاع. فأضلاعه وزواياه المتقابلة متساوية، زا مثل ب ه. وكذلك ب ه مثل ج ط. ب ه مثل زا. و زا مثل ج ط. وكذلك ا ج مثل ط ز. و ا ج مثل ح ك. ط ز مثل ح ك. وبمثل ذلك ج[2] مثل ط ك. فسطح ز ط ك ح متساوي الأضلاع قائم الزوايا. و ه ب مواز لخط زا. ووقع عليهما خط ب ز. والزاويتان اللتان من ه ب ز ومن ا ز ب مثل زاويتين قائمتين. والتي من ه ب ز قائمة. وبقيت التي من ا ز ب قائمة. ومن أجل ذلك الزوايا التي عليها ط وك وح وز قائمة. فسطح ز ط ك ح مربع متساوي الأضلاع قائم الزوايا. وقد عمل على دائرة ا ب ج د مربعا يحيط بها.

وذلك ما أردنا بيانه.

[1]שוות] +⟨⟨נ״א [..] על [..] [..]נצבות⟩⟩ [2]לח״כ] +⟨⟨ד״[ת] וב״ד שוה לא״ג לפי ששני ⟨⟨[..]⟩⟩ קטרי העגול ואם כן ז״ח וט״כ שוים נכוחיים לבד ושוים [..] [הם] שוים לא״ג⟩⟩

[1]قوائم] قايمة ⟨*⟩ [2]ج] ا ج

نخطّ زح يَماسُّ الدَّائِرَةَ. وَقَدْ أُخْرِجَ مِنْ حَيْثُ يَماسُّها خَطُّ اه إلى المَرْكَزِ. فَهُوَ عَمُودٌ على زح. وَزاوِيَتا زاه ەاح قائِمَتانِ. وَكَذَلِكَ تَكونُ الزَّوايا الَّتي عِنْدَ نقط ب ج د قائِمَةً. فَزاوِيَتا زاه ب ەا مثل قائِمَتَيْنِ. نخطّا ەب زا مُتَوازِيانِ. وَكَذَلِكَ يَكونُ خَطّا ب ز ەا مُتَوازِيَيْنِ. فَسَطْح ز ب ەا مُتوازي الأَضْلاعِ. فأَضلاعُهُ وَزَواياهُ المُتَقابِلَةُ مُتَساوِيَةٌ. فَضِلْع زا مثل ضلع ب ە. وكَذَلِكَ يَكونُ ب ە مثل ج ط، وا ح مثل ە د[1] ومثل ج ك[2]^. فسطح ز ط ك ح مُتساوي الأضلاع. و ب ە يُوازي زا، وَقَدْ وَقَعَ عَلَيْهِما خَطُّ ب ز. فَزاوِيَتا ە ب ز ا ز ب الدَّاخِلَتَيْنِ مثل قائِمَتَيْنِ. وزاوية ە ب ز قائِمَةٌ. فتَبْقَى زاويةُ ا ز ب قائِمَةً. ولِذَلِكَ تَكونُ الزَّوايا^ الَّتي عِنْدَ نُقَطِ ط ك ح د قائِمَةً. فَسَطْحُ ز ط ك ح مُرَبَّعٌ. وَهُوَ مَعْمُولٌ على دائِرَةِ ا ب ج د.

وذَلِكَ ما أَرَدْنا أَنْ نُبَيِّنَ.

הנה ז״ח ימשש העגולה. וכבר הוצא ממקום אשר ימששה קו א״ה אל המרכז. אם כן הוא עמוד על ז״ח. ושתי זויות זא״ה הא״ח נצבות[1]. וכן תהיינה הזויות אצל[2] נקודות[3] ב׳ ג׳ ד׳ נצבות. אם כן שתי זויות זא״ה בה״א כמו[4] שתי נצבות. אם כן שני קוי ה״ב ז״א נכחיים[5]. וכן יהיו שני קוי ז״ב א״ה נכחיים. הנה שטח זבה״א נכחי הצלעות. הנה צלעותיו וזויותיו המתנגדים שוות. אם כן צלע ז״א כמו צלע ב״ה[6]. וכן יהיה ב״ה כמו ט״ג, וא״ח כמו ה״ד וכמו ג״כ. הנה ב״ד כמו ז״ח[7] וכמו ט״כ[8]. ולכן[9] א״ג כמו ט״ז וכמו ח״ב. הנה שטח זטכ״ח שוה הצלעות. וב״ה ינגד א״ז, וכבר נפל עליהם קו ב״ז. אם כן שתי זויות הב״ז אז״ב הפנימיות[10] כמו שתי נצבות. וזוית הב״ז נצבת. הנה נשארה[11] זוית אז״ב נצבת. וכן תהיה כל אחת מן הזויות אשר אצל נקודות[12] ט׳ כ׳ ח׳ נצבות. אם כן שטח זטכ״ח מרובע. והוא עשוי על עגולת אבג״ד[13].

וזה מה שרצינו לבאר.

---

[1] ە د] ط ز MPR1R2 [2] ج ك] ح ك MPR1R2

---

[1] נצבות] ⟨⟨*⟩⟩ **ב** [2] אצל] אשר+ **טפק**
[3] נקודות] נקודת **ספ** [4] כמו] ליתא **ק**
[5] נכחיים] נכחיי⟨⟨ם⟩⟩ ~~הצלעות~~ **ק** [6] ב״ה] ב״א **ב** [7] ז״ח] ~~ט״ח~~+ **פ** [8] וכמו ט״כ] ⟨⟨*⟩⟩ **ק**
[9] ולכן] +יהיה **ס** [10] הפנימיות] הפנימי⟨ו⟩ת **ז** [11] נשארה] נשאר⟨ה⟩~~ת~~ **ב** [12] נקודות] נקודת **ס** [13] אבג״ד] +⟨⟨הידועה⟩⟩ **ק**

[Proposition IV.8]

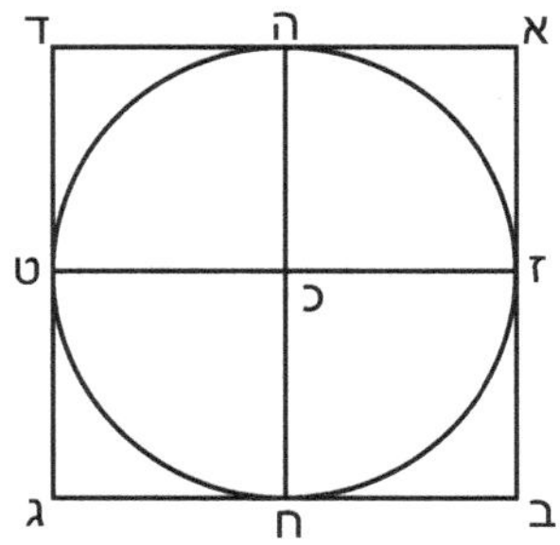

נרצה לעשות בתוך מרובע ידוע עגול יקיף בו המרובע.

نُرِيدُ أَنْ نُبَيِّنَ كَيْفَ نَعْمَلُ فِي مُرَبَّعٍ مَعْلومٍ دائِرَةً يُحِيطُ بِها.

ונשים המרובע הידוע מרובע אבג״ד. וכאשר רצינו לעשות בתוכו עגול יקיף בו המרובע, נחלוק כל אחד מקוי א״ד וא״ב לחציים על ה׳ ועל ז׳. ונוציא מהם שני קוים על זויות נצבות והם ה״ח וז״ט.

مِثالُهُ: أَنَّا نَجْعَلُ المُرَبَّعَ المَعْلومَ مُرَبَّعَ ا ب ج د. وَنُرِيدُ أَنْ نَعْمَلَ فيهِ دائِرَةً يُحِيطُ بِها. فَيَقْطَعُ كُلُّ واحِدٍ مِنْ خَطَّي ا د ا ب بِنِصْفَيْنِ عَلَى ه و ز. وَنُخْرِجُ مِنْها خَطَّيْنِ عَلَى زوايا قائِمَةٍ، وَهُما خَطّا ه ح ز ط، يَتَقاطَعانِ عَلَى نُقْطَةِ كَ. فَأَقولُ إِنَّ عَلَى مَرْكَزِ كَ وَبِبُعْدِ ك ه ك ز ك ح ك ط نَعْمَلُ الدّائِرَةَ الَّتي يُحِيطُ بِها مُرَبَّعُ ا ب ج د.

نُرِيدُ أَنْ نَعْمَلَ فِي مُرَبَّعٍ مَعْلُومٍ دائِرَةً يُحِيطُ بِها.

נרצה לעשות במרובע[1] ידוע עגולה[2] יקיף בה.

فَنَجْعَلَ المُرَبَّعَ[1] ا ب ج د. وَنَقْسِمَ[2] كُلَّ واحِدٍ مِنْ خَطَّيْ ا د ا ب بِنِصْفَيْنِ على نُقْطَتَيْ ه ز. وَنُخْرِجَ مِنْ نُقْطَتَيْ ه ز خَطَّيْ ه ح ز ط على زَوايا قائِمَةٍ.

הנה נשים המרובע אבג״ד. ונחלק כל אחד מקוי א״ד א״ב בשני חציים על שתי נקודות ה׳ ז׳. ונוציא משתי נקודות ה׳ ז׳ שני קוי ה״ח ז״ט על זויות נצבות.

---

[1]المربع] U +المعلوم [2]ونقسم] מכאן ועד סוף ההוכחה יש הבדלים חשובים רבים בין הנוסח בכתבי היד הערביים ממשפחה A-1 ובכתב יד U לנוסחים הנמסרים בכתבי היד האחרים; תרגום אבן תיבון נסמך בעיקרו על הנוסח בכתבי היד ממשפחה A-1

שרטוט—ס: הריבוע חופף חלקית למעגל
[1]במרובע] ⟨ב⟩מרובע **ט** [2]עגולה] בעגולה **ט**

וא״ה שוה לה״ד. וא״ד כלו כפל א״ה. וא״ב כפל א״ז. ואם כן א״ה שוה לא״ז, לפי שא״ד[1] שוה לא״ב. וא״ה נכוחי לז״כ. וא״ז נכוחי לה״כ. ואם כן שטח אזכ״ה נכוחי הצלעות. ואם כן כל צלעיו וזויותיו שוות[2]. ואם[3] כן (ז״א)[4] שוה לכ״ה. וז״כ שוה לא״ה. וה״ח שוה לא״ב. וא״ז שוה לה״כ. ומזה הצד יתבאר שכ״ה שוה לכ״ט וכ״ט לכ״ח [!]. ואם כן הקוים הארבעה כ״ה וכ״ז וכ״ח וכ״ט שוים. ונקיף על מרכז כ׳ ובמרחק ז׳ וה׳ וט׳ וח׳ עגול אשר תוך מרובע אבג״ד. והמרובע יקיף בו, והוא עגול הזח״ט.

بُرْهانُهُ: أَنَّ ا ه مِثْلُ ه د. وَجَمِيعُ ا د مِثْلا ا ه. وَكَذَلِكَ ا ب مِثْلا ا ز. ا د مِثْل ا ب. ا ه مِثْل ا ز. و ا ه مُوازٍ لِخَطِّ ز ك. و ا ز مُوازٍ لِخَطِّ ه ك. فَسَطْحُ ا ز ك ه مُتَوازي الأَضْلاعِ. فَأَضْلاعُهُ وَزَواياهُ المُتَقابِلَةُ بَعْضُها لِبَعْضٍ مُتَساوِيَةٌ، ز ا مِثْل ك ه. و ز ك مِثْلُ ا ه. وَلَكِنَّ ك ز مِثْلُ ب ح. ب ح مِثْلُ ه ا. ه ح مِثْل ا ب. ا ز مِثْلُ ه ك. وَلَكِنَّ جَمِيعَ ا ب مِثْلا ا ز. و ح ه مِثْلا ه ك. ه ك مِثْل ك ح. وَكَذَلِكَ ز ك مِثْل ك ط. ا ه مُوازٍ لِخَطِّ ك ز. وَوَقَعَ عَلَيْهِما خَطُّ ز ا. فَالزَّاوِيَتانِ الدَّاخِلَتانِ مِثْل زَاوِيَتَيْنِ قائِمَتَيْنِ. وَالزَّاوِيَةُ الَّتي مِنْ ا ز ك قائِمَةٌ. وبَقِيَتِ الَّتي من ز ا ه قائِمَةٌ. وَالَّتي مِنْ ا ه ك قائِمَةٌ. وَالباقِيَةُ الَّتي من ه ك ز قائِمَةٌ. فَسَطْحُ ا ك ه ز زَواياهُ قائِمَةٌ. وَسَطْحُ ا ز ك ه مُتَساوي الأَضْلاعِ قائِمِ الزَّوايا. ه ك مِثْل ك ز. وَكَذَلِكَ ك ط و ك ح مِثْل ز ك و ك ه. والأَضْلاعُ الأَرْبَعَةُ ك ه و ك ح و ك ط و ك ز مُتَساوِيَةٌ. فَعَلى مَرْكَزِ ك وَبِبُعْدِ ز و ه و ح و ط [يُ]حيطُ الدَّائِرَةَ الَّتي في مُرَبَّعِ ا ب ج د ويُحيطُ بِها، وَهِيَ دائِرَةُ ه ز ح ط.

[1]לפי שא״ד] ⟨⟨לפי ש⟩⟩וא״ד [2]שוות] שו⟨ו⟩ת
[3]ואם] ⟨⟨[..]⟩⟩+ [4]ז״א] ז״כ

فكُلّ واحدٍ مِنْ سُطُوحِ ا ب ح ه و ط ز ب ج مُتَوازي الأَضلاعِ. فكُلّ واحِدَةٍ مِنَ^ الزَّوايا الَّتي عِنْدَ ح ط قائِمَةٌ، لأَنّها تُقابلُ[1] الزَّوايا القائِمَةَ.^ والزَّوايا الَّتي عِنْدَ نُقْطَتَيْ ه ز قَوائِم. وَلأَنَّ خطّ ا ب مثل خطّ ا د، يكونُ خطُّ ا ز مثل خطِّ ا ه، لأَنَّ خَطَّيْ ا ز ا ه نِصْفا خَطَّيْ ا د ا ب. وخطّ ا ه مثل خطّ ب ح، لأَنَّهُ يُقابِلُهُ. وخطّ ز ك مثل خطِّ ا ه أيضًا. وخطّ ز ا مثل خطِّ ك ه ومثل خطِّ ط د. فكَذَلِكَ خطّ ه د مثل خطِّ ك ط ومثل خطِّ ج ح. وخطّ ز ب مثل خطِّ ك ح ومثل خطِّ ط ج. فسُطُوح ا ك ه ط ك ب ك ج الأَرْبَعَة مُرَبَّعات مُتَساوِيَة. فتكونُ خُطوطُ ك ز ك ه ك ط ك ح الأَرْبَعَةُ مُتَساوِيَةً. فَقَدْ تَبَيَّنَ أَنَّ الزَّوايا المُخرَجَةَ مِنْ أَطْرافِ هَذِهِ الخُطوطِ قائِمَةٌ. فَإذا جَعَلْنا نُقْطَةَ ك مَرْكَزًا، وَأَدَرْنا دائِرَةً بِبُعْدِ كُلِّ واحِدٍ مِنْ خُطوطِ ك ز ك ه ك ح ك ط، مَرَّتِ الدَّائِرَةُ بِنُقَطِ ز ه ط ح، وَماسَّتْ أَضْلاعَ مُرَبَّعِ^ ا ب ج د، لأَنَّ خَطَّيْ ز ط ه ح قُطْرا الدَّائِرَةِ. وَالزَّوايا الَّتي عِنْدَ أَطْرافِها، كُلّ واحِدَةٍ مِنْها^، قائِمَة. فَأَضْلاعِ مُرَبَّعِ ا ب ج د إذا خَطَطْنا هَذِهِ الدَّائِرَةَ—تَماسُّها. فلنَخُطّ هَذِهِ^ الدَّائِرَةَ في المُرَبَّعِ المَفروضِ، فتكونُ مُماسَّةً لأَضْلاعِ ا ب ب ج ج د د ا، لأَنَّ كُلّ واحِدَةٍ مِنَ^ الزَّوايا الَّتي عِنْدَ نُقْطَةِ ه ز ح ط قائِمَةٌ.

[1] تقابل] תמאתל P

הנה כל אחד משטחי אחה״ג[1] אטז״ג נכחי הצלעות. אם כן הזויות אשר אצל ח׳ ט׳ נצבות, כי הם יקבילו הזויות הנצבות אשר אצל נקודות[2] ב׳ א׳ ד׳ מן המרובע הידוע. והזויות אשר אצל ה׳ ז׳ נצבות. ומפני כי א״ב כמו א״ד, יהיה א״ז כמו א״ה, מפני כי א״ז וא״ה חציי א״ד א״ב. וקו א״ה כמו ב״ח, כי הוא יקבילהו. וז״כ כמו א״ה גם כן. וז״א כמו[3] כ״ה וד״ט. וכן ה״ד כמו כ״ט וג״ח[4]. וז״ב כמו כ״ח וט״ג. אם כן שטחי א״כ ה״ט כ״ב כ״ג הארבעה מרובעים שוים[5]. אם כן יהיו[6] קוי כ״ז כ״ה כ״ט כ״ח הארבעה שוים. וכבר התבאר כי הזויות המוצאות מקצוי אלו הקוים נצבות. אם כן כאשר שמנו נקודת כ׳ מרכז, וסבבנו במרחק כל אחד מקוי כ״ז כ״ה כ״ח כ״ט עגולה, הלכה העגולה בנקודות ז׳ ה׳ ט׳ ח׳, ותמשש צלעות אבג״ד[7], מפני כי שני קוי ז״ט ה״ח שני קוטרי העגולה. והזויות אשר אצל קצויהם נצבות. אם כן צלעות מרובע אבג״ד, כאשר קוינו העגולה הזאת—תמששם[8]. אם כן הנה נקוה העגולה[9] במרובע המונח, ותהיה ממששת[10] לצלעות א״ב ב״ג ג״ד א״ד, מפני כי הזויות אשר אצל נקודות ה׳ ז׳ ח׳ ט׳ נצבות.

[1] אחה״ג] ~~א״ג~~+ ק [2] נקודות] נקודת ס [3] כמו] ליתא ב [4] וג״ח] וג״ט בו ח״ג ס [5] שוים] +והם בוס [6] יהיו] ליתא ב [7] אבג״ד] מרובע+ טק [8] תמששם] תתמששה ב תמששה וס [9] העגולה] +היא בו +הזאת טק [10] ממששת] המששת פ

וזה מה שרצינו.

وَذَلِكَ ما أَرَدْنا بَيانَهُ.

وذَلِكَ ما أَرَدْنا أَنْ نُبَيِّنَ.

וזה מה שרצינו לבאר.

[Proposition IV.9]

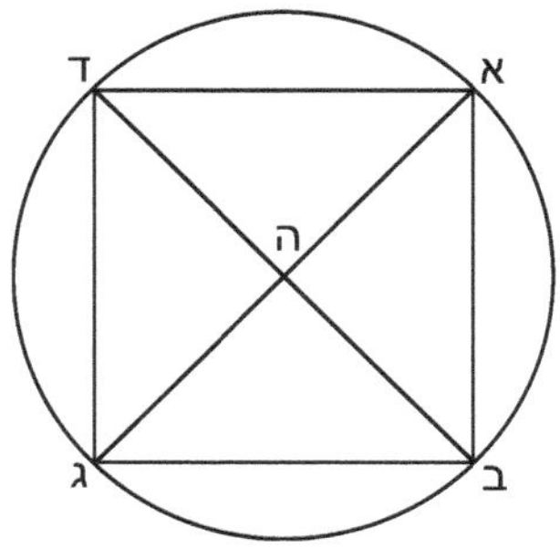

נרצה שנעשה על מרובע ידוע עגול מקיף בו.

نُرِيدُ أَنْ نُبيِّنَ كَيْفَ نَعْمَلُ عَلى مُرَبَّعٍ مَعْلومٍ دائِرَةً تُحِيطُ بِهِ.

ונשים המרובע הידוע מרובע אבג״ד. וכאשר רצינו לעשות עליו עגול מקיף בו, נדביק א׳ בג׳ וב׳ בד׳.

مِثالُهُ: أَنَّا نَجْعَلُ المُرَبَّعَ المَعْلومَ مُرَبَّعَ ا ب ج د. وَنُرِيدُ أَنْ نَعْمَلَ عَلَيْهِ دائِرَةً تُحِيطُ بِهِ. فَنَصِلُ ا ج و ب د. فَأَقولُ إِنَّ عَلى مَرْكَزِهِ وَبِبُعْدِ ه ا و ه ب ه ج و ه د بِخَطِّ الدّائِرَةِ عَلى م مُرَبَّعِ ا ب ج د المَعْلوم.

نُرِيدُ أَنْ نَعْمَلَ على مُرَبَّعٍ مَعْلومٍ دائِرَةً تُحيطُ بِهِ.

נרצה לעשות על מרובע ידוע עגולה תקיף בו.

فَنَجْعَلَ المُرَبَّعَ المَعْلومَ مُرَبَّعَ ا ب ج د. ونُرِيدُ أَنْ نَعْمَلَ عَلَيْهِ دائِرَةً تُحيطُ بِهِ. فَنُخْرِجَ خَطَّيْ ا ج ب د.

הנה נשים המרובע הידוע מרובע אבג״ד. ונרצה לעשות עליו עגולה תקיף בו. הנה נוציא שני קוי[1] א״ג ב״ד.

שרטוט—**ס**: אדג״ב הוא מלבן [1]קוי] ליתא **ב**

ולפי שא״ב שוה לא״ד, וזוית בא״ד נצבה, תהיין זוית [!] אב״ד ואד״ב כל אחת מהן חצי נצבה. וכמו[1] כן התבאר ששתי זויות דא״ג ובא״ג[2] כל אחת מהן חצי נצבה. ואם כן זוית אד״ב שוה לזוית דא״ג. וצלע ה״א שוה לצלע ה״ד. וכמו כן יתבאר שה״ג וה״ב שוים לא״ה וה״ד. ואם כן בארבעה [!] קוים ה״א וה״ד וה״ב וה״ג הם שוים. ונקיף עגול על מרכז ה׳ ובמרחק א׳ וב׳ וג׳ וד׳, ויקיף במרובע אבג״ד.

וזה מה שרצינו.

بُرْهانُهُ: أَنَّ ا ب مِثْل ا د. وَالزَّاوِيَةُ الَّتِي مِنْ ب ا د قائِمَةٌ. فَالزَّاوِيَتانِ اللَّتانِ مِنْ ا ب د ومِنْ ا د ب، كُلّ واحِد مِنْهُما، نِصْفُ قائِمَة. وَكَذَلِكَ الزَّاوِيَتانِ اللَّتانِ مِنْ د ا ج وَمِنْ ب ا ج، كُلُّ واحِدَةٍ نِصْف قائِمَةٍ. فَالزَّاوِيَةُ الَّتِي من ا د ب مِثْل الَّتِي من د ا ج. وضِلْعُ ه ا مِثْل ضِلْعِ ه د. وكَذَلِكَ ه ج و ه ب مِثْل ا ه و ه د. وَالأَضْلاعُ الأَرْبَعَةُ ه ا و ه د و ه ب و ه ج مُتَساوِيَةٌ. فَعَلى مَرْكَزِ ه وَبِبُعْدِ ا و ب و ج و د نَخُطّ الدَّائِرَةَ الَّتِي تُحِيطُ بِمُرَبَّعِ ا ب ج د، وتَجُوزُ عَلى سائِرِ الخُطُوطِ، وَهِيَ دائِرَةُ ا ب ج د.

وَذَلِكَ ما أَرَدْنا بَيانَهُ.

[1] וכמו] ~~ואם כן זוית~~+ [2] ובא״ג] +⟨⟨ודג״א⟩⟩

ف اب مثل اد. وزاوية ب اد قائمة. فزاويتا اب د ا د ب — كلّ واحدة[1] نصف قائمة. وكذلك تكون زاويتا د اج د ج ا — كلّ واحدة نصف قائمة. فزاوية ا د ب مثل زاوية د اج. فضلع اه مثل ضلع ه د. وكذلك يكون ب ه مثل ه ج. و ه ج مثل د ه[2]. فخطوط^ ه د ه ا ب ه ه ج متساوية. فعلى مركزه وببعدا ب ج د^ نخطّ الدّائرة المحيطة[3] بالمربّع ا ب ج د.

הנה א״ב כמו א״ד. וזוית בא״ד נצבת. אם כן שתי זויות אב״ד אד״ב—כל אחת[1] חצי נצבת. וכן יהיו שתי זויות דא״ג דג״א—כל אחת חצי נצבת. אם כן זוית אד״ב כמו זוית דא״ג. אם כן צלע א״ה כמו צלע ה״ד. וכן יהיה ב״ה כמו ה״ג. וה״ג כמו ד״ה. אם כן ה״ד ה״א ה״ב ה״ג שוים. הנה על מרכז ה׳ ובמרחק ה״ג הקפנו עגולה מקפת במרובע אבג״ד.

وذلك ما أردنا أن نبيّن

וזה מה שרצינו לבאר.

[1] واحدة] +منهما CamMO2PR1R2

[2] وكذلك ... د ه] MPR1R2 وكذلك يكون خطي ب ه ه ج مثل خطي ا ه ه د [3] الدّائرة المحيطة] MPR1R2 دائرة يحيط

[1] אחת] +~~צב״ד~~ ב

[Proposition IV.10]

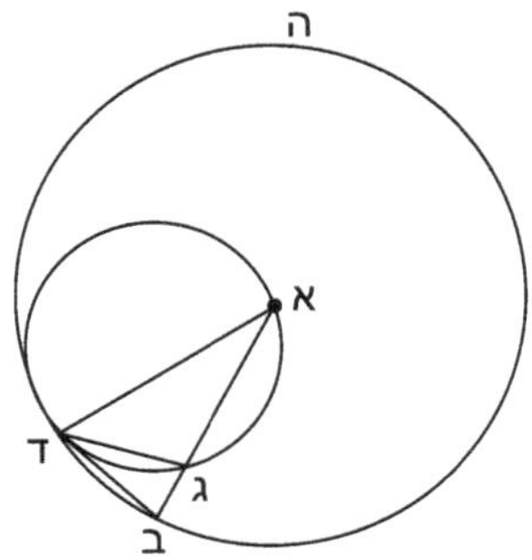

נרצה לעשות משולש שוה השוקים ותהיה כל אחת מזויותיו אשר אצל התושבת כפל הזוית הנשארת.

ונרשום קו א״ב. ונחלוק אותו על נקודת ג׳ בדרך שיהיה העולה מא״ב בב״ג, שוה לעולה מא״ג בעצמו. ונקיף על מרכז א׳ במרחק (א״ב)[1] עגול בד״ה. ונשים בעיגול זה קו שנוציא מנקדת ב׳ שוה לקו א״ג, והוא קו ב״ד. ונדביק ג׳ בד׳ וא׳ בד׳. ונקיף על משולש אג״ד עגול אג״ד.

نُرِيدُ أَنْ نَعْمَلَ مُثَلَّثًا مُتَساوِيَ الأَضْلاعِ السَّاقَيْنِ (تكونُ)[1] كُلُّ واحِدَةٍ مِنْ زاوِيَتَيْهِ اللَّتَيْنِ عَلى القاعِدَةِ مِثْلَيِ الزَّاوِيَةِ الباقِيَةِ.

مثالُه: أَنَّا نَخُطُّ خَطًّا وَنَرْسُمُ عَلَيْهِ ا ب. وَنَقْسِمُهُ عَلى نُقْطَةِ ج حَتَّى يَصِيرَ ما يَكونُ مِنْ ضَرْبِ ا ب في ب ج مِثْلَ الَّذي يَكونُ مِنْ ضَرْبِ ا ج في نَفْسِهِ. فَنَخُطُّ عَلى مَرْكَزِ ا وَبِبُعْدِ ا ب دائِرَةَ ب د ه. وَنَجْعَلُ في هَذِهِ الدَّائِرَةِ خَطًّا مِثْلَ خَطِّ ا ج، وَهُوَ ب د. وَنَصِلُ ا د. فَأَقولُ إِنَّ مُثَلَّثَ (ا ج د)[2] كَما أَرَدْنا.

[1] א״ב] ב׳

[1] تكون] يكون [2] ا ج د] ا ب د

نُرِيدُ أَنْ نَعْمَلَ مُثَلَّثًا مُتساوي السَّاقَيْنِ تَكونُ كُلُّ واحِدَةٍ مِنْ زاوِيَتَيْهِ اللَّتَيْنِ على القاعِدَةِ مِثْلَيِ الزَّاوِيَةِ الباقِيَةِ.

فنخُطُّ خطّ ا ب. وَنَقْسِمَهُ على ج قِسْمَةً يَصيرُ بِها ا ب ب ج[1] مثل ا ج في نَفْسِهِ^. وَنَخُطُّ على مَرْكَزِ ا وبِبُعْدِ ا ب دائِرَةَ ب د ه. ونُخْرِجَ مِنْ ب وتر^ ب د مثل خطّ ا ج. ونُخْرِجَ خَطَّيْ ج د د ا. ونَخُطَّ على مُثَلَّثِ ا ج د دائِرَةً تُحيطُ بِهِ وَهِيَ دائِرَةُ ا ج د.

נרצה לעשות משולש שוה השוקים תהיה כל אחת מזויותיו אשר על התושבת כפל הזוית הנשארת.

הנה נקוה[1] קו א״ב. ונחלקהו על ג׳ חלוקה יהיה בה א״ב ב״ג כמו א״ג בעצמו. ונשים נקודת א׳ מרכז. ונקוה[2] על מרכז א׳ ובמרחק א״ב[3] עגולת[4] הב״ד. ונוציא מן ב׳ מיתר יהיה שוה אל א״ג, והוא ב״ד. ונגיע קוי א״ד ג״ד. ונקוה על משולש אג״ד עגולה תקיף בו, והיא עגולת אג״ד.

[1] ب ج] T2 في+

שרטוט—**בוסטפק**: $ [1]נקוה] ⟨נ⟩קוה **ב** [2]ונקוה] +גם כן **ט** +⟨⟨גם כן⟩⟩ **פ** [3]א״ב] ב׳ **בו** [4]עגולת] עגולה **טספ**

بُرْهانُهُ: أَنْ نَصِلَ ج د ه. وَنَخُطَّ عَلَى مُثَلَّثِ ا ج د دائِرَةً عَلَيْها ا ج د. فالَّذي يَكونُ مِنْ ضَرْبِ ا ب في ج ب مِثْل الَّذي يَكونُ مِنْ ضَرْبِ ا ج في نَفْسِهِ. و ج ا مِثْل ب د. فالَّذي يَكونُ مِنْ ضَرْبِ ا ب في ب ج مِثْل الَّذي يَكونُ مِنْ ضَرْبِ (ب د)[1] في نَفْسه. ونُقْطَة ب خارِجَة من دائِرَةِ ا ج د. وَقَد أُخْرِجَ مِنْ ب خَطّانِ إِلى دائِرَةِ (ا ج د)[2]، أَحَدُهُما يَقْطَعُ الدَّائِرَةَ، وَهُوَ ب ا، والآخَرُ يَنْتَهي إِلَيْها، وَهُوَ ب د. والَّتي يَكونُ مِنْ ضَرْبِ ا ب في ب ج مِثْل ما يَكونُ مِنْ ضَرْبِ ب د في نَفْسِهِ. ب د يَماسُّ دائِرَةَ ا ج د. وَنُخْرِجُ مِنَ المَوْضِعِ الَّذي يَماسُّها فيهِ خَطُّ د ج فَقَطَعَ الدَّائِرَةَ. فَالزَّاوِيَتانِ اللَّتانِ عَنْ جَنْبَتَيْ د ج مِثْل الزَّاوِيَتَيْنِ اللَّتَيْنِ تَقَعانِ في قِطْعَتَي الدَّائِرَةِ المُتَبادَلَتَيْنِ الَّتي مِنْ ج د ب مِثْل الَّتي مِنْ ج ا د. وَنَجْعَلُ[3] الَّتي من ج د ا مُشْتَرَكَةً. فَجَميع الَّتي من ب د ا مِثْل الَّتي من ج د ا ومن د ا ج. وَلٰكِنَّ اللَّتانِ من ج د ا ومن د ا ج مِثْلهما جَميعًا الَّتي من خارِجِ المُثَلَّثِ من ب ج د. فَالَّتي من ب ج د مِثْل الَّتي من ب د ا. والَّتي من ب د ا مِثْل الَّتي من د ب ا. والَّتي من د ب ا مِثْل الَّتي من ب ج د. فَضِلْع ب د مِثْل ضِلْع ج د. و ب د مِثْل ا ج. ا ج مِثْل ج د. والزَّاوِيَةُ الَّتي من ج ا د مِثْل الَّتي من ج د ا. والَّتي من خارِجِ المُثَلَّثِ

[1] ب د] ب ه [2] ا ج د] ا ج ك [3] ونجعل] ونجعل التي من ج د ا+

ולפי שהעולה מא״ב בב״ג שוה לעולה מא״ג בעצמו, וג״א שוה לב״ד, יהיה העולה מא״ב בב״ג שוה לעולה מב״ד בעצמו. ונקודת ב׳ היא חיצונה לעיגול אג״ד. וכבר יצאו מב׳ שני קוים אל עיגול אג״ד—אחד מהם יחתוך אותו, והוא ב״א, והאחר יכלה אליו, והוא ב״ד. והעולה מא״ב בב״ג שוה לעולה בב״ד בעצמו. ואם כן ב״ד ימשש עיגול אג״ד. וכבר יצא ממקום מישושו אליו קו ד״ג חותך העגול. ואם כן שתי הזויות אשר אצל צדי קו ד״ג שוות לשתי הזויות אשר תפלנה בשתי חתיכות העגול המתחלפות להן. ואם כן תהיה זוית גד״ב שוה לזוית גא״ד. ונשים זוית גד״א משותפת. ותהיה זוית בד״א שוה[1] לזוית גד״א וזוית דא״ג. וזוית החצונה למשולש, והיא זוית בג״ד, שוה לשתיהן יחד. ואם כן זוית בג״ד שוה לזוית בד״א. וזוית בד״א שוה לזוית דב״א. ואם כן זוית דב״א שוה לזוית דג״ב. ואם כן צלע ב״ד שוה לצלע ג״ד. ותהיה זוית גא״ד שוה לזוית גד״א, ושתיהן יחד הן כפל לזוית (גא״ד)[2]. והזוית החצונה, והיא זוית בג״ד, שוה לזוית גא״ד וזוית גד״א יחד. ואם כן זוית בג״ד כפל זוית דא״ג. וזוית בג״ד שוה לזוית אב״ד ולזוית אד״ב. ואם כן כל אחת מזויות אב״ד ואד״ב הוא כפל זוית בא״ד. וכבר עשינו משולש שוה השוקים והוא שעליו אב״ד ושתי הזויות אשר אצל התושבת כל אחת מהן כפל הזוית הנשארת.

[1] שוה] ~~בלתי~~+ [2] גא״ד] גה״ד

فيكونُ اب في ب ج مثل اج في نَفْسِهِ. و ج ا مثل ب د. ف اب في ب ج مثل ب د في نَفْسِهِ. ونُقْطَة ب خارجة عن دائِرَةِ اج د. وَقَدْ خَرَجَ مِنْها إلى دائِرَةِ اج د خَطّانِ— أَحَدُهُما يَقْطَعُها، وَهُوَ اب، وَالآخَرُ يَنْتَهي إِلَيْها، وهو ب د[1]. والَّذي يكونُ مِنْ اب في ب ج مثل ب د في نَفْسِهِ. ف ب د يَماسُّ دائِرَةَ اج د. وَقَدْ خَرَجَ مِنْ حَيْثُ يَماسُّها خَطُّ د ج فَقَطَعَ الدَّائِرَةَ. فالزَّاوِيَتانِ اللَّتانِ عَنْ جَنْبَتَيْهِ مثل اللَّتَيْنِ تَقَعانِ في قِطْعَتَيْ الدَّائِرَةِ المُتَبادلَتَيْنِ لَهُما. فزاوية ج د ب مثل زاوية ج ا د، وزاوية ج د ا مشتركة. فجميع زاوية ب د ا مثل زاوِيَتَيْ ج د ا د ا ج. ولكنّ زاوِيَتَيْ ج د ا د ا ج^ مِثْلُهما زاوية ب ج د الخارجة مِنَ المُثَلَّث. فزاوية ب ج د مثل زاوية ب د ا، وزاوية ب د ا مثل زاوية د ب ا. فزاوية د ب ا مثل زاوية ب ج د. فضلع ب د مثل ضلع ج د. و ب د مثل اج. ف اج مثل ج د. فزاوية ج ا د مثل زاوية ج د ا. وَزاوِيَتا ج ا د ج د ا مثلًا زاوية ج ا د. فزاوية ب ج د الخارِجَة مِنَ المُثَلَّثِ مثل زاوِيَتَيْ ج ا د ج د ا جَمِيعًا. فزاوية ب ج د مِثْلا زاوية د ا ج. وزاوية ب ج د مثل زاوية اب د ومثل زاوية اد ب. فكُلّ واحِدَةٍ مِنْ زاوِيَتَيْ اب د اد ب مِثْلا زاوية ب ا د.

[1] ب د] U

הנה יהיה א״ב בב״ג כמו א״ג בעצמו. וג״א כמו ב״ד. אם כן יהיה א״ב בב״ג כמו ב״ד בעצמו. ונקודת ב׳ חוץ מעגולת אג״ד. וכבר יצאו ממנה אל עגולת אג״ד שני קוים—אחד מהם יחתכה, והוא א״ב, והאחר יכלה אליה, והוא ב״ד. ואשר יהיה[1] מן א״ב בב״ג כמו ב״ד בעצמו. אם כן ב״ד ימשש עגולת אג״ד. וכבר יצא מהמקום שימששה קו ד״ג וחתך העגולה[2]. ושתי הזויות אשר משני צדדיו כמו[3] אשר יפלו בשתי חתיכות העגולה המומרות לשתיהן. אם כן זוית גד״ב[4] כמו זוית גא״ד, וזוית גד״א משותפת. אם כן כל[5] זוית בד״א כמו[6] שתי זויות גד״א דא״ג. אבל שתי זויות גד״א דא״ג[7] שתיהן יחד כמו זוית[8] בג״ד החיצונה מן המשולש[9]. אם כן זוית בג״ד כמו זוית בד״א, וזוית בד״א כמו זוית דב״א. אם כן[10] זוית דב״א[11] כמו זוית בג״ד. אם כן צלע ב״ד כמו צלע ג״ד. וב״ד כמו א״ג. אם כן[12] א״ג כמו ג״ד. אם כן זוית גא״ד כמו זוית גד״א. ושתי זויות גא״ד גד״א כפל זוית[13] גא״ד. וזוית[14] בג״ד החיצונה מן המשולש כמו שתי זויות גא״ד גד״א[15] יחד. אם כן זוית בג״ד כפל זוית דא״ג. וזוית בג״ד כמו זוית אב״ד[16] וכמו זוית אד״ב. הנה כל אחת משתי זויות אב״ד אד״ב כפל זוית בא״ד.

[1] יהיה] ליתא **בוס** [2] העגולה] עגולה **ב** +על זולת המרכז **ס** [3] כמו] +שתי זויות **ס** [4] גד״ב] ⟨⟨*⟩⟩ **ב** [5] כל] ליתא **בפ** [6] כמו] +כל **וק** [7] דא״ג] ⟨⟨*⟩⟩ **ב** [8] זוית] זויות **פ** [9] המשולש] המשולשים **ב** [10] כן] ליתא **פ** [11] זוית דב״א] ⟨⟨*⟩⟩ **ב** [12] כן] ליתא **ב** [13] זוית] זויות **בס** [14] וזוית] אם כן זוית **ק** [15] גד״א] ליתא **ס** [16] אב״ד] אד״ב **ב** אב״ה **ט**

الَّتي من ب ج د مِثْل اللّتينِ من ج ا د ومن ج د ا جَميعًا. والَّتي من ب ج د مِثْلا الَّتي من د ا ج. والَّتي من ب ج د مِثْل الَّتي من (ا ب د)[1] مِثْل الَّتي من ا د ب. فَكُلّ واحِدَة مِنَ اللّتَيْنِ من ا ب د ومن (ا د ب)[2] هي مِثْلا الَّتي من ب ا د. فَقَدْ عَمِلْنَا مُثَلَّثًا مُتَساوي السَّاقَيْنِ عَلَيْه ا ب د وزاوِيَتاهُ اللّتانِ عَنْ ب د كُلّ واحِدَةٍ مِنْهُا مِثْلا الزَّاوِيَة الباقِيَة.

وَذَلِكَ ما أَرَدْنا بَيانَهُ.

וזה מה שרצינו.

[1] ا ب د] ب ا د [2] ا د ب] د ب ا

فَقَدْ عَمِلْنا مُثَلَّثًا مُتَساوي السَّاقَيْنِ عَلَيْهِ ا ب د، تكونُ كُلُّ واحِدَةٍ مِنْ زاوِيَتَيْهِ اللَّتَيْنِ على قاعِدَةِ[1] ب د مِثْلَي الزَّاوِيَةِ الباقيَةِ.

הנה כבר עשינו משולש שוה השוקים עליו אב״ד, תהיה כל אחת מזויותיו אשר על תושבת[1] ב״ד כפל הזוית[2] הנשארת[3].

وذلك ما أَرَدْنا أَنْ نُبيِّنَ.

וזה מה שרצינו לבאר[4].

---

[1] قاعِدَةِ] O2U التي هي

---

[1] תושבת] התושבת **ק** [2] הזוית] הזויות **ב** [3] הנשארת] הנשאר **בס** [4] וזה ... לבאר] ונשלם ביאורו **טס**

[Proposition IV.11]

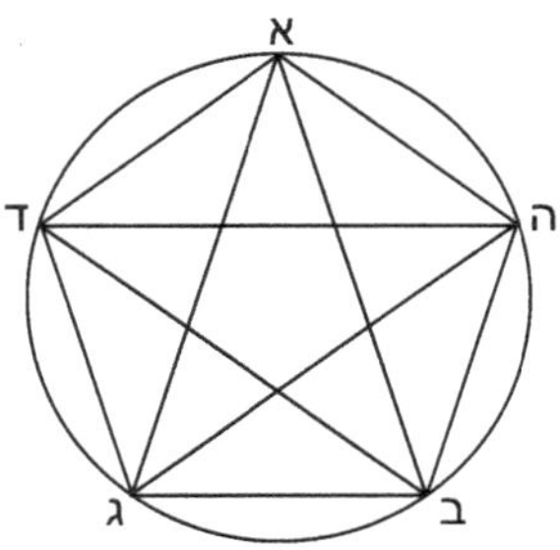

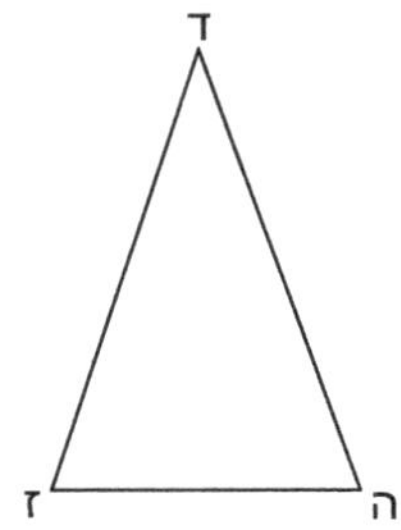

נרצה לעשות בתוך עגול ידוע מחומש שוה הצלעות והזויות יקיף בו העגול.

ונשים העגול הידוע עגול אב״ג. וכאשר רצינו שנעשה בתוכו מחומש שוה הצלעות והזויות שיקיף בו העגול, נעשה[1] משולש שוה השוקים, היה כל[2] אחת מזויותיו שאצל התושבת כפל הזוית הנשארת. ותהיה זה משולש דה״ז. ונעשה בתוך עגול אב״ג משולש אב״ג[3] שוה הזויות לזויות משולש דה״ז. ונשים כל אחת מזויות אב״ג ואג״ב כפל זוית בא״ג הנשארת. ונחתוך זוית ב׳ לחציים בקו ב״ד, וזוית ג׳ גם כן לחציים בקו ג״ה. ונדביק א׳ ב״ה׳, וה׳ בב׳, וא׳ בד׳, וד׳ בג׳. ואומר שכבר עשינו בתוך עגול אב״ג מחומש שוה הצלעות, והזויות עליו אהבג״ד.

نُرِيدُ أَنْ نُبَيِّنَ كَيْفَ نَعْمَلُ مُخَمَّسًا مُتَساوي الأَضْلاعِ وَالزَّوايا في دائِرَةٍ مَعْلومَةٍ تُحيطُ بِهِ.

مِثالُهُ: أَنَّا نَجْعَلُ الدَّائِرَةَ المَعْلومَةَ دائِرَةَ ا ب ج. وَنُرِيدُ أَنْ نَعْمَلَ فيها مُخَمَّسًا مُتَساوِيَةَ الأَضْلاعِ وَالزَّوايا تُحيطُ به الدَّائِرَةُ. فَنَعْمَلُ مُثَلَّثًا مُتَساوي السَّاقَيْنِ تَكونُ كُلُّ زاوِيَةٍ مِنْ زَواياهُ اللَّتَيْنِ عَلَى القاعِدَةِ مِثْلَ الزَّاوِيَةِ الباقِيَةِ. وَنَرْسُمُ عَلى المُثَلَّثِ د ه ز. وَنَعْمَلُ في دائِرَةِ ا ب ج مُثَلَّثًا عَلَيْهِ ا ب ج مُساوِيَةً زَواياهُ لِزَوايا مُثَلَّثِ د ه ز. وَنَجْعَلُ كُلَّ واحِدَةٍ مِنْ زاوِيَتَيْ مُثَلَّثِ ا ب ج اللَّتَيْنِ مِنْ ا ب ج ومن ا ج ب مِثْلَيِ الزَّاوِيَةِ الَّتِي من ب ا ج. وَنَقْطَعُ زاوِيَةَ ب بِنِصْفَيْنِ بِخَطِّ ب د، وَزاوِيَةَ ج أَيْضًا نِصْفَيْنِ بِخَطِّ ج ه. وَنَصِلُ ا ه ه ب ا د د ج. فَأَقولُ إِنَّا قَدْ عَمِلْنا في دائِرَةِ ا ه مُخَمَّسًا مُتَساوي الأَضْلاعِ وَالزَّوايا عَلَيْهِ (ا ه ب ج د)[1].

שרטוט—B: ה׳ על המחומש ← ط, שלא כנכתב

[1]נעשה] ונעשה [2]כל] +~~הזויות~~ [3]אב״ג] ⟨⟨*⟩⟩

[1]ا ه ب ج د] ا ب ج د

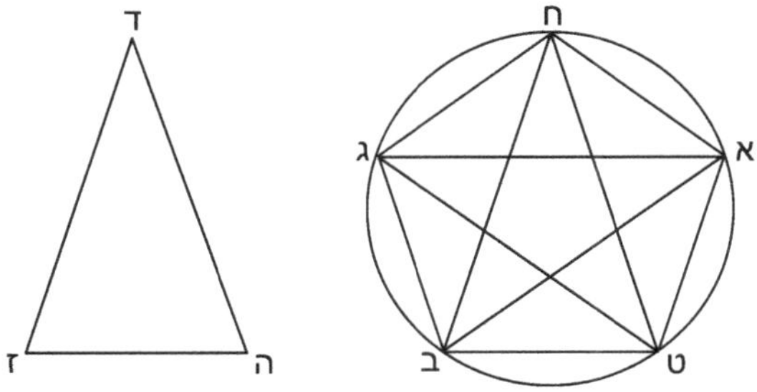

نُرِيدُ أَنْ نَعْمَلَ في دائِرَةٍ مَعْلومةٍ مُخَمَّسًا مُتَساوي الأَضْلاعِ والزَّوايا تُحيطُ بِهِ.

נרצה לעשות בעגולה ידועה מחומש שוה הצלעות והזויות תקיף בו.

فَنَجْعَلَ الدَّائِرَةَ المَعْلومَةَ دائِرَةَ ا ب ج. ونُرِيدُ أَنْ نَعْمَلَ فيها مُخَمَّسًا مُتَساوي الأَضْلاعِ وَالزَّوايا تُحيطُ بِهِ. فنَعْمَلَ مُثَلَّثًا مُتَساوي السّاقَيْنِ، تكونُ كُلُّ واحِدَةٍ مِنْ زاوِيَتَيْهِ اللَّتَيْنِ على القاعِدَةِ مِثْلَيِ الزَّاوِيَةِ الباقِيَةِ، وَهُوَ مُثَلَّثُ د ه ز. ونَعْمَلُ في دائِرَةِ ا ب ج مُثَلَّثَ ا ب ج مُساوِيَة زواياهُ لِزَوايا مُثَلَّثِ د ه ز.

הנה נשים העגולה הידועה עגולת אב״ג. ונרצה לעשות בה מחומש שוה הצלעות והזויות תקיף בו. הנה נעשה משולש שוה השוקים, תהיה כל אחת מזויותיו אשר אצל התושבת כפל הזוית[1] הנשארת, והוא משולש דה״ז. ונעשה בעגולת אב״ג משולש אב״ג, שוה [!] זויותיו לזויות משולש דה״ז[2].

שרטוט—**ס**: חסר ט״ח [1]הזוית] הזויות **ב**
[2]ונעשה ... דה״ז] ליתא **ב**

מופת זה: לפי שכל אחת מזויות ב׳ וג׳ כפל זוית א׳, וכבר נחלקה לחציים כל אחת מזויות ב׳ וג׳, תהיין הזויות החמש, והן זוית בא״ג ואג״ה והג״ב וגב״ד ודב״א, שוות. והחמש קשתות שעליהן א״ה וה״ב וב״ג [!] וג״ד וד״א שוות, ומיתריהם שוים. ואם כן מחומש אבגד״ה שוה הצלעות. והוא גם כן שוה הזויות, לפי שקשת ה״ב שוה לקשת ג״ד. ונשים הא״ד משותפת. יהיה בהא״ד שוה לגדא״ה. ואמנם זוית בג״ד היא על קשת בהא״ד, וזוית[1] גב״ה היא על קשת גדא״ה. ואם כן זוית גב״ה שוה לזוית בג״ד. ומצד זה תהיין זויות גד״א ודא״ה ואה״ב שוות לזויות בג״ד וגב״ה. ואם כן מחומש אהבג״ד שוה הצלעות והזויות. וכבר נעשה בתוך עגול אב״ג והעגול מקיף בו.

וזה מה שרצינו.

[1] וזוית] וקשת⟨זוית⟩

بُرْهانُهُ: أَنَّ كُلَّ واحِدَةٍ مِنْ زاوِيَتَيْ بَ جَ[1] مِثْلا زاوِيَةِ آ. فَقَدْ قُطِعَتْ كُلُّ واحِدَةٍ مِنْهُما بِنِصْفَيْنِ. فَالزَّوايا الخَمْسُ اللَّواتي مِنْ ب ا ج ومن ا ج ه ومن ه ج ب ومن ج ب د ومن د ب ا مُتَساوِيَةٌ. وَالقِسِيُّ الخَمس اللَّواتي عَلَيْها ا ه و ه ب و ب ج وج د و د ا مُتَساوِيَةٌ. فَمُخَمَّسُ ا ب ج د ه مُتَساوي الأَضْلاعِ وَالزَّوايا. وقَوْسُ ه ب مِثْلِ قَوْسِ ج د. ونَجْعَلُ قَوْسَ ه ا د مُشْتَرَكَةً. فَجَميعُ قَوْسِ ب ه ا د مِثْل جَميع قَوْسِ (ج د ا ه)[2]. وَلكِنَّ الزّاوِيَةَ الَّتي مِنْ ب ج د على قَوْسِ ب ه ا د. وَالزّاوِيَةُ الَّتي من ج ب ه وَهِيَ الَّتي عَلى قَوْسِ ج د ا ه. والَّتي من ج ب ه مِثْل الَّتي من ب ج د. وَمِنْ أَجْلِ ذَلِكَ اللَّواتي مِنْ ج د ا ومن د ا ه ومن ا ه ب مِثْل اللَّتَيْنِ من د ه ب وَمِنْ ج ب ه. فَمُخَمَّسُ ا ه ب ج د قَدِ اسْتَبانَ أَنَّهُ مُتَساوي الأَضْلاعِ وَالزَّوايا وَقَدْ عُمِلَ في دائِرَةِ ا ب ج.

وَذَلِكَ ما أَرَدْنا بَيانَهُ.

[1] ب ج] + د [2] ج د ا ه] د ا ه

فكُلّ واحِدَةٍ مِنْ زاوِيَتَيْ ا ب ج ا ج ب مِثْلا زاوية ب ا ج. فنَقْسِم زاويةَ ا ب ج بِنِصْفَيْنِ بخطّ ب ح. وزاوية ا ج ب بِنِصْفَيْن بخطِ ج ط. ونُخْرِجَ خُطوطَ ا ج ط ب ا ح ح ج. فكُلّ واحدةٍ مِنْ زاوِيَتَيْ ا ب ج ا ج ب مِثْلا زاوية ب ا ج. وَقَدْ قُسِمَتْ كُلّ واحدةٍ مِنْهما بِنِصْفَيْن. فزَوَايا ب ا ج ا ج ط ط ج ب ب ح ب ج ح ب ا الخَمْس—مُتَساوِيَة. فقُسِيّ ا ط ط ب ب ج ج ح ح ا الخَمْس—مُتَساوِيَة. فمُخَمَّس ا ط ب ج ح متساوي الأَضلاع. وقَوْس ط ب مثل قَوْس ج ح. ونَجْعَلُ قَوْسَ ط ا ح مُشْتَرَكًا. فجميع قَوْسِ ب ط ا ح مثل جميع قَوْسِ ج ح ا ط. وزاوية[1] ب ج ح هي على قَوْس ب ط ا ح. وزاوية ج ب ط على قَوْسِ ج ح ا ط.^ فزَاوِيَة ج ب ط مثل زاوية ب ج ح. وكَذَلكَ تكونُ زوايا ح ب ا ح ا ط ا ط ب مثل كُلّ واحدةٍ مِنْ زاوِيَتَيْ ب ج ح ج ب ط. فمُخَمّس ا ط ب ج ح مُتَساويِ الأَضلاعِ والزَّوايا، وَقَدْ عُمِلَ في دائِرَةِ ا ب ج[2].

وذَلِكَ ما أَرَدْنا أَنْ نُبيِّنَ.

הנה כל אחת משתי זויות אב״ג אג״ב כפל זוית בא״ג. ונחלק זוית אב״ג בשני חציים בקו ב״ח, וזוית אג״ב בשני חצאים[1] בקו ג״ט. ונוציא קוי א״ט ט״ב[2] א״ח ח״ג. הנה כל אחת משתי זויות אב״ג אג״ב כפל זוית בא״ג. וכבר נחלקה כל אחת מהן בשני חציים. אם כן זויות[3] בא״ג אג״ט טג״ב חב״ג[4] חב״א החמש—שוות. אם כן קשתות א״ט ט״ב ב״ג[5] ג״ח ח״א החמשה—שוים. אם כן מחומש אטבג״ח שוה הצלעות. וקשת ב״ט כמו קשת ג״ח. ונשים קשת טא״ח משותף. אם כן[6] כל קשת בטא״ח כמו כל קשת גחא״ט. וזוית גב״ט על קשת גחא״ט[7]. וזוית בג״ח[8] על קשת חאט״ב. אם כן זוית גב״ט כמו זוית בג״ח. וכן יהיו זויות גח״א חא״ט אט״ב כמו כל אחת משתי זויות בג״ח גב״ט. אם כן מחומש[9] אטבג״ח שוה הצלעות והזויות כבר נעשה בעגולת אב״ג[10].

וזה מה שרצינו לבאר.

---

[1] وزاوية] MPR1R2 ولكن زاوية [2] ا ب ج] +المعلومة OST2T4

---

[1] בשני חצאים] ליתא **בוס** [2] ט״ב] +ב״ג **ו** [3] זויות] זוית **בס** [4] חב״ג] חג״ב **ס** [5] ב״ג] ליתא **ב** [6] כן] ליתא **ב** [7] על ... גחא״ט] ליתא **בו** ⟨על קשת⟩ גחא״ט **ט** +וזוית בג״א על קשת חאט״ב **ס** +~~וזוית גב״ט על קשת גחא״ט~~ **פ** [8] וזוית בג״ח] ליתא **בו** [9] מחומש] +שוה **ק** [10] אב״ג] +⟨⟨הידועה⟩⟩ **ק**

[Proposition IV.12]

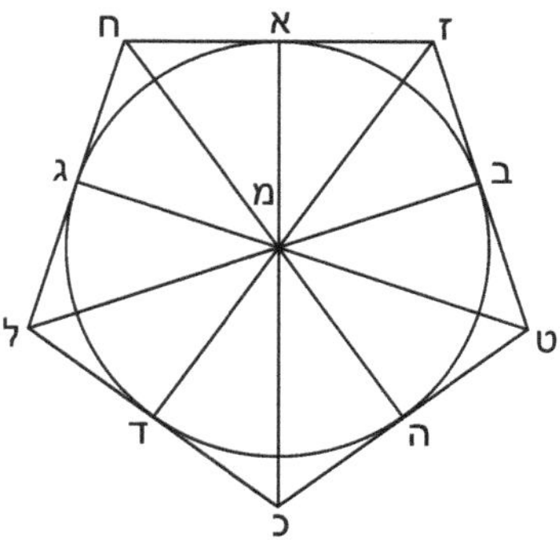

נרצה לעשות על עגול ידוע מחומש שוה הצלעות והזויות מקיף על[1] העגול.

نُرِيدُ أَنْ نُبَيِّنَ كَيْفَ نَعْمَلُ عَلَى دائِرَةٍ مَعْلومَةٍ مُخَمَّسًا يُحيطُ بِها مُتَساوِي الأَضْلاعِ وَالزَّوايا.

ונשים העגול הידוע עגול אב״ג. וכאשר רצינו לעשות עליו מחומש שוה הצלעות והזויות, נרשום על העגול נקודות א׳ וב׳ וה׳ וד׳ וג׳. ונוציא מאילו [!] הנקודות קוים ממששים העגול, והם ח״ז וז״ט וט״כ וכ״ל ול״ח. ויהיה מרכז העגול מ׳. ונדביק מ׳ בא׳ ובב׳ ובז׳ ובט׳, ובאותיות כלם. וא׳ וב׳ וה׳ וד׳ וג׳ הם נקודות המחומש שעשינו תחלה בתוך עגול אב״ג שוה הצלעות והזויות.

مِثالُهُ: أَنْ نَجْعَلَ الدَّائِرَةَ المَعْلومَةَ دائِرَةَ ا ب ج. وَنُرِيدُ أَنْ نَعْمَلَ عَلَيْها مُخَمَّسًا يُحيطُ بِها مُتَساوِي الأَضْلاعِ وَالزَّوايا. فَنُعَلِّمُ عَلى الدَّائِرَةِ نُقَطًا، وَهِيَ ا و ب و ه و د و ج. وَنُخْرِجُ مِنْ هَذِهِ النُّقَطَةِ خُطوطًا مُماسَّةً لِلدَّائِرَةِ، وَهِيَ خُطوطَ ح ز و ز ط و ط ك و ك ل و ل ح. فَأَقولُ إِنَّا قَدْ عَمِلْنا عَلى دائِرَةِ ا ب ج مُخَمَّسًا يُحيطُ بِها مُتَساوِي الأَضْلاعِ وَالزَّوايا، وَهُوَ مُخَمَّسُ ح ز ط ك ل.

[1] על] +~~אצל~~

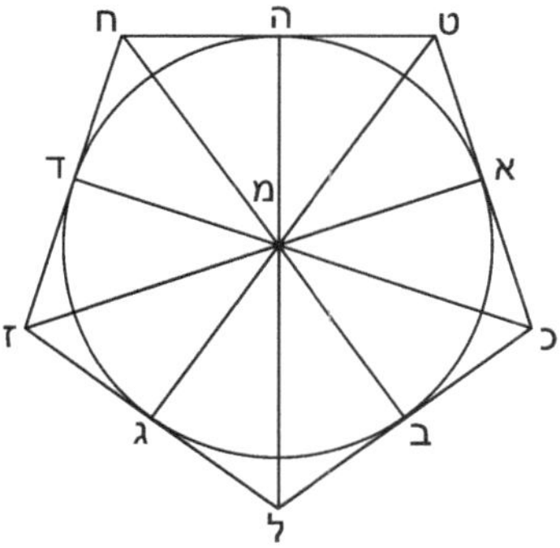

نُرِيدُ أَن نَعْمَلَ على دائِرَةٍ مَعْلومَةٍ مُخَمَّسًا مُتساوي الأَضْلاعِ والزَّوايا يُحيطُ بِها.

فَنَجْعَلَ الدَّائِرَةَ المَعْلومَةَ دائِرَةَ ا ب ج د ه. وَنُرِيدُ أَنْ نَعْمَلَ عَلَيها مُخَمَّسًا متساوي الأَضلاع والزَّوايا يُحيطُ بها. فنَخُطَّ في دائِرَةَ ا ب ج د ه مُخَمَّسًا مُتساوي الأَضلاعِ والزَّوايا تُحيطُ بِهِ. ولْتَكُنْ نُقَطُ زَوايا المُخَمَّسِ نُقَط ا ب ج د ه. ونُجيزَ على هذه النُّقَطِ خُطوطًا مُماسَّةً للدائِرَةِ عَلَيها ك ل ل ز ز ح ح ط ط ك. ونَجْعَلَ مَرْكَزَ الدَّائِرَةِ نُقْطَةَ م. ونَصِلَ خُطوطَ م ك م ب م ل م ج م ز م د م ح م ه م ط م ا.

נרצה לעשות על עגולה ידועה[1] מחומש שוה הצלעות והזויות יקיף בה.

הנה נשים[2] העגולה הידועה עגולת אבגד״ה. ונרצה לעשות עליה מחומש שוה הצלעות והזויות יקיף בה. הנה נקוה בעגולת אבגד״ה מחומש שוה הצלעות והזויות תקיף בו[3] [4]. ויהיו נקודות זויות המחומש נקודות א׳ ב׳ ג׳ ד׳ ה׳. ונעביר על אלו הנקודות קוים ממששים לעגולה, עליהם כ״ל ל״ז ז״ח ח״ט ט״כ. ונשים מרכז העגולה נקודת מ׳[5]. ונגיע קוי מ״כ מ״ב מ״ל מ״ג מ״ז מ״ד מ״ח מ״ה מ״ט מ״א.

שרטוט—**וט**: חסר המחומש הפנימי; **ס**: חסרות ה׳ ג׳ ב׳ כ׳ א׳ [1]ידועה] ידוע **ט** [2]נשים] ליתא **ב** [3]הנה ... בו] ליתא **ב** [4]תקיף בו] ליתא **וט** תקיף בה **ס** [5]נקודת מ׳] ליתא **ב** מ׳ **וס**

ואם כן קשת ב״א שוה לקשת א״ג. וזוית במ״א שוה לזוית אמ״ג. וכבר יצאו מנקודת[1] ז׳ שני קוים אל העגול ממששים בו, והם ז״ב וז״א. ואם כן ז״ב שוה לז״א. וז״מ משותף. ואם כן שני קוי א״ז וז״מ שוים לשני קוי ב״ז וז״מ. ותושבת א״מ שוה לתושבת ב״מ. ואם כן זוית אז״מ שוה לזוית בז״מ. ומשולש אז״מ שוה למשולש בז״מ. ושתי הזויות הנשארות, והם זוית זא״מ וזוית אמ״ז, שוות לשתי זויות הנשארות האחרות, והן זוית זב״מ וזוית זמ״ב, כל אחת לדומה לה—זוית במ״ז שוה לזוית זמ״א. ולזה זוית אמ״ח שוה לזוית חמ״ג, וזוית אמ״ב שוה לזוית אמ״ג. ואמנם זוית מא״ז היא נצבה, והיא שוה לזוית מא״ח הנצבה. ושתי זויות משלש זא״מ, והן זוית זמ״א וזוית זא״מ, שוות לשתי זויות משולש חא״מ, והן זוית חא״מ וזוית אמ״ח. וצלע אחד משותף אשר אצל צדיו שתי הזויות הנצבות, והוא מ״א. ואם כן שני צלעי המשולש שוים לשני צלעי המשולש האחר הנשארים, כל אחד לדומה לו. ואם כן יהיה צלע ז״מ שוה לצלע מ״ח. וזוית מז״א שוה לזוית מח״א. וז״א שוה לא״ח. וז״ח כלו כפל ח״א. ובזה ל״ח כפל ג״ח. וא״ח שוה לח״ג. ואם כן ז״ח שוה לח״ל. ובזה ל״ב וכ״ט וט״ז שוים לח״ל וז״ח. ואם כן מחומש חזטכל״ח [!] שוה הצלעות.

[1] מנקודת] מ⟨נ⟩קודת

بُرهانه: أَنَّ مَرْكَزَ الدَّائِرَةِ م. فَنَصِلُ م ا م ز م ب م ط. وَبِالحُروفِ كُلِّها ا و ب و ه و د و ج هِيَ نُقَطُ المُخَمَّسِ الَّتي عَمِلْنا قَبْلُ في دائِرَةِ ا ب ج المُتَساوي الأَضْلاعِ وَالزَّوايا. فَقَوْسُ ب ا مِثْل قَوْسِ ا ج. والزَّاوِيَةُ الَّتي من ب م ا مِثْل الَّتي من ا م ج. وَقَدْ خَرَجَ من نُقْطَةِ ز خَطَّانِ إِلى الدَّائِرَةِ مُماسَّانِها ز ب و ز ا، ز ب مِثْل ز ا. و ز م مُشْتَرَكٌ. فَكِلا خَطَّيْ ا ز و ز م مِثْل كِلَيْ خَطَّيْ ب ز و ز م، كُلّ واحِد مِثْل نَظيرِهِ. وَقاعِدَةُ ا م مِثْل قاعِدَةِ ب م. وَالزَّاوِيَةُ الَّتي من ا ز م مِثْل الَّتي من ب ز م. وَمُثَلَّثُ ا ز م مِثْل مُثَلَّثِ ب ز م. فَالزَّاوِيَتانِ الباقِيَتانِ اللَّتانِ مِنْ ز ا م وَمِنْ ا م ز مِثْل الزَّاوِيَتَيْنِ (الباقِيَتَيْنِ)[1] اللَّتَيْنِ من ز ب م ومن ز م ب، كُلّ واحِدَةٍ مِثْل نَظيرَتِها—الَّتي مِنْ ب م ز مِثْل الَّتي مِنْ ز م ا. وَكَذَلِكَ الَّتي مِنْ ا م ح مِثْل الَّتي مِنْ ح م ج. وَالَّتي مِنْ ب م ز مِثْل الَّتي مِنْ ا م ز. وَالَّتي مِنْ ب م ا مِثْلا الَّتي مِنْ ز م ا. وَكَذَلِكَ الَّتي مِنْ ج م ا مِثْلا الَّتي من م ح ا. وَالَّتي مِنْ ب م ا مِثْل الَّتي مِنْ ا م ج. وَالَّتي مِنْ ز م ا مِثْل الَّتي مِنْ ا م ح. وَلَكِنَّ الَّتي مِنْ م ا ز قائِمَةٌ، وَهِيَ مِثْل الَّتي مِنْ م ا ح القائِمَةِ. فَزاوِيَتا مُثَلَّثِ ا ز م اللَّتانِ مِنْ ز م ا وَمِنْ (ز ا م)[2] مِثْل زاوِيَتَيْ مُثَلَّثِ ح ا م اللَّتَيْنِ مِنْ ح ا م وَمِنْ ا م ح. وَضِلْعٌ واحِدٌ مُشْتَرَكٌ لِلزَّوايا المُتَساوِيَةِ

[1] الباقيتين] الناقيتين [2] زام] ازم

فلأنَّ خطَّي ج ز ز د قد خرَجا مِنْ نُقْطَةِ ز، فمَاسَّا دائِرَةَ ا ب ج د ه، يكونُ خطُّ ج ز مُساوِيًا لِخَطِّ ز د. وخطّ ز م مُشْتَرَكٌ. فكِلا خَطَّيْ ج ز ز م مُسَاوِيانِ لكِلَيْ خَطَّيْ د ز ز م، كُلّ واحدٍ لِنظيرِه. وقاعدَةُ ج م مُساوِيَةٌ لقاعدَةِ م د، لأنَّهما خَرَجا مِنْ مَرْكَزِ الدَّائِرَةِ إلى المُحِيطِ بِها. فزاوِيةُ ج ز م مُساوِيَةٌ لِزاوِيَةِ د ز م. فخطّ م ز قد قَسَمَ زاويةَ ج ز د بِنِصْفَيْنِ. وكذلكَ أيضًا تَبَيَّنَ أَنَّ زوايا^ ه ح د ا ط ه ب ك ا ب ل ج قَدْ قَسَمَتْها خُطوطُ م ح م ط م ك م ل بِنِصْفَيْنِ نِصْفَيْنِ.^ وأيْضًا فإنَّ خطّ ج م مُساوٍ لِخَطِّ م د. وخطّ م ز مُشْتَرَكٌ. فكِلا خَطَّيْ ج م م ز مُساوِيانِ لِكِلَيْ خَطَّيْ ز م م د، كُلّ واحدٍ لِنظيرِه. وقاعِدَةُ ج ز مُساوِيَةٌ لقاعِدَةِ ز د. فزاوِيةُ ج م ز مُساوِيَةٌ لِزاوِيَةِ ز م د. فزاوِيةُ ج م د قد قُسِمَتْ بِنِصْفَيْنِ بخطّ م ز. وكذلكَ أيضًا تَبَيَّنَ أَنَّ زوايا د م ه ه م ا ا م ب ب م ج قد قَسَمَتْها خُطوطُ م ح م ط م ك م ل بِنِصْفَيْنِ نِصْفَيْنِ.^ ولأنَّ قَوْسَ د ه مُساوِيَةٌ لقَوْسِ ج د، إذْ كان يُوتِرُهُما ضِلْعا مُخَمَّسٍ، تكونُ زاويةُ ج م د مُساوِيَةٌ لِزاوِيَةِ د م ه. فأمّا زاويةُ ج م د فَهِيَ مِثْلا زاويةِ د م ز. وأمّا زاويةُ ه م د فهي مِثْلا زاويةِ د م ح. فزاويةُ ز م د مُساوِيَةٌ لِزاوِيَةِ د م ح. وزاويةُ م د ز^ قائِمَةٌ، لأنَّ خطّ د م الّذي يمرُّ بالْمَرْكَزِ قَدْ خَرَجَ

הנה מפני כי שני קוי ג״ז ז״ד כבר יצאו מנקודת ז׳, ומששו עגולת[1] אבגד״ה, יהיה קו ג״ז שוה לקו ז״ד. וקו ז״מ משותף. הנה כל שני קוי ג״ז[2] ז״מ שוים לכל שני קוי ד״ז ז״מ, כל אחד לדומה לו. ותושבת ג״מ שוה לתושבת מ״ד, מפני כי שתיהם יוצאות ממרכז העגולה אל הקו המקיף בהם. אם כן זוית גז״מ שוה לזוית דז״מ. הנה קו מ״ז כבר חלק זוית גז״ד בשני חציים. וכן גם כן התבאר כי זוית [!] הח״ד אט״ה בכ״א בל״ג כבר חלקום קוי מ״ח מ״ט מ״כ מ״ל בשני חציים. וגם כן[3] הנה ג״מ שוה לקו מ״ד. וקו מ״ז משותף. אם כן כל שני קוי ג״מ מ״ז שוים לכל שני קוי ז״מ מ״ד, כל אחד לדומה לו. ותושבת ג״ז שוה לתושבת ז״ד. אם כן זוית גמ״ז שוה לזוית זמ״ד. אם כן זוית גמ״ד כבר נחלקה בשני חציים בקו מ״ז. וכן גם כן התבאר כי זויות דמ״ה המ״א אמ״ב במ״ג כבר חלקום קוי מ״ח מ״ט מ״כ מ״ל בשני חציים. ומפני כי קשת ד״ה שוה לקשת ג״ד, כי היה מיתרם[4] צלעות מחומש[5], תהיה זוית גמ״ד שוה לזוית דמ״ה. ואולם זוית גמ״ד הנה היא כפל זוית[6] דמ״ז. ואולם זוית המ״ד הנה היא כפל דמ״ח. אם כן זוית זמ״ד שוה לזוית דמ״ח. וזוית מד״ח נצבת, מפני כי קו ד״מ אשר יעבור במרכז כבר יצא ממקום המשוש. ולכן תהיה זוית מד״ז נצבת. אם כן שתי זויות זמ״ד מד״ז ממשולש זמ״ד שוות לשתי זויות דמ״ח חד״מ ממשולש[7] מד״ח, כל (אחת)[8] לדומה[9] לה. וקו מ״ד משותף בין שתיהן. אם כן הצלעות הנשארות[10] שוות לצלעות הנשארות, כל אחת לדומה לה. אם כן קו ז״ד שוה לד״ח.

---

[1]עגולת] ליתא **בו** [2]ג״ז] ז״מ **ו** [3]וגם כן] וכן גם כן **בו** [4]מיתרם] מיתרים **ב** [5]מחומש] +תהיה **פ** [6]זוית] ליתא **ס** [7]ממשולש] ⟨*⟩ **ט** [8]אחת] אחד **בוטספק** [9]לדומה] על דומה **ק** [10]הצלעות הנשארות] צלעות הנשארות **ק**

مِنْ جَنَبَتَيْهِ الزَّاوِيَتَانِ القَائِمَتَانِ، وَهُوَ م ا. وَضِلْعَا المُثَلَّثِ البَاقِيَانِ مِثْل ضِلْعَي المُثَلَّثِ الآخَرِ البَاقِيَيْنِ، كُلٌّ واحِدٍ مِثْل نَظِيرِهِ. ضِلْعُ زم مِثْل م ح. والزَّاوِيَةُ الَّتِي مِنْ م ز ا مِثْل الَّتِي مِنْ (م ح ا)[1]. ز ا مِثْل ا ح. وَجَمِيعُ ز ح مِثْلا ح ا. وَكَذَلِكَ (ل ح)[2] مِثْلا ج ح. ا ح مِثْل ح ج. ز ح مِثْل ح ل. وَكَذَلِكَ ل ك و ك ط و ط ز مِثْل ز ح و ح ل. فَمُخَمَّسُ ح ز ط ك ل مُتَسَاوِي الأَضْلاعِ.

---

1 م ح ا] ج ح ا 2 ل ح] ا ح

مِنْ مَوْضِعِ التَّماسِّ. وَلِذَلِكَ تكونُ زاويةُ م د ح^ قائمةً. فزاوِيتا ز م د م د ز مِنْ مُثَلَّثِ ز م د مُساوِيتانِ لِزاوِيَتَيْ د م ح ح د م مِنْ مُثَلَّثِ م د ح، كُلّ واحِدَةٍ لِنَظيرَتِها. وخطّ م د مُشْتَرَك بينهما[1]. فالأَضْلاعُ الباقِيَةُ مُساوِيَةٌ للأَضْلاعِ الباقِيَةِ، كُلّ واحدٍ لِنَظيرِه. فخطّ ز د مُساوٍ لخطّ د ح. وزاوية م ز د الباقية مُساوِيَةٌ لِزاوِيَةِ م ح د الباقية^. وَلأَنَّ خطّ ج ز مُساوٍ لخطّ ز د، وضِعْفُ ج ز هو ز ل، وضِعْفُ ز د هو ز ح، يكونُ خطُّ ز ح مُساوِيًا لخطّ ز ل. وكذلكَ أيضًا تَبَيَّنَ أنَّ خطَّ ز ح مُساوٍ لخطّ ح ط، وأنَّ خطّ ح ط مُساوٍ لخطّ ط ك، وأنَّ خطَّ ط ك مُساوٍ لخطّ ك ل، وأنَّ خطَّ ك ل مُساوٍ لخطّ ل ز، فخُطوط ك ل ل ز ز ح ح ط ط ك مُتَساوِيَةٌ، وَمُخَمَّس ز ح ط ك ل مُتَساوي الأَضْلاعِ.

וזוית מז״ד הנשארת שוה לזוית מח״ד הנשארת. וכן התבאר כי קו ל״ג שוה לקו ג״ז. ומפני כי קו ג״ז שוה לקו ז״ד, וכפל ג״ז הוא ז״ל, וכפל ז״ד הוא ז״ח, יהיה קו ז״ח שוה לקו ז״ל. וכן גם כן התבאר כי קו ז״ח שוה לקו ח״ט, ושקו ח״ט שוה לקו ט״כ, ושקו ט״כ שוה לקו כ״ל, ושקו כ״ל שוה לקו ל״ז. אם כן קוי כ״ל ל״ז ז״ח ח״ט ט״כ שוים, ומחומש זחטכ״ל[1] שוה[2] הצלעות.

---

[1] بينهما] MR1R2 لهما

---

[1] זחטכ״ל] ליתא **בו** ⟨*⟩ **פ** [2] שוה] שוים **בופ**

והוא גם כן שוה הזויות. לפי שזוית מז״ב שוה לזוית מז״א. וזוית בז״א כלה כפל זוית מז״א. וכמו כן זוית (אח״ג)[1] כפל זוית מח״א. ואם כן זוית בז״א שוה לזוית אח״ג. וכמו כן יתבאר שכל אחת מזויות גל״ד ודכ״ה (והט״ב)[2] שוות לזויות ז׳ וח׳[3]. ואם כן מחומש זטכל״ח שוה צלעות והזויות. וכבר נעשה על עגול אב״ג הידוע.

וזה מה שרצינו.

وَالزَّاوِيَةُ الَّتِي من م ز ب مِثْل الَّتِي من م ز ا. وَجَميعُ الَّتي من ب ز ا مِثْلا الَّتِي من م ز ا. وَكَذَلِكَ الَّتي مِنْ ا ح ج مِثْلا الَّتي مِنْ م ح ا. وَالَّتي مِنْ م ز ا مِثْل الَّتي مِنْ م ح ا. وَالَّتي مِنْ ب ز ا مِثْل الَّتي مِنْ ا ح ج. وَكَذَلِكَ تَبَيَّنَ أَنَّ كُلَّ واحِدَةٍ مِنَ الزَّوايا اللَّواتي من ج ل د ومن د ك ه ومن ه ط ب مِثْل كُلِّ واحِدَةٍ مِنْ زاوِيَتَيْهِ ب ز ا ا ح ج. فَمُخَمَّسُ (ز ط ك ل ح)[1] مُتَساوي الأَضْلاعِ وَالزَّوايا. وَقَدْ عُمِلَ عَلَى دائِرَةِ ا ب ج المَعْلومَة.

وَذَلِكَ ما أَرَدْنا بَيانَهُ.

[1] אח״ג] אח״מ [2] והט״ב] והא״ב [3] וח׳] ⟨ו⟩ח׳

[1] ز ط ك ل ح] ز ط ك ل م

وَأَقولُ إِنّهُ مُتَساوي الزَّوايا. فَلأَنَّ زاوِيةَ م ز د مُساوِيَةٌ لِزَاوِيَةِ م ز د^، ومِثْلا زاوية م ز د هي زاوية ج ز د، ومِثْلا زاوية م ز د هي زاوية ه ح د، فزاوية ج ز د مُساوِيَة لِزَاوِيَةِ ه ح د. وكَذلكَ أيضًا تَبيَّنَ أَنَّ زاوِيَةَ ز ح ط مُسَاوِيَةٌ لِزَاوِيَةِ ك ط ح، وأَنَّ زاوِيَةَ ك ط ح مُسَاوِيَةٌ لِزَاوِيَةِ ل ك ط، وأَنَّ زاوِيَةَ ل ك ط مُسَاوِيَةٌ لِزَاوِيَةِ ك ل ز، وزاوية ك ل ز لِزَاوِيَةِ^ ل ز ح. فالزَّوايا الخَمْس^ الَّتي عَلَيْها ل ك ط ك ل ز ل ز ح ز ح ط ح ط ك مُتَساوِيَة. فَخُمَّس ز ح ط ك ل مُتَساوي الزَّوايا. وَقَدْ كُنَّا بَيَّنَّا أَنّهُ مُتَساوي الأضلاع، وهو يُحيطُ[1] بدائِرَةِ ا ب ج د ه.

وذَلِكَ ما أَرَدْنا أَنْ نُبيِّنَ.

[1] يُحيط] MPR1R2 محيط

ואומר שהוא שוה[1] הזויות[2]. ומפני כי כפל זוית מז״ד היא זוית גז״ד, וכפל זוית מח״ד היא זוית הח״ד, תהיה זוית גז״ד שוה[3] לזוית הח״ד. וכן גם כן התבאר כי זוית זח״ט שוה לזוית כט״ח, ושזוית כט״ח שוה לזוית לכ״ט, ושזוית לכ״ט שוה לזוית[4] לז״ח. אם כן זויות המחומש אשר עליהם לכ״ט כל״ז לז״ח זח״ט חט״כ שוות. אם כן מחומש זחטכ״ל שוה הזויות. וכבר בארנו שהוא שוה הצלעות, והוא יקיף בעגולת אבגד״ה.

וזה מה שרצינו לבאר.

[1] שוה] ליתא ס [2] הזויות] +⟨⟨מפני שזוית מז״ד שוה לזוית מח״ד⟩⟩ ק [3] שוה] ~~שוה גז״ד~~+ פ [4] לזוית] +⟨⟨כל״ז ושזוית כל״ז שוה לזוית⟩⟩ ק

[Proposition IV.13]

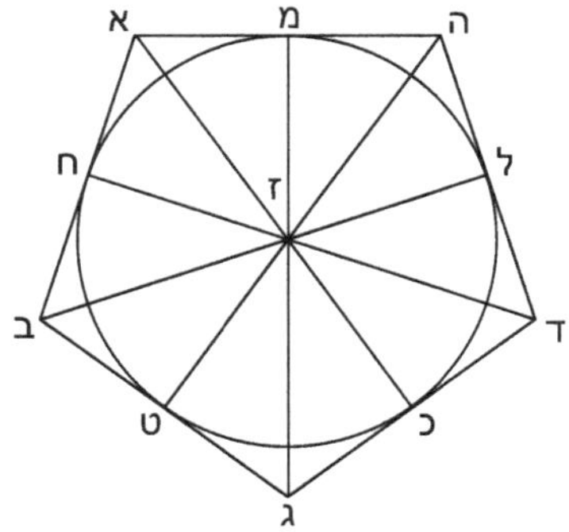

نُرِيدُ أَنْ نُبَيِّنَ كَيْفَ نَعْمَلُ فِي مُخَمَّسٍ مَعْلومٍ مُتَساوِي الأَضْلاعِ وَالزَّوايا دائِرَةً يُحيطُ بِها.

مِثالُهُ: أَنَّا نَجْعَلُ المُخَمَّسَ المَعْلومَ مُخَمَّسَ ا ب ج د ه. ونُرِيدُ أَنْ نَعْمَلَ فيهِ دائِرَةً يُحيطُ بِها. فَنَقْطَعُ زاوِيَةَ ج بِنِصْفَيْنِ بِخَطِّ ج ز، وَزاوِيَةَ د بِنِصْفَيْنِ بِخَطِّ د ز. وَيَلْتَقِيانِ عَلَى ز. فَنُخْرِجُ من ز أَعْمِدَةً إِلى خَطِّ ا ب وَخَطِّ ب ج وَخَطِّ ج د وَخَطِّ د ه وَخَطِّ ه ا، وَهِيَ ز ح و ز ط و ز ك و ز ل و ز م[1]. فَأَقولُ إِنَّ عَلَى مَرْكَزِ ز وَبِبُعْدِ كُلِّ واحِدٍ مِنْ ح و ط و ك و ل (و م)[2] بِخَطِّ الدَّائِرَةِ فِي مُخَمَّسِ ا ب ج د ه المَعْلومِ.

נרצה לעשות בתוך מחומש ידוע שוה הצלעות והזויות עגול שיקיף המחומש.

ונשים המחומש הידוע מחומש אבגד״ה. וכאשר רצינו לעשות בתוכו עגול שיקיף בו המחומש, נחלק זוית ג׳ לחציים בקו ג״ז, וזוית ד׳ לחציים בקו ד״ז. ויפגשו על ז׳. ונוציא מז׳ עמודים על קו א״ב וקו ב״ג וקו ג״ד וקו ד״ה וקו ה״א, והם ז״ח וז״ט וז״כ וז״ל וז״מ.

---

1 زم] زل ⟨م⟩ 2 م] ج

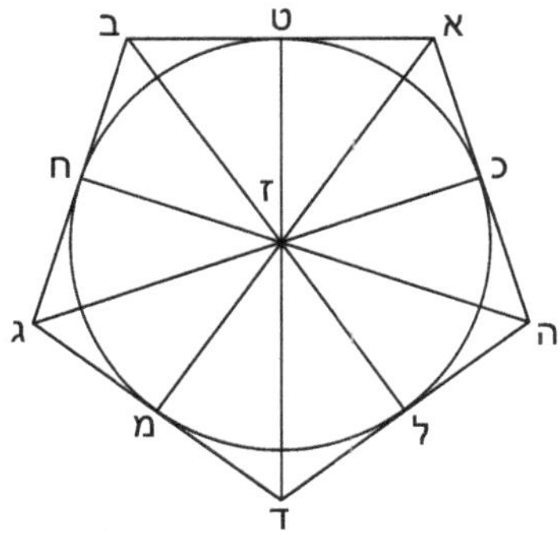

نُرِيدُ أَنْ نَعْمَلَ فِي مُخَمَّسٍ مَعْلومٍ مُتَساوي الأَضْلاعِ والزَّوايا دائِرَةً يُحيطُ بِها.

נרצה לעשות במחומש ידוע[1] שוה הצלעות והזויות עגולה יקיף בה.

فَنَجْعلَ المُخَمَّسَ المَعْلومَ المُتَساوي الأَضْلاعِ والزَّوايا مُخَمَّسَ^ ا ب ج د ه. ونُريدُ أَنْ نَعْمَلَ فيهِ دائِرَةً يُحيطُ بِها. فنَقْسِمَ زاوِيَتَيْ ب ج د ج د ه بِنِصْفَيْنِ نِصْفَيْنِ^ بِخَطَّيْ زج ز د. وَنَصِلَ خُطوطَ ا ز ز ب ه ز. ونُخْرِجَ مِنْ نُقْطَةِ زَ إلى خُطوطِ ا ب ب ج ج د د ه ا ه أَعْمِدَةَ زح زط زك زل زم.

הנה נשים המחומש הידוע השוה[2] הצלעות והזויות אבגד״ה. ונרצה לעשות בו עגולה יקיף בה. הנה נחלק שתי זויות בג״ד גד״ה כל אחד בשני[3] חציים בשני קווי[4] ז״ג ז״ד. ונגיע קוי א״ז ז״ב ה״ז. ונוציא מנקודת ז׳ אל קוי א״ב[5] ב״ג ג״ד ה״ד א״ה עמודים ז״ח ז״ט ז״כ ז״ל ז״מ.

שרטוט—השרטוט זהה לשרטוט בעותק **O2**
[1]ידוע] ⟨*⟩ **ק** [2]השוה] ושוה **ב** השוה **ט**
[3]בשני] לשני **ק** [4]בשני קוי] קו **ב** [5]א״ב]
+~~ג׳ ד׳ ה׳~~ **פ**

ונדביק ב׳ בז׳ וז׳ בה׳ ובא׳. וזוית בג״ז שוה לזוית דג״ז. וצלע ב״ג שוה לצלע ג״ד. וכאשר שמנו צלע ג״ז משותף, היו צלעי ב״ג ג״ז שוים לצלעי ד״ג ג״ז. וזוית בג״ז שוה לזוית דג״ז. ואם כן תושבת ב״ז שוה לתושבת ד״ז. וזוית זב״ג שוה לזוית זד״ג. אמנם זוית גז״ה[1] כפל זוית גז״ב. אם כן זוית אב״ג כפל זוית זב״ג. ואם כן כבר נחלקה זוית ב׳ לחציים. ובזה יתבאר שכל אחת מזויות א׳ וה׳ נחלקה גם כן לחציים. ועוד זוית טג״ז שוה לזוית כג״ז. אמנם זוית גט״ז נצבה, והיא שוה לזוית גכ״ז. אם כן משלש זט״ג שתי זויותיו[2], והן זוית טג״ז וזוית גט״ז, שוות לשתי זויות משולש זכ״ג, והן זוית כג״ז וזוית (גכ״ז)[3], כל אחת לדומה לה—זוית טג״ז שוה לזוית כג״ז, וזוית גט״ז שוה לזוית גכ״ז. ושתי הזויות השוות—צלע אחד משותף הוא מיתר להן יחד, והוא ג״ז. ואם כן שני צלעי המשולש הנשארים שוות [!] לשני צלעי המשולש האחר הנשארים, כל אחד לדומה לו. ואם כן ט״ז שוה לכ״ז. וכמו כן יתבאר שז״מ וז״ל וז״ח שוים לז״ט וז״כ. והקוים החמשה ז״ח וז״כ וז״ט וז״ל וז״מ שוים. ונקיף עגול על מרכז ז׳ ובמרחק ז״ח, ויעבר על שאר הנקודות, וימשש צלעי א״ב וב״ג וג״ד וד״ה וה״א על זויות ח׳ וט׳ וכ׳ ול׳ ומ׳ הנצבות. וכבר עשינו בתוך מחומש אבגד״ה שוה הצלעות והזויות הידוע עגול, והוא עגול חטכל״מ. והמחומש שמקיף בו.

וזה מה שרצינו.

برهانه: أن نصل ب ز ج ز د ز ز ه ز ا. فالزاوية التي من ط ج ز مثل الزاوية التي من ك ج ز. ولكنّ التي من (ج ط ز)[1] قائمة، وهي مثل التي من ج ك ز. فمثلث ز ط ج زاويتاه اللتان من ط ج ز ومن ج ط ز مثل زاويتي مثلث ز ك ج اللتين من ك ج ز ومن ج ك ز، كلّ واحدة مثل نظيرتها [..] التي من ك ج ز والتي من ج ط ز مثل التي من ج ك ز. والزاويتان المتساويتان— ضلع واحد يوترهما جميعًا، وهو ج ز. فضلعا المثلث الباقيان مثل ضلعي المثلث الآخر الباقيين، كلّ واحد مثل نظيره. ط ز مثل ك ز. وكذلك ز ل و ز م و ز ح مثل ز ط و ز ك. فالأضلاع الخمسة ز ح و ز ط و ز ك و ز ل و ز م متساوية. فعلى مركز ز وببعد ز ح نخطّ الدائرة، وتجوز على سائر النقط. وهي تماسّ الأضلاع ا ب و ب ج و ج د و د ه و ه ا. وزوايا ح و ط و ك و ل و م قائمة. والمخمّس يحيط بها.

وذلك ما أردنا بيانه.

[1] גז״ה] גז״[..] ⟨*⟩ [2] זויותיו] +~~והן זוית~~ [3] וזוית גכ״ז] ⟨⟨וזוית כג״ז⟩⟩

[1] ج ط ز] ط ج ز

فلأنّ ضلع ب ج مساوٍ لضلع ج د، وذلك أنّ المخمّس متساوي الأضلاع، وخطّ ج ز مشترك، يكون كلا خطّي ب ج ج ز مساويين لكلّي خطّي ج ز ج د، كلّ واحد لنظيره. وزاوية ب ج ز مساوية لزاوية ز ج د. فقاعدة ب ز مساوية لقاعدة ز د، ومثلّث ب ج ز مساوٍ لمثلّث ز د ج، وسائر الزوايا مساوية[1] لسائر الزوايا التي توترها الأضلاع المتساوية. فزاوية ج ب ز مساوية لزاوية ز د ج. وتبقى زاوية ا ب ز مساويةً لزاوية ز د ه. ولكنّ زاوية ز د ه مساويةٌ لزاوية ز د ج. فزاوية ا ب ز مساوية لزاوية ز د ج. وقد كانت زاوية ز د ج مساويةً لزاوية ز ب ج. فزاوية ا ب ز مساوية لزاوية ز ب ج. فزاوية ا ب ج قد قسمت بنصفين بخطّ ب ز. وكذلك تبيّن أنّ كلّ واحدة من زاويتي ب ا ه ا ه د قد قسمت بنصفين نصفين بخطّي ا ز ز ه. ولأنّ زاوية ب ج ز مساويةٌ لزاوية ز ج د، وزاوية ز ح ج قائمة مساوية لزاوية ز م ج، يكون كلا زاويتي ز م ج ز ج م مساويتين لكلّي زاويتي ز ج ح ز ح ج، كلّ واحدة لنظيرتها.

הנה מפני כי צלע ב״ג שוה לצלע ג״ד, כי[1] המחומש שוה הצלעות, וקו ז״ג משותף, יהיו כל שני קוי ב״ג ג״ז שוים[2] לכל שני קוי ג״ד ג״ז, כל אחד לנכחי לו. וזוית בג״ז שוה לזוית זג״ד. אם כן תושבת ב״ז שוה לתושבת ז״ד, ומשולש בג״ז שוה למשולש זד״ג, ושאר הזויות שוה לשאר הזויות אשר יהיו מיתריהם הצלעות השוות. אם כן זוית גב״ז שוה לזוית זד״ג. ונשארה זוית אב״ז שוה לזוית זד״ה. ויהיה זוית זד״ה שוה לזוית זד״ג. הנה זוית אב״ז שוה לזוית זד״ג. וכבר היה זוית זד״ג שוה לזוית[3] זב״ג. אם כן זוית אב״ז[4] שוה לזוית זב״ג[5]. אם כן זוית אב״ג כבר חולקה בשני חציים[6] בקו ב״ז. וכן התבאר כי כל אחת משתי זויות בא״ה אה״ד כבר נחלקה כל אחת בשני חציים בשני קוי א״ז ז״ה. ומפני כי זוית בג״ז שוה לזוית זג״ד, וזוית זח״ג נצבת שוה לזוית זמ״ג, יהיו כל שתי זויות זמ״ג זג״מ שוות לכל שתי זויות זח״ג זג״ח, כל אחת לנכחית[7] לה.

[1] مساوية] **CamR1**

[1] כי] +* **פ**   [2] שוים] שוי **ב** שוה **ו**   [3] לזוית] ליתא **ב**   [4] אב״ז] ~~אב״ג~~ ⟨⟨*⟩⟩ **ק**   [5] וכבר ... זב״ג] ⟨*⟩ **ס**   [6] חציים] +כל אחת **בס**   [7] לנכחית] לנכחי **בוס**

وخطّ زج مشْتَرَكٌ لِلْمُثَلَّثَيْنِ جَميعًا—فَالضِّلعانِ الباقِيانِ مُسَاوِيانِ لِلضِّلْعَيْنِ الباقِيَيْنِ، كُلّ واحدٍ لِنَظيرِه. فخطّ م ز مساوٍ لخطّ ز ح. وكذلك[1] تَبَيَّنَ[2] أنَّ خطَّ ز ح مُساوٍ لخطّ ز ط، وخطّ ز ط لخطّ ز ك، وخطّ ز ك لخطّ ز ل، وخطّ زل لخطّ ز م، فالخُطوط الخَمْسَة الّتي عَلَيها ح ز ط ز ك ز ل ز م مُتَساوِيَةٌ. فَإذا جَعَلْنا نُقْطَةَ ز مَرْكَزًا، وَأَدَرْنا بِبُعْدِ إحْدَى نُقَطِ ح ط ك ل م دائِرَةً مَرَّتِ الدَّائِرَةُ بِسائِرِ النُّقَطِ وَماسَّتْ أَضْلاعَ مُخَمَّسِ ا ب ج د ه، لِأَنَّ الزَّوايا الَّتي عِنْدَ نُقَطِ ح ط ك ل م قائِمَة. فَنَخُطّ هذه الدَّائِرَة، وهي دائِرَةُ ا ب ج د ه، فَقَدْ عَمِلْنا في مُخَمَّس ا ب ج د ه دائِرَةً يُحيطُ بِها وَهِيَ دائِرَة ح ط ك ل م.

وذلك ما اردنا ان نبين.

וקו ז״ג משותף לשני המשולשים יחד—יהיו[1] הצלעות הנשארות שוות לשתי הצלעות[2] הנשארות, כל אחת לנכחית[3] לה. אם כן קו מ״ז שוה לקו ז״ח. וכן התבאר כי קו ח״ז שוה לקו ז״ט, וקו ז״ט לקו ז״כ, וקו ז״כ לקו ז״ל, וקו ז״ל[4] גם כן לקו ז״מ, הנה הקוים החמשה אשר עליהם[5] ח״ז ז״ט ז״כ ז״ל ז״מ שוים. וכאשר שמנו נקודת ז׳ מרכז, וסבבנו במרחק אחת מנקודות[6] ח׳ ט׳ כ׳ ל׳ מ׳ עגולה[7], עברה העגולה על שאר הנקודות[8] ומששה צלעות מחומש אבגד״ה, מפני כי הזויות[9] אשר אצל נקודות[10] ח׳ ט׳ כ׳ ל׳ מ׳ נצבות. ונקוה העגולה הזאת, והיא עגולת חטכל״מ. הנה כבר עשינו במחומש אבגד״ה עגולה יקיף בה, והיא עגולת חטכל״מ.

וזה מה שרצינו לבאר[11].

---

1 وكذلك] MPR1R2 +أيضا
2 تبين] CO1T1T3T4 يتبين

---

1 יהיו] +⟨⟨שתי⟩⟩ ק 2 הצלעות] צלעות פ 3 אחת לנכחית] לנכחית ב אחד לנכחי ו 4 לקו ... ז״ל] ליתא ב לקו ז״ל וס 5 עליהם] +~~ח״ט ז״כ~~ ק 6 אחת מנקודות] אחת מנקודת בפ אחד מנקודות ו 7 עגולה] עגולת בספ 8 הנקודות] +ח״ט ב 9 הזויות] הזוית ס 10 נקודות] נקודת ס 11 לבאר] לבארו בפ ביאורו ס

[Proposition IV.14]

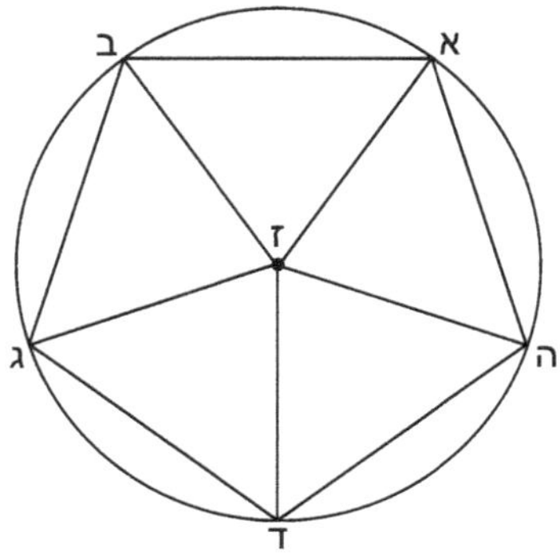

نُرِيدُ أَنْ نُبَيِّنَ كَيْفَ نَعْمَلُ عَلَى مُخَمَّسٍ مَعْلُومٍ مُتَساوِي الأَضْلاعِ وَالزَّوايا دائِرَةً تُحِيطُ بِهِ.

مِثالُهُ: أَنَّا نَجْعَلُ المُخَمَّسَ المَعْلُومَ المُتَساوِي الأَضْلاعِ وَالزَّوايا مُخَمَّسَ ا ب ج د ه. ونُرِيدُ أَنْ نَعْمَلَ عَلَيْهِ دائِرَةً تُحِيطُ بِهِ. فَنَقْطَعُ الزَّاوِيَةَ الَّتِي ب ج د بِنِصْفَيْنِ بِخَطِّ ج ز. وَالَّتِي من ج د ه بِنِصْفَيْنِ بِخَطِّ د ز. وَيَلْتَقِيانِ عَلَى نُقْطَةِ ز. فَنَصِلُ ز ب ز ا ز ه. فَعَلَى مَرْكَزِ ز وَبِبُعْدِ ا و ب وج و د و ه نَخُطُّ الدَّائِرَةَ الَّتِي تُحِيطُ بِمُخَمَّسِ ا ب ج د ه المُتَساوِي الأَضْلاعِ وَالزَّوايا المَعْلُوم.

נרצה לעשות על מחומש ידוע שוה הצלעות והזויות עגול מקיף בו.

ונשים המחומש הידוע מחומש אבגד״ה. וכאשר רצינו לעשות עליו עגול מקיף בו, נחלק זוית בג״ד לחציים בקו ג״ז. וזוית גד״ה לחציים בקו ד״ז. ויפגשו על ז׳. ונדביק ז׳ בב׳ ובא׳ ובה׳.

نُرِيدُ أَنْ نَعْمَلَ على مُخَمَّسٍ مَعْلومٍ مُتَساوي الأَضْلاعِ والزَّوايا دائِرَةً تُحيطُ بِهِ.

فَنَجْعَلَ المُخَمَّسَ المَعْلومَ مُخَمَّس ا ب ج د ه. وَنُرِيدُ أَن نَعْمَلَ عَلَيْهِ دائِرَةً تُحيطُ بِهِ. فَنَقْسِمَ زاوِيَةَ ب ج د بِنِصْفَيْنِ بِخَطِّ ج ز. وَزاوِيَةَ ج د ه بِنِصْفَيْنِ بِخَطِّ د ز. وَلْيَلْتَقِيا على نُقْطَةِ ز. وَنُخْرِجَ خُطوطَ ز ب ز ا ز ه.

נרצה לעשות על מחומש ידוע[1] שוה הצלעות והזויות[2] עגולה תקיף בו.

הנה נשים המחומש הידוע מחומש אבגד״ה. ונרצה לעשות עליו[3] עגולה תקיף בו[4]. הנה נחלק זוית בג״ד בשני חציים בקו ג״ז. וזוית גד״ה בשני חציים בקו ד״ז. ויפגשו על נקודת ז׳. ונוציא קוי ז״ב ז״א ז״ה.

[1] ידוע] ⟨*⟩ ק [2] והזויות] ~~והעגולות~~ ⟨⟨*⟩⟩ פ [3] עליו] ליתא ס [4] הנה ... בו] ⟨⟨*⟩⟩ ק

וב״ג[1] שוה לג״ד. וג״ז משותף. ואם כן שני קוי ב״ג וג״ז שוים לשני קוי ג״ד[2] וג״ז, כל אחד לדומה לו. וזוית בג״ז שוה לזוית דג״ז. ואם כן תושבת ב״ז שוה לתושבת ז״ד. ומשלש זב״ג שוה למשלש זד״ג. ושתי הזויות הנשארות, והן זוית זב״ג וזוית בז״ג, שוות לשתי הזויות הנשארות, והן זוית זד״ג וזוית דז״ג, כל אחת לדומה לה. ואם כן זוית זב״ג שוה לזוית זד״ג. וזוית זד״ג שוה לזוית זד״ה. וזוית גד״ה היא כפל זוית זד״ג. וזוית גד״ה שוה לזוית גב״א. וזוית זד״ג שוה לזוית זב״ג. אם כן זוית גב״א כפל זוית גב״ז[3]. וזוית אב״ז גם כן שוה לזוית גב״ז. ועוד קו א״ב שוה לקו ב״ג. וב״ז משותף. ואם כן שני קוי א״ב וב״ז שוים לשני קוי ג״ב וב״ז, כל אחד לדומה לו: זוית אב״ז שוה לזוית גב״ז. ואם כן תושבת ז״א שוה לתושבת ז״ג. וכמו כן ז״ה וז״ד שוים לא״ז וז״ג. והקוים החמשה ז״א וז״ב וז״ג (וז״ד)[4] וז״ה שוים. ונקיף עגול על מרכז ז׳ וברוחק א׳, ויקיף במחומש אבגד״ה שוה הצלעות והזויות הידוע.

بُرْهانُهُ: أَنَّ ب ج مِثْل ج د. و ج ز مُشْتَرَكٌ. فَكِلَيْ خَطَّيْ ب ج ج ز مِثْل كِلَيْ خَطَّيْ ج د و ج ز، كُلّ واحِد مِثْل نَظيرِهِ. وَالزَّاوِيَةُ الَّتي مِنْ ب ج ز مِثْل الَّتي من د ج ز. فَقاعِدَةُ ب ز مِثْل قاعِدَة ز د. وَمُثَلَّثُ ز ب ج مِثْلُ مُثَلَّثِ ز د ج. فَالزَّاوِيَتانِ الباقِيَتانِ اللَّتانِ من ب ز ج وَمِنْ ز ب ج مِثْل الزَّاوِيَتَيْنِ الباقِيَتَيْنِ اللَّتَيْنِ مِنْ ز د ج ومن د ز ج، كُلّ واحِدَة مِثْل نَظيرَتِها الَّتي توتِرُها الأَضْلاعُ المُتَساوِيَةُ. فَالَّتي مِنْ ز ب ج مِثْل الَّتي مِنْ (ز د ج)[1]. والَّتي من ز د ج مِثْل الَّتي من د ز ه. والَّتي من ج د ه مِثْلا الَّتي من ز د ج. والَّتي من ج د ه مِثْلا الَّتي من ج ب ا. والَّتي من ز د ج مِثْل الَّتي من ز ب ج. فَالَّتي من ج ب ا مِثْلا الَّتي من ز ب ج. والَّتي من ا ب ز مِثْل الَّتي من ج ب ز. وَأَيْضًا خَطُّ ا ب مِثْل خَطّ ب ج. و ب ز مُشْتَرَكٌ. فَكِلَيْ خَطَّيْ ا ب ب ز مِثْل كِلَيْ خَطَّيْ (ج ب)[2] و ب ز، كُلّ واحِد مِثْل نَظيرِهِ. وَالزَّاوِيَةُ الَّتي من (ز ب ا)[3] مِثْل الَّتي من ج ب ز. وقاعِدَةُ ز ا مِثْل قاعِدَةِ ز ج. وَكَذَلِكَ ز ه و ز د مِثْل ا ز و ب ز و ز ج. فَالأَضْلاعُ الخَمْسَةُ ز ا و ز ب و ز ج و ز د و ز ه مُتَساوِيَةٌ. وَعَلى مَرْكَزِ ز وَبِبُعْدِ ا نَخُطُّ الدَّائِرَةَ الَّتي تُحيطُ مُخَمَّس ا ب ج د ه المُتَساوي الأَضْلاعِ وَالزَّوايا المَعْلوم.

[1] וב״ג] ~~[..]~~ ⟨⟨*⟩⟩ [2] ג״ד] ~~ב״ג וג״ז~~+ [3] גב״ז] ~~אב״ג~~+
[4] וז״ד] וז״ב

[1] ز د ج] د ز ج [2] ج ب] ج ز [3] ز ب ا] ب ز ا

فخطّ ب ج مثل خطّ ج د. و ج ز مُشْتَرَكٌ. فخطّا ب ج ج ز مثل خَطَّيْ ج د ج ز. وزاوية ب ج ز مثل زاوية د ج ز. فقاعِدَة ب ز مثل قاعِدَة ز د. فمُثَلَّث ز ب ج مثل مُثَلَّث ز د ج. وزاوِيَتا ز ب ج ب ز ج الباقِيَتانِ مثل زاوِيَتَيّ ز د ج د ز ج، كُلّ واحِدَةٍ مثل نَظيرَتِها. فزاوية ز ب ج مثل زاوية ز د ج. وزاوية ز د ج نِصْف زاوية ج د ه. وزاوية ج د ه مثل زاوية ج ب ا. فزاوية ز ب ج نِصْف زاوية ج ب ا. فزاوية ز ب ج مثل زاوية ج ب ا. وخطّ ا ب مثل خطّ ب ج. وخطّ ب ز مُشْتَرَك. فخطّا ا ب ب ز مثل خَطَّيْ ج ب ب ز. وزاوية ا ب ز مثل زاوية ج ب ز. فقاعِدَة ز ا مثل قاعِدَة ز ج. وكذَلِكَ أَيْضًا تبيَّنَ[1] أَنَّ كُلَّ واحِدٍ مِنْ ز ه ز د مثل كُلِّ واحِدٍ مِنْ ز ا ز ب ز ج. فخُطُوط ز ا ز ب ز ج ز د ز ه الخَمْسَة مُتَساوِيَةٌ. فإذا جَعَلْنا نُقْطَةَ ز مَرْكَزًا وَأَدَرْنَا بِبُعْدِ ا دائِرَةً، مَرَّتْ بِنُقَطِ ب ج د ه[2]، وَأَحاطَتْ بالمُخَمَّسِ المُتَساوي الأَضْلاعِ وَالزَّوايا المَعْلوم، وهي دائِرَة ا ب ج د ه.

הנה קו[1] ב״ג כמו קו ג״ד. וג״ז משותף. אם כן שני קוי ב״ג ג״ז[2] כמו שני קוי ג״ד ג״ז. וזוית בג״ז כמו זוית דג״ז. אם כן תושבת ב״ז[3] כמו תושבת ז״ד. ומשולש זב״ג כמו משולש זד״ג[4]. ושתי זויות זב״ג בז״ג הנשארות כמו[5] שתי זויות זד״ג דז״ג, כל אחת לנכחית לה. אם כן זוית זב״ג כמו זוית זד״ג. וזוית זד״ג חצי זוית[6] גד״ה[7]. וזוית גד״ה כמו זוית גב״א. אם כן זוית זב״ג[8] חצי זוית גב״א. אם כן זוית זב״ג כמו זוית זב״א. וקו א״ב כמו קו[9] ב״ג. וקו ב״ז משותף. אם כן שני קוי א״ב ב״ז כמו שני קוי[10] ג״ב ב״ז. וזוית אב״ז כמו זוית[11] גב״ז. אם כן תושבת א״ז כמו תושבת ז״ג. וכן גם כן התבאר כי כל אחד מן ז״ה ז״ד כמו כל אחד מן א״ז ז״ב ז״ג[12]. אם כן קוי א״ז ז״ב ז״ג ז״ד ז״ה החמשה שוים. הנה כאשר שמנו נקודת ז׳ מרכז והקפנו במרחק א׳ עגולה, הלכה בנקודות ב׳ ג׳ ד׳ ה׳, והקותה במחומש השוה הצלעות והזויות הידוע, והיא עגולת[13] אבגד״ה.

1 تبين] CO1T1T3

2 ه] +ا CCamO1T1T2T3T4

1 קו] קוי **ב** 2 ג״ז] ב⟨ג⟩״ז **פ** 3 ב״ז] +~~ז״ד~~ **ק** 4 ומשולש ... זד״ג] ומשותף **ב** 5 כמו] +כל **ו** 6 זוית זד״ג ... חצי זוית] ליתא **ו** 7 זוית גד״ה] ⟨*⟩ **ב** 8 זב״ג] ~~גב״א~~+ **ס** 9 קו] ליתא **ס** 10 קוי] ליתא **בפ** 11 אב״ז ... זוית] ליתא **ב** 12 ז״ג] +שוים **ס** 13 עגולת] עגולה **בוספ**

וזה מה שרצינו.

وَذَلِكَ ما أَرَدْنا بَيَانَهُ.

وذلك ما اردنا ان نعمل.

וזה מה שרצינו לבאר[1].

[1] וזה ... לבאר] ונשלם ביאורו **וט**

[Proposition IV.15]

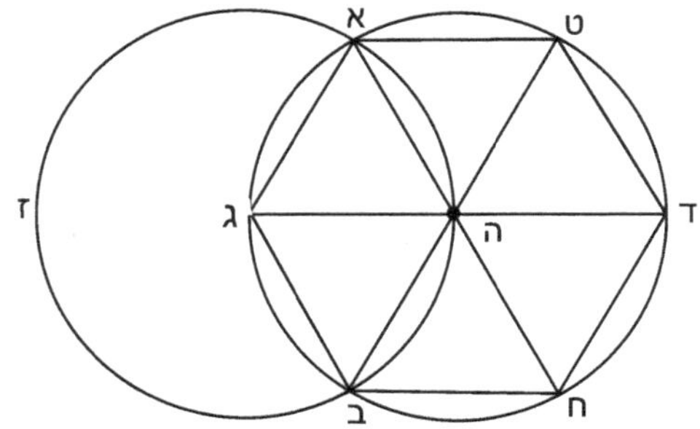

נֻرِيدُ أَنْ نُبيِّنَ[1] كَيْفَ نَعْمَلُ فِي دائِرَةٍ مَعْلومَةٍ مُسَدَّسًا مُتَساوِيَ الأَضْلاعِ وَالزَّوايا (تُحيطُ)[2] بِهِ الدّائِرَةُ.

נרצה לעשות בתוך עגול ידוע משושה שוה הצלעות והזויות.

مِثالُهُ: أَنْ نَجْعَلَ الدّائِرَةَ المَعْلومَةَ دائِرَةَ ا ب ج. وَقُطْرُها د ج. وَمَرْكَزُها ه. وَنُرِيدُ أَنْ نَعْمَلَ فيها مُسَدَّسًا مُتَساوِيَ الأَضْلاعِ وَالزَّوايا تُحيطُ بِهِ دائِرَةُ ا ب ج. فَنَخُطُّ عَلى مَرْكَزِ ج وَبِبُعْدِ ه دائِرَةً عَلَيْها (ه ب ز)[3]. وَنَصِلُ ا ه ه ب بِخَطَّيْنِ يُخْرِجُهُما عَلى اسْتِقامَةٍ إِلى ط و ح مِنَ الخَطِّ المُحيطِ. وَنَصِلُ ا ج ج ب ب ح ح د د ط ط ا. فَأَقولُ إِنّا قد عَمِلْنا فِي دائِرَةِ ا ب ج مُسَدَّسًا مُتَساوِيَ الأَضْلاعِ وَالزَّوايا تُحيطُ به دائِرَةُ ا ب ج، وَهُوَ مُسَدَّسُ ا ط د ح ب ج.

ונשים העגול הידוע אבג״ד, וקטרו ד״ג, ומרכזו נקודת ה׳. וכאשר רצינו לעשות בתוכו משושה שוה הצלעות והזויות, נקיף בו על מרכז ג׳ ובשיעור מרחק מה שבין ג׳ וה׳— עגול עליו הב״ז. ונדביק א׳ בה׳ וב׳ בה׳ בשני קוים. ונוציאם עד ח׳ וט׳ על יושר. ונדביק א׳ בג׳ וג׳ בב׳ וב׳ בח׳ וח׳ בד׳ וד׳ בט׳ וט׳ בא׳. ואומר שכבר עשינו בתוך עגול אבג״ד משושה שוה הצלעות והזויות עליו אטדחב״ג.

---

[1] أن نبين] +* [2] تحيط] يحيط [3] ه ب ز] ه ب ز ا

שרטוט—$

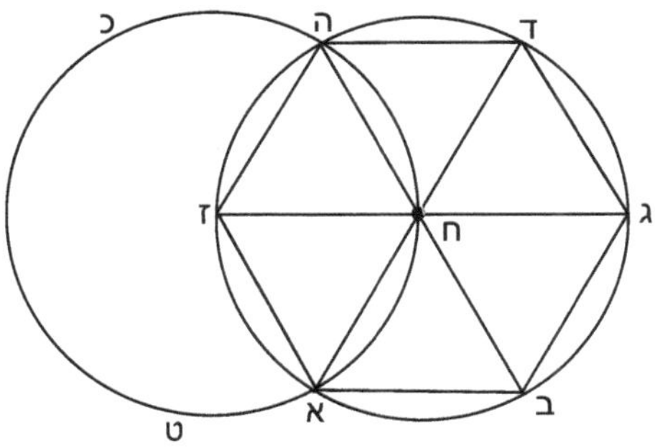

نُرِيدُ أَنْ نَعْمَلَ في دائِرَةٍ مَعلومَةٍ مُسَدَّسًا مُتَساوي الأَضْلاعِ والزَّوايا.

נרצה לעשות בעגולה ידועה משושה[1] שוה הצלעות והזויות.

فَلْتَكُنِ الدَّائِرَةُ المَعْلومَةُ دائِرَةَ ا ب ج د ه ز. ونُرِيدُ أَنْ نَعْمَلَ فيها مُسَدَّسًا مُتَساوي الأَضْلاعِ والزَّوايا. فَنُخْرِجَ قُطْرَ الدَّائِرَةِ، وَهُوَ ج ز. وَلْيَكُنِ المَرْكَزُ نُقْطَةَ ح. ونَجْعَلَ نُقْطَةَ ز مَرْكَزًا. ونُدِيرَ بِبُعْدِ ز ح دائِرَةً عليها ح ا ط ك ه. وَنَصِلَ خُطوطَ ا ز ا ح ه ح ه ز. ونُخْرِجَ خَطَّيْ ا ح ه ح إلى نُقْطَتَيْ د ب. وَنَصِلَ خُطوطَ ا ب ج ب ج د د ه.

תהיה העגולה הידועה עגולת אבגדה״ז. ונרצה לעשות בה משושה[2] שוה הצלעות והזויות[3]. הנה נוציא קוטר העגולה, והוא ג״ז. ויהיה המרכז נקודת ח׳[4]. ונשים נקודת ז׳ מרכז. ונקיף במרחק ז״ח עגולה עליה חאטכ״ה. ונגיע קוי א״ז א״ח[5] ח״ה ה״ז[6]. ונוציא שני קוי א״ח ה״ח אל שתי נקודות ד׳ ב׳. ונגיע קוי א״ב ג״ב ג״ד ד״ה[7].

שרטוט ראשון—**פ**: $ [1]משושה] משושש **ו** [2]משושה] משושת **בופ** [3]תהיה ... והזויות] ⟨*⟩ **ט** [4]המרכז ... ח׳] נקודת ח׳ המרכז **ב** [5]א״ח] +~~א״ה~~ **ק** [6]ה״ז] ליתא **ב** [7]ד״ה] ⟨*⟩ **ט**

מופת זה: לפי שה׳ מרכז עגול אבג״ד, יהיה א״ה שוה לה״ג. ועוד, לפי שנקדת ג׳ מרכז עגול ב״א[1], יהיה א״ג שוה לג״ה. ואם כן משולש אה״ג שוה הצלעות והזויות. ותהיה זוית אה״ג שני שלישי זוית נצבה. וכמו כן זוית גה״ב שני שלישי זוית נצבה. וזוית אה״ב כלה היא זוית נצבה ושליש. וזוית טה״א וזוית אה״ב שוות לשתי זויות נצבות. וזוית אה״ב היא נצבה ושליש. ונשארה זוית אה״ט שני שלישי זוית נצבה. וזוית אה״ט שוה לזוית בה״ח. וזוית אה״ג שוה לזוית דה״ח. וזוית גה״ב שוה לזוית טה״ד. ואם כן כל שש הזויות אשר עליהן ה׳—כל אחת מהן שני שלישי זוית נצבה, והן שוות. ואם כן השש קשתות גם כן הן שוות, והן א״ט וט״ד וד״ח וח״ב וב״ג וג״א. ומיתריהם שוים. ואם כן משושה אטדחב״ג שוה הצלעות. והוא גם כן שוה[2] הזויות, לפי שקשת ד״ח שוה לקשת ב״ג. וקשת דטא״ג משותפת. ואם כן קשת דטאג״ב כלה שוה לקשת גאטד״ה כלה. ואמנם קשת דטאג״ב עליה זוית דח״ב. ועל קשת גאטד״ה זוית חב״ג. אם כן זוית חב״ג שוה לזוית דח״ב. ומצד זה הזויות אשר על ד׳ וט׳ וא׳ וג׳ שוות לזויות דח״ב וחב״ג. וכבר התבאר שזה[3] המשושה שוה הצלעות והזויות, ונעשה בתוך עיגול[4] אבג״ד.

بُرْهانُهُ: أَنَّ ه مَرْكَزُ دائِرَةِ ا د ب ج. ا ه مِثْل ه ج. وَأَيْضًا نُقْطَةُ ج مَرْكَزُ دائِرَةِ ب ا. ا ج مِثْل ج ه. فَمُثَلَّثُ ا ه ج مُتَساوِي الأَضْلاعِ وَالزَّوايا. وَالزَّاوِيَةُ الَّتِي مِنْ ا ه ج هِيَ ثُلُثا زاوِيَةٍ قائِمَةٍ. وَأَيْضًا الَّتِي مِنْ ج ه ب ثُلُثا قائِمَةٍ. وَلٰكِنَّ جَمِيعَ الَّتِي مِنْ ا ه ب هِيَ زاوِيَةٌ قائِمَةٌ وَثُلُثُ قائِمَةٍ. وَالَّتِي من ط ه ا وَالَّتِي مِنْ ا ه ب مِثْل زاوِيَتَيْنِ قائِمَتَيْنِ. وَالزَّاوِيَةُ الَّتِي مِنْ ا ه ط مِثْل الَّتِي مِنْ ب ه ح. وَالَّتِي مِنْ ا ه ج مِثْل الَّتِي من د ه ح. وَالَّتِي من ج ه ب مِثْل الَّتِي من ط ه د. فَجَمِيعُ الزَّوايا السِتّ الَّتِي عَلَى ه كُلُّ واحِدَةٍ (مِنْها)[1] ثُلُثا زاوِيَةٍ قائِمَةٍ، وَهِيَ مُتَساوِيَةٌ. وَالقِسِيُّ مُتَساوِيَةٌ، وَهِيَ ا ط ط د د ح ح ب ب ج ج ا. فَمُسَدَّسُ ا ط د ح ب ج مُتَساوِي الأَضْلاعِ. وَقَوْسُ د ح مِثْل قَوْسِ ب ج. وَقَوْسُ د ط ا ج مُشْتَرَكَةٌ. فَجَمِيعُ قَوْسِ د ط ا ج ب مِثْل جَمِيع قَوْسِ ج ا ط د ح. وَلٰكِنَّ قَوْسَ (د ط ا ج ب)[2] عَلَيْها الزَّاوِيَةُ الَّتِي مِنْ د ح ب. وَعَلَى قَوْسِ ج ا ط د ح الزَّاوِيَةُ الَّتِي مِنْ ح ب ج. وَالَّتِي مِنْ ح ب ج مِثْل الَّتِي مِنْ (ب ح د)[3]. وَكَذٰلِكَ الَّتِي عَلَى د و ط و ا و ج مُساوِيَةٌ الَّتِي من د ح ب وَالَّتِي من ح ب ج. فَقَد اسْتَبانَ أَنَّ المُسَدَّسَ مُتَساوِيَ الأَضْلاعِ وَالزَّوايا مَعْمُولٌ في دائِرَةِ ا ب ج.

[1] ב״א] +⟨⟨הבא״ז⟩⟩ [2] שוה] ~~נצב~~+ [3] שזה] ש⟨⟨זה⟩⟩ [4] עיגול] ~~ה~~עיגול

[1] منها] منهي [2] د ط ا ج ب] ج ا ج ب [3] ب ح د] ب ج د

فلأنّ نقطة ح مركز دائرة اج ه، يكون خطّ اح مساويًا لخطّ ح ه^. ولأنّ نقطة ز مركز دائرة اط ك ه، يكون خطّ از مساويًا لخطّ ز ه. وكلّ واحد من مثلّثي اح ز ح ه ز متساوي الأضلاع. وخطّ اح مساوٍ لخطّ ح ه. وخطّ از مساوٍ لخطّ ز ه. وخطّ ز ح مشترك. فكلّا خطّي اح ح ز مساويان لكلّا خطّي ه ح و ح ز، كلّ واحد لنظيره. وقاعدة از مساوية لقاعدة ز ه. فزاوية اح ز مساوية لزاوية ز ح ه. وزاوية ز ح ه مساوية لزاوية ب ح ج[1]. ولكنّ زاوية اح ز مساوية لزاوية ج ح د. فزوايا ب ح ج ج ح د اح ز ز ح ه الأربع متساوية. ولأنّ خطّ اح مساوٍ لخطّ ح ز، إذ هما يخرجان من مركز الدائرة إلى الخطّ المحيط بها، تكون زاوية ح ا ز مساوية لزاوية ا ز ح. فكلّا زاويتي ح ا ز ا ز ح مثلا زاوية ح ا ز. ولكنّ زاويتي ح ا ز ا ز ح مساويتان لزاوية اح ج الخارجة. فزاوية اح ج الخارجة مثلا زاوية ح ا ز. وزاوية ح ا ز مساوية لزاوية اح ز، لأنّ المثلّث متساوي الأضلاع.

[1]وزاوية ز ح ه ... ب ح ج] CamMPR1R2

הנה מפני כי נקודת ח׳ מרכז עגולת אג״ה, יהיה קו א״ח שוה לקו ז״ח[1]. ומפני כי נקודת ז׳ גם כן מרכז עגולת אטכ״ה[2], יהיה קו א״ז שוה לקו ז״ח[3]. וכל[4] אחד ממשולשי[5] אח״ז[6] חה״ז[7] שוה הצלעות. וקו א״ח שוה לקו ח״ה. וקו א״ז שוה לקו ז״ה. וקו ח״ז משותף. הנה כל שני קוי א״ח ח״ז שוים לכל שני קוי ה״ח ח״ז, כל אחד לנכחי לו[8]. ותושבת א״ז שוה לתושבת ז״ה. אם כן[9] זוית אח״ז שוה לזוית זח״ה[10]. אבל זוית אח״ז שוה לזוית[11] גח״ד. וזוית בח״ג שוה לזוית זח״ה[12]. אם כן זויות בח״ג גח״ד אח״ז זח״ה הארבע[13] שוות. ומפני כי קו א״ח שוה לקו ח״ז, כי הם יוצאים ממרכז העגולה אל הקו המקיף בהם, תהיה זוית חא״ז[14] שוה לזוית אז״ח. אם כן[15] כל שתי זויות[16] חא״ז אז״ח[17] כפל זוית חא״ז. אבל שתי זויות חא״ז אז״ח שוות לזוית אח״ג החיצונה. אם כן זוית[18] אח״ג החיצונה[19] כפל זוית חא״ז. וזוית חא״ז שוה לזוית אח״ז[20], מפני כי המשולש שוה הצלעות.

[1]ז״ח] ח״ה ⟨⟨*⟩⟩ ק [2]אטכ״ה] ~~אג״ה~~+ ק [3]ז״ח] ~~ז״ח~~ ⟨⟨*⟩⟩ ק [4]וכל] * ⟨⟨נ״א הנה⟩⟩ ק [5]ממשולשי] ממשולת **ב** ממשולש **ו** [6]אח״ז] אה״ז **ס** [7]חה״ז] הח״ז **בוס** [8]לו] ליתא **ב** [9]כן] +~~תושבת~~ **ט** [10]זח״ה] +אבל זוית אח״ז שוה לזויות זח״ה **ב** [11]לזוית] +~~גב״ד~~ **ס** [12]זח״ה] ~~בח״ה~~+ **פ** [13]הארבע] הארבעה **ו** [14]חא״ז] הח״ז **ב** אז״ח **ו** [15]כן] ליתא **ב** [16]זויות] ⟨⟨*⟩⟩ ק [17]אז״ח] +~~שוות לזוית אח״ג החיצונה~~ ק [18]זוית] ⟨*⟩ **ס** [19]אם ... החיצונה] ⟨⟨*⟩⟩ **ו** [20]אח״ז] ~~אח״ד~~+ **פ**

وزاوية اح ز مُسَاوِيَة لِزَاوِيَةِ ب ح ج. فزاوية
اح ج مِثْلا زاوية ب ح ج. فزاوية ب ح ج
مُسَاوِيَة لِزَاوِيَةِ ب ح ا[1]. فزاوية ب ح ج
مُسَاوِيَة لِزَاوِيَةِ ب ح ا. وَلٰكِنَّ زاوية ب ح ا
مُسَاوِيَة لِزَاوِيَةِ د ح ه. وزاوية ب ح ج مُسَاوِيَة
لِكُلِّ واحِدَةٍ مِنْ زوايا اح ز زح ه ج ح د.
فكُلّ واحِدَةٍ مِنْ زوايا ب ح ج ج ح د اح ز
زح ه مُسَاوِيَة لِكُلِّ واحِدَةٍ مِنْ زاوِيَتَيْ ب ح ا
و د ح ه. فالزَّوايا السِتّ الَّتِي عِنْدَ نُقْطَة ح
مُساوٍ بَعْضُها لِبَعْض. والزَّوايا المُتَساوِيَة يوتِرُها
قِسِيٌّ مُتَساوِيَةٌ. فَقُسِيّ اب ب ج ج د د ه ه ز
زا السِتّ مُتَساوِيَة. فخُطوط اب ب ج ج د
د ه ه ز زا السِتّة إِذَنْ مُتَساوِيَة. لأَنَّ القُسِيّ
المُتَساوِيَة توتِرُها خُطوطٌ مُتَساوِيَةٌ. فَمُسَدَّس
اب ج د ه ز مُتَساوِي الأَضلاع. وأَقولُ إِنَّهُ
مُتَساوِي الزَّوايا. فَلأَنَّ قَوْسَ ا ز مُسَاوِيَةٌ لقَوْسِ
ب ج، وقَوْس ج د ه ز مُشْتَرَكَة، فَجَمِيع قَوْس
ب ج د ه ز مُسَاوِيَة لجميع قَوْسِ ا ز ه د ج.
وزاوية ب ا ز على قَوْس ب ج د ه ز. وزاوية
ا ب ج على قَوْس ا ز ه د ج. والزَّوايا الَّتي تكونُ
على القُسِيّ المُتَساوِيَةِ هِيَ مُتَساوِيَةٌ. فزاوية ب ا ز
مُسَاوِيَةٌ لِزَاوِيَةِ ج ب ا. وَكَذٰلِكَ أَيْضًا يَتَبَيَّنُ
أَنَّ زَاوِيَةَ ج ب ا مُسَاوِيَةٌ لِزَاوِيَةِ ب ج د.
وزاوية ب ج د مُسَاوِيَة لِزَاوِيَةِ ج د ه، وزاوية
ج د ه لِزَاوِيَةِ د ه ز، وزاوية د ه ز لِزَاوِيَةِ ا ز ه.

[1] فزاوية اح ج ... ب ح ا] **CamO2QU**

וזוית אח״ז שוה לזוית בח״ג. אם כן זוית בח״ג
שוה לזוית חא״ז. אם כן זוית אח״ג כפל זוית
בח״ג. אם כן זוית בח״ג שוה לזוית בח״א. אבל
זוית בח״א[1] שוה לזוית[2] דח״ה[3]. וזוית בח״ג
שוה לכל אחת מזויות[4] אח״ז זח״ה גח״ד.
הנה[5] כל אחת מזויות בח״ג גח״ד אח״ז זח״ה[6]
שוות לכל אחת משתי זויות בח״א דח״ה. אם
כן הזויות השש[7] אשר[8] אצל נקודת ח׳ שוות
קצתם אל קצת. והזויות השוות יהיו מיתריהם
קשתות שוות. אם כן קשתות א״ב ב״ג ג״ד[9]
ד״ה ה״ז ז״א השש[10] שוות. הנה קוי א״ב ב״ג
ג״ד ד״ה ה״ז ז״א הששה[11] אם כן שוים[12],
מפני כי הקשתות השוות יהיו מיתריהם
קוים[13] שוים. אם כן משושת אבגדה״ז שוה
הצלעות. ואומר שהוא שוה הזויות. הנה[14]
מפני כי קשת א״ז שוה לקשת ב״ג, וקשת
גדה״ז משותפים[15], הנה כל קשת בגדה״ז
שוה לכל קשת אזהד״ג. וזוית בא״ז על קשת
בגדה״ז. וזוית אב״ג על קשת אזהד״ג. והזויות
אשר תהיינה על הקשתות השוות[16] הם[17]
שוות. אם כן זוית בא״ז שוה לזוית גב״א.
וכן גם[18] כן יתבאר כי זוית גב״א שוה לזוית
בג״ד, וזוית בג״ד שוה לזוית גד״ה, וזוית
גד״ה לזוית דה״ז, וזוית דה״ז לזוית אז״ה.

[1] אבל ... בח״א] +* **ב** [2] אבל ... לזוית] ליתא **ז** [3] לזוית דח״ה] לזוית בח״ג **ב** לדח״ה **ס** ⟨⟨לזוית⟩⟩ לדח״ה **פ** [4] מזויות] מזויותיו **ס** [5] הנה] ~~אם כן~~+ **ק** [6] זח״ה] ⟨⟨זה״ח⟩⟩ **ב** [7] השש] ⟨*⟩ **פ** [8] אשר] ליתא **ב** [9] ג״ד] ליתא **ס** [10] השש] +קצוות **ס** [11] הנה ... שוים] ליתא **ז** [12] הששה] השש שוות הנה קו ח״ב ב״ג ג״ד ד״ה ה״ז ז״א הששה **ס** [13] קוים] +ישרים **ק** [14] הנה] כי+ **ק** [15] משותפים] ⟨⟨משותף⟩⟩ **ק** * [16] הקשתות השוות] קשתות שוות **ק** [17] הם] +גם כן **ז** [18] וכן גם] וגם **ס**

וכבר אפשר לנו לעשות על עגול ידוע משושה שוה הצלעות והזויות מקיף בעיגול, ולעשות בתוך משושה שוה הצלעות והזויות ידוע עגול מקיף בו, כמו שבארנו במחומש. והנה התבאר לך הנה כי חצי קוטר בעיגול ישלים מיתרי העגול בששה פעמים, ושצלע המשושה הוא חצי קוטר העגול.

وَذَلِكَ ما أَرَدْنا أَنْ نَعْمَلَ. وَنَعْمَلُ عَلى دائِرَةٍ مَعْلومَةٍ مُسَدَّسًا مُتَساوِيَ الأَضْلاعِ وَالزَّوايا يُحِيطُ بِها. وَنَعْمَلُ في مُسَدَّسٍ مَعْلومٍ مُتَساوي الأَضْلاعِ دائِرَةً تُحِيطُ بِهِ كَنَحْوِ ما فَسَّرْنا في المُخَمَّسِ. فَقَدِ اسْتَبانَ أَنَّا قَدْ عَمِلْنا في دائِرَةٍ مَعْلومَةٍ مُسَدَّسًا مُتَساوِيَ الأَضْلاعِ وَالزَّوايا تُحِيطُ بِهِ، فَإِنَّ ضِلْعَ المُسَدَّسِ مِثْلُ نِصْفِ القُطْرِ.

وَذَلِكَ ما أَرَدْنا بَيانَهُ.

فالزَّوايا السِتّ الَّتِي^ عِنْدَ نُقْطَةِ ا ب ج د ه ز مُتَساوِيَةٌ. فَمُسَدَّس ا ب ج د ه ز مُتَساوي الزَّوايا. وَقَدْ كُنَّا بَيَّنا أَنَّهُ مُتَساوي الأَضْلاعِ، وَهُوَ مَعْمُولٌ في دائِرَةِ ا ب ج د ه ز.

אם כן הזויות השש אצל[1] נקודות א׳ ב׳ ג׳ ד׳ ה׳ ז׳ שוות. אם כן משושה אבגדה״ז שוה הזויות. וכבר בארנו שהוא שוה הצלעות[2], והוא עשוי בעגולת אבגדה״ז.

وذَلِكَ ما أَرَدْنا أَنْ نُبَيِّنَ.

וזה מה שרצינו לבאר[3].

وَقَدْ يمْكِنُنا أَنْ نَعْمَلَ على دائِرَةٍ مَعْلومَةٍ مُسَدَّسًا مُتَساوي الأَضْلاعِ وَالزَّوايا، وَأَنْ نَعْمَلَ في مُسَدَّسٍ مَعْلومٍ مُتَساوي الأَضْلاعِ وَالزَّوايا دائِرَةً^ يُحيطُ بِها، وَأَنْ نَعْمَلَ عَلَيْهِ دائِرَةً تُحيطُ بِها، على مِثْلِ ما وَصَفْنا في المُخَمَّسِ.

וכבר היה אפשר לנו שנעשה על עגולה[4] ידועה משושה שוה הצלעות והזויות יקיף בה, ושנעשה עליו עגולה[5] תקיף בו, על דמיון מה שספרנו במחומש[6].

وهُنالِكَ اسْتَبَانَ أَنَّ نِصفَ قُطْرِ الدَّائِرَةِ يوتِرُ الخَطَّ المُحيطَ بِها في سِتِّ مَرَّاتٍ. وَأَنَّ ضِلْعَ المُسَدَّسِ مُساوٍ لِنِصْفِ قُطْرِ الدَّائِرَةِ.

ובכאן התבאר כי חצי קוטר העגולה יהיה מיתר הקו המקיף בה בששה פעמים. כי[7] צלע המשושה[8] שוה לחצי קוטר העגולה.

---

[1] אצל] אשר+ **בס** 〈〈אשר〉〉+ **ק** [2] הצלעות] ~~הצלעות הזויות~~+ **ס** [3] לבאר] לבארו **בטפ**
[4] עגולה] עגולת **ב** [5] ידועה ... עגולה] +* **ס**
[6] במחומש] בו מחומש **ס** [7] כי] +הנה **טפק**
[8] המשושה] **משושה ס**

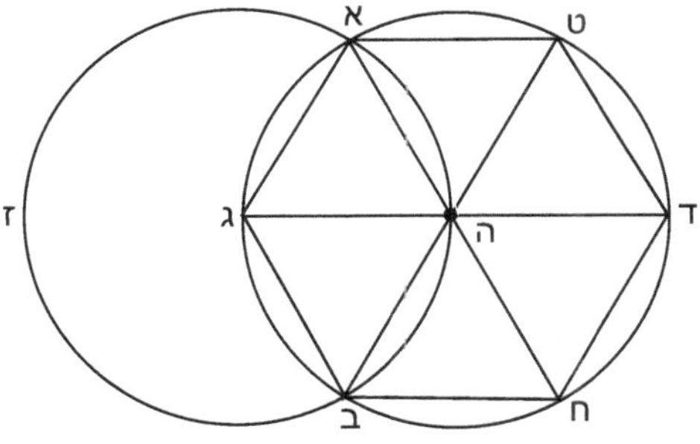

وَوَجَدْنا في نُسْخَةٍ أُخرى هذا الشَّكلَ مُعادًا بِنَحْوٍ آخرَ على ما تَرى[1]:

نُريدُ أَن نَعْمَلَ في دائِرَةٍ مَعْلومَةٍ مُسَدَّسًا مُتَساوي الأَضْلاعِ وَالزَّوايا. فَنَجْعَلَ الدّائِرَةَ المَعْلومَةَ دائِرَةَ ا ب ج، وَقُطْرَها د ج، ومَرْكَزَها ه. ونُريدُ أَنْ نَعْمَلَ فيها مُسَدَّسًا مُتَساوي الأَضْلاعِ والزَّوايا يُحيطُ بِهِ. فَنَخُطَّ على مَرْكَزِ ج. وَبِبُعْدِ ه دائِرَةَ ه ب ز. ونُخْرِجَ خَطَّيْ ا ه ب ه. ونُخْرِجَها إلى ح ط على اسْتِقامَةٍ. ونُخْرِجَ خُطوطَ ا ج ج ب ب ح ح د د ط ط ا. فَمَرْكَزُ دائِرَةِ ا ب ج د. نُقْطَةُ ه. فَخَطُّ ا ه مثل خطِّ ه ج. وَأَيْضًا فَإِنَّ مَرْكَزَ دائِرَةِ ه ب ز نُقْطَةُ ج. فَخَطُّ ا ج مثل خطِّ ج ه. ومُثَلَّثُ ا ه ج مُتَساوي الأَضلاعِ والزَّوايا. وكَذَلِكَ أيضًا^ تَبَيَّنَ[2] أَنَّ مُثَلَّثَ ج ه ب مُتَساوي الأَضلاعِ والزَّوايا. فزاويةُ ا ه ج ثُلُثا قائِمَةٍ. وزاوية

[1]ووجدنا ... ترى] זה הנוסח בעד נוסח U והוא הקרוב ביותר לנוסח שיש בתרגום מא"ת [2]تبين] M

מצאנו התמונה הזאת בנסחא אחרת במין אחר לפי מה שתראה:

נרצה שנעשה בעגולה ידועה משושה שוה הצלעות והזויות. הנה נשים העגולה אב"ג, והקוטר[1] שלה ד"ג, ומרכזה[2] ה׳. ונרצה לעשות בה משושה[3] שוה הצלעות והזויות תקיף בו[4]. הנה נקוה על מרכז ג׳ ובמרחק ה׳ עגולת הב"ז. ונוציא א"ה ב"ה[5]. ונוציאם[6] אל ח׳ ט׳ על יושר. ונוציא קוי[7] א"ג ג"ב ב"ח ח"ד ד"ט א"ט. הנה מרכז עגולת אבג"ד[8] נקודת ה׳. אם כן קו א"ה כמו קו ה"ג. וגם כן[9] הנה מרכז עגולת הב"ז נקודת ג׳[10]. אם כן קו א"ג כמו קו[11] ג"ה. ומשולש אה"ג שוה[12] הצלעות והזויות. וכן התבאר שמשולש גה"ב שוה הצלעות והזויות. אם כן זוית אה"ג שני שלישי נצבת. וזוית גה"ב שני שלישי נצבת

[1]והקוטר] וקוטר ב [2]ומרכזה] ומרכ⟨ז⟩ה ב [3]משושה] משושת ט [4]בו] ליתא ב [5]ב"ה] ב"ט ס [6]ונוציאם] ונוציאיהם ב [7]קוי] ליתא ב [8]עגולת אבג"ד] ⟨⟨*⟩⟩ ק [9]כן] ליתא ב [10]נקודת ג׳] ליתא ס [11]א"ג ... קו] ליתא ס [12]שוה] +~~למשולש~~ ק

ج ه ب ثُلُثا قائِمَةٍ أَيْضًا[1]. فجميع زاوية ا ه ب قائِمَة وثُلُث قائِمَة. وزاوية ط ه ا وزاوية ا ه ب مِثْل قائِمَتَيْنِ. وزاوية ا ه ب قائِمَة وثُلُث قائِمَة. بَقِيَتْ زاوِيةُ ا ه ط ثُلُثَيّ قائِمَة^. وزاوية ا ه ط مثل زاوية ب ه ح. وزاوية ا ه ج مثل زاوية د ه ح. وزاوية ج ه ب مثل زاوية ط ه د. فكُلّ واحدةٍ مِنَ الزَّوايا السِتّ الَّتي عند^ نُقْطَة ه ثُلُثا قائِمَة. فهي مُتَساوِيَة. والقُسِيّ السِتّ^ الَّتي عليها ا ط ط د د ح ح ب ب ج ا ج مُتَساوِيَةٌ. ومُسَدَّس ا ط د ح ب ج مُتَساوي الأَضلاع. وقَوْس د ح مثل قَوْسِ^ ب ج. وقَوْس د ط ا ج مشتركة. فجميع قَوْس د ط ا ج ب مثل جَميع قَوْسِ ج ا ط د ح. لكنَّ قَوْسَ د ط ا ج ب عَلَيها زاوِيَةُ د ح ب. وعلى قَوْسِ ج ا ط د ح زاوية ح ب ج. فزاوية ح ب ج مثل زاوِيَةِ ب ح د. وكَذَلِكَ تَكونُ الزَّوايا الَّتي عِنْدَ نُقْطَةِ د ط ا ج مُساوِيَةً لِزاوِيَتَيْ د ح ب ح ب ج. فَقَدِ اسْتَبانَ أَنَّ المُسَدَّسَ مُتَساوي الأَضلاع والزَّوايا، وَهُوَ مَعْمُولٌ في دائِرَةِ ا ب ج.

גם כן. אם כן כל זוית אה״ב[1] נצבת ושליש[2] נצבת. וזוית טה״א וזוית[3] אה״ב כמו שתי נצבות[4]. וזוית אה״ב נצבת ושליש נצבת. הנה ישאר זוית אה״ט שני שלישי נצבת גם כן. וזוית אה״ט כמו זוית בה״ח. וזוית אה״ג כמו זוית דה״ח. וזוית גה״ב כמו זוית טה״ד. הנה כל אחת מן הזויות השש אשר [!] נקודת[5] ה׳ שני שלישי נצבת. אם כן הן שוות. והקשתות[6] אשר עליהם א״ט ט״ד ד״ח ח״ב ב״ג א״ג שוות. ומשושת[7] אטדחב״ג שוה הצלעות. וקשת ד״ח כמו ב״ג[8]. וקשת דטא״ג משותף. אם כן כל קשת דטאג״ב כמו[9] כל קשת גאטד״ח. אבל קשת דטאג״ב[10] עליהם [!] זוית דח״ב. ועל קשת גאטד״ח זוית[11] חב״ג. אם כן[12] זוית חב״ג[13] כמו זוית בח״ד. וכן יהיו[14] הזוויות אשר אצל נקודת ד׳ ט׳ א׳ ג׳ שוות לשתי זויות דח״ב חב״ג. הנה[15] כבר התבאר כי המשושה[16] שוה הצלעות והזויות, והוא עשוי בעגולת אב״ג,

[1] أيضا] O2U

[1] אה״ב] ~~גה״ב~~+ ק [2] ושליש] ושלישי ב [3] וזוית] חצי זוית ב [4] כמו ... נצבות] ליתא ב [5] נקודת] על+ ב ~~על~~+ ק [6] והקשתות] +השש טק [7] ומשושת] ומשושה ו ומשושה ~~א״ג~~ ק [8] ב״ג] קשת+ ט [9] כמו] כל ⟨⟨*⟩⟩ ק [10] כמו ... דטאג״ב] ליתא בו [11] זוית] זויות בוס [12] כן] ליתא ב [13] אם ... חב״ג] ליתא ס [14] יהיו] ליתא בו [15] הנה] ליתא ו [16] המשושה] המשושת בוט המשוששת פ

وَذَلِكَ ما أَرَدْنا أَنْ نُبَيِّنَ[1].

وזה מה שרצינו לבאר.

وَهُنالِكَ اسْتَبَانَ أَنَّهُ إِنْ عُمِلَ في دائِرَةٍ مَعْلومة مُسَدَّسًا مُتَساوي الأَضْلاعِ والزَّوايا، فَإِنَّ ضِلْعَهُ الوَاحِدَ مُساوٍ لِنِصْفِ قُطْرِ الدَّائِرَةِ.

ובכאן התבאר כי אם נעשה בעגולה משושת שוה[1] הצלעות והזויות, הנה צלעו שוה לחצי קוטר העגולה.

[1] نبين] O1

[1] שוה] ⟨*⟩ ק

[Proposition IV.16]

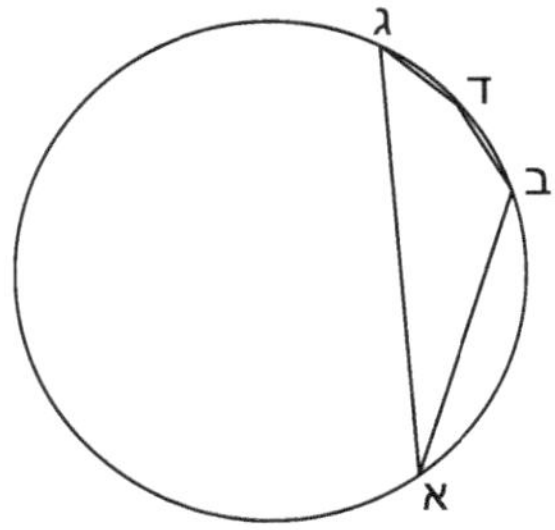

נרצה לעשות בתוך עגול ידוע תמונה בעלת חמש עשרה זויות שוות הצלעות והזויות מקיף בה העגול.

نُرِيدُ أَنْ نُبَيِّنَ كَيْفَ نَعْمَلُ فِي دائِرَةٍ شَكْلًا ذا خَمْس عَشْرَةَ زاوِيَةً مُتَساوي الأَضْلاعِ وَالزَّوايا تُحيطُ بِهِ الدَّائِرَةُ.

نُرِيدُ أَنْ نَعْمَلَ فِي دائِرَةٍ مَعْلومَةٍ شَكْلًا لَهُ خَمْسَ عَشْرَةَ زاوِيَةً مُتَساوِي الأَضلاعِ والزَّوايا تُحِيطُ بِهِ الدَّائِرَةُ.

נרצה לעשות בעגולה ידועה תמונה יש[1] לה חמשה עשר זויות[2] שוות הצלעות והזויות תקיף בה העגולה.

שרטוט—**בק**: קשת א״ב נחלקת לשלושה חלקים והוספו עליהם מיתרים ללא אותיות נוספות, וקו א״ג הושלם למשולש שוה צלעות חסום במעגל שקדקדו השלישית על ה׳; **ט**: קשת א״ב נחלקת לשלושה חלקים והוספו עליהם מיתרים ללא אותיות נוספות; **ס**: קשת א״ב נחלקת לשני חלקים והוספו עליהם מיתרים ללא אות נוספת; **פ**: המצולע בעל חמש עשרה הצלעות חסום במעגל, וקו א״ג הושלם למשולש שוה צלעות

[1] יש] ⟨*⟩ **ב**

[2] זויות] +⟨⟨צלעות⟩⟩ **ק**

ונשים העגול הידוע עגול אב״ג. וכאשר רצינו לעשות בתוכו תמונה בעלת חמש עשרה זויות שוות הצלעות והזויות יקיף בה העגול, נרשום בעגול צלע משולש שוה הצלעות והזויות. ונרשום בקצותיו א׳ וג׳. ונוציא מא׳ קו אל צלע מחומש שוה הצלעות והזויות מורשם בתוך עגול בקשת ג״א, והוא א״ב. וכאשר נחלק הקו המקיף בעגול בחמשה עשר חלקים, נפל קו א״ג על חמשה חלקים מהם, ונפל קו א״ב שהוא צלע מחומש על שלשה חלקים. ונשארו שני חלקים, והם קשת ב״ג. ונחלק אותה לחציים על ד׳. ונדביק ב׳ בד׳ וד׳ בג׳.

مِثالُهُ: أَنْ نَجْعَلَ الزَّاوِيَةَ المَعْلومَةَ دائِرَةَ ا ب ج. وَنُرِيدُ أَنْ نَعْمَلَ فيها شَكْلًا ذا خَمْسَ عَشْرَةَ زاوِيَةً مُتَساوِي الأَضْلاعِ وَالزَّوايا تُحِيطُ بِهِ الدَّائِرَةُ. فَنَخُطُّ في الدَّائِرَةِ ضِلْعَ مُثَلَّثٍ مُتَساوِي الأَضْلاعِ[1] وَالزَّوايا. وَنَرْسُمُ على طَرَفَيْهِ ا ج. وَنُخْرِجُ مِنْ ا ضِلْعَ مُخَمَّسٍ عَلَى مُتَساوِي الأَضْلاعِ وَالزَّوايا مَخْطوط في الدَّائِرَةِ في قَوْسِ ا ج، وَهُوَ ا ب. فَإِذا قَسَمْنَا الخَطَّ المُحِيطَ بالدَّائِرَةِ بِخَمْسَةَ عَشَرَ جُزءًا، وَقَعَ خَطُّ ا ج على خَمْسَةِ أَجْزاءٍ. وَيَقَعُ خَطُّ ا ب الَّذي هُوَ ضِلْعُ[2] مُخَمَّسٍ عَلَى (ثَلاثَةِ)[3] أَجْزاء أو (جُزْأَيْنِ)[4]، وَهُما قَوْسُ ب ج. فَنَقْسِمُهُما بِنِصْفَيْنِ عَلَى نُقْطَةِ د. وَنَصِلُ ب د د ج.

---

[1] الأضلاع] +‹الساقين› [2] متساوي ... ضلع] ‹*›
[3] ثلاثة] ثلثة [4] جزأين] جزؤين

فَنَجْعَلَ الدَّائِرَةَ المَعْلومَةَ دائِرَةَ ا ب ج. ونُريدُ أَن نَعْمَلَ فيها شَكْلًا لَهُ خَمْس عَشَرَةَ زاوية مُتَساوي الأَضْلاعِ والزَّوايا تُحيطُ بِهِ الدَّائِرَةُ. فَلْنَخُطّ في الدَّائِرَةِ وَتَرًا يكونُ ضِلْعَ مُثَلَّث مُتَساوي الأَضْلاعِ والزَّوايا، وَهُوَ^ وتَرُ ا ج. ونُخْرِجُ مِنْ نُقْطَةِ آ وَتَرًا يَكونُ ضِلْعَ مُخَمَّسٍ مُتَساوي الأَضْلاع والزَّوايا في قَوْسِ ا ج، وَهُوَ وَتَرُ ا ب. فَإِذا قُسِمَ الخَطُّ المُحيطُ بِخَمْسة عشر جُز[ءً]ا وَقَعَ وَتَرُ ا ج على خَمْسَةِ أجزاءٍ مِنْها. ويَقَعُ وَتَرُ ا ب على ثلاثةِ أجزاءٍ مِنها. وَيَبْقَى جُزْءانِ، وَهُما قَوْسُ ب ج. فَنَقْسِمها بِنِصْفَيْنِ على د. ونُخْرِجُ وَتَرَيْ ب د د ج.

הנה נשים העגולה הידועה עגולת אב״ג. ונרצה שנעשה בה תמונה בעלת חמשה עשר זויות שוות הצלעות והזויות תקיף בה העגולה[1]. הנה נקוה בעגולה מיתר יהיה צלע משולש[2] שוה הצלעות והזויות, מיתר[3] א״ג. ונוציא מנקודת[4] א׳ מיתר יהיה צלע מחומש שוה[5] הצלעות והזויות בקשת א״ג, והוא מיתר א״ב. וכאשר חולק הקו המקיף בחמשה עשר חלק [!] יפול מיתר א״ג על חמשה חלקים מהם. ויפול מיתר א״ב על שלשה חלקים מהם[6]. וישארו שני חלקים, והם קשת ב״ג. ונחלקם בשני חציים על ד׳. ויצאו שני מיתרי ב״ד וד״ג.

---

[1] הנה ... העגולה] ‹‹*›› ס [2] משולש] ~~מחומש~~ ‹‹*›› ב [3] מיתר] והוא+ ס [4] מנקודת] מנק‹ו›דת ב [5] מחומש שוה] +* ס [6] מהם] ‹*› ו

ולפי שקשת ב״ד[1] שוה לקשת ד״ג, יהיה מיתר ב״ד[2] שוה למיתר ד״ג. וכאשר חלקנו כל הקו המקיף בעיגול בכמות קשת ד״ג, ושמנו על כל קשת מהם מיתר, כבר עשינו בתוך העיגול תמונה [אשר] לה חמש עשרה זויות שוות הצלעות והזויות.

فقَوْسُ ب د مِثْل قَوْسِ د ج. وَوَتَرُ (ب د)[1] مِثْل وَتَرِ د ج. فَإذا قَسَمْنا جَميعَ الخَطِّ المُحيطِ بالدَّائِرَةِ مِثْل قَوْسِ د ج وَجَعَلْنا عَلى كُلِّ قَوْسٍ وَتَرًا، فَقَدْ خَطَطْنا شَكْلًا ذا خَمْس عَشْرَةَ زاوِيَةً مُتَساوي الأَضْلاعِ وَالزَّوايا في دائِرَةِ ا ب ج المَعْلومَةِ (تُحيطُ)[2] بِهِ الدَّائِرَةُ.

וזה מה שרצינו.

וכבר אפשר לנו לעשות על עגול ידוע תמונה אשר לה חמש עשרה זויות מקיפות בו, ולעשות בתוך תמונה ידועה אשר לה חמש עשרה זויות עגול מקפת בו, ולעשות עליה עיגול מקיף בה—כדרך שהראינו במחומש.

ثُمَّ نَخُطُّ عَلى الدَّائِرَةِ المَعْلومَةِ شَكْلًا لَهُ خَمْس عَشْرَةَ زاوِيَةً مُتَساوي الأَضْلاعِ وَالزَّوايا يُحيطُ بالدَّائِرَةِ، وَفي شَكْلٍ مَعْلومٍ ذي خَمْس عَشْرَةَ زاوِيَةً مُتَساوي الأَضْلاعِ وَالزَّوايا، نَعْمَلُ دائِرَةً يُحيطُ بِها. وَعَلى شَكْلٍ ذي خَمْس عَشْرَةَ زاوِيَةً مُتَساوي الأَضْلاعِ وَالزَّوايا مَعْلوم نَخُطُّ دائِرَةً (تُحيطُ)[3] بِهِ كَنَحْوِ ما فَسَّرْنا في المُخَمَّسِ.

وَذَلِكَ ما أَرَدْنا بَيانَهُ.

נשלם המאמר הרביעי בכח השם ועזרו ולו לבדו יאתה תהילה.

[1] ב״ד] ~~מדיז~~+ [2] ב״ד] +~~שויתר~~

[1] ب د] ح د [2] تحيط] يحيط [3] تحيط] يحيط

فَقَوْس ب د مثل قَوْسِ د ج. فَوَتَر ب د مثل وَتَرِ د ج. فَإِذا قَسَمْنَا جميعَ الخَطِّ المُحِيطِ بِمِثْلِ قَوْس د ج. وجَعَلْنا على كُلِّ قَوْسٍ وَتَرًا. فَقَدْ عَمِلْنا في الدَّائِرَةِ شَكْلًا لَهُ خَمْسَ عَشْرَةَ زاوِيَةً مُتَساوي الأَضلاعِ والزَّوايا.

אם כן קשת ב״ד כמו קשת ד״ג. אם כן מיתר ב״ד כמו מיתר[1] ד״ג. הנה אם כן[2] חלקנו כל הקו המקיף כמו קשת ד״ג, ושמנו על כל קשת מיתר. הנה כבר עשינו בעגולה תמונה יש לה חמשה עשר זויות שוות הצלעות והזויות.

وذَلِكَ ما أَرَدْنا أَنْ نُبَيِّنَ[1].

וזה מה שרצינו לבאר[3].

وَقَدْ يُمْكِنُنَا[2] أَنْ نَعْمَلَ على دائِرَةٍ مَعْلومَةٍ شَكْلًا لَهُ خَمْسَ عَشْرَةَ زاوِيَةً مُتَساوي الأَضْلاعِ والزَّوايا، وأَنْ نَعْمَلَ في شَكْلٍ مَعْلومٍ لَهُ خَمْسَ عَشْرَةَ زاوِيَةً دائِرَةً يُحِيطُ بِها^، وأَنْ نَعْمَلَ عَلَيْهِ دائِرَةً تُحِيطُ بِهِ على مِثْلِ ما وَصَفْنا في المُخَمَّسِ.

והנה אפשר לנו שנעשה על עגולה ידועה תמונה יש לה חמש עשרה[4] זויות שוות הצלעות והזויות, ושנעשה עליה עגולה תקיף בה, על דמיון מה שספרנו במחומש.

وذَلِكَ ما أَرَدْنا أَنْ نُبَيِّنَ[3].

וזה מה שרצינו לבאר.

تَمَّتِ المَقالَةُ الرَّابِعَةُ.

נשלם המאמר הרביעי[5].

---

[1] נبين] M [2] يمكننا] CamO1R1ST4 +أيضا
[3] وذلك ... نبين] MO1

[1] כמו מיתר] ליתא ו [2] כן] ליתא ו [3] לבאר] לבארו **בספק** [4] חמש עשרה] חמשה עשרה **בס** חמשה עשר **ט** [5] הרביעי] +שבח לאל נותן כח ⟨ו⟩עצמה **ב** +מספר אקלידס החכם תל״ח **ו** +מספר אקלידס החכם ת״ל **ס** +מספר אקלידס שבח לאל נותן כח ועצמה **פ**

[Book v]

(המאמר החמישי)[1]

[1] השיעור הקטן הוא חלק מן השיעור הגדול כאשר ישער הגדול.

[2] וקצת הוא אשר לא יכלול הגדול כשישערהו.

[3] [H2] ויהיה הגדול כפלים לקטן כאשר יפל עליו השיעור בקטן.

[4] [H3] היחס הוא הצטרפות מה בשיעור בין שני שעורים ממין אחד.

[5] ההתיחס הוא הדמות היחסים [!].

[6] [H4] השיעורים אשר יאמר שבין קצתם ובין קצת יחס הם אשר יתכן כי כשיכפלו יעדף קצתם על קצת.

[1] המאמר החמישי] כותרת המאמר הושמטה

المَقالَةُ الخامِسَةُ مِنْ كِتابِ الأَركان

[1] الجُزءُ مِقْدارُ الأَصْغَرِ مِنَ الأَعْظَمِ إِذا كانَ بَعَدِ الأَعْظَمِ.

[2] ذو الأَضْعافِ هُوَ الأَعْظَمُ مِنَ الأَصْغَرِ إِذا كانَ [بِ]قَدْرِهِ الأَصْغَرِ.

[3] النِّسْبَةُ هِيَ [أَيِّيَة][1] قَدْر مِقْدارَيْنِ مُتَجانِسَيْنِ كُلّ واحِدٍ مِنْهُما مِنَ الآخَرِ أي قَدْرٍ كانَ.

[4] المُناسَبَةُ هِيَ اشْتِباهُ النِّسَبِ.

[5] الأَقْدارُ الَّتي تُسَمَّى [ب]أن لِبَعْضِها نِسْبَة إِلى بَعْض هِيَ الَّتي تكونُ تَضاعيفُها يَزيدُ بَعضُها عَلى بَعْضٍ.

[1] أَيِّية] אולי: أَيَّة

المَقالَةُ الخامِسَةُ مِنْ كِتابِ أُقْليدِس

המאמר החמישי מספר אקלידס

[1] [E1] المِقْدارُ الأَصْغَرُ يَكونُ جُز[ءً]ا[1] مِنَ المِقْدارِ الأَعْظَمِ مَتَى كانَ يُقَدِّرُ الأَعْظَمَ.

[1] השיעור הקטן הוא חלק מן השיעור הגדול כאשר ישוער הגדול.

[2] [E1a] والبَعْضُ هُوَ الَّذي لا يَسْتَغْرِقُ الأَعْظَمَ إذا قُدِّرَ بِهِ.

[2] וקצת [!] הוא אשר לא יכלול הגדול כאשר ישערהו.

[3] [E2] وَيَكونُ الأَكْبَرُ[2] أَضْعافًا لِلأَصْغَرِ مَتَى كانَ يَقَعُ عَلَيْهِ التَّقديرُ بالأَصْغَرِ.

[3] [H2] ויהיה הגדול כפלים לקטן כאשר יפול עליו השיעור בקטן.

[4] [E3] النِّسْبَةُ هي إِضافةٌ ما في القَدْرِ بَيْنَ مِقْدارَيْنِ مِنْ جِنْسٍ واحِدٍ.

[4] [H3] היחס הוא הצטרפות מה בשיעור בין שני שיעורים מסוג אחד.

[5] [E3a] التَّناسُبُ هو تَشابُهُ النِّسَبِ.

[5] ההתיחס הוא הדמות היחסים.

[6] [E4] المَقاديرُ الَّتي يُقالُ إِنَّ بَيْنَ بَعْضِها وَبَيْنَ بَعْضٍ[3] نِسْبَةً هِيَ الَّتي يُمْكِنُ[4] إذا ضُوعِفَتْ أَنْ يَفْضُلَ بَعْضُها على بَعْضٍ.

[6] [H4] השיעורים אשר יאמר כי בין קצתם ובין קצתם יחס הם אשר אפשר כי כשיכפלו שיעדוף קצתם על קצת.

---

[1] جُز[ءً]ا] M جزؤا   [2] الأكبر] M الأعظم

[3] وبين بعض] M وبعض   [4] يمكن] CopM +قد

[7] [H5] יאמר בשיעורים שהם על יחס אחד, הראשון אל השני והשלישי אל הרביעי, כשיהיו כפלי הראשון והשלישי שוי הפעמים, אם שיעדפו יחד על כפלי השני והרביעי שוי הפעמים, כמה שהיו, ואם שישוו אליהם, ואם שיגרעו מהם יחד כשיוקשו בסדר קצתם לקצת.

[6] الأَقدارُ الَّتي تُسَمَّى أَنَّها عَلى نِسْبَةٍ واحِدَة، الأَوَّل إِلى الثّاني والثّالث إِلى الرّابِع، هِيَ الَّتي إِذا كانَتْ أَضْعافُ الأَوَّلِ والثّالث المُتَساوِيَة، إِمّا زائِدَة معًا عَلى أَضْعافِ الثّاني والرّابِع المُتَساوِيَة، أي التَّضاعيفِ كانَتْ، وَإِمّا مُساوِيَة لَها وَإِمّا ناقِصَة عَنْها مَعًا إِذا أُخِذَتْ تَتْلو بَعْضُها بَعْضا.

[8] [H6] ויקראו השיעורים אשר יחסם יחס אחד בעצמו המתייחסים.

[7] وَالأَقدارُ الَّتي عَلى نِسْبَةٍ واحِدَةٍ فَلْتُسَمَّ المُتَناسِبَةَ.

[9] [H7] וכשיהיו הכפלים שוי הפעמים, וכפלי הראשון מהם יעדפו על כפלי השני, וכפלי השלישי לא יעדפו על כפלי הרביעי, אז יחס הראשון עם זה אל השני יאמר לו כי הוא גדול מיחס השלישי אל הרביעי.

[8] وَإِذا كانَتِ الأَضْعافُ المُتَساوِيَةُ أَضْعاف الأَوَّلِ زائِدَة عَلى أَضْعافِ الثّاني، وَأَضْعاف الثّالِث غَيْر زائِدَة عَلى أَضْعافِ الرّابِع، فَإِنَّ الأَوَّلَ يُسَمَّى بِأَنَّهُ أَكْبَرُ نِسْبَةً إِلى الثّاني مِنَ الثّالِثِ إِلى الرّابِع.

[10] [H8] והמעט אשר יהיה בו ההתיחס[1] שלשה גבולים.

[9] وَأَقَلُّ ما تَكونُ المُناسَبَةُ في (ثَلاثَة)[1] حُدودٍ.

[1] ההתיחס] ⟨⟨ה⟩⟩התיחס

[1] ثلاثة] ثلثة

[7] [E5] وَيُقالُ في المَقاديرِ إنَّها في نِسْبَةٍ واحِدَةٍ، الأَوَّلِ إلى الثَّاني والثَّالِثِ إلى الرَّابِعِ، مَتَى كانَتْ أضعافُ الأَوَّلِ والثَّالِثِ المُتَساوِيَةُ المرَّاتِ، أيّ أَضْعافٍ^ كانَتْ[1]، إمَّا أَنْ تَفْضُلَ مَعًا على أَضعافِ الثَّاني والرَّابعِ المُتَساوِيَةِ المرَّاتِ، أيّ أَضْعافٍ كانَتْ، وَإمَّا أَنْ تُساوِيَها مَعًا، وَإمَّا أَنْ تَنْقُصَ عَنْها مَعًا إذا قِيسَتْ على التَّوالي بعضها بِبَعْضٍ. وبِالْعَكْسِ مَتَى كانَتِ المَقاديرُ في نِسْبَةٍ واحِدَةٍ بِعَيْنِها على التَّوالي—فَإنَّ أَضْعافَ الأَوَّلِ والثَّلاثِ تَكونُ إمَّا زائِدَةً على أَضْعافِ الثَّاني وَالرَّابِعِ، وَإمَّا ناقِصَةً مَعًا مِنْهُما، وَإمَّا مُساوِيَةً مَعًا لَهُما[2].

[7] [H5] יאמר בשיעורים שהם על יחס אחד, הראשון אל השני והשלישי אל הרביעי, כשיהיו כפלי הראשון והשלישי שווי הפעמים, אי זה מין שיהיו, אם שיעדפו יחד על כפלי השני והרביעי שווי הפעמים, אי זה כפל שיהיו, ואם שישוו אליהם יחד, ואם שיגרעו מהם יחד כשיוקשו בסדר קצתם לקצת. ובהפך כשיהיו השיעורים ביחס אחד בעצמו על הסדר—הנה כפלי הראשון והשלישי יהיו אם נוספים על כפלי השני והרביעי, ואם חסרים יחד מהם, ואם שוים יחד לשניהם.

[8] [E6] وَلْتُسَمَّ المَقاديرُ الَّتي نِسْبَتُها واحِدَةٌ بِعَيْنِها المُتَناسِبَةَ.

[8] [H6] ויקראו השיעורים אשר יחסם[1] אחד בעצמו המתיחסים.

[9] [E7] وَمَتَى كانَتِ الأَضْعافُ المُتَساوِيَةُ المرَّاتِ، إمَّا أَضعافُ الأَوَّلِ مِنْهُما[3] فَتَفْضُلُ على أَضْعافِ الثَّاني، وَإمَّا أَضعافُ الثَّالِثِ فلا تَفْضُلُ على أَضعافِ الرَّابِعِ، فإنَّ[4] نِسْبَةَ^ الأَوَّلِ إلى الثَّاني^ أَعْظَمُ مِنْ نِسْبَةِ الثَّالِثِ إلى الرَّابِعِ.

[9] [H7] וכאשר היו הכפלים שווי הפעמים, וכפלי הראשון מהם יעדפו על כפלי השני, וכפלי השלישי לא יעדפו על כפלי הרביעי, הנה הראשון אז אל השני יאמר שהוא[2] גדול מיחס השלישי אל הרביעי.

[10] [E8] وَأَقَلُّ ما يَكونُ التَّناسُبُ في ثَلاثَةِ حُدودٍ.

[10] [H8] והמעט שיהיה בו ההתיחס שלשה גבולים.

---

[1] أي ... كانت] M [2] وبالعكس ... لهما] MPR2 [3] منهما] M [4] فان] M

---

[1] יחסם] +יחד **בפ** +יחס **וטק** [2] יאמר שהוא] נראה שאבן תיבון השלים את הנוסח כאן, מדעתו או לפי הערה בכתב היד הערבי שהשתמש בו (ראו האפראט הערבי)

[11] [H9] וכאשר יהיו שלשה שיעורים מתיחסים נאמר כי יחס הראשון אל השלישי[1] כפל יחסו אל השני, רצוני לומר נשנה בהשנות.

[10] وَإذا كانَتْ ثَلاثَةُ أَقْدارٍ نِسْبَةً واحِدَةً، فَإِنَّ نِسْبَةَ الأَوَّلِ إِلى الثَّالِثِ كَنِسْبَتِهِ إِلى الثَّاني مُثَنَّاةً بِالتَّكْرِيرِ.

[12] [H10] וכאשר היו ארבעה שיעורים מתיחסים נאמר כי יחס הראשון אל הרביעי שלשה כפלי יחסו אל השני, רצוני לומ׳ משולש. ועל זה המשל ילך מה שימשך לזה.

[13] [H11] יאמר בשיעורים שהם מסודרים ביחס כשהוקשו הראשונים עם הראשונים והנמשכים עם הנמשכים.

[11] الأَقْدارُ المُتَشابِهَةُ النِّسْبَةِ المُقَدَّماتُ إِلى التَّوالي وَالمُقَدَّماتِ إِلى التَّوالي وَأَنَّها أَيْضًا مِنْ خِلافٍ كَما التَّالي إِلى المُقَدَّمِ، وَكَذَلِكَ التَّالي إِلى المُقَدَّمِ. وَعَلى التَّبادُلِ أَيْضًا كَما المُقَدَّمُ إِلى المُقَدَّمِ، كَذَلِكَ التَّالي إِلى التَّالي.

[15] [H12] חלוף היחס הוא לקיחת הראשון אצל הראשון והנמשך אצל הנמשך.

[14] [H13] והפך היחס הוא כשילקח הנמשך במדרגת הראשון והראשון במדרגת הנמשך.

[16] [H14] הרכבת היחס היא לקיחת הראשון עם הנמשך במדרגת דבר אחד אצל הנמשך.

[12] تَرْكِيبُ النِّسْبَةِ هُوَ كَما المُقَدَّمِ وَالتَّالي جَميعًا إِلى التَّالي، كَذَلِكَ المُقَدَّمُ وَالتَّالي جَميعًا إِلى التَّالي.

[17] [H15] חלוק היחס הוא לקיחת מותר הראשון על הנמשך אצל הנמשך.

[13] تَفْصِيلُ النِّسْبَةِ هُوَ نِسْبَةُ زِيادَةِ المُقَدَّمِ عَلى التَّالي إِلى التَّالي.

---

[1] השלישי] ~~הרביעי~~+

[11] [E9] وإذا كانَتْ ثلاثةُ مَقاديرَ مُتَنَاسِبَةً قِيلَ إنَّ نِسْبَةَ الأَوَّلِ إلى الثَّالِثِ ضِعْفُ نِسْبَتِهِ إلى الثَّاني أي مُثَنَّاة.

[11] [H9] וכאשר יהיו שלשה שיעורים מתיחסים נאמר כי יחס הראשון אל השלישי כפל יחסו אל השני, כלומר נשנה בהשנות.

[12] [E10] وإذا كانَتْ أَرْبَعَةُ مقاديرَ مُتَنَاسِبَةً قِيلَ إنَّ نِسْبَةَ الأَوَّلِ إلى الرَّابِعِ ثلاثة أَضعافِ نِسْبَتِهِ إلى الثَّاني أي مُثَلَّثَة. وعلى هذا المِثَالِ يَجْرِي ما يَتْلُو ذَلِكَ.

[12] [H10] וכאשר היו ארבעה שיעורים מתיחסים נאמר כי יחס הראשון אל הרביעי שלשה כפלי יחסו אל השני כלומר משולש. ועל זה המשל ילך מה שימשך לזה.

[13] [E11] يُقالُ في المَقاديرِ إنّها مُتَّسِقَةٌ في النِّسْبَةِ وَنَظيرَةٌ^ إذا قِيسَتِ المُقَدّماتُ مَعَ المُقَدّماتِ وَالتَّوَالي مَعَ التَّوَالي.

[13] [H11] ויאמר בשיעורים שהם מסודרים ביחס ושהוא נכחי לו כאשר הוקשו הראשונים עם הראשונים והנמשכים עם הנמשכים.

[15] [E12] تَبديلُ النِّسْبَةِ هُوَ أَخذُ المُقَدَّمِ عِنْدَ المُقَدَّمِ وَالتَّالي عِنْدَ التَّالي.

[14] [H12] תמורת היחס הוא לקיחת הראשון אצל הראשון והנמשך אצל הנמשך[1].

[14] [E13] عَكْسُ النِّسْبَةِ هُوَ أَخذُ التَّالي بِمَنْزِلَةِ المُقَدَّمِ عِنْدَ المُقَدَّمِ^ وَالمُقَدَّمِ[1] بِمَنْزِلَةِ التَّالي.

[15] [H13] והפך היחס הוא כשילקח הנמשך במדרגת הראשון והראשון במדרגת הנמשך.

[16] [E14] تَرْكيبُ النِّسْبَةِ هو أَخذُ المُقَدَّمِ مَعَ التَّالي بِمَنْزِلَةِ شَيْءٍ واحِدٍ عِنْدَ التَّالي.

[16] [H14] הרכבת היחס הוא לקיחת הראשון עם הנמשך במדרגת דבר אחד אצל הנמשך.

[17] [E15] تَفْصيلُ النِّسْبَةِ هُوَ أَخذُ فَضْلِ المُقَدَّمِ على التَّالي عِنْدَ التَّالي.

[17] [H15] הבדל היחס הוא לקיחת מותר הראשון על הנמשך אצל הנמשך.

[1] תמורת ... הנמשך] בכתבי היד הערביים PR2T2T3 מקומה של הגדרה זו כמו בתרגום מא״ת

[1] والمقدم] C

[18] [H16] התהפכות היחס הוא לקיחת הראשון אצל מותרו על הנמשך.

[14] قَلْبُ النِّسْبَةِ هُوَ نِسْبَةُ المُقَدَّمِ إِلى زِيادَةٍ عَلى التَّالي.

[19] [H17] יחס השווי הוא כשיהיו כמה שיעורים שיהיו ושיעורים אחרים כפי מספרם, וכשנלקחו שנים מן האחרים מהם היו על יחס שנים מן האחרים ונלקחו הקצוות מלבד מה שביניהם.

[15] نِسْبَةُ المُساواةِ هِيَ نِسْبَةُ الأَطْرافِ بَعْضُها إِلى بَعْضٍ إِذا كانَتِ الأَقدارُ أَكْثَرَ مِنْ قَدْرَيْنِ مُقارَنَةَ الأَقدارِ أُخَرَ عَلى عِدَّتِها [وبرحت] العِدَّة الواسطة [باسير] فَإِنَّها مُتَناسِبَةً.

[20] היחס ההולך על סדר הוא כשהראשון אצל הנמשך כראשון אצל הנמשך והנמשך אצל דבר אחר כנמשך אצל דבר אחר.

[21] [H18] והיחס הנפתל במערכת הוא כשיהיה הראשון אצל הנמשך כראשון אצל הנמשך והנמשך אצל דבר אחר כדבר אחר אצל הראשון.

[18] [E16] قَلْبُ النِّسْبَةِ هُوَ أَخْذُ المُقَدَّمِ عِنْدَ فَضْلِهِ عَلَى التَّالي.

[18] [H16] התהפכות היחס הוא לקיחת הראשון אצל מותרו על הנמשך.

[19] [E17] نِسْبَةُ المُساواةِ تكونُ مَتَى كانَتْ أيَّ مقادير كانَتْ ومقادير أُخَر على عِدَّتِها وكانَتْ إذا أُخِذَ اثْنانِ مِنْ أَحَدَيْهِما كانَتْ على نِسْبَةِ اثْنَيْنِ مِنَ الآخَرِ فَأُخِذَتِ الأَطرافُ دونَ ما بَيْنَهَا.

[19] [H17] יחס השווי כשיהיו כמה שיעורים שיהיו ושעורים אחרים כפי מספרם, וכשילקחו שנים מאחד משניהם יהיו על יחס שנים מן האחר ונלקחו הקצוות מלבד מה שביניהם.

وعلى جِهَةٍ أُخْرى: إذا كانَتْ عِدَّةُ مَقادير وَمَقادير أُخَر عَلَى عِدَّتِها فَإِنَّ نِسْبَةَ المُساواةِ هِيَ نِسْبَةُ الأَطْرافِ إِذا أُسْقِطَتِ المَقاديرُ الَّتي في الوَسَطِ[1].

ועל פנים אחרים: כשיהיו שיעורים ושעורים אחרים על מספרם הנה יחס השווי הוא יחס הקצוות כשהפילו השעורים אשר באמצע.

[20] [E17b] المُناسَبَةُ المُنْتَظِمَةُ هي مَتَى كانَ المُقَدَّمُ عِنْدَ التَّالي كَالمُقَدَّمِ عِنْدَ التَّالي وَالتَّالي عِنْدَ شَيْءٍ آخَر كَالتَّالي عِنْدَ شَيْءٍ آخَر.

[20] היחס ההולך על סדר הוא כשהראשון אצל הנמשך כראשון אצל הנמשך והנמשך אצל דבר אחר כנמשך אצל דבר אחר.

[21] [E18] المُناسَبَةُ المُضْطَرِبَةُ هِيَ مَتَى كانَ المُقَدَّمُ عِنْدَ التَّالي كالمُقَدَّمِ عِنْدَ التَّالي والتَّالي عِنْدَ شَيْءٍ آخَر كَشَيْءٍ آخر عِنْدَ المُقَدَّمِ.

[21] [H18] והיחס הנפתל במערכת הוא כשיהיה הראשון אצל הנמשך כראשון אצל הנמשך והנמשך אצל דבר אחר כדבר אחר אצל הראשון.

[1] وعلى ... الوسط] **CamO1O2U**

[Proposition v.1]

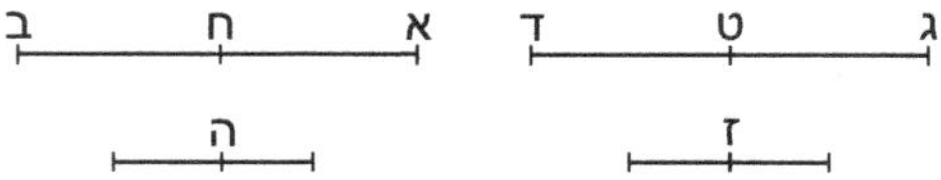

כשיהיו שיעורים בהם כפלי שיעורים אחרים מחוברים להם כפי מספרם, וכפליהם שוים, יהיה באחד[1] מכפלו המחובר לו כמה שיש בכל מכפלי הכל.

إذا كانَتْ أَقْدارٌ كَمْ كانَتْ فيها أَضْعافُ أَقْدارٍ أُخَر مُقارِنَةً لَها عَلى عِدَّتِها، وَأَضْعافُها مُتَساوِيَةٌ، فَإِنَّ ما في أَحَدِ القَدْرَيْنِ مِنْ أَضْعافِ قَرينِهِ مِثْل ما في الجَميعِ مِنْ أَضْعافِ الجَميعِ.

המשל בו: ששיעורי א״ב וג״ד בהם כפלי שני שיעורים אחרים, והם שיעורי ה׳ ז׳, וכפליהם שוים—מה שבא״ב[2] מכפלי ה׳ כמו אשר בג״ד מכפלי ז׳. ואומר כי מה שבא״ב מכפלי ה׳ כמו מה שבא״ב וג״ד מכפלי ה׳ ז׳ יחד.

مِثالُهُ: أَنّا نَضَعُ قَدْرَيْنِ عَلَيْهِما ا ب ج د، فيهِما أَضْعافُ قَدْرَيْنِ آخَرَيْنِ عَلَيْهِما ه ز، وأَضْعافُهُما مُتَساوِيَةٌ—ما في ا ب مِنْ أَضْعافِ ه مِثْل ما في ج د مِنْ أَضْعافِ ز [..] مِثْل ما في ا ب وج د جَميعًا مِنْ أَضْعافِ ه و ز جَميعًا.

[1] באחד] ~~כפלים~~ ⟨⟨ב⟩⟩אחד ⟨⟨שב⟩⟩א״ב

[2] שבא״ב] ~~א״ב~~

إذا كانَتْ مَقاديرُ فيها أَضعافُ مَقاديرَ أُخَرَ مُقارِبَةً لَها على عِدَّتِها، وَأَضعافها مُتَساوِيَةٌ، فإنَّ ما في الواحِدِ مِنْ أَضْعافِ قَريبِهِ مِثْل ما في الْجَميعِ مِنْ أَضعافِ الْجَميعِ.

كאשר יהיו שיעורים בהם כפלי[1] שיעורים אחרים[2] קרובים[3] להם על מספרם, וכפליהם שוים, הנה מה שבאחד מכפל קרובו כמו מה שבכל מכפל הכל[4].

مِثالُهُ: أَنَّ في مِقْدارَيْ اب جد أَضعافُ مُسَاوِيَةَ لِمِقْدارَيْ ه ز ما في اب مِنْ أَضعافِ ه، مِثْل ما في جد مِنْ أَضعاف ز. فَأَقولُ إِنَّ ما في اب مِنْ أَضعافِ ه مِثْل ما في اب جد جَميعًا مِنْ أَضعافِ ه ز جَميعًا.

המשל בו: כי בשיעורי א״ב ג״ד כפל שוה[5] לשיעורי ה׳ וז׳[6] וה׳ בא״ב מכפל מה[7], כמו מה שבג״ד מכפלי[8] ז׳[9]. הנה אומר כי מה שבא״ב מכפל ה׳ כמו מה[10] שבא״ב וג״ד מקובצים מכפלי[11] ה׳ ז׳ יחד.

שרטוט—**בוטספק**: גט״ד וז׳ משמאל לאח״ב וה׳ [1]כפלי] כפל ⟨⟨נ״א כפל לשעורים אחרי׳ כגיל{ם} על מספרם וכו׳⟩⟩ **ס** [2]אחרים] ליתא **בוס** [3]קרובים] ⟨⟨נ׳ נגדיים⟩⟩ **פ** ⟨נ׳ נגדיים⟩ **ק** [4]הכל] הכליי **פ** הכלי **ק** [5]כפל שוה] * ⟨⟨כפלים שוים⟩⟩ **ק** [6]וז׳] ליתא **ב** [7]וה׳ ... מה] ⟨⟨ומה⟩⟩ שבא״ב מכפל ה׳ **פ** * ⟨⟨ומה שא״ב מכפלי⟩⟩ **ק** [8]מכפלי] מכפל **בו** [9]ז׳] +⟨נ״א לשיעורי ה״ז ומה שבא״ב כפלי ה׳ כמו שבג״ד מכפלי ז׳⟩ **ס** [10]מה] ליתא **בופ** [11]מכפלי] ⟨*⟩ **ט**

بُرْهانُهُ: أَنَّا نَقْسِمُ اب بِقَدْرِ ه. فَيَكُونُ اح وح ب. وَنَقْسِمُ ج د بِقَدْرِ ز. فَيَكُونُ ج ط وط د. فَعِدَّةُ اح وح ب مِثْلُ عِدَّةِ ط د وط ج. اح مِثْلُ ه. (وج ط)[1] مِثْلُ ز. وَكَذٰلِكَ ح ب (وط د)[2] مِثْل ه وز. فَمَا في اب مِنْ أَضْعافِ ه مِثْل ما في اب وج د جَميعًا مِنْ أَضْعافِ ه وز جَميعًا.

מופת זה: שנחלק א״ב בשיעור ה׳. ויהיו חלקיו א״ח וח״ב. ונחלק ג״ד בשיעור ז׳, ויהיו חלקיו ג״ט ט״ד. ומספר א״ח וח״ב במספר ג״ט וט״ד. וא״ח כמו ה׳, וג״ט כמו ז׳. וא״ח וג״ט כמו ה׳ וז׳. וכמו כן ח״ב וט״ד כמו ה׳ וז׳. אם כן יהיה מה שבא״ב מכפלי ה׳ כמו מה שבא״ב וג״ד יחד מכפלי ה׳ וז׳ יחד.

وَذٰلِكَ ما أَرَدْنا بَيانَهُ.

וזה מה שרצינו לבאר.

[1]وج ط] ود ط [2]وط د] وط ج

بُرْهانُه: نَقْسِمُ اب بِقَدْرِ ه. فتكونُ أَقسامُهُ اح ح ب. ونَقْسِمُ ج د بقَدْرِ ز. فتكونُ أَقسامُهُ ج ط ط د. فَعِدَّةُ اح ح ب مِثْل عِدَّةِ ج ط ط د^. فَقَدْرُ اح مِثْل ه، و ج ط مِثْل ز. فَجميعُ اح ج ط مِثْل ه ز جَميعًا. وكذلِكَ جَميعًا ح ب ط د مِثْل ه ز جَميعًا. فما في اب مِنْ أَضعافِ ه مُساوٍ لِعدَّةِ[1] ما في اب ج د جَميعًا مِنْ أَضعافِ ه ز جَميعًا.

وذَلِكَ ما أَرَدْنا أَنْ نُبيِّنَ.

המופת: אנחנו נחלק א״ב בשיעור[1] ה׳. ויהיו חלקיו א״ח וח״ב. ונחלק ג״ד בשיעור ז׳. ויהיו[2] חלקיו ג״ט ט״ד. אם כן שיעור א״ח כמו ח״ב. ושיעור ג״ט כמו ט״ד. וא״ח כמו ה׳. וג״ט כמו ז׳. אם כן כל א״ח וג״ט כמו ה׳ ז׳[3] יחד. וכן כל ח״ב וט״ד כמו ה׳ ז׳ יחד. הנה מה שבא״ב[4] מכפל ה׳ שוה למספר מה שבא״ב[5] וג״ד יחד מכפל ה׳ ז׳ יחד.

וזה מה שרצינו לבאר.

[1] لعدة] F

[1] בשיעור] בשיעורי ס [2] ויהיו] היו ס יהיו פ
[3] כמו ... ז׳] ⟨⟨*⟩⟩ ק [4] שבא״ב] שא״ב ו
[5] שבא״ב] ש⟨ב⟩א״ב ט

[Proposition V.2]

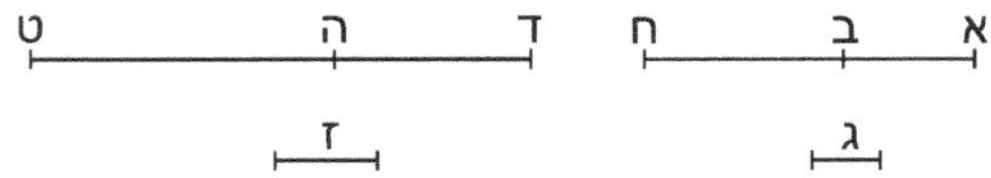

כשיהיו שיעורים, והיו בראשון מהם מכפלי השני כמו אשר בשלישי מכפלי הרביעי, ובחמשי מכפלי השני כמו אשר בששי מכפלי הרביעי, יהיה מה שבראשון והחמשי יחד מכפלי השני כמו שיש בשלישי ובששי יחד מכפלי הרביעי.

إِذا كانَتْ أَقْدارٌ، وَفي الأَوَّلِ مِنْ أَضْعافِ الثَّاني مِثْلُ ما في الثَّالِثِ مِنْ أَضْعافِ الرَّابِعِ، وَفي الخامِسِ مِنْ أَضْعافِ الثَّاني مِثْلُ ما في السَّادِسِ مِنْ أَضْعافِ الرَّابِعِ، فَإِنَّ جَميعَ ما في الأَوَّلِ وَالخامِسِ مِنْ أَضْعافِ الثَّاني مِثْلُ جَميعِ ما في الثَّالِثِ وَالسَّادِسِ مِنْ أَضْعافِ الرَّابِعِ.

המשל בזה: כי הראשון א״ב. ובו מכפלי השני, והוא ג׳, כמו שיש בשלישי, והוא ד״ה, מכפלי הרביעי, והוא ז׳. ובחמשי, והוא ב״ח, מכפלי השני, והוא ג׳, כמו שיש בששי, והוא ה״ט, מכפלי הרביעי, והוא ז׳. ואומר כי יחד מה שבראשון ובחמשי, והם א״ח, מכפלי השני, כמו מה שבשלישי ובשישי יחד, והם ד״ט, מכפלי הרביעי.

مِثالُهُ: أَنَّ الأَوَّلَ اب. وَفيهِ مِنْ أَضْعافِ الثَّاني، وَهُوَ ج [..] مِثْلُ ما في السَّادِسِ، وَهُوَ ه ط، مِنْ أَضْعافِ الرَّابِعِ، وَهُوَ ز. فَأَقولُ إِنَّ جَميعَ ما في الأَوَّلِ وَالخامِسِ، وَهُما اح، مِنْ أَضْعافِ الثَّاني، مِثْلُ ما في جَميعِ الثَّالِثِ وَالسَّادِسِ، وَهُما د ط، مِنْ أَضْعافِ الرَّابِعِ.

إِذَا كَانَتْ مَقَادِيرُ فِي الأَوَّلِ[1] مِنْ أَضْعافِ الثَّانِي مِثْل مَا فِي الثَّالِثِ مِنْ أَضْعافِ الرَّابِعِ، وَفِي الْخَامِسِ مِنْ أَضْعافِ الثَّانِي مِثْل مَا فِي السَّادِسِ مِنْ أَضْعافِ الرَّابِعِ، فَإِنَّ مَا فِي جَمِيعِ الأَوَّلِ وَالْخَامِسِ مِنْ أَضْعافِ الثَّانِي مِثْل مَا فِي جَمِيعِ الثَّالِثِ وَالسَّادِسِ مِنْ أَضْعافِ الرَّابِعِ.

مِثَالُهُ: أَنَّ فِي الأَوَّلِ، وهُوَ ا ب، مِنْ أَضْعافِ الثَّانِي، وهُوَ ج، مثل ما فِي الثَّالِثِ، وهُوَ د ه، مِنْ أَضْعافِ الرَّابِعِ، وهُوَ ز. وفِي الخامِسِ، وهُوَ ب ح، مِنْ أَضْعافِ الثَّانِي، وهُوَ ج، مثل ما فِي السَّادِسِ، وهُوَ ه ط، مِنْ أَضْعاف الرَّابِعِ، وهُوَ ز. فَأَقولُ إنّ ما فِي جميعِ ^ الأَوَّل والخامِس، وَهُما ا ح، مِنْ أَضْعافِ الثَّانِي، وهُوَ ج، مِثْل ما فِي جميعِ الثَّالِثِ والسَّادِسِ، وَهُما د ط، مِنْ أَضْعَافِ الرَّابِعِ، وهُوَ ز.

כשיהיו שיעורים בראשון מכפלי השני כמו מה שבשלישי מכפלי הרביעי, ובחמשי מכפלי השני[1] כמו מה שבששי מכפלי[2] הרביעי, הנה מה[3] שבכל הראשון והחמישי מכפלי[4] השני כמו מה שבכל השלישי[5] והששי מכפלי הרביעי[6].

המשל בו: כי[7] בראשון, והוא א״ב, מכפלי השני, והוא ג׳[8], כמו מה שבשלישי[9], והוא ד״ה[10], מכפלי הרביעי, והוא ז׳[11]. ובחמישי, והוא ב״ח, מכפלי השני, והוא ג׳, כמו מה שבששי, והוא ט״ה, מכפלי הרביעי, והוא ז׳. הנה אומר כי מה שבראשון והחמישי, והוא א״ח, מכפלי השני, והוא ג׳, כמו מה[12] שבכל השלישי והששי, והם ד״ט, מכפלי הרביעי, והוא ז׳[13].

---

[1]الأول] Q +منها

שרטוט—**פ**: א״ב מסומן ב״ראשון״, ג׳ מסומן ב״שני״, ב״ח מסומן ב״חמשי״, ד״ה מסומן ב״שלישי״, ז׳ מסומן ב״רביעי״, וה״ט מסומן ב״ששי״; **ק**: סימונים כמו ב**פ** חוץ מסימון ה״ט

[1]השני] שני **ס** [2]מכפלי] מכפל **ס** [3]מה] +שנכפל **ס** [4]מכפלי] מכפל **ו** [5]שבכל השלישי] שבשלישי **ו** שבכפל+ **ס** [6]הנה ... הרביעי] ליתא **ב** [7]כי] ליתא **בו** [8]והוא ג׳] ⟨⟨*⟩⟩ **ט** [9]שבשלישי] +~~והשיני והשישי~~ **ב** שבב~~כל השלישי~~ **ט** [10]והוא ד״ה] ⟨*⟩ **ב** [11]כמו ... ז׳] ⟨⟨*⟩⟩ **פ** [12]מה] ליתא **בו** [13]ז׳] ⟨*⟩ **פ**

מופת זה: כי מה שבא״ב מכפלי ג׳ כמו מה שיש בד״ה מכפלי ז׳. ומה שבב״ח מכפלי ג׳, כמו מה שבה״ט מכפלי ז׳—יהיה כל מה שבא״ח מכפלי ג׳ כמו כל מה שבד״ט מכפלי ז׳.

بُرْهانُهُ: أَنَّ عِدَّةَ ما في ا ب مِنْ أَضْعافِ جَ مِثْل عِدَّةِ ما في د ه مِنْ أَضْعافِ زَ. وَعِدَّة ما في ح ب مِنْ أَضْعافِ جَ مِثْلُ عِدَّةِ ما في ه ط مِنْ أَضْعافِ زَ. وكُلّ ما في ا ح مِنْ أَضْعافِ جَ مِثْلُ ما في د ط مِنْ أَضْعافِ زَ.

וזה מה שרצינו לבאר.

وَذَلِكَ ما أَرَدْنا بيانَهُ.

بُرْهانُهُ: أَنَّ ما في اب مِنْ أَضعافِ ج مثل ما في د ه مِنْ أَضْعاف ز. فَعِدَّة ما في اب مِنَ الأَقْدار المُسَاوِيَة ل ج مثل عِدَّة ما في د ه مِنَ الأَقْدار المُسَاوِيَة ل ز. وما في ب ح مِنْ أَضْعاف ج مثل ما في ه ط مِنْ أَضْعاف ز. فَعِدَّة ما في ب ح مِنَ الأَقْدار المُسَاوِيَة ل ج مثل عِدَّة ما في ه ط مِنَ الأَقْدار المُسَاوِيَة ل ز. فَعِدَّة ما في اح إِذًا مِنَ الأَقْدار المُسَاوِيَة ل ج مثل عِدَّة ما في د ط مِنَ الأَقْدارِ المُسَاوِيَةِ ل ز. فَمَا في كُلِّ اح مِنْ أَضْعافِ ج مِثْل ما في كُلِّ د ط مِنْ أَضْعافِ ز.

وذَلِكَ ما أَرَدْنا أَنْ نُبَيِّنَ.

מופתו: כי מה שבא״ב מכפלי ג׳ כמו מה[1] שבד״ה מכפלי ז׳. הנה מספר מה[2] שבא״ב מן השיעורים השוים[3] אל ג׳ כמו מספר מה[4] שבד״ה מן השיעורים השוים אל ז׳. ומה שבב״ח[5] מכפלי ג׳ כמו[6] מה[7] שבה״ט מכפלי ז׳. הנה מספר מה שבב״ח מן השיעורים השוים אל ג׳ כמו מספר מה שבה״ט מן השיעורים השוים אל ז׳. הנה מספר מה שבא״ח אם כן מן השיעורים השוים אל ג׳ כמו מספר מה שבד״ט מן השיעורים השוים אל ז׳. אם כן מה שבכל א״ח מכפלי ג׳ כמו מה שבכל ד״ט מכפלי ז׳[8].

וזה מה שרצינו לבאר[9].

[1] מה] ליתא **ס** +* **ק** [2] מה] ליתא **בוס**
[3] השוים] שוים **בו** [4] מה] ליתא **בוטסס**
[5] שבב״ח] שב״ח **בו** ש⟨ב⟩ב״ח **פק** [6] כמו]
+מספר **ב** [7] מה] ⟨*⟩ **ט** [8] ז׳] ליתא **ב**
[9] לבאר] לבארו **טפק**

[Proposition v.3]

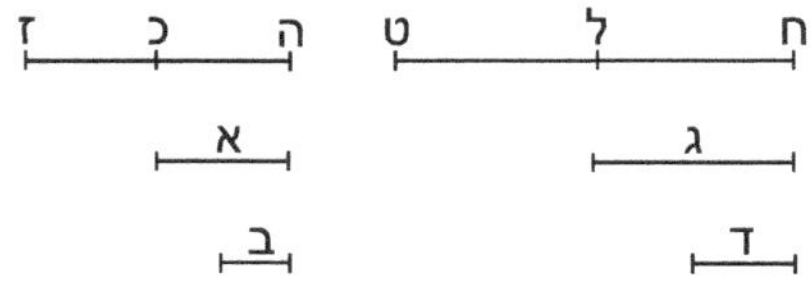

כשיהיה בראשון מכפלי השני כמו שיש בשלישי מכפלי הרביעי, ונלקחו לראשון ולשלישי כפלים שוי הפעמים כמה שהם, יהיה מה שבכפלי הראשון הלקוחים מכפלי השני כמו מה שבכפלי השלישי הלקוחים מכפלי הרביעי.

إِذا كانَ فِي الأَوَّلِ مِنْ أَضْعافِ الثَّاني مِثْل ما فِي الثَّالِثِ مِنْ أَضْعافِ الرَّابِعِ، وَوُضِعَ لِلأَوَّلِ أَضْعافٌ وَلِلثَّالِثِ أَضْعافٌ مِثْلها، فَإِنَّ ما فِي المَوْضُوعِ الَّذي هُوَ أَضْعافٌ لِلأَوَّلِ مِنْ أَضْعافِ الثَّاني، مِثْل ما فِي المَوْضُوعِ الَّذي هُوَ أَضْعافٌ لِلثَّالِثِ أَضْعافُ الرَّابِعِ.

إذا كانَ في الأَوَّلِ مِنْ أَضْعاف الثَّاني مِثْل ما في الثَّالِثِ مِنْ أَضْعافِ الرَّابِعِ، وَأُخِذَ لِلأَوَّلِ والثَّالِثِ أَضْعافٌ مُتَساوِيَةُ المَرَّاتِ، ما كانَتْ، فإنَّ ما في أَضْعافِ الأَوَّلِ المأْخوذَةِ مِنْ أَضْعافِ الثَّاني مِثْل ما في أَضْعاف^ الثَّالِثِ المأْخوذَةِ مِنْ أَضْعافِ الرَّابِعِ.

כשיהיה בראשון מכפלי[1] השני כמו מה שבשלישי מכפלי הרביעי, ונלקח לראשון והשלישי[2] כפלים שווי[3] הפעמים, אי זה מספר שיהיו, הנה מה שבכפלי[4] הראשון הלקוח מכפלי השני כמו[5] מה[6] שבשלישי[7] הלקוח מכפלי הרביעי.

שרטוט—**בוטספק**: חל״ט, ג׳ וד׳ משמאל להכ״ז, א׳ וב׳; **ס**: הכ״ז, א׳ וב׳ מסומנים ב״נ״א״, ושלושה קווים אחרים ובאורך שווה מסומנים בהתאמה באותיות כ׳ ח׳ וב׳—קו כ׳ מסומן ב״שני״ וקו ב׳ מסומן ב״ראשון״; **פ**: ה״כ וא׳ מסומנים ב״ראשון״, ב׳ מסומן ב״שני״, ח״ל וג׳ מסומנים ב״שלישי״, ד׳ מסומן ב״רביעי״, כ״ז מסומן ב״חמישי״, ול״ט מסומן ב״ששי״ [1]מכפלי] מכפל **בו** [2]לראשון והשלישי] הראשון והשלישי **בס** לראשון ולשלישי **טפ** ⟨⟨אל⟩⟩ הראשון והשלישי **ק** [3]שווי] שוה **ט** [4]שבכפלי] שבכ⟨פ⟩לי **פ** [5]כמו] והוא+ **פ** [6]מה] ליתא **ס** [7]שבשלישי] +⟨⟨שבכפלי השלישי⟩⟩ **פ**

המשל בזה: כי הראשון א׳. ובו כפלי השני, והוא ב׳, כמו מה שבשלישי, והוא ג׳, מכפלי הרביעי, והוא ד׳. ולוקחו לא׳ כפלים והם ה״ז, ולג׳ כפלים שוים להם, והם ח״ט. ואומר כי מה שבה״ז מכפלי ב׳, כמו שיש בח״ט מכפלי ד׳.

مِثالُهُ: أَنَّ الأَوَّلَ آ. فِيهِ مِنْ أَضْعافِ الثَّاني، وَهُوَ بَ، مِثْل ما في الثَّالِث، وَهُوَ جَ، مِنْ أَضْعافِ الرَّابِعِ، وَهُوَ دَ. فَالمَوْضُوعُ الَّذي هُوَ أَضْعافُ الأَوَّلِ هُوَ ه ز. وَالْمَوْضُوعُ الَّذي هُوَ أَضْعافُ الثَّالِثِ هُوَ ح ط. وَما في ه ز مِنْ أَضْعافِ الأَوَّلِ، وَهُوَ آ، مِثْل ما في ح ط مِنْ أَضْعافِ الثَّالِثِ، وَهُوَ جَ. فَأَقولُ إِنَّ ما في ه ز مِنْ أَضْعافِ بَ مِثْل ما في ح ط مِنْ أَضْعافِ دَ.

مِثَالُهُ: أنَّ الأَوَّل آ، وفيهِ مِنْ أَضْعافِ الثَّاني، وهُوَ بَ، مثل ما في الثَّالث، وهُوَ جَ، مِنْ أَضْعافِ الرَّابِعِ، وهُوَ دَ. وَقَدْ أُخِذَتْ أَضْعافُ لِقَدْرِ آ، وهي ه ز، وأَضْعافُ لِقَدْرِ جَ، مُسَاوِيَةً^ لها في عَدَدِ المَرَّاتِ، وهي ح ط. فَأَقولُ إنَّ ما في ه ز مِنْ أَضْعافِ بَ، مثل ما في ح ط مِنْ أَضْعافِ دَ.

המשל בו: כי הראשון א׳, ובו מכפלי השני, והוא ב׳, כמו מה שבשלישי, והוא ג׳, מכפלי הרביעי, והוא ד׳. וכבר נלקחו כפלים לשיעור א׳, והם ה״ז, וכפלים לשיעור[1] ג׳, והם שוים להם במספר הפעמים, והם ח״ט. הנה אומר כי מה שבה״ז מכפלי ב׳, כמו מה[2] שבח״ט[3] מכפלי ד׳[4].

1 לשיעור] לשעורים **ב** 2 מה] ליתא **בו** 3 שבח״ט] שבה״ט **ס ב** 4 ד׳] ג׳ ⟨נ״א ד׳⟩

מופת זה: שנחלק ה״ז בשיעור א׳, וחלקיו ה״כ וכ״ז. ונחלק ח״ט בשיעור ג׳, וחלקיו ח״ל ול״ט. ויהיה מספר מה שבה״ז מכפלי א׳, והם ה״כ וכ״ז, כמספר מה שבח״ט מכפלי ג׳, והם ח״ל ול״ט. ומה שבא׳ מכפלי ב׳ כמו מה שבג׳ מכפלי ד׳. וא׳ כמו ה״כ. וג׳ כמו ח״ל. ואם כן יהיה מה שבה״כ מדמיוני ב׳ כמו מה שבח״ל מכפלי ד׳. ולזה מה שבכ״ז[1] מדמיוני ב׳ כמו מה שבל״ט מדמיוני ד׳. ונשים הראשון ה״כ. ובו מכפלי ב׳ כמו שיש בשלישי, והוא ח״ל, מכפלי ד׳. והחמישי כ״ז[2]. ובו מכפלי ב׳ כמו שיש בששי, והוא ל״ט, מכפלי ד׳. וכאשר חוברו הראשון (והחמשי)[3], והם ה״ז, היה כפל בם[4] יחד מכפלי ב׳, כמו שיש בששי ובשלישי כשיחוברו, והם ח״ט, מכפלי ד׳.

بُرْهانُهُ: أَنْ نَقْسِمَ ه ز بِقَدْرِ ا. وَأَقْسامُهُ ه ك ك ز. وَنَقْسِمُ ح ط بِقَدْرِ ج، وَأَقْسامُهُ ح ل ل ط. فَعِدَّةُ ما في ه ز مِنْ أَضْعافِ ا، وَهُوَ ه ك ك ز، مِثْلُ عِدَّةِ ما في ح ط مِنْ أَضْعافِ ج، وَهُوَ ح ل ل ط. وَما في ا مِنْ أَضْعافِ ب مِثْلُ ما في ج مِنْ أَضْعافِ د. ا مِثْلُ ه ك. ج مِثْلُ ح ل. وَما في ه ك مِنْ أَمْثالِ ب مِثْلُ ما في ح ل مِنْ أَمْثالِ د. وَكَذَلِكَ ما في ك ز مِنْ أَمْثالِ ب مِثْلُ ما في ل ط مِنْ أَمْثالِ د. فَنَجْعَلُ الأَوَّلَ ه ك. فِيهِ مِنْ أَضْعافِ الثَّاني، وَهُوَ ب، مِثْلُ ما في الثَّالِثِ، وَهُوَ ح ل، مِنْ أَضْعافِ الرَّابِعِ، وَهُوَ د. وَالخامِسُ ك ز. فيهِ مِنْ أَضْعافِ الثَّاني، وَهُوَ ب، مِثْلُ ما في السَّادِسِ، وَهُوَ ل ط، مِنْ أَضْعافِ الرَّابِـعِ، وَهُوَ د. فَإِذا جُمِـعَ الأَوَّلُ وَالخامِسُ، وَهُما ه ز، كانَ فيهِما جَميعًا مِنْ أَضْعافِ الثَّاني، وَهُوَ ب، مِثْلُ ما في الثَّالِثِ وَالسَّادِسِ إِذا جُمِعا، وَهُما ح ط، مِنْ أَضْعافِ الرَّابِـعِ، وَهُوَ د. فَما في ه ز مِنْ أَضْعافِ ب مِثْلُ ما في ح ط مِنْ أَضْعافِ د.

וזה מה שרצינו לבאר.

وَذَلِكَ ما أَرَدْنا بَيانَهُ.

[1]שבכ״ז] שבב~~כ״ה~~⟨כ״ז⟩ [2]כ״ז] ⟨⟨*⟩⟩ [3]והחמשי] והששי [4]כפל בם] כפל~~יהם~~⟨⟨בם⟩⟩ ~~בה״ז~~

بُرْهانُه: أَنَّ ما في ه ز مِنْ أَضْعافِ آ مِثْل ما في ح ط مِنْ أَضْعافِ ج. فَعِدَّة ما في ه ز مِنَ الأقدارِ المُسَاوِيَةِ لِقَدْرِ آ، مِثْل عِدَّة ما في ح ط مِنَ الأَقْدارِ المُسَاوِيَةِ لِقَدْرِ ج. فنَقْسِمُ ه ز بِقَدْرِ آ، فتكونُ أَقسامُه ه ك ك ز. ونَقْسِمُ ح ط بِقَدْرِ ج. فتكونُ أَقسامُهُ ح ل ل ط. فَعِدَّة ه ك ك ز مِثْل عِدَّة ح ل ل ط. وما في آ مِنْ أَضْعاف ب مِثْل ما في ج مِنْ أَضْعافِ د. و آ مِثْل ه ك. وج مِثْل ح ل. فَمَا في ه ك مِنْ أَضْعافِ ب مِثْل ما في ل ط مِنْ أَضْعافِ د. وكذَلكَ ما في ك ز مِنْ أَضْعافِ ب مِثْل ما في ل ط مِنْ أَضْعافِ د. فَنَجْعَلُ الأَوَّلَ ه ك، وفيه مِنْ أَضْعافِ^ ب[1]، مِثْل ما في الثَّالِثِ، وهو ح ل، مِنْ أَضْعافِ د[2]. والخامس ك ز، وفيه مِنْ أَضْعاف^ ب[3]، مِثْل ما في السَّادِسِ، وهو ل ط، مِنْ أَضْعافِ د[4]. فإذا جُمِعَ الأَوَّلُ والخامِسُ، وَهُما ه ز، كانَ فيهما مِنْ أَضْعافِ ب، مِثْل ما في الثَّالِثِ والسَّادِسِ، وَهُما ح ط، مِنْ أَضْعافِ د. فَمَا في ه ز مِنْ أَضْعافِ ب مِثْل ما في ح ط مِنْ أَضْعافِ د.

وذَلِكَ ما أَرَدْنا أَنْ نُبَيِّنَ.

המופת: כי מה שבה״ז מכפלי א׳ כמו מה שבח״ט מכפלי ג׳. אם כן מספר מה שבה״ז מן השיעורים השוים לשיעור א׳, כמו מספר מה שבח״ט מן השיעורים השוים לשיעור ג׳. הנה נחלק ה״ז בשיעור א׳, ויהיו חלקיו ה״כ וכ״ז. ונחלק ח״ט בשיעור ג׳. ויהיו חלקיו ח״ל ול״ט. אם כן מספר ה״כ וכ״ז כמו מספר ח״ל ול״ט. ומה שבא׳ מכפלי ב׳ כמו מה שבג׳ מכפלי ד׳. וא׳ כמו ה״כ. וג׳ כמו ח״ל. אם כן מה שבה״כ מכפלי ב׳ כמו מה שבח״ל מכפלי ד׳. וכן מה שבכ״ז מכפלי ב׳ כמו מה שבל״ט מכפלי ד׳. ונשים הראשון ה״כ, ובו כפלי ב׳ כמו מה שבשלישי, והוא ח״ל, מכפלי ד׳, והחמישי כ״ז[1], ובו כפלי ב׳ כמו[2] מה שבששי, והוא ל״ט, מכפלי ד׳[3]. וכאשר קובץ הראשון והחמישי, והם ה״ז, היה בשניהם מכפלי ב׳ כמו מה שבשלישי והששי, והם ח״ט, מכפלי ד׳. אם כן מה שבה״ז מכפלי ב׳ כמו מה שבח״ט מכפלי ד׳.

וזה מה[4] שרצינו לבאר.

1 ب] CopFM +وهو الثاني   2 د] M +وهو الرابع   3 ب] CopFM +وهو الثاني   4 د] CopFM +وهو الرابع

1 כ״ז] והוא+ ק   2 כמו] ליתא ב   3 ד׳] ~~+והחמישי כ״ז בו כפלי ב׳ כמו מה שבששי והוא ל״ט מכפלי ד׳~~ ט   4 מה] ⟨*⟩ ז

[Proposition v.4]

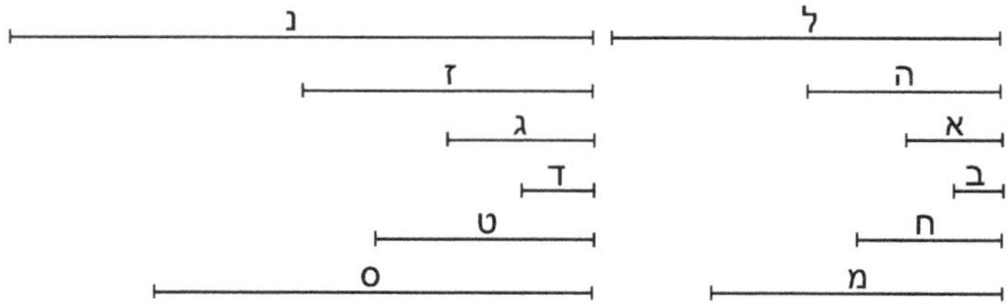

כשיהיה יחס הראשון אל השני כיחס השלישי אל הרביעי, ונלקחו לראשון ולשלישי כפלים שוי הפעמים כמו שהם, ולשני ולרביעי כפלים שוי הפעמים כמה שהם, יהיה יחס כפלי הראשון הנלקחים אל כפלי השני הוא יחס כפלי השלישי הנלקחים אל כפלי הרביעי.

إِذا كانَتْ نِسْبَةُ الأَوَّلِ إِلى الثَّاني كَنِسْبَةِ الثَّالِثِ إِلى الرَّابِعِ، وَوُضِعَ لِلأَوَّلِ أَضْعافٌ وَلِلثَّالِثِ أَضْعافٌ مِثْلها، وَلِلثَّاني أَضْعافٌ أُخَر، وإِلى الرَّابِعِ أُخَر مِثْلها، فَإِنَّ أَضْعاف الأَوَّلِ، وَالثَّالِثِ عِنْدَ أَضْعافِ الثَّاني، وَالرَّابِعِ كَذَلِكَ، يَكونُ أَيَّ التَّضاعيفِ كانَتْ، إِذا أُخِذَتْ يَتْلُو بَعْضُها بَعْضًا.

המשל בזה: שהראשון א׳. ויחסו אל השני, והוא ב׳, כיחס השלישי, והוא ג׳, אל הרביעי, והוא ד׳. ונלקחו לא׳ ולג׳ כפלים שוי הפעמים, והם ה׳ וז׳. ולב׳ וד׳ כמו כן כפלים שוים אחרים, והם ח׳ ט׳. ואומר כי יחס ה׳ אל ח׳ כיחס ז׳ אל ט׳.

مِثالُهُ: أَنَّ اَ هُوَ الأَوَّلُ. وَقَدْرُهُ عِنْدَ الثَّاني، وَهُوَ بَ، كَقَدْرِ الثَّالِثِ، وَهُوَ جَ، عِنْدَ الرَّابِعِ، وَهُوَ دَ. وَنَضَعُ جَ وَ اَ أَضْعافًا مُتَساوِيَةً، وَهُوَ زَ وَ هَ. وَنَضَعُ لِحَطَّيْ دَ وَ بَ أَضْعافًا مُتَساوِيَةً، وَهِيَ حَ وَ طَ. فَأَقولُ إِنَّ قَدْرَ هَ عِنْدَ حَ كَقَدْرِ زَ عِنْدَ طَ.

إذا كانَتْ نِسْبَةُ الأَوَّلِ إلى الثَّاني هِيَ نِسْبَةُ الثَّالِثِ إلى الرَّابِعِ، وَأُخِذَ لِلأَوَّلِ والثَّالِثِ أَضْعافٌ مُتَساوِيَةُ المَرَّاتِ، ما كانَتْ، وَلِلثَّاني وَالرَّابِعِ أَضْعافٌ مُتَساوِيَةُ المَرَّاتِ، ما كانَتْ، فإنَّ نِسْبَةَ أَضْعافِ الأَوَّلِ المَأْخوذَةُ إلى أَضْعافِ الثَّاني هِيَ نِسْبَةُ أَضْعافِ الثَّالِثِ المَأْخوذَةُ إلى أَضْعافِ الرَّابِعِ.

כאשר היה יחס הראשון אל השני הוא יחס השלישי אל הרביעי, ונלקח[1] לראשון ולשלישי[2] כפלים שווי[3] הפעמים, אי זה מספר שיהיה, ולשני והרביעי[4] כפלים שווי[5] הפעמים, אי זה מספר שיהיה, הנה יחס כפלי הראשון הלקוחים אל כפלי השני[6] הם יחס כפלי השלישי[7] הלקוחים אל כפלי הרביעי.

مِثالُهُ: أَنَّ نِسْبَةَ الأَوَّلِ، وَهُوَ ا، إلى الثَّاني، وَهُوَ ب، هِيَ نِسْبَةُ الثَّالِثِ، وَهُوَ ج، إلى الرَّابِعِ، وَهُوَ د. وَقَدْ أُخِذَ لِقَدْرَيْ ا ج، أَضْعافٌ مُتَساوِيَةُ المَرَّاتِ، وَهُوَ ه ز، وِلِقَدْرَيْ ب د أَضْعافٌ مُتَساوِيَةُ المَرَّاتِ، وَهِيَ ح ط. فَأَقولُ إنَّ نِسْبَةَ ه إلى ح هِيَ نِسْبَةُ ز إلى ط.

המשל בו: כי יחס הראשון, והוא א׳, אל השני, והוא ב׳, הוא יחס השלישי, והוא[8] ג׳, אל הרביעי, והוא ד׳. וכבר נלקח לשיעורי א׳ ג׳ כפלים שווי[9] הפעמים, והם ה׳ ז׳, ולשיעורי ב׳ ד׳ כפלים שווי[10] הפעמים, והם ח׳ ט׳. הנה אומר כי יחס ה׳ אל ח׳ הוא יחס ז׳ אל ט׳.

שרטוט—**ספ**: א׳ מסומן ב״ראשון״, ב׳ מסומן ב״שני״, ג׳ מסומן ב״שלישי״ וד׳ מסומן ב״רביעי״ [1]ונלקח] ולקח **ס** [2]ולשלישי] והשלישי **טק** [3]שווי] שווה ⟨שוי⟩ **ט** שוה **ס** [4]ולשני והרביעי] והשני והרביעי **בטס** ~~הנה יחס כפלי~~ והשני והרביעי **פ** ו⟨⟨אל⟩⟩ השני והרביעי **ק** [5]שווי] שווה ⟨שוי⟩ **ט** [6]השני] ⟨⟨*⟩⟩ **פ** [7]השלישי] השלי⟨⟨שי⟩⟩ **ק** [8]והוא] והוא יחס השלישי+ **ב** [9]שווי] שווה ⟨שוי⟩ **ט** שוים **ק** [10]שווי] שווה ⟨שוי⟩ **ט** שוה **פ**

מופת זה: שנקח לה׳ ולז׳ עוד כפלים שוים והם ל׳ ונ׳. ולח׳ ולט׳ כפלים שוים, והם מ׳ וס׳. ויהיה שיעור ה׳ אצל א׳ כשיעור ז׳ אצל ג׳. והנה ל׳ ונ׳ כפלי ה׳ וז׳. ויהיה שיעור ל׳ אצל א׳ כשיעור נ׳ אצל ג׳. ולזה שיעור מ׳ אצל ב׳ כשיעור ס׳ אצל ד׳. ושיעור א׳ אצל ב׳ כשיעור ג׳ אצל ד׳. וכבר נלקחו לא׳ ולג׳ כפלים שוים, והם ל׳ ונ׳. ולב׳ ולד׳ כפלים שוים, והם מ׳ וס׳. ואם כן יהיה שיעור ל׳ אצל מ׳ כשעור נ׳ אצל ס׳. ואם היו שוים—גם הם שוים. ואם היו [!] אחד מהם קטן מן האחר—גם הם יהיו אחד מהם[1] קטן מן האחר בזה השיעור. וכפלי ל׳ ונ׳ לה׳ ולז׳ ככפלי מ׳ וס׳ לח׳ ולט׳. ואם כן יהיה שיעור ה׳ אצל ח׳ כשיעור ז׳ אצל ט׳.

بُرْهانُه: أنَّا نَضَعُ ه و ز أَضْعافًا مُتَساوِيَةً، وَهِيَ ن و ل. و ط و ح أَضْعافًا مُتَساوِيَةً، وَهِيَ م س. فَقَدْرُ ه عِنْدَ ا كَقَدْرِ ز عِنْدَ ج. و ن و ل أَضْعاف ز و ه. فَقَدْرُ ل عِنْدَ ا كَقَدْرِ ن عِنْدَ ج. وَكَذَلِكَ قَدْرُ م عِنْدَ ب كَقَدْرِ س عِنْدَ د. وَقَدْرِ ا عِنْدَ ب كَقَدْرِ ج عِنْدَ د. وَقَدْ وُضِعَتْ لِحَطَّيْ ا و ج أَضْعافٌ، وَهِيَ ن و ل. و د و ب أَضْعافٌ، وَهِيَ م و س. فَقَدْرُ ل عِنْدَ م كَقَدْرِ ن عِنْدَ س إِنْ كانَا مُتَساوِيَيْنِ فَهُما مُتَساوِيانِ، وَإِنْ كانَ أَحَدُهُما أَقَلَّ مِنَ الآخَرِ وَأَحَدُهُما أَيضًا أَقَلُّ مِنَ الآخَرِ بِذَلِكَ القَدْرِ. وَقَدْرُ ن و ل عِنْدَ ز و ه كَقَدْرِ س و م عِنْدَ ط و ح. فَقَدْرُ ه عِنْدَ ح كَقَدْرِ ز عِنْدَ ط.

וזה מה שרצינו ביאורו.

وَذَلِكَ ما أَرَدْنا بَيانَهُ.

[1]מן ... מהם] ⟨⟨*⟩⟩

بُرهانُهُ: أَنَّا نَأخُذُ لِقَدْرَيْ ه ز أَضْعافًا مُتَساوِيَةَ المَرَّاتِ وهي ل ن، ولِقَدْرَيْ ح ط أَضْعافًا مُتَساوِيَةَ المَرَّاتِ، وهي م س. فأَضْعاف ل لِقَدْرِ ا كأَضْعافِ ن لِقَدْرِ ج. وكَذَلِكَ أَضْعاف م لِقَدْرِ ب كأَضْعافِ س. لِقَدْرِ د. ونِسْبَةَ ا إلى ب هي نِسْبَةُ ج إلى د. وَقَدْ أُخِذَ لِقَدْرَيْ ا ج أَضْعافٌ مُتَساوِيَةُ المَرَّاتِ، وهي ل ن، ولِقَدْرَيْ ب د أَضْعافٌ مُتَساوِيَةُ المَرَّاتِ، وهي م س. فَقَدْرا^ ل ن—إِمّا زائِدانِ مَعًا عَلَى قَدْرَيْ م س، وإِمَّا مُساوِيانِ مَعًا لَهُمَا، وإِمَّا ناقِصانِ مَعًا مِنْهُما. وَقَدْرا^ ل ن أَضْعاف مُتَساوِيَة لِقَدْرَيْ ه ز وَقَدْرَا^ م س أَضْعاف مُتَساوِيَة لِقَدْرَيْ ح ط فَنِسْبَةُ ه إلى ح هي نِسْبَةُ ز إلى ط.

وذَلِكَ ما أَرَدْنا أَنْ نُبَيِّنَ.

מופתו: אנחנו נקח לשני שיעורי ה׳ ז׳ כפלים שווי[1] הפעמים, והם ל׳ נ׳, ולשני שיעורי ח׳ ט׳ כפלים שווי הפעמים, והם מ׳ ס׳. הנה כפלי ל׳ לשיעור[2] א׳ כמו כפלי[3] נ׳ לשיעור ג׳. וכן כפלי מ׳ לשיעור ב׳ כמו כפלי ס׳ לשיעור ד׳[4]. ויחס א׳[5] אל ב׳ הוא יחס[6] ג׳ אל ד׳[7]. וכבר[8] נלקח לשני שיעורי א׳ ג׳ כפלים שווי[9] הפעמים, והם ל׳[10] נ׳, ולשני שיעורי ב׳ ד׳ כפלים שווי הפעמים והם מ׳ ס׳. אם כן שיעור [!] ל׳ נ׳[11]— אם מוסיפים יחד על שיעור[12] מ׳ ס׳[13], ואם שוים יחד לשניהם, ואם גורעים יחד מהם. ושיעור [!] ל׳ נ׳[14] כפלים שוים לשני שיעורי ה׳ ז׳. ושיעור [!] מ׳ ס׳[15] כפלים שוים לשני שיעורי ח׳ ט׳. אם כן יחס ה׳ אל ח׳ הוא יחס[16] ז׳ אל ט׳.

וזה מה שרצינו לבאר.

---

[1] שווי] שוה **פ** [2] לשיעור] לשיעורי **ב** [3] כפלי] +ל׳ לשיעור א׳ כמו כפלי **ב** [4] ד׳] ⟨*⟩ **ט** [5] א׳] ⟨*⟩ **ק** [6] הוא יחס] כיחס **ס** [7] ד׳] א״ד **ב** [8] וכבר] +* **ק** [9] שווי] שווה ⟨שוי⟩ **ט** [10] ל׳] מ׳ ס׳ **ב** [11] ל׳ נ׳] ל״נ **וספק** [12] שיעור] ליתא **בו** [13] מ׳ ס׳] מ״ס **סק** [14] ל׳ נ׳] ל״נ **וספק** [15] מ׳ ס׳] מ״ס **וספק** [16] יחס] ליתא **ס**

[Proposition v.5]

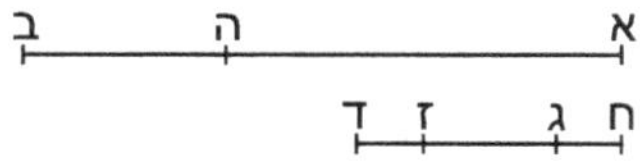

כשיהיו שני שיעורים—אחד מהם כפלים לאחר, וחוסרו מהם שני שיעורים, והיה במחוסר מכפלי המחוסר כמו שיש בכל מכפלי הכל, יהיה מה שבנשאר מכפלי הנשאר כמו שיש בכל מכפלי הכל.

إذا كانَ قَدْرانِ أَحَدُهُما جُزْءٌ مِنَ الآخَرِ، وَنَقَصَ مِنْ كُلٍّ واحِدٍ مِنْهُما ذَلِكَ الجُزْءُ، فَإِنَّ الباقِيَ مِنَ الباقِي مِثْلُ الكُلِّ مِنْ الكُلِّ.

המשל בזה: שג״ד חלק מא״ב. והמחוסר מהם א״ה וג״ז. והיה בא״ה מכפלי ג״ז כמו שבא״ב מכפלי ג״ד. ואומר כי מה שבה״ב הנשאר מכפלי ז״ד הנשאר[1], כמו שבא״ב מכפלי ג״ד.

مِثالُهُ: أَنَّ ج د جُزْءٌ مِنْ ا ب. وَالمَنْقوصُ مِنْهُما ج ز و ا ه. وَفِي ا ه مِنْ أَضْعافِ (ج ز)[1] مِثْلُ ما في ا ب مِنْ أَضْعافِ ج د. فَأَقولُ إِنَّ أَضْعافَ (ه ب)[2] الباقي ل ز د الباقي كَقَدْرِ أَضْعافِ ج د ل ا ب.

[1] הנשאר] להנשאר

[1] ج ز] ج [2] ه ب] ه ز

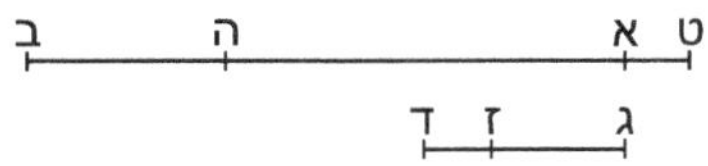

إذا كانَ مِقْدارانِ، أَحَدُهُما أَضْعافٌ لِلآخَرِ، ونَقَصَ مِنْهُما مِقْدارانِ، فَكَانَ ما[1] في المَنْقُوصِ مِنْ أَضْعافِ المَنْقُوصِ مِثْل ما في الكُلِّ مِنْ أَضْعافِ الكُلِّ، فإِنَّ ما في الباقِي مِنْ أَضْعافِ الباقِي مِثْل ما في الكُلِّ مِنْ أَضْعافِ الكُلِّ.

כאשר יהיו שני שיעורים, אחד משניהם כפלים[1] לאחר, וחוסר משניהם שני שיעורים, והיה מה שבמחוסר מכפלי המחוסר כמו מה שבכל מכפלי[2] הכל, הנה מה שבנשאר[3] מכפלי הנשאר כמו מה שבכל מכפלי הכל.

مِثالُهُ: أَنّ مِقْدارَ اب أَضْعافٌ لِمِقْدارِ ج د. والمَنْقُوصانِ مِنْهُما اه ج ز. وما في[2] اه مِنْ أَضْعافِ ج ز مِثْل ما في اب مِنْ أَضْعافِ ج د. فَأَقُولُ إِنَّ ما[3] في ه ب الباقِي مِنْ أَضْعاف ز د الباقِي مِثْل ما في اب مِنْ أَضْعافِ ج د.

המשל בו: כי שיעור א״ב כפלים לשיעור ג״ד. והמחוסרים משניהם א״ה וג״ז. ומה שבא״ה מכפלי ג״ז כמו מה שבא״ב[4] מכפלי ג״ד. הנה אומר כי מה שבה״ב הנשאר מכפלי ז״ד[5] הנשאר כמו מה שבא״ב מכפלי ג״ד.

שרטוט—ס: ב ← ט   [1]שיעורים ... כפלים] ⟨*⟩ ט   [2]מכפלי] +~~המחוסר כמו ש מה שבכל~~ ב   [3]שבנשאר] שנשאר ו   [4]שבא״ב] שא״ב ט   [5]ז״ד] ~~[..]~~ ⟨*⟩ ק

[1]ما] R2   [2]وما في] M وفي   [3]ما] R2T3

מופת זה: שנשים כפלי ה״ב לג״ח[1] ככפלי א״ה לג״ז[2]. ואם כן יהיו כפלי א״ה לג״ז ככפלי א״ב לח״ז. ושיעור מה שהיו כפלי[3] א״ה לג״ז כשיעור א״ב לג״ד. ושיעור כפלי א״ה לג״ז כשיעור כפלי א״ב לח״ז. ואם כן ח״ז[4] כמו ד״ג. וכשנשליך ג״ז המשותף, ישאר ח״ג כמו ז״ד. ויהיה שיעור כפלי א״ה לג״ז כשיעור כפלי ה״ב לז״ד. ואם כן שיעור כפלי ה״ב לז״ד [!] כשיעור כפלי א״ב לג״ד.

וכבר נשאר ה״ב בו מכפלי ז״ד כשיעור מה שבא״ב מכפלי ג״ד.

וזה מה שרצינו לבאר.

بُرْهانُه: أَنَّا نَجْعَلُ في ه ب مِنْ أَضْعافِ ج ح مِثْلَ ما في ا ه مِنْ أَضْعافِ ج ز. فَتَصيرُ أَضْعافُ ج ز ل ا ه كَقَدْرِ أَضْعافِ ا ب ل (ح ز)[1]. وَقَدْر ما كانَ مِنْ أَضْعافِ ا ه ل ج ز كَقَدْرِ أَضْعافِ ا ب ل ج د. ح ز مِثْل د ج. فَتُلْقِي ج ز المُشْتَرَكَ. فَيَبْقَى ح ج مِثْل ز د. فَقَدْر أَضْعافِ ج ز ل ا ه كَقَدْرِ أَضْعافِ ج ح و ج ح ل ه ب مِثْل ز د. فَقَدْر أَضْعافِ ج ز ل ا ه كَقَدْرِ أَضْعافِ ه ب ل ز د. وَقَدْر أَضْعافِ ا ه ج ز كَقَدْرِ أَضْعافِ ا ب ل ج د. فبَقِيَ ه ب فِيهِ مِنْ أَضْعافِ ز د كَقَدْرِ ما هِيَ في ا ب مِنْ أَضْعافِ ج د.

وَذَلِكَ ما أَرَدْنَا بَيانَهُ.

[1] לג״ח] ⟨⟨*⟩⟩ [2] לג״ז] ⟨⟨*⟩⟩ [3] כפלי] ⟨⟨*⟩⟩
[4] ואם כן ח״ז] ⟨⟨*⟩⟩

[1] ح ز] ج ز

بُرهانُه: أَنَّا نَجْعَلُ في ا ط مِنْ أَضْعافِ ز د مِثْلَ ما في ا ه مِنْ أَضْعافِ ج ز. فيكونُ ما في ه ط كُلِّهِ[1] مِنْ أَضْعافِ ج د مِثْلَ ما في ا ه مِنْ أَضْعافِ ج ز. وَقَدْ كانَ ما في ا ب مِنْ أَضْعافِ ج د مِثْلَ ما في ا ه مِنْ أَضْعافِ ج ز^. فما في ه ط مِنْ أَضْعافِ ج د مثل ما في ا ب مِنْ أَضْعافِ ج د. ف ه ط مثل ا ب. فتُلْقِي^ ا ه المُشْتَرَكَ. فيَبْقَى ا ط مثل ه ب. وَقَدْ كانَ في[2] ا ط مِنْ أَضْعافِ ز د مثل ما في ه ا مِنْ أَضْعافِ ج ز. فما في ه ب مِنْ أَضْعافِ ز د مثل ما في ه ا مِنْ أَضْعافِ ج ز. وَقَدْ كانَ ما في ا ب كُلِّهِ مِنْ أَضْعافِ ج د مثل ما في ه ا مِنْ أَضْعافِ ج ز. فَمَا في ه ب إِذًا مِنْ أَضْعافِ ز د مثل ما في ا ب كُلّه مِنْ أَضْعافِ ج د.

وذَلِكَ ما أَرَدْنا أَنْ نُبَيِّنَ.

מופתו: אנחנו נשים בא"ט מכפלי ז"ד כמו מה[1] שבא"ה מכפלי ג"ז.[2] אם כן מה שבה"ט כולו מכפלי ג"ד כמו מה שבא"ה[3] מכפלי ג"ז[4]. אם כן מה[5] שבה"ט מכפלי ג"ד כמו מה שבא"ב[6] מכפלי ג"ד. אם כן ה"ט כמו א"ב. ותשליך א"ה המשותף. וישאר א"ט כמו ה"ב. וכבר היה בא"ט[7] מכפלי ז"ד[8] כמו מה שבא"ה מכפלי ג"ז. אם כן מה שבה"ב מכפלי ז"ד כמו מה שבא"ה מכפלי ג"ז[9]. וכבר היה מה שבא"ב כולו מכפלי ג"ד כמו מה שבא"ה מכפלי ג"ז[10]. אם כן מה שבה"ב מכפלי ז"ד[11] כמו מה[12] שבא"ב[13] כולו[14] מכפלי ג"ד[15].

וזה מה שרצינו לבאר.

---

1 كله] F   2 في] CopFMT1 +ما

1 מה] ליתא **ס**   2 ג"ז] ג"ד כמו מה שבא"ה מכפלי ז' ז"ג **ב** +וכבר היה בא"ב מכפלי ג"ד כמו מה שבא"ה מכפלי ג"ז **ז** +⟨⟨וכבר היה מה שבא"ב מכפלי ג"ד כמו שבא"[..] מכפלי ג"ז⟩⟩ **פ**   3 מה שבא"ה] שבא"ה **ספ** מה ~~שבא"ב~~ שבא"ה **ק**   4 ג"ז] +⟨⟨וכבר היה מה שבא"ב מכפלי ג"ד כמו מה שבא"ה מכפלי ג"ז⟩⟩ **ק**   5 מה] ⟨*⟩ **ט**   6 מה שבא"ב] מה שא"ב **ט** שבא"ב **ספ**   7 מכפלי ... בא"ט] ליתא **ב**   8 ז"ד] ~~ג"ד~~+ **ק**   9 אם ... ג"ז] ליתא **וטפ** ⟨⟨*⟩⟩ **ק**   10 וכבר ... ג"ז] ליתא **פ**   11 ז"ד] ~~ג"ז~~+ **ק**   12 מה] ליתא **ס**   13 שבא"ב] שבא"ה **פ**   14 כולו] ליתא **בזפ** ⟨*⟩ **ט**   15 ג"ד] ג"ז ~~וכבר היה מה שבא"ב כלו מכפלי ג,ד כמו שבא"ה מכפלי ב"ז אם כן מה שבה"ב מכפלי ז"ד~~ כמו שבא"ב כלו מכפלי ג"ד **פ**

[Proposition v.6]

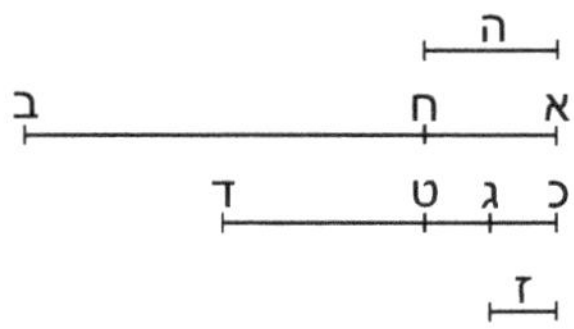

כשיהיו שני שיעורים אשר בהם כפלים שוים לשני שיעורים אחרים, וחוסרו מן הגדולים כפלים שוים לקטנים, יהיו הנשארים אם שהם שוים לקטנים או שוי כפלים לקטנים.

إِذا كانَ قَدْرانِ فيهِما أَضْعافُ قَدْرَيْنِ آخَرَيْنِ، وكانَتْ أَضْعافُهُما مُتَساوِيَةً، وَنُزِعَ الأَصْغَرانِ مِنَ الأَكْبَرَيْنِ، كُلٌّ واحِدٍ مِنَ الَّذي هُوَ أَضْعافُهُ، فَإِنَّ الَّذي يَكونُ في الباقِيَيْنِ مِنْ أَضْعافِ الأَصْغَرَيْنِ أَيْضًا مُتَساوِيَةً.

המשל בזה: כי א״ב וג״ד כפלים לה׳ ולז׳. ומה שבא״ב מכפלי ה׳ כמו מה שיש בג״ד מכפלי ז׳. וחוסרו מא״ב וג״ד דמיוני ה׳ וז׳ והם א״ח וג״ט. ואומר כי מה שבח״ב וט״ד הנשארים מכפלי ה׳ וז׳ שוים.

مِثالُهُ: أَنَّ في جـ د و ا ب أَضْعافُ ل ز و ه. ما في ا ب مِنْ أَضْعافِ ه مِثْلَ ما في جـ د مِنْ أَضْعافِ ز. وَنُزِعَ مِنْ ا ب وَمِنْ جـ د ا ح و جـ ط. وَهُما مِثْلُ ه و ز. فَأَقولُ إِنَّ ما في ط د وح ب الباقِيَيْنِ مِنْ أَضْعافِ ز و ه مُتَساوِيَةٌ.

إذا كانَ مِقْدارانِ فِيهِما أَضْعافٌ مُتَساوِيَةٌ لِمِقْدارَيْنِ آخَرَيْنِ، وَنَقَصَ مِنَ الأَكْبَرَيْنِ أَضْعافٌ مُتَساوِيَةٌ لِلأَصْغَرَيْنِ، فَإِنَّ الباقِيَيْنِ إِمّا مُساوِيانِ لِلأَصْغَرَيْنِ وَإِمّا أَضْعافٌ لَهُما مُتَساوِيَةٌ.

כאשר היו שני שיעורים בשניהם כפלים שוים[1] לשני שיעורים אחרים, וחוסר מהשנים הגדולים כפלים שוים לקטנים, הנה השנים הנשארים אם שוים לשנים הקטנים[2] ואם כפלים לשניהם שוים.

مِثالُهُ: أَنَّ أَضْعافَ اب لِقَدْرِ ه وَأَضْعافَ ج د لِقَدْرِ ز مُتَساوِيَةٌ. وَقَدْ نَقَصَ مِنْ اب وَ ج د أَضْعافٌ مُتَساوِيَةٌ لِقَدْرَيْ ه وَ ز، وَهُما اح وَ ج ط. فَأَقُولُ إِنَّ ب ح ط د الباقِيَيْنِ إِمّا مُساوِيانِ مَعًا لِقَدْرَيْ ه ز، وَإِمّا أَضْعافٌ لَهُما مُتَساوِيَةٌ. فَلْيَكُنْ ب ح أَوَّلًا أَضْعافًا لِقَدْرِ ه. فَأَقُولُ إِنَّ ط د أَضْعافُ مِثْلِها لِقَدْرِ ز.

המשל בו: כי כפלי[3] א״ב לשיעור ה׳ וכפלי ג״ד לשיעור[4] ז׳ שוים[5]. וכבר חוסר[6] מן א״ב וג״ד[7] כפלים שוים לשיעור ה׳ וז׳, והם א״ח וג״ט. הנה אומר כי ב״ח וט״ד הנשארים אם שוים יחד לשני שיעורי ה׳ ז׳, ואם כפלים לשניהם שוים. ויהיה ב״ח תחלה כפלים לשיעור ה׳. הנה אומר כי ט״ד כפלים כמו הם[8] לשיעור ז׳.

שרטוט—**ס**: ג ↔ ט, ט׳ מסומנת גם בקטע ח״ב, א״ח מסומן ב״ראשון״, ג״ד מסומן ב״שני״, ט״ג מסומן ב״שלישי״, ז׳ מסומן ב״רביעי״, ט״ב (?) מסומן ב״חמישי״, ג״ד מסומן ב״ששי״; **פ**: א״ח מסומן ב״ראשון״, ה׳ מסומן ב״שני״, ג״ט מסומן ב״שלישי״, ז׳ מסומן ב״רביעי״, ח״ב מסומן ב״רביעי״, כ״ג מסומן ב״ששי״

[1]שוים] ליתא **ס** [2]הקטנים] קטנים **ט**
[3]כפלי] כפלים **ב** [4]לשיעור] +ה׳ וכפלי ב׳ **ב** [5]שוים] ליתא **ב** [6]חוסר] חסר **ב** ליתא **ז**
[7]וג״ד] ⟨ו⟩ג״ד **ט** [8]הם] כן+ **ס** ה⟨ם⟩ **פ**

מופת זה: שנשים ראשונה ח״ב כמו ה׳[1]. ואומר כי ט״ד כמו ז׳. ונשים ג״כ כמו ז׳. ויהיו כפלי א״ב לה׳ כשיעור כפלי ג״ד לז׳. וה׳ כמו ב״ח. וז׳ כמו ג״כ. ואם כן שיעור כפלי א״ב לה׳ כשיעור כפלי כ״ט לז׳. ושיעור כפלי א״ב לה׳ כשיעור כפלי ג״ד לז׳. ואם כן כ״ט כמו ג״ד. וכשנשליך ג״ט המשותף ישאר כ״ג כמו ט״ד. ואמנם כ״ג הוא כמו ז׳. ויהיה ט״ד גם כן כמו ז׳. ובזה יתבאר אם הוא בח״ב כפלים הרבה מן ה׳, היה בט״ד כמו אלו הכפלים מז׳. ושעור מה שבח״ב מכפלי ה׳ כשיעור מה שבט״ד מכפלי ז׳.

بُرْهانُهُ: أَنَّا نَجْعَلُ ح ب أَوَّلًا مِثْلَ ه. وَأَقولُ إِنَّ ط د مِثْلُ ز. وَنَجْعَلُ ج ك مِثْلَ ز. فَقَدْرُ أَضْعافِ ا ب لل ه [..] كَقَدْرِ أَضْعافِ ك ط لل ز. وَقَدْرُ أَضْعافِ ا ب لل ه كَقَدْرِ ج د لل ز. فَقَدْرُ أَضْعافِ ز ل ك ط كَقَدْرِ أَضْعافِ ز ل ج د. ك ط مِثْلُ ج د. فَتَنْزَعُ ج ط المُشْتَرَكَ. فَيَبْقَى ك ج مِثْلَ ط د أَيضًا. وَلَكِنَّ ك ج مِثْلُ ز. و ط د مِثْلُ ز. وَكَذَلِكَ [تَ]بيَّنَ أَيْضًا أَنَّهُ لَوْ كانَ في ح ب أَضْعافٌ كَثيرَةٌ مِثْلَ ه، لَكانَ في د ط أَيضًا مِثْلَ تِلْكَ الأَضْعافِ مِنْ ز. فَقَدْرُ ما في ح ب مِنْ أَضْعافِ ه كَقَدْرِ ما في ط د مِنْ أَضْعافِ ز.

[1] ה׳] ~~ח׳~~+

بُرهانُهُ: أنَّا نَجْعَلُ ما في ج ك مِنْ أَضْعاف ز مثل ما في ح ب مِنْ أَضْعاف ه. فما في ا ح، وهو الأَوَّل، مِنْ أَضْعاف ه، وهو الثَّاني، مثل ما في ج ط، وهو الثَّالِث، مِنْ أَضْعاف ز، وهو الرَّابِع. وما في ح ب، وهو الخامس، مِنْ أَضْعاف ه الثَّاني[1]، مثل ما في ج ك، وهو السَّادِس، مِنْ أَضْعاف ز الرَّابِع[2]. فَمَا في ا ب، وهو الأَوَّل والخامس جَميعًا، مِنْ أَضْعاف ه الثَّاني[3]، مثل ما في ك ط، وهو^ الثَّالِثُ والسَّادِسُ جَميعًا، مِنْ أَضْعاف ز الرَّابِع[4]. وَقَدْ كانَ[5] في ا ب مِنْ أَضْعاف ه مثل ما في ج د مِنْ أَضْعاف ز. ف ك ط مثل ج د. فنُلْقِي ج ط المُشْتَرَك. فيَبْقَى ك ج مثل^ ط د. وَقَدْ كانَ ما في ح ب مِنْ أَضْعاف ه مثل ما في ك ج مِنْ أَضْعاف ز. فَمَا في ط د مِنْ أَضْعاف ز مثل ما في ح ب مِنْ أَضْعاف ه.

מופתו: אנחנו נשים מה שבג״כ מכפלי ז׳ כמו מה שבח״ב מכפלי ה׳. אם כן מה שבא״ח, והוא הראשון, מכפלי ה׳, והוא השני, כמו מה שבג״ט, והוא השלישי, מכפלי ז׳, והוא הרביעי. ומה שבח״ב[1], והוא החמשי, מכפלי ה׳ השני, כמו מה[2] שבג״כ, והוא הששי, מכפלי ז׳ הרביעי[3]. אם כן מה שבא״ב, והוא הראשון והחמשי יחד, מכפלי ה׳[4] השני, כמו מה שבכ״ט השלישי[5] והששי יחד, מכפלי ז׳ הרביעי. וכבר היה בא״ב מכפלי ה׳ כמו מה[6] שבג״ד מכפלי ז׳. אם כן כ״ט כמו ג״ד. ונשליך ג״ט המשותף. וישאר כ״ג כמו מה[7] שבט״ד. וכבר היה מה שבח״ב מכפלי ה׳ כמו מה שבכ״ג מכפלי ז׳. אם כן מה שבט״ד מכפלי ז׳[8] כמו מה שבח״ב מכפלי ה׳.

وَلْيَكُنْ أيضًا ح ب مُسَاوِيَةً ل ه. فَأَقولُ إنَّ ط د مُساوٍ ل ز.

ויהיה גם כן ח״ב שוה אל ה׳. הנה אומר כי[9] ט״ד שוה אל ז׳[10].

بُرهانُهُ: أنَّا نَجْعَلُ ج ك مُساوِيًا ل ز. فَبِمِثْلِ العَمَلِ الأَوَّل نُبيِّنُ[6] أَنَّ ك ط مثل ج د. فإذا أُلْقِيَ ج ط المُشْتَرَك، بَقِيَ ط د مثل ك ج. وَلَكِنَّ^ ك ج مثل ز. ف ط د إذًا مثل ز.

המופת: אנחנו נשים ג״כ שוה אל ז׳. הנה בכמו המעשה הראשון נבאר כי כ״ט כמו ג״ד. הנה כאשר הושלך ג״ט המשותף, ישאר ט״ד כמו כ״ג. ומפני שכ״ג כמו ז׳, הנה ט״ד אם כן כמו ז׳.

1 الثاني] **FM** +وهو 2 الرابع] **F** +وهو 3 الثاني] **F** +وهو 4 الرابع] **T2** +وهو 5 كان] **M** +ما 6 نبين] **T2T3** تبين

1 שבח״ב] +~~מכפלי~~ **ק** 2 מה] ליתא **טפ** 3 הרביעי] והוא+ **ק** 4 ה׳] ליתא **ס** 5 השלישי] והוא+ **ס** 6 מה] ליתא **בופ** 7 מה] ליתא **ט** 8 אם ... ז׳] ליתא **ז** 9 כי] ליתא **ס** 10 ז׳] ה״ז **ס**

وَذَلِكَ ما أَرَدْنا بَيَانَهُ. וזה מה שרצינו שנבאר.

وَذَلِكَ ما أَرَدْنا أَنْ نُبَيِّنَ.

וזה מה שרצינו לבאר.

[Proposition v.7]

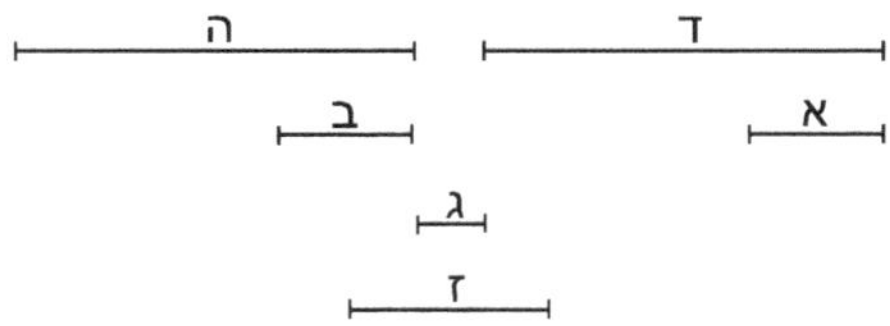

השיעורים השוים יחסם אל שיעור אחר אחד, ויחסו גם כן אליהם אחד.

الأَقْدارُ المُتَساوِيَةُ[1] إِذا نُسِبَتْ إِلى قَدْرٍ آخَرَ كانَتْ نِسْبَتُها إِلَيْهِ واحِدَةً، وَإِذا نُسِبَتْ إِلَيْها أَيْضًا كانَتْ نِسْبَتُهُ إِلَيْها واحِدَةً.

המשל בזה: כי שעורי א׳ וב׳ שוים, ושעור ג׳ שעור אחר. ואומר כי יחס כל אחד מא׳ וב׳ אל ג׳ כמו יחס האחר, ויחס ג׳ גם כן אל כל אחד מהם כמו יחסו אל האחר.

مِثالُهُ: أَنَّ قَدْرَ بَ وَ اَ مُتَساوِيانِ. وَقَدْرُ جَ قَدْرٌ آخَرَ. فَأَقولُ إِنَّ نِسْبَةَ كُلِّ واحِدٍ مِنْ بَ وَ اَ إِلى جَ مِثْلُ نِسْبَةِ الآخَرِ إِلَيْهِ، وَنِسْبَةُ جَ أَيْضًا إِلى كُلِّ واحِدٍ مِنْ بَ وَ اَ مِثْلُ نِسْبَتِهِ إِلى الآخَرِ.

---

[1] المتساوية] ~~المتناسبة~~ ⟨⟨*⟩⟩

المَقاديرُ المُتَساوِيَةُ نِسْبَتُها[1] إلى قَدْرٍ آخَرَ واحِدَةٌ، وَنِسْبَتُهُ^ هُوَ أيضًا إِلَيْها واحِدَةٌ.

השעורים השוים יחסם[1] אל שעור[2] אחר אחד, ויחס הוא גם כן אליהם[3] אחד.

مِثالُهُ: أَنَّ قَدْرَيْ آ بَ مُتَساوِيان، ومِقْدارَ جَ قَدْرٌ آخَر. فَأَقولُ إنّ نِسْبَةَ آ إلى جَ كَنِسْبَةِ بَ إلى جَ، ونِسْبَةَ جَ أيضًا إلى آ كَنِسْبَتِهِ إلى بَ.

המשל בו: כי שני שעורי א׳ ב׳ שוים, ושיעור ג׳ שיעור אחר[4]. הנה אומר כי יחס א׳ אל ג׳ כיחס ב׳ אל ג׳, ויחס ג׳ גם כן אל א׳ כיחסו אל ב׳.

---

[1] نسبتها] نسبته #

---

[1] יחסם] ⟨⟨הנה⟩⟩+ ק [2] שעור] השעור ז
[3] אליהם] עליהם ס [4] אחר] +~~הנה אומר כי יחס א״ל א׳ אל ג׳ שוים ושיעור ג׳ שיעור אחר~~ פ

המופת בזה: שנקח לא׳ ולב׳ כפלים שוים והם ד׳ וה׳. ונקח לג׳ כפלים אחרים והם ז׳. ומה שבד׳ מכפלי א׳ כמו מה שבה׳ מכפלי ב׳. וא׳ שוה לב׳—יהיה אם כן ד׳ שוה לה׳. ושם שעור אחר כמו שהוא, והוא ז׳—יהיה שעור תוספת ד׳ על ז׳ כשעור תוספת ה׳ על ז׳. ואם היה שוה לו, גם הוא שוה לו. ואם היה חסר ממנו, גם הוא חסר ממנו בזה השיעור. וד׳ וה׳ הם כפלים שוים לא׳ ולב׳. וז׳ הוא כפלי ג׳, והוא השעור האחר[1]. ונמצא יחס א׳ וב׳ אל ג׳ אחד. ואומר כי יחס ג׳ אל כל אחד מא׳ וב׳ כמו יחסו אל האחר. כי הנהגתם אחת. וכמו כן יתבאר כי ד׳ שוה לה׳. ושעור תוספת ז׳ על ד׳ כשעור תוספתו על ה׳. וכמו כן אם היה שוה לו או חסר ממנו. וז׳ כפלי ג׳. וד׳ וה׳ כפלי א׳ וב׳—נמצא יחס ג׳ אל כל אחת מא׳ וב׳ כמו יחסו אל האחר.

וזה מה שרצינו שנבאר.

[1] האחר] ⟨⟨ה⟩⟩אחר

بُرْهانُهُ: أَنَّا نَأْخُذُ لِ ا و ب أَضْعافًا مُتَساوِيَةً، وَهِيَ د و ه. وَنَأْخُذُ لِ ج أَضْعافًا أُخَرَ، كَيْفَما كانَ، وَهُوَ ز. فَمَا في د مِنْ أَضْعافِ ا مِثْلُ ما في ه مِنْ أَضْعافِ ب. و ا مُساوٍ لِ ب و د مُساوٍ لِ ه. وَقَدْرٌ آخَرُ، كَيْفَما كانَ، وَهُوَ ز. فَقَدْ[رُ] زِيادَةِ د على ز كَقَدْرِ زِيادَةِ ه على ز. فَإِنْ كانَ مُساوِيًا لَهُ، فَهُوَ مُساوٍ لَهُ. وَإِنْ كانَ ناقِصًا مِنْهُ، فَهُوَ ناقِصٌ مِنْهُ بِذَلِكَ القَدْرِ. و ه و د هُما أَضْعافٌ مُتَساوِيَةٌ لِ ب و ا. و ز هُوَ أَضْعافُ ج. وَهُوَ قَدْرٌ آخَرُ. فَكُلُّ واحِدٍ مِنْ ا و ب نِسْبَتُهُ إِلى ج مِثْلُ نِسْبَتِهِ لِلآخَرِ. وَأَقُولُ إِنَّ نِسْبَةَ ج إِلى كُلِّ واحِدٍ مِنْ ب و ا مِثْلُ نِسْبَتِهِ إِلى الآخَرِ. بُرْهانُهُ: أَنَّ تَدْبِيرَهُما واحِدٌ. وَكَذَلِكَ (تَبَيَّنَ)[1] أَنَّ (د)[2] مُساوٍ لِ ه. وَقَدْرُ زِيادَةِ ز على د كَقَدْرِ زِيادَتِهِ على ه. وَكَذَلِكَ إِنْ كانَ مُساوِيًا لَهُ أَو ناقِصًا عَنْهُ [!]. و ز أَضْعافُ ج. و د و ه أَضْعافٌ مُتَساوِيَةٌ لِ ا و ب. فَنِسْبَةُ ج إِلى كُلِّ واحِدٍ مِنْ ا و ب مِثْلُ نِسْبَتِهِ إِلى الآخَرِ.

وَذَلِكَ ما أَرَدْنا بَيانَهُ.

[1] تبين] نبين [2] د] ه

بُرهانُ ذَلِكَ: أَنَّا نَأخُذُ لِقَدْرَيْ ا ب أَضْعافًا مُتَساوِيَةً، وهي د ه. ولِقَدْرِ ج أَضْعافًا ما كانَتْ، وهي ز. فأَضْعاف د لِقَدْرِ ا كأَضْعاف ه لِقَدْرِ ب. و ا مثل ب. ف د مثل ه. وَقَدْرُ ز قَدْرٌ آخَر ما كانَ. فقَدْرا د ه إِمّا مُساوِيانِ مَعًا لِقَدْرِ ز، وَإِمّا زائِدانِ مَعًا عَلَيْه، وَإِمّا ناقِصانِ مَعًا مِنْهُ. وَهُما أَضْعافٌ مُتَساوِيَةٌ لِقَدْرَيْ ا ب. و ز أَضْعافُ لِقَدْرِ ج. فنِسْبَةُ ا إلى ج كنِسْبَةِ ب إلى ج. وأَقولُ إِنَّ نِسْبَةَ ج إلى ا كنِسْبَتِهِ إلى ب، لِأَنَّ تَدْبيرَهُما واحِدٌ. وَقَدْ تَبَيَّنَ أَنَّ د ه مُتَساوِيانِ، وأَنَّ ز إِمّا مُساوٍ لَهُما مَعًا^، وَإِمّا زائِدٌ عَلَيْهِما مَعًا^، وَإِمّا ناقِصٌ مِنْهُما مَعًا. و ز أَضْعافُ لِقَدْرِ ج. و د ه أَضْعافٌ مُتَساوِيَةٌ لِقَدْرَيْ ا ب. فنِسْبَةُ ج إلى ا كَنِسْبَتِهِ إلى ب.

وذَلِكَ ما أَرَدْنا أَنْ نُبَيِّنَ.

מופתו: אנחנו נקח לשני שיעורי א׳ ב׳ כפלים שוים, והם ד׳ ה׳. ולשיעור[1] ג׳ כפלים[2] מה שיהיו, והם ז׳. והנה כפלי ד׳ לשיעור א׳ ככפלי ה׳[3] לשיעור ב׳. וא׳ כמו ב׳. אם כן ד׳ כמו ה׳. ושיעור ז׳ שיעור אחר אי זה שיהיה. אם כן שני שיעורי ה׳ ד׳ אם שוים יחד לשעור ז׳[4], ואם נוספים יחד עליו, ואם חסרים יחד ממנו. ושניהם כפלים שוים לשני שעורי א׳ ב׳. וז׳ כפלים לשיעור ג׳. אם כן יחס[5] א׳ אל ג׳ כיחס ב׳ אל ג׳. ואומר כי[6] יחס ג׳ אל א׳ כיחסו אל ב׳, מפני כי הנהגת שניהם אחת. וכבר התבאר כי ד׳ ה׳ שוים, וכי ז׳ אם שוה לשניהם, ואם נוסף על שניהם, ואם חסר משניהם יחד. וז׳ כפלים לשיעור ג׳. וד׳ ה׳ כפלים שוים לשני שעורי א׳ ב׳. אם כן יחס ג׳ אל א׳ כיחסו אל ב׳.

וזה[7] מה שרצינו לבאר.

---

1 ולשיעור] לשעור ס   2 כפלים] +~~שוים~~ ק

3 ג׳ ... ה׳] ליתא ב   4 לשעור ז׳] לשעורי ח׳ ס

5 יחס] +* ב +~~אל~~ ק   6 כי] +~~יחסו אל~~ ק

7 וזה] +* ב

[Proposition v.8]

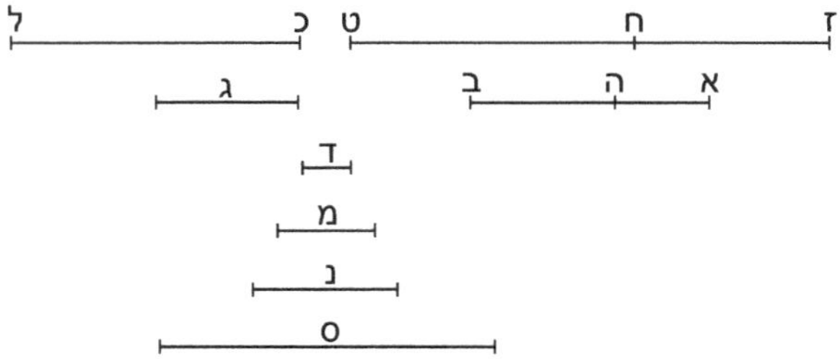

השיעורים המתחלפים כשיוחסו אל שעור אחר, היה הגדול רב היחס אליו מן הקטן. וכשייוחס הוא אליהם יהיה יחסו אל הקטן רב מיחסו אל הגדול.

الأَقْدارُ المُخْتَلِفَةُ إذا نُسِبَتْ إلى قَدْرٍ آخَرَ، فَإِنَّ أَعْظَمَها هُوَ أَعْظَمُها إِلَيْهِ نِسْبَةً مِمَّا هُوَ أَصْغَرُ مِنْهُ. وَكَذَلِكَ القَدْرُ الآخَرُ إذا نُسِبَتْ إلى أَصْغَرِ الأَقْدارِ المُخْتَلِفَةِ كانَتْ نِسْبَتُهُ إِلَيْهِ أَعْظَمَ مِنْ نِسْبَتِهِ إلى ما هُوَ أَكْبَرُ مِنْهُ.

המשל בזה: ששיעורי א״ב ושעור ג׳[1] מתחלפים, א״ב גדול מג׳, ושעור ד׳ שעור אחר. ואומר כי א״ב רב היחס אל ד׳[2] מג׳ אל ד׳. וד׳ רב היחס אל ג׳ ממה שהוא אל א״ב.

مِثالُهُ: أَنَّ قَدْرَا اب وقَدْرَ ج مُخْتَلِفانِ، اب أَعْظَمُ مِنْ ج. وَقَدْرُ د قَدْرٌ آخَرُ. فَأَقُولُ إِنَّ اب أَعْظَمُ نِسْبَةً إلى د مِنْ ج إلى د. وأَنَّ د أَعْظَمُ نِسْبَةً إلى ج مِنْهُ إلى اب.

[1] ג׳] ~~ג״ח~~+ [2] אל ד׳] ⟨⟨*⟩⟩

الأَقْدارُ المُخْتَلِفَةُ إذا نُسِبَتْ إلى قَدْرٍ آخَرَ، فَالأَعْظَم أَكْبَرُ ^ نِسْبَةً إِلَيْهِ مِنَ الأَصْغَرِ. وإذا نُسِبَ هُوَ إِلَيْها فَنِسْبَتُهُ إلى الأَصْغَرِ أَكْبَرُ مِنْ ^ نِسْبَتِهِ إلى الأَعْظَمِ.

השעורים המתחלפים כאשר נתיחסו[1] אל שעור אחר, הנה הגדול—יותר יחסו אליו מן הקטן. וכאשר נתיחס הוא[2] אליהם הנה יחסו אל הקטן יותר מיחסו אל הגדול.

مِثالُهُ: أَنَّ^ قَدْرَيْ ا ب ج مُخْتَلِفانِ، ا ب أعظم مِنْ ج. وَقَدْر د قَدْرٌ[1] آخر. فَأَقولُ إنّ ا ب أَكبر نِسْبَةً^ إلى د مِنْ ج إلى د، وإنَّ د أَكبر نِسْبَةً^ إلى ج مِنْهُ إلى ا ب.

המשל בו[3]: שני שעורי א״ב וג׳ מתחלפים, א״ב[4] יותר גדול מן ג׳. ושעור ד׳ שיעור[5] אחר. הנה אומר כי א״ב יותר גדול יחסו אל ד׳[6] מן ג׳ אל ד׳, ושד׳ יותר גדול[7] יחסו אל ג׳ מאשר הוא אל א״ב.

[1] قدر] وقدر #

שרטוט—**בס**: ה׳ נוספת סומנה על כ״ל
[1] נתיחסו] התיחסו **ק** [2] הוא] ⟨*⟩ **ט** [3] בו] +⟨כי⟩ **ט** +כי **ס** [4] א״ב] ⟨*⟩ **ט** וא״ב **ס** [5] ד׳ שעור] ליתא **ב** [6] ד׳] +~~מיחסו~~ **ק** [7] גדול] ליתא **ס**

מופת זה: שנשים ב״ה כמו ג׳. וא״ה קטן מה״ב. וא״ה כשיכפל יהיה גדול מד׳. ונקח לא״ה ולה״ב ולג׳ כפלים שוים, והם ז״ח וח״ט וכ״ל[1]. ונשים מ׳ שני כפלי ד׳, ונ׳ שלשת כפליו. וכל אחד ימשוך אחר האחר עד שישיג אל שיעור ראשון מכפליו שיהיה גדול מל״כ. ויהיה ס׳[2] הוא ראשון כפלי ד׳ שיהיה גדול מכ״ל. ויהיה שעור מה שבז״ח מכפלי א״ה כמו מה שבח״ט מכפלי ה״ב. ואם כן שבז״ח [!] מכפלי א״ה כמו מה שבז״ט מכפלי א״ב. ומה שבז״ח מכפלי א״ה כמו מה שבכ״ל מכפלי ג׳[3]. וה״ב כמו ג׳. ואם כן ח״ט כמו כ״ל. וס׳ גדול מכ״ל. וכ״ל איננו קטן מנ׳. וכ״ל כמו ח״ט. ואם כן ח״ט איננו קטן מנ׳. וז״ח גדול מד׳. ואם כן ז״ט גדול מד׳ ונ׳ יחד. וס׳ כמו ד׳ ונ׳. ואם כן ז״ט גדול מס׳. וכ״ל איננו גדול מס׳. וז״ט וכ״ל כפליהם לא״ב ולג׳ שוים. וס׳ הוא כפל ד׳. ויהיה אם כן[4] א״ב רב היחס אל ד׳ מג׳ אל ד׳.

بُرْهانُه: أَنْ نَجْعَلَ ب ه مِثْلَ ج. ا ه أَصْغَرُ مِنْ د. و ا ه إِذا ضُوعِفَتْ قَدْ تَصِيرُ أَعْظَمَ مِنْ د. فَتُضَعَّفُ ا ه حَتَّى تَصيرَ أَعْظَمَ مِنْ د مُجْتَمِعٌ مِنْهُ ز ح. وَنَجْعَلُ في ح مِنْ أَضْعافِ ه ب مِثْلَ ما في ز ح مِنْ أَضْعافِ ا ه. وَكَذَلِكَ نَجْعَلُ في ك ل مِنْ أَضْعافِ ج تِلْكَ العِدَّةَ نَفْسَها. وَنَجْعَلُ م ضِعْفَي د، و ن ثَلاثَةَ[1] أَضْعافِه. وَكُلَّ واحد يَتْلُو الآخَرَ حَتَّى يَبْلُغَ أَوَّلَ أَضْعافِ د وَتَكونُ أَعْظَمُ مِنْ ك ل. فَلْيَكُنْ س، وَهُوَ أَوَّلُ أَضْعافِ د، أَعْظَمَ مِنْ ك ل. فَقَدْرُ ما في ز ح مِنْ أَضْعافِ ا ه مِثْلُ ما في ح ط مِنْ أَضْعافِ ه ب. وَما في ز ح مِنْ أَضْعافِ ا ه مِثْلُ ما في ك ل مِنْ أَضْعافِ ج. لِأَنَّ أَضْعافَهُما هِيَ مُتَساوِيَةٌ. وَأَيْضًا ما في ح ط مِنْ أَضْعافِ ه ب مِثْلُ ما في ك ل مِنْ أَضْعافِ ج. ه ب مِثْل ج. ح ط مِثْل ك ل. و س أَعْظَم مِنْ ك ل. ك ل لَيْسَ بِأَصْغَرَ مِنْ ن. ك ل مِثْلُ ح ط. ح ط لَيْسَ بِأَصْغَرَ مِنْ ن. و ز ح أَعْظَمُ من د. ز ط أَعْظَمُ مِنْ د و ن. و س مِثْلُ د ن. ز ط أَعْظَمُ مِنْ س. و ك ل لَيْسَ بِزائِد عَلَى س. ز ط و ك ل أَضْعافهما ل ا ب و ج مُتَساوِيَة. و س أَضْعافُ د. ا ب أَعْظَمُ نِسْبَةً إِلى د مِنْ ج إِلى د.

[1]וכ״ל] ⟨⟨*⟩⟩ [2]ס׳] [..יחס]+ [3]ג׳] +⟨⟨אם כן מה שבז״[ט] ר״ל מכפלי א״ב וג׳ הם שוים וכמו כן מה שבח״ט מכפלי ה״ב כמו מה ש[כ]״ל מכפלי ג׳ וה״ב כמו ג׳ ואם כן⟩⟩ [4]כן] ⟨*⟩

[1]ثلاثة] ثلثة

بُرهانُه: أنَّا نَفْصِلُ مِنْ ا ب مثل ج، وهو ب ه. فأَصْغَر مِقْدارَيْ ا ه ه ب قد يُمْكِنُ أَنْ يُضاعَفَ حتَّى يَصيرَ أَعْظَمَ مِنْ د. فَلْيَكُنْ ا ه أصغرَهُما. وَلْيَكُنْ أَضْعافُهُ الزَّائدة على مِقْدارِ د أَضْعاف ز ح. ونَأخُذ لِقَدْرَيْ ه ب ج[1] أَضْعافًا مُتَساوِيَةً لأَضْعافِ ز ح، وهي ح ط ك ل. ونَجْعَلُ م ضِعْفَيْ د. و ن ثلاثة أَضْعافِهِ. ثُمَّ لا نَزالُ نَأْخُذُ أَضْعافَ د على الوَلاءِ إلى أَنْ يَنْتَهِيَ إلى أَوَّلِ أَضْعافٍ لَهُ يكونُ أعظم مِنْ ك ل. فَلْيَكُنْ س الَّتي بعد ن^ أَوَّل أَضْعاف د، الَّتي تكونُ أعظم مِنْ ك ل. فأَضْعاف ز ح لِقَدْرِ ا ه كأَضْعافِ ح ط لِقَدْرِه ب. فأَضْعاف ز ح لِقَدْرِ ا ه كأَضْعافِ ز ط لِقَدْرِ ا ب. وأَضْعاف ز ح لِمِقْدارِ ا ه كأَضْعافِ ك ل لِقَدْرِ ج. فأَضْعاف ز ط و ك ل لِقَدْرَيْ ا ب و ج مُتَساوِيَةٌ. وأيضًا فإنَّ أَضْعافَ ح ط لِقَدْرِ ه ب كأَضْعافِ ك ل لِقَدْرِ ج. و ه ب مثل ج. ف ح ط مثل ك ل. و س أعظم مِنْ ك ل. و ك ل لَيْسَ بأَصْغَرَ مِنْ ن. و ك ل مثل ح ط. ف ح ط لَيْسَ بأَصْغَرَ مِنْ ن. و ز ح أعظم مِنْ د. ف ز ط أعظم مِنْ د و ن جَميعًا. و س مثل د و ن جَميعًا. و ز ط أعظم مِنْ س. فَهُوَ زائِدٌ عَلَيْه. و ك ل ليس بزائد على س. و ز ط و ك ل أَضْعاف مُتَساوِيَة لِقَدْرَيْ ا ب و

1 ه ب ج] ه ب ج #

מופתו: שנבדיל מן א״ב[1] כמו ג׳, והוא ב״ה. הנה היותר קטן משני שעורי א״ה ה״ב— הנה איפשר שיכפל עד שיהיה יותר גדול מן ד׳. ויהיה א״ה היותר קטן משניהם. ויהיה כפלו הנוסף על שעור ד׳ כפל ז״ח. ונקח לשני שעורי ה״ב וג׳ כפלים שוים לכפלי ז״ח נוסף[2], והוא ח״ט וכ״ל[3]. ונשים מ׳ שני כפלי ד׳. ונ׳ שלשה כפליו[4]. אחר כן לא תסור לקחת[5] כפלי ד׳ על הסדר עד שיכלה אל הראשון מכפליו, עד שיהיה יותר גדול מן כ״ל. ויהיה ס׳[6] הוא[7] ראשון מכפלי ד׳ אשר הוא יותר גדול מן כ״ל. הנה כפלי ז״ח לשיעור א״ה ככפלי ח״ט לשעור[8] ה״ב. אם כן כפלי ז״ח לשיעור א״ה ככפלי[9] ז״ט לשיעור א״ב. וכפלי[10] ז״ח לשיעור א״ה ככפלי כ״ל לשיעור ג׳. וכפלי ז״ט וכ״ל לשני שעורי א״ב וג׳ שוה. וגם כן הנה כפלי ח״ט לשעור ה״ב ככפלי כ״ל לשעור ג׳. וה״ב כמו ג׳. אם כן ח״ט כמו כ״ל. וס׳[11] יותר גדול מכ״ל. וכ״ל אינו יותר קטן מנ׳. וכ״ל כמו ח״ט. אם כן ח״ט אינו יותר קטן מן נ׳. וז״ח יותר גדול מן ד׳. אם כן ז״ט יותר גדול מן ד׳ ונ׳ יחד[12]. וס׳ כמו ד׳ ונ׳[13] יחד. אם כן ז״ט יותר גדול מס׳. אם כן הוא נוסף עליו. וכ״ל בלתי נוסף על[14] ס׳. וז״ט וכ״ל כפלים שוים לשעורי א״ב וג׳. וס׳ כפלים[15] לשעור ד׳[16]. אם כן שיעור[17] א״ב יותר גדול היחס אל ד׳ מן ג׳ אל ד׳.

1 א״ב] +א״ג **ס** 2 ז״ח נוסף] ז״ח **ט** נוסף ⟨נ׳ ז״ח⟩ **ק** 3 וכ״ל] ליתא **ב** 4 כפליו] כפלו **בפ** כפלי **וס** 5 לקחת] נקח **בו** ליקח **ט** לקח **פ** 6 ס׳] ס״ה **ס** 7 הוא] והוא **ט** 8 לשעור] +~~ח״ב~~ **ק** 9 ח״ט ... ככפלי] ⟨⟨*⟩⟩ **פ** 10 וכפלי] ושעור **בו** 11 וס׳] וה׳ **ס** 12 יחד] ליתא **ב** 13 ונ׳] ליתא **ב** 14 על] עליו **ב** 15 כפלים] +שוים **בו** 16 ד׳] +מהפתיחה **ב** 17 שיעור] שעורי **ס**

ואומר עוד שד׳ רב היחס אל ג׳ ממה שהוא אל א״ב, כי הנהגתם אחת. וכמו כן יתבאר כי ס׳ מוסיף על כ״ל ואינו מוסיף על ז״ט. וס׳ בו כפלי ד׳. וכ״ל וז״ט כפליהם שוים לג׳ ולא״ב. ויהיה אם כן יחס ד׳ אל ג׳ רב מיחסו אל א״ב.

وأَقولُ أَيضًا إِنَّ د أَعْظَمُ نِسْبَةً إِلى ج مِنْهُ إِلى ا ب. بُرْهانُه: أَنَّ تَدْبيرَهُما واحِدٌ. وَكَذَلِكَ تَبَيَّنَ أَنَّ س زائِدٌ عَلَى كـ ل. وَلَيْسَ بِزائِدٍ عَلَى ز ط. و س فِيهِ أَضْعافُ د. و كـ ل و ز ط أَضْعافُهُما ل ج ول ا ب مُتَساوِيَةٌ. فَنِسْبَةُ د إِلى ج أَعْظَمُ مِنْ نِسْبَتِهِ إِلى ا ب.

וזה מה שרצינו לבאר.

وَذَلِكَ ما أَرَدْنا بَيانَهُ.

جَ. و سَ أَضْعاف ل دَ. فقَدْرُ ا ب أَكْبَر نِسْبَةً إلى دَ مِنْ جَ إلى دَ. وأَقولُ أيْضًا إنَّ دَ أَكْبَر نِسْبَةً إلى جَ مِنْه إلى ا ب، لِأَنَّ تَدْبِيرَهُما واحِدٌ. فتَبَيَّنَ أَنَّ سَ زائِدٌ على كـ ل، ولَيْسَ بِزائد على ز ط. و سَ أَضْعاف دَ. و ز ط و كـ ل أَضْعاف مُتَساوِيَة لِقَدْرَيْ ا ب و جَ. فنِسْبَة دَ إلى جَ أَكْبَر مِنْ نِسْبَتِهِ إلى ا ب.

ואומר גם כן כי ד׳ יותר גדול היחס אל ג׳ מאשר הוא אל א״ב[1], מפני כי הנהגתם אחת. והתבאר כי ס׳ נוסף על כ״ל, ואינו נוסף על ז״ט. וס׳ כפלי ד׳[2]. וכ״ל וז״ט כפלים שוים לשיעורי א״ב וג׳. אם כן יחס ד׳ אל ג׳ יותר גדול מיחסו אל א״ב[3].

وذلك ما اردنا ان نبين.

וזה מה שרצינו לבאר.

[1] א״ב] ב׳ **בו** [2] כפלי ד׳] כפל ד׳ **ב** כפל ⟨ד׳⟩ ד״ט **ו** [3] א״ב] ב׳+ **ב** ב׳ ⟨*⟩ **ו** ב׳ **ט**

[Proposition v.9]

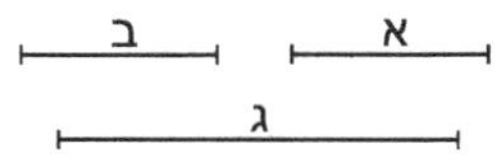

השיעורים אשר יחסם אל שעור אחר—אחד, הם שוים. ואם יחס שיעור אל שיעורים אחרים אחד[1], השעורים הם שוים.

الأَقْدارُ الَّتي تَكونُ نِسْبَتُها إِلى قَدْرٍ آخَرَ—واحِدَةً، فَهِيَ مُتَساوِيَةٌ. وَإِنْ كانَ قَدْرٌ نِسْبَتُهُ إِلى أَقْدارٍ أُخَرَ واحِدَةً، فَالأَقْدارُ مُتَساوِيَةٌ.

המשל בזה: ששעור א׳ ושעור ב׳ יחסם אל שעור ג׳—אחד. ואומר כי א׳ שוה לב׳.

مِثالُهُ: أَنَّ قَدْرَ آ وَقَدْرَ بَ نِسْبَتُهُما إِلى قَدْرِ جَ واحِدَةٌ. فَأَقولُ إِنَّ آ مُساوٍ لِ بَ.

---

[1] אחד] ⟨⟨*⟩⟩

الأَقْدَارُ الَّتِي نِسْبَتُهَا إلى قَدْرٍ آخَر واحِدَةٌ، فهِيَ مُتَساوِيَةٌ. وإذا كانَ قَدْرُ نِسْبَتِهِ إلى أَقْدارٍ أُخرى واحِدَة، فَالأَقْدارُ مُتَساوِيَةٌ.

השעורים אשר יחסם אל שעור אחר—אחד, הם[1] שוים. ואם היה שעור יחסו אל שעורים אחרים[2]—אחד, הנה[3] השעורים שוים.

مِثالُهُ: أَنَّ قَدْرَيْ اَ بَ نِسْبَتُهُما إلى قَدْرِ جَ واحِدَةٌ. فَأَقولُ إنَّ اَ مِثْل بَ.

המשל בו: כי שני שעורי א׳ ב׳ יחסם אל שעור ג׳—אחד. הנה אומר[4] כי[5] א׳ כמו ב׳.

---

[1]הם] הנה+ **ק** [2]אחרים] ליתא **בוס** [3]אחד הנה] ~~הנה~~ אחד 〈〈הנה〉〉 **פ** [4]אומר] +~~בו~~ **פ** [5]כי] +~~שני שעורי א״ב~~ **פ**

כי אם לא היה שוה לו, הוא גדול או קטן ממנו. ואי אפשר להיות א׳ גדול מב׳, שאם היה גדול ממנו היה רב היחס אל ג׳ מב׳ אל ג׳. ואין יחסו אל ג׳ גדול. אם כן אין א׳ גדול מב׳. וכן אי אפשר שיהיה א׳ קטן מב׳. שאם היה קטן ממנו, היה יחסו אל ג׳ קטן מיחס ב׳ אל ג׳. ואיננו כן. אם כן א׳ איננו גדול מב׳ ולא קטן ממנו. אמנם הוא שוה לו. ועוד, יחס ג׳ אל א׳ ואל ב׳—אחד. ואומר שא׳ שוה אל ב׳. מופת זה: שאם איננו שוה, הוא גדול או קטן ממנו. ואין א׳ גדול מב׳. שאם היה גדול ממנו, היה יחס ג׳ אליו קטן מיחסו אל ב׳, ואיננו כן. ואין א׳ גם כן קטן מב׳. שאם היה קטן ממנו, היה יחס ג׳ אליו גדול מיחסו אל ב׳, ואיננו כן. אם כן אי אפשר להיות א׳ גדול או קטן מב׳. אמנם הוא שוה לו.

וזה[1] מה שרצינו לבאר.

بُرْهانُهُ: إِنْ لَمْ يَكُنْ كَذَلِكَ مُساوِيًا لَهُ، فَهُوَ أَكْبَرُ مِنْهُ أَوْ أَصْغَرُ مِنْهُ. فَلَيْسَ ا بِأَكْبَرَ مِنْ ب، لِأَنَّهُ لَوْ كانَ أَكْبَرَ مِنْهُ، لَكانَ أَعْظَمَ نِسْبَةً إِلى ج مِنْ ب إِلى ج. وَلَيْسَ نِسْبَتُهُ إِلى ج بِأَعْظَمَ. وَلَيْسَ ا أَيْضًا بِأَصْغَرَ مِنْ ب، لِأَنَّهُ لَوْ كانَ أَصْغَرَ، لَكانَتْ نِسْبَتُهُ إِلى ج أَصْغَرَ مِنْ نِسْبَةِ ب إِلى ج. وَلَيْسَتْ كَذَلِكَ. فَلَيْسَ ا بِأَكْبَرَ وَأَصْغَرَ مِنْ ب، وَلَكِنَّهُ مُساوٍ لَهُ. وَأَيْضًا نِسْبَةُ ج إِلى ا و ب واحِدَةٌ. فَأَقُولُ إِنَّ ا مُساوٍ لِ ب. بُرْهانُهُ: إِنْ لَمْ يَكُنْ مُساوِيًا لَهُ، فَهُوَ أَكْبَرُ أَوْ أَصْغَرُ مِنْهُ. فَلَيْسَ ا بِأَعْظَمَ مِنْ ب، لِأَنَّهُ لَوْ كانَ أَعْظَمَ مِنْهُ لَكانَتْ نِسْبَةُ ج إِلَيْهِ أَصْغَرَ مِنْها إِلى ب. وَلَيْسَتْ كَذَلِكَ. وَلَيْسَ ا أَيْضًا بِأَصْغَرَ مِنْ ب، لِأَنَّهُ لَوْ كانَ أَصْغَرَ مِنْهُ لَكانَتْ نِسْبَةُ ج إِلَيْهِ أَعْظَمَ. وَلَيْسَتْ كَذَلِكَ. فَلَيْسَ ا بِأَعْظَمَ وَلا بِأَصْغَرَ مِنْ ب. ا مُساوٍ لِ ب.

وَذَلِكَ ما أَرَدْنا بَيانَهُ.

[1]וזה] ⟨ו⟩זה

فَإِنْ لم يَكُنْ مِثْلَهُ، فَهُوَ أعظم مِنْهُ أَو أَصْغَر. وَلَوْ كانَ أعظم، لكانَتْ نِسْبَتُهُ إلى ج أَكْبَر. ولو كانَ أَصْغَر، لكانَتْ نِسْبَتُهُ إلى ج أَصْغَر، وَلَيْسَتْ كَذَلِكَ. فَلَيْسَ آ بأَعْظَمَ وَلا^ أَصْغَرَ مِنْ ب، وَلٰكِنَّهُ مُساوٍ لَهُ. وَأَيْضًا فإِنَّ نِسْبَةَ ج إلى آ و ب واحِدَة. فَأَقُولُ إنَّ آ مِثْل ب. فَإِنْ لم يَكُنْ مِثْلَهُ فَهُوَ أَعْظَم أَو أَصْغَر مِنْهُ. وَلَوْ كانَ أَعْظَم لَكانَتْ نِسْبَةُ ج إِلَيْهِ أَصْغَر. وَلَوْ كانَ أَصْغَر لَكانَتْ نِسْبَةُ ج إِلَيْهِ أَعْظَم. فَلَيْسَ آ بأَعْظَمَ وَلا أَصْغَرَ مِنْ ب، وَلٰكِنَّهُ مُساوٍ لَهُ.

وذَلِكَ ما أَرَدْنا أَنْ نُبَيِّنَ.

שאם לא יהיה כמוהו, הנה הוא יותר גדול ממנו או יותר קטן. ואלו היה יותר גדול[1], יהיה יחסו אל ג׳ יותר גדול. ואלו היה[2] יותר קטן, היה יחסו אל ג׳ יותר קטן, ואינו כן. אם כן[3] אין א׳[4] יותר גדול ויותר קטן מן ב׳, אבל הוא שוה לו. וגם כן הנה יחס ג׳ אל א׳ וב׳—אחד. הנה אומר כי א׳ כמו ב׳. שאם לא יהיה כמוהו, הנה הוא יותר גדול או יותר קטן ממנו. ואלו היה יותר גדול, היה יחס ג׳ אליו יותר קטן. ואלו היה יותר קטן, היה יחס ג׳ אליו יותר גדול[5]. אם כן אין א׳ יותר גדול ולא יותר[6] קטן מן ב׳, אבל הוא שוה לו.

וזה מה שרצינו לבאר.

---

[1]גדול] ⟨*⟩ **ב** [2]היה] היתה **ט** [3]כן] ליתא **ב** [4]אין א׳] אינו **ו** [5]גדול] ~~קטן~~+ **פ** [6]ולא יותר] ויותר **ט**

[Proposition V.10]

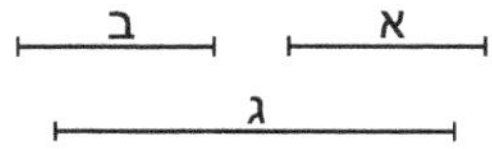

أَعْظَمُ الأَقْدارِ نِسْبَتُهُ إِلى قَدْرٍ آخَرَ هُوَ أَكْبَرُها. وَالَّذي نِسْبَةُ القَدْرِ إِلَيْهِ أَعْظَمُ هُوَ أَصْغَرُ الأَقْدارِ.

مِثالُهُ: أَنَّ قَدْرَ ا أَعْظَمُ نِسْبَةً إِلى قَدْرِ ج مِنْ قَدْرِ ب. فَأَقولُ إِنَّ ا أَكْبَرُ مِنْ ب.

רב היחס מן השעורים אל שעור אחר הוא הגדול מהם. ואשר יחס השעור אליו רב הוא הקטן מהם.

המשל בזה: ששעור א׳ רב היחס אל שעור ג׳ משעור ב׳. ואומר כי א׳ גדול מב׳.

أَكْبَرُ الأَقْدارِ نِسْبَةً إلى قَدْرٍ آخَر فَهُوَ أَعْظَمُهَا. وَالَّذي نِسْبَةُ القَدْرِ إِلَيْهِ أَكْبَرُ فَهُوَ أَصْغَرُها.

مِثَالُهُ: أَنَّ ا أَكْبَرَ نِسْبَةً إلى جَ مِنْ بَ. فَأَقولُ إِنَّ اَ أَعْظَمَ مِنْ بَ.

גדול השעורים יחס[1] אל שעור הוא היותר[2] גדול מהם. ואשר יחס השעור אליו יותר גדול הוא היותר קטן מהם.

המשל בו: כי א׳ יותר[3] גדול יחסו אל ג׳ מן ב׳[4]. הנה אומר כי א׳ יותר גדול[5] מן ב׳.

שרטוט—**פ**: א׳ משמאל לב׳ [1]גדול ... יחס] היותר גדול מהשעורים יחסו **ב** [2]היותר] ⟨ה⟩יותר **ק** [3]כי ... יותר] ליתא **ב** [4]מן ב׳] ליתא **בז** [5]גדול] +יחסו **ז**

שאם לא היה גדול ממנו, יהיה כמהו או קטן ממנו. ואין א׳ כמו ב׳. שאם היה כמהו, היה יחסם אל ג׳ אחד. ואיננו אחד. אם כן אין א׳ כמו ב׳. וגם אין א׳ קטן מב׳. שאם היה קטן ממנו, היה יחסו אל ג׳ קטן מיחס ב׳ אל ג׳. ואיננו כן. כבר התבאר שאין א׳ שוה לב׳ ולא קטן ממנו. אם כן הוא גדול מב׳. ועוד נשים יחס ג׳ אל ב׳ רב מיחסו אל א׳. ואומר כי א׳ גדול מב׳. שאם לא היה גדול ממנו, הוא שוה לו או קטן ממנו. ואין א׳ שוה לב׳. שאם היה שוה לו, היה יחס ג׳ אליהם אחד. ואיננו כן. ואין א׳ גם כן קטן מג. שאם היה קטן ממנו, היה יחס ג׳ אליו רב (ממה)[1] שהוא אל ב׳. ואיננו רב ממנו. אם כן אין א׳ קטן מב׳ ולא שוה לו. אמנם א׳ גדול מב׳.

بُرْهانُهُ: إِنْ لَمْ يَكُنْ أَكْبَرَ مِنْهُ، فَهُوَ مِثْلُهُ أَوْ أَصْغَرَ مِنْهُ. فَلَيْسَ ا مِثْلَ ب، لِأَنَّهُ لَوْ كانَ مِثْلَهُ لَكانَتْ نِسْبَتُهُما إِلى ج واحِدَةً. وَلَيْسَتْ واحِدَةً. فَلَيْسَ ا مُساوِيًا لِ ب. وَلَيْسَ ا أَيْضًا بِأَصْغَرَ مِنْ ب، لِأَنَّهُ لَوْ كانَ أَصْغَرَ مِنْهُ لَكانَتْ نِسْبَتُهُ إِلى ج أَصْغَرَ مِنْ نِسْبَةِ ب إِلى ج. وَلَيْسَتْ بِأَصْغَرَ. فَقَدِ اسْتَبانَ أَنَّ ا لَيْسَ مِثْلَ ب وَلا أَصْغَرَ مِنْهُ. ا أَعْظَمُ مِنْهُ. وَأَيْضًا نَجْعَلُ نِسْبَةَ ج إِلى ب أَعْظَمَ مِنْ نِسْبَتِهِ إِلى ا. فَأَقُولُ إِنَّ ا أَكْبَرُ مِنْ ب. بُرْهانُهُ: إِنْ لَمْ يَكُنْ أَكْبَرَ مِنْهُ، فَهُوَ مِثْلُهُ أَوْ أَصْغَرُ مِنْهُ. وَلَيْسَ ا بِمِثْلِ ب، لِأَنَّهُ لَوْ كانَ مِثْلَهُ لَكانَتْ نِسْبَةُ ج إِلَيْهِما واحِدَةً. وَلَيْسَتْ بِواحِدَةٍ. وَلَيْسَ ا أَيْضًا بِأَصْغَرَ مِنْ ب، لِأَنَّهُ لَوْ كانَ أَصْغَرَ مِنْهُ لَكانَتْ نِسْبَةُ ج إِلَيْهِ أَعْظَمَ مِنْها إِلى ب. وَلَيْسَتْ بِأَصْغَرَ. فَلَيْسَ ا بِمِثْلِ ب وَلا أَصْغَرَ مِنْهُ. ف ا أَكْبَرُ مِنْ ب.

וזה מה שרצינו לבאר. ר״ל[2] המופת הוא שאם דבר החולק יצא שקר הנגזר תחלה הוא אמת.

وَذلِكَ ما أَرَدْنا بَيانَهُ.

---

[1]ממה] ממ[נה]; המילה תוקנה לא בבירור
[2]ר״ל] ⟨⟨*⟩⟩

فَإِنْ لَمْ يَكُنْ أعظم مِنْهُ، فَهُوَ مثله أَو أَصْغَر مِنْهُ. وَلَوْ كانَ مِثْلهُ، لكانَتْ نِسْبَتُهما إلى ج واحِدَة. وَلَوْ كانَ أَصْغَر مِنْه، لكانَتْ نِسْبَتهُ إلى ج أَصْغَر. ولَيْسَ كَذٰلِكَ. فَقَدِ اسْتَبانَ أَنَّ ا لَيْسَ مثل ب ولا أَصْغَر مِنْهُ. فَهُوَ إِذَنْ[1] أَعظَم مِنْه. وأيضًا فإِنَّ نِسْبَةَ ج إلى ب أَكْبَر مِنْ نِسْبَتِهِ إلى ا. فَأَقولُ إِنَّ ا أَعظم مِنْ ب. فإِنْ لم يَكُنْ أَعْظَم مِنْهُ، فَهُوَ مِثْلُه أَو أَصْغَر مِنْهُ. وَلَوْ كانَ مثله، لَكانَتْ نِسْبَةُ ج إِلَيْه وَإلى ب واحِدَةً. ولَوْ كانَ أَصْغَر مِنْهُ، لَكانَتْ نِسْبَةُ ج إِلَيْه أَكْبَر. وَلَيْسَتْ كَذٰلِكَ. فَلَيْسَ ا مثل ب وَلا أَصْغَر مِنْهُ، لَكِنَّهُ أَعْظَم مِنْهُ.

وذٰلِكَ ما أَرَدْنا أَنْ نُبَيِّنَ.

[1] اذن] M اذا

שאם לא יהיה יותר גדול ממנו, הנה הוא כמוהו או[1] קטן ממנו. ואלו היה כמוהו, היה יחס[2] שניהם אל ג׳ אחד[3]. ואלו היה יותר קטן ממנו, היה יחסו אל ג׳ יותר קטן. ואינו כן. הנה[4] כבר התבאר כי א׳[5] אינו כמו ב׳ ולא קטן ממנו. הנה הוא אם כן[6] יותר גדול ממנו. וגם כן הנה יחס ג׳ אל ב׳ יותר גדול מיחסו אל א׳. הנה אומר כי א׳ יותר גדול מן ב׳. שאם לא יהיה יותר גדול ממנו, הנה הוא כמוהו או יותר קטן ממנו. ואלו היה כמוהו היה יחס ג׳ אליו ואל ב׳ אחד. ואלו היה יותר קטן ממנו, היה יחס ג׳ אליו יותר גדול. ואינו כן. אם כן[7] אין א׳[8] כמו ב׳ ולא קטן ממנו, אבל הוא יותר גדול ממנו.

וזה מה שרצינו לבאר.

[1] או] +יותר **ז** [2] יחס] יחסם **ב** [3] אחד] ~~אחד~~+ **ב** [4] הנה] ליתא **ט** [5] א׳] +יותר גדול **ב** [6] הנה ... כן] אם כן הוא **ז** אם כן הנה הוא **ק** [7] אם כן] ליתא **ב** [8] אין א׳] אינו **ז**

[Proposition V.11]

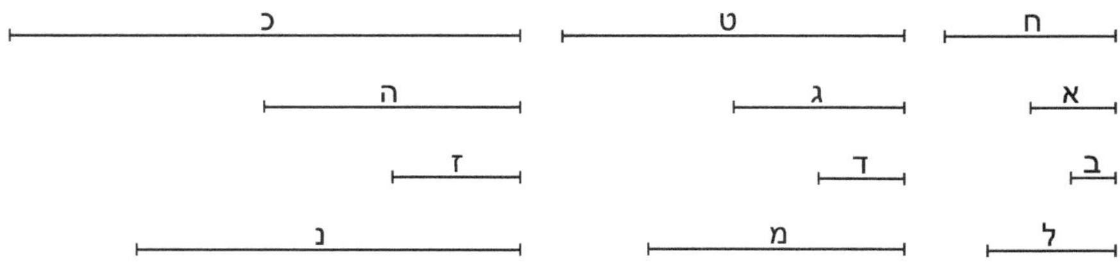

השיעורים אשר יחסם שוה ליחס אחד יחסם גם כן שוה.

الأَقْدارُ الَّتي نِسْبَتُها مُساوِيَةٌ لِنِسْبَتِهِ فَإِنَّ نِسَبها مُتَساوِيَةٌ.

המשל בזה: שיחס א׳ אל ב׳ כיחס ג׳ אל ד׳, ויחס ג׳ אל ד׳ כיחס ה׳ אל ז׳. ואומר שיחס א׳ אל ב׳ כיחס ה׳ אל ז׳.

مِثالُهُ: أَنَّ نِسْبَةَ آ إِلى بَ كَنِسْبَةِ جَ إِلى دَ. وَنِسْبَةُ جَ إِلى دَ كَنِسْبَةِ هَ إِلى زَ. فَأَقولُ إِنَّ نِسْبَةَ آ إِلى بَ كَنِسْبَةِ هَ إِلى زَ.

שרטוט—B: د ← ز

الأَقْدارُ الَّتي نِسْبَتُها مُسَاوِيَةٌ لِنِسْبَةٍ واحِدَةٍ[1] فإنَّ نِسْبَتَها مُتَساوِيَةٌ.

مِثالُهُ: أَنَّ نِسْبَةَ آ إلى بَ كَنِسْبَةِ جَ إلى دَ. ونِسْبَةَ هَ إلى زَ كَنِسْبَةِ جَ إلى دَ. فأَقولُ إنَّ نِسْبَةَ آ إلى بَ كَنِسْبَةِ هَ إلى زَ.

השעורים אשר יחסם שוים ליחס אחד הנה יחסם[1] שוים.

המשל בו: כי יחס א׳ אל ב׳ כיחס ג׳ אל ד׳. ויחס ה׳ אל ז׳ כיחס ג׳ אל ד׳. הנה אומר כי יחס א׳[2] אל ב׳ כיחס ה׳[3] אל ז׳.

[1] واحدة] Q

[1] יחסם] ליתא ב   [2] א׳] ד ⟨*⟩ ט   [3] ה׳] ד׳ פ

מופת זה: שנקח לא׳ ולג׳ ולה׳ כפלים שוים והם ח׳ וט׳ וכ׳. ונקח לב׳ ולד׳ ולז׳ כפלים שוים והם ל׳[1] ומ׳ ונ׳. ושיעור א׳ אצל ב׳ כשעור ג׳ אצל ד׳. וכבר לוקח לא׳ ולג׳ כפלים שוים, והם ח׳ וט׳. ולב׳ ולד׳ כפלים שוים, והם ל׳ ומ׳. ואם כן שעור תוספת ח׳ על ל׳ כשעור תוספת ט׳ על מ׳. ואם היה ח׳ שוה לל׳, יהיה ט׳ שוה למ׳. ואם ח׳ חסר מל׳, יהיה ט׳ חסר ממ׳. וכמו כן שעור ג׳ אל ד׳ כשעור ה׳ אצל ז׳. וכבר לוקח לג׳ ולה׳ כפלים שוים והם ט׳ וכ׳. ולד׳ ולז׳ כפלים שוים, והם מ׳ ונ׳. ואם כן שעור תוספת ט׳ על מ׳ כשעור תוספת כ׳ על נ׳. ואם היה ט׳ שוה למ׳, יהיה כ׳ שוה לנ׳. ואם ט׳ חסר ממ׳, יהיה כ׳ חסר מנ׳ בזה השעור. יהיה אם כן שיעור תוספת ח׳ על ל׳ כשיעור תוספת כ׳ על נ׳. ואם היה ח׳ שוה לל׳, יהיה כ׳ שוה לנ׳. ואם ח׳ חסר מל׳, גם כ׳ חסר מנ׳ בזה החסרון. וח׳ וכ׳ כפלים שוים לא׳ ולה׳. ול׳ ונ׳ כפלים שוים לב׳ ולז׳. אם כן שיעור א׳ אצל ב׳ כשיעור ה׳ אצל ז׳.

וזה מה שרצינו לבאר.

بُرْهانُهُ: أَنّا نَأْخُذُ ا و ج و ه أَضْعافًا مُتَساوِيَةً، وَهِيَ ح و ط و ك. وَنَأْخُذُ ل ب و د و ز أَضْعافًا مُتَساوِيَةً، وَهِيَ ل و م ن. فَقَدْرُ ا عِنْدَ ب كَقَدْرِ ج عِنْدَ د. (وَقَدْ أُخِذَ)[1] ا و ج أَضْعافٌ مُتَساوِيَةٌ، وَهِيَ ح ط. ولل ب و د أَضْعافٌ مُتَساوِيَةٌ، وَهِيَ ل و م. فَقَدْرُ زِيادَةِ ح عَلَى ل كَقَدْرِ زِيادَةِ ط عَلَى م. وَإِنْ كانَ ح مُساوِيًا لل ط مُساوٍ ل م. فَإِنْ كانَ ح أَنْقَص مِنْ ل، فَبِقَدْرِ ذَلِكَ يَكونُ ط أَنْقَص مِنْ م. وَأَيضًا قَدْرُ ج عِنْدَ د كَقَدْرِ ه عِنْدَ ز. وَقَد أُخِذَ ل ج و ه أَضْعافٌ مُتَساوِيَةٌ [..] وَهِيَ م و ن. فَقَدْرُ زِيادَةِ ط عَلَى م كَقَدْرِ زِيادَةِ ك عَلَى ن. وَإِنْ كانَ ط مُساوِيًا ل م. فَإِنَّ (ك)[2] مُساوِيًا ل ن. وَإِنْ كانَ ط أَنْقَص مِنْ م، فَبِقَدْرِ ذَلِكَ يَكونُ ك أَنْقَص مِنْ ن. وَقَدْرِ زِيادَةِ ح عَلَى ل كَقَدْرِ زِيادَةِ ك عَلَى ن. وَإِنْ كانَ مُساوِيًا ل ل، فَإِنَّ ك مُساوٍ ل ن. وَإِنْ كانَ ح أَنْقَص مِنْ ل، فَبِقَدْرِ ذَلِكَ يَكونُ ك أَنْقَص مِنْ ن. و ح و ك هُما أَضْعافٌ مُتَساوِيَةٌ ا و ه. و ل و ن أَضْعافٌ مُتَساوِيَةٌ ل ب و ز فَقَدْرُ ا عِنْدَ (ب)[3] كَقَدْرِ ه عِنْدَ ز.

وَذَلِكَ ما أَرَدْنا بَيانَهُ.

---

[1] ל׳] ~~ח׳ וט׳ וכ׳~~ +

---

[1] وقد أخذ] وقدر اخر [2] ك] كان [3] ب] ליתא

بُرهانُه: أَنَّا نَأْخُذُ الأَقْدارَ اَ جَ هَ أَضْعافًا مُتَساوِيَةً، وهي حَ طَ كَ، وَلأَقْدارِ بَ دَ زَ أَضْعافًا مُتَساوِيَةً، وهي لَ مَ نَ. فَنِسْبَةُ اَ إلى بَ كَنِسْبَةِ جَ إلى دَ. وَقَدْ أُخِذَ لِقَدْرَيْ اَ جَ أَضْعافٌ مُتَساوِيَةٌ، وهي حَ طَ، وَلِقَدْرَيْ بَ دَ أَضْعاف مُتَساوِيَة، وهي لَ مَ. فَقَدْرا حَ طَ إِمَّا زائِدانِ مَعًا عَلى قَدْرَيّ لَ مَ، وَإمّا مُساوِيانِ مَعًا لَهُما، وَإِمّا ناقِصانِ مَعًا مِنْهُما. وَأَيْضًا فإِنَّ نِسْبَةَ جَ إلى دَ كَنِسْبَةِ هَ إلى زَ. وَقَدْ أُخِذَ لِقَدْرَيْ جَ هَ أَضْعافٌ مُتَساوِيَةٌ، وَهِيَ طَ كَ، وَلِقَدْرَيْ دَ زَ أَضْعافٌ مُتَساوِيَةٌ، وَهِيَ مَ نَ. فَقَدْرا طَ كَ إِمّا زائِدانِ مَعًا عَلى قَدْرَيْ مَ نَ، وَإمّا مُساوِيَانِ^ لَهُما، وَإِمّا ناقِصانِ مَعًا مِنْها[1]. وحَ كَ أَضْعافٌ مُتَساوِيَةٌ لِقَدْرَيْ اَ هَ. ولَ نَ أَضْعافٌ مُتَساوِيَةٌ لِقَدْرَيْ بَ زَ. فَنِسْبَةُ اَ إلى بَ كَنِسْبَةِ هَ إلى زَ.

מופתו: אנחנו[1] נקח לשעורי א׳ ג׳ ה׳ כפלים שוים, והם ח׳ ט׳ כ׳, ולשיעורי ב׳ ד׳ ז׳ כפלים שוים, והם ל׳ מ׳ נ׳. הנה יחס[2] א׳ אל ב׳ כיחס ג׳ אל ד׳. וכבר לוקח לשיעורי א׳ ג׳ כפלים שוים, והם ח׳ ט׳, ולשיעורי ב׳ ד׳ כפלים שוים, והם ל׳ מ׳. הנה שני שעורי ח׳ ט׳ אם נוספים יחד על שני שעורי ל׳ מ׳, ואם שוים לשניהם יחד, ואם חסרים משניהם יחד. וגם כן הנה יחס ג׳ אל ד׳ כיחס ה׳ אל ז׳. וכבר נלקח לשני שעורי ג׳ ה׳ כפלים שוים, והם ט׳ כ׳, ולשני שעורי ד׳ ז׳ כפלים שוים, והם מ׳ נ׳. אם כן שני שעורי ט׳ כ׳ אם נוספים יחד על שני שעורי מ׳ נ׳, ואם שוים יחד לשניהם, ואם חסרים יחד משניהם. וח׳ כ׳ כפלים שוים לשני שעורי א׳ ה׳. ול׳ נ׳[3] כפלים שוים לשני שעורי ב׳ ז׳. אם כן יחס[4] א׳ אל ב׳ כיחס ה׳ אל ז׳.

وذَلِكَ ما أَرَدْنا أَنْ نُبَيِّنَ.

וזה מה שרצינו לבאר.

[1]منها] CopMT1 +وقدرا حَ كَ إما زائدان معا على قدري لَ نَ وإما مساويان لهما وإما ناقصان معا منهما

[1]אנחנו] +~~נשים~~ ט [2]יחס] +~~א״נ~~ פ [3]ול׳ נ׳] הו״ל נ׳ ב [4]יחס] +אל פ

[Proposition V.12]

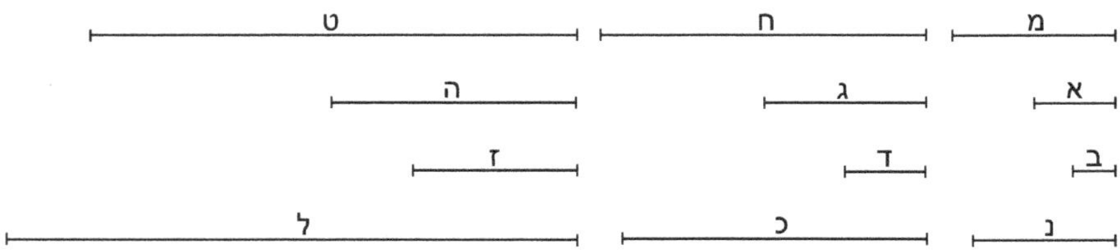

כשיהיה יחס הראשון אל השני כיחס השלישי אל הרביעי, ויחס השלישי אל הרביעי רב מיחס החמשי אל השישי, יהיה יחס הראשון אל השני גדול מיחס החמשי אל השישי.

إذا كانَتْ نِسْبَةُ الأَوَّلِ إلى الثَّاني كَنِسْبَةِ الثَّالِثِ إلى الرَّابِعِ، وَنِسْبَةُ الثَّالِثِ إلى الرَّابِعِ أَعْظَمُ مِنْ نِسْبَةِ الخامِسِ إلى السَّادِسِ، فَإِنَّ نِسْبَةَ الأَوَّلِ إلى الثَّاني أَعْظَمُ مِنْ نِسْبَةِ الخامِسِ إلى السَّادِسِ.

המשל בזה: שיחס א׳ אל ב׳ כיחס ג׳ אל ד׳. ויחס ג׳ אל ד׳ רב מיחס ה׳ אל ז׳. ואומר שיחס א׳ אל ב׳ רב מיחס ה׳ אל ז׳.

مِثالُهُ: أَنَّ نِسْبَةَ ا إلى ب كَنِسْبَةِ ج إلى د. وَنِسْبة ج إلى د أَعْظَمُ مِنْ نِسْبَةِ ه إلى ز. فَأَقولُ إِنَّ نِسْبَةَ ا إلى ب أَعْظَمُ مِنْ نِسْبَةِ ه إلى ز.

إذا كانَتْ نِسْبَةُ الأَوَّلِ إلى الثَّاني كنِسْبَةِ الثَّالِثِ إلى الرَّابِعِ ونِسْبَةُ الثَّالِثِ إلى الرَّابِعِ أَكْبَرَ مِنْ نِسْبَةِ الخَامِسِ إلى السَّادِسِ فَإِنَّ نِسْبَةَ الأَوَّلِ إلى الثَّاني أَكْبَرُ مِنْ نِسْبَةِ الخَامِسِ إلى السَّادِسِ.

مِثَالُهُ: أَنَّ نِسْبَةَ اَ إلى بَ كنِسْبَةِ جَ إلى دَ. ونِسْبَةَ جَ إلى دَ أَكْبَرَ مِنْ نِسْبَةِ هَ إلى زَ. فأَقُولُ إنَّ نِسْبَةَ اَ إلى بَ أَكْبَرُ مِنْ نِسْبَةِ هَ إلى زَ.

כאשר היה יחס הראשון אל השני כיחס השלישי אל הרביעי, ויחס השלישי[1] אל הרביעי יותר גדול מיחס החמישי אל[2] הששי, הנה יחס הראשון אל השני יותר גדול מיחס החמישי אל הששי.

המשל בו: כי יחס א׳ אל[3] ב׳ כיחס ג׳ אל ד׳. ויחס ג׳ אל ד׳[4] יותר גדול[5] מיחס ה׳ אל ז׳. הנה אומר כי יחס[6] א׳ אל ב׳ יותר גדול מיחס ה׳ אל ז׳.

[1] אל ... השלישי] ליתא **ס** [2] אל] +~~השני~~ **ב** על **ז** [3] כי ... אל] כיחס ח׳ **ב** [4] ג׳ ... ד׳] ~~ג׳ אל ד׳ ויחס~~ **ב** [5] גדול] ⟨ה⟩גדול **ב** [6] כי יחס] כיחס **ב**

מופת זה: כי יחס ג׳ אל ד׳ רב מיחס ה׳ אל ז׳. ונקח לג׳ ולה׳ כפלים שוים, ולד׳ ולז׳ כפלים שוים אחרים. ויהיו כפלי ג׳ עודפים על כפלי ד׳. וכפלי ה׳ אינם עודפים על כפלי ז׳. וכפלי ג׳ וה׳ השוים הם ח׳ וט׳. וכפלי ד׳ וז׳ האחרים השוים הם כ׳ ול׳. וח׳ יעדיף על כ׳. וט׳ לא יעדיף על ל׳. ונשים כפלי מ׳ לא׳ ככפלי ח׳ לג׳. וכפלי נ׳ לב׳ ככפלי כ׳ לד׳. ושעור א׳ אצל שעור ב׳ כשעור ג׳ אצל ד׳. וכבר לוקח לא׳ ולג׳ כפלים שוים, והם מ׳ וח׳. ולוקח לב׳ ולד׳ כפלים שוים, והם נ׳ וכ׳—יהיה שיעור תוספת מ׳ על נ׳ כשעור תוספת ח׳ על כ׳. ואם היה מ׳ שוה לנ׳, יהיה ח׳ שוה לכ׳. ואם מ׳ חסר מנ׳, גם ח׳ חסר מכ׳ בשעור זה. וכבר התבאר כי ח׳ יעדף על כ׳. וט׳ לא יעדף על ל׳. ומ׳ כמו כן יעדף על נ׳. וט׳ לא יעדף על ל׳. וט׳ ומ׳ הם כפלים שוים לא׳ ולה׳. ונ׳ ול׳ הם כפלים שוים לב׳ ולז׳. אם כן א׳ רב היחס אל ב׳ מיחס ה׳ אל ז׳.

بُرْهانُه: أَنَّ نِسْبَةَ ج إِلى د أَعْظَمُ مِنْ نِسْبَةِ ه إِلى ز. ونأخُذُ ل ج و ه أَضْعافًا مُتَساوِيَةً، وَهِيَ ح و ط. و د و ز أَضْعافًا مُتَساوِيَةً أُخَر، وَهِيَ ك و ل. وَأَضْعافُ ج تَزيدُ عَلى أَضْعافِ د. وأَضْعافُ ه لَيْسَتْ تَزيدُ على أَضْعافِ ز. ح يَزيدُ على ك. و ط لَيْسَ يَزيدُ عَلى ل. ونَجْعَلُ أَضْعافَ (م)[1] (ل)[2] ا كَأَضْعافِ ح ل ج. وَأَضْعافُ ن ل ب كأَضْعافِ ك ل د. فَقَدْرُ ا عِنْدَ ب كَقَدْرِ ج عِنْدَ د. وَقَدْ أُخِذَ ل ا و ج أَضْعافٌ مُتَساوِيَةٌ، وَهِيَ م وح. وَأُخِذَ ل د و ب أَضْعافٌ مُتَساوِيَةٌ، وَهِيَ ن وك. فَقَدْ[رُ] زِيادَةِ م عَلى ن كَقَدْرِ زِيادَةِ ح على ك. وَإِنْ كانَ م مُساوِيًا ل ن، فَإِنَّ ح مُساوٍ ل ك. وَإِنْ كانَ م ناقِصًا مِنْ ن، فَبِقَدْرِ ذَلِكَ يَكونُ ح ناقِصًا مِنْ ك. وَقَدِ اسْتَبانَ أَنَّ ح يَزيدُ عَلى ك. وَأَنَّ ط لا تَزيدُ عَلى ل. وم و ط هُما أَضْعافٌ مُتَساوِيَةٌ ل ا و ه. و ن و ل وَهُما أَضْعافٌ مُتَساوِيَةٌ ل ب و ز. و ا أَعْظَمُ نِسْبَةً إِلى ب مِنْ ه إِلى ز.

וזה מה שרצינו לבאר.

وَذَلِكَ ما أَرَدْنا بَيانَه.

הגהה: ר״ל[1] אחר שתוציא להם כפלים שוים, ותמצא כפלי[2] א׳ וב׳ יעדיפו זה על זה. וכפלי ה׳ וז׳ לא יעדפו זה על זה. אם כן א׳ רב היחס אל ב׳ מיחס ה׳ אל ז׳, כפי מה שאמרנו בפתיחת השע[ר] הזה[3].

---

[1] ר״ל] ⟨⟨*⟩⟩ [2] כפלי] ⟨*⟩ [3] הזה] ⟨⟨*⟩⟩

[1] م] ج [2] ل] لا

بُرهانُه: أَنَّ نِسْبَةَ ج إلى د أَكْبَرُ مِنْ نِسْبَةِ ه إلى ز. فَقَدْ^ يكونُ ل ج و ه أَضْعاف مُتَساوِيَة، ول د و ز أَضْعاف مُتَساوِيَة. ويكونُ أَضْعاف ج زائِدَة على أَضْعافِ د. وأَضْعاف ه غير زائِدَة على أَضْعافِ ز. فَلْيَكُنْ أَضْعاف ج و ه المُتَساوِيَة وأَضْعاف د و ز المُتَساوِيَة الَّتي هَذِهِ حالُها إِمَّا أَضْعاف ج ه ف ح ط، وإِمّا أَضْعاف د ز ف ك ل. وَلْيَكُنْ ما في م مِنْ أَضْعاف ا مثل ما في ح مِنْ أَضْعاف ج. وما في ن مِنْ أَضْعاف ب مثل ما في ك مِنْ أَضْعاف د. فنِسْبَةَ ا إلى ب كنِسْبَةِ ج إلى د. وقَدْ أُخِذَ لِقَدْرَيْ ا ج أَضْعافٌ مُتَساوِيَةٌ، وهي م ح، ولِقَدْرَيْ ب د أَضْعاف مُتَساوِيَة، وهي ن ك. فَقَدْرَيْ م ح إِمّا زائِدانِ مَعًا عَلى قَدْرَيْ ن ك، وإِمّا مُساوِيانِ مَعًا لَهُما، وإِمّا ناقِصانِ مَعًا مِنْهُما. ولكنَّ^ ح يَزِيدُ على ك، ف م يَزِيدُ على ن. و ط لَيْسَ يَزِيدُ على ل. ف م ط أَضْعاف مُتَساوِيَة لِقَدْرَيْ ا ه. و ن ل أَضْعاف مُتَساوِيَة لِقَدْرَيْ ب ز. فنِسْبَةَ ا إلى ب أَكْبَرُ مِنْ نِسْبَةِ ه إلى ز.

وَذَلِكَ ما أَرَدْنا أَنْ نُبيِّنَ.

מופתו: כי יחס[1] ג׳ אל ד׳ יותר גדול מיחס ה׳ אל ז׳. הנה יהיה אל ג׳ וה׳ כפלים שוים, ואל ד׳ וז׳ כפלים שוים[2]. ויהיו כפלי ג׳ נוספים על כפלי ד׳. וכפלי ה׳ בלתי נוספים על כפלי ז׳. הנה יהיו כפלי ג׳ וה׳ השוים וכפלי ד׳ וז׳ השוים אשר אלו[3] ענינם[4], אולם[5] כפלי ג׳ ה׳—הנה ח׳ ט׳, ואולם כפלי ד׳ ז׳—הנה כ׳ ל׳. ויהיה מה שבמ׳ מכפלי א׳ כמו מה שבח׳ מכפלי ג׳, ומה שבנ׳ מכפלי ב׳ כמו מה שבכ׳ מכפלי ד׳. הנה יחס א׳ אל ב׳ כיחס ג׳ אל ד׳. וכבר הוסף[6] לשני שעורי א׳ ג׳ כפלים שוים, והם מ׳ ח׳, ולשני שעורי ב׳ ד׳ כפלים שוים, והם נ׳ כ׳. הנה שני שעורי מ׳ ח׳ אם נוספים יחד על שני שעורי נ׳ כ׳, ואם שוים יחד לשניהם, ואם חסרים יחד משניהם. אבל אם היה נוסף ח׳[7] על כ׳, הנה[8] מ׳ נוסף על נ׳. וט׳ בלתי נוסף על ל׳. הנה מ׳ ט׳ כפלים שוים לשני שעורי א׳ ה׳. ונ׳ ל׳ כפלים שוים לשני שעורי ב׳ ז׳. אם כן יחס א׳ אל ב׳ יותר גדול מיחס ה׳ אל ז׳.

וזה מה שרצינו לבאר.

---

[1] כי יחס] כיחס **ב** מפני+ **ז**   [2] ואל ... שוים] ליתא **ב**   [3] אלו] לו **ב**   [4] ענינם] * ענינים **ס**   [5] אולם] ליתא **בוספ** ⟨*⟩ **ק**   [6] הוסף] לוקח **ס**   [7] נוסף ח׳] ח׳ נוסף **פ**   [8] הנה] +מ׳ נוסף על נ׳ **ס**

[Proposition V.13]

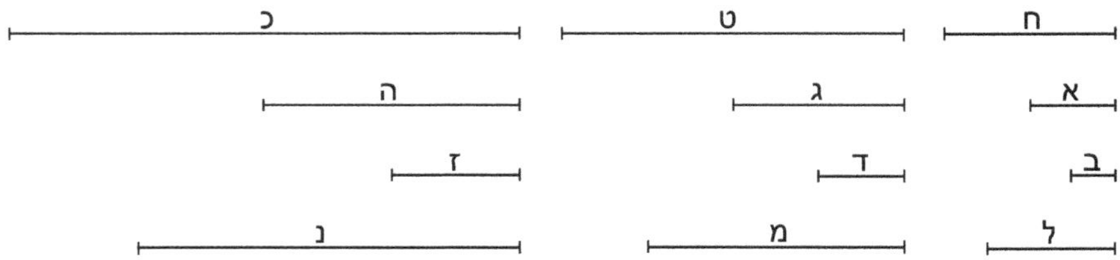

השעורים אשר יחסם אל שעורים אחרים על מספרם—אחד, כמה שהיו, יהיה יחס האחד[1] אל האח[ר] חברו כיחס הכל אל הכל.

המשל בזה: שיחס א׳ לב׳ וג׳ לד׳ וה׳ לז׳—אחד. ואומר שיחס א׳ אצל ב׳ כיחס א׳ וג׳ וה׳ יחד אצל ב׳ וד׳ וז׳ יחד.

الأَقْدارُ الَّتي نِسْبَتُها إِلى أَقْدُرٍ أُخَرَ على عِدَّتِها—واحِدَةٌ، كَمْ كانَتْ، فَإِنَّ نِسْبَةَ الواحِدِ إِلى الواحِدِ كَنِسْبَةِ الجَميعِ إِلى الجَميعِ.

مِثالُهُ: أَنَّ نِسْبَةَ ا و ب وج و د و ه و ز واحِدَةٌ—قَدْرُ ا عِنْدَ ب كَقَدْرِ ج عِنْدَ د، وكَقَدْرِ ه عِنْدَ ز. فَأَقولُ إِنَّ قَدْرَ ا عِنْدَ ب كَقَدْرِ ا وج و ه جَميعًا عِنْدَ ب و د و ز جَميعًا.

[1]האחד] ⟨⟨*⟩⟩

الأَقْدارُ الَّتِي نِسْبَتُها إلى أَقْدارٍ أُخَرَ مُقارِنَةً لَهَا عَلى عِدَّتِها كَمْ كانَتْ—واحِد، فَإِنَّ نِسْبَةَ الواحِدِ إلى قَرِينِهِ كَنِسْبَةِ الْجَمِيعِ إلى الْجَمِيعِ.

مِثالُهُ: أَنَّ نِسْبَةَ ا إلى ب و ج إلى د و ه إلى ز واحِدَةٌ. فَأَقُولُ إِنَّ نِسْبَةَ ا إلى ب كَنِسْبَةِ ا ج ه جَمِيعًا إلى ب د ز جَمِيعًا.

השעורים אשר יחסם אל שיעורים[1] אחרים קרובים להם על מספרם כמה שהיו[2]—אחד, הנה יחס האחד אל קרובו כיחס הכל אל הכל.

המשל בו: כי יחס[3] א׳ אל ב׳ וג׳ אל ד׳ וה׳ אל ז׳—אחד. הנה אומר כי יחס א׳ אל ב׳ כיחס א׳ ג׳ ה׳ יחד אל ב׳ ד׳ ז׳ יחד.

1 שיעורים] שעורי **ס** 2 שהיו] שיהיו **וספ** 3 כי יחס] כיחס **ב** +~~אל~~ **ק**

מופת זה: שנקח לא׳ ולג׳ ולה׳ כפלים שוים, והם ח׳ וט׳ וכ׳. ונקח לב׳ ולד׳ וז׳ כפלים שוים, והם ל׳ ומ׳ ונ׳. ושעור א׳ אצל ב׳ כשעור ג׳ אצל ד׳ וכשעור ה׳ אצל ז׳. וכבר לקחנו לא׳ ולג׳ ולה׳ כפלים שוים, והם ח׳ וט׳ וכ׳. ולב׳ ולד׳ ולז׳ כפלים שוים, והם ל׳ ומ׳ ונ׳—יהיה (שעור)[1] ח׳ על ל׳ כשיעור ט׳ על מ׳, וכ׳ על נ׳. ואם היה שוה לו יהיו הם שוים להם. ואם היה חסר ממנו יהיו בזה השעור חסרים מהם. ומה שיעדף א׳ על ל׳ כפי מה שיעדפו ח׳ וט׳ וכ׳ על ל׳ ומ׳ ונ׳. ואם היה שוה לו הם שוים להם. ואם היה חסר ממנו גם הם יחסרו מהם בזה השיעור. וכפלי ח׳ לא׳ ככפלי ח׳ וט׳ וכ׳ לא׳ ולג׳ ולה׳. וכפלי ל׳ לב׳ ככפלי ל׳ ומ׳ ונ׳ לב׳ ולד׳ ולז׳. נמצא שעור א׳ אצל ב׳ כשעור א׳ וג׳ וה׳ יחד אצל ב׳ וד׳ וז׳ יחד.

وזה מה שרצינו לבאר.

بُرْهانُهُ: أَنَّا نَأْخُذُ لِ ا وَ ج وَ ه أَضْعافًا مُتَساوِيَةً، وَهِيَ ح وَ ط وَ ك. وَنَأْخُذُ لِ ب وَ د وَ ز أَضْعافًا مُتَساوِيَةً، وَهِيَ ل وَ م وَ ن. فَقَدْرُ ا عِنْدَ ب كَقَدْرِ ج عِنْدَ د. وَكَقَدْرِ ه عِنْدَ ز. وَقَدْ أُخِذَ لِ ا وَ ج وَ ه أَضْعافٌ مُتَساوِيَةٌ، وَهِيَ ح وَ ط وَ ك. وَلِ ب وَ د وَ ز أَضْعافٌ مُتَساوِيَةٌ، وَهِيَ ل وَ م وَ ن. فَقَدْرُ زِيادَةِ ح عَلى ل كَقَدْرِ زِيادَةِ ط عَلى م، وَ ك عَلى ن. وَإِنْ كانَ مُساوِيًا لَهُ، فَهُوَ مُساوٍ لَها. وَإِنْ كانَ ناقِصًا مِنْهُ بِذَلِكَ القَدْرِ يَنْقُصُ مِنْها. وَأَضْعافُ ح لِ ا كَأَضْعافِ ح وَ ط وَ ك لِ ا وَ ج وَ ه. وَأَضْعافُ ل لِ ب كَأَضْعافِ ل وَ م وَ ن لِ ب وَ د وَ ز. فَقَدْرُ ا عِنْدَ ب كَقَدْرِ ا وَ ج وَ ه جَميعًا عِنْدَ ب وَ د وَ ز جَميعًا.

وَذَلِكَ ما أَرَدْنا بَيانَهُ.

[1]שעור] +תוספת ⟨⟨ר״ל⟩⟩ ~~מן הנודע מסגלת כפלים מתיחסים~~ ⟨⟨ע״כ⟩⟩

بُرهانُه: أنّا نأخُذُ لأَقْدارِ ا ج ه أَضْعافًا مُتَساوِيَةً، وهي ح ط ك، ولِأَقْدار ب د ز أَضْعافًا مُتَساوِيَةً، وهي ل م ن. فنِسْبَةُ ا إلى ب كَنِسْبَةِ ج إلى د، وكَنِسْبَةِ ه إلى ز. وح ط ك أَضْعاف مُتَساوِيَة لأَقْدارِ ا ج ه. و ل م ن أَضْعاف مُتَساوِيَة لأَقْدارِ ب د ز. ف ح ط ك إمّا زائدة معًا على ل م ن، وإمّا مُساوِيَة لها معًا، وإمّا ناقِصَة مِنْها معًا^. فإنْ كانَ ح زائدًا على ل، فإنَّ ح ط ك مجموعة زائدة على ل م ن مجموعة. وإنْ كانَ ناقِصًا مِنْهُ، فَهِيَ ناقِصَةٌ مِنْها. وإنْ كانَ مُساوِيًا لَهُ، فَهِيَ مُساوِيَةٌ لها. فأَضْعاف ح لِقَدْرِ ا كأَضْعافِ ح ط ك مجموعة لأَقْدارِ ا ج ه مجموعة. وأَضْعاف ل لِقَدْرِ ب كأَضْعافِ ل م ن مجموعة لأَقْدارِ ب د ز مجموعة. فنِسْبَةُ ا إلى ب كَنِسْبَةِ ا ج ه جَميعًا[1] إلى ب د ز جَميعًا.

وَذَلِكَ ما أَرَدْنا أَنْ نُبَيِّنَ.

מופתו: אנחנו נקח לשיעורי א׳ ג׳ ה׳ כפלים שוים, והם ח׳ ט׳ כ׳[1], ולשיעורי ב׳ ד׳ ז׳ כפלים שוים, והם ל׳ מ׳ נ׳[2]. הנה יחס א׳ אל ב׳ כיחס ג׳ אל ד׳, וכיחס ה׳ אל ז׳. וח׳ וט׳ וכ׳ כפלים שוים לשעורי א׳ ג׳ ה׳. ול׳ מ׳ נ׳ כפלים שוים לשעורי ב׳ ד׳ ז׳. אם כן ח׳ ט׳ כ׳ אם נוספים יחד על ל׳ מ׳ נ׳, ואם חסרים מהם יחד, ואם שוים להם יחד. ואם היה ח׳ נוסף על ל׳[3], הנה ח׳ ט׳ כ׳ מקובצים נוספים על ל׳ מ׳ נ׳ מקובצים[4]. ואם היה חסר ממנו, הנה הם חסרים מהם. ואם היה שוה לו[5], הנה הם שוים להם. וכפלי ח׳ לשעור א׳ ככפלי[6] ח׳ ט׳ כ׳ מקובצים לשעורי א׳ ג׳ ה׳ מקובצים. וכפלי ל׳ לשעור ב׳ ככפלי ל׳ מ׳ נ׳ לשעורי ב׳ ד׳ ז׳ מקובצים. אם כן יחס א׳ אל ב׳ כיחס א׳ ג׳ ה׳ מקובצים אל ב׳ ד׳ ז׳ יחד[7].

וזה מה שרצינו לבאר.

1 جميعا] R1R2T2T3

1 כ׳] ⟨*⟩ פ 2 נ׳] +~~ה״נ~~ פ 3 ל׳] ל״כ ס 4 מקובצים] ליתא ז נוספים+ פ 5 שוה לו] שוים ב 6 ח׳ ... ככפלי] ליתא ב 7 יחד] ליתא ספ

[Proposition v.14]

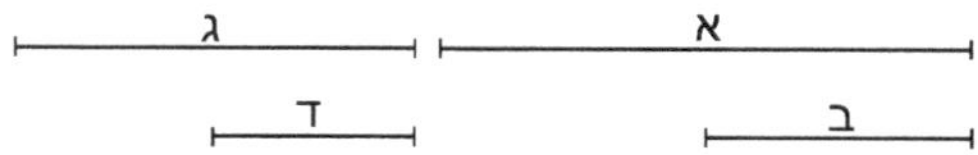

כשהיו ארבעה שעורים מתיחסים, והראשון גדול מן השלישי, יהיה השני גדול מן הרביעי. ואם היה שוה לו, גם הוא שוה לו. ואם היה קטן ממנו גם הוא קטן ממנו.

إذا كانَتْ أَرْبَعَةُ أَقْدارٍ مُتَناسِبَةً، وَكانَ الأَوَّلُ أَعْظَمَ مِنَ الثَّالِثِ، فَإِنَّ الثَّاني أَعْظَمُ مِنَ الرَّابِعِ. وَإِنْ كانَ مُساوِيًا لَهُ، فَهُوَ مُساوٍ لَهُ، وَإِنْ كانَ أَصْغَرَ مِنْهُ، فَهُوَ أَصْغَرُ مِنْهُ.

המשל בזה: ארבעה [!] שיעורים מתיחסים, והם א׳ וב׳ וג׳ וד׳. יחס א׳ אל ב׳ כיחס ג׳ אל ד׳. וא׳ גדול מג׳. ואומר שב׳ גדול מד׳.

مِثالُهُ: أَنَّ أَرْبَعَةَ أَقْدارٍ مُتَناسِبَةً ا ب وج د. نِسْبَةُ ا عِنْدَ ب كَنِسْبَةِ ج عِنْدَ د. ا أَعْظَمُ مِنْ ج. فَأَقولُ إِنَّ ب أَعْظَمُ مِنْ د.

מופת זה: לפי שא׳ גדול מג׳, יהיה א׳ רב היחס אל ב׳ מג׳ אל ב׳. ואמנם שעור א׳ אצל ב׳ כשעור ג׳ אצל ד׳. ויהיה ג׳ רב היחס אל ד׳ מיחסו אל ב׳. ואשר יהיה היחס אליו רב הוא הקטן. אם כן ד׳ קטן מב׳. וב׳ גדול מד׳. ולזה אם היה א׳ שוה לג׳ היה ב׳ שוה לד׳. ואם היה א׳ קטן מג׳ היה ב׳ קטן מד׳.

بُرْهانُهُ: أَنَّ ا أَعْظَمُ مِنْ ج. ا أَعْظَمُ نِسْبَةً إِلى ب[1] مِنْ ج إِلى (ب)[2]. وَلَكِنَّ قَدْرَ ا عِنْدَ ب كَقَدْرِ ج عِنْدَ د. وج أَعْظَمُ نِسْبَةً إِلى د مِنْ ج إِلى ب. وَالَّذي تَكونُ النِّسْبَةُ إِلَيْهِ أَعْظَمَ هُوَ أَصْغَرُ. د أَصْغَرُ مِنْ ب. وب أَعْظَمُ مِنْ د. وَكَذَلِكَ (تَبَيَّنَ)[3] أَنَّ ا لَوْ كانَ مِثْلَ ج لَكانَ ب مِثْلَ د. وَلَوْ كانَ ا أَصْغَرَ مِنْ ج لَكانَ ب أَصْغَرَ مِنْ د.

וזה מה שרצינו באורו.

وَذَلِكَ ما أَرَدْنا بَيانَهُ.

---

[1] نسبة ... ب] ⟨*⟩ [2] ب] اب [3] تبين] نبين

إذا كانَتْ أَرْبَعَةُ أَقْدارٍ مُتَنَاسِبَةً، وَكانَ الأَوَّلُ أَعْظَمَ مِنَ الثَّالِثِ، فَإِنَّ الثَّاني أَعْظَمَ مِنَ الرَّابِعِ. وَإِنْ كانَ مُساوِيًا لَهُ، فَهُوَ مُساوٍ لَهُ. وَإِنْ كانَ أَصْغَرَ مِنْهُ، فَهُوَ أَصْغَرُ مِنْهُ.

כאשר היו ארבעה שעורים[1] מתיחסים, והיה הראשון יותר גדול מן השלישי, הנה השני[2] יותר גדול מן הרביעי. ואם היה שוה לו, הנה הוא שוה לו. ואם היה יותר קטן ממנו, הנה הוא יותר קטן ממנו[3].

مِثالُهُ: أَنَّ أَقْدارَ ا ب ج د الأَرْبَعَةَ مُتَنَاسِبَةٌ— نِسْبَةُ ا إِلى ب كَنِسْبَةِ ج إِلى د. وَا أَعْظَمُ مِنْ ج. فَأَقُولُ إِنَّ ب أَعْظَمُ مِنْ د.

המשל בו: כי שעורי א׳ ב׳ ג׳ ד׳ הארבעה מתיחסים—יחס א׳ אל ב׳ כיחס ג׳ אל ד׳. וא׳ יותר גדול מן ג׳. הנה אומר כי ב׳ יותר גדול מד׳.

بُرْهانُهُ: أَنَّ ا أَعْظَمُ مِنْ ج. فَا أَكْبَرُ نِسْبَةً إِلى ب مِنْ ج إِلى ب. وَلٰكِنَّ نِسْبَةَ ا إِلى ب كَنِسْبَةِ ج إِلى د. فَنِسْبَةُ ج إِلى د أَعْظَمُ مِنْ نِسْبَتِهِ إِلى ب. وَالَّذي تَكُونُ النِّسْبَةُ إِلَيْهِ أَكْبَرَ، فَهُوَ أَصْغَرُ. فَ د أَصْغَرُ مِنْ ب. فَ ب أَعْظَمُ مِنْ د. وَكَذٰلِكَ يَتَبَيَّنُ أَنَّ ا لَوْ كانَ مِثْلَ ج، لَكانَ ب مِثْلَ د. وَلَوْ كانَ أَصْغَرَ مِنْ ج، لَكانَ ب أَصْغَرَ مِنْ د.

מופתו: כי א׳ יותר גדול מן ג׳[4]. אם כן א׳ יותר גדול היחס אל ב׳ מג׳ אל ב׳. אבל יחס א׳ אל ב׳ כיחס ג׳ אל ד׳. אם כן יחס ג׳ אל ד׳ יותר גדול מיחסו אל ב׳. ואשר[5] יהיה היחס אליו יותר גדול, הנה הוא יותר קטן. אם כן ד׳ יותר קטן מן ב׳. אם כן ב׳ יותר גדול מן ד׳. ובן התבאר כי א׳ אלו[6] היה כמו ג׳, היה ב׳ כמו ד׳. ואלו היה[7] יותר קטן מן ג׳, היה ב׳ יותר קטן מן[8] ד׳.

وَذٰلِكَ ما أَرَدْنا أَنْ نُبَيِّنَ.

וזה מה שרצינו לבאר.

שרטוט—ו: ב ↔ ג [1]שעורים] שעורי ב [2]השני] ⟨ה⟩שני ב [3]ואם ... ממנו] ⟨⟨*⟩⟩ ק [4]ג׳] ⟨*⟩ ט [5]ואשר] וכאשר ס [6]כי ... אלו] אלו היה א׳ ו [7]היה] +א׳ ו [8]מן] ליתא ס

[Proposition v.15]

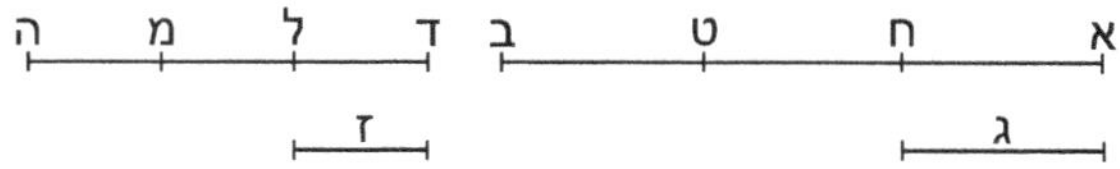

الأَجْزاءُ الَّتي أَضْعافُها مُتَساوِيَةٌ فَإِنَّ نِسْبَتَها واحِدَةٌ إِذا أُخذت يَتْلو بَعْضُها بَعْضًا.

החלקים אשר כפליהם שוים יהיה יחס קצתם אל קצת כיחס כפליהם קצתם אל קצת.

مِثالُه: أَنَّ في ا ب مِنْ أَضْعافِ ج مِثْلَ ما في د ه مِنْ أَضْعافِ ز. فَأَقولُ إِنَّ قَدْرَ ا ب عِنْدَ د ه كَقَدْرِ ز عِنْدَ ج.

והמשל בזה: כי בא״ב מכפלי ג׳ כמו שיש בד״ה מכפלי ז׳. ואומר כי שעור א״ב אצל ד״ה כשעור ג׳ אצל ז׳.

بُرْهانُه: أَنَّ في ا ب مِنْ أَضْعافِ ج مِثْلَ ما في د ه مِنْ أَضْعافِ ز. فَنَقْسِمُ ا ب بِقَدْرِ ج. فتكونُ أَقْسامُهُ ا ح وح ط و ط ب. ونَقْسِمُ د ه بِقَدْرِ ز. فتكونُ د ل ول م وم ه. وا ح وح ط وط ب مُتَساوِيَةً. ود ل ول م وم ه مُتَساوِيَة. وَعِدَّةُ ا ح وح ط و ط ب مِثْلُ عِدَّةِ د ل و ل م و م ه. فَقَدْرُ ا ح عِنْدَ د ل كَقَدْرِ ح ط [..] عِنْدَ التّالي. ثُمَّ كَذٰلِكَ هِيَ فَقَدْرُ ا ح عِنْدَ د ل كَقَدْرِ ا ب عِنْدَ د ه. ا ح مِثْلُ ج. و د ل مِثْلُ ز. فَقَدْرُ ا ب عِنْدَ د ه كَقَدْرِ[1] ج عِنْدَ ز.

מופת זה: כי מה שיש בא״ב מכפלי ג׳ כמו שיש בד״ה מכפלי ז׳. ונחלק א״ב כשיעור ג׳. ויהיו חלקיו א״ח וח״ט וט״ב. ונחלק ד״ה בשעור ז׳. ויהיו חלקיו ד״ל ל״מ מ״ה. וא״ח וח״ט וט״ב שוים. וד״ל ול״מ ומ״ה שוים. ומספר א״ח וח״ט וט״ב כמספר ד״ל ל״מ[1] מ״ה. ושעור א״ח אצל ד״ל כשיעור ח״ט אצל ל״מ וכשעור ט״ב אצל מ״ה. ושעור אחד מן הראשונים אצל אחד מן הנמשכים אחריהם כשעור כל הראשונים אצל כל הנמשכים. ואם כן יהיה שיעור א״ח אצל ד״ל כשיעור א״ב אצל ד״ה. וא״ח כמו ג׳. וד״ל כמו ז׳[2]. ואם כן שיעור א״ב אצל ד״ה כשיעור ג׳ אצל ז׳.

وَذٰلِكَ ما أَرَدْنا بَيانَه.

וזה מה שרצינו לבאר.

---

[1] كقدر] ⟨⟨ك⟩⟩قدر

שרטוט—B: ה ← ט [1] ל״מ] ⟨*⟩ [2] ז׳] א+

الأَجْزاءُ الَّتِي أَضْعافُها مُتَساوِيَةٌ فَإِنَّ نِسْبَةَ بَعْضِها إِلى بَعْضٍ كَنِسْبَةِ أَضْعافِها بعضها إِلى بَعْضٍ.

החלקים אשר כפליהם שוים הנה יחס קצתם אל קצת כיחס כפלי קצתם אל קצת.

مِثالُهُ: أَنَّ أَضْعاف ا ب لِقَدْرِ ج كأَضْعافِ د ه لِقَدْرِ ز. فَأَقولُ إِنَّ نِسْبَةَ ا ب إِلى د ه كَنِسْبَةِ ج إِلى ز.

המשל בו: כי[1] כפלי א״ב אל שעור ג׳ ככפלי[2] ד״ה לשעור ז׳. הנה אומר כי יחס א״ב אל ד״ה כיחס ג׳ אל ז׳[3].

بُرهانُهُ: أَنَّا نَقْسِمُ ا ب بِقَدْرِ ج فأقسامُهُ ا ح ح ط ط ب، وهي مُتَساوِيَةٌ. وَنَقْسِمُ د ه بِقَدْرِ ز. فأقسامُهُ د ل ل م م ه، وهي مُتَساوِيَةٌ. وعِدَّةُ ا ح ح ط ط ب كَعِدَّةِ د ل ل م م ه. فَنِسْبَةُ ا ح إِلى د ل كَنِسْبَةِ ح ط إِلى ل م، وكَنِسْبَةِ ط ب إِلى م ه. ونِسْبَةُ واحِدٍ إِلى قَرِينِهِ كَنِسْبَةِ الجَمِيعِ إِلى الجَمِيعِ. فَنِسْبَةُ ا ح إِلى د ل كَنِسْبَةِ ا ب إِلى د ه. وا ح مِثْل ج. و د ل مِثْل ز. فَنِسْبَةُ ا ب إِلى د ه كَنِسْبَةِ ج إِلى ز.

המופת: אנחנו נחלק א״ב בשעור ג׳. וחלקיו א״ח וח״ט וט״ב, והם שוים. ונחלק ד״ה בשיעור ז׳. וחלקיו ד״ל ול״מ ומ״ה, והם שוים. ומספר א״ח וח״ט וט״ב[4] כמספר ד״ל ול״מ ומ״ה. אם כן יחס א״ח אל ד״ל כיחס ח״ט אל ל״מ, וכיחס ט״ב אל מ״ה. ויחס האחד אל קרובו כיחס הכל אל הכל. אם כן יחס א״ח אל ד״ל כיחס א״ב אל ד״ה. וא״ח כמו ג׳. וד״ל כמו ז׳. אם כן יחס א״ב אל ד״ה כיחס ג׳ אל ז׳[5].

وذَلِكَ ما أَرَدْنا أَنْ نُبَيِّنَ.

וזה מה שרצינו לבאר[6].

---

[1]כי] +כפלו ב [2]ככפלי] כשיעור ז [3]ז׳] +⟨ד״ת כי יחס ג׳ אל ז׳ כיחס א״ב אל ד״ה⟩ טק [4]וט״ב] ליתא ב [5]ז׳] +⟨ד״ת יחס ג׳ אל ז׳ כיחס א״ב אל ד״ה⟩ טק [6]לבאר] לבארו טק

[Proposition V.16]

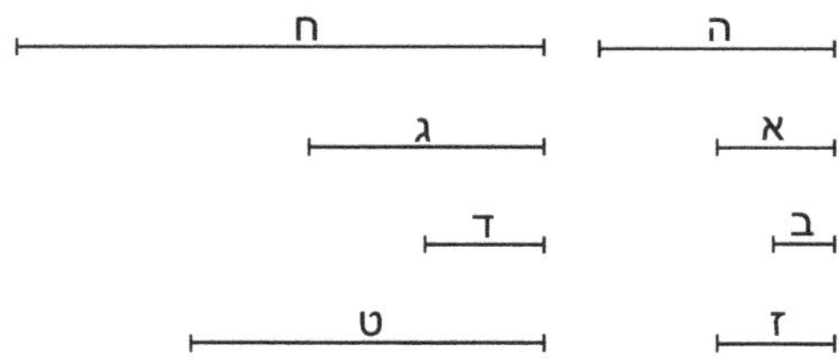

כשיהיו ארבעה שעורים מתיחסים, כשיתחלפו יהיו גם כן מתיחסים.

המשל בזה: שארבעה שיעורים מתיחסים, א׳ וב׳ וג׳ וד׳, שיעור א׳ אצל ב׳ כשיעור ג׳ אצל ד׳. ואומר שאם יתחלפו יהיו מתיחסים, ויהיה[1] שיעור א׳ אצל ג׳ כשיעור ב׳ אצל ד׳.

إِذا كَانَتْ أَرْبَعَةُ أَقْدارٍ مُتَناسِبَةً، فَإِنَّها إِذا بُدِّلَتْ تَكونُ مُتَساوِيَةً.

مِثالُهُ: أَنَّ أَرْبَعَةَ أَقْدارٍ[1] مُتَناسِبَةٌ، ا ب وج د. قَدْرُ ا عِنْدَ ب كَقَدْرِ ج عِنْدَ د. فَأَقولُ إِنَّها إِذا بُدِّلَتْ كانَتْ مُتَناسِبَةً، يَكُونُ قَدْرُ ا عِنْدَ ج كَقَدْرِ ب عِنْدَ د.

[1] ويهיה] ויהי⟨ה⟩

[1] أقدار] ~~اعداد~~ ⟨*⟩

إذا كَانَتْ أَرْبَعَةُ أَقْدارٍ مُتَنَاسِبَةً، فَإِنَّها إذا بُدِّلَتْ تَكُونُ مُتَنَاسِبَةً.

כאשר היו ארבעה[1] שעורים מתיחסים, הנה הם יהיו, כאשר הומרו, מתיחסים.

مِثَالُهُ: أَنَّ أَقْدارَ ا ب ج د الأَرْبَعَةَ مُتَنَاسِبَةٌ، نِسْبَةَ ا إلى ب كنِسْبَةِ ج إلى د. فَأَقولُ إنَّ نِسْبَةَ ا إلى ج كنِسْبَةِ ب إلى د على التَّبْدِيلِ.

המשל בו: כי שעורי א׳ ב׳ ג׳ ד׳ הארבעה מתיחסים, יחס א׳ אל ב׳ כיחס ג׳ אל ד׳. הנה אומר כי יחס א׳ אל ג׳ כיחס ב׳ אל ד׳ על התמורה.

[1]ארבעה] ליתא **בו**

מופת זה: שנקח לא׳ ולב׳ כפלים שוים, והם ה׳ וז׳, ולג׳ ולד׳ כפלים שוים, והם ח׳ וט׳. וה׳ וז׳ הם כפלים שוים לא׳ ולב׳. והחלקים אשר כפליהן שוים הם מתיחסים, ויחסם אחד. ויהיה שיעור א׳ אצל ב׳ כשיעור ה׳ אצל ז׳. ושיעור א׳ אצל ב׳ כשיעור ג׳ אצל ד׳. ואם כן יהיה שיעור ה׳ אצל ז׳ כשיעור ג׳ אל ד׳. ועוד ח׳ וט׳ כפלים שוים לג׳ ולד׳. ויהיה שיעור ג׳ אצל ד׳ כשעור ח׳ אצל ט׳. ואמנם שיעור ג׳ אצל ד׳ כשעור ה׳ אצל ז׳. ויהיה שעור ה׳ אצל ז׳ כשיעור ח׳ אצל ט׳. ושעור תוספת ה׳ על ח׳ כשעור תוספת ז׳ על ט׳. ואם היה ה׳ שוה לח׳, יהיה ז׳ שוה לט׳. ואם היה ה׳ חסר כשעור זה, יהיה חסרון ז׳ מט׳. וה׳ וז׳ כפלים שוים לא׳ ולב׳. וח׳ וט׳ כפלים שוים לג׳ ולד׳. נמצא שעור א׳ אצל ג׳ כשיעור ב׳ אצל ד׳.

بُرْهانُه: أَنَّا نَأْخُذُ ا و ب أَضْعافًا مُتَساوِيَةً، وَهِيَ ه و ز، وج و د أَضْعافًا مُتَساوِيَةً، وَهِيَ ح و ط. ه و ز هُما أَضْعافُ مُتَساوِيَةٌ ا و ب. (وَالأَجْزاءُ)[1] الَّتي أَضْعافُها مُتَساوِيَةٌ، وَهِيَ مُتَناسِبَةٌ. فَقَدْرُ ا عِنْدَ ب كَقَدْرِ ه عِنْدَ ز. وَأَيْضًا ح و ط أَضْعافُ مُتَساوِيَةٌ ل ج و د. فَقَدْرُ ج عِنْدَ د كَقَدْرِ ح عِنْدَ ط. وَلَكِنَّ قَدْرَ ج عِنْدَ د كَقَدْرِ ه عِنْدَ ز. وَقَدْرُ ه عِنْدَ ز كَقَدْرِ ح عِنْدَ ط. فَقَدْرُ زِيادَةِ ه عَلى ح كَقَدْرِ زِيادَةِ ز عَلى ط. وَإِنْ كانَ ه مُساوِيًا ل ح، فَإِنَّ ز مُساوٍ ل ط. وَإِنْ كانَ ه ناقِصًا مِنْ ح فَبِقَدْرِ ذَلِكَ يَكونُ نُقْصانُ ز من ط. و ه و ز أَضْعافُ مُتَساوِيَةٌ ا و ب. وح و ط هُما أَضْعافُ مُتَساوِيَةٌ ل ج و د. فَقَدْرُ ا عِنْدَ ج كَقَدْرِ ب عِنْدَ د.

וזה מה שרצינו לבאר.

وَذَلِكَ ما أَرَدْنا بَيانَهُ.

[1] والاجزاء] والاخر

بُرهانُهُ: أَنَّا نَأْخُذُ لِقَدْرَيْ ا ب أَضْعافًا مُتَساوِيَةً، وَهِيَ ه ز، وَلِقَدْرَيْ ج د أَضْعافًا مُتَساوِيَةً، وَهِيَ ح ط. والأَجزاء الَّتي أَضْعافُها مُتَساوِيَةٌ، نِسْبَةُ بَعْضِها إلى بَعْض كَنِسْبَةِ أَضْعافِها بعضها إلى بعض. و ه ز أَضْعاف مُتَساوِيَةٌ^ لِقَدْرَيّ ا ب. فَنِسْبَةَ ا إلى ب كَنِسْبَةِ ه إلى ز. وأيضًا فإنَّ ح ط أَضْعافُ مُتَساوِيَةٌ لِقَدْرَيْ ج د. فَنِسْبَةَ ج إلى د كَنِسْبَةِ ح إلى ط. ولكنَّ نِسْبَةَ ا إلى ب كَنِسْبَةِ ه إلى ز^. فَنِسْبَةَ ه إلى ز[1] كَنِسْبَةِ ح إلى ط. فَإِنْ كانَ ه يَزِيدُ على ح، فإنَّ ز يَزِيدُ على ط. وإنْ كان يُساوِيهِ، فَهُوَ يُساوِيهِ. وإنْ كانَ يَنْقُصُ عَنْهُ، فَهُوَ يَنْقُصُ عَنْهُ. فَقَدْرَا ه ز إِمّا زائِدانِ مَعًا على قَدْرَيْ ح ط، وَإِمّا مُساوِيانِ مَعًا لَهُما، وَإِمّا ناقِصانِ مَعًا مِنْهُما. و ه ز أَضْعاف مُتَساوِيَة لِقَدْرَيْ ا ب. و ح ط أَضْعاف مُتَساوِيَة لِقَدْرَيْ ج د. فَنِسْبَةَ ا إلى ج كَنِسْبَةِ ب إلى د.

وَذلِكَ ما أَرَدْنا أَنْ نُبَيِّنَ.

מופתו: אנחנו נקח לשעורי א׳ ב׳ כפלים שוים, והם[1] ה׳ ז׳, ולשעורי ג׳ ד׳ כפלים שוים, והם ח׳ ט׳. והחלקים אשר כפליהם שוים[2], יחס קצתם אל קצת[3] כיחס כפליהם קצתם אל קצת. וה׳ ז׳ כפלים לשעורי א׳ ב׳. אם כן יחס א׳[4] אל ב׳ כיחס ה׳ אל ז׳. וגם כן הנה ח׳ וט׳ כפלים שוים לשיעורי ג׳ וד׳. אם כן יחס[5] ג׳ אל ד׳ כיחס ח׳ אל ט׳. אבל יחס[6] ג׳ אל ד׳ כיחס א׳ אל ב׳ וכיחס ה׳ אל ז׳. אם כן יחס ה׳ אל ז׳ כיחס ח׳ אל ט׳[7]. ואם היה ה׳ מוסיף על ח׳, הנה ז׳ יוסיף על ט׳. ואם היה שוה לו[8], הנה הוא שוה לו, ואם היה שיחסר ממנו, הנה יחסר ממנו. אם כן שני שעורי ה׳ וז׳ אם נוספים יחד על שני שעורי ח׳ וט׳, ואם[9] שוים יחד להם, ואם חסרים יחד מהם. וה׳ ז׳ כפלים שוים[10] לשני שעורי א׳ וב׳. וח׳ וט׳ כפלים שוים לשני שעורי ג׳ וד׳. אם כן יחס א׳ אל ג׳ כיחס[11] ב׳ אל ד׳.

וזה מה שרצינו לבאר[12].

---

[1] فنسبة ه ... ز] PR1R2

---

[1] והם] ⟨*⟩ **ס** [2] שוים] +הם **ק** [3] קצת] קצתם **ז** [4] א׳] +ג׳ **פ** [5] יחס] ליתא **ז** [6] יחס] ליתא **ז** [7] אבל ... ט׳] ליתא **פ** [8] לו] ליתא **ז** [9] ואם] +נוספים יחד על שני שעורי ח׳ וט׳ אם **ב** [10] שוים] ליתא **ט** [11] כיחס] כי יחסו **ב** [12] וזה ... לבאר] ונשלם באורו **ט** זה מה שרצינו לבארו **ק**

[Proposition v.17]

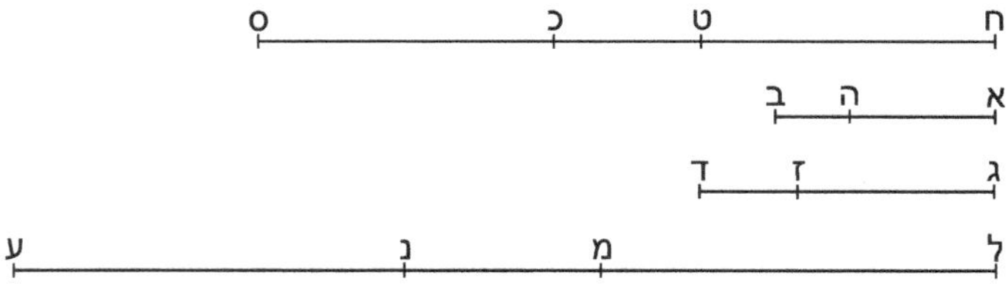

כשיהיו שעורים מורכבים מתיחסים, אם יתחלקו יהיו גם כן מתיחסים.

إِذا كانَتْ أَقْدارٌ مُرَكَّبَةٌ مُتَناسِبَةً [..].

המשל בזה: ששעורי א״ב וב״ה וג״ד וד״ז מורכבים מתיחסים—שיעור א״ב אצל ב״ה כשיעור ג״ד אצל ד״ז. ואומר שהם אם יתחלקו יהיו מתיחסים, ויהיה שעור א״ה אצל ה״ב[1] כשיעור ג״ז אצל ז״ד.

مثالُهُ: أَنَّ قَدْرَا اب وب ه وجـ د و د ز المُرَكَّبَةَ مُتَناسِبَةٌ—قَدْرُ اب عِنْدَ ب ه كَقَدْرِ جـ د عِنْدَ ز د. فَأَقولُ إِنَّها إِذا فُصِلَتْ تَكونُ مُتَناسِبَةً، يَكونُ قَدْرُ ا ه عِنْدَ ه ب كَقَدْرِ جـ ز عِنْدَ ز د.

[1] ה״ב] ~~א״ב~~ ⟨⟨ה״ב⟩⟩

إذا كانَتْ أَقْدارٌ مُرَكَّبةٌ مُتَنَاسِبَةً، فَإنَّها إذا فُصِلَتْ تَكُونُ مُتَنَاسِبَةً.

مِثالُهُ: أَنَّ أَقْدارَ اب ب ه ج د د ز مُرَكَّبةً مُتَنَاسِبَةٌ—نِسْبَةُ اب إلى ب ه كَنِسْبَةِ ج د إلى د ز. فَأَقولُ إِنَّها إذا فُصِلَتْ تكونُ مُتَنَاسِبَةً، فَتَكونُ نِسْبَةُ اه إلى ه ب كَنِسْبَةِ ج ز إلى ز د.

כאשר היו שעורים מורכבים מתיחסים, הנה הם כאשר נבדלו יהיו מתיחסים.

המשל בו: כי שעורי א״ב וב״ה וג״ד וד״ז מורכבים מתיחסים—יחס א״ב אל ב״ה[1] כיחס ג״ד אל ד״ז. הנה אומר שהם כאשר נבדלו יהיו מתיחסים, ויהיה יחס א״ה אל ה״ב כיחס ג״ז אל ז״ד.

שרטוט—**פ**: ט״כ מסומן ב״ראשון״, ה״ב מסומן ב״שני״, מ״נ מסומן ב״שלישי״, ז״ד מסומן ב״רביעי״, כ״ס מסומן ב״חמישי״, נ״ע מסומן ב״ששי״ [1]ב״ה] ד״ה **ס**

מופת זה: שנקח לא״ה וה״ב וג״ז וז״ד כפלים שוים, והם ח״ט וט״כ ול״מ ומ״נ. ויהיה שעור כפלי ח״ט לא״ה כשיעור כפלי ט״כ לה״ב. ואם כן שיעור כפלי ח״ט לא״ה כשעור כפלי ח״כ לא״ב. ושעור כפלי ח״ט לא״ה כשעור כפלי ל״מ לג״ז. ואם כן שעור כפלי ח״כ לא״ב כשעור כפלי ל״מ לג״ז. ועוד שעור כפלי ל״מ לג״ז כשיעור כפלי מ״נ לז״ד. ואם כן שיעור כפלי ל״מ לג״ז כשיעור כפלי ל״נ לג״ד. וכבר התבאר כי שעור כפלי ל״מ לג״ז כשעור כפלי ח״כ לא״ב. ואם כן שעור כפלי ח״כ לא״ב כשעור כפלי ל״נ לג״ד. וחט״כ למ״נ הם כפלים שוים לא״ב וג״ד. ונקח לה״ב וז״ד כפלים שוים אחרים, והם כ״ס ונ״ע. ויהיה שעור כפלי הראשון, והוא ט״כ, לשני, והוא ה״ב, כשיעור כפלי השלישי, והוא מ״נ, לרביעי, והוא ז״ד. וכמו כן כפלי החמשי, והוא כ״ס, לשני, והוא ה״ב, ככפלי הששי, והוא נ״ע, לרביעי, והוא ז״ד. וכאשר יתחברו הראשון והחמשי, והם ט״כ כ״ס[1], יהיו כפליהם לשני, והוא ה״ב, כשעור כפלי השלישי והששי כשיתחברו, והם מ״ע, לרביעי, והוא ז״ד. ושעור א״ב אצל ב״ה כשיעור ג״ד אצל ד״ז. וכבר לוקח לא״ב ולג״ד כפלים שוים, והם ח״כ ול״נ. ולה״ב וז״ד כפלים שוים אחרים כמו שהם, והם ט״ס ומ״ע. ויהיה שיעור תוספת ח״כ על ט״ס כשיעור תוספת ל״נ על מ״ע.

بُرْهانُهُ: أَنَّا نَأْخُذُ ل ا ه و ه ب و ج ز و ز د أَضْعافًا مُتَساوِيَةً، وَهِيَ ح ط و ط ك و ل م و م ن. فَقَدْرُ أَضْعافِ ح ط ل ا ه كَقَدْرِ أَضْعافِ ك ط ل ه ب. فَقَدْرُ أَضْعافِ ح ط ل ا ه كَقَدْرِ أَضْعافِ ح ك ل ا ب. وَقَدْرُ أَضْعاف ح ط ل ا ه كَقَدْرِ أَضْعافِ م ل ل ج ز. فَقَدْرُ أَضْعافِ ح ك ل ا ب كَقَدْرِ أَضْعافِ ل م ل ج ز. وَقَدْرُ أَضْعافِ ل م ل ج ز كَقَدْرِ أَضْعافِ م ن ل ز د. فَقَدْرُ أَضْعافِ م ل ل ج ز كَقَدْرِ أَضْعافِ ل ن ل ج د. وَقَدْ كانَ اسْتَبانَ أَنَّ قَدْرَ أَضْعافِ ل م ل ج ز كَقَدْرِ أَضْعافِ ح ك ل ا ب. فَقَدْرُ أَضْعافِ ح ك ل ا ب كَقَدْرِ أَضْعافِ ل ن ل ج د. ح ط ك ل م ن هِيَ أَضْعافٌ مُتَساوِيَةٌ ل ا ب و ج د. وَنَأْخُذُ ل ب ه و ز د أَضْعافًا مُتَساوِيَةً أُخْرَى، وَهِيَ ك س ن ع. فَقَدْرُ أَضْعافِ الأَوَّلِ، وَهُوَ ط ك، لِلثَّانِي، وَهُوَ ه ب، كَقَدْرِ أَضْعافِ الثَّالِثِ، وَهُوَ م ن، لِلرَّابِعِ، وَهُوَ ز د. وَقَدْرُ أَضْعافِ الخامِسِ، وَهُوَ ك س، لِلثَّانِي، وَهُوَ ه ب، كَقَدْرِ أَضْعافِ السَّادِسِ، وَهُوَ ن ع، لِلرَّابِعِ، وَهُوَ ز د. وَإِذا جَمَعْنا الأَوَّلَ وَالخامِسَ، وَهُما ط س، يَكُونُ قَدْرُ أَضْعافِهِ لِلثَّانِي، وَهُوَ ه ب، كَقَدْرِ أَضْعافِ الثَّالِثِ وَالسَّادِسِ إِذا جُمِعا، وَهُما م ع، لِلرَّابِعِ،

---

[1] כ״ס] ⟨⟨*⟩⟩ נראה שצ״ל: ט״ס

بُرْهانُ ذَلِكَ: أَنَّا نَأخُذُ لأَقْدارِ ا ه ه ب ج ز ز د أَضْعافَ مُتَساوِيَةً، وهي ح ط ط ك ل م م ن. فأَضْعافُ ح ط لِقَدْرِ ا ه كأَضْعافِ ط ك لِقَدْرِ ه ب. فأَضْعافُ ح ط لِقَدْرِ ا ه كأَضْعافِ ح ك لِقَدْرِ ا ب. وأَضْعافُ ح ط لِقَدْرِ ا ه كأَضْعافِ ل م لِقَدْرِ ج ز. فأَضْعافُ ح ك لِقَدْرِ ا ب كأَضْعافِ ل م لِقَدْرِ ج ز. وأيضًا فإنَّ أَضْعافَ ل م لِقَدْرِ ج ز كأَضْعافِ م ن لِقَدْرِ ز د. فأَضْعافُ ل م لِقَدْرِ ج ز كأَضْعافِ ل ن لِقَدْرِ ج ز. وَقَدْ كانَ اسْتَبانَ أَنَّ أَضْعافَ ل م لِقَدْرِ ج ز كأَضْعافِ ح ك لِقَدْرِ ا ب. فأَضْعافُ ح ك لِقَدْرِ ا ب كأَضْعافِ ل ن لِقَدْرِ ج د[1]. ونأخُذُ لِقَدْرَيْ ه ب ز د أَضْعافًا مُتَساوِيَةً، وهي ك س ن ع. فأَضْعافُ ط ك الأَوَّل لِقَدْرِ ه ب الثَّاني كأَضْعافِ م ن الثَّالث لِقَدْرِ ز د الرَّابِع. وكذلكَ أَضْعافُ ك س الخامس لِقَدْرِ ه ب الثَّاني كأَضْعافِ ن ع السَّادِس لِقَدْرِ ز د الرَّابِع. فأَضْعافُ الأَوَّل والخامس مجْموعَيْن، وَهُما ط س، لِقَدْرِ ه ب الثَّاني، كأَضْعافِ الثَّالث والسَّادِس مجْموعَيْن، وَهُما م ع، لِقَدْرِ ز د الرَّابِع. ونِسْبَة ا ب إلى ب ه كنِسْبَة ج د إلى د ز. وَقَدْرا ح ك ل ن أَضْعافٌ مُتَساوِيَةٌ لِقَدْرَيْ ا ب ج د. وَقَدْرا ط س م ع أَضْعافٌ مُتَساوِيَةٌ

[1] ج د] F +فقدرا ح ك ل ن هي أضعاف متساوية لقدري ا ب ج د

מופתו: אנחנו נקח לשיעורי א״ה וה״ב וג״ז וז״ד כפלים שוים, והם ח״ט וט״כ ול״מ ומ״נ. אם כן כפלי ח״ט לשיעור[1] א״ה ככפלי ט״כ לשיעור ה״ב. אם כן כפלי ח״ט לשיעור א״ה[2] ככפלי ח״כ[3] לשעור א״ב. וכפלי ח״ט לשעור א״ה ככפלי ל״מ לשעור ג״ז, אם כן כפלי ח״כ לשעור א״ב ככפלי ל״מ לשעור ג״ז. וגם כן הנה כפלי ל״מ לשעור ג״ז ככפלי מ״נ לשעור ז״ד. אם כן כפלי ל״מ[4] לשעור ג״ז ככפלי ל״נ לשעור ג״ז. וכבר התבאר כי כפלי ל״מ לשיעור ג״ז[5] ככפלי ח״כ[6] לשיעור א״ב. אם כן[7] כפלי ח״כ לשעור א״ב ככפלי ל״נ לשעור ג״ד. ונקח לשעורי ה״ב וז״ד כפלים שוים[8] והם כ״ס ונ״ע. הנה כפלי ט״כ הראשון לשעור ה״ב השני ככפלי מ״נ השלישי לשעור ז״ד הרביעי. וכן כפלי[9] כ״ס החמישי לשיעור ה״ב השני ככפלי נ״ע הששי לשעור ז״ד הרביעי. אם כן כפלי הראשון והחמישי מקובצים, והם[10] ט״ס[11], לשעור ה״ב השני, ככפלי[12] השלישי והששי מקובצים, והם מ״ע, לשעור ז״ד הרביעי[13]. ויחס א״ב אל ב״ה כיחס ג״ד אל ד״ז. ושני שעורי ח״כ[14] ול״נ כפלים שוים לשני שעורי א״ב וג״ד. ושני שעורי ט״ס[15] מ״ע כפלים שוים לשני שעורי ה״ב וז״ד. אם כן שני[16] שעורי ח״כ ול״נ אם נוספים יחד על שני שעורי ט״ס ומ״ע, ואם שוים יחד לשניהם, ואם חסרים יחד משניהם.

[1] לשיעור] לשעורי ב [2] ככפלי ... א״ה] ליתא בוטפ ⟨⟨*⟩⟩ ק [3] ח״כ] כ׳ ב [4] ל״מ] ~~מ״נ~~+ פ [5] ל״נ ... ג״ז] ליתא ו [6] ח״כ] ס״ב; אולי מחוקה [7] כן] ליתא ו [8] שוים] ⟨*⟩ ט [9] כפלי] ⟨*⟩ ק [10] והם] ליתא ו [11] ט״ס] ט״ח ס [12] ככפלי] כשעור כפלי ק [13] הרביעי] ~~החמישי~~+ פ [14] ח״כ] מ״ב ס [15] ט״ס] ט״ו ב [16] שני] ליתא ס

وَهُوَ ز د. فَقَدْرُ ا ب عِنْدَ ب ه كَقَدْرِ ج د عِنْدَ د ز. وَقَدْ أُخِذَ ل ا ب ج د أَضْعافٌ مُتَساوِيَةٌ، وَهِيَ ح ك و ل ن. ول ه ب و ز د أَضْعافٌ مُتَساوِيَةٌ أُخَرُ، كَيْفَما كانَتْ، وَهِيَ ط س و م ع. فَقَدْرُ زِيادَةِ ح ك عَلَى ط س كَقَدْرِ زِيادَةِ ل ن عَلى م ع.

وَإِذا نَزَعْنا ط ك و م ن المُشْتَرَكَيْنِ، فَقَدْرُ ما يَزِيدُ ح ط عَلى ك س كَقَدْرِ ما يَزِيدُ ل م عَلى ن ع. وَإِنْ كانَ مُساوِيًا لَهُ، وَإِنْ كانَ ناقِصًا مِنْهُ. فَبِذَلِكَ يُنْقَصُ مِنْهُ. وح ط و ل م هُما أَضْعافٌ مُتَساوِيَةٌ ل ا ه و ج ز. و ك س و ن ع هُما أَضْعافٌ مُتَساوِيَةٌ ل ه ب و ز د. فَقَدْرُ ا ه عِنْدَ ب ه كَقَدْرِ ج ز عِنْدَ ز د.

וכאשר הסירונו ט״כ ומ״נ המשותפים, יהיה שיעור מה שיוסיף ח״ט על כ״ס כשעור מה שיוסיף ל״מ על נ״ע. ואם היה שוה לו גם הוא שוה לו, ואם היה חסר ממנו, גם הוא יחסר בזה השעור ממנו. וח״ט ול״מ הם כפלים שוים לא״ה וג״ז. וכ״ס ונ״ע הם כפלים שוים לה״ב וז״ד. ואם כן שיעור א״ה אצל ה״ב כשעור ג״ז אצל ז״ד.

وَذَلِكَ ما أَرَدْنا بَيانَهُ.

וזה מה שרצינו לבארו.

لِقَدْرَيْ ه ب ز د. فَقَدْرا ح ك ل ن إِمّا زائِدانِ مَعًا على قَدْرَيْ ط س م ع، وإِمّا مُساوِيانِ مَعًا لَهُما، وإِمّا ناقِصانِ مَعًا مِنْهُما.

فإِنْ نَقَّصْنا ط ك و م ن المُشْتَرَكَيْنِ، فَقَدْرا ح ط ل م إِمّا زائِدانِ مَعًا على قَدْرَيْ ك س ن ع، وإِمّا مُساوِيانِ مَعًا لَهُما، وإِمّا ناقِصان مَعًا مِنْهُما. وح ط و ل م أَضْعاف مُتَساوِيَة لِقَدْرَيْ ا ه ج ز. و ك س ن ع أَضْعاف مُتَساوِيَة لِقَدْرَيْ ه ب ز د. فنِسْبَة ا ه إلى ه ب كَنِسْبَةِ ج ز إلى ز د.

ואם חסרנו ט״כ ומ״נ המשותפים, הנה שני שעורי ח״ט ול״מ אם נוספים יחד על שני שעורי כ״ס ונ״ע, ואם שוים יחד לשניהם[1], ואם חסרים יחד משניהם. וח״ט ול״מ כפלים שוים לשני שעורי א״ה[2] וג״ז. וכ״ס ונ״ע כפלים שוים לשני שעורי ה״ב וז״ד. אם כן יחס א״ה אל ה״ב כיחס ג״ז אל ז״ד.

وَذَلِكَ ما أَرَدْنا أَنْ نُبَيِّنَ.

וזה מה שרצינו לבאר.

---

1 לשניהם] +~~ואם חסרים יחד לשניהם~~ ס

2 א״ה] ליתא ב

[Proposition v.18]

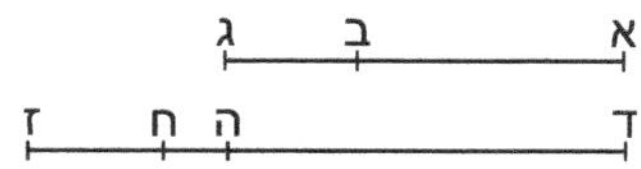

כשיהיו שעורים נחלקים מתיחסים, אם יורכבו יהיו גם הם מתיחסים.

إِذا كانَتْ أَقْدارٌ مُفَصَّلَةٌ مُتَناسِبَةً، فَإِنَّها إِذا رُكِّبَتْ تَكونُ مُتَناسِبَةً.

המשל בזה: ששיעורי א״ב וב״ג וד״ה וה״ז נחלקים מתיחסים—שיעור א״ב אצל ב״ג כשיעור ד״ה אצל ה״ז. ואומר שאם יורכבו יהיו מתיחסים, ויהיה שיעור א״ג אצל ב״ג כשעור ד״ז אצל ה״ז.

مِثالُهُ: أَنَّ أَقْدارَ ا ب و ب ج و د ه و ه ز مُفَصَّلَةٌ مُتَناسِبَةٌ—قَدْرُ ا ب عِنْدَ ب ج كَقَدْرِ د ه عِنْدَ ه ز. فَأَقولُ إِنَّها إِنْ رُكِّبَتْ تَكونُ مُتَناسِبَةً، قَدْرُ ا ج عِنْدَ ج ب كَقَدْرِ د ز عِنْدَ ز ه.

إذا كانَتْ أَقْدارٌ مُفَصَّلَةٌ مُتَناسِبَةً، فَإِنَّهُ إذا رُكِّبَتْ تكونُ مُتَناسِبَةً.

مَثَالُهُ: أَنَّ أَقْدارَ ا ب ب ج د ه ه ز مُتَنَاسِبَةٌ[1]— نِسْبَةُ ا ب إلى ب ج كَنِسْبَةِ د ه إلى ه ز. فَأَقولُ إنَّها إذَا رُكِّبَتْ تكونُ مُتَنَاسِبَةً، فَتَكونُ^ نِسْبَةُ ا ج إلى ج ب كَنِسْبَةِ د ز إلى ز ه.

כאשר היו שעורים נבדלים מתיחסים, הנה הם כאשר הורכבו יהיו מתיחסים.

המשל בו: כי שעורי א״ב וב״ג וד״ה וה״ז מתיחסים[1]—יחס א״ב אל ב״ג[2] כיחס ד״ה אל ה״ז. הנה אומר שהם כאשר[3] הורכבו יהיו מתיחסים, יחס[4] א״ג אל ג״ב כיחס ד״ז אל ז״ה.

[1] متناسبة] R1R2 +مفصلة

[1] מתיחסים] יהיו+ **ק** [2] ב״ג] ~~א״ג~~+ **ס**

[3] כאשר] ⟨*⟩ **ק** [4] יחס] ליתא **בוטספ**

מופת זה: שאם לא יהיה שעור א״ג אצל ב״ג כשעור ד״ז[1] אצל ה״ז, יהיה שיעור א״ג אצל ב״ג כשעור ד״ז אצל שעור שהוא קטן או גדול מה״ז. ויהיה ראשונה אצל מי שהוא קטן ממנו, אם הוא אפשר, והוא ז״ח. ויהיה שיעור א״ג אצל ג״ב כשעור ד״ז אצל ז״ח. וכאשר חלקנו היה שעור א״ב אצל ב״ג כשעור ד״ח אצל ח״ז. ואם כן שיעור ד״ח[2] אצל ח״ז כשעור ד״ה אצל ה״ז. וכאשר החלפנו היה שיעור ח״ד אצל ד״ה כשעור ח״ז אצל ז״ה. וח״ד גדול מד״ה—יהיה ח״ז גדול מז״ה, הקטן רב מן הגדול, זה חלוף. אם כן אין שיעור א״ג אצל ב״ג כשעור ד״ז אצל שעור שהוא קטן מז״ה. וכמו כן יתבאר שלא הוא אצל גדול ממנו. ואם כן שעור א״ג אצל ג״ב כשעור ד״ז אצל ז״ה.

וזה מה שרצינו לבאר.

بُرْهانُهُ: أَنَّهُ إِنْ لَمْ يَكُنْ قَدْرُ ا ج عِنْدَ ج ب كَقَدْرِ د ز عِنْدَ ز ه، فَإِنَّ قَدْرَ ا ج عِنْدَ ج ب كَقَدْرِ ز د عِنْدَ ما هُوَ أَصْغَرُ أَوْ أَكْبَرُ مِنْ ز ه. فَلْيَكُنْ أَوَّلًا ما هُوَ أَصْغَرُ، إِنْ كانَ يُمْكِنُ، وَهُوَ ز ح. فَقَدْرُ ا ج عِنْدَ ج ب كَقَدْرِ د ز عِنْدَ ز ح. وَإِذا فَصَلْنا كانَ قَدْرُ ا ب عِنْدَ ب ج كَقَدْرِ د ح عِنْدَ ح ز. فَقَدْرُ د ح عِنْدَ ح ز كَقَدْرِ د ه عِنْدَ ز ه. فَإِذا بَدَّلْنا كانَتْ قَدْرُ د ح عِنْدَ د ه كَقَدْرِ ح ز عِنْدَ ز ه. ح د أَعْظَمُ مِنْ د ه. ح ز أَعْظَمُ مِنْ ز ه، الأَصْغَرُ أَعْظَمُ مِنَ الأَكْبَرِ، هَذا خُلْفٌ لا يُمْكِنُ. فَلَيْسَ قَدْرُ ا ج عِنْدَ ج ب كَقَدْرِ د ز عِنْدَ قَدْرٍ هُوَ أَصْغَرُ مِنْ (ز ه)[1]. وَكَذَلِكَ (تَبَيَّنَ)[2] وَلا عِنْدَ أَكْبَرَ مِنْهُ. فَقَدْرُ ا ج عِنْدَ ج ب كَقَدْرِ د ز عِنْدَ ز ه.

وَذَلِكَ ما أَرَدْنا بَيانَهُ.

---

[1]ד״ז] ~~ב״ז~~ ⟨*⟩ [2]ד״ח] ⟨⟨*⟩⟩

[1]ز ه] د ه [2]تبين] نبين

بُرهانُه: أنَّهُ إنْ لم تَكُنْ نِسْبَةُ اج إلى ج ب كَنِسبَةِ د ز إلى ز ه، فإنَّ نِسْبَةَ اج إلى ج ب كَنِسْبَةِ د ز إلى ما هُوَ أَصْغَرُ أو أَكبَرُ مِنْ ز ه. فَلْيَكُنْ أوَّلًا إلى ما هُوَ أَصْغَرُ مِنْهُ، وهُوَ ز ح. فَنِسْبَةُ اج إلى ج ب كَنِسْبَةِ د ز إلى ز ح. وإذا فَصَلْنَا، فَنِسْبَةُ اب إلى ب ج كَنِسبَةِ د ح إلى ح ز. وَقَدْ كانَتْ نِسبَةُ اب إلى ب ج كَنِسْبَةِ د ه إلى ه ز. فَنِسْبَةُ د ح إلى ح ز كَنِسْبَةِ د ه إلى ه ز. وح د الأَوَّل أعظم مِنْ د ه الثَّالِث. ف ح ز الثَّاني أعظم مِنْ ز ه الرَّابِع. وَقَدْ كانَ ز ه أَعْظَم مِنْ ز ح. فالأَصْغَر إذًا أَعْظَم مِنَ الأَعْظَم[1]—هذا خُلْفٌ. فَلَيْسَتْ نِسْبَةُ اج إلى ج ب كَنِسْبَةِ د ز إلى قَدْرٍ هُوَ أَصْغَر مِنْ ز ه. وكَذَلِكَ يَتَبَيَّنُ أنه لَيْسَتْ نِسْبَةُ اج إلى ج ب كَنِسْبَةِ د ز إلى قَدْرٍ هُوَ أَعْظَم مِنْ ز ه.

وَذَلِكَ ما أَرَدْنا أَنْ نُبَيِّنَ.

מופתו: אם לא יהיה יחס א״ג אל ג״ב כיחס ד״ז אל ז״ה, הנה יחס א״ג[1] אל ג״ב כיחס ד״ז אל[2] מה שהוא יותר[3] קטן[4] או יותר גדול מן ז״ה. ויהיה תחלה אל מה שהוא יותר קטן ממנו, והוא ז״ח. הנה יחס א״ג אל ג״ב כיחס ד״ז[5] אל ז״ח. וכאשר הבדלנו, הנה[6] יחס א״ב אל ב״ג כיחס ד״ח אל ח״ז. וכבר היה יחס א״ב אל ב״ג כיחס ד״ה אל ה״ז. אם כן[7] יחס ד״ח אל ח״ז כיחס ד״ה אל ה״ז. וח״ד הראשון יותר גדול מן ד״ה השלישי. אם כן ח״ז השני יותר גדול מן ז״ה הרביעי. וכבר היה ז״ה[8] יותר גדול מן ז״ח. הנה הקטן אם כן[9] יותר גדול מן הגדול—זה שקר. אם כן אין יחס א״ג אל ג״ב כיחס ד״ז אל שעור הוא[10] יותר קטן מן ז״ה. וכן יתבאר כי יחס[11] א״ג אל ג״ב אינו כיחס ד״ז אל שעור הוא[12] יותר גדול מן ז״ה.

וזה מה שרצינו לבאר.

[1] من الأعظم] PR$_1$R$_2$T$_2$T$_3$

[1] א״ג] ~~ד״ז~~+ פ [2] אל ... אל] ⟨⟨(*)⟩⟩ ק [3] יותר] ליתא ס [4] קטן] +~~ממנו~~ ב +ממנו ס [5] ד״ז] ד״ה ס [6] הנה] יהיה ק [7] כן] ⟨(*)⟩ ו [8] היה ז״ה] ז״ה היה ס [9] אם כן] ⟨(*)⟩ ט [10] הוא] ליתא ס [11] כי יחס] כיחס ב [12] הוא] ליתא ט

[Proposition v.19]

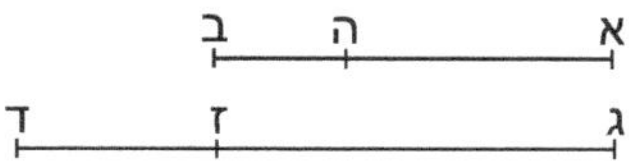

כשחוסרו שני שיעורים משני שיעורים מכל אחד מהם שיעור, והיה יחס המחוסר אצל המחוסר כיחס הכל אל הכל, יהיה יחס הנשאר אל הנשאר כיחס הכל אל הכל.

إذا كانَ قَدْرانِ، وَنَقَصَ مِنْ كُلِّ واحِدٍ مِنْهُما قَدْرٌ، وَكانَ الكُلُّ إِلى الكُلِّ كَقَدْرِ المَنْقوصِ مِنَ المَنْقوصِ، فَإِنَّ قَدْرَ الباقي مِنَ الباقي كَقَدْرِ الكُلِّ مِنَ الكُلِّ.

המשל בזה: ששעור א״ב חוסר ממנו א״ה, ושעור ג״ד חוסר ממנו ג״ז. והיה שעור א״ב אצל שעור ג״ד כשעור א״ה אצל ג״ז. ואומר שיהיה שעור ב״ה הנשאר אצל ז״ד הנשאר כשעור כל א״ב אצל כל ג״ד.

مِثالُهُ: أَنَّ قَدْرَ ا ب نَقَصَ مِنْهُ ا ه. وَقَدْرُ ج د نَقَصَ مِنْهُ ج ز. وَقَدْرُ ا ب عِنْدَ ج د كَقَدْرِ ا ه عِنْدَ ج ز. فَأَقولُ إِنَّ قَدْرَ ه ب الباقي عِنْدَ ز د الباقي كَقَدْرِ ا ب عِنْدَ ج د.

إذا نَقَصَ مِنْ مقدارَيْنِ[1] مِنْ كُلِّ واحِدٍ مِنْهُما قَدْرٌ، فَكَانَتْ نِسْبَةُ المَنْقُوصِ إلى المَنْقُوصِ كَنِسْبَةِ الكُلِّ إلى الكُلِّ، فَإِنَّ نِسْبَةَ البَاقِي إلى البَاقِي كَنِسْبَةِ الكُلِّ إلى الكُلِّ.

כאשר חוסר משני שעורים מכל אחד משניהם שעור, והיה יחס המחוסר אל המחוסר כיחס הכל אל הכל, הנה יחס[1] הנשאר אל הנשאר כיחס הכל אל הכל.

مِثَالُهُ: أَنَّ ا ب نَقَصَ مِنْهُ ا ه، ج د نَقَصَ مِنْهُ ج ز. وِنِسْبَةَ ا ه إلى ج ز كَنِسْبَةِ ا ب إلى ج د^[2]. فَأَقُولُ إِنَّ نِسْبَةَ ه ب الباقي إلى ز د الباقي كَنِسْبَةِ ا ب الكُلِّ إلى ج د الكُلِّ.

המשל בו: כי א״ב[2] חוסר ממנו א״ה. וג״ד חוסר ממנו ג״ז. ויחס א״ה אל ג״ז כיחס א״ב הכל אל ג״ד הכל. הנה אומר כי יחס ה״ב הנשאר אל ז״ד הנשאר[3] כיחס א״ב הכל אל ג״ד הכל[4].

[1] مقدارين] Q قدرين [2] ونسبة ... ج د] F ونسبة ا ب الى ج د كنسبة ا ه الى ج د

[1] יחס] ~~יחס~~+ **פ** [2] כי א״ב] +* **ס** [3] אל ... הנשאר] ליתא **ב** [4] הכל] ליתא **בוס**

מופת זה: כי שיעור א״ב אצל ג״ד כשיעור א״ה אצל ג״ז. וכאשר החלפנו היה שיעור א״ב אצל א״ה כשעור ג״ד אצל ג״ז. וכאשר חלקנו היה שעור ב״ה אצל א״ה כשעור ד״ז אצל ז״ג. ואם החלפנו היה שעור ב״ה אצל ז״ד כשעור א״ה אצל ג״ז. ושעור א״ה אצל ג״ז היה כשעור א״ב אצל ג״ד. וכבר נשאר שיעור ה״ב אצל שיעור ז״ד הנשאר כשיעור כל א״ב אצל כל ג״ד.

וזה מה שרצינו שנבאר.

بُرْهانُهُ: أَنَّ قَدْرَ ا ب عِنْدَ ج د كَقَدْرِ ا ه عِنْدَ ج ز. وَإِذا بَدَّلْنَا كانَ قَدْرُ ب ا عِنْدَ ا ه كَقَدْرِ ج د عِنْدَ ج ز. وَإِنْ فَصَلْنا كانَ قَدْرُ ب ه عِنْدَ ا ه كَقَدْرِ د ز عِنْدَ ج ز. وَإِنْ بَدَّلْنا يكونُ قَدْرُ ب ه عِنْدَ ز د كَقَدْرِ ا ه عِنْدَ ج ز. وَقَدْرُ ا ه عِنْدَ ج ز كَذَلِكَ قَدْرُ ا ب عِنْدَ ج د. فَقَدْرُ ب ه عِنْدَ د ز كَقَدْرِ ا ب عِنْدَ ج د. وَقَدْرُ ا ه عِنْدَ ج ز كَقَدْرِ ا ب عِنْدَ ج د، وَهِيَ قَدْرُ ب ه عِنْدَ ز د الباقي كَقَدْرِ الكُلِّ، وَهُوَ ا ب، عِنْدَ الكُلِّ، وَهُوَ ج د.

وَذَلِكَ ما أَرَدْنا بَيانَهُ.

بُرْهانُ ذَلِكَ: أَنَّ نِسْبَةَ ا ب إلى ج د كَنِسْبَةِ ا ه إلى ج ز. وإذا بَدَّلْنا، فَنِسْبَةُ ب ا إلى ا ه كَنِسْبَة د ج إلى ج ز. وإذا فَصَلْنَا، فَنِسْبَة ب ه إلى ه ا كَنِسْبَةِ د ز إلى ز ج. وإذا بَدَّلْنا، فَنِسْبَة ب ه إلى ز د كَنِسْبَةِ ا ه إلى ج ز. وَقَدْ كانَتْ نِسْبَةُ ا ه إلى ج ز كَنِسْبَةِ ا ب إلى ج د. فَنِسْبَة ب ه إلى ز د كَنِسْبَةِ ا ب إلى ج د.

מופתו: כי יחס א״ב אל ג״ד כיחס א״ה אל ג״ז. וכאשר המירונו, הנה יחס א״ב אל א״ה כיחס ד״ג אל ג״ז. וכאשר הבדלנו, הנה יחס ב״ה אל א״ה כיחס ד״ז אל ז״ג. וכאשר המירונו, הנה יחס ב״ה אל[1] ז״ד כיחס א״ה אל ג״ז. וכבר היה יחס[2] א״ה אל ג״ז כיחס א״ב אל ג״ד. אם כן יחס ב״ה אל ז״ד כיחס א״ב אל ג״ד[3].

وَذَلِكَ ما أَرَدْنا أَنْ نُبَيِّنَ.

וזה מה שרצינו לבאר.

[1] א״ה ... אל] ליתא ז [2] יחס] ליתא ז [3] אם ... ג״ד] ⟨*⟩ ס

[Proposition v.20]

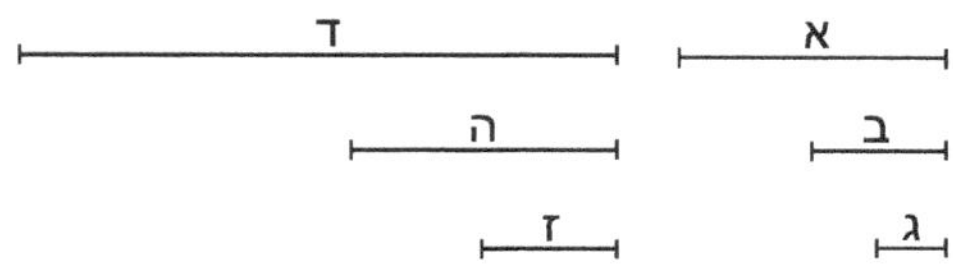

כשיהיו שעורים[1] מה ושעורים אחרים על מספרם, כל שנים מן הראשונים על יחס שנים מן האחרים, אם יהיה הראשון מן הראשונים ביחס השווי גדול מן האחרון, יהיה הראשון מן האחרים גדול מן האחרון. ואם היה שוה לו, גם הוא שוה לו. ואם היה קטן ממנו, גם הוא קטן ממנו.

לשון אחר בהעתקה אחרת דוק: אם היו שלשה שיעורים ואחרים כפי מספרם, וכל שנים שיוקחו מן השלשה הם על יחס[2] שנים מן האחרים, אם יהיה הראשון ביחס השווי גדול מן (השלישי)[3], יהיה (השני)[4] גדול מן הרביעי. ואם היו שוים יהיו גם הם שוים. ואם היה קטן ממנו גם הוא יהיה קטן ממנו.

إِذا كانَتْ (ثَلاثَةُ)[1] أَقْدارٍ وَأُخَرُ على عِدَّتِها، وَكُلُّ اثْنَيْنِ يُؤْخَذانِ مِنَ (الثَّلاثَةِ)[2] على نِسْبَةِ اثْنَيْنِ مِنَ الثَّلاثَةِ الأُخَرِ، فَإِنَّ الأَوَّلَ في نِسْبَةِ المُساواةِ إِنْ كانَ أَعْظَمَ مِنَ الثَّالِثِ، فَإِنَّ الثَّاني أَعْظَمُ مِنَ الرَّابِعِ. فَإِنْ كانَا مُتَساوِيَيْنِ، فَهُما مُتَساوِيانِ. فَإِنْ كانَ أَصْغَرَ مِنْهُ، فَهُوَ أَصْغَرُ مِنْهُ.

---

[1] שעורים] ⟨⟨*⟩⟩ [2] יחס] [∵] ⟨*⟩ [3] השלישי] השני [4] השני] השלישי

---

[1] ثلاثة] ثلث [2] الثلاثة] الثلثة

إذا كانَتْ أَقْدارٌ ما كانَتْ وأَقْدارٌ أُخَر عَلَى عِدَّتِها، كُلُّ اثْنَيْنِ مِنَ الأَوَّلِ عَلَى نِسْبَةِ اثْنَيْنِ مِنَ الأُخْرَى[1]، فَإِنَّ الأَوَّلَ مِنَ الأُولى في نِسْبَةِ المُساواةِ—إِنْ كانَ أَعْظَم مِنَ الأَخيرِ، فَإِنَّ الأَوَّلَ مِنَ الأُخْرى أَعْظَمُ مِنَ الأَخيرِ. وَإِنْ كانَ مُساوِيًا لَهُ[2]، فَهُوَ مُساوٍ لَهُ. وَإِنْ كانَ[3] أَصْغَرَ مِنْهُ، فَهُوَ أَصْغَرُ مِنْهُ.

כאשר היו שעורים מה שיהיו ושעורים אחרים על מספרם, כל שנים מן הראשון על יחס שנים מן[1] האחר, הנה הראשון[2] מן הראשון ביחס השווי—אם היה יותר גדול מן האחרון, הנה הראשון מן האחר יותר גדול מן האחרון[3]. ואם היה שוה לו, הנה הוא שוה לו. ואם היה יותר קטן ממנו, הנה הוא יותר קטן ממנו.

[1] الاخرى] CRamT1 +وانتظمت النسب [2] له] QR1R2T2T3 [3] كان] CMRamT1 +آ

[1] מן] כל **ב** [2] הראשון] ⟨ה⟩ראשון **ב** [3] הנה ... האחרון] ⟨⟨*⟩⟩ **סק**

המשל בזה: ששלשה שיעורים, עליהם א׳ וב׳ וג׳, ושלשה שעורים אחרים, עליהם ד׳ וה׳ וז׳—כל שנים שילקחו מהם, הם עליהם שוים מן השלשה הראשונים. ויהיה שעור א׳ אצל ב׳ כשעור ד׳ אצל ה׳. ושעור ב׳ אצל ג׳ כשעור ה׳ אצל ז׳. ונשים א׳ גדול מג׳. ואומר שד׳ גדול מז׳.

مثالُهُ: أنّا نَضَعُ (ثلاثَةَ)[1] أَقْدارٍ عَلَيْها ا و ب و ج. ونَضَعُ (ثلاثَةَ)[2] أَقْدارٍ أُخَر، كُلّ اثْنَيْنِ يُؤخَذانِ مِنْها، وَهُما عَلى نِسْبَةِ اثْنَيْنِ مِنْ تِلْكَ (الثَّلاثَةِ)[3] الأُوَلِ، عَلَيْها د و ه و ز. وَقَدْرُ ا عِنْدَ ب كَقَدْرِ د عِنْدَ ه. وَقَدْرُ ب عِنْدَ ج كَقَدْرِ ه عِنْدَ ز. ونَجْعَلُ ا أَعْظَمَ مِنْ ج. فَأَقولُ إِنَّ د أَعْظَمُ من ز.

מופת זה: לפי שא׳ גדול מג׳, ושעור ב׳ שעור אחר, יהיה א׳ רב היחס אצל ב׳ מג׳ אצל ב׳. ואמנם שעור א׳ אצל ב׳ כשעור ד׳ אצל ה׳. ושעור ג׳ אצל ב׳ כשעור ז׳ אצל ה׳. ואם כן יהיה ד׳ רב היחס אצל ה׳ מז׳ אצל ה׳. ואשר יחסו רב[1] הוא הגדול. אם כן ד׳ גדול מז׳. וכמו כן יתבאר שאם היה א׳ שוה לג׳, יהיה ד׳ שוה לז׳. ואם א׳ קטן מג׳, יהיה ד׳ קטן מז׳.

بُرْهانُهُ: أَنَّ ا أَعْظَمُ مِنْ ج. وَقَدْرُ ب قَدْرٌ آخَرُ أَكْبَرُ نِسْبَةً إِلى ب مِنْ[4] ج إِلى ب. وَلَكِنَّ قَدْرَ ا عِنْدَ ب كَقَدْرِ د عِنْدَ ه. وَقَدْرُ ج عِنْدَ ب كَقَدْرِ ز عِنْدَ ه. د أَكْبَرُ نِسْبَةً إِلى ه مِنْ ز إِلى ه. وَالَّذي نِسْبَتُهُ أَكْبَرُ هُوَ أَعْظَمُ. د أَعْظَمُ مِنْ ز. وَكَذَلِكَ (تَبَيَّنَ)[5] أَنَّ ا لو كانَ مُساوِيًا ل ج لَكانَ د مُساوِيًا ل ز. وَلَوْ كانَ ا أَصْغَرَ مِنْ ج لَكانَ د أَصْغَرَ مِنْ د.

וזה מה שרצינו לבאר.

وَذَلِكَ ما أَرَدْنا بَيانَهُ.

[1] רב] [ꢀ]+

[1] ثلاثة] ثلثة [2] ثلاثة] ثلثة [3] الثلاثة] الثلثة [4] ب من] ⟨*⟩ [5] تبين] نبين

مِثَالُهُ: أَنَّ أَقْدَارَ ا ب ج على عِدَّةِ أَقْدَارِ د ه ز. وكُلّ اثْنَيْنِ مِنْهَا على نِسْبَةِ اثْنَيْنِ مِنْهَا—نِسْبَةُ ا إلى ب كَنِسْبَةِ د إلى ه، ونِسْبَةَ ب إلى ج كَنِسْبَةِ ه إلى ز. ونَجْعَلُ الأَوَّلَ مِنَ الأُولى، وَهُوَ ا، أعظم مِنَ الأَخيرِ، وَهُوَ ج. فَأَقولُ إنّ الأَوَّلَ من الأُخرى، وَهُوَ د، أعظم مِنَ الأَخيرِ، وَهُوَ ز.

המשל בו: כי שעורי א׳ ב׳ ג׳ על מספר שעורי ד׳ ה׳ ז׳. וכל שנים מהם על יחס שנים מהם—יחס א׳ אל ב׳ כיחס ד׳ אל ה׳, ויחס ב׳ אל ג׳ כיחס ה׳ אל ז׳. ונשים הראשון מן הראשון, והוא א׳, יותר גדול מן האחרון, והוא ג׳. הנה אומר כי הראשון מן האחר, והוא ד׳, יותר גדול מן האחרון והוא ז׳.

بُرْهانُ ذَلِكَ: أَنَّ ا أعظم مِنْ ج، وَقَدْرَ ب قَدْرٌ آخر. فَقَدْرُ ا أَكْبَرَ نِسْبَةً إلى ب مِنْ ج إلى ب. وَلَكِنَّ نِسْبَةَ ا إلى ب كَنِسْبَةِ د إلى ه. ونِسْبَةَ ج إلى ب كَنِسْبَةِ ز إلى ه. فَقَدْرُ د أَكْبَرَ نِسْبَةَ إلى ه مِنْ ز إلى ه. والَّذي نِسْبَتُهُ أَكْبَرَ فَهُوَ أعظم. فَقَدْرُ د أعظم مِنْ ز. وكذَلِكَ تَبَيَّنَ أَنَّ ا لَوْ كانَ مُساوِيًا ل ج، لَكانَ د مُساوِيًا ل ز، ولَوْ كانَ أَصْغَرَ مِنْ ج، لَكانَ د أَصْغَرَ مِنْ ز.

מופתו: כי א׳ יותר גדול מן ג׳, ושעור ב׳ שעור אחר. אם כן שיעור[1] א׳ יותר גדול היחס אל ב׳ מן ג׳ אל ב׳. אבל יחס א׳ אל ב׳ כיחס ד׳ אל ה׳[2]. ויחס ג׳ אל ב׳ כיחס ז׳ אל ה׳. אם כן שעור ד׳ יותר גדול יחס[3] אל ה׳ מן[4] ז׳ אל ה׳. ואשר יחסו יותר גדול הנה הוא יותר גדול. אם כן שעור ד׳ יותר גדול מן ז׳. וכן התבאר כי א׳ אלו היה שוה אל ג׳, היה ד׳[5] שוה אל ז׳, ואלו היה יותר קטן מן ג׳, היה ד׳ יותר קטן מז׳.

وَذَلِكَ ما أَرَدْنا أَنْ نُبَيِّنَ.

וזה מה שרצינו לבאר[6].

[1]אם ... שיעור] אם שעור כן **ב** [2]ה׳] +אם כן יחס ד׳ אל ה׳ גדול היחס מן ג׳ אל ב׳ **ס** [3]יחס] היחס **ז** ⟨*⟩ **ק** [4]מן] +אל ה׳ מן **ב** [5]ד׳] ⟨*⟩ **פ** [6]וזה ... לבאר] ונשלם ביאורו **ס** ליתא **ק**

[Proposition V.21]

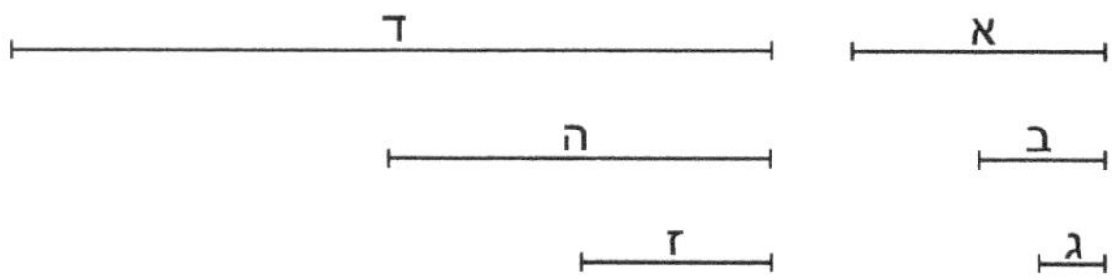

כשיהיו שעורים מה ושעורים אחרים כמספרם, כל שנים מן הראשונים עליהם שנים מן האחרים, והשתוה היחס במערכה: אם יהיה הראשון[1] מן הראשונים כיחס השווי גדול מן האחרון, יהיה הראשון מן האחרים גדול מן האחרון[2], ואם היה שוה לו גם הוא יהיה שוה לו, ואם היה קטן ממנו גם הוא קטן ממנו.

לשון אחר דוק: כשהיו שלשה שעורים ואחרים כפי מספרם, כל שנים שילקחו מהם הם עליהם שוים מן האחרים, והתחלף היחס בשיעורים בקדימה ובאחור: אם הראשון מן הראשונים ביחס השווי הוא גדול מן השלישי יהיה השני גדול מן הרביעי, ואם היה שוה לו גם הוא שוה לו, ואם היה קטן ממנו גם הוא קטן ממנו.

إِذا كانَتْ (ثَلاثَةُ)[1] أَقْدارٍ وَأُخَرُ على عِدَّتِها، كُلُّ اثْنَيْنِ يُؤْخَذانِ مِنَ (الثَّلاثَةِ)[2]، فَهُما عَلى نِسْبَةِ اثْنَيْنِ مِنَ (الثَّلاثَةِ)[3] الأُوَلِ، وَاخْتَلَفَتِ النِّسَبُ في الأَقْدارِ بِالتَّقْديمِ وَالتَّأْخيرِ. فَإِنَّ الأَوَّلَ في نِسْبَةِ المُساواةِ إِنْ كانَ أَعْظَمَ مِنَ الثَّالِثِ، فَإِنَّ الثَّاني أَعْظَمُ مِنَ الرَّابِعِ. وَإِنْ كانَ مُساوِيًا لَهُ، فَهُوَ مُساوٍ لَهُ. وَإِنْ كانَ أَصْغَرَ مِنْهُ، فَهُوَ أَصْغَرُ مِنْهُ.

[1]הראשון] ~~הראשונים~~+ [2]יהיה ... האחרון] ⟨⟨*⟩⟩

[1]ثلاثة] ثلثة [2]الثلاثة] الثلثة [3]الثلاثة] الثلثة

إِذا كَانَتْ أَقْدارٌ ما وَأَقْدارٌ أُخَرُ عَلى عِدَّتِها، كُلُّ اثْنَيْنِ مِنَ الأُولى عَلى نِسْبَةِ اثْنَيْنِ مِنَ الأُخْرى، وَاضْطَرَبَتِ النِّسْبَةُ، فَإِنَّ الأَوَّلَ مِنَ الأُولى فِي نِسْبَةِ المُساوَاة: إِنْ كَانَ أعظم مِنَ الأَخيرِ، فَإِنَّ الأَوَّل مِنَ الأُخرى أعظم مِنَ الأَخيرِ، وَإِنْ كَانَ مُساوِيًا لَهُ فَهُوَ مُساوٍ لَهُ، وَإِنْ كَانَ أَصْغَرَ مِنْهُ فَهُوَ أَصْغَرَ مِنْهُ.

כאשר היו שעורים מה[1] ושעורים אחרים על מספרם, כל שנים מן הראשון על יחס שנים מן האחר, ונתבלבל היחס[2], הנה הראשון[3] מן הראשון כיחס השווי: אם היה יותר גדול[4] מן האחרון, הנה הראשון מן האחר יותר גדול מן האחרון, ואם היה שוה לו הנה הוא שוה לו, ואם היה[5] יותר קטן ממנו הנה הוא יותר קטן ממנו.

שרטוט—**בוטספק**: הקטעים מונחים בשתי שורות: בעליונה א׳ ב׳ ג׳, בסדר זה, ובתחתונה ד׳ ה׳ ז׳, בסדר זה [1]מה] +שיהיו **ב** [2]ונתבלבל היחס] ליתא **ב** [3]הראשון] ⟨ה⟩ראשון **ב** [4]גדול] ⟨*⟩ **פ** [5]היה] ליתא **פ**

המשל בזה: כי שלשה שיעורים עליהם א׳ ב׳ ג׳, ושלשה אחרים עליהם ד׳ וה׳ וז׳, כל שנים שיוקחו מהם הם על יחס שנים מן האחרים, והיחס מתחלף בשיעורים בקדימה ובאחור: שיעור א׳ אצל ב׳ כשיעור ה׳ אצל ז׳, ושיעור ב׳ אצל ג׳ כשיעור ד׳ אצל ה׳. ונשים א׳ גדול מג׳. ואומר שד׳ גדול מז׳.

مِثالُهُ: أَنّا نَضَعُ (ثَلاثَةَ)[1] أَقْدارٍ عَلَيْها ا بَ وجَ، (وَثَلاثَةَ)[2] أُخَر، كُلُّ اثْنَيْنِ يُؤْخَذانِ مِنْها، فَهُما عَلى نِسْبَةِ اثْنَيْنِ مِنَ (الثَّلاثَةِ)[3] الأُوَلِ، عَلَيْها دَ وهَ وزَ. وَالنِّسَبُ مُخْتَلِفَةٌ في الأَقْدارِ بِالتَّقْديمِ وَالتَّأْخيرِ: قَدْرُ ا عِنْدَ بَ كَقَدْرِ هَ عِنْدَ زَ، وَقَدْرُ بَ عِنْدَ جَ كَقَدْرِ دَ عِنْدَ هَ. وَنَجْعَلُ ا أَعْظَمَ مِنْ جَ. فَأَقولُ إِنَّ دَ أَعْظَمُ مِنْ زَ.

מופת זה: לפי שא׳ גדול מג׳, ושיעור ב׳ שיעור אחר, אם כן א׳ רב היחס אצל ב׳ מג׳ אצל ב׳. ואמנם שיעור א׳ אצל ב׳ כשעור ה׳ אצל ז׳. ושיעור ג׳ אצל[1] ב׳ כשיעור ה׳ אצל ד׳. ואם כן יהיה ה׳ רב היחס אצל ז׳ מה׳ אצל ד׳. ואשר היחס אליו רב הוא הקטן. אם כן ז׳ קטן מד׳, וד׳ גדול מז׳. וכמו כן יתבאר שאם היה א׳[2] שוה לג׳, היה ד׳ שוה לז׳. ואם היה א׳ קטן מג׳, היה ד׳ קטן מז׳.

بُرْهانُهُ: أَنَّ ا أَعْظَمُ مِنْ جَ. وَقَدْرُ بَ قَدْرٌ آخَرُ أَكْبَرُ نِسْبَةً إِلى بَ مِنْ جَ إِلى بَ. وَلٰكِنَّ قَدْرَ ا عِنْدَ بَ كَقَدْرِ هَ عِنْدَ زَ. وَقَدْرُ جَ عِنْدَ بَ كَقَدْرِ هَ عِنْدَ دَ. هَ أَكْبَرُ نِسْبَةً إِلى زَ مِنْ هَ إِلى دَ. وَالَّذي نِسْبَتُهُ إِلَيْهِ أَكْبَرُ، هُوَ أَصْغَرُ. زَ أَصْغَرُ مِنْ دَ. وَ دَ أَعْظَمُ مِنْ زَ. وَكَذٰلِكَ (تَبَيَّنَ)[4] أَنَّ ا لو كانَ مِثْلَ جَ، لَكانَ دَ مِثْلَ[5] زَ. ولو كانَ ا أَصْغَرَ مِنْ جَ لَكانَ دَ أَصْغَرَ مِنْ زَ.

וזה מה שרצינו שנבאר.

وَذٰلِكَ ما أَرَدْنا بَيانَهُ.

1 אצל] ~~אצל ג׳~~+ 2 א׳] ~~ב׳~~+

1 ثلاثة] ثلثة 2 وثلاثة] وثلثة 3 الثلاثة] الثلثة 4 تبين] نبين 5 مثل] +*

مِثالُهُ: أَنَّ أَقْدارَ ا ب ج وأَقْدار د ه ز على عِدَّةٍ واحدَةٍ^. وكُلّ اثْنَيْنِ مِنَ الأُولى على نِسْبَةِ[1] اثْنَيْنِ مِنَ الأُخْرى. والنِّسْبَة مُضْطَرِبَةٌ: نِسْبَةُ ا إلى ب كنِسْبَةِ ه إلى ز، ونِسْبَة ب إلى ج كَنِسبة د إلى ه. ونَجْعَلُ ا أعظم مِنْ ج. فَأَقولُ إنَّ د أعظم مِنْ ز.

המשל בו: כי שעורי א׳ ב׳ ג׳ על מספר שעורי ד׳ ה׳ ז׳. וכל שנים מן הראשון[1] על יחס שנים מן האחר. והיחס מתבלבל[2]: יחס א׳ אל ב׳ כיחס ה׳ אל ז׳[3], ויחס ב׳ אל ג׳ כיחס ד׳ אל ה׳. ונשים א׳ יותר גדול מג׳. הנה אומר כי ד׳ יותר גדול מן ז׳[4].

بُرهانُهُ: أَنَّ ا أعظم مِنْ ج. و ب قَدْرٌ آخر، فَقَدْر ا أَكْبَر نِسْبَةً إلى ب مِنْ ج إلى ب. ولكنَّ نِسْبَةَ ا إلى ب كنِسْبَةِ ه إلى ز. فنِسْبَة ه إلى ز أَكْبَر مِنْ نِسْبَة ج إلى ب. ونِسْبَة ج إلى ب كنِسْبَةِ ه إلى د. فنِسْبَة ه إلى ز أعظم مِنْ نِسْبَة ه إلى د[2]. والَّذي تَكونُ النِّسْبَةُ إِلَيْهِ أَكْبَر فَهُوَ أَصْغَر. فَقَدْر ز أَصْغَر مِنْ د. ف د أعظم مِنْ ز. وكَذَلِكَ تَبَيَّنَ أَنَّ ا لَوْ كانَ مُساوِيًا لِقَدْرِ ج لكانَ د مُساوِيًا ل ز. وَلَوْ كانَ أَصْغَر مِنْ ج لَكانَ د أَصْغَر مِنْ ز.

وَذَلِكَ ما أَرَدْنا أَنْ نُبَيِّنَ.

מופתו: כי א׳ יותר גדול מן ג׳. וב׳[5] שיעור אחר. הנה שעור א׳ יותר גדול היחס אל ב׳ מן ג׳ אל ב׳. אבל יחס א׳ אל ב׳ כיחס ה׳ אל ז׳. אם כן יחס ה׳ אל ז׳ יותר גדול מיחס ג׳ אל ב׳. ויחס ג׳ אל ב׳ כיחס ה׳ אל ד׳. אם כן יחס ה׳ אל ז׳ יותר גדול מיחס ה׳ אל ד׳. ואשר יהיה היחס אליו יותר גדול הנה הוא יותר קטן. אם כן שעור ז׳ יותר קטן[6] מן ד׳. אם כן ד׳[7] יותר גדול[8] מן ז׳. וכן התבאר כי א׳ אלו היה שוה לשיעור ג׳ היה ד׳ שוה לז׳[9]. ואלו היה יותר קטן מן ג׳ היה ד׳ יותר קטן מן ז׳.

וזה מה שרצינו לבאר.

---

[1]نسبة] CRamMT1 +بالعكس [2]د] CopMRam

---

[1]מן הראשון] מהם **ב** [2]מתבלבל] מהם **ב** [3]ה׳ ... ז׳] ד׳ אל ה׳ **ב** [4]כיחס ד׳ ... ז׳] ה׳ אל ז׳ ונשים הראשון מן הראשון והוא א׳ יותר גדול מן האחרון והוא ג׳ הנה אומר כי הראשון מן האחד והוא ד׳ יותר גדול מן האחרון והוא ז׳ **ב** [5]וב׳] ושעור ב׳ **ב** [6]ה׳ אל ז׳ אם ... ז׳ יותר קטן] ד׳ אל ה׳ ויחס ג׳ אל ב׳ כיחס ז׳ אל ה׳ אם כן שעור ד׳ יותר גדול יחס אל ה׳ מן אל ה׳ מן ז׳ אל ה׳ ואשר יחסו יותר גדול הנה הוא יותר גדול **ב** +~~יותר קטן~~ **ס** [7]ד׳] שעור+ **ב** [8]גדול] ליתא **ק** [9]לשיעור ... לז׳] אל ג׳ היה ד׳ שוה אל ז׳ **ב**

[Proposition V.22]

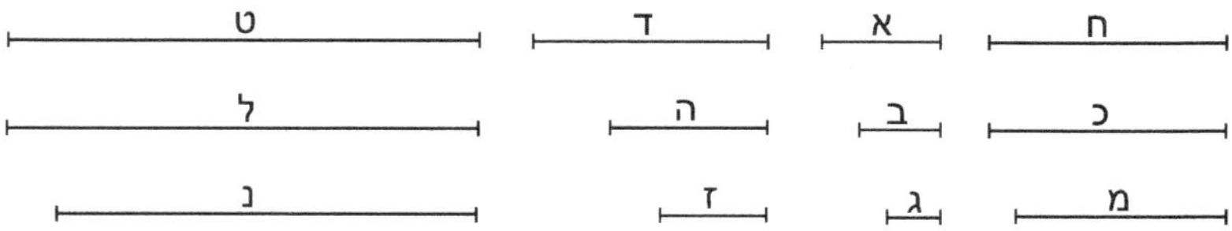

כשיהיו שעורים מה וכו׳, בלשון אחד.

לשון אחר: כשיהיו שלשה שיעורים ואחרים כפי מספרם, כל שנים שיוקחו מהם הם עליהם שנים מן השלשה הראשונים[1], יהיו ביחס השווי גם כן על יחסם.

إذا كانَتْ (ثلاثَةُ)[1] أَقْدارٍ، وَأُخَرُ عَلى عِدَّتِها، كُلُّ اثْنَيْنِ يُؤخَذانِ مِنْها فَهُما عَلى نِسْبَةِ اثْنَيْنِ مِنَ (الثَّلاثَةِ)[2] الأُوَلِ، فَإِنَّها في نِسْبَةِ المُساواةِ يَكونُ عَلى نِسْبَتِها.

המשל בזה: ששלשה שיעורים עליהם א׳ וב׳ וג׳, ושלשה שעורים אחרים ועליהם[2] ד׳ וה׳ וז׳, כל שנים יוקחו מהם הם על יחס שנים מן השלשה הראשונים—שיעור א׳ אצל ב׳ כשעור ד׳ אצל ה׳, ושעור ב׳ אצל ג׳ כשעור ה׳ אצל ז׳. ואומר כי ביחס השווי יהיה שיעור א׳ אצל ג׳ כשעור ד׳ אצל ז׳.

مِثالُهُ: أَنَّ (ثَلاثَةَ)[3] أَقْدارٍ عَلَيْها آ و ب و ج. (وَثَلاثَةَ)[4] أَقْدارٍ أُخَر كُلُّ اثْنَيْنِ يُؤخَذانِ مِنْها، فَهُما عَلى نِسْبَةِ اثْنَيْنِ مِنَ (الثَّلاثَةِ)[5] الأُوَلِ. عَلَيْها د و ه و ز—وَقَدْرُ آ عِنْدَ ب كَقَدْرِ د عِنْدَ ه، وَقَدْرُ ب عِنْدَ ج كَقَدْرِ ه عِنْدَ ز. فَأَقولُ إِنَّ بِالمُساواةِ يَكونُ قَدْرُ آ عِنْدَ ج كَقَدْرِ د عِنْدَ ز.

[1]הראשונים] ⟨ה⟩ראשונים ⟨⟨*⟩⟩ [2]א׳ ... ועליהם]

[1]ثلاثة] ثلثة [2]الثلاثة] الثلثة [3]ثلاثة] ثلثة [4]وثلاثة] و ثلثة [5]الثلاثة] الثلثة

إِذا كانَتْ أَقْدارٌ ما وَأُخَرَ[1] على عِدَّتِها، كُلّ قَدْرَيْنِ مِنَ الأُولى على نِسْبَةِ قَدْرَيْنِ مِنَ الأُخْرى[2]. فَإِنَّها في نِسْبَةِ المُساواةِ تكونُ في نِسْبَتِها.

כאשר היו שעורים מה[1] ואחרים על מספרם, כל שני שעורים מן הראשון על יחס שני שיעורים מן האחר. הנה הם ביחס השווי יהיו ביחסם[2].

مِثالُهُ: أَنَّ أَقْدارَ ا ب ج وأَقْدار د ه ز على عِدَّةٍ واحِدَةٍ. وكُلّ اثْنَيْنِ مِنَ الأُولى على نِسْبَةِ اثْنَيْنِ مِنَ الأُخرى—ونِسْبَةَ ا إلى ب كنِسْبَةِ د إلى ه، ونِسْبَةَ ب إلى ج كنِسْبَةِ ه إلى ز. فَأَقولُ إنّ نِسْبَةَ ا إلى ج كنِسْبَةِ د إلى ز.

המשל בו[3]: כי שעורי א׳ ב׳ ג׳ ושעורי ד׳ ה׳ ז׳ על מספר אחד. וכל שנים מן הראשון על יחס שנים מן האחר—יחס א׳ אל ב׳ כיחס ד׳ אל ה׳, ויחס ב׳ אל ג׳ כיחס ה׳ אל ז׳. הנה אומר כי יחס א׳ אל ג׳ כיחס ד׳ אל ז׳.

---

[1]واخر] R1 واقدار اخر [2]الاخرى] CFRamT1 +وانتظمت النسبة

שרטוט—**ב**: שורת הקטעים הראשונה והשורה השניה $ [1]מה] +~~ושעורים~~ **פ** [2]ביחס] ביחס **ב** כיחסם **ס** יחסם **ק** [3]בו] ליתא **ט**

מופת זה: שנקח לא׳ ולד׳ כפלים שוים, והם ח׳ וט׳. ונקח לב׳ ולה׳ כפלים שוים אחרים כמה שהם, והם כ׳ ול׳. ונקח לג׳ ולז׳ כפלים אחרים שוים כמה שהם, והם מ׳ ונ׳. ושעור א׳ אצל ב׳ כשעור ד׳ אצל ה׳. וכבר לוקחו לא׳ ולד׳ כפלים שוים, והם ח׳ וט׳. ולוקח לב׳ ולה׳ כפלים שוים והם כ׳ ול׳, יהיה שיעור (ח׳)[1] אצל כ׳ כשעור ט׳ אצל ל׳. ועוד שיעור ב׳ אצל ג׳ כשיעור ה׳ אצל ז׳. וכבר לוקח לב׳ ולה׳ כפלים שוים והם כ׳ ול׳. ולוקח לג׳ ולז׳ כפלים שוים אחרים, והם מ׳ ונ׳, יהיה שיעור כ׳ אצל מ׳ כשיעור ל׳ אצל נ׳. וכבר התבאר כי שיעור ח׳ אצל כ׳ כשיעור ט׳ אצל ל׳. ואם כן יהיה שיעור תוספת ח׳ על מ׳ כשיעור תוספת ט׳ על נ׳. ואם היה שוה לו גם הוא שוה לו. ואם היה קטן ממנו גם הוא קטן ממנו בזה השעור. וח׳ וט׳ הם כפלי א׳ וד׳ ומ׳ ונ׳, ויש להם סגלת כפלים משעור היחס הם כפלים אחרים לג׳ ולז׳. אם כן שיעור א׳ אצל ג׳ כשעור ד׳ אצל ז׳.

וזה מה שרצינו שנבאר.

بُرْهانُهُ: أَنَّا نَأْخُذُ ا و د أَضْعافًا مُتَساوِيَةً، وَهِيَ ح و ط. وَنَأْخُذُ لِ ب و ه أَضْعافًا مُتَساوِيَةً أُخَرَ، كَيْفَما كانَتْ، وَهِيَ ك و ل. وَنَأْخُذُ لِ ج و ز أَضْعافًا مُتَساوِيَةً، وَهِيَ م و ن. وَقَدْرُ ا عِنْدَ ب كَقَدْرِ د عِنْدَ ه. وَقَدْ أَخَذَ لِ ا و د أَضْعافُ مُتَساوِيَةٌ، وَهِيَ ح و ط. وَأُخِذَ لِ ه و ب أَضْعافٌ مُتَساوِيَةٌ، وَهِيَ ك و ل. فَقَدْرُ ح عِنْدَ ك كَقَدْرِ ط عِنْدَ ل. وَأَيْضًا قَدْرُ ب عِنْدَ ج كَقَدْرِ ه عِنْدَ ز. وَقَدْ أُخِذَ لِ ب و ه أَضْعافٌ مُتَساوِيَةٌ، وَهِيَ ك و ل. فَقَدْرُ ح عِنْدَ ك كَقَدْرِ ط عِنْدَ ل. وَأَيْضًا قَدْرُ ب عِنْدَ ج كَقَدْرِ ه عِنْدَ ز. وَقَدْ أُخِذَ لِ ب و ه أَضْعافٌ مُتَساوِيَةٌ، وَهِيَ م و ن. وَأُخِذَ لِ ج و ز أَضْعافٌ مُتَساوِيَةٌ، وَهِيَ م و ن. فَقَدْرُ ك عِنْدَ م كَقَدْرِ ل عِنْدَ ن. وَقَدِ اسْتَبانَ أَنَّ قَدْرَ ح عِنْدَ ك كَقَدْرِ ط عِنْدَ ل. فَقَدْرُ زِيادَةِ ح عَلى م كَقَدْرِ زِيادَةِ ط عَلى ن. فَإِنْ كانَ مُساوِيًا لَهُ، فَهُوَ مُساوٍ لَهُ، وَإِنْ كانَ ناقِصًا مِنْهُ، فَهُوَ ناقِصٌ مِنْهُ بِذَلِكَ القَدْرِ. وح و ط هُما أَضْعافُ لِ ا و د. و م و ن هُما أَضْعافُ أُخَرُ لِ ج و ز. فَقَدْرُ ا عِنْدَ ج كَقَدْرِ د عِنْدَ ز.

وَذَلِكَ ما أَرَدْنا بَيانَهُ.

[1] ח׳] [א׳] א׳

بُرهانُه: أَنَّا نَأخُذُ لِقَدْرَيْ ا د أَضْعافًا مُتَساوِيَةً، وهي ح ط، ولِقَدْرَيْ ب ه أَضْعافًا مُتَساوِيَةً، وهي ك ل. ولِقَدْرَيْ ج ز أَضْعافًا مُتَساوِيَةً، وهي م ن. فنِسْبَةَ ا إلى ب كنِسْبَةِ د إلى ه. وح ط أَضْعاف مُتَساوِيَة لِقَدْرَيْ ا د. وك ل أَضْعاف مُتَساوِيَة لِقَدْرَيْ ب ه. فنِسْبَة ح إلى ك كنِسْبَة ط إلى ل. وأيضًا فإنَّ نِسْبَةَ ب إلى ج كنِسْبَةِ ه إلى ز. وك ل أَضْعاف مُتَساوِيَة لِقَدْرَيْ ب ه. وم ن أَضْعاف مُتَساوِيَة لِقَدْرَيْ ج ز. فنِسْبَة ك إلى م كنِسْبَةِ ل إلى ن. وقَدْ تَبَيَّنَ أَنَّ نِسْبَةَ ح إلى ك كنِسْبَةِ ط إلى ل. فقَدْرَيْ ح ط إِمّا زائِدانِ مَعًا على قَدْرَيْ م ن، وإِمّا مُساوِيانِ مَعًا لَهُما، وإِمّا ناقِصانِ مَعًا مِنْهُما. وح ط أَضْعاف مُتَساوِيَة لِقَدْرَيِّ ا د. وم ن أَضْعاف مُساوِيَة لِقَدْرَيْ ج ز. فنِسْبَة ا إلى ج كنِسْبَةِ د إلى ز.

وذَلِكَ ما أَرَدْنا أَنْ نُبَيِّنَ.

מופתו: אנחנו נקח[1] לשני שעורי א׳ ד׳ כפלים שוים, והם ח׳ ט׳, ולשני שעורי ב׳ ה׳ כפלים שוים[2], והם כ׳ ל׳, ולשני שעורי ג׳ ז׳ כפלים שוים, והם מ׳ נ׳[3]. הנה יחס א׳ אל ב׳ כיחס ד׳ אל ה׳. וח׳ ט׳ כפלים שוים לשני שעורי א׳ ד׳. וכ׳ ל׳ כפלים שוים לשני שעורי ב׳ ה׳. אם כן יחס ח׳[4] אל כ׳ כיחס ט׳ אל ל׳. וגם כן[5] הנה יחס ב׳ אל ג׳ כיחס ה׳ אל ז׳. וכ׳ ל׳ כפלים[6] שוים לשני שעורי ב׳ ה׳. ומ׳ נ׳ כפלים שוים לשני שעורי ג׳ ז׳. אם כן יחס כ׳ אל מ׳ כיחס ל׳ אל נ׳. וכבר התבאר כי יחס ח׳ אל כ׳ כיחס ט׳ אל ל׳. אם כן שני שעורי ח׳ ט׳ אם נוספים יחד על שני שעורי מ׳ נ׳, ואם שוים יחד לשניהם, ואם חסרים יחד משניהם. וח׳ ט׳ כפלים שוים לשעורי א׳ ד׳. ומ׳ נ׳ כפלים שוים לשני שעורי[7] ג׳ ז׳. אם כן יחס א׳ אל ג׳ כיחס ד׳ אל ז׳.

וזה מה שרצינו לבאר.

[1] נקח] ~~נשים~~ ⟨⟨*⟩⟩ **ט** נשים **ס**
[2] והם ... שוים] ⟨⟨*⟩⟩ **פ**
[3] מ׳ נ׳] ליתא **ב**
[4] ח׳] א׳ **ס**
[5] כן] ליתא **ב**
[6] כפלים] כפלי **ב**
[7] לשני שעורי] לשעורי א״ד ומ״נ כפלים שוים+ **ט** לשם **ס**

[Proposition v.23]

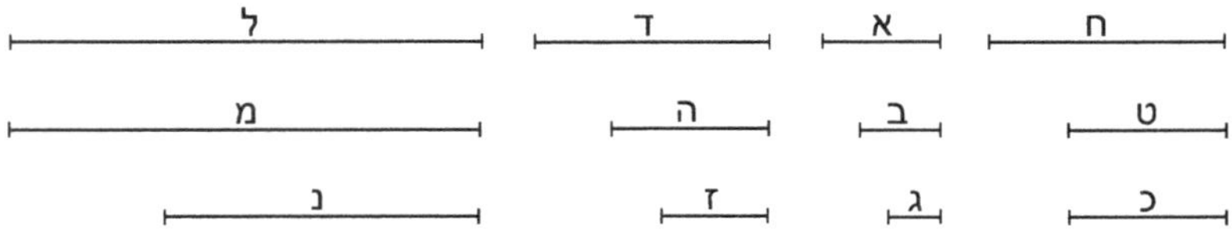

כשיהיו שעורים מה וכו׳, בלשון אחד.

לשון אחר: כשיהיו שלשה שעורים ואחרים על מספרם, כל שנים שיוקחו מהם הם על יחס שנים מן האחרים הראשונים, והתחלף היחס בשעורים בקדימה ובאחור, יהיו גם כן כיחס השווי על יחסם.

إذا كانَتْ (ثَلاثَةُ)[1] أَقْدارٍ وَأُخَرُ عَلى عِدَّتِها، كُلّ اثْنَيْنِ يؤُخَذانِ مِنْها، فَهُما عَلى اثْنَيْنِ مِنَ (الثَّلاثَةِ)[2] الأُوَلِ، وَاخْتَلَفَتِ النِّسَبُ في الأَقْدارِ بِالتَّقْديمِ وَالتَّأْخيرِ، فَإِنَّها بِالْمُساواةِ تَكونُ عَلى نِسْبَتِها.

המשל בזה: כי שלשה שיעורים עליהם א׳ וב׳ וג׳, ושלשה אחרים עליהם ד׳ וה׳ וז׳, כל שנים שיוקחו מהם הם על יחס שנים מן השלשה הראשונים. והיחס המתחלף בשעורים בקדימה ובאחור—שעור א׳ אצל ב׳ כשעור ה׳ אצל ז׳, ושעור ב׳ אצל ג׳ כשעור ד׳ אצל ה׳. ואומר כי כיחס השווי יהיה שעור א׳ אצל ג׳ כשעור ד׳ אצל ז׳.

مِثالُهُ: أَنَّ (ثَلاثَةَ)[3] أَقْدارٍ عَلَيْها آ و بَ و جَ. (وَثَلاثَةُ)[4] أَقْدارٍ أُخَرُ، كُلُّ اثْنَيْنِ يؤُخَذانِ مِنْها عَلى نِسْبَةِ اثْنَيْنِ مِنَ (الثَّلاثَةِ)[5] الأُوَلِ، عَلَيْها دَ و هَ و زَ. فَالنِّسْبَةُ مُخْتَلِفَةٌ في الأَقْدارِ بِالتَّقْديمِ والتَّأْخيرِ، قَدْرُ آ عِنْدَ بَ كَقَدْرِ هَ عِنْدَ بَ. وَقَدْرُ بَ عِنْدَ جَ كَقَدْرِ دَ عِنْدَ هَ. فَأَقولُ إِنَّ بِالْمُساواةِ يَكونُ قَدْرُ آ عِنْدَ جَ كَقَدْرِ دَ عِنْدَ زَ.

[1] ثلاثة] ثلثة [2] الثلاثة] الثلثة [3] ثلاثة] ثلثة [4] وثلاثة] وثلثة [5] الثلاثة] الثلثة

إِذا كانَتْ أَقْدارٌ كَمْ كانَتْ وَأَقْدارٌ أُخَر على عِدَّتِها، كُلّ اثْنَيْنِ مِنَ الأُولى على نِسْبَةِ اثْنَيْنِ مِنَ الأُخْرَى، وَاضْطَرَبَتِ النِّسْبَةُ، فَإِنَّها في نِسْبَةِ المُسَاواةِ تكونُ على نِسْبَتِها.

כאשר היו שעורים[1] כמה שיהיו ושעורים אחרים על מספרם, כל שנים מן הראשון על יחס שנים מן האחר, ונתבלבל היחס, הנה הם ביחס השווי יהיו על יחסם.

مِثالُهُ: أَنَّ أَقْدارَ ا ب ج وأَقْدارَ د ه ز على عِدَّةٍ واحِدَةٍ. وَكُلّ اثْنَيْنِ مِنَ الأُولى على نِسْبَةِ اثْنَيْنِ مِنَ الأُخرى. وَالنِّسْبَة مُضْطَرِبَةٌ: نِسْبَةُ ا إلى ب كنِسْبَةِ ه إلى ز، ونِسْبَةِ ب إلى ج كِنِسْبَةِ د إلى ه. فَأَقولُ إنَّ نِسْبَةَ ا إلى ج كِنِسْبَةِ د إلى ز.

המשל בו: כי שעורי א׳ ב׳ ג׳ ושיעורי ד׳ ה׳ ז׳ על מספר אחד. וכל שנים[2] מן הראשון על יחס שנים מן האחרון. והיחס מבולבל: יחס א׳ אל ב׳ כיחס ה׳ אל ז׳, ויחס ב׳ אל ג׳ כיחס ד׳ אל ה׳. הנה אומר כי יחס[3] א׳ אל ג׳ כיחס ד׳ אל ז׳.

שרטוט—**ב**: $ [1]שעורים] שני * ~~או~~ **ט**
[2]שנים] ~~אחד~~+ **ס** [3]יחס] ליתא **ס**

مופת זה: שנקח לא׳ ולב׳ ולד׳ כפלים שוים, והם ח׳ וט׳ ול׳. ולה׳ ולז׳ ולג׳ כפלים שוים אחרים כמה שהיו, והם מ׳ ונ׳ וכ׳. ושיעור כפלי ח׳ לא׳ כשיעור כפלי ט׳ לב׳. והחלקים אשר כפליהם שוים יהיו [!] יחסם אחד. ואם כן יהיה שעור א׳ אצל ב׳ כשעור ח׳ אצל ט׳. ואמנם שעור א׳ אצל ב׳ כשעור ה׳ אצל ז׳. ושעור ה׳ אצל ז׳ כשעור מ׳ אצל נ׳. וכבר היה שעור ה׳ אצל ז׳ כשעור ח׳ אצל ט׳. אם כן שיעור ח׳ אצל ט׳ כשעור מ׳ אצל נ׳. ועוד שיעור ב׳ אצל ג׳ כשעור ד׳ אצל ה׳. וכבר לוקח לב׳ ולד׳ כפלים שוים, והם ט׳ ול׳. ולוקח לג׳ ולה׳ כפלים שוים אחרים כמה שהם, והם כ׳ ומ׳. ויהיה שעור ט׳ אצל כ׳ כשעור ל׳ אצל מ׳. וכבר התבאר כי שעור ח׳ אצל ט׳ כשעור מ׳ אצל נ׳. ואם כן יהיה שיעור תוספת ח׳ על כ׳ כשעור תוספת ל׳ על נ׳. ואם היה שוה לו גם הוא שוה לו. ואם היה חסר ממנו גם הוא חסר ממנו[1] בזה השעור. וח׳ ול׳ הם כפלים שוים לא׳ ולד׳. וכ׳ ונ׳ הם כפלים שוים אחרים לג׳ ולז׳. ויאמר על שיעורים שהם מתיחסים על יחס אחד—הראשון אל השני כשלישי אל הרביעי—כשהיו כפלי הראשון והשלישי השוים אצל כפלי[2] השני והרביעי השוים אצל[3] האחרים, אם שהם עודפים יחד, אם שהם שוים יחד, ואם שהם חסרים יחד. אם כן התבאר ששיעור א׳ אצל ג׳ כשעור ד׳ אצל ז׳.

וזה מה שרצינו לבאר.

بُرْهانه: أَنَّا نَأْخُذُ ل ا و ب و د أَضْعافًا مُتَساوِيَةً، وَهِيَ ح و ط و ل. وَنَأْخُذُ ل ه و ز و ج أَضْعافًا أُخَرَ، كَيْفَما كانَتْ، مُتَساوِيَةً، وَهِيَ م و ن و ك. فَقَدْرُ أَضْعافِ ح ل ا كَقَدْرِ أَضْعافِ ط ل ب. وَالأُخَرُ الَّتي أَضْعافُها مُتَساوِيَةٌ، فَإِنَّ نِسْبَتَها واحِدَةٌ. فَقَدْرُ ا عِنْدَ ب كَقَدْرِ ح عِنْدَ ط. وَلَكِنَّ قَدْرَ ا عِنْدَ ب كَقَدْرِ ه عِنْدَ ز. وَقَدْرُ ه عِنْدَ ز كَقَدْرِ ح عِنْدَ ط. وَأَيْضًا أَضْعافُ م ل ه كَأَضْعافِ ن ل ز. وَقَدْرُ ه عِنْدَ ز كَقَدْرِ م عِنْدَ ن. وَقَدْ[1] كانَ قَدْرُ ه عِنْدَ (ز)[2] كَقَدْرِ ح عِنْدَ ط. فَقَدْرُ ح عِنْدَ ط كَقَدْرِ م عِنْدَ ن. وَأَيْضًا قَدْرُ ب عِنْدَ ج كَقَدْرِ د عِنْدَ ه. وَقَدْ أُخِذَ ل ب و د أَضْعافٌ مُتَساوِيَةٌ، وَهِيَ ط و ل. وَأُخِذَ ل ج و ه أَضْعافٌ مُتَساوِيَةٌ أُخَرُ، كَيْفَما كانَتْ، وَهِيَ ك و م. فَقَدْرُ ط عِنْدَ ك كَقَدْرِ ل عِنْدَ م. وَقَدِ اسْتَبانَ أَنَّ قَدْرَ ح عِنْدَ ط كَقَدْرِ م عِنْدَ ن. فَقَدْرُ زِيادَةِ ح عَلى ك كَقَدْرِ زِيادَةِ ل عَلى ن. وَإِنْ كانَ مُساوِيًا لَهُ، فَهُوَ مُساوٍ لَهُ. وَإِنْ كانَ أَقَلَّ مِنْهُ، فَهُوَ أَقَلُّ مِنْهُ. وح و ل هُما أَضْعافٌ مُتَساوِيَةٌ ل ا و د. و ك و ن هُما أَضْعافٌ أُخَرُ مُتَساوِيَةٌ ل ج و ز. فَقَدْرُ ا عِنْدَ ج كَقَدْرِ د عِنْدَ ز.

وَذَلِكَ ما أَرَدْنا بَيانَهُ.

[1]גם ... ממנו] +* [2]כפלי] ⟨*⟩ [3]אצל] א⟨צ⟩ל

[1]وقد] ⟨*⟩ [2]ز] ب

بُرهانُهُ: أنَّا نأخُذُ لأَقْدارِ ا ب د أَضْعافًا مُتَساوِيَةً وهي ح ط ل، ولأَقْدارِ ه ز ج أَضْعافًا مُتَساوِيَةً، وهي م ن ك. فأَضْعاف ح لِقَدْرِ ا كأَضْعافِ ط لِقَدْرِ ب. الأَجزاء الَّتي أَضْعافُها مُتَساوِيَةٌ[1] فنِسْبَة بَعْضِها إلى بَعْضٍ كَنِسْبَةِ أَضْعافِها بعضِها إلى بعض. فنِسْبَة ا إلى ب كَنِسْبَة ح إلى ط. ولكِنَّ نِسْبَةَ ا إلى ب كَنِسْبَة ه إلى ز. فنِسْبَة ه إلى ز كَنِسْبَة ح إلى ط. وأَيْضًا فإِنَّ أَضْعاف م لِقَدْرِ ه كأَضْعافِ ن لِقَدْرِ ز. فنِسْبَة ه إلى ز كَنِسْبَة م إلى ن. وَلكِنَّ نِسْبَة ه إلى ز كَنِسْبَة ح إلى ط. فنِسْبَة م إلى ن كَنِسْبَة ح إلى ط. وَأَيْضًا فَإِنَّ نِسْبَة ب إلى ج كَنِسْبَة د إلى ه. وَقَدْ أُخِذَ لِقَدْرَيْ ب د أَضْعافٌ مُتَساوِيَةٌ وهي ط ل. ولِقَدْرَيْ ج ه أَضْعاف مُتَساوِيَة، وَهِيَ ك م. فنِسْبَة ط إلى ك كَنِسْبَة ل إلى م. وَقَدْ كانَ تَبَيَّنَ أَنَّ نِسْبَةَ ح إلى ط كَنِسْبَةِ م إلى ن. فَقَدْر ح ل إِمّا زائِدانِ مَعًا عَلى قَدْرَيْ ك ن، وَإِمّا مُساوِيانِ مَعًا لَهُما، وَإِمّا ناقِصانِ مَعًا مِنْهُما. و ح ل أَضْعاف مُساوِيَة لِقَدْرَيّ ا د. و ك ن أَضْعاف مُساوِيَة لِقَدْرَيْ ج ز. فنِسْبَة ا إلى ج كَنِسْبَة د إلى ز.

وَذلِكَ ما أَرَدْنا أَنْ نُبَيِّنَ.

[1] مُتَساوِيَة] R1R2T2T3 المتساوية

מופתו: אנחנו נקח לשיעורי א׳ ב׳ ד׳ כפלים שוים והם ח׳ ט׳ ל׳, ולשיעורי ה׳ ז׳ ג׳ כפלים שוים, והם מ׳ נ׳ כ׳. הנה כפלי ח׳ לשעור א׳ ככפלי ט׳ לשיעור ב׳. והחלקים אשר כפליהם[1] שוים הנה יחס קצתם אל קצת כיחס כפליהם קצתם אל קצת. אם כן יחס א׳ אל ב׳ כיחס ח׳ אל ט׳. אבל יחס א׳ אל ב׳ כיחס ה׳ אל ז׳. אם כן יחס ה׳ אל ז׳ כיחס ח׳ אל ט׳. וגם כן הנה כפלי מ׳[2] לשיעור ה׳ ככפלי נ׳ לשיעור ז׳. אם כן יחס ה׳ אל ז׳ כיחס מ׳ אל נ׳. אבל יחס ה׳ אל ז׳ כיחס ח׳ אל ט׳[3]. אם כן יחס מ׳ אל נ׳ כיחס ח׳[4] אל ט׳. וגם כן הנה יחס ב׳ אל ג׳ כיחס ד׳ אל ה׳. וכבר נלקח לשני שעורי ב׳ ד׳ כפלים שוים והם ט׳ ל׳. ולשני שעורי ג׳ ה׳ כפלים שוים והם כ׳ מ׳. אם כן[5] יחס ט׳ אל כ׳ כיחס ל׳ אל מ׳. וכבר התבאר כי יחס ח׳ אל ט׳ כיחס מ׳ אל נ׳. אם כן שני שעורי[6] ח׳ ל׳ אם נוספים יחד על שני שעורי[7] כ׳ נ׳, ואם שוים יחד להם, ואם חסרים יחד משניהם. וח׳ ל׳ כפלים שוים לשני שעורי א׳ ד׳. וכ׳ נ׳ כפלים שוים[8] לשני שעורי ג׳ ז׳. אם כן יחס א׳ אל ג׳ כיחס ד׳ אל ז׳.

וזה מה שרצינו לבאר[9].

[1]אשר כפליהם] ז׳ **ב** [2]מ׳] ט׳ **ב** [3]וגם ... ט׳] ⟨* וגם כן יחס ב׳ אל ג׳ כיחס ד׳ אל ה׳ וכבר נלקח⟩ **ס** [4]כיחס ח׳] יחס ה׳ **ב** [5]כן] ליתא **ב** [6]שעורי] ⟨*⟩ **ט** [7]שעורי] +~~כ׳ ל׳~~ **ק** [8]כפלים שוים] ליתא **ו** [9]וזה ... לבאר] ונשלם ביאורו **בוספק**

[Proposition V.24]

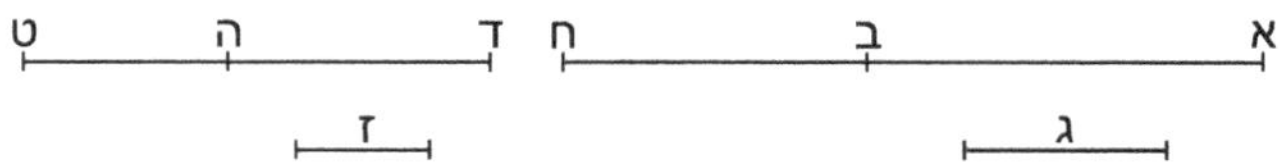

כשיהיה יחס הראשון אל השני כיחס השלישי אל הרביעי, ויחס החמשי אל השני כיחס הששי אל הרביעי, יהיה יחס הראשון והחמשי יחד אל השני כיחס השלישי והששי יחד אל הרביעי.

اذا كانت نسبة الاول الى الثاني كنسبة الثالث الى الرابع, ونسبة الخامس الى الثاني كنسبة السادس الى الرابع, فان نسبة الاول والخامس مركبين الى الثاني كنسبة الثالث والسادس مركبين الى الرابع.

המשל בזה: כי הראשון א״ב, ויחסו אל השני, והוא ג׳, כיחס השלישי, והוא ד״ה, אל הרביעי, והוא ז׳. ויחס החמישי, והוא ב״ח, אל השני, והוא ג׳, כיחס הששי, והוא ה״ט, אל הרביעי, והוא ז׳. ואומר כי יחס הראשון והחמשי מורכבים, והם א״ח, אל השני, והוא ג׳, כיחס השלישי והרביעי מורכבים, והם ד״ט, אל הרביעי, והוא ז׳.

مثالُهُ: أَنَّ الأَوَّلَ اب وَنِسْبَتُهُ إلى الثَّاني، وَهُوَ ج، كَنِسْبَةِ الثَّالِثِ، وَهُوَ (د ه)[1] إلى الرَّابِعِ، وَهُوَ ز. وَنِسْبَةُ الخامِسِ، وَهُوَ ب ح، إلى الثَّاني، وَهُوَ ج، كَنِسْبَةِ السَّادِسِ، وَهُوَ ه ط، إلى الرَّابِعِ، وَهُوَ ز. فَأَقولُ نِسْبَةُ الخامِسِ وَالأَوَّلِ مُرَكَّبَيْنِ، وَهُما اح، إلى الثَّاني، وَهُوَ ج، كَنِسْبَةِ الثَّالِثِ وَالسَّادِسِ مُرَكَّبَيْنِ، وَهُما ط د، إلى الرَّابِعِ، وَهُوَ ز.

[1] د ه] د

إذا كانَتْ نِسْبَةُ الأَوَّلِ إلى الثَّاني كَنِسْبَةِ الثَّالِثِ إلى الرَّابِعِ، ونِسْبَةُ الخامسِ إلى الثَّاني كَنِسْبَةِ السَّادِسِ إلى الرَّابِعِ، فَإِنَّ نِسْبَةَ الأَوَّلِ والخامسِ مَجْموعَيْنِ إلى الثَّاني، كَنِسْبَةِ الثَّالِثِ والسَّادِس مَجْمُوعَيْنِ إلى الرَّابِعِ.

כאשר היה יחס[1] הראשון אל השני כיחס השלישי אל הרביעי, ויחס החמישי אל השני כיחס הששי אל הרביעי, הנה יחס הראשון והחמישי מקובצים אל השני, כיחס השלישי והששי מקובצים אל הרביעי.

مِثالُهُ: أَنَّ نِسْبَةَ الأَوَّلِ، وهو اب، إلى الثَّاني، وهو ج، كَنِسْبَةِ الثَّالِثِ، وهو ده، إلى الرَّابِعِ، وهو ز. ونِسْبَةَ الخامسِ، وهو[1] ب ح، إلى الثَّاني، وهو ج، كَنِسْبَةِ السَّادِسِ، وهو ه ط، إلى الرَّابِعِ، وهو ز. فَأَقولُ إنَّ نِسْبَةَ الأَوَّلِ والخامسِ مَجْموعَيْنِ، وَهُما اح، إلى الثَّاني، وهو ج، كَنِسْبَةِ الثَّالِثِ والسَّادِسِ مَجْموعَيْنِ، وَهُما دط، إلى الرَّابِعِ، وهو ز.

המשל בו: כי יחס[2] הראשון, והוא א״ב, אל השני, והוא ג׳, כיחס השלישי, והוא ד״ה, אל הרביעי, והוא ז׳. ויחס החמישי, והוא ב״ח[3], אל השני, והוא ג׳, כיחס הששי, והוא ה״ט, אל הרביעי, והוא ז׳. הנה אומר כי יחס[4] הראשון והחמישי מקובצים, והם א״ח, אל השני, והוא ג׳, כיחס השלישי והששי מקובצים, והם ד״ט, אל הרביעי, והוא ז׳.

---

[1] وهو] R2T2T3

שרטוט—**בוטספק**: אב״ח, ג׳, דה״ט וז׳ מונחים זה מתחת לזה [1] יחס] ליתא **ס** [2] כי יחס] כיחס **ב** [3] ב״ח] ב״ז **ס** [4] כי יחס] כיחס **ב**

بُرْهانُهُ: أَنَّ قَدْرَ اب عِنْدَ ج كَقَدْرِ ده عِنْدَ ز. وَلَكِنَّ قَدْرَ ج عِنْدَ ب ح كَقَدْرِ ز عِنْدَ ه ط. فَبِالمُساواةِ يكونُ عَلى قَدْرِ (اب)[1] عِنْدَ ب ح كَقَدْرِ ده عِنْدَ ه ط. وَإِذا رَكَّبْنَا كانَ قَدْرُ اح عِنْدَ ح ب كَقَدْرِ د ط عِنْدَ ط ه. وَلَكِنَّ قَدْرَ ح ب عِنْدَ ج كَقَدْرِ ه ط عِنْدَ ز. فَبِالمُساواةِ يَكونُ قَدْرُ اح عِنْدَ ج كَقَدْرِ د ط عِنْدَ ز.

وَذَلِكَ ما أَرَدْنا بَيانَهُ.

מופת זה: כי שעור א״ב אל ג׳ כשעור ד״ה אצל ז׳. ואמנם שיעור ג׳ אצל ב״ח כשיעור ז׳ אצל ה״ט. וביחס השווי יהיה שעור א״ב אצל ב״ח כשיעור ד״ה אצל ה״ט. וכאשר הרכבנו יהיה יחס א״ח אצל ח״ב כשיעור ד״ט אצל ט״ה. ואמנם שיעור ח״ב אצל ג׳ כשיעור ה״ט אצל ז׳. ובשווי יהיה שיעור א״ח אצל ג׳ כשעור ד״ט אצל ז׳.

וזה מה שרצינו לבאר.

[1] اب] اح

بُرهانُه: أَنَّ نِسْبَةَ اب إلى ج كنِسْبَةِ د ه إلى ز. وَلٰكِنَّ نِسْبَةَ ج إلى ب ح كنِسْبَةِ[1] ز إلى ه ط. فَبِالْمُساواةِ تكونُ نِسْبَةُ اب إلى ب ج كنِسْبَةِ د ه إلى ه ط. وَإِذا رَكَّبْنا فَنِسْبَةُ اح إلى ح ب كنِسْبَةِ د ط إلى ط ه. وَلٰكِنَّ نِسْبَةَ ح ب إلى ج كنِسْبَةِ ه ط إلى ز. فَبِالْمُساواةِ تكونُ نِسْبَةُ اح إلى ج كنِسْبَةِ د ط إلى ز.

وَذٰلِكَ ما أَرَدْنا أَنْ نُبَيِّنَ.

מופתו: כי יחס א״ב אל ג׳[1] כיחס ד״ה אל ז׳. אבל יחס ג׳ אל ב״ח כיחס ז׳ אל ה״ט[2]. הנה[3] השווי יהיה יחס א״ב אל ב״ח כיחס ד״ה אל ה״ט[4]. וכאשר הרכבנו[5] הנה יחס א״ח אל ח״ב כיחס ד״ט אל ט״ה. אבל יחס ח״ב אל ג׳ כיחס ה״ט אל ז׳. הנה בשווי יהיה יחס א״ח אל ג׳ כיחס ד״ט אל ז׳.

וזה מה שרצינו לבאר.

1 كنسبة] CopM بالعكس+

1 כי ... ג׳] ליתא **ב** 2 ה״ט] ~~ח״ט~~ ה״[ט] **ב**
3 הנה] +כיחס **ס** 4 הנה ... ה״ט] ⟨*⟩ **ט**
5 הרכבנו] הורכבנו **ב** הורכבו **ס**

[Proposition v.25]

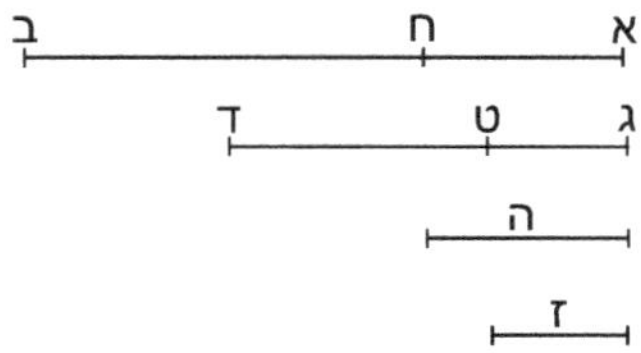

כשיהיו ארבעה שיעורים מתיחסים, והיה הראשון מהם הגדול מהם, והאחרון הקטן מהם, יהיו שניהם מחוברים גדולים מן השנים הנשארים מחוברים.

إذا كانَتْ أَرْبَعَةُ أَقْدارٍ مُتَناسِبَةً، وَكانَ الأَوَّلُ أَعْظَمَها، وَالآخِرُ أَصْغَرَها، فَإِنَّ الأَوَّلَ وَالآخِرَ مُرَكَّبَيْنِ أَعْظَمُ مِنَ الباقِيَيْنِ.

המשל בזה: ארבעה שיעורים מתיחסים עליהם א״ב וג״ד וה׳ וז׳. ושעור א״ב אצל ג״ד כשעור ה׳ אצל ז׳. והגדול שבשיעורים (א״ב)[1], והקטן שבהם ז׳. ואומר כי א״ב וז׳ מחוברים גדולים מג״ד וה׳ מחוברים.

مِثالُهُ: أَرْبَعَةُ أَقْدارٍ مُتَناسِبَةٌ عَلَيْها ا ب و ج د و ه و (ز)[1] وَقَدْرُ ا ب عِنْدَ ج د كَقَدْرِ ه عِنْدَ ز. وأَعْظَمُ الأَقْدارِ ا ب، وَأَصْغَرُها ز. فَأَقولُ إِنَّ ا ب و ز مُرَكَّبَيْنِ أَعْظَمُ مِنْ ج د و ه.

---

[1] א״ב] א׳

[1] ز] ب

إِذَا كَانَتْ أَرْبَعَةُ أَقْدَارٍ مُتَنَاسِبَةً، وَكَانَ الْأَوَّلُ أَعْظَمَهَا، وَالْأَخِيرُ أَصْغَرَهَا، فَإِنَّهُمَا مَجْمُوعَيْنِ أَعْظَمَ مِنَ الْبَاقِيَيْنِ مَجْمُوعَيْنِ.

מִثَالُهُ: أَنَّ أَرْبَعَةَ أَقْدَارٍ اب ج د ه ز مُتَنَاسِبَةٌ، نِسْبَةُ اب إلى ج د كَنِسْبَةِ ه إلى ز — اب أَعْظَمَهَا، و ز أَصْغَرَهَا. فَأَقُولُ إِنَّ اب و ز مَجْمُوعَيْنِ أَعْظَمَ مِنْ ج د و ه مَجْمُوعَيْنِ.

כאשר יהיו ארבעה שעורים מתיחסים, והיה הראשון היותר גדול מהם, והאחרון היותר קטן מהם, הנה שניהם מקובצים[1] יותר גדולים מן הנשארים מקובצים[2].

המשל בו: כי ארבעה[3] שעורי א״ב ג״ד ה׳ ז׳ מתיחסים, יחס א״ב אל ג״ד כיחס ה׳ אל ז׳— וא״ב היותר גדול מהם, וז׳ היותר קטן מהם. הנה אומר כי א״ב וז׳ מקובצים יותר גדולים מג״ד וה׳ מקובצים.

[1] מקובצים] ליתא ט [2] מקובצים] ליתא פ
[3] ארבעה] ליתא ק

מופת זה: שנקח מא״ב כשעור ה׳, והוא א״ח. ונקח מג״ד כשעור ז׳, והוא ג״ט. ושיעור א״ב אצל ג״ד כשעור ה׳ אצל ז׳. וה׳ שוה לא״ח. וז׳ שוה לג״ט. אם כן יהיה שיעור א״ב אצל ג״ד כשעור א״ח אצל ג״ט. ואם כן שיעור א״ב אצל ג״ד כשיעור ב״ח אצל ט״ד. וא״ב גדול מג״ד. אם כן ב״ח גדול מט״ד. ונשים א״ח וג״ט משתתפים—יהיו א״ב וג״ט גדולים מא״ח וג״ד. וג״ט שוה לז׳, וא״ח שוה לה׳. אם כן התבאר שא״ב וז׳ מחוברים גדולים מג״ד וה׳ מחוברים.

وزה מה שרצינו לבאר.

نשלם המאמר החמשי בכח השם ובעזרו.

بُرْهانُهُ: أَنَّا نَأْخُذُ مِنْ اب مِثْلَ ه، وَهُوَ اح. وَنأْخُذُ مِنْ ج د مِثْلَ ز، وَهُوَ ج ط. فَقَدْرُ اب عِنْدَ ج د كَقَدْرِ ه عِنْدَ ز. و ه مِثْلُ اح. و ز مِثْلُ ج ط. وَقَدْرُ اب عِنْدَ ج د كَقَدْرِ اح عِنْدَ ج ط. فَيَبْقَى ح ب أَعْظَم مِنْ ط د الباقي. وَنَجْعَلُ اح و ط ج إِذا جُمِعا فَإِنَّهُما قَدْرٌ واحِدٌ مُشْتَرَكَيْنِ—اب و ط ج هُما أَعْظَمُ مِنْ اح وج د. وج ط مِثْل ز. اح مِثْل ه. اب و ز هُما أَعْظَمُ من ج د و ه.

وَذَلِكَ ما أَرَدْنا بَيانَهُ.

بُرهانُهُ: أَنَّا نَفْصِلُ مِنْ ا ب مثل ه، وهو ا ح، ومِنْ ج د مثل ز، وهو ج ط. فنِسْبَةُ ا ب إلى ج د كنِسْبَةِ ه إلى ز. و ه مثل ا ح. و ز مثل ج ط. فنِسْبَةُ ا ب إلى ج د كنِسْبَةِ ا ح إلى ج ط. فنِسْبَةُ ا ب إلى ج د كنِسْبَةِ ح ب الباقي إلى ط د الباقي. وإذا بَدَّلْنَا، كانَتْ نِسْبَةُ ا ب إلى ب ح كنِسْبَةِ ج د إلى د ط. و ا ب الأَوَّل أعظم مِنْ ج د الثَّالِث. ف ب ح الثَّاني أعظم مِنْ ط د الرَّابِع. وَنَجْعَلُ ا ح ط ج مُشْتَرَكَيْن. فَقَدْرا ا ب ط ج أعظم مِنْ ا ح ج د. و ج ط مثل ز. و ا ح مثل ه. ف ا ب ز مُجْمُوعَيْن أعظم مِنْ ج د ه مُجْمُوعَيْن.

מופתו: אנחנו נבדיל מן א״ב כמו ה׳, והוא א״ח, ומן ג״ד כמו ז׳, והוא ג״ט. הנה יחס א״ב אל ג״ד כיחס ה׳ אל ז׳. וה׳ כמו א״ח. וז׳ כמו ג״ט. אם כן יחס א״ב אל ג״ד כיחס א״ח אל ג״ט. אם כן יחס א״ב אל ג״ד[1] כיחס ח״ב[2] הנשאר אל[3] ט״ד הנשאר. וכאשר המירונו, היה יחס א״ב אל ב״ח כיחס ג״ד אל ד״ט. וא״ב הראשון יותר גדול מן ג״ד השלישי. אם כן ב״ח השני יותר גדול מן ט״ד הרביעי. ונשים א״ח וט״ג משותפים. אם כן שני שעורי א״ב וט״ג[4] יותר גדולים מן א״ח וג״ד. וג״ט כמו ז׳. וא״ח כמו ה׳. אם כן א״ב וז׳ מקובצים יותר גדולים מן ג״ד וה׳ מקובצים.

وَذَلِكَ ما أَرَدْنَا أَنْ نُبَيِّنَ.

וזה מה שרצינו לבאר.

تَمَّتِ الْمَقَالَةُ الْخامِسَةُ مِنْ كِتَابِ أُوقليدِس^.

נשלם המאמר החמישי מספר אקלידס החכם[5].

[1] ג״ד] ג״ט **ס** [2] ה׳ ... ח״ב] ליתא **ב** [3] הנשאר אל] הנשאם אל ה׳ **ב** [4] א״ב וט״ג] ~~א״ב וט״ג משותפים~~ אם כן שני שעורי א״ב וט״ג **פ** [5] נשלם ... החכם] +בע״ה **ב** +תל״ח **ו** ליתא **ט** +ת״ל **ס** נשלם המאמר החמישי בעזרת י״י **ק**

[Book VI]

המאמר הששי

[1] השטחים המתדמים הם שזויותיהם שוות וצלעותם המקיפות בזויותיהם השוות מתיחסות.

[2] והשטחים המתהפכים הצלעות הם אשר צלעותם מתיחסות בקדימה ובאחור.

[3] [H4] יאמר הגבה בתמונה העמוד היוצא מנקודת ראשו אל תושבתו.

المَقالَةُ السَّادِسَةُ مِنْ كِتابِ الأَرْكانِ

[2] وَالسُّطوحُ المُستَقيمَةُ الخُطوطِ المُتَشابِهَةُ هِيَ الَّتي زَوايا كُلِّ سَطْحٍ مِنْها مُساوِيَةٌ لِزَوايا السَّطْحِ الآخَرِ، وَالأَضْلاعُ المُحيطَةُ بِالزَّوايا المُتَساوِيَةِ مُتَناسِبَةٌ.

[3] وَالسُّطوحُ المُستَقيمَةُ الخُطوطِ تَكونُ مُتَكافِئَةَ الأَضْلاعِ مَتَى كانَتْ في كُلِّ واحِدٍ مِنَ السَّطْحَيْنِ نِسْبَةً مُقَدَّمَةً وَتالِيَةً عَلَى التَّقديمِ وَالتَّأخيرِ.

[1] الارتِفاعُ في كُلِّ شَكْلٍ هُوَ العَمودُ المُخْرَجُ مِنْ رَأْسِهِ إِلى قاعِدَتِهِ.

المَقَالَةُ السَّادِسَةُ مِنْ كِتَابِ أُقْلِيدِس^

[1] السُّطُوحُ المُتَشَابِهَةُ هِيَ الَّتِي زَوَايَاهَا مُتَسَاوِيَةٌ وَأَضْلَاعُهَا المُحِيطَةُ بِالزَّوَايَا المُتَسَاوِيَةِ مُتَنَاسِبَةٌ.

[2] والسُّطُوحُ المُتَكَافِئَةُ الأَضْلَاعِ هِيَ الَّتِي أَضْلَاعُهَا مُتَنَاسِبَةٌ على التَّقْدِيمِ وَالتَّأْخِيرِ. وَوَجَدْنَا فِي نُسْخَةٍ أُخرى: المُتَكَافِئَةُ هِيَ[1] الَّتِي فِي كُلِّ وَاحِدٍ مِنْهَا مُقَدَّمَاتٌ وَتَوَالٍ.

[3] الارْتِفَاعُ فِي الشَّكْلِ هُوَ العَمُودُ المُخْرَجُ مِنْ رَأْسِهِ^ إلى قاعِدَتِهِ. وَوَجَدْنَا فِي بَعْضِ النُّسَخِ: الارتِفاعُ^ هُوَ أَعْظَمُ الأَعْمِدَةِ الواقِعَةِ مِنْ نُقْطَةٍ ما مِنَ النُّقَطِ الَّتِي على مُحِيطِ شَكْلٍ مِنَ الأَشْكَالِ أَيِّ شَكْلٍ كانَ على قاعِدَتِهِ^ أو على الخَطِّ^ الَّذي على اسْتِقامَتِها[2].

[1]هي] MPR2 [2]ووجدنا ... استقامتها] ההגדרה כולה נמצאת רק בעדי הנוסח של משפחה B חוץ מעד נוסח Cam

המאמר הששי מספר אקלידס החכם[1]

[1] השטחים המתדמים הם אשר זויותיהם שוות וצלעותיהם המקיפות בזויות השוות[2] מתיחסות.

[2] והשטחים המספיקות[3] הצלעות הם אשר צלעותיהם מתיחסות על ההקדמה והאיחור. נמצא בנסחא אחרת: המספיקות הם אשר בכל אחד מהם הקדמות ונמשכות.

[3] [H4] הגובה בתמונה הוא העמוד המוצא מנקודת ראשה[4] אל תושבתה. ומצאנו בקצת הנסחאות: הגובה בתמונה הוא[5] היותר גדול שבעמודים הנופלים מנקודת[6] מה מן הנקודות אשר על מקיף תמונה מן התמונות אי זה תמונה שתהיה על תושבת או על קו אשר[7] על ישרו.

[1]מספר ... החכם] ליתא ס [2]השוות] ששוות ס [3]המספיקות] המקיפות ז [4]ראשה] ראשו וספ [5]הוא] ליתא ב [6]מנקודת] מנקודה פק [7]אשר] ליתא ס

[4] [H3] ויאמר בקו הישר שהוא על [!] יחס בעל עצם אמצע ושתי קצוות, כשיהיה יחס הקו בעצמו אל גדול חלקיו כיחס גדול חלקיו אל קטנם.

[4] يُقالُ في الخطّ إنَّهُ قُسِمَ عَلى نِسْبَةٍ ذات وَسَطٍ وَطَرَفَيْنِ مَتَى كانَتْ نِسْبَةُ الخطّ بِأَسْرِهِ إِلى أَعْظَمِ قِسْمَيْهِ كَنِسْبَةِ أَعْظَمِ قِسْمَيْهِ إِلى أَصْغَرِهِما.

[5] יאמר שהיחס מחובר מיחסים כששעורי היחסים נכפלו בעצמם ועשו יחס מה.

[5] يُقالُ إِنَّ نِسْبَةً مُؤَلَّفَةٌ مِنْ نِسَبٍ مَتَى كانَتْ أَقْدارُ النِّسَبِ إِذا ضُوعِفَتْ بِأَنْفُسِها فَعَلَتْ نِسْبَةً ما.

[4] يُقَالُ فِي الخَطِّ المُسْتَقِيمِ إنَّهُ قُسِمَ[1] على نِسْبَةٍ ذاتِ وَسَطٍ وطَرَفَيْنِ، مَتَى كانَتْ نِسْبَةُ الخَطِّ بأَسْرِهِ إلى أَعْظَمِ قِسْمَيْهِ كَنِسْبَةِ أَعْظَمِ قِسْمَيْهِ إلى أَصْغَرِهِما.

[4] [H3] ויאמר בקו הישר שהוא חולק על יחס בעל אמצע[1] ושתי קצוות, כאשר היה יחס הקו בכללו אל היותר גדול שבחלקיו כיחס היותר גדול שבחלקיו אל היותר קטן משניהם.

[5] يُقَالُ إنَّ النِّسْبَةَ مُؤَلَّفَةٌ مِنْ نِسَبٍ مَتَى كانَتْ أَقْدارُ النِّسَبِ إذا ضُوعِفَتْ بأَنْفُسِها فَعَلَتْ نِسْبَةً ما. وُجِدَ[2] في نُسْخَةٍ أُخْرَى: يُقَالُ^ إنَّ النِّسْبَةَ المُؤَلَّفَةَ[3] تَنْقَسِمُ بِنِسَبٍ إذا كانَتِ النِّسَبُ^ مَتَى^ جُزِّئَتْ بَعْضُها على بَعْضٍ أَحْدَثَتْ نِسْبَةً ما^.

[5] יאמר כי היחס מחובר[2] מיחסים כאשר היו[3] שעורי היחסים כאשר נכפלו בעצמיהם[4] עשו יחס מה. נמצא בנסחא אחרת: כי היחס המחובר יחלק ליחסים כאשר היו שעורי היחסים מן חלוקת[5] קצתם על קצת חדשו יחס[6] אחד.

[1]قسم] قد+ CCopT1T2T3 [2]وجد] CopT2 وجدنا [3]المؤلفة] R2Q

[1]אמצע] אמצעי **בסק**; אפשר שהנוסח הנכון הוא "אמצעי" [2]מחובר] מחוברים **ס** [3]כאשר היו] ⟨נ"א כאשר היו שעורי' הן⟩ **ס** [4]בעצמיהם] בעצמ⟨ה⟩ם **ב** [5]חלוקת] ליתא **בוס** [6]יחס] דבר **פ**

[Proposition VI.1]

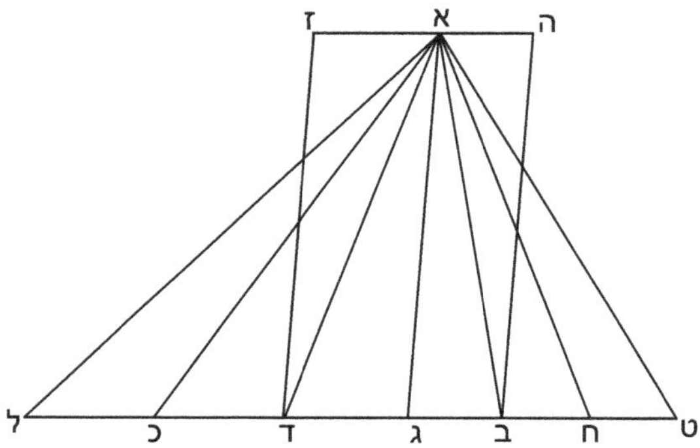

השטחים נכוחי הצלעות והמשולשים כשיהיה גבהם בשיעור אחד, יהיה יחס קצתם אל קצת כיחס תושבותם קצתם אל קצת.

السُّطوحُ المُتَوازِيَةُ الأَضْلاعِ والمُثَلَّثات إِذا ضُوعِفَتْ كانَ ارتِفاعُها بِقَدْرٍ واحِدٍ، فَإِنَّ قَدْرَ بَعْضِها عِنْدَ بَعْضٍ كَقَدْرِ قَواعِدِها بَعْضِها عِنْدَ بَعْضٍ.

המשל בזה: שנצייר שני משלשים על שיעור אחד בגובה, עליהם אב״ג ואג״ד, ושני שטחים נכוחי הצלעות על שעור גבה המשולשים[1], והם עליהם ה״ג וג״ז. ואומר כי שיעור תושבת ב״ג אצל תושבת ג״ד כשיעור משולש אב״ג אצל משולש אג״ד, וכשיעור שטח ה״ג נכוחי הצלעות אצל שטח ג״ז נכוחי הצלעות.

مِثالُهُ: أَنَّا نَخُطُّ مُثَلَّثَيْنِ عَلى قَدْرٍ واحِدٍ في الارْتِفاعِ، عَلى أَحَدِهِما ا ب ج وَعَلى الآخَرِ ا ج د. وسَطْحَيْنِ مُتَوازِيَي الأَضْلاعِ على ارْتِفاعِ المُثَلَّثَيْنِ عَلَيْهِما ه ج وج ز. فَأَقولُ إِنَّ قَدْرَ قاعِدَة ب ج عِنْدَ قاعِدَةِ ج د كَقَدْرِ مُثَلَّث ا ب ج عِنْدَ مُثَلَّث ا ج د. وكَقَدْرِ ه ج المُتَوازي الأَضْلاعِ عِنْدَ سَطْحِ ج ز المُتَوازي الأَضْلاعِ.

[1] המשולשים] המש⟨ו⟩לשים

السُّطُوحُ المُتَوازِيَةُ الأَضْلاعِ وَالمُثَلَّثاتُ إذا كانَ ارتِفاعُها بِقَدْرٍ واحِدٍ فإنَّ نِسْبَةَ بَعْضِها إلى بَعْضٍ كَنِسْبَةِ قَواعِدِها بَعْضِها إلى بَعْضٍ.

השטחים הנכחי הצלעות והמשולשים כאשר היה גבהותם בשעור אחד, הנה יחס[1] קצתם אל קצת כיחס תושבותיהם קצתם אל קצת.

مِثالُهُ: أَنَّ سَطْحَيْ ه ج ج ز المُتَوازي الأَضْلاعِ، ومُثَلَّثَيْ ا ب ج ا ج د على قَدْرٍ واحِدٍ في الارتِفاعِ. فَأَقولُ إنَّ نِسْبَةَ قاعِدَةِ ب ج إلى قاعِدَةِ ج د كَنِسْبَةِ مُثَلَّثِ ا ب ج إلى مُثَلَّثِ ا ج د وكَنِسْبَةِ سَطْحِ ه ج المُتَوازي الأَضْلاعِ إلى سَطْحِ ج ز المُتَوازي الأَضْلاعِ.

המשל בו: כי שני[2] שטחי ה״ג וג״ז נכחי הצלעות, ושני משולשי אב״ג ואג״ד על שעור אחד בגובה. הנה אומר כי יחס[3] תושבת ב״ג אל תושבת ג״ד כיחס משולש אב״ג אל משולש אג״ד וכיחס שטח ה״ג[4] הנכחי הצלעות אל שטח[5] ג״ז הנכחי הצלעות[6].

---

[1] הנה יחס] הני יחסו **ב** הנה **פ** [2] שני] ⟨*⟩ **ט** [3] כי יחס] כיחס **ב** [4] ה״ג] ג״ז **ב** [5] שטח] יחס **ז** [6] אל ... הצלעות] ליתא **ב**

מופת זה: שנוציא קו ב״ד לשני הצדדים. ונחלוק ממנו בשיעור ב״ג כמו שנרצה, והוא ב״ח וח״ט. ונחתוך ממנו בשעור ג״ד כמו שנרצה, והוא ד״כ וכ״ל. ונדביק א׳ בט׳ ובח׳ ובכ׳ ובל׳. ותושבות ט״ח וח״ב וב״ג שוות. ואם כן משולשי אט״ח אח״ב אב״ג שוים גם כן. ושעור כפלי תושבת ט״ג לתושבת ב״ג [כ]שעור כפלי משולש אג״ט למשולש אב״ג. ולזה גם כן שעור כפלי תושבת ג״ל לתושבת ג״ד כשעור כפלי משולש אג״ל למשולש אג״ד. וכשיעור מה שתוסיף תוספת[1] ג״ט על תוספת[2] ג״ל, יוסיף משלש אג״ט על משלש אג״ל[3]. ואם היתה שוה לו גם הוא שוה לו, ואם היתה חסרה ממנה גם הוא חסר ממנו בזה השעור. והנה השעורים ארבעה—שתי תושבות ב״ג וג״ד ושני משולשי אב״ג ואג״ד. וכבר לוקח לתושבת ג״ב ולמשולש אג״ב כפלים שוים, והם תושבת ט״ג ומשולש אג״ט. ולוקח לתושבת ג״ד ולמשלש אג״ד כפלים שוים אחרים, והם תושבת ג״ל ומשלש אג״ל. וכבר התבאר כי שיעור מה שתוסיף תושבת ט״ג על תושבת ג״ל הוא כשיעור מה שיוסיף משולש אט״ג על משלש אג״ל. ואם היתה שוה לה גם הוא שוה לו, ואם היתה חסרה ממנה גם הוא חסר ממנה בזה השיעור. אם כן שעור תושבת ב״ג אצל תושבת ג״ד כשעור משלש אב״ג אצל משולש אג״ד. ולזה יתבאר כי כמו כן הוא שטח (ה״ג)[4] נכוחי הצלעות אצל שטח ג״ז נכוחי הצלעות.

بُرْهانُهُ: أَنَّا نُخْرِجُ خَطَّ (ب د)[1] في كِلْتَيِ الجِهَتَيْنِ. وَنَفْصِلُ مِنْهُ مِثْلَ ب ج كَمْ شِئْنا، وَهُوَ ب ح وَهُوَ ح ط. وَنَفْصِلُ مِنْهُ في الجِهَةِ الأُخْرى مِثْلَ ج د وَعَلى تِلْكَ العِدَّةِ، وَهُوَ د ك و ك ل. وَنَصِلُ ا ط ا ح ا ك بِخُطوط ا ط و ا ح و ا ك و ا ل. فَقَواعِدُ ط ح و ح ب و ب ج مُتَساوِيَةٌ. ومُثَلَّثاتُ ا ط ح و ا ح ب و ا ب ج أَيضًا مُتَساوِيَةٌ. فَقَدْرُ أَضْعافِ قاعِدَةِ ط ج كَقاعِدَةِ ب ج كَقَدْرِ أَضْعافِ مُثَلَّثِ ا ط ج لِمُثَلَّثِ ا ب ج. وكَذَلِكَ قَدْرُ أَضْعافِ قاعِدَةِ ج ل كَقاعِدَةِ ج د كَقَدْرِ أَضْعافِ مُثَلَّثِ ا ج ل لِمُثَلَّثِ ا ج د. فَإِنْ كانَتْ قاعِدَةُ ج ط تَزيدُ عَلى قاعِدَةِ ج ل. فَإِنَّ مُثَلَّثَ (ا ج ط)[2] يَزيدُ عَلى مُثَلَّثِ (ا ج ل)[3]. فَإِنْ كانَتْ مُساوِيَةً لَها، فَهُوَ مُساوٍ لَهُ، وَإِنْ كانَتْ بِأَقْصَرَ مِنْها، فَهُوَ ناقِصٌ مِنْهُ بِذَلِكَ القَدْرِ. فَالأَقْدارُ الأَرْبَعَةُ قاعِدَتا ب ج و ج د وَمُثَلَّثا ا ب ج و ا ج د. وَقَدْ أُخِذَ لِقاعِدَةِ ب ج وَلِمُثَلَّثِ ا ب ج أَضْعافٌ مُتَساوِيَةٌ، وَهِيَ قاعِدَةُ ط ج وَمُثَلَّثُ ا ط ج. وَأُخِذَ لِقاعِدَةِ ج د وَلِمُثَلَّثِ ا ج د أَضْعافٌ مُتَساوِيَةٌ أُخَرُ، وَهِيَ قاعِدَةُ ج ل

[1]תוספת] נראה שצ״ל: תושבת [2]תוספת] נראה שצ״ל: תושבת [3]אג״ל] ~~אב״ג~~ ⟨⟨*⟩⟩ [4]ה״ג] ב״ג

[1]ب د] ب ج [2]ا ج ط] ا ب ج [3]ا ج ل] ا ج د

بُرهانُهُ: أَنَّا نُخْرِجُ خَطَّ ب د في كِلْتَيّ^ الجِهَتَيْنِ. ونَفْصِلُ مِنْهُ مثل ب ج كَمْ شِئْنا، وَهُوَ ب ح ح ط، ومثل ج د أيضًا كَمْ شِئْنا، وَهُوَ د ك ك ل. ونُخْرِجُ خُطوطَ ا ط وا ح وا ك وا ل. فَقَواعِد ط ح ح ب و ب ج مُتَساوِيَة. فمُثَلَّثات ا ط ح و ا ح ب و ا ب ج أَيْضًا مُتَساوِيَة. فأَضْعاف قاعِدَة ط ج لقاعِدَة ب ج كأَضْعافِ مُثَلَّثِ ا ط ج لمُثَلَّثِ ا ب ج. وكَذَلِكَ أَضْعاف قاعِدَةِ ج ل لقاعِدَةِ ج د كأَضْعافِ مُثَلَّثِ ا ج ل لمُثَلَّثِ ا ج د. فَإِنْ كانَتْ قاعِدَةُ ج ط تَزيدُ على قاعِدَةِ ج ل، فَإِنَّ مُثَلَّثَ ا ج ط يَزيدُ على مُثَلَّثِ ا ج ل. وإنْ كانَتْ تُساويها فَهُوَ يُساويهِ، وَإِنْ كانَتْ تَنْقُصُ مِنْها فَهُوَ يَنْقُصُ مِنْهُ. فالأَقْدارُ الأَرْبَعَةُ—قاعِدَتا ب ج و ج د وَمُثَلَّثا ا ب ج و ا ج د. وَقَدْ أَخَذْنا[1] لِقاعِدَةِ ب ج ومُثَلَّثِ[2] ا ب ج أَضْعافًا مُتَساوِيَةً، وهِيَ قاعِدَةُ ط ج ومُثَلَّث ا ط ج. وَأَخَذْنا[3] لِقاعِدَةِ ج د ومُثَلَّثِ ا ج د أَضْعافًا مُتَساوِيَةً، وَهِيَ قاعِدَةُ ج ل ومُثَلَّث ا ج ل. فاسْتَبانَ أَنَّ قاعِدَةَ ط ج ومُثَلَّثَ ا ط ج إِمَّا زائِدانِ مَعًا على قاعِدَةِ ج ل ومُثَلَّثِ ا ج ل،

مופתו: אנחנו נוציא קו ב״ד[1] בשני הצדדים. ונבדיל ממנו כמו ב״ג כמה שנרצה, והוא ב״ח וח״ט, וכמו ג״ד גם כן כמה שנרצה, והוא ד״כ וכ״ל. ונוציא קוים א״ט וא״ח וא״כ וא״ל[2]. הנה תושבות[3] ט״ח[4] וח״ב וב״ג שוות. אם כן משולשי[5] אט״ח ואח״ב ואב״ג גם כן שוים. אם כן כפלי[6] תושבת ט״ג לתושבת ב״ג ככפלי משולש אט״ג למשולש אב״ג. וכן כפלי תושבת ג״ל לתושבת ג״ד ככפלי משולש אג״ל למשולש אג״ד. הנה אם[7] היתה תושבת ג״ט מוסיפה על תושבת[8] ג״ל, הנה משולש אג״ט[9] יוסיף על משולש אג״ל. ואם היתה שוה לה הנה הוא ישוה אליו[10], ואם היתה חסרה ממנה הנה הוא[11] יחסר ממנו. והשעורים הארבעה—שתי[12] תושבות ב״ג וג״ד ושני משולשי אב״ג ואג״ד. וכבר לקחנו לתושבת ב״ג ומשולש אב״ג כפלים שוים, והם תושבת ט״ג ומשולש אט״ג. ולקחנו[13] לתושבת ג״ד ומשולש אג״ד כפלים שוים, והם תושבת ג״ל ומשולש אג״ל. והתבאר כי תושבת ט״ג ומשולש אט״ג אם נוספים יחד על תושבת ג״ל ומשולש אג״ל, אם שוים יחד לשניהם, ואם חסרים יחד לשניהם[14] [15]. אם כן יחס תושבת ב״ג אל תושבת ג״ד כיחס משולש אג״ב אל משולש אג״ד. ושטח ה״ג[16] כפל משולש אב״ג. ושטח ז״ג כפל משולש אד״ג. אם כן יחס שטח ה״ג אל שטח[17] ג״ז כיחס משולש אב״ג אל משולש אג״ד, מפני

[1] اخذنا] اخذ **MPR2** [2] ومثلث] **MPR2**

[3] واخذنا] اخذ **MPR2**

[1] ב״ד] ב״⟨ד⟩ **ט** [2] וא״ל] ליתא **ז** [3] תושבות] תושבת **וס** [4] ט״ח] ס״ב+ **ס** [5] משולשי] המשולשים **ב** משולש **וס** [6] כפלי] ⟨*⟩ **ט** ⟨⟨*⟩⟩ **פ** [7] אם] +כן **ב** [8] תושבת] ליתא **ס** [9] אג״ט] אג״ד **ס** [10] ישוה אליו] שוה לה **ב** שוה אליו **ט** [11] הוא] +* **ט** [12] שתי] שני **ט** ⟨⟨להם⟩⟩+ **ק** [13] ולקחנו] נלקחנו **ס** [14] ואם ... לשניהם] ⟨*⟩ **ז** [15] לשניהם] משניהם **טק** [16] משולש ... ה״ג] ⟨*⟩ **ט** [17] שטח] ליתא **ב**

وَمُثَلَّثُ ا ج ل. فَقَدِ اسْتَبانَ أَنَّ قاعِدَةَ ط ج ومُثَلَّثَ ا ط ج إِمَّا زائِدانِ مَعًا عَلى قاعِدَةِ ج ل وَمُثَلَّثِ ا ج ل. وَإِمَّا مُتَساوِيَتانِ مَعًا لَهُما. وَإِمّا ناقِصانِ مَعًا مِنْهُما. فَنِسْبَةُ قاعِدَةِ ب ج إِلى قاعِدَةِ ج د كَنِسْبَةِ مُثَلَّثِ ا ب ج إِلى مُثَلَّثِ ا ج د. وَقَدِ اسْتَبانَ أَنَّهُ كَذَلِكَ نِسْبَةُ سَطْحِ ه ج إِلى سَطْحِ ج ز المُتَوازِي الأَضْلاعِ.

וזה מה שרצינו לבאר.

وَذَلِكَ ما أَرَدْنا بَيانَهُ.

وَإِمّا مُسَاوِيانِ مَعًا لَهُما، وَإِمّا ناقِصانِ مَعًا مِنْهُما. فَنِسْبَةُ قاعِدَةِ ب ج إلى قاعِدَةِ ج د كَنِسْبَةِ مُثَلَّث ا ب ج إلى مُثَلَّث ا ج د. وسَطْح ه ج ضِعْف مُثَلَّث ا ب ج. وسَطْح ز ج ضِعْف مُثَلَّث ا د ج. فَنِسْبَة سَطْح ه ج إلى سَطْح ز ج كَنِسْبَة مُثَلَّث ا ب ج إلى مُثَلَّث ا ج د^. ونِسْبَة مُثَلَّث ا ب ج إلى مُثَلَّث ا ج د كَنِسْبَة خطّ ب ج إلى خطّ ج د. فَنِسْبَة سَطْح ه ج إلى سَطْح ج ز كَنِسْبَة خَطّ ب ج إلى خطّ ج د.

وَذَلِكَ ما أَرَدْنا أَنْ نُبَيِّنَ.

כי החלקים אשר כפליהם שוים יחס קצתם אל קצת כיחס כפליהם קצתם אל קצת. ויחס[1] משולש אב״ג אל משולש אג״ד כיחס קו ב״ג אל קו[2] ג״ד. אם כן יחס שטח ה״ג אל שטח ג״ז כיחס קו[3] ב״ג אל קו ג״ד.

וזה מה שרצינו לבאר.

[1]ויחס] אבל יחס ט [2]קו] ליתא ס +ב״ד ק
[3]קו] +* ב

[Proposition VI.2]

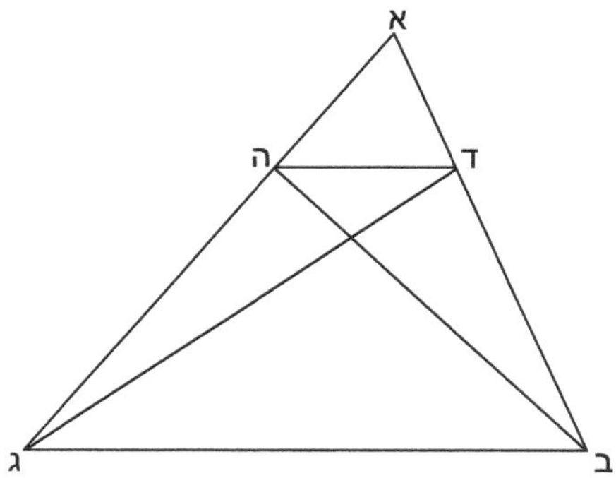

כל משלש שיצא מצלע מצלעיו קו אל צלע אחר ממנו, אם יהיה הקו היוצא נכוחי לצלע המשולש הנשארת, יחתוך שתי הצלעות על יחס אחד, ואם חתך שתי הצלעות על יחס אחד, יהיה נכוחי לצלע המשולש הנשארת.

كُلُّ مُثَلَّثٍ يَخْرُجُ مِنْ ضِلْعٍ مِنْ أَضْلاعِهِ خَطٌّ إِلى ضِلْعٍ آخَرَ، فَإِنَّ الخَطَّ المُخْرَجُ إِن كانَ مُوازِيًا لِضِلْعِ المُثَلَّثِ الباقي، فَإِنَّه يَقْطَعُ الضِّلْعَيْنِ عَلى نِسْبَةٍ واحِدَةٍ. فَإِنْ قَطَعَ الضِّلْعَيْنِ عَلى نِسْبَةٍ واحِدَةٍ، فَإِنَّهُ مُوازٍ لِضِلْعِ المُثَلَّثِ.

המשל בזה: שנצייר משולש עליו אב"ג. ונוציא מצלע א"ב אל צלע א"ג קו נכוחי לצלע ב"ג, והוא קו ד"ה. ואומר כי קו ד"ה חותך צלעות א"ב וא"ג על יחס אחד, והוא ששיעור ב"ד אצל ד"א כשעור ג"ה אצל ה"א.

مِثالُهُ: مُثَلَّثُ ا ب ج. وَقَدْ خَرَجَ مِنْ ضِلْعِ ا ج خط د ه مُوازِيًا لِضِلْعِ ب ج. وَأَقولُ إِنَّ خَطَّ د ه قَدْ قَطَعَ ا ب و ا ج عَلى نِسْبَةٍ واحِدَةٍ، قَدْرُ ب د عِنْدَ د ا كَقَدْرِ ج ه عِنْدَ ه ا.

كُلُّ مُثَلَّثٍ يَخْرُجُ مِنْ ضِلْعٍ مِنْ أَضْلاعِهِ خَطٌّ إلى ضِلْعٍ مِنْهُ آخَرَ، فَإِنَّ الخَطَّ المُخْرَجَ إِنْ كانَ مُوازِيًا لِضِلْعِ المُثَلَّثِ الباقي، فَإِنَّهُ يَقْطَعُ الضِّلْعَيْنِ على نِسْبَةٍ واحِدَةٍ. وَإِنْ قَطَعَ الضِّلْعَيْنِ على نِسْبَةٍ واحِدَةٍ، فَإِنَّهُ مُوازٍ لِضِلْعِ المُثَلَّثِ.

כל משולש יצא מצלע[1] מצלעותיו קו אל צלע אחר ממנו[2], הנה הקו המוצא אם היה נכחי לצלע המשולש הנשאר, הנה הוא יחתוך שתי הצלעות על יחס אחד. ואם חתך שתי הצלעות על יחס אחד, הנה הוא נכחי לצלע המשולש[3].

مِثالُهُ: أَنَّا نُخْرِجُ مِنْ ضِلْعِ ا ب مِنْ مُثَلَّثِ ا ب ج خَطًّا إلى ضِلْعِ ا ج مُوازِيًا لِضِلْعِ ب ج، وَهُوَ د ه. فَأَقُولُ إِنَّ د ه قَدْ قَطَعَ ا ب و ا ج على نِسْبَةٍ واحِدَةٍ، فَصارَتْ نِسْبَةُ ب د إلى ا د كَنِسْبَةِ ج ه إلى ه ا.

המשל בו: שנוציא מצלע א״ב ממשולש אב״ג קו אל צלע א״ג נכחי לצלע ב״ג, והוא ד״ה. הנה אומר כי ד״ה כבר חתך א״ב וא״ג על יחס אחד, והיה יחס ב״ד אל א״ד כיחס ג״ה אל א״ה.

שרטוט—**בוטספק**: משולש אב״ג הוא שווה שוקיים   1 מצלע] ליתא **וטפ** צלע **ס** ⟨מ⟩צלע **ק**   2 ממנו] ליתא **ז**   3 המשולש] +הנשאר **ס**

מופת זה: שנדביק ד׳ בג׳ וה׳ בב׳. ויהיה משלש בד״ה שוה למשלש גד״ה, כי הם על תושבת אחת, והוא ד״ה, ובין שני קוים נכוחיים ד״ה וב״ג. והשעורים השוים יחסם אל שיעור אחר הוא אחד. ואם כן שיעור משלש בד״ה אצל משלש אד״ה כשעור משלש גד״ה אצל משלש אד״ה. ושיעור משלש בד״ה אצל משלש אד״ה כשיעור תושבת ב״ד אצל ד״א. ושעור משולש גד״ה אצל משולש אד״ה כשעור ג״ה אצל ה״א. אם כן שעור ב״ד אצל ד״א כשיעור ג״ה אצל ה״א. עוד נשים שיעור ב״ד אצל ד״א כשעור ג״ה אצל ה״א. ואומר כי ד״ה נכוחי לב״[..]. כי הנהגתם אחת, לפי [..]ב״ד אצל ד״[א] כשעור [..] אצל ה״א[1]. ואמנם שעור ב״ד אצל ד״א כשיעור משלש בד״ה אצל משלש אד״ה. ושיעור ג״ה אצל ה״א כשעור משלש גד״ה אצל משלש אד״ה. ואם כן משלש בד״ה הוא שוה למשלש גד״ה, והם על תושבת ד״ה. אם כן קו ד״ה נכוחי לקו ב״ג.

بُرْهانُه: أنَّا نَصِلُ د ب ج و ه ب ب بِخَطَّيْ د ج ه ب. فَمُثَلَّثُ د ب ه مِثْلُ مُثَلَّثِ ج د ه، لِأَنَّهُما عَلى قاعِدَةٍ واحِدَةٍ، وَهِيَ د ه، وبَيْنَ خَطَّيْ د ه وب ج مُتَوازِيَيْنِ. وَالأَشْياءُ المُتَساوِيَةُ نِسْبَتُها إِلى الشَّيْءِ الواحِدِ—واحِدَةٌ. فَقَدْرُ (مُثَلَّث)[1] ب د ه عِنْدَ مُثَلَّثِ ا د ه كَقَدْرِ مُثَلَّثِ ج د ه عِنْدَ مُثَلَّثِ ا د ه. وَقَدْرُ مُثَلَّثِ ب د ه عِنْدَ مُثَلَّثِ ا ه د كَقَدْرِ ب د عِنْدَ د ا. وَقَدْرُ مُثَلَّثِ ج د ه عِنْدَ مُثَلَّثِ د ه ا كَقَدْرِ ج ه عِنْدَ ا ه. فَقَدْرُ ب د عِنْدَ د ا كَقَدْرِ ج ه عِنْدَ ه ا. ثُمَّ نَجْعَلُ قَدْرَ ب د عِنْدَ د ا كَقَدْرِ ج ه عِنْدَ ه ا. فَأَقولُ إِنَّ د ه موازٍ ل ب ج. بُرهانُه: أَنَّ قَدْرَ (ب د)[2] عِنْدَ د ا كَقَدْرِ مُثَلَّثِ ب د ه عِنْدَ مُثَلَّثِ ا د ه. وَقَدْرُ ج ه عِنْدَ ه ا كَقَدْرِ مُثَلَّثِ ج د ه عِنْدَ مُثَلَّثِ ا د ه. فَمُثَلَّثُ ب د ه مِثْلُ مُثَلَّثِ ج د ه. وَهُما عَلى قاعِدَةٍ واحِدَةٍ موازٍ ل ب ج.

---

[1] עוד ... ה״א] ⟨*⟩

[1] مثلث] مثل [2] ب د] ج د

بُرهانُهُ: أَنَّا نُخْرِجُ خَطَّي د ج و ه ب. فَمُثَلَّثُ ب د ه مثل مُثَلَّثِ ج د ه، لأَنَّهُما على قاعِدَةٍ واحدةٍ، وهي د ه، وَبَيْنَ خَطَّيْنِ مُتَوازِيَيْنِ، وَهُما د ه و ب ج. والمُتَساوِيَةُ نِسْبَتُها إلى شيءٍ واحدٍ بِعَيْنِهِ واحدة. فَنِسْبَةُ مُثَلَّثِ ب د ه إلى مُثَلَّثِ ا د ه كَنِسْبَةِ مُثَلَّثِ ج د ه إلى مُثَلَّثِ ا د ه، ونِسْبَةَ مُثَلَّثِ ب د ه إلى مُثَلَّثِ ا د ه كَنِسْبَةِ ب د إلى د ا، لِأَنَّ ارتفاعَهُما واحِدٌ. وكَذَلِكَ نِسْبَةُ مُثَلَّثِ ج د ه إلى مُثَلَّثِ ا د ه كَنِسْبَةِ ج ه إلى ه ا. فنِسْبَةُ ب د إلى د ا كَنِسْبَةِ ج ه إلى ه ا. ثُمَّ لِتَكُنْ نِسْبَةُ ب د إلى د ا كَنِسْبَةِ ج ه إلى ه ا. فَأَقولُ إنّ د ه يُوازي ب ج، لِأَنَّ تَدْبيرَهُما واحِدٌ. فَنِسْبَةُ ب د إلى د ا كَنِسْبَةِ ج ه إلى ه ا. لكنّ نِسْبَةَ ب د إلى د ا كَنِسْبَةِ مُثَلَّثِ ب د ه إلى مُثَلَّثِ ا د ه. ونِسْبَةَ ج ه إلى ه ا كَنِسْبَةِ مُثَلَّثِ ج د ه إلى مُثَلَّثِ ا د ه. فَمُثَلَّثُ ب د ه مِثْلُ مُثَلَّثِ ج د ه، وَهُما على قاعِدَةٍ واحِدَةٍ، وهي قاعِدَةُ[1] د ه. فخَطّ د ه مُوازٍ لِخَطِّ ب ج.

מופתו: אנחנו נוציא שני קוי ד״ג וה״ב. הנה משולש בד״ה כמו משולש גד״ה, מפני כי שניהם על תושבת אחת, והיא ד״ה, ובין שני קוים נכחיים, והם ד״ה וב״ג. והדברים השוים יחסם אל[1] דבר אחד בעינו אחד. אם כן יחס משולש בד״ה אל משולש אד״ה כיחס משולש גד״ה אל משולש אד״ה. ויחס משולש בד״ה אל משולש אד״ה[2] כיחס ב״ד אל א״ד, מפני כי גבהותם אחד. וכן יחס משולש גד״ה אל משולש אד״ה כיחס ג״ה אל א״ה. הנה יהיה[3] יחס ב״ד אל א״ד כיחס ג״ה[4] אל א״ה. עוד יהיה[5] יחס ב״ד אל א״ד כיחס ג״ה אל א״ה. הנה אומר כי ד״ה[6] נכחי אל ב״ג, מפני כי הנהגת[7] שניהם אחת. הנה יחס ב״ד אל א״ד כיחס ג״ה אל א״ה. אבל יחס ב״ד אל א״ד כיחס משולש[8] בד״ה אל משולש אד״ה. ויחס ג״ה אל א״ה כיחס משולש גד״ה[9] אל משולש אה״ד. אם כן[10] משולש[11] בד״ה כמו משולש[12] גד״ה, ושניהם על תושבת אחת, והיא ד״ה. אם כן קו ד״ה נכחי לקו ב״ג.

[1] واحدة ... قاعدة] PMR2

[1] אל] +משולש גד״ה אל אד״ה ויחס משולש בד״ה אל משולש ~~אד״ה כיחס~~ **פ** [2] כיחס ... אד״ה] ליתא **בו** ⟨⟨*⟩⟩ **פ** [3] יהיה] ליתא **ס** [4] ג״ה] ~~ב״ה~~ ⟨⟨*⟩⟩ **פ** [5] יהיה] +* **ס** [6] ד״ה] ⟨*⟩ **פ** [7] הנהגת] ⟨ה⟩נהגת **ב** [8] משולש] +אד״ה **ק** [9] גד״ה] +~~אל משולש הד״ג~~ **פ** [10] אה״ד ... כן] הד״ג אל משולש+ **ב** [11] משולש אה״ד ... משולש] ⟨⟨משולש אד״ה אם כן יחס משולש בד״ה אל משולש אד״ה כיחס משולש⟩⟩ משולש הד״ג אל משולש אה״ד אם כן משולש בד״ה כמו משולש **ק** [12] משולש] +* **ב**

وَذَلِكَ ما أَرَدْنا بيانَهُ.	וזה מה שרצינו לבאר.

وذَلِكَ ما أَرَدْنا أَنْ نُبيِّنَ.

וזה מה שרצינו לבאר[1].

[1]וזה ... לבאר] ונשלם ביאורו **בטספק**

[Proposition VI.3]

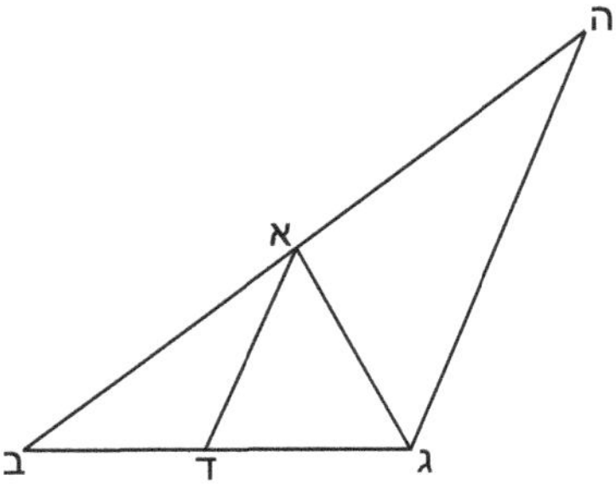

כל משלש שיצא מזויתו קו אל התושבת, ויחלוק הזוית לחציים, יהיה יחס[1] אחד מחלקי התושבת אל האחר כיחס צלעות המשולש הנשארות אחת מהן אל האחרת. ואם היו חלקי התושבת יחס אחד מהם אל האחר כיחס[2] צלעות המשלש הנשארות אחת מהן אל האחרת, אז הקו יחלוק הזוית לחציים.

كُلُّ مُثَلَّثٍ يَخْرُجُ مِنْ زاوِيَةٍ مِنْ زَواياهُ خَطٌّ إلى القاعِدَةِ، فَيَقْسِمُ الزَّاوِيَةَ بِنِصْفَيْنِ. فَإِنَّ قِسْمَي القاعِدَةِ تكونُ نِسْبَةُ أَحَدِهما إلى الآخَرِ كَنِسْبَةِ ضِلْعَي المُثَلَّثِ الباقِيَيْنِ أَحَدُهُما إلى الآخَرِ. وَإِنْ كانَتْ نِسْبَةُ قِسْمَي القاعِدَةِ أَحَدُهُما إلى الآخَرِ كَنِسْبَةِ الضِّلْعَيْنِ الباقِيَيْنِ أَحَدُهُما إلى الآخَرِ، فَإِنَّ الخَطَّ يَقْسِمُ الزَّاوِيَةَ بِنِصْفَيْنِ.

המשל בזה: כי משלש [!] אב״ג יצא מזוית בא״ג קו אל תושבת ב״ג, והוא א״ד, ויחלוק הזוית לחציים[3]. ואומר ששיעור ב״ד אצל ד״ג כשיעור ב״א אצל א״ג.

مِثالُهُ: أَنَّ مُثَلَّثَ ا ب ج. قَدْ خَرَجَ مِنْ زاوِيَةِ ب ا ج مِنْهُ خَطُّ ا د إلى قاعِدَةِ ب ج، فَقَطَعَ الزَّاوِيَةَ بِنِصْفَيْنِ. فَأَقولُ إِنَّ قَدْرَ ب د عِنْدَ د ج كَقَدْرِ ب ا عِنْدَ ا ج.

שרטוט—א״ד ה״ג ⊥ [1]יחס] ~~זוית~~ ⟨*⟩ [2]כיחס] ~~ל~~⟨⟨כ⟩⟩יחס [3]לחציים] ⟨⟨*⟩⟩

كُلّ مُثَلَّثٍ يَخْرُجُ مِنْ زاوِيَتِهِ[1] خَطٌّ إلى القاعِدَةِ، فيَقْسِمُ الزَّاوِيَةَ بِنِصْفَيْنِ، فَإِنَّ قِسْمَيِ القاعِدَةِ تكونُ نِسْبَةَ^ أحَدِهما إلى الآخَرِ كنِسْبَةِ ضِلْعَيِ المُثَلَّثِ الباقِيَيْنِ أحَدُهُما إلى الآخر. وإنْ كانَتْ نِسْبَةُ قِسْمَيِ القاعِدَةِ أحَدُهُما إلى الآخَرِ كنِسْبَةِ الضِّلْعَيْنِ الباقِيَيْنِ أحَدُهُما إلى الآخر، فَإِنَّ الخطّ يَقْسِمُ الزَّاوِيَةَ بِنِصْفَيْنِ.

כל משולש יצא מזויתו[1] קו[2] אל התושבת, ויחלק הזוית בשני חציים, הנה שני חלקי התושבת יהיה יחס שני חלקי התושבת אחד משניהם אל האחר כיחס שתי צלעות המשולש הנשארות אחת משתיהן[3] אל האחרת. ואם היה[4] יחס שני חלקי התושבת אחד אל האחר כיחס שתי הצלעות[5] הנשארות אחת משתיהן אל האחרת[6], הנה הקו יחלק הזוית בשני חציים.

مِثالُه: أنّهُ قد أُخْرِجَ مِنْ زاوِيَةِ ب ا ج مِنْ مُثَلَّث ا ب ج خطُّ ا د إلى قاعِدَة ب ج، فقَسَمَ الزَّاوِيَةَ بِنِصْفَيْنِ. فَأَقولُ إنَّ نِسْبَةَ ب د إلى د ج كنِسْبَةِ ب ا إلى ا ج.

המשל בו: כי כבר יצא מזוית בא״ג ממשולש אב״ג קו א״ד אל תושבת ב״ג, וחתך הזוית בשני חציים. הנה אומר כי יחס ב״ד אל ד״ג כיחס א״ב אל א״ג.

שרטוט—**בוטספק**: $; **וסק**: א״ד ה״ג ⊥
[1]מזויתו] מזויותיו **ב** מזויותו **ט** [2]קו] ⟨*⟩ **ב** [3]משתיהן] משתיהם **ט** [4]שני חלקי התושבת יהיה ... היה] ליתא **ס** [5]הצלעות] צלעות המשולש **ס** [6]ואם ... האחרת] ⟨*⟩ **ט** +* **ס**

[1]زاويته] **MPR2** زاوية من زواياه

המופת בזה: שנוציא מנקודת ג׳ קו נכוחי לקו א״ד, והוא קו ג״ה. ונוציא ב״א עד שיפגוש קו ג״ה על ה׳ חוץ למשולש. וקו א״ד נכוחי לקו ה״ג. ונפל עליהם קו ה״ב. אם כן הזוית החיצונה, שהיא לבא״ד, שוה לפנימית שהיא כנגדה, והיא לאה״ג. וכמו כן קו א״ד נכוחי לקו ה״ג. ונפל עליהם קו א״ג. אם כן שתי זויות המתחלפות, שהן מדא״ג ומ[ה]ג״א הן שוות. ואמנם הזוית שהיא מבא״ד שוה לזוית שהיא מדא״ג. ואם כן אשר היא מאה״ג שוה לאשר היא מאג״ה. אם כן צלע א״ג שוה לצלע א״ה. ומשלש בג״ה כבר יצא מצלע מצלעיו קו נכוחי לצלע ג״ה, והוא קו ד״א. ואם כן שעור ב״ד אצל ד״ג כשעור ב״א אצל א״ה. וא״ה שוה לא״ג. א״כ שיעור ב״ד אצל ד״ג כשעור ב״א אצל א״ג. וכמו כן נדביק א׳ בד׳. ואומר כי זוית בא״ד שוה לזוית דא״ג,

بُرْهانُهُ: أَنَّا نُخْرِجُ مِنْ نُقْطَةِ ج خَطَّ ج ه مُوازِيًا لِخَطِّ ا د. وَنُخْرِجُ خَطَّ ب ا حَتَّى يَلْقَى خَطَّ ج ه عَلَى نُقْطَةِ ه مِنْ خارِجِ المُثَلَّثِ. فَخَطُّ ا د مُوازٍ لِخَطِّ ه ج. ووَقَعَ عَلَيْهِما خَطُّ ب ه. فَزاوِيَةُ ب ا د الخارِجَةُ مِثْلُ زاوِيَةِ ا ه ج الخارِجَةِ المُقابِلَةِ لَها. وَأَيضًا خَطُّ ا د مُوازٍ لِخَطِّ ه ج. ووَقَعَ عَلَيْهِما خَطُّ ا ج. فَزاوِيَتا د ا ج و ا ج ه المُتَبادَلَتانِ مُتَساوِيَتانِ. وَلَكِنَّ زاوِيَةَ ب ا د مِثْلُ د ا ج. فَزاوِيَةُ ا ه ج مِثْلُ زاوِيَةِ ا ج ه. فَضِلْعُ ا ج مِثْلُ ضِلْعِ ا ه. وَمُثَلَّثُ ب ج ه قَدْ خَرَجَ مِنْ ضِلْعٍ مِنْ أَضْلاعِهِ خَطٌّ مُوازٍ لِقاعِدَةِ ه ج، وَهُوَ خَطُّ د ا، مُوازِيًا لِضِلْعِ ج ه. فَقَدْرُ ب د عِنْدَ د ج كَقَدْرِ ا ب عِنْدَ ا ه. و ا ه مِثْلُ ا ج. فَقَدْرُ ب د عِنْدَ د ج كَقَدْرِ ا ب عِنْدَ ا ج. وَأَيضًا نَجْعَلُ قَدْرَ ب د عِنْدَ د ج كَقَدْرِ ا ب عِنْدَ ا ج، وَنَصِلُ ا د بِخَطِّ ا د. فَأَقُولُ إِنَّ زاوِيَةَ ب ا د مِثْلُ زاوِيَةِ د ا ج.

بُرْهانُهُ: أَنّا نُخْرِجُ مِنْ نُقْطَةِ ج خَطًّا يُوازي ا د، وهو ج ه. ونُخْرِجُ ب ا حتَّى يَلْقَى ج ه على نُقْطَة ه. فَخَطّ ا د يُوازي ج ه، وَقَدْ وَقَعَ عَلَيْهِما ب ه. فزاوية ب ا د الخارجة^ مثل زاوية ا ه ج الدّاخِلَة^. وَأَيضًا فَإِنَّ خطّ ا د يُوازي ه ج. وَقَدْ وَقَعَ عَلَيْهِما ا ج. فَزاوِيَتا د ا ج ه ج ا المُتَبادَلَتانِ مُتَساوِيَتانِ. لَكِنَّ زاويةَ ب ا د مثل زاويةِ د ا ج. فزاوية ا ه ج مثل زاويةِ ا ج ه. فضلع ا ج مثل ضلع ا ه. فمُثَلَّث ب ج ه قد أُخْرِجَ مِنْ ضِلْعِ ب ه^ مِنْهُ خطٌّ يُوازي خَطَّ ج ه، وهو خَطُّ د ا. فنِسْبَة ب د إلى د ج كنِسْبَةِ ب ا إلى ا ه. و ا ه مثل ا ج. فنِسْبَة ب د إلى د ج كنِسْبَةِ ا ب إلى ا ج. وَأَيْضًا^ فَلْتَكُنْ نِسْبَةُ ب د إلى د ج كنِسْبَةِ ا ب إلى ا ج. فَأَقولُ إنَّ زاويةَ ب ا ج قد انْقَسَمَتْ بِنِصْفَيْنِ، فَصارَتْ زاويةُ ب ا د مِثْل زاويةِ د ا ج.

מופתו: אנחנו נוציא מן נקודת ג׳ קו יהיה נכחי אל א״ד, והוא ג״ה. ונוציא א״ב עד שיפגוש ג״ה על נקודת ה׳. הנה א״ד נכחי לג״ה, וכבר נפל על שניהם ב״ה. אם כן זוית בא״ד שוה לזוית אה״ג[1]. וגם כן קו א״ד[2] נכחי אל ה״ג[3]. וכבר נפל על שניהם א״ג. אם כן[4] שתי זויות דא״ג והג״א המומרות שוות. אבל זוית בא״ד כמו זוית דא״ג. אם כן זוית אה״ג כמו[5] זוית אג״ה. אם כן צלע א״ג[6] כמו צלע א״ה[7]. ומשולש בג״ה כבר הוצא מצלע ב״ג ממנו קו נכחי לקו ג״ה, והוא א״ד. אם כן יחס ב״ד אל ד״ג כיחס א״ב אל א״ה. וא״ה כמו א״ג. אם כן יחס ב״ד אל ד״ג כיחס א״ב אל א״ג[8]. ואם היה יחס ב״ד אל ד״ג[9] כיחס א״ב אל א״ג, הנה אומר כי זוית בא״ג כבר נחלקה בשני חציים, והיתה זוית בא״ד כמו זוית דא״ג.

---

[1] אה״ג] *+ **ס** [2] א״ד] ~~א״ה~~ ⟨⟨*⟩⟩ **פ** [3] אל ה״ג] ה״ג **וטפ** [4] זוית ... אם כן] ליתא **ב** [5] כמו] אם כן **ב** [6] א״ג] א״ה **ו** [7] א״ה] א״ג+ **ב** א״ג **ו** [8] כיחס ... א״ג] ⟨*⟩ **פ** [9] ב״ד ... ד״ג] ⟨ב״ד⟩ אל א״ג **ב** ב״ד אל ~~א״ד~~ ד״ג **ס**

כי הנהגתם אחת: ששיעור ב״ד אצל ד״ג כשיעור ב״א אצל א״ג, ושעור ב״ד אצל ד״ג כשעור ב״ה אצל א״ה[1]. ושיעור א״ב אצל א״ג ואצל א״ה יחסו אליהם אחד. אם כן א״ג שוה לא״ה. וזוית אג״ה שוה לזוית אה״ג. וזוית אג״ה שוה לזוית דא״ג. וזוית אה״ג שוה לזוית בא״ד. ואם כן זוית בא״ד שוה לזוית דא״ג.

بُرْهانُهُ: أَنَّ تَدْبِيرَهُما واحِدٌ. فَقَدْرُ ب د عِنْدَ د ج كَقَدْرِ ب ا عِنْدَ ا ج. وَلٰكِنَّ قَدْرَ ب د عِنْدَ د ج كَقَدْرِ ب ه عِنْدَ ا ه. ا ب عِنْدَ ا ج وَعِنْدَ ا ه نِسْبَتُهُ إِلَيْهِما واحِدَةٌ. ا ج مِثْلُ ا ه. وَزاوِيَةُ ا ج ه مِثْلُ زاوِيَةِ ا ه ج. وَزاوِيَةُ ا ج ه مِثْلُ زاوِيَةِ د ا ج. وَزاوِيَةُ ا ه ج مِثْلُ زاوِيَةِ ب ا د. فَزاوِيَةُ ب ا د مِثْلُ زاوِيَةِ د ا ج.

וזה מה שרצינו לבאר.

وَذٰلِكَ ما أَرَدْنا بَيانَهُ.

[1] ושעור ... ש א״ה] ⟨⟨*⟩⟩

بُرْهانُه[1]: أَنَّا إذا أَخْرَجْنا خَطَّ ج ه مُوازِيًا لِخَطِّ د ا، وَأَخْرَجْنا خَطَّ ب ا حتّى يَلْقَى خَطَّ ج ه على نُقْطَةِ ه، كما فَعَلْنا قَبْلُ، كانَتْ نِسْبَةُ ب د إلى د ج كَنِسْبَةِ ب ا إلى ا ه. لكِنَّ نِسْبَةَ ب د إلى د ج كَنِسْبَةِ ب ا إلى ا ج. فَنِسْبَةَ ب ا إلى كُلِّ واحِدٍ مِنْ خَطَّيْ ا ج ا ه—واحِدَةٌ. ف ا ج و ا ه مُتَساوِيان. فَزاوِيَتا ا ه ج و ا ج ه مُتَساوِيَتانِ. لكِنَّ زاوِيَةَ ا ج ه مثل زاوِيَةِ د ا ج المُبادِلَةَ لها. وزاوية ا ه ج الدَّاخِلَةَ مثل زاوية ب ا د الخارِجَة. فَزاوِيَتا د ا ب و د ا ج مُتَساوِيَتانِ.

وذَلِكَ ما أَرَدْنا أَنْ نُبَيِّنَ.

מופתו: אנחנו כאשר הוצאנו[1] קו ג״ה נכחי לקו א״ד, והוצאנו[2] א״ב עד שיפגש קו ג״ה על נקודת ה׳, כמו שעשינו קודם, יהיה יחס ב״ד אל ד״ג כיחס א״ב אל א״ה. אבל[3] יחס ב״ד אל ד״ג כיחס ב״א אל א״ג. אם כן[4] יחס א״ב אל כל אחד משני[5] קוי א״ג וא״ה—אחד. אם כן א״ג וא״ה[6] שוים. אם כן שתי זויות אה״ג ואג״ה שוות. אבל זוית אג״ה כמו זוית דא״ג המומרת[7] לה. וזוית אה״ג[8] הפנימית כמו זוית בא״ד החיצונה. אם כן שתי זויות דא״ב[9] ודא״ג שוות.

וזה מה שרצינו לבאר[10].

[1] برهانه] MPR2

[1] הוצאנו] הוצי[נ]ו ב הוציאנו ספ [2] והוצאנו] והוציאנו ב [3] אבל] ⟨⟨*⟩⟩ אז ז [4] כן] ⟨*⟩ ט [5] משני] משניה ק [6] אחד ... וא״ה] ליתא ס [7] המומרת] המומרות ספ [8] המומרת ... אה״ג] ליתא ב [9] דא״ב] +החיצונה ב [10] וזה ... לבאר] ונשלם ביאורו ס

[Proposition VI.4]

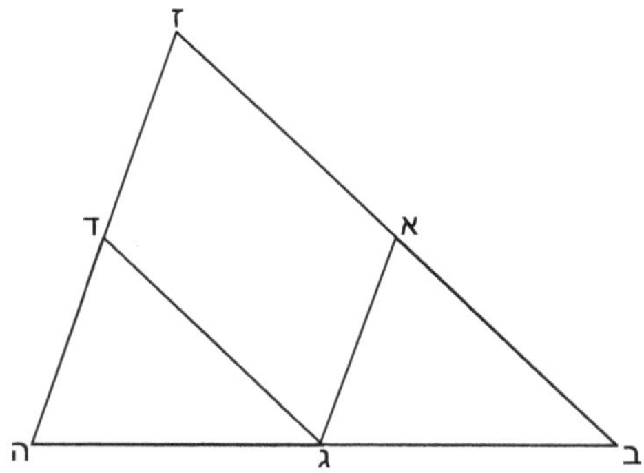

כל שני משולשים שזויות[1] אחת מהם שוות לזויות האחר, יהיו מיתרי הזויות השוות מתיחסים.

كُلُّ مُثَلَّثٍ تكونُ زَوايا أَحَدِهما مُساوِيَةً لِزَوايا الآخَرِ، فَإِنَّ الأَضْلاعَ الَّتي تُوتِرُ الزَّوايا المُتَساوِيَةَ مُتَناسِبَةٌ.

המשל בזה: שנצייר שני משולשים על קו אחד, עליהם אב״ג ודג״ה. ונשים זוית אב״ג שוה לזוית דג״ה. וזוית אג״ב שוה לזוית דה״ג. וזוית בא״ג שוה לזוית גד״ה. ואומר כי צלעות המשולשים שהם מיתרי הזויות השוות— יחסם אחד, שיעור ב״א אצל ד״ג כשעור ב״ג אצל ג״ה, וכשיעור א״ג אצל ד״ה.

مِثالُهُ: مُثَلَّثُ ا ب ج و ج د ه. وزاوية ا ب ج مِثْل زاوية د ج ه. وزاوية ب ا ج مِثْل زاوية ج د ه. وزاوية ب ج ا مِثْل زاوية ج ه د. فَأَقولُ إِنَّ الأَضْلاعَ الَّتي توتِرُ الزَّوايا المُتَساوِيَةَ مُتَناسِبَةٌ، قَدْرُ ا ب عِنْدَ ج د كَقَدْرِ ب ج عِنْدَ ج ه وكَقَدْرِ ا ج عِنْدَ د ه.

[1] שזויות] שזוי⟨⟨ו⟩⟩ת

كُلُّ مُثَلَّثَيْنِ تُسَاوِي زَوَايَا أَحَدِهِمَا زَوَايَا الْآخَرِ، فَإِنَّ أَوْتَارَ الزَّوَايَا الْمُتَسَاوِيَةِ مُتَنَاسِبَةٌ.

כל שני משולשים תשתוינה זויות אחד משניהם לזויות האחר, הנה מיתרי הזויות השוות מתיחסות.

مِثَالُهُ: أَنَّ زوايا مُثَلَّثَيْ ا ب ج ج د ه مُتَسَاوِيَةٌ— زاوية ا ب ج مثل زاوية د ج ه، وزاوية ا ج ب مثل زاوية د ه ج، وزاوية ب ا ج مثل زاوية ج د ه. فَأَقولُ إِنَّ أَوْتارَ الزَّوايا المُتَساوِيَةِ مُتَنَاسِبَةٌ—نِسْبَةَ ب ا إلى د ج كَنِسْبَةِ ب ج إلى ج ه وكَنِسْبَةِ ا ج إلى د ه.

המשל בו: כי זויות שני משולשי[1] אב״ג גד״ה שוות—זוית אב״ג כמו זוית דג״ה, וזוית אג״ב כמו זוית דה״ג, וזוית בא״ג כמו זוית גד״ה. הנה אומר כי מיתרי הזויות השוות מתיחסים—יחס א״ב אל ד״ג[2] כיחס ב״ג אל ג״ה וכיחס א״ג אל ד״ה.

שרטוט—**בופק**: משולשים בה״ז, אב״ג וגד״ה שווי שוקיים [1]שני משולשי] ליתא **ופ** ⟨⟨*⟩⟩ **ק** [2]ד״ג] ~~א״ג~~+ **פ**

מופת זה: שנוציא קוי ב״א וה״ד. ויפגשו זה את זה על נקדת ז׳. וזוית אב״ג שוה לזוית דג״ה. אם כן קוי א״ג וה״ז נכוחיים. ומזה הצד ג״ד וב״ז נכוחיים. ואם כן צלע מצלעיו קו נכוחי לצלע ה״ז[1], והוא קו א״ג. יהיה שיעור ב״א אצל א״ז כשעור ב״ג אצל ג״ה. וא״ז שוה לג״ד. ואם כן שעור ב״א אצל ג״ד כשעור ב״ג אצל ג״ה[2]. ועוד משלש בה״ז כבר יצא[3] מאחד מצלעיו קו נכוחי לצלע ב״ז, והוא קו ג״ד. אם כן שיעור ד״ז אל ד״ה כשיעור ב״ג אל ג״ה. וד״ז שוה לא״ג. אם כן שיעור א״ג אל ד״ה כשעור ב״ג אל ג״ה. ושעור ב״ג אל ג״ה כשעור א״ב אל ד״ג. ואם כן שעור א״ב אל ד״ג כשיעור א״ג אל ד״ה, וכשעור ב״ג אל ג״ה.

وזה מה שרצינו לבאר.

بُرْهانُهُ: أَنَّا نُخْرِجُ خَطَّيْ ب ا و ه د، حَتَّى يَلْتَقِيا عَلى نُقْطَةِ ز. فَزاوِيَةُ ا ب ج مِثْلُ زاوِيَةِ د ج ه. فَخَطّا ز ب و د ج مُتَوازِيانِ. وَأَيْضًا زاوِيَةُ ج ه د مِثْلُ زاوِيَةِ ب ج ا. فَخَطّا ج ا و ه ز مُتَوازِيانِ. فَسَطْحُ ا ج د ز مُتَوازِي الأَضْلاعِ، ا ج مِثْلُ د ز، و ج د مِثْلُ ا ز. فَمُثَلَّثُ ب ه ز قَدْ خَرَجَ مِنْ ضِلْعٍ مِنْ أَضْلاعِهِ خَطُّ ا ج مُوازِيًا لِضِلْعِ ه ز. فَقَدْرُ ب ا عِنْدَ ا ز كَقَدْرِ (ب ج)[1] عِنْدَ ج ه. و ا ز مِثْلُ ج د. فَقَدْرُ ب ا عِنْدَ ج د كَقَدْرِ ب ج عِنْدَ ج ه. وَأَيْضًا مُثَلَّثُ ب ه ز قَدْ خَرَجَ مِنْ ضِلْعٍ لَهُ خَطُّ ج د مُوازِيًا لِضِلْعِ ب ز. فَقَدْرُ ز د عِنْدَ د ه كَقَدْرِ ب ج عِنْدَ ج ه. و د ز مِثْلُ ا ج. فَقَدْرُ ا ج عِنْدَ د ه كَقَدْرِ ب ج عِنْدَ ج ه. [..] كَقَدْرِ ا ب عِنْدَ د ج. [..] كَقَدْرِ ا ج عِنْدَ ه د وكَقَدْرِ ب ج عِنْدَ ج ه.

وَذلِكَ ما أَرَدْنا بَيانَهُ.

[1] ואם ... ה״ז] ⟨⟨ד״ך ואם כן אגד״ז נכוחי הצלעות שצלעותיו המקבילות שוות ומשלש זב״ה יצא מצלע מצלעיו קו נכוחי לצלע ה״ז וכו׳⟩⟩ [2] וא״ז ... ג״ה] ⟨⟨*⟩⟩ [3] כבר יצא] ~~כש~~⟨כבר⟩ יצא

[1] ب ج] ب د

بُرْهانُهُ: أَنَّا نُنْزِلُ ب ج مُتَّصلًا ب ج ه على اسْتِقَامَةٍ. ونُخْرِجُ ب ا ه د حتّى يَلْتَقِيا على نُقْطَةِ ز. فزاوية ا ب ج مثل زاويةِ د ج ه. فخطّا ب ز د ج مُتوازِيانِ. وأَيْضًا فَإِنَّ زاويَةَ ا ج ب مثل زاويةِ د ه ج. فَخَطّا ج ا ه ز مُتوازِيانِ. فَسَطْح ا ج د ز مُتَوازي الأَضْلاعِ. ف ا ج مثل د ز. و ج د مثل ا ز. فَمُثَلَّث ب ه ز قد خَرَجَ مِنْ ضِلْعِ ب ه مِنْهُ خَطُّ ا ج يُوازي ه ز. فَنِسْبَة ب ا إلى ا ز كَنِسْبَةِ ب ج إلى ج ه. و ا ز مثل ج د. ونِسْبَة ا ب إلى ج د كَنِسْبَةِ ب ج إلى ج ه. وأيضًا فَإِنَّ مُثَلَّثَ ب ه ز قد خَرَجَ مِنْ ضِلْعِ ب ه مِنْهُ خطّ ج د يُوازي ب ز. فَنِسْبَةَ ز د إلى د ه كَنِسْبَة ب ج إلى ج ه. و ز د مثل ا ج. فَنِسْبَةَ ا ج إلى د ه كَنِسْبَةَ ب ج إلى ج ه. ونِسْبَة ب ج إلى ج ه كَنِسْبَةَ ا ب إلى د ج. فنِسبة ا ب إلى د ج كَنِسْبَةَ ا ج إلى د ه وكَنِسْبَةَ ب ج إلى ج ه.

وذَلِكَ ما أَرَدْنا أَنْ نُبَيِّنَ.

מופתו: אנחנו נניח ב״ג מדובק בג״ה על יושר. ונוציא א״ב ה״ד עד שיפגשו על נקודת ז׳. הנה זוית אב״ג כמו זוית דג״ה. אם כן שני קוי ב״ז ד״ג נכחיים. וגם כן הנה זוית אג״ב[1] כמו זוית דה״ג. אם כן שני קוי א״ג ה״ז נכחיים. אם כן שטח אגד״ז נכחי[2] הצלעות. אם כן א״ג כמו ד״ז. וג״ד כמו א״ז. אם כן משולש בה״ז כבר יצא מצלע ב״ה[3] ממנו קו א״ג נכחי אל ה״ז. אם כן יחס א״ב אל א״ז כיחס ב״ג אל ג״ה. וא״ז כמו ג״ד. אם כן יחס[4] א״ב[5] אל ג״ד[6] כיחס ב״ג אל ג״ה. וגם כן הנה משולש בה״ז כבר יצא מצלע ב״ה ממנו קו ג״ד נכחי אל ב״ז. אם כן יחס ז״ד אל ד״ה כיחס ב״ג אל ג״ה. וז״ד כמו א״ג. אם כן יחס[7] א״ג אל ד״ה כיחס ב״ג אל ג״ה. ויחס ב״ג אל ג״ה כיחס א״ב אל ד״ג. אם כן יחס א״ב אל ד״ג כיחס א״ג אל[8] ד״ה[9] וכיחס ב״ג אל ג״ה.

וזה מה שרצינו לבאר.

---

[1]אג״ב] ~~אב״ג~~ ⟨⟨*⟩⟩ **פ** [2]אגד״ז נכחי] נכחי גד״ז **ב** [3]ב״ה] +וגם כן הנה משולש בה״ז כבר יצא מצלע ב״ה ממנו קו ג״ד נכחי אל ב״ז אם כן יחס ז״ד אל ד״ה כיחס ב״ג אל ג״ה וז״ד כמו א״ג אם ⟨⟨כן⟩⟩ יחס א״ג אל ד״ה כיחס ב״ג אל ג״ה ויחס ב״ג אל ג״ה כיחס **ז** ליתא **בטספ** ⟨*⟩ **ק** [4]אם ... יחס] +* **ב** [5]א״ב] ⟨*⟩ **ק** [6]ג״ד] ג״ב+ **ב** [7]יחס] ⟨*⟩ **ס** [8]ג״ה וגם ... אל] ליתא **ז** [9]ד״ה] ~~ד״ג~~+ **ס**

[Proposition VI.5]

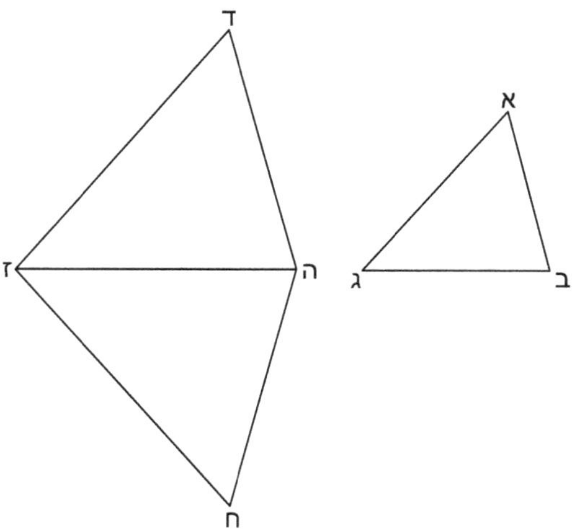

כל שני משולשים אשר צלעותם מתיחסות, הזויות אשר הצלעות המתיחסות הן להן מיתרים—תהיינה שוות.

המשל בזה: שנצייר שני משולשים עליהם אב״ג ודה״ז. ונשים יחס צלעות אחד מהם אל צלעות האחר כל צלע לדומה לו אחד: שיעור א״ב אצל ד״ה כשיעור א״ג אצל ד״ז וכשיעור ב״ג אצל ה״ז. ואומר כי זויות משולש[1] אב״ג שוות לזויות משלש דה״ז: זוית בא״ג שוה לזוית הד״ז, וזוית גב״א שוה לזוית זה״ד וזוית בג״א שוה לזוית הז״ד[2].

كُلُّ مُثَلَّثَيْنِ تكونُ أَضْلاعُهُما مُتَناسِبَةً، فَإِنَّ الزَّوايا الَّتي تُوتِرُها الأَضْلاعُ المُتَناسِبَةُ هِيَ مُتَساوِيَةٌ.

مثالُهُ: أَنَّ مُثَلَّثَيْ ا ب ج و د ه ز، أَضْلاعُهُما مُتَناسِبَةٌ. قَدْرُ ا ب عِنْدَ د ه كَقَدْرِ ا ج عِنْدَ د ز وكَقَدْرِ ب ج عِنْدَ ه ز. فَأَقولُ إِنَّ زَوايا مُثَلَّثِ ا ب ج مُساوِيَةٌ لِزَوايا مُثَلَّثِ د ه ز. زاوِيَةُ ب ا ج مِثْلُ زاوِيَةِ ه د ز، وَزاوِيَةُ ا ج ب مِثْلُ زاوِيَةِ د ز ه، وَزاوِيَةُ ج ب ا مِثْلُ زاوِيَةِ ز ه د.

[1] משולש] ⟨⟨*⟩⟩ [2] הז״ד] ~~ה[..ד]~~ ⟨⟨*⟩⟩

كُلُّ مُثَلَّثَيْنِ أَضْلَاعُهُمَا مُتَنَاسِبَةٌ، فَإِنَّ زَوَايَاهُمَا الَّتِي تُوتِرُهَا الْأَضْلَاعُ الْمُتَنَاسِبَةُ—مُتَسَاوِيَةٌ.

مِثَالُهُ: أَنَّ أَضْلَاعَ مُثَلَّثَيْ ا ب ج د ه ز مُتَنَاسِبَةٌ—نِسْبَةُ ا ب إلى د ه كَنِسْبَةِ ا ج إلى د ز وَكَنِسْبَةِ ب ج إلى ه ز. فَأَقُولُ إِنَّ زوايا مُثَلَّثِ ا ب ج مُسَاوِيَةٌ لِزَوَايَا مُثَلَّثِ د ه ز، كُلّ واحِدَةٍ الَّتِي يُوتِرُها الضِّلعُ المُنَاسِبُ لِوَتَرِها: فزاوية ب ا ج مثل زاوية ه د ز، وزاوية ا ج ب مثل زاوية د ز ه، وزاوية ج ب ا مثل زاوية ز ه د.

כל שני משולשים צלעות שניהם מתיחסות, הנה זויות שניהם[1] אשר יהיו מיתריהם הצלעות המתיחסות—שוות.

המשל בו: כי צלעות שני משולשי אב״ג דה״ז מתיחסות—יחס א״ב אל ד״ה כיחס א״ג אל ד״ז וכיחס ב״ג אל ה״ז. הנה אומר כי זויות[2] משולש אב״ג שוות לזויות משולש דה״ז, כל אחת אשר יהיה מיתרה הצלע המתיחס למיתרה: זוית[3] בא״ג כמו זוית הד״ז, וזוית אג״ב כמו זוית[4] דז״ה, וזוית גב״א כמו זוית זה״ד.

שרטוט—**בוסק**: כל המשולשים שווי שוקיים [1]שניהם] משניהם **ס** [2]זויות] זוית **וס** [3]זוית] הזוית **פ** [4]זוית] ליתא **ב**

מופת זה: שנעמיד על קו ה״ז, על נקודת ה׳, זוית שוה לזוית אב״ג, והיא זוית זה״ח. ונעמיד עוד על קו ה״ז, על נקודת ז׳, זוית שוה לזוית אג״ב, והיא זוית הז״ח. ותשאר זוית הח״ז שוה לזוית בא״ג. אם כן זויות משלש אב״ג שוות לזויות משלש הח״ז. ויהיה שעור ב״ג אל ה״ז כשעור א״ב אל ה״ח. וכבר היה שעור ב״ג אל ה״ז כשיעור א״ב אל ד״ה. אם כן שעור א״ב אל ד״ה כשעור ב״א אל ה״ח. אם כן ד״ה שוה לה״ח[1]. וכמו כן יתבאר שז״ד שוה לז״ח, וד״ה שוה לה״ח. וה״ז משותף. ושני קוי ד״ה וה״ז שוים לשני קוי ה״ח וה״ז, כל אחד לדומה לו. ותושבת ז״ד שוה לתושבת ז״ח. ואם כן זוית דה״ז שוה לזוית חה״ז. וכמו כן יתבאר כי זוית דז״ה שוה לזוית חז״ה. וזוית חה״ז שוה לזוית אב״ג. ואם כן[2] זוית דה״ז שוה לזוית אב״ג. וכמו כן יתבאר שזוית דז״ה שוה לזוית אג״ב. ונשארה זוית הד״ז שוה לזוית בא״ג. אם כן זויות משלש אב״ג שוות לזויות משלש דה״ז.

وזה מה שרצינו לבאר.

بُرْهانُهُ: أَنَّا نُقِيمُ عَلَى نُقْطَةِ ه مِنْ خَطِّ ه ز زاوِيَةَ ز ه ح مِثْلَ زاوِيَةَ ا ب ج. وَنُقِيمُ أَيْضًا على نُقْطَةِ ز مِنْ خَطِّ ه ز زاوِيَةَ ه ز ح مِثْلَ زاوِيَةِ (ا ج ب)[1]. فَزَوايا مُثَلَّثِ ا ج ب مُساوِيَةٌ لِزَوايا مُثَلَّثِ ه ح ز. فَقَدْرُ ب ج عِنْدَ ه ز كَقَدْرِ ا ب عِنْدَ ه ح. وَقَدْ كانَ قَدْرُ ج ب عِنْدَ ه ز كَقَدْرِ ا ب عِنْدَ د ه. فَقَدْرُ ا ب عِنْدَ د ه كَقَدْرِ (ا ب عِنْدَ)[2] ه ح. د ه مِثْلُ ه ح. وَكَذَلِكَ تَبَيَّنَ أَنَّ ز د مِثْلُ ز ح، و د ه مِثْلُ ه ح. ه ز مُشْتَرَكٌ. فَكِلَا خَطَّيْ د ه و ه ز مِثْلَيْ كِلَيْ خَطَّيْ ه ح و ه ز، كُلّ واحِدٍ مِثْل نَظِيرِهِ. وَقاعِدَةُ ز د مِثْلُ قاعِدَةِ ز ح. فَزاوِيَةُ د ه ز مِثْلُ زاوِيَةِ ح ه ز. وَكَذَلِكَ (نُبَيِّنُ)[3] أَنَّ زاوِيَةَ د ز ه مِثْلُ زاوِيَةِ ح ز ه. وَزاوِيَةُ ح ه ز مِثْلُ زاوِيَةِ ا ب ج. وَزاوِيَةُ ح ز ه مِثْلُ زاوِيَةِ ا ج ب. فَزاوِيَةُ د ه ز مِثْلُ زاوِيَةِ ا ب ج. وَزاوِيَةُ د ز ه مِثْلُ زاوِيَةِ ا ج ب. وبَقِيَتْ زاوِيَةُ ه د ز مِثْلُ زاوِيَةِ ب ا ج. فَزَوايا مُثَلَّثِ ا ب ج مُساوِيَةٌ[4] لِزَوايا مُثَلَّثِ د ه ز.

وَذَلِكَ ما أَرَدْنا بَيانَهُ.

[1] לה״ח] + ~~יחסו אחד אם כן ד״ה שוה לה״ח~~ [2] ואם כן] ו⟨⟨אם כן⟩⟩

[1] ا ج ب] ب ا ح [2] ا ب عند] ליתא [3] تبين] نبين
[4] مساوية] متساوية+

بُرهانُه: أنّا نَعمَلُ على نُقْطَةِ ه مِنْ خطِّ ه ز زاوِيةً مثل زاوِيةِ ا ب ج، وهي زاوِيةُ ز ه ح، وعلى نُقْطَةِ ز مِنْ خطِّ ه ز زاوية مثل زاوية ا ج ب، وهي زاوية ه ز ح. فَتَبْقَى زاوِيَةُ ح مِنْ مُثَلَّثِ ه ز ح الحادث^ مُساوِيَةً لِزاوِيَةِ ا. فَزَوايا مُثَلَّثِ ا ب ج مُسَاوِيَةٌ لزوايا مُثَلَّثِ ه ز ح. وأَضلاعُهُمَا المُوتِرَاتِ لِزَوَاياهما المُتَساوِيَةَ—مُتَنَاسِبَة. فنِسْبَة ب ج إلى ه ز كَنِسْبَة ا ب إلى ه ح. وَقَدْ كانَتْ نِسْبَةُ خطِّ ب ج إلى خطِّ ه ز كَنِسْبَةِ خطِّ ا ب إلى د ه. فنِسْبَة خطّ ا ب إلى كُلّ واحِدٍ مِنْ خَطَّيْ د ه و ه ح—واحدة[1]. ف د ه مثل ه ح. وكذَلِكَ تَبَيَّنَ أَنَّ د ز مثل ز ح. و د ه مثل ه ح. و ه ز مُشْتَرَك. فكِلا د ه ه ز مثل كِلا ه ح ه ز. وقاعِدَة ز د مثل قاعِدَة ز ح. فزاوية د ه ز مثل زاوية ز ه ح. وزاوية^[2] د ز ه مثل زاوية ه ز ح. فزاوية ه د ز مثل زاوية ه ح ز. فزاويا مُثَلَّث د ه ز مُسَاوِيَة لزوايا مُثَلَّث ه ح ز، كُلّ واحِدَةٍ لِنَظيرَتِها. وزاوية مُثَلَّث ا ب ج مُسَاوِيَة لزوايا مُثَلَّث ه ح ز، كُلّ واحِدَة لِنَظيرَتِها[3]. فزوايا مُثَلَّث ا ب ج مُسَاوِيَة لزوايا مُثَلَّث د ه ز، كُلّ واحِدَةٍ لِنَظيرَتِها.

وذَلِكَ ما أَرَدْنا أَنْ نُبَيِّنَ.

[1] خط ب ج ... واحدة] MPR2 ا ب الى د ه كنسبة ب ج الى ه ز; P: השמטה עד د ه
[2] وزاوية] ComPMR2T2 وكذلك
[3] وزاوية ... لنظيرتها] PT2

מופתו: אנחנו נשים על נקודת ה׳ מקו ה״ז זוית כמו זוית אב״ג, והיא זוית זה״ח, ועל נקודת ז׳ מקו ה״ז זוית כמו זוית אג״ב, והיא זוית[1] הז״ח. ותשאר זוית ח׳ ממשולש הז״ח כמו זוית א׳. אם כן זויות משולש אב״ג שוות לזויות משולש הז״ח. וצלעות שניהם העושות מיתרים לזויות[2] שניהם השוות[3]—מתיחסות. אם כן יחס ב״ג אל ה״ז כיחס א״ב אל ה״ח. וכבר היה יחס ב״ג אל ה״ז כיחס א״ב אל ד״ה. אם כן יחס א״ב אל כל אחד משני קוי ד״ה ה״ח—אחד. אם כן ד״ה כמו ה״ח. וכן התבאר כי ד״ז כמו ז״ח. וד״ה כמו ה״ח. וה״ז משותף. אם כן כל ד״ה ה״ז כמו כל ה״ח ה״ז. ותושבת ז״ד כמו תושבת ז״ח. אם כן זוית דה״ז כמו זוית זה״ח. ולכן זוית דז״ה כמו זוית הז״ח. אם[4] כן זוית הד״ז כמו זוית הח״ז. אם כן זויות[5] משולש דה״ז שוות לזויות משולש הח״ז, כל אחת לנכחית[6] לה. וזויות משולש אב״ג שוות לזויות משולש הח״ז, כל אחת לנכחית[7] לה. אם כן[8] זויות[9] משולש אב״ג שוות לזויות דה״ז[10], כל אחת לנכחית[11] לה.

וזה מה שרצינו לבאר.

[1] ועל ... זוית] ליתא **ב** [2] לזויות] לזוית **ב** [3] השוות] ⟨⟨*⟩⟩ **ק** [4] אם] +* **ק**
[5] זויות] שתי+ **ט** [6] לנכחית] לנכחי **בוטספ**
[7] לנכחית] לנכחי **בט** [8] וזויות ... כן] ליתא **ק**
[9] זויות] זוי⟨ו⟩ת **ב** וזויות **ק** [10] כל ... דה״ז] ליתא **ס** [11] לנכחית] לנכחי **בוטפק**

[Proposition VI.6]

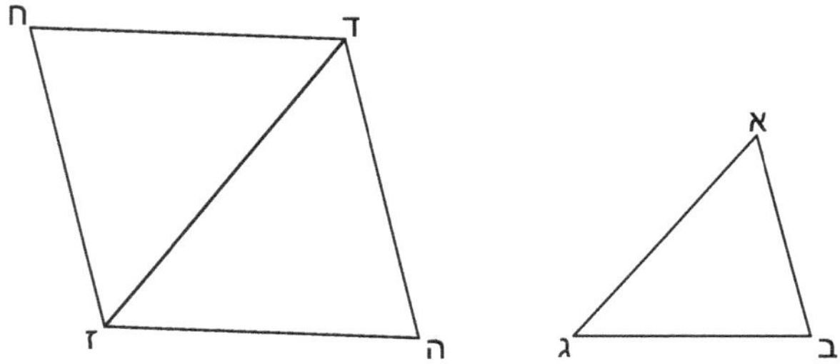

כשתהיינה שתי זויות משני משולשים שוות, והצלעות המקיפות בהן מתיחסות, תהיינה זויות המשולש האחד שוות לזויות המשולש האחר.

המשל בזה: שזוית משלש אב״ג שעליה א׳ שוה לזוית משלש דה״ז אשר עליה ד׳. והצלעות המקיפות בהן מתיחסות: שיעור א״ב אל ד״ה כשעור א״ג אל ד״ז[1]. ואומר כי זויות משלש אב״ג שוות לזויות דה״ז.

كُلُّ مُثَلَّثَيْنِ يَكونُ في أَحَدِهما زاوِيَةٌ مِثْلَ زاوِيَةٍ في الآخَرِ، وَتَكونُ الأَضْلاعُ الَّتي تُحيطُ بِالزَّاوِيَتَيْنِ المُتَساوِيَتَيْنِ مُتَناسِبَةً، فَإِنَّ زَوايا المُثَلَّثَيْنِ مُتَساوِيَةٌ.

مِثالُهُ: أَنَّ زاوِيَةَ ا مِنْ مُثَلَّثِ ا ب ج مِثْلُ زاوِيَةِ د مِنْ مُثَلَّثِ د ز ه وَالأَضْلاعُ الَّتي تُحيطُ بِهِما مُتَناسِبَةٌ: قَدْرُ ا ب عِنْدَ د ه كَقَدْرِ ا ج عِنْدَ د ز. فَأَقولُ إِنَّ زَوايا مُثَلَّثَيْ ا ب ج و د ه ز مُتَساوِيَةٌ.

---

[1] ד״ז] ~~ד״ה~~ ⟨⟨*⟩⟩

إِذَا تَسَاوَتْ زَاوِيَتَانِ مِنْ مُثَلَّثَيْنِ، وَتَنَاسَبَتْ أَضْلَاعُهُمَا الْمُحِيطَةُ بِهِمَا، فَزَوَايَا أَحَدِهِمَا مُسَاوِيَةٌ لِزَوَايَا الْآخَرِ[1].

مِثَالُهُ: أَنَّ زَاوِيَتَيْ ا د مِنْ مُثَلَّثَيْ ا ب ج د ه ز مُتَسَاوِيَتَانِ. وَأَضْلَاعُهُمَا الْمُحِيطَةُ بِهِمَا مُتَنَاسِبَةٌ: نِسْبَةُ ا ب إِلَى د ه كَنِسْبَةِ ا ج إِلَى د ز. فَأَقُولُ إِنَّ زَوَايَا مُثَلَّثِ ا ب ج مُسَاوِيَةٌ لِزَوَايَا مُثَلَّثِ د ه ز، كُلُّ وَاحِدَةٍ لِنَظِيرَتِهَا.

כאשר תשתוינה שתי זויות משני משולשים, ויתיחסו[1] צלעות שניהם המקיפות בשניהם, הנה זויות אחד משניהם שוות[2] לזויות האחר.

המשל בו: כי שתי זויות א׳ ד׳ משני משולשי אב״ג דה״ז שוות. וצלעות שניהם המקיפות בהם מתיחסות: יחס א״ב אל ד״ה כיחס א״ג אל ד״ז. הנה אומר כי זויות משולש אב״ג שוות לזויות משולש דה״ז, כל אחת לנבחית לה.

---

[1] الأخر] + كل واحدة لنظيرتها CQT1T3

שרטוט—**בוסק**: כל המשולשים שווי שוקיים [1] ויתיחסו] ורתיחסו **ב** [2] שוות] ~~שוות~~+ **ב**

מופת זה: שנעמיד על קו ד״ז על נקודת ד׳ זוית שוה לזוית א׳, והיא זוית זד״ח. ונעמיד עוד על קו ד״ז, על נקודת ז׳, זוית שוה לזוית ג׳, והיא זוית דז״ח. ותשאר זוית ב׳ שוה לזוית ח׳. ואם כן זויות משלש אב״ג שוות לזויות משלש דז״ח. ואם כן שיעור א״ג אל ד״ז כשעור א״ב אל ד״ח. וכבר היה שיעור א״ג אל ד״ז כשיעור א״ב אל ד״ה. וא״ב אל ד״ה ואל ד״ח יחסו אחד. אם כן ד״ה שוה לד״ח. וז״ד משותף. אם כן שני קוי ה״ד וז״ד שוים לשני קוי ז״ד וד״ח, כל אחד לדומה לו. וזוית חד״ז שוה לזוית הד״ז. ואם כן תהיה תושבת ה״ז שוה לתושבת ז״ח. ומשולש דה״ז שוה למשולש דז״ח. והזויות הנשארות שוות לזויות הנשארות שהצלעות השוות הן להן מיתרים—זוית הז״ד שוה לזוית דז״ח. וזוית ה׳ שוה (לח׳)[1]. ואמנם זוית דז״ח שוה לזוית ג׳, וזוית ח׳ שוה לזוית ב׳. ואם כן זוית דז״ה שוה לזוית ג׳. וזוית ה׳ שוה לזוית ב׳. וזוית ד׳ שוה לזוית א׳. אם כן זויות משלש אב״ג שוות לזויות משלש דה״ז.

[1] לח׳] לז׳ ח׳

بُرْهانُهُ: أَنَّا نُقيمُ عَلى نُقْطَةِ د مِنْ خَطِّ د ز زاوِيَةَ (ز د ح)[1] مِثْلَ زاوِيَةِ ا. وَنُقيمُ أَيْضًا عَلى نُقْطَةِ (ز)[2] مِنْ خَطِّ د ز زاوِيَةَ د ز ح مِثْلَ زاوِيَةِ ج. وَبَقِيَتْ زاوِيَةُ ب مِثْلَ زاوِيَةِ ح. فَزَوايا مُثَلَّثِ ا ب ج مُساوِيَةٌ لِزَوايا مُثَلَّثِ ز د ح. فَقَدْرُ ا ج عِنْدَ د ز كَقَدْرِ ا ب عِنْدَ د ح. وَقَدْ كانَ قَدْرُ ا ج عِنْدَ د ز كَقَدْرِ ا ب عِنْدَ د ه. [..] وَعِنْدَ د ح وَنِسْبَتُهُ إِلَيْهِما واحِدَةٌ. د ه مِثْلُ د ح. د ز مُشْتَرَكٌ. فَكِلَيْ خَطَّيْ د ه د ز مِثْلُ كِلَيْ خَطَّيْ ز د و د ح، كُلُّ واحِدٍ مِثْلُ نَظيرِهِ. وَزاوِيَةُ ح د ز مِثْلُ زاوِيَةِ ه د ز. فَقاعِدَةُ ه ز مِثْلُ قاعِدَةِ ز ح. وَمُثَلَّثُ د ه ز مِثْلُ مُثَلَّثِ د ز ح. فَالزَّوايا الباقِيَةُ مِثْلُ الزَّوايا الباقِيَةِ الَّتي تُوتِرُها الأَضْلاعُ المُتَساوِيَةُ—زاوِيَةُ ه ز د مِثْلُ زاوِيَةِ د ز ح. فَزاوِيَةُ ه مِثْلُ زاوِيَةِ ح. وَلٰكِنَّ زاوِيَةَ د ز ح مِثْلُ زاوِيَةِ ج. وَزاوِيَةُ ح مِثْلُ زاوِيَةِ ب. فَزاوِيَةُ د ز ه مِثْلُ زاوِيَةِ ج. وَزاوِيَةُ ه مِثْلُ زاوِيَةِ ب. وَزاوِيَةُ د مِثْلُ زاوِيَةِ ا. فَزَوايا مُثَلَّثِ ا ب ج مِثْلُ زَوايا مُثَلَّثِ د ه ز.

[1] ز د ح] د ز ح [2] ز] د

بُرهانُهُ: أَنَّا نُقيمُ على نُقْطَةِ دَ مِنْ خطّ د ز زاويةً مِثْل زاويةِ ا، وهي زاويةُ ز د ح، وعلى نُقْطَة زَ مِنْ خطّ د ز أَيْضًا، زاوية^ مثل زاوية ج، وهي زاوية د ز ح. فتكونُ زاويةُ ال ح مثل زاويةِ ال ب. فزوايا مُثَلَّثِ ا ب ج مُساوِيَةٌ لِزَوايا مُثَلَّثِ د ز ح. فِنسبَة أَوْتارِ الزَّوايا المُسَاوِيَة بعضها إلى بَعْضٍ—واحدة. فنِسْبَة ا ب إلى د ح كنِسْبَة ا ج إلى د ز. وَقَدْ كانَتْ نِسْبَةُ ا ب إلى د ه كنِسْبَةِ ا ج إلى د ز. فنِسْبَة ا ب إلى د ه وإلى د ح—واحدة. ف د ه مثل د ح. و ز د مُشْتَرَكٌ. فكلا ه د د ز مثل كِلا ح د د ز. وزاويتا دَ مُتَساوِيَتانِ، لِأَنَّ كُلّ واحِدَةٍ مِنْهُما مُسَاوِيَةٌ لِزَاوِيَةِ ا. فقاعِدَة ه ز مثل قاعِدَةِ ز ح. ومُثَلَّث د ه ز مثل مُثَلَّث د ح ز. وسائر الزَّوايا مُسَاوِيَة لِسائِرِ[1] الزَّوايا، كُلّ واحدة مُسَاوِيَة^[2] لِنَظيرَتِها الَّتي يُوتِرُها الضِّلْعُ المُساوي لِلضِّلْع الَّذي يُوتِرُ الأُولى. فَزاوِيَتا ح د ز و ز د ه مُتَساوِيَتانِ. وزاوية ال ج مثل زاوية د ز ح. فزاوية ال ج إِذَنْ مثل زاوية د ز ه. وَقَدْ كانَتْ زاوِيَةُ ال ا مثل زاوِيَةِ ه د ز. تَبْقَى زاوِيَةُ ال ه مثل زاوِيَةِ ال ب. فزوايا مُثَلَّثِ ا ب ج و د ه ز مُتَساوِيَةٌ كُلّ واحِدَةٍ مثل نظيرَتِها[3].

---

[1]مساوية لسائر] **MPR2** مثل سائر [2]مساوية] مثل **MPR2** [3]للضلع ... نظيرتها] **MPR2**

מופתו: אנחנו נעמיד על נקודת ד׳ מקו ד״ז זוית[1] כמו זוית א׳[2], והיא זוית זד״ח, ועל נקודת ז׳ מקו[3] ד״ז גם כן[4], כמו זוית ג׳[5], והיא[6] זוית דז״ח[7] [8]. ותהיה זוית ח׳[9] כמו זוית ב׳. הנה זויות משולש אב״ג שוות לזויות משולש דח״ז[10]. ויחס מיתרי הזויות השוות קצתם אל קצת—אחד. אם כן יחס א״ב אל ד״ח[11] כיחס א״ג אל ד״ז. וכבר היה יחס א״ב אל ד״ה כיחס א״ג אל ד״ז[12]. אם כן[13] יחס א״ב אל ד״ה ואל ד״ח[14]—אחד. אם כן ד״ה כמו ד״ח[15]. וז״ד משותף. הנה כל ה״ד ד״ז כמו כל ח״ד[16] ד״ז. וזויות[17] ד׳[18] שוות, מפני[19] כי כל אחת משתיהן[20] שוות לזוית[21] א׳. אם כן תושבת ה״ז כמו תושבת ז״ח. ומשולש דה״ז כמו משולש דז״ח. ושאר הזויות שוות לשאר הזויות, כל אחת לנכחית[22] לה אשר יהיה מיתרה הצלע השוה לצלע אשר יהיה מיתרו[23] הראשון. אם כן שתי זויות ז׳ שוות. וזוית[24] ג׳ כמו זוית דז״ח[25]. אם כן זוית ג׳ כמו זוית דז״ה. וכבר היתה זוית א׳ כמו זוית הד״ז. ותשאר זוית ה׳ כמו זוית ב׳. אם כן זויות[26] משולשי אב״ג דה״ז שוות כל אחת כמו הנכחית[27] לה.

---

[1]זוית] כמו זוית ~~ב׳~~ ד״ח מקו ד״ז **ב** ליתא **ו** [2]א׳] ~~כמו זוית~~ **ט** [3]מקו] מנקודת **ו** [4]גם כן] ליתא **ב** +זוית **ו** [5]ג׳] אג״ב **ס** [6]והיא] והוא **ו** [7]זד״ח ... דז״ח] ליתא **ט** [8]מקו ... דז״ח] ⟨⟨זוית כמו זוית ג׳⟩⟩ **פ** [9]ח׳] א׳ **ס** [10]דח״ז] דא״ז **ס** [11]ד״ח] ~~א״ח~~+ **ב** [12]וכבר ... ד״ז] ⟨*⟩ **ו** [13]כן] ⟨*⟩ **ט** [14]ד״ח] ד״א **ס** [15]ד״ח] ד״א **ס** [16]ח״ד] א״ד **ס** [17]וזויות] זויות **ב** וזוי⟨ו⟩ת **פ** [18]ד׳] +⟨⟨שתיהן⟩⟩ **ק** [19]מפני] ליתא **ב** [20]משתיהן] משתיהם **בטספ** [21]לזוית] לזויות **בוס** [22]לנכחית] לנכחי **בס** [23]הצלע ... מיתרו] ליתא **ב** [24]וזוית] וזויות **ב** [25]דז״ח] דז״א **ס** [26]ב׳ ... זויות] זוית **ב** [27]כמו הנכחית] כמו הנכחי **בספ** לנכחית **ו**

וזה מה שרצינו לבאר.

وَذَلِكَ ما أَرَدْنا بَيانَهُ.

وذَلِكَ ما أَرَدْنا أَنْ نُبيِّنَ.

וזה מה שרצינו לבאר.

[Proposition VI.7]

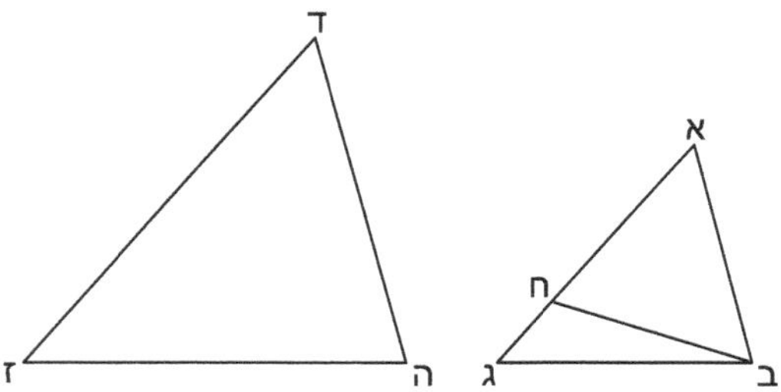

כשתהיינה שוות שתי זויות משני משלשים, ויתיחסו צלעות שתי זויות אחרות, והיתה כל אחת משתי זויות הנשארות מהם קטנה או גדולה מנצבה, תהיינה זויותהם הדומות שוות.

كُلُّ مُثَلَّثَيْنِ يكونُ في أَحَدِهِما زاوِيَةٌ مِثْلَ زاوِيَةٍ في الآخَرِ، وَتَكونُ إِحْدى الزّاوِيَتَيْنِ الباقِيَتَيْنِ مِنْ كُلِّ مُثَلَّثٍ يُحيطُ بِهِما الأَضْلاعُ مُتَناسِبَةً، فَإِنَّ الزّاوِيَتَيْنِ الباقِيَتَيْنِ مِنَ المُثَلَّثَيْنِ إِنْ كانَتْ كُلُّ واحِدَةٍ مِنْهُما حادَّةً أَو مُنْفَرِجَةً، فَإِنَّ زَوايا كُلِّ واحِدٍ مِنْهُما مُساوِيَةٌ لِزَوايا الآخَرِ.

המשל בזה: כי משולשי אב״ג ודה״ז באחד מהם זוית שוה לזוית האחר, והן זויות א׳ וד׳. והצלעות המקיפות בזויות ב׳ וה׳ מתיחסות: שעור א״ב אצל ד״ה כשיעור ב״ג אצל ז״ה[1]. וכל אחת מזויות ג׳ וז׳ קטנה או גדולה מנצבה. ואומר כי זויות משלש אב״ג שוות לזויות משלש דה״ז.

مِثالُهُ: مُثَلَّثا ا ب ج و د ه ز فيهِما زاوِيَتا ا و د مُتَساوِيَتانِ. وَالأَضْلاعُ الَّتي تُحيطُ بِزاوِيَتَيْ ب و ه مُتَناسِبَةٌ: قَدْرُ ا ب عِنْدَ ه د كَقَدْرِ ب ج عِنْدَ ه ز. وَكُلُّ واحِدَةٍ مِنْ زاوِيَتَيْ ج و ز حادَّةٌ أَو مُنْفَرِجَةٌ. فَأَقولُ إِنَّ زَوايا مُثَلَّثِ ا ب ج مُساوِيَةٌ لِزَوايا مُثَلَّثِ د ه ز.

---

[1] ז״ה] ~~ג״ה~~+

إذا تَساوَتْ زاوِيَتانِ مِنْ مُثَلَّثَيْنِ، وتَنَاسَبَتْ أَضْلاعُ زاوِيَتَيْنِ أُخْرَيَيْنِ مِنْهُما^، وكانَتْ كُلُّ واحِدَةٍ مِنَ الزَّاوِيَتَيْنِ الباقِيَتَيْنِ مِنْهُما أَصْغَرَ أَو أَكْبَرَ مِنْ قائِمَةٍ، فَإِنَّ زواياهُما النَّظائِرَ مُتَساوِيَةٌ.

כאשר ישתוו שתי זויות משני משולשים, ויתיחסו צלעות[1] שתי זויות[2] אחרות, והיתה כל אחת משתי הזויות הנשארות משניהם יותר קטנה או יותר גדולה מנצבת, הנה[3] זויות שניהם הנכחיות שוות.

مِثالُهُ: أَنَّ زاوِيَتَيْ ا د مِنْ مُثَلَّثَيْ ا ب ج د ه ز مُتَساوِيَتانِ. والأَضلاع المُحِيطَة بزاوِيَتَيْ ب ه مُتَنَاسِبَةٌ: نِسْبَةُ ا ب إلى د ه كَنِسْبَةِ ب ج إلى ه ز. وكُلّ واحدةٍ مِنْ زاوِيَتَيْ ج ز أَكْبَر أَو أَصْغَر[1] مِنْ قائِمَةٍ. فَأَقولُ إِنَّ زوايا مُثَلَّث ا ب ج مُساوِيَةٌ لِزَوايا مُثَلَّثِ د ه ز كُلّ واحدةٍ لِنَظيرَتِها.

המשל בו: כי שתי זויות א׳ ד׳ משני משולשי אב״ג ודה״ז שוות. והצלעות המקיפות שתי זויות[4] ב׳ ה׳ מתיחסות: יחס א״ב אל ה״ד כיחס ב״ג אל ה״ז. וכל אחת משתי זויות ג׳ ז׳ יותר גדולה או יותר קטנה מנצבת. הנה אומר כי זויות משולש אב״ג שוות לזויות משולש דה״ז, כל אחת לנכחית לה[5].

[1] أكبر ... أصغر] **MPR2**

שרטוט—**בוטספק**: משולשים אב״ג ודה״ז שווי שוקיים; **טפ**: משולש דה״ז משמאל למשולש אב״ג [1] צלעות] +* **ט** [2] זויות] זוית **ב** [3] הנה] +* **פ** [4] זויות] הזויות **ק** [5] לה] +~~ותהיה~~ **ק**

ונשים תחילה כל אחת מהן גדולה מנצבה. ואומר כי זוית ב׳ שוה לזוית ה׳, לא יתכן כי אם זה. ואם היה אפשר שתהיה אחת מהן גדולה מן האחרת, תהיה זוית ב׳ גדולה מזוית ה׳. ונעמיד על קו א״ב, על נקודת ב׳, זוית שוה לזוית ה׳, והיא זוית אב״ח. וזוית א׳ היא שוה לזוית ד׳—תשאר זוית ח׳ שוה לזוית ז׳. ותהיינה זויות משלש אח״ב שוות לזויות משלש דה״ז. ויהיה שעור א״ב אל ד״ה כשיעור ב״ח אל ה״ז. ושיעור א״ב אל ד״ה כשיעור ב״ג אל ה״ז. אם כן שיעור ב״ח אל ה״ז כשיעור ב״ג אל ה״ז. ואם כן יחס ב״ג וב״ח אל ה״ז—אחד. ואם כן ב״ח שוה לב״ג. וזוית ח׳ שוה לזוית ג׳. וזוית ג׳ אינה קטנה מנצבה, ולא זוית ח׳ קטנה מנצבה. אם כן שתי זויות משלש בח״ג אשר עליהן ח׳ ג׳ אינן קטנות משתי זויות נצבות—חלוף [!] ודבר בטל. ואם כן אין זוית ב׳ גדולה מזוית ה׳. וכמו כן יתבאר שאינה קטנה ממנה. וזוית א׳ שוה לזוית ד׳. ותשאר זוית ג׳ שוה לזוית ז׳.

فَلْيَكُنْ أَوَّلًا كُلُّ واحِدَةٍ مِنْهُما مُنْفَرِجَةً. فَأَقولُ إِنَّ زاوِيَةَ ب مِثْلُ زاوِيَةِ ه. بُرْهانُهُ: أَنَّهُ لا أَمْكَنَ إِلّا ذلِكَ. فَإِنْ أَمْكَنَ، فَلْتَكُنْ زاوِيَةُ ب أَعْظَمَ مِنْ زاوِيَةِ ه. وَنُقيمُ عَلى نُقْطَةِ ب مِنْ خَطِّ ا ب زاوِيَةَ ا ب ح مِثْلَ زاوِيَةِ ه. وَزاوِيَةُ ا مِثْلُ زاوِيَةِ د. وَبَقِيَتْ زاوِيَةُ ح مِثْلَ زاوِيَةِ ز. فَزَوايا مُثَلَّثِ (ا ب ح)[1] مُساوِيَةٌ لِزَوايا مُثَلَّثِ د ه ز. فَقَدْرُ ا ب عِنْدَ د ه كَقَدْرِ ب ح عِنْدَ ه ز. وَقَدْرُ ا ب عِنْدَ د ه كَقَدْرِ ب ج عِنْدَ ه ز. فَنِسْبَةُ ب ج و ب ح إِلى ه ز واحِدَةٌ. ب ح مِثْلُ (ب ج)[2]. وَزاوِيَةُ ح مِثْلُ زاوِيَةِ ج. وَزاوِيَةُ ج مُنْفَرِجَةٌ. وَزاوِيَةُ ح مُنْفَرِجَةٌ. فَزاوِيَتا ح وج مِنْ مُثَلَّثِ (ب ح ج)[3] مُنْفَرِجَتانِ—هذا خُلْفٌ لا يُمْكِنُ. فَلَيْسَتْ زاوِيَةُ ه بِأَصْغَرَ مِنْ زاوِيَةِ ب. وَكَذلِكَ تَبَيَّنَ أَنَّها لَيْسَتْ بِأَعْظَمَ مِنْها. وَزاوِيَةُ ا مِثْلُ زاوِيَةِ د. فَبَقِيَتْ زاوِيَةُ ج مِثْلَ زاوِيَةِ ز. فَزَوايا مُثَلَّثِ ا ب ج مُساوِيَةٌ لِزَوايا مُثَلَّثِ د ه ز.

[1] ا ب ح] ا ب ج [2] ب ج] ه د [3] ب ح ج] ب ح ه

ותהיה כל אחת משתי זויות ג׳ ז׳ תחלה יותר גדולה מנצבת. הנה[1] אומר כי[2] זוית[3] ב׳ שוה לזוית ה׳, אי אפשר זולתו. שאם היה אפשר[4], נאמר שתהיה זוית ב׳ יותר גדולה. ונעמיד על נקודת ב׳ מקו א״ב זוית כמו זוית ה׳, והיא זוית אב״ח. אם כן זוית אב״ח כמו זוית ה׳[5]. וזוית[6] א׳ כמו זוית ד׳. הנה נשארה[7] זוית אח״ב כמו זוית ז׳. אם כן זויות משולש אב״ח שוות לזויות משולש דה״ז, כל אחת לנכחית לה. אם כן יחס א״ב אל ד״ה כיחס[8] ב״ח אל ה״ז. וכבר היה יחס א״ב אל ד״ה כיחס ב״ג אל ה״ז. אם כן יחס ב״ח וב״ג אל ה״ז—אחד. אם כן ב״ח כמו ב״ג[9]. אם[10] כן זוית בח״ג כמו זוית בג״ח. וזוית ג׳ יותר גדולה מנצבת. אם כן זוית בח״ג[11] יותר גדולה מנצבת[12]. הנה במשולש בג״ח אם כן שתי זויות יותר גדולות משתי נצבות—זה שקר. אם כן אין זוית ב׳[13] יותר גדולה מזוית ה׳. וכן גם[14] כן התבאר שהיא[15] אינה קטנה ממנה—הנה היא אם כן[16] שוה[17]. וזוית א׳ כמו זוית ד׳. ונשארה זוית ג׳ כמו זוית ז׳[18]. אם כן זויות משולש אב״ג שוות לזויות משולש דה״ז.

1 הנה] +* פ 2 זויות משולש ... כי] ליתא ב 3 זוית] +~~משולש אב״ג שוות לזויות משולש דה״ז כל אחת לנכחית לה~~ פ 4 אפשר] +~~זולתו~~ ב 5 והיא ... ה׳] ליתא ז 6 ה׳ וזוית] ליתא ב 7 נשארה] נשארת ב 8 כיחס] +* ק 9 ב״ח ... ב״ג] יותר גדולה מנצבת פ 10 אם] ~~אם כן זוית בח״ג כמו זוית ב״ג~~+ פ 11 יותר ... בח״ג] ליתא ופ 12 אם ... מנצבת] ⟨*⟩ ט 13 ב׳] ליתא ב 14 וכן גם] וגם ז 15 שהיא] ליתא ב 16 היא ... כן] ליתא ז 17 שוה] +לה ס 18 כמו ... ז׳] ליתא ז

فَلْتَكُنْ كُلُّ واحِدَةٍ مِنْ زاوِيَتَيْ ج ز أَوَّلًا أَكْبَرَ مِنْ قائِمَةٍ. فَأَقولُ إِنَّ زاوِيَةَ ب مُساوِيَةٌ[1] لِزاوِيَةِ ه، لا يُمْكِنُ غيره. فَإِنْ أَمْكَنَ، فَلْتَكُنْ^ زاوِيَةُ ب أعظم. وَنُقيمُ على نُقْطَةِ ب مِنْ ا ب زاوِيَةً مثل زاوية ه، وهي زاوية ا ب ح. فزاوية ا ب ح مثل زاوية ه. وزاوية ا مثل زاوية د. فَبَقِيَتْ زاوِيَةُ ا ح ب مثل زاوية ز. فزوايا مُثَلَّث ا ب ح مُساوِيَةٌ لزوايا مُثَلَّث د ه ز، كُلّ واحِدَةٍ لِنَظيرَتِها. فنِسْبَةُ ا ب إلى د ه كَنِسْبَةِ ب ح إلى ه ز. وَقَدْ كانَتْ نِسْبَةُ[2] ا ب إلى د ه كَنِسْبَةِ ب ج إلى ه ز. فنِسْبَةُ ب ح و ب ج إلى ه ز—واحدة. ف ب ح مثل ب ج. فزاوية ب ح ج مثل زاوية ب ج ح. وزاوية ج أَكْبَرُ مِنْ قائِمَةٍ. فزاوية ب ح ج أَكْبَرُ مِنْ قائِمَةٍ. فَفي مُثَلَّثِ ب ج ح إِذَنْ[3] زاويتان أَكْبَرُ مِنْ قائِمَتَيْنِ—هذا خُلْفٌ. فَلَيْسَتْ زاوية ب بِأَعْظَمَ مِنْ زاوية ه. وكَذَلِكَ تَبَيَّنَ أَنَّها لَيْسَتْ بِأَصْغَرَ مِنْها—فهي إِذَنْ مُساوِيَةٌ لها. وزاوية ا مثل زاوية د. وبَقِيَتْ زاوِيَةُ ج مثل زاوِيَةِ د. فَزَوايا مُثَلَّثِ ا ب ج مُساوِيَةٌ لِزَوايا مُثَلَّثِ د ه ز، كُلّ واحِدَةٍ لِنَظيرَتِها^.

1 مساوية] MPR2 مثل 2 وقد ... نسبة] MR2
3 إذن] P

עוד כל אחת מזויות ג׳ וז׳ קטנה מנצבה. ואומר כי זוית ב׳ שוה לזוית ה׳. כי הנהגתם אחת. ויהיה ב״ח שוה לב״ג. וזוית ג׳ שוה לזוית ח׳. וזוית ג׳ קטנה מנצבה. אם כן זוית ח׳ קטנה מנצבה. ותהיה זוית ח׳ האחרת גדולה מנצבה, והיא שוה לזוית ז׳. אם כן זוית ז׳ גדולה מנצבה—זה חלוף לא יתכן. כי היא היתה קטנה מנצבה. אם כן אין זוית ב׳ בלתי שוה לזוית ה׳. וזוית א׳ שוה לזוית ד׳. ונשארה זוית ג׳ שוה לזוית ז׳. ואם כן זווית משלש אב״ג[1] שוות לזויות דה״ז.

وَنَجْعَلُ أَيضًا كُلَّ واحِدَةٍ مِنْ زاوِيَتَيْ ج و ز حادَّةً. فَأَقولُ إِنَّ زاوِيَةَ ب مِثْلُ زاوِيَةِ ه. بُرْهانُه: أَنَّ تَدْبيرَهُما واحِدٌ. وَزاوِيَةُ ج مِثْلُ زاوِيَةِ ح. وَزاوِيَةُ ج حادَّةٌ. وَزاوِيَةُ ح حادَّةٌ. فَزاوِيَةُ ح الأُخرى مُنْفَرِجَةٌ، وَهِيَ مِثْلُ زاوِيَةِ (ز)[1]. فَزاوِيَةُ ز إِذًا مُنْفَرِجَةٌ—هذا خُلْفٌ لا يُمْكِنُ، لِأَنَّها كانَت حادَّةً. فَلَيْسَتْ زاوِيَةُ ب بِغَيْرِ مُساوِيَةٍ لِزاوِيَةِ ه. وَزاوِيَةُ ا مِثْلُ زاوِيَةِ د. وبَقِيَتْ زاوِيَةُ ج مِثْلُ زاوِيَةِ ز. فَزَوايا مُثَلَّثِ ا ب ج مُساوِيَةٌ لِزَوايا مُثَلَّثِ د ه ز.

וזה מה שרצינו לבאר.

وَذلِكَ ما أَرَدْنا بَيانَهُ.

---

[1]אב״ג] ⟨*⟩

[1] ز] ه

ثُمَّ لِتَكُنْ^ أيضًا كُلُّ واحِدَةٍ مِنْ زاوِيَتَيْ جَ زَ أَصْغَرَ مِنْ قائِمَةٍ. فَأَقولُ إنَّ زاويةَ بَ مثل زاويةِ هَ. والتَّدبير واحِدٌ. ف ب ح مثل ب ج. فزاوية جَ مثل زاوية ب ح ج. وزاوية جَ أَصْغَر مِنْ قائِمَةٍ. فزاوية ب ح ج أَصْغَر مِنْ قائِمَةٍ. فزاوية ب ح ا أعظم مِنْ قائِمَةٍ. وهي مثل زاوية زَ. فزاوية زَ أعظم مِنْ قائِمَةٍ. وَقَدْ كانَتْ أَصْغَرَ. فزاوية ال زَ أَكْبَر وأَصْغَر مِنْ شيءٍ واحِدٍ بِعَيْنِهِ[1]—هذا خُلْفٌ. فزاوية بَ مثل زاوية هَ. وزاوية اَ مثل زاوية دَ. فَبَقِيَتْ زاوِيَةُ جَ مثل زاوية زَ. فزوايا مُثَلَّث ا ب ج مُسَاوِيَة لزوايا مُثَلَّث د ه ز، كُلّ واحِدَةٍ لِنَظِيرَتِها.

وذَلِكَ ما أَرَدْنا أَنْ نُبَيِّنَ.

[1] فزاوية ... بعينه] MPR2

עוד נאמר גם כן שתהיה כל אחת משתי זויות ג׳ ז׳ יותר קטנה מנצבת. הנה אומר כי זוית ה׳ כמו זוית ב׳. וההנהגה אחת. אם כן ב״ח כמו ב״ג. אם כן זוית ג׳ כמו זוית בח״ג. וזוית ג׳ יותר קטנה מנצבת. אם כן זוית בח״ג[1] יותר קטנה מנצבת. אם כן[2] זוית בח״א יותר גדולה מנצבת, והיא כמו זוית ז׳. אם כן זוית ז׳[3] יותר גדולה מנצבת[4]. וכבר[5] היתה יותר קטנה. אם כן זוית ז׳ יותר גדולה ויותר קטנה מדבר אחד בעינו—זה שקר[6]. אם כן זוית ב׳ כמו זוית ה׳. וזוית א׳ כמו זוית ד׳. ונשארה זוית[7] ג׳ כמו זוית ז׳. אם כן זויות משולש אב״ג שוות[8] לזויות משולש דה״ז, כל אחת לנכחית לה.

וזה מה שרצינו לבאר.

[1] מנצבת ... בח״ג] ⟨*⟩ **ו** [2] זוית ... כן] ⟨*⟩ **ט** [3] ז׳] ⟨*⟩ **ט** [4] והיא ... מנצבת] ליתא **ו** [5] וכבר] +~~התבאר~~ **פ** [6] שקר] +אי אפשר **ט** [7] ונשארה זוית] ונשאר הזויות **ב** [8] שוות] שוה **בוטפק**

[Proposition VI.8]

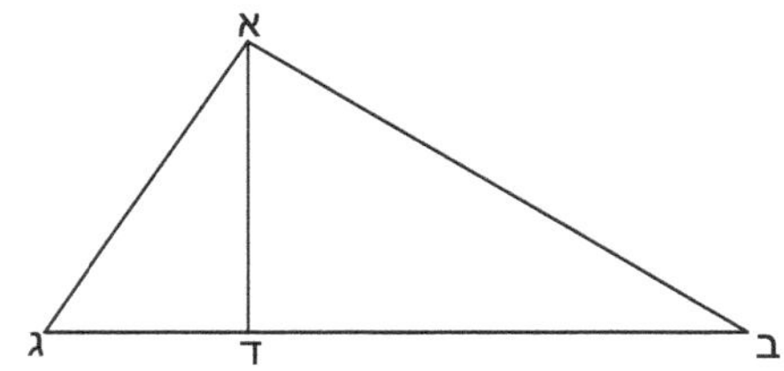

כל זוית נצבה ממשלש יצא ממנה עמוד על התושבת, יהיו בצדי העמוד שני משולשים מתדמים וידמו למשלש הגדול.

והנה יתבאר לך כי כל זוית נצבה יצא ממנה עמוד אל התושבת, יהיה העמוד אמצעי ביחס בין חלקי התושבת, ושכל אחד מצלעות המשולש אמצעי ביחס בין התושבת ואחד מחלקיה הנמשך לצלע ההיא.

המשל בזה: כי משלש אב״ג נצב (הזוית)[1], והיא זוית א׳. וכבר יצא ממנה עמוד על התושבת ב״ג, והוא א״ד. ואומר כי משולשי אב״ד ואד״ג מתדמים והם דומים לאב״ג[2].

كُلُّ مُثَلَّثٍ قائِمِ الزّاوِيَةِ إذا أُخْرِجَ مِنْ زاوِيَتِهِ القائِمَةِ عَمودٌ[1] إلى القاعِدَةِ، فَالمُثَلَّثانِ اللَّذانِ عَنْ جَنْبَتَيِ العَمودِ يُشْبِهانِ المُثَلَّثَ الأَعْظَمَ، وَيُشْبِهُ كُلُّ واحِدٍ مِنْهُما الآخَرَ.

مِثالُهُ: أَنَّ زاوِيَةَ اَ مِنْ مُثَلَّثِ ا ب ج قائِمَةٌ، وقَدْ خَرَجَ مِنْها إلى قاعِدَةِ ب ج عَمودُ ا ه. فَأَقولُ مُثَلَّثَي ا ب د وا د ج يُشْبِهانِ مُثَلَّثَ ا ب ج، وَيُشْبِهُ كُلُّ واحِدٍ مِنْهُما الآخَرَ.

---

[1]הזוית] הזויות [2]לאב״ג] ⟨⟨לא⟩⟩ב״ג

[1]عمود] عمودا

كُلُّ زاوِيَةٍ قائِمَةٍ مِنْ مُثَلَّثٍ يَخْرُجُ مِنْها عَمُودٌ إلى القاعِدَةِ، فَعَنْ جَنْبَتَيِ العَمُودِ مُثَلَّثانِ مُتَشابِهانِ، وَيُشْبِهانِ المُثَلَّثَ الأَعْظَمَ.

כל זוית נצבת ממשולש יצא ממנה עמוד אל התושבת, הנה משני צדדי העמוד שני משולשים דומים, ושניהם ידמו המשולש הגדול.

مِثالُهُ: أَنَّ زاوِيَةَ اَ مِنْ مُثَلَّثِ ا ب ج قائِمَةٌ. وَقَدْ أُخْرِجَ[1] مِنْها عَمُودُ ا د إلى قاعِدَةِ ب ج. فَأَقُولُ إنَّ مُثَلَّثَيْ ا ب د ا د ج، يَتَشابَهانِ وَيُشْبِهانِ مُثَلَّثَ ا ب ج.

המשל בו: כי משולש אב״ג זוית א׳[1] ממנו נצבת. וכבר הוצא ממנה עמוד א״ד אל תושבת ב״ג. הנה אומר כי משולש אב״ד ידמה משולש אג״ד, ושניהם ידמו משולש[2] אב״ג.

[1]أخرج] MPR2

שרטוט—**בוטסק**: $; **ס**: א״ד ∡ [1]זוית א׳] ליתא **ב** [2]משולש] ליתא **ס**

מופת זה: כי זוית א׳ נצבה, והיא שוה לזוית [ד׳][1]. וזוית ב׳ משותפת למשולש אב״ג ולמשלש אב״ד. ונשארה זוית בא״ד שוה לזוית ג׳. ויהיו זויות משלש אב״ג שוות לזויות משלש אב״ד. אם כן יהיה שעור ב״ג אצל ב״א כשיעור א״ב אצל ב״ד וכשעור ג״א אצל א״ד. ונמצאו משלשי [!] אב״ג ומשולש אב״ד מתדמים.

بُرْهانُهُ: أَنَّ زاوِيَةَ ا قائِمَةٌ، وَمُساوِيَةٌ لِكُلِّ واحِدَةٍ مِنْ زاوِيَتَيْ د. وزاوِيَةُ ب مُشْتَرَكَةٌ لِمُثَلَّثَيْ ا ب ج و ا ب د. وبَقِيَتْ مِنَ الزّاوِيَتَيْنِ القائِمَتَيْنِ زاوِيَةُ ب ا د مِثْلُ زاوِيَةِ ب ج ا مِنَ المُثَلَّثِ الأَعْظَمِ الباقِيَةَ مِنَ الزّاوِيَتَيْنِ القائِمَتَيْنِ. وَزَوايا مُثَلَّثِ ا ب ج مُساوِيَةٌ لِزَوايا مُثَلَّثِ ا ب د. فَقَدْرُ ب ج مِنَ المُثَلَّثِ الأَعْظَمِ عِنْدَ ب ا مِنَ المُثَلَّثِ الأَصْغَرِ كَقَدْرِ ا ب مِنَ المُثَلَّثِ الأَعْظَمِ عِنْدَ ب د مِنَ المُثَلَّثِ الأَصْغَرِ وَكَقَدْرِ ا ج مِنَ المُثَلَّثِ الأَعْظَمِ عِنْدَ ا د مِنَ المُثَلَّثِ الأَصْغَرِ، لِأَنَّ الأَضْلاعَ الَّتِي تُوتِرُ زَوايا مُتَساوِيَةً—مُتَناسِبَةٌ. وَالمُثَلَّثاتُ جِنْسُها واحِدٌ. فَمُثَلَّثا ا ب ج و ا ب د مُتَشابِهانِ.

וכמו כן יתבאר כי[2] משלש אד״ג ומשלש אב״ג מתדמים. ואומר כי משלש אב״ד ידמה למשלש אד״ג.

وَكَذَلِكَ مُثَلَّثُ ا د ج و ا ب ج مُتَشابِهانِ. فَأَقُولُ إِنَّ مُثَلَّثَ ا ب د يُشْبِهُ مُثَلَّثَ ا د ج.

[1] ד׳] +⟨ממשלש אב״ד⟩ [2] כי] ⟨⟨*⟩⟩

بُرهانُهُ: أَنَّ زاوِيَتَيْ ب ا ج ب د ا مُتَساوِيَتانِ، لأَنَّهُما قائِمَتانِ. وزاوية ب مُشْتَرَكَة لِمُثَلَّثَيْ ا ب ج و ا ب د. فَبَقِيَتْ زاويةُ ب ا د مثل زاوية ج. فزوايا مُثَلَّث ا ب ج مُساوِيَةٌ لِزَوايا مُثَلَّث ا ب د. فأَضلاعهما المُوتِرات لِزَواياهما المُتَساوِيَةِ— مُتَناسِبَةٌ. فِنسْبَةَ ب ج إلى ب ا كَنِسْبَةِ ا ب إلى ب د وكَنِسْبَةِ ا ج إلى ا د. فمُثَلَّثا ا ب ج و ا ب د مُتَشابِهانِ. ولِذَلِكَ يكونُ[1] مُثَلَّثا ا ب ج و ا د ج مُتَشابِهانِ.

מופתו: כי שתי זויות בא״ג אד״ב שוות, מפני שהם נצבות. וזוית ב׳ משותפת לשני משולשי אב״ג ואב״ד. ונשארה[1] זוית בא״ד כמו זוית ג׳. אם כן זויות[2] משולש אב״ג שוות לזויות משולש אב״ד. אם כן צלעותיהם שהם מיתרים לזויות שניהם השוות—מתיחסות. אם כן יחס ב״ג אל א״ב כיחס א״ב[3] אל ב״ד וכיחס א״ג אל א״ד. אם כן שני משולשי אב״ג ואב״ד מתדמים. ולכן יהיה [!] אב״ג[4] ואד״ג דומים.

وَأَيْضًا فَإِنَّ[2] مُثَلَّثَيْ ا ب د و ا د ج مُتَشابِهانِ.

וגם כן הנה שני[5] משולשי אב״ד ואד״ג דומים.

[1] يكون] **MPR2** [2] فإن] **MPR2**

[1] ואב״ד ונשארה] ונשארת **ב** ואב״ד ונשאר **ס** [2] זויות] זוית **בספ** [3] א״ב] ~~ב״ד~~+ **ק** [4] אב״ג] אב״ד **ט** [5] שני] ליתא **ו**

מופת זה: כי זוית ד׳ נצבה, והיא שוה לזוית ד׳ האחרת. וזוית ב׳ שוה לזוית דא״ג. וזוית בא״ד שוה לזוית ג׳. אם כן זויות משלש אב״ד שוות לזויות משולש אד״ג. ויהיה שעור א״ג אצל א״ב כשעור ג״ד אצל א״ד וכשעור א״ד אצל ד״ב. ונמצאו משלש אב״ד ומשלש אד״ג מתדמים. והמה ידמו למשלש אב״ג.

بُرْهانُهُ: أَنَّ زاوِيَةَ د قائِمَةٌ. وَهِيَ مِثْلُ زاوِيَةِ د الأُخْرى. وَزاوِيَةُ ب مِثْلُ زاوِيَةِ د ا ج. وَزاوِيَةُ ب ا د مِثْلُ زاوِيَةِ ج. فَزَوايا مُثَلَّثِ ا ب د مُساوِيَةٌ لِزَوايا مُثَلَّثِ ا د ج. فَقَدْرُ ا ج عِنْدَ ا ب كَقَدْرِ ج د عِنْدَ د ا. وَكَقَدْرِ ا د عِنْدَ د ب. فَمُثَلَّثا ا ب د و ا د ج مُتَشابِهانِ. وَهُما يُشْبِهانِ مُثَلَّثَ ا ب ج.

وَهُناكَ اسْتَبانَ: كُلُّ [!] مُثَلَّث قائِمِ الزَّاوِيَةِ، إِذا أُخْرِجَ مِنْ زاوِيَتِهِ القائِمَةِ عَمودٌ إِلى القاعِدَةِ، فَإِنَّهُ يَكونُ مُناسِباً فيما بَيْنَهُما. وَأَنَّ كُلَّ واحِدٍ مِنَ الضِّلْعَيْنِ المُحيطَيْنِ بِالزَّاوِيَةِ القائِمَةِ مِنَ المُثَلَّثِ الأَعْظَمِ يَكونُ مُناسِباً لِكُلِّ القاعِدَةِ وَلِأَحَدِ قِسْمَيْها فيما بَيْنَهُما مِنَ المُثَلَّثِ الأَعْظَمِ.

והנה יתבאר לך וכו׳[1].

وَذٰلِكَ ما أَرَدْنا بَيانَهُ.

---

[1] וכו׳] +⟨⟨כמו שכת׳ למעלה⟩⟩

بُرهانُهُ: أَنَّ زاوِيَتَيْ دَ مُتَساوِيَتانِ، لأَنَّهُما قائِمَتانِ. وَقَدْ كانَ تَبَيَّنَ أَنَّ زاويةَ ب ا د مثل زاوية ال ج. فَزاوِيَتا ا د ب و ب ا د مِنْ مُثَلَّثِ ا د ب مثل زاوِيَتَيْ ا د ج و د ج ا مِنْ مُثَلَّثِ ا د ج. تَبْقَى[1] زاويةُ ب مثل زاويةِ د ا ج. فَزَوايا مُثَلَّثَيْ ا ب د ا ج د مُتَساوِيَةٌ. فأَضْلاعُهُما المُوتِراتُ لِزَواياهُما المُتَساوِيَةِ—مُتَناسِبَةٌ. فَنِسْبَةُ ا ج إلى ا ب كَنِسْبَةِ ج د إلى د ا وكَنِسْبَةِ ا د إلى د ب. فَمُثَلَّثا ا ب د ا د ج مُتَشابِهانِ، وَيُشْبِهانِ مُثَلَّثَ ا ب ج.

وذَلِكَ ما أَرَدْنا أَنْ نُبَيِّنَ.

وَهُنالِكَ اسْتَبانَ أَنَّ كُلَّ زاويةٍ قائِمَةٍ في مُثَلَّثٍ يَخْرُجُ مِنْها عَمُودٌ إلى القاعِدَةِ، فَإِنَّ العَمودَ وَسَطٌ في النِّسْبَةِ بَيْنَ^ قِسْمَيِ القاعِدَةِ، وأَنَّ كُلَّ واحدٍ مِنْ ضِلْعَيِ المُثَلَّثِ وَسَطٌ في النِّسْبَةِ بَيْنَ القاعِدَةِ وَأَحَدِ قِسْمَيْها الَّذي يَلِيهِ.

وذَلِكَ ما أَرَدْنا أَنْ نُبَيِّنَ.

המופת כי שתי זויות ד׳ שוות, מפני כי שתיהן נצבות. והנה כבר התבאר[1] כי זוית בא״ד כמו זוית ג׳. אם כן שתי זויות אד״ב ובא״ד ממשולש אד״ב כמו שתי זויות אד״ג ודג״א ממשולש אד״ג. ותשאר זוית[2] דא״ג כמו זוית אב״ד[3]. אם כן זויות[4] שני משולשי אב״ד אג״ד שוים. אם כן צלעותיהם שהן[5] מיתרים לזויות שניהם השוות—מתיחסות. אם כן יחס א״ג אל א״ב כיחס ג״ד אל א״ד וכיחס א״ד אל ד״ב. אם כן שני משולשי אב״ד אד״ג דומים, וידמו משולש אב״ג.

וזה מה שרצינו לבאר[6].

ובכאן התבאר כי כל זוית נצבת במשולש יצא ממנה עמוד אל התושבת, הנה העמוד אמצע ביחס משני חלקי התושבת. והנה כל אחת משתי צלעות המשולש אמצע ביחס בין התושבת ואחד מחלקיו הנלוה אליו.

וזה מה שרצינו לבאר[7].

[1]وقد ... تبقى] **MPR2**

[1]התבאר] התבארו **ס** [2]זוית] ליתא **ט** [3]אב״ד] ~~אג״ד~~+ **ק** [4]זויות] זוי⟨ו⟩ת **ב** [5]שהן] שהם **בק** [6]מה ... לבאר] ליתא **ז** מה שרצינו **ס** [7]לבאר] לבארו **בפ**

[Proposition VI.9]

[H13]

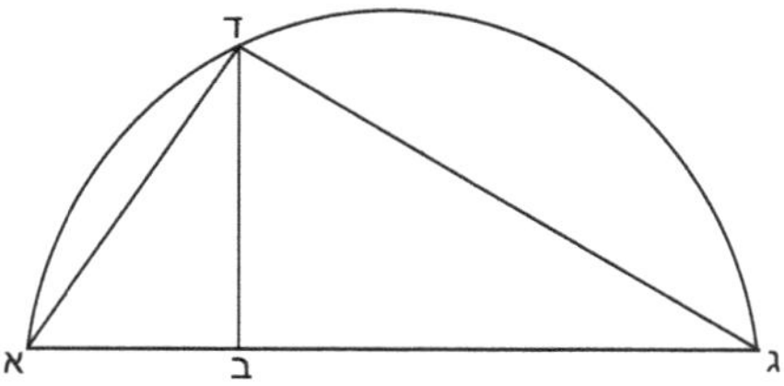

נרצה שנמצא קו שלישי מתיחס לשני קוים ידועים שיהיה ביניהם.

ונשים הקוים הידועים קוי א״ב וב״ג. וכשנרצה למצא קו שיהיה ביניהם אשר לו יחס אליהם, נחבר אחד מהם באחר על יושר. ונצייר על קו א״ג חצי עגלה עליה אד״ג. ונוציא מקו א״ג, מנקדת ב׳, קו על זוית נצבה, והוא ב״ד.

نُرِيدُ أَنْ نُبَيِّنَ كَيْفَ نَجِدُ خَطًّا مُناسِباً بِخَطَّيْنِ مَعْلومَيْنِ فيما بَيْنَهُما.

مِثالُهُ: خَطّا ا ب و ب ج. وَنُرِيدُ أَنْ نَجِدَ خَطًّا فيما بَيْنَهُما مُناسِباً لَهُما. فَنَصِلُ أَحَدَ الخَطَّيْنِ بِالآخَرِ عَلى اسْتِقامَةٍ. وَنَخُطُّ عَلى ا ج نِصْفَ دائِرَةِ ا د ج. وَنُخْرِجُ مِنْ نُقْطَةِ ب مِنْ خَطِّ ا ج عَمودًا[1] ب د. فَأَقولُ إِنَّ خَطَّ ب د كَما أَرَدْنا.

[1] عمودا] عمود

نُرِيدُ أَنْ نَجِدَ خَطًّا مُنَاسِبًا لخَطَّيْنِ مَعْلومَيْنِ فيما بَيْنَهُما.

נרצה שנמצא קו מתיחס לשני קוים ידועים במה שבין שניהם[1].

فَنَجْعَلَ الخَطَّيْنِ المَعْلومَيْنِ ا ب ب ج. ونُرِيدُ أَنْ نَجِدَ بَيْنَهُما خَطًّا مُنَاسِبًا لَهُما. فَنَصِلَ أَحَدَهُما بالآخَرِ على اسْتِقَامَةٍ[1]. ونَخُطَّ عَلَيْهِما مَوْصُولَيْنِ، وَهُما ا ج، نِصفَ دائِرَةٍ ا د ج. وَنُخْرِجَ مِنْ ا ج، مِنْ نُقْطَةِ ب^، عَمُودَ ب د. وَنُخْرِجَ خَطَّيْ د ا د ج.

הנה נשים השני קוים הידועים א״ב וב״ג. ונרצה שנמצא בין שניהם קו מתיחס לשניהם. הנה נגיע אחד משניהם באחר על יושר. ונקוה על שניהם המגיעים, והם א״ג, חצי עגולה אד״ג. ונוציא מן א״ג, מנקודת ב׳ ממנו, עמוד ב״ד. ונוציא שני קוי א״ד וד״ג.

---

[1] على استقامة] MPR2

שרטוט—**ט**: ב״ד ⊥

שביניהם ט

[1] שבין שניהם]

ונדביק א׳ בד׳, וד׳ בג׳. ולפי שבחצי העגלה זוית נצבה, והיא זוית אד״ג, וכבר יצא ממנה עמוד ב״ד אל תושבת משלש אד״ג, והיא תושבת א״ג, יתבאר כי קוי א״ב וב״ג, והם חלקי התושבות, כבר נמצא ביניהם קו שיש לו אליהם יחס, והוא קו ב״ד.

[!]

بُرْهانُهُ: أَنْ نَصِلَ د ا د ج بِخَطَّيْ د ا د ج. فَزاوِيَةُ (ا د ج)[1] قائِمَةٌ لِأَنَّها في نِصْفِ دائِرَةٍ. وَقَدْ خَرَجَ[2] مِنْها عَمودُ د ب إِلى قاعِدَةِ جا مِنْ مُثَلَّثْ ا د ج. فَخَطّا ا ب و ب ج هُما قِسْما القاعِدَةِ. وَقَدْ وَجَدْنا بَيْنَهُما خَطَّ ب د مُناسِباً لَهُما.

وَذَلِكَ ما أَرَدْنا بَيانَهُ.

---

[1] ا د ج] د ا ج [2] وقد خرج] +*

فزاوية ا د ج قائمَة، لأنّها في نِصْفِ دائِرَة، وَقَدْ أُخْرِجَ مِنْها عَمُودُ د ب إلى قاعِدَةِ ا ج[1]. نخَطّا ا ب ب ج هُما قِسْما القاعِدَةِ. وَقَدْ وَجَدْنا بَيْنَهُما خَطًّا مُنَاسِباً لهُمَا[2].

הנה זוית אד״ג נצבת, מפני שהיא (בחצי)[1] עגולה. וכבר יצא ממנה עמוד ד״ב אל תושבת[2] א״ג. אם כן שני קוי א״ב וב״ג הם שני חלקי התושבת. וכבר[3] מצאנו בין שניהם קו מתיחס לשניהם[4].

وذَلِكَ ما أَرَدْنا أَنْ نبيِّنَ.

ונשלם ביאורו.

[1] ا ج] MPR2 +من مثلث ا د ج CT1T3 +وهو خط ب د [2] لهما]

[1] בחצי] חצי **בוטספק** [2] תושבת] התושבת **ס** [3] וכבר] ליתא **ב** [4] לשניהם] +⟨והוא ד״ב⟩ **ק**

[Proposition VI.10]

[H11]

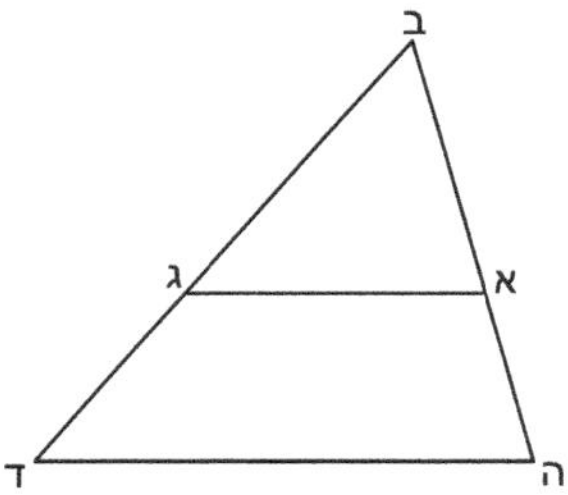

נרצה שנמצא קו שלישי מתיחס לשני קוים ידועים.

ונשים הקוים הידועים א״ב וב״ג. וכאשר רצינו שנמצא קו שלישי שיהיה לו יחס אליהם, נדביק א׳ בג׳, ונוציא א״ב וב״ג אל ה׳ וד׳. ונשים א״ה שוה לב״ג. ונוציא מן ה׳ קו נכוחי לקו א״ג והוא ה״ד.

نُرِيدُ أَنْ نُبَيِّنَ كَيْفَ نَجِدُ خَطًّا ثالِثًا مُناسِبًا لِخَطَّيْنِ مَعْلومَيْنِ.

مِثالُهُ: خَطّا اب وب ج. وَنُرِيدُ أَنْ نَجِدَ خَطًّا ثالِثًا مُناسِبًا. فَنَصِلُ اَجَ بِخَطِّ اج. وَنُخْرِجُ خَطَّيْ اب وب ج إِلى ه و د. وَنَجْعَلُ اه مِثْلَ ب ج. وَنُخْرِجُ مِنْ ه خَطَّ ه د مُوازِيًا لِخَطِّ اج. فَأَقولُ إِنَّ خَطَّ ج د كَما أَرَدْنا.

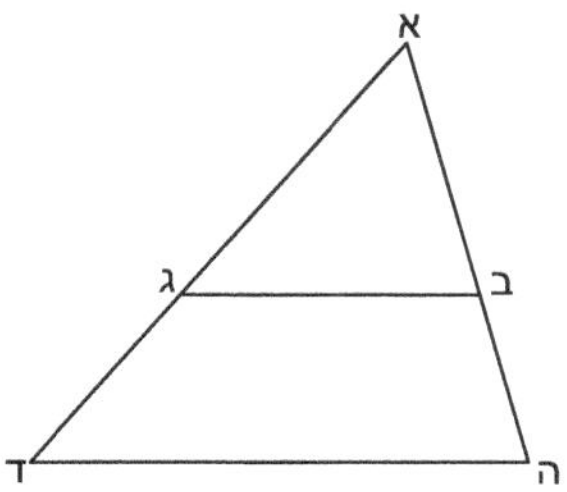

نُرِيدُ أَنْ نَجِدَ خَطًّا ثَالِثًا مُنَاسِبًا لِخَطَّيْنِ مَعْلُومَيْنِ.

נרצה שנמצא קו שלישי מתיחס[1] לשני קוים ידועים.

فَنَجْعَلَ الخَطَّيْنِ المَعْلُومَيْنِ ا ب و ا ج. وَنَجْعَلُهُما يُحِيطانِ بِزاوِيَةِ ا. وَنُرِيدُ أَنْ نَجِدَ خَطًّا ثَالِثًا مُنَاسِبًا لَهُما. فَنُخْرِجَ خَطَّ ب ج. وَنُخْرِجَ خَطَّيْ ا ب و ا ج إلى ه د. وَنَجْعَلَ خَطَّ ب ه مثل ا ج. وَنُخْرِجَ ه د يُوازِي ب ج.

הנה נשים השני קוים הידועים א״ב וא״ג. ונשים שניהם יקיפו בזוית א׳. ונרצה שנמצא קו שלישי מתיחס[2] לשניהם. ונוציא קו ב״ג. ונוציא שני קוי א״ב וא״ג אל ה״ד. ונשים קו ב״ה כמו א״ג. ונוציא קו ה״ד נכחי לקו ב״ג.

שרטוט—**בוטספק**: משולש אד״ה שווה שוקיים [1]מתיחס] מתיחסם **ב** [2]מתיחס] מתיחסם **ב**

ויהיה משלש בה״ד כבר יצא מצלע מצלעיו[1] קו נכוחי לקו ה״ד, והוא קו א״ג. ויהיה שעור ב״א אל א״ה כשיעור ב״ג אל ג״ד. וכבר מצאנו קו שלישי שהוא מתיחס לקוי א״ב וב״ג הידועים, והוא קו ג״ד.

بُرْهانُهُ: أَنَّ مُثَلَّثَ ب ه د قَدْ خَرَجَ مِنْ ضِلْعٍ لَهُ خَطُّ ا ج مُوازِيًا لِخَطِّ ه د. فَقَدْرُ ب ا عِنْدَ ا ه كَقَدْرِ ب ج عِنْدَ ج د. و ا ه مِثْلُ ب ج. فَقَدْرُ ب ا عِنْدَ ب ج كَقَدْرِ ب ج عِنْدَ ج د. فَقَدْ وَجَدْنا لِخَطَّيْنِ ا ب و ب ج خَطًّا ثالِثًا مُناسِبًا لَهُما، وَهُوَ ج د.

וזה מה שרצינו לבאר.

وَذَلِكَ ما أَرَدْنا بَيانَهُ.

[1] מצלע מצלעיו] 〈〈*〉〉

فنِسْبَةَ اب إلى ب ه كنِسْبَةَ اج إلى ج د. و ب ه مثل اج. فنِسْبَةَ اب إلى اج كنِسْبَةَ اج إلى ج د. فقد وَجَدْنا خطًّا ثالثًا مُناسِبًا لخطَّيْن مَعْلومَيْن، وَهُما اب و اج، وهو ج د.

הנה יחס א״ב אל ב״ה כיחס א״ג אל ג״ד. וב״ה כמו א״ג. אם כן יחס א״ב אל א״ג כיחס[1] א״ג אל ג״ד. הנה מצאנו קו שלישי מתיחס[2] לשני קוים ידועים, והם א״ב וא״ג, והוא ג״ד.

وذلِكَ ما أَرَدْنا أَنْ نُبَيِّنَ.

וזה מה שרצינו לבאר[3].

[1] כיחס] +~~ג״ד~~ **ק**   [2] מתיחס] מתיחסם **ב**
[3] וזה ... לבאר] ונשלם ביאורו **בוספק**

[Proposition VI.11]

[H12]

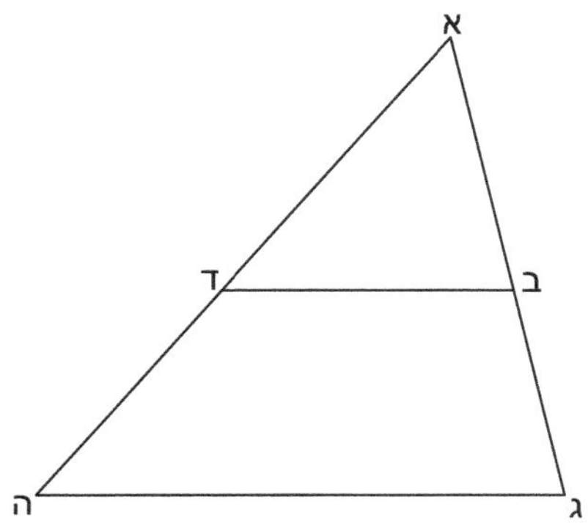

נרצה שנמצא קו רביעי מתיחס לשלשה קוים ידועים.

ונשים הקוים הידועים א״ב ב״ג א״ד. וכאשר רצינו למצא קו רביעי מתיחס להם, נדביק ב׳ בד׳. ונוציא מנקודת ג׳ קו נכוחי לב״ד, והוא קו ג״ה. ונוציא קו א״ד עד ה׳.

שרטוט—משולש אג״ה שווה שוקיים

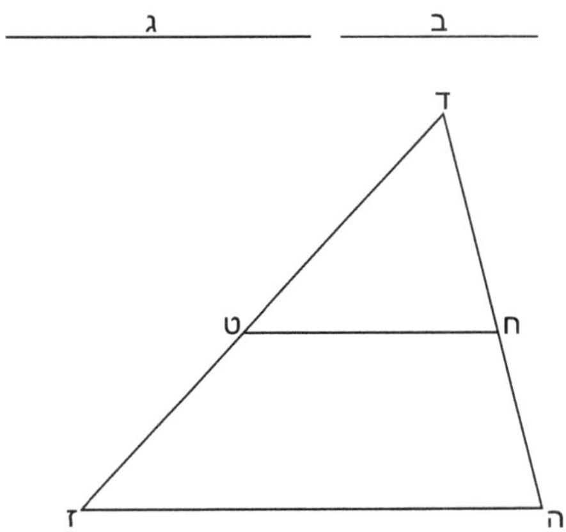

نُرِيدُ أَنْ نَجِدَ خَطًّا رابِعًا مُنَاسِبًا لِثَلاثَةِ خُطوطٍ مَعْلومَةٍ.

נרצה למצוא קו רביעי מתיחס[1] לשלשה קוים ידועים.

فَلْتَكُنِ الخُطوطُ الثَّلاثَةُ خُطوطَ اَ بَ جَ. ونُرِيدُ أَنْ نَجِدَ^ خَطًّا رابِعًا مُنَاسِبًا لِخُطوطِ اَ بَ جَ.

ויהיו הקוים השלשה קוי א׳ ב׳ ג׳. ונרצה שנקוה[2] קו רביעי מתיחס לקוי א׳ ב׳ ג׳.

שרטוט—**בוטספק**: משולש דה״ז שווה שוקיים [1]מתיחס] מיתיחסם **ב** [2]שנקוה] שנקו⟨ה⟩ **ו**

ומשלש אג״ה כבר יצא מצלע מצלעיו קו נכוחי לקו ג״ה, והוא קו ב״ד. ואם כן יהיה שיעור א״ב אל ב״ג כשיעור א״ד אל ד״ה. וכבר מצאנו קו רביעי מתיחס לשלשה קוים ידועים.

וזה מה שרצינו לבאר.

فَنَخُطُّ خَطَّيْنِ مُسْتَقيمَيْنِ يُحيطانِ بِزاوِيَةٍ، وَهُما د ه د ز. وَلْيَكُنْ د ه مثل ا و ب جَميعًا. ونَفْصِلُ مِنْهُ مثل ا، وهو د ح. فَيَبْقَى ح ه مثل ب^. وَلْيَكُنْ د ز أعظم مِنْ ج^. ونَفْصِلُ مِنْه مثل خطّ ال ج[1]، وهو د ط. وَنُخْرِجُ خَطَّ ح ط. وَنُخْرِجُ مِنْ نُقْطَةِ ه خَطًّا موازيًا لِخَطِّ ح ط، وهو ه ز. فَلِأَنَّهُ قَدْ خَرَجَ في^ مُثَلَّثِ د ه ز مِنْ ضِلْعٍ مِنْ أَضْلاعِهِ خَطُّ ح ط يُوازي القاعِدَةَ، وهي ه ز، تكونُ نِسْبَةُ د ح إلى ح ه مثل نِسْبَةِ د ط إلى ط ز. لكنَّ د ح مثل ا. وح ه مثل ب. و د ط مثل ج. فنِسْبَةُ ا إلى ب كَنِسْبَةِ ج إلى ط ز. فقد وَجَدْنا خَطًّا رابِعًا، وهو ط ز، مُناسِبًا لِخُطوطِ ا ب ج الثَّلاثَةِ[2].

وذَلِكَ ما أَرَدْنا أَنْ نُبَيِّنَ.

הנה נקוה שני קוים ישרים לא יקיפו בזוית, והם ד״ח וח״ה. ונשימם דבקים על יושר. ויהיה ד״ח כמו א׳[1], וח״ה כמו ב׳. ויהיה ד״ז יותר[2] גדול מן ג׳. ויקיף עם ד״ה בזוית ד׳[3]. ונבדיל ממנו כמו ג׳[4], והוא ד״ט. ונוציא קו ח״ט. ונוציא מנקודת ה׳ קו נכחי לקו ח״ט, והוא ה״ז. הנה מפני שהוא כבר יצא ממשולש דה״ז מצלע מצלעותיו קו ח״ט נכחי לתושבת, והוא ה״ז, יהיה יחס ד״ח אל ח״ה כמו יחס[5] ד״ט אל ט״ז. אבל ד״ח כמו א׳. וח״ה כמו ב׳. וט״ד כמו ג׳. הנה יחס א׳ אל ב׳ כיחס ג׳ אל ט״ז. הנה כבר מצאנו קו רביעי, והוא ט״ז, מתיחס לקוי א׳ ב׳ ג׳ השלשה.

וזה מה שרצינו לבאר.

1 مثل ... ال ج] PMR2 مثله CopT2 2 الثلاثة]

1 א׳] ~~וא׳~~ ב 2 יותר] ויותר ב 3 ד׳] ליתא בוטפ ⟨*⟩ ק 4 ג׳] +~~ונוציא~~ ק 5 ח״ה ... יחס] ~~ח״ה~~ ה״ח כמו ס

[Proposition VI.12]

[H9, B11]

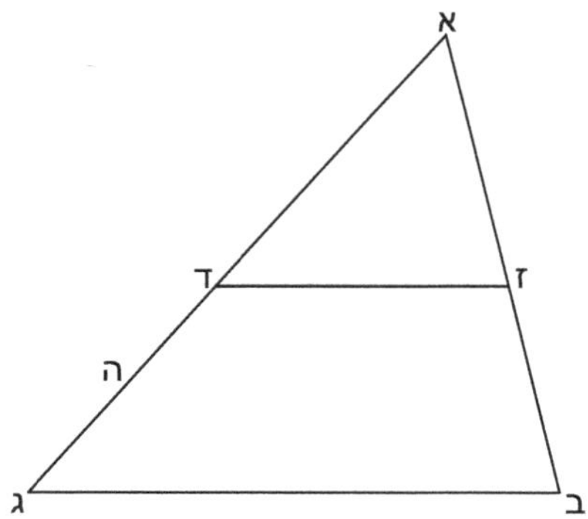

נרצה שנחתוך מקו ידוע איזה חלק שנרצה.

نُرِيدُ أَنْ نُبَيِّنَ كَيْفَ نَفْصِلُ مِنْ خَطٍّ مَعْلُومٍ أَيَّ جُزْءٍ شِئْنَا.

ונשים החלק[1] השליש והקו הידוע קו א״ב. וכשנרצה שנחתוך מקו א״ב שלישו,

مثالُهُ: خَطُّ ا ب الْمَعْلوم وَالجُزْ[ء] الثُّلث. وَنُرِيدُ أَنْ نَفْصِلَ مِنْ ا ب ثُلُثَهُ.

[1] החלק] קו+

نُرِيدُ أَنْ نَفْصِلَ مِنْ خَطٍّ مَعْلومٍ أَيَّ جُز[ءٍ] شِئْنا[1].

נרצה שנבדיל מקו ידוע אי זה חלק שנרצה.

فَنَجْعَلَ الخطَّ المَعْلومَ ا بَ والجُزْءَ الثُّلُثَ. ونُرِيدُ أَنْ نَفْصِلَ مِنْ ا ب ثُلُثَهُ.

הנה נשים הקו הידוע א״ב והחלק השלישי[1]. ונרצה שנבדיל[2] מן א״ב שלישיתו[3].

[1] شينا] أردنا **Q**

שרטוט—**בוטספק**: משולש אב״ג שווה שוקיים [1] השלישי] השליש **טק** [2] שנבדיל] +שלישיתו **ט** [3] שלישיתו] ליתא **ט**

נוציא מנקודת א׳ קו אחר, והוא א״ג. ויהיו א״ב וא״ג מקיפים בזוית א׳. ויהיו בא״ג שלשה חלקים שוים, והם א״ד וד״ה וה״ג. ונדביק ב׳ בג׳. ונוציא מקו א״ג, מנקודת ד׳, קו נכוחי לקו ב״ג, והוא ד״ז. וכבר יצא ממשלש אב״ג, מצלעו האחת, קו נכוחי לצלע ב״ג, והוא ד״ז. אם כן שיעור ג״ד אצל ד״א כשיעור ב״ז אצל א״ז. וג״ד כפל א״ד—יהיה ב״ז כפל א״ז. וכל ב״א שלשה פעמים א״ז. והתבאר שא״ז שליש א״ב. וכבר חתכנו מקו א״ב הידוע החלק שרצינו והוא השליש.

וזה מה שרצינו לבאר.

فنُخْرِجُ مِنْ نُقْطَةِ ا خَطًّا آخَرَ، وَهُوَ ا ج. ا ب (ا ج)[1] يُحيطانِ بِزاوِيَةِ ا. ونَفْصِلُ مِنْ ا ج (ثَلاثَةَ)[2] أَجْزا[ءٍ] مُتَساوِيَةً، وَهِيَ ا د و د ه و ه ج. وَنَصِلُ ب ج بِخَطِّ ب ج. ونُخْرِجُ مِنْ نُقْطَةِ د مِنْ خَطِّ ا ج خَطَّ د ز مُوازِيًا لِخَطِّ ب ج. فَأَقولُ إِنَّ ا ز ثُلُثُ ا ب.

بُرهانُهُ: أَنَّ مُثَلَّثَ ا ب ج قَدْ خَرَجَ مِنْ ضِلْعٍ لَهُ خَطُّ د ز مُوازِيًا لِضِلْعِ ب ج. وَقَدْرُ ج د عِنْدَ د ا كَقَدْرِ ب ز عِنْدَ ز ا. و ج د مِثْلا د ا. ب ز مِثْلا ز ا. فَكُلُّ ب ا (ثَلاثَةُ)[3] أَمْثالِ ز ا. و ا ز ثُلُثُ ا ب. فَقَدْ فَصَلْنا مِنْ خَطِّ ا ب المَعْلوم الجُزْءَ الَّذي أَرَدْناهُ، وَهُوَ الثُّلُثُ، وَهُوَ ا ز.

وَذَلِكَ ما أَرَدْنا بَيانَهُ.

[1] ا ج] ליתא [2] ثلاثة] ثلثة [3] ثلاثة] ثلثة

فَنُخْرِجَ خَطَّ ا د، مُحيطًا مع خَطِّ ا ب بِزاوِيَةِ ا. وَنَزيدَ في ا د مِثْلَهُ، وهو د ه، وفي د ه مِثْلَهُ، وهو ه ج. ونُخْرِجَ خَطَّ ب ج. وَنُخْرِجَ د ز يُوازي ب ج. فنِسْبَةَ خَطِّ ج د إلى خطِّ د ا كنِسْبَةِ ب ز إلى ز ا. وج د مِثْلا د ا. ف ب ز مِثْلا ز ا. فَكُلُّ ب ا ثَلاثَةَ أَمْثال ز ا. ف ا ز ثُلُثُ ا ب. فقد فَصَلْنا مِنْ ا ب المَعْلومِ^ الجُزْ[ءَ] الَّذي أَرَدْنا، وهو ا ز، ومِقْدارُهُ الثُّلُثُ.

הנה נוציא קו א״ד, מקיף עם קו א״ב בזוית א׳. ונוסיף בא״ד כמוהו, והוא ד״ה, ובד״ה כמוהו, והוא ה״ג. ונוציא[1] קו ב״ג. ונוציא ד״ז נכחי אל ב״ג. הנה יחס ג״ד אל א״ד כיחס ב״ז אל[2] א״ז. וג״ד כפל א״ד. אם כן ב״ז[3] כפל א״ז. אם כן כל א״ב שלשה כפלים[4] א״ז. אם כן א״ז שליש א״ב. הנה כבר הבדלנו מקו א״ב הידוע אי זה חלק שרצינו, והוא א״ז, והוא שיעור השליש.

وذَلِكَ ما أَرَدْنا أَنْ نُبيِّنَ.

וזה מה שרצינו לבאר.

[1] ונוציא] ליתא ס [2] אל] +~~א״ד~~ ק [3] ב״ז] קו+ פ [4] כפלים] +⟨נ׳ פעמים כמו א״ז⟩ ק

[Proposition VI.13]

[H10, B12]

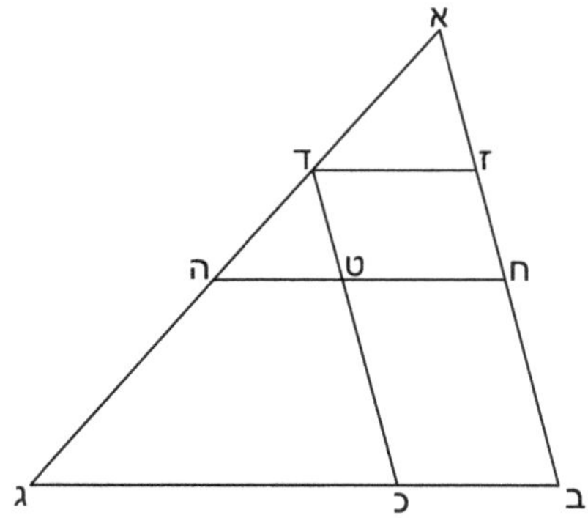

נרצה שנחלק קו ידוע ביחס קו אחר בכמו חלקיו.

نُرِيدُ أَنْ نُبَيِّنَ كَيْفَ نَقْسِمُ خَطًّا مَعْلومًا غَيْرَ مَقْسومٍ كَمَا قُسِمَ خَطٌّ آخَرَ مَقْسوم بِمِثْلِ أَجْزائِهِ.

ונשים שני הקוים הידועים א״ב וא״ג. ואשר נרצה לחלק א״ב, והקו שנחלק א״ג על שתי נקודות ד׳ וה׳. וכאשר נרצה לחלק קו א״ב, כמו שנחלק[1] א״ג בכמו חלקיו.

مِثالُهُ: خَطُّ اب و اج. وَالَّذي لَمْ يُقْسَمْ اب، والَّذي قُسِمَ اج عَلى نُقْطَتَيَ د و ه. وَنُرِيدُ أَنْ نَقْسِمَ خَطَّ اب كَمَا قُسِمَ خَطُّ اج بِمِثْلِ أَجْزائِهِ.

---

[1] שנחלק] ⟨⟨ש⟩⟩נחלק

نُرِيدُ أَنْ نَقِّسِمَ خَطًّا مَعْلُومًا بِقِسْمَةِ خَطٍّ آخَرَ[1] بِمِثْلِ أَقْسامِهِ.

فَنَجْعَلَ الخَطَّيْنِ المَعْلومَيْنِ ب ا ا جَ. وَلْيَكُونا مُحيطَيْنِ بِزاوِيَةِ ال آ[2]. والمَقَّسُوم مِنْهُما ا جَ على نُقْطَتَيْ د ه. ونُرِيدُ أَنْ نَقِّسِمَ ا ب كَقِسْمَةِ ا ج^ بِمِثْلِ أَقْسامِهِ.

נרצה לחלק קו ידוע בחלוקת[1] קו אחר כמו[2] חלקיו.

הנה[3] נשים שני הקוים הידועים[4] א״ב וא״ג. ויהיו מקיפים בזוית א׳. והמחולק משניהם[5] א״ג על שתי נקודות ד׳ ה׳. ונרצה לחלק בחלוקה[6] ההיא א״ב כמו חלקיו.

---

[1]أخر] MPR2 +معلوم [2]ال آ] MPR2

שרטוט—**בוטספק**: משולש אב״ג שווה שוקיים [1]בחלוקת] בחלוק **ב** +⟨⟨נ״א ביחס⟩⟩ **ק** [2]כמו] +~~חלקי~~ **ב** [3]הנה] ליתא **ס** [4]הידועים] ליתא **בוס** [5]משניהם] לשניהם **ט** [6]בחלוקה] ⟨נ׳ ביחס⟩⟩ **ק**

נדביק ב׳ בג׳. ונוציא מנקדת ד׳ וה׳ שני קוים נכוחיים לקו ב״ג, והם קוי ד״ז ה״ח. ונוציא מד׳ קו נכוחי לקו א״ב, והוא ד״כ. ויהיו שטחי ז״ט וח״כ נכוחי הצלעות. אם כן ז״ח שוה לד״ט. וח״כ שוה לט״ב. ומשלש דב״ג כבר יצא מצלעו קו נכוחי לצלע ב״ג, והוא קו ט״ה. אם כן שיעור ג״ה אצל ה״ד כשיעור כ״ט אצל ט״ד. וכבר היה כ״ט שוה לב״ח. וט״ד שוה לח״ז. אם כן שיעור ג״ה אצל ה״ד כשעור ב״ח אצל ח״ז. ועוד משלש אח״ה כבר יצא מצלעו קו נכוחי לצלע ה״ח, והוא קו ד״ז. אם כן שיעור ה״ד אצל ד״א כשעור ח״ז אצל ז״א. וכבר היה שיעור ג״ה אצל ה״ד כשעור ב״ח אצל ח״ז. והתבאר כי קו א״ב כבר נחלק כמו שנחלק[1] קו א״ג ובכמו חלקיו.

וזה מה שרצינו לבאר.

فَنَصِلُ ب ج بِخَطِّ ب ج. وَنُخْرِجُ مِنْ نُقْطَتَيْ د وه خَطَّيْ د و ز. وح ه مُوازِيَيْنِ لِخَطِّ ب ج. فَأَقولُ إِنّا قَدْ قَسَمْنا خَطَّ ا ب بِعَدَدِ أَقْسامِ خَطِّ ا ج بِمِثْلِ أَجْزائِهِ على نُقْطَتَيْ ز وح.

بُرْهانُهُ: أَنْ نُخْرِجَ مِنْ د خَطَّ د ك مُوازِيًا لِخَطِّ ا ب. فَسَطْحا ز ط و ح ك مُتَوازِيا الأَضْلاعِ. ز ح مِثْلُ ط د. و ح ب مِثْلُ ط ك. فَمُثَلَّثُ د ك ج قَدْ خَرَجَ مِنْ ضِلْعٍ لَهُ خَطُّ ط ه مُوازِيًا لِضِلْعِ ك ج. فَقَدْرُ ج ه عِنْدَ ه د كَقَدْرِ ك ط عِنْدَ ط د. وَقَدْ كانَ ك ط مِثْلَ ب ح. و ط د مِثْل ح ز. فَقَدْرُ ج ه عِنْدَ ه د كَقَدْرِ ب ح عِنْدَ ح ز. وَأَيْضًا مُثَلَّثُ ح ا ه قَدْ خَرَجَ مِنْ ضِلْعٍ لَهُ خَطُّ د ز مُوازِيًا لِضِلْعِ ه ح. فَقَدْرُ ه د عِنْدَ د ا كَقَدْرِ ح ز عِنْدَ ز ا. وَقَدْ كانَ قَدْرُ ج ه عِنْدَ ه د كَقَدْرِ ب ح عِنْدَ ح ز. فَخَطُّ ا ب قَدْ قُسِمَ عَلَى نُقْطَتَيْ ز ح بِعَدَدِ أَقْسامِ خَطِّ ا ج بِمِثْلِ أَجْزائِهِ.

وَذَلِكَ ما أَرَدْنا بَيانَهُ.

[1] כמו שנחלק] ~~כאשר נח~~ ⟨⟨כמו ש⟩⟩נחלק

فَنَجْعَلُهُما يُحيطانِ بِزاوِيَةِ ا. ونُخْرِجُ خَطَّ ب ج. ونُخْرِجُ خَطَّيْ د ز و ه ح يُوازِيان ب ج. ونُخْرِجُ خَطَّ د ط ك يُوازي ا ب. فَسَطْحَا ز ط ك ح مُتَوازِيا الأَضْلاعِ. ف ز ح مثل د ط. و ب ح مثل ك ط. فَنِسْبَةُ ك ط إلى ط د كَنِسْبَةِ ب ح إلى ح ز. وفي مُثَلَّثِ ك د ج خطّ ط ه يُوازي ك ج. فَنِسْبَةُ ج ه إلى ه د كَنِسْبَةِ ك ط إلى ط د. وَقَدْ كانَ تَبَيَّنَ أَنَّ نِسْبَةَ ك ط إلى ط د كَنِسْبَةِ ب ح إلى ح ز. فَنِسْبَةُ ج ه إلى ه د كَنِسْبَةِ ب ح إلى ح ز. وَأَيْضًا فَإِنَّ في مُثَلَّثِ ح ا ه خَطّ د ز يُوازي ه ح. فَنِسْبَةُ ه د إلى د ا كَنِسْبَةِ ح ز إلى ز ا. وَقَدْ كانَتْ نِسْبَةُ[1] ج ه إلى ه د كَنِسْبَةِ ب ح إلى ح ز. وَنِسْبَةُ خطّ ه د إلى خطّ د ا كَنِسْبَةِ خطّ ح ز إلى خطّ ز ا[2]. نَخَطّ ا ب قَدْ قُسِمَ على نُقْطَتَيْ ز ح كَقِسْمَةِ خطّ ا ج بِمِثْلِ^ أَقْسامِهِ.

وَذَلِكَ ما أَرَدْنا أَنْ نُبَيِّنَ.

[1] وقد ... نسبة] MPR2 [2] ونسبة ... ز ا] MPR2

ונשימם מקיפים בזוית א׳. ונוציא קו ב״ג. ונוציא שני קוי ד״ז וה״ח[1] נכחיים אל ב״ג. ונוציא קו דט״כ נכחי א״ב[2]. הנה שני שטחי ז״ט ח״כ נכחיי הצלעות. אם כן ז״ח כמו ד״ט. וח״ב כמו ט״כ[3]. אם כן יחס ט״כ אל ט״ד— יחס[4] ב״ח אל ח״ז. ובמשולש כד״ג קו ט״ה נכחי אל כ״ג[5]. אם כן יחס ג״ה אל ה״ד כיחס[6] ט״כ אל ט״ד. וכבר התבאר כי יחס כ״ט אל ט״ד כיחס ב״ח אל ח״ז. אם כן יחס ג״ה אל ה״ד כיחס ב״ח אל ח״ז. וגם כן הנה במשולש חא״ה קו ד״ז נכחי אל ה״ח. אם כן יחס ה״ד אל א״ד כיחס ח״ז אל א״ז. וכבר היה יחס ג״ה אל ה״ד[7] כיחס ב״ח אל ח״ז[8]. ויחס ה״ד אל א״ד כיחס ח״ז אל א״ז. אם כן קו א״ב כבר חולק על שתי נקודות ז׳ ח׳ כמו חלוקת קו א״ג ויחס חלקיו.

ונשלם ביאורו.

[1] וה״ח] ה״ח ויהיו **ס** [2] א״ב] לקו+ **ס** אל+ **ק** [3] ט״כ] ט״ד **ס** [4] יחס] כיחס **ס** [5] כ״ג] קו+ **וס** [6] כיחס] +~~ב״ח אל ח״ז וגם כן הנה במשולש חא״ה קו ד״ז נכח אל~~ **ב** [7] ה״ד] א״ד+ **ק** [8] ח״ז] +אם כן יחס ג״ה אל ה״ד כיחס ב״ח אל ח״ז **בטפק**

[Proposition VI.14]

[B13]

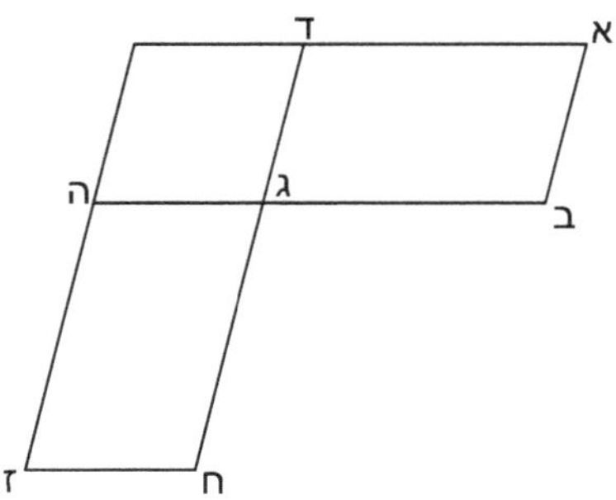

כשיהיו שני שטחים נכוחי צלעות, ושתי זויות מהם תהיינה שוות, ויהיו השטחים שוים, תהיינה הצלעות המקיפות באלו הזויות מתהפכות ביחס. ואם תהיינה הצלעות המקיפות בשתי הזויות מתהפכות, יהיו השטחים שוים.

كُلُّ سَطْحَيْنِ مُتَوازِيَيِ الأَضْلاعِ في أَحَدِهِما زاوِيَةٌ مِثْلُ زاوِيَةٍ في الآخَرِ، فَإِنَّ السَّطْحَيْنِ إِنْ كانا مُتَساوِيَيْنِ فَإِنَّ أَضْلاعَهُما الَّتي تُحيطُ بِالزّاوِيَتَيْنِ المُساوِيَتَيْنِ مُتَكافِئَةٌ. وَإِنْ كانَتْ أَضْلاعُهُما الَّتي تُحيطُ بِالزّاوِيَتَيْنِ المُتَساوِيَتَيْنِ مُتَكافِئَةً، فَالسَّطْحانِ مُتَساوِيانِ.

המשל בזה: ששטחי א״ג וג״ז נכוחי הצלעות, ובאחד מהם זוית שוה לזוית אשר באחר, והם אשר עליהם ג׳. ונשים קו ב״ג מחובר בקו ג״ה על יושר. והשטחים שוים. ואומר כי הצלעות[1] המקיפות בשתי הזויות השוות מתהפכות ביחס, יהיה שיעור ב״ג אצל ג״ה כשיעור ח״ג אצל ג״ד.

مِثالُهُ: أَنَّ سَطْحَيْ (ا ج)[1] و ج ز مُتَوازِيَيِ الأَضْلاعِ، في أَحَدِهِما زاوِيَةٌ في الآخَرِ، وَهُما زاوِيَتا ج. وَنَجْعَلُ خَطَّ ب ج مُتَّصِلًا بِخَطِّ ج ه عَلى اسْتِقامَةٍ. وَالسَّطْحانِ مُتَساوِيانِ. فَأَقولُ إِنَّ الأَضْلاعَ الَّتي تُحيطُ بِالزّاوِيَتَيْنِ المُتَساوِيَتَيْنِ مُتَكافِئَةٌ، فَقَدْرُ ب ج عِنْدَ ج ه كَقَدْرِ ح ج عِنْدَ ج د.

[1] הצלעות] קו+

[1] ا ج] ه ا ج

إِذا تَوازَتْ أَضْلاعُ سَطْحَيْنِ، وَتَساوَتْ زاوِيَتانِ مِنْهُما، وَكانَا مُتَساوِيَيْنِ، فَإِنَّ الأَضْلاعَ المُحِيطَةَ بِتَيْنِكَ الزَّاوِيَتَيْنِ مُتَكافِئَةٌ. وَإِنْ كانَتِ الأَضْلاعُ المُحِيطَةُ بِالزَّاوِيَتَيْنِ مُتَكافِئَةً، فَإِنَّ السَّطْحَيْنِ مُتَساوِيانِ.

כאשר יתנגדו צלעות שני שטחים, ותשתוינה שתי זויות מהם, והיו[1] שוים, הנה הצלעות המקיפות בשתי הזויות ההם מספיקות. ואם היו הצלעות המקיפות בשתי הזויות[2] מספיקות[3], הנה שני השטחים שוים.

مِثالُهُ: أَنَّ أَضْلاعَ سَطْحَيْ ا ج و ج ز مُتَوازِيَةٌ. وَزاوِيَتا ج مِنْهُما مُتَساوِيَتانِ. وَالسَّطْحانِ مُتَساوِيانِ. فَأَقُولُ إِنَّ الأَضْلاعَ المُحِيطَةَ بِزاوِيَتَيْ ج مُتَكافِئَةٌ—نِسْبَةُ ب ج إلى ج ه كَنِسْبَةِ ح ج إلى ج د.

המשל בו: כי צלעות שטח א״ג וג״ז נכחיים. ושתי זויות ג׳[4] משניהם שוות. והשני שטחים שוים. הנה אומר כי הצלעות[5] המקיפות בשתי זויות ג׳ מספיקות—יחס ב״ג אל ג״ה כיחס ח״ג אל ג״ד.

שרטוט פ: $ [1]והיו] +⟨⟨השטחים⟩⟩ ק [2]הזויות] +ההם ס [3]ואם ... מספיקות] ⟨⟨*⟩⟩ פ [4]ג׳] ליתא ו [5]הצלעות] השתי צלעות ק

המופת בו: שנשלים שטח ד״ה. וא״ג הוא שוה לג״ז. והשיעורים השוים יחסם אל שיעור אחר—אחד. והיה שעור ג״ז אצל ד״ה כשעור א״ג אצל ד״ה. ושעור א״ג אצל ד״ה כשעור ב״ג אצל ג״ה. ושעור ג״ז אצל ד״ה כשעור ח״ג אצל ג״ד. ואם כן יהיה שעור ב״ג אצל ג״ה כשעור ח״ג אצל ג״ד.

بُرْهانُهُ: أَنَّا نُتِمُّ سَطْحَ د ه. ا ج مِثْلُ ج ز. وَالْمُتَساوِيَةُ نِسْبَتُها إِلى الشَّيْءِ الواحِدِ—واحِدَةٌ. فَقَدْرُ ج ز عِنْدَ ه د كَقَدْرِ ا ج عِنْدَ د ه[1]. وَقَدْرُ ا ج عِنْدَ ه د كَقَدْرِ ب ج عِنْدَ ج ه. وَقَدْرُ ج ز عِنْدَ د ه كَقَدْرِ ج ح عِنْدَ ج د. فَقَدْرُ ب ج عِنْدَ ج ه كَقَدْرِ ح ج عِنْدَ ج د.

עוד נשים שיעור ב״ג אצל ג״ה כשיעור ח״ג אצל ג״ד. ועוד הצלעות המקיפות בשתי הזויות מתהפכות ביחס. והיה שעור ב״ג אצל ג״ה כשעור ח״ג אצל ג״ד[1]. ואומר ששטחי א״ג וג״ז שוים.

فَأَقولُ إِنَّ سَطْحَيْ ا ج و ج ز مُتَوازِيانِ مُتَساوِيانِ.

מופת זה: כי שיעור ב״ג אצל ג״ה כשיעור ח״ג אצל ג״ד. ושיעור ב״ג אצל ג״ה כשיעור א״ג אצל ד״ה. ושיעור ח״ג אצל ג״ד כשיעור ג״ז אצל ד״ה. ואם כן יהיו א״ג וג״ז יחסם אצל ד״ה אחד. אם כן הם שוים.

بُرْهانُهُ: أَنَّ قَدْرَ ب ج عِنْدَ ج ه كَقَدْرِ ح ج عِنْدَ د ج. فَقَدْرُ ب ج عِنْدَ ج ه كَقَدْرِ ا ج عِنْدَ د ه. وَقَدْرُ ح ج عِنْدَ (ج د)[2] كَقَدْرِ ج ز عِنْدَ ه د. ا ج و ج ز نِسْبَتُهُما إِلى د ه واحِدَةٌ فَهُما مُتَساوِيانِ.

[!]

وَذلِكَ ما أَرَدْنا بَيانَهُ.

---

[1] ועוד ... ג״ד] ⟨⟨*⟩⟩

[1] د ه] +⟨⟨إلى سطح د ه⟩⟩ [2] ج د] د

بُرهانُه: أَنَّا نُنْزِلُ[1] ب ج مُتَّصِل ب ج ه على اسْتِقَامَةٍ. فَيصيرُ د ج مُتَّصِلًا ب ج ح على اسْتِقَامَةٍ. ونُتِمُّ سَطْحَ د ه. ف ا ج مثل ج ز. فَنِسْبَتُهُما إلى د ه واحِدَةٌ.^ فنِسْبَةَ ا ج إلى د ه كنِسْبَةِ ب ج إلى ج ه. فنِسْبَةَ سَطْحِ^ ج ز إلى د ه[2] كنِسْبَةِ خَطِّ ب ج إلى خطّ ج ه. ولكِنّ[3] ج ز إلى د ه كنِسْبَةِ ح ج إلى ج د. فنِسْبَة ب ج إلى ج ه كنِسْبَةِ ح ج إلى ج د.

המופת: אנחנו נניח ב״ג מדובק בג״ה על יושר. ויהיה ד״ג דבוק בג״ח על יושר[1]. ונשלים[2] שטח ד״ה. הנה א״ג[3] כמו ג״ז. אם כן יחסם אל ד״ה אחד. אם כן יחס א״ג אל ד״ה כיחס ז״ג אל ד״ה. ויחס א״ג אל ד״ה כיחס[4] ב״ג אל ג״ה. אם כן יחס ג״ז אל ד״ה כיחס ב״ג אל ג״ה. אבל יחס ג״ז אל ד״ה כיחס ח״ג אל ג״ד. אם כן[5] יחס ב״ג אל ג״ה כיחס ח״ג אל ג״ד.

ثُمَّ نَجْعَلُ نِسْبَةَ ح ج إلى ج د كنِسْبَة ب ج إلى ج ه. فَأَقولُ إنَّ سَطْحَيْ ا ج ج ز مُتَساوِيانِ.

עוד נשים יחס ח״ג אל ג״ד כיחס ב״ג אל ג״ה. הנה אומר כי שטחי[6] א״ג וג״ז שוים.

بُرهانُه: أَنَّ التَّدْبيرَ واحِدٌ. فنِسْبَةَ ب ج إلى ج ه كنِسْبَةَ ح ج إلى ج د. ونِسْبَة ب ج إلى ج ه كنِسْبَة ا ج إلى د ه. فنِسْبَة سَطْحِ ا ج إلى سَطْحِ د ه كنِسْبَةِ[4] خَطِّ ح ج إلى ج د، ونِسْبَة خطّ ح ج إلى خطّ ج د[5] كنِسْبَة سطح ج ز إلى د ه. فنِسْبَةَ ا ج و ج ز إلى د ه واحِدَةٌ. فَهُما مُتَساوِيان.

מופתו: כי ההנהגה אחת. הנה יחס ב״ג אל ג״ה כיחס ח״ג אל ג״ד. ויחס ב״ג אל ג״ה כיחס א״ג אל ד״ה. אם כן יחס א״ג[7] אל ד״ה כיחס ח״ג אל ג״ד[8]. ויחס ח״ג[9] אל ג״ד כיחס ג״ז אל ד״ה[10]. אם כן יחס א״ג וג״ז אל ד״ה אחד. אם כן שניהם שוים[11].

وذَلِكَ ما أَرَدْنا أَنْ نُبَيِّنَ.

ונשלם ביאורו.

---

[1]ننزل] MPR2 [2]كنسبة ... د ه] M [3]ولكن] M [4]سطح ... كنسبة] R2 [5]ونسبة ... ج د] M

[1]ויהיה ... יושר] ליתא בוטפק [2]ונשלים] +יושר ס [3]א״ג] שטח+ ס [4]אל ד״ה ויחס ... כיחס] ליתא ב [5]כן] ליתא ב [6]שטחי] ⟨שני⟩+ ק שני בוטספ [7]א״ג] +ח״ג ב ⟨*⟩ ס [8]ח״ג ... ג״ד] ד״ג אל ה״ג ס [9]ח״ג] ג״ז ס [10]ד״ה] ג״ד ס [11]שניהם שוים] +* ט

[Proposition VI.15]

[B14]

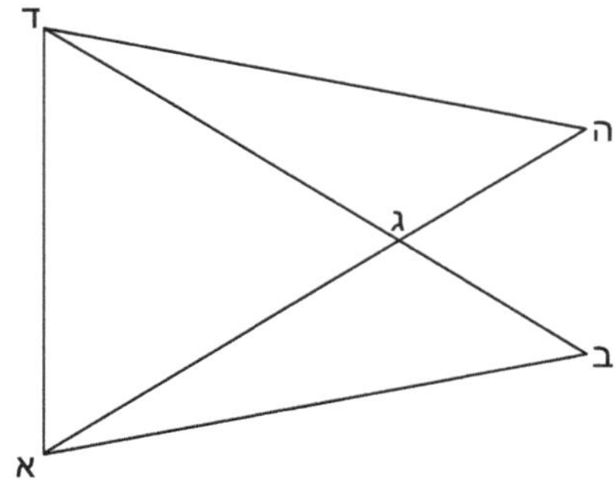

كُلُّ مُثَلَّثَيْنِ يَكُونُ فِي أَحَدِهِمَا زَاوِيَةٌ مِثْلَ زَاوِيَةِ الآخَرِ، وَيَكُونُ المُثَلَّثَانِ مُتَسَاوِيَيْنِ، فَإِنَّ الأَضْلاعَ الَّتِي تُحِيطُ بِالزَّاوِيَتَيْنِ المُتَسَاوِيَتَيْنِ مُتَكَافِئَةٌ. وَإِنْ كَانَتِ الأَضْلاعُ الَّتِي تُحِيطُ بِالزَّاوِيَتَيْنِ مُتَكَافِئَةً، فَإِنَّ المُثَلَّثَيْنِ مُتَسَاوِيَةٌ.

כשתהיין שתי זויות משני משולשים שוות, והיו המשולשים שוים, תהיין הצלעות המקיפות בשתי הזויות מתהפכות ביחס. ואם היו הצלעות המקיפות בשתי הזויות מתהפכות ביחס[1], יהיו המשולשים שוים.

مِثالُهُ: أَنَّ مُثَلَّثَيْ ا ب ج و ج د ه، فِي أَحَدِهِمَا زَاوِيَةٌ مِثْلَ زَاوِيَةٍ فِي الآخَرِ، وَهُمَا اللَّتَيْنِ عَلَيْهِمَا ج. وَالمُثَلَّثَانِ مُتَسَاوِيَانِ. فَأَقُولُ إِنَّ الأَضْلاعَ الَّتِي تُحِيطُ بِالزَّاوِيَتَيْنِ المُتَسَاوِيَتَيْنِ مُتَكَافِئَةٌ، قَدْرُ ب ج عِنْدَ ج د كَقَدْرِ ه ج عِنْدَ ج ا.

המשל בזה: כי משולשי אב״ג וגד״ה שוים. ובאחד מהם זוית שוה לזוית כאשר באחר, והן אשר עליהן ג׳. ואומר כי הצלעות המקיפות בזויות השוות מתהפכות, ויהיה שעור ב״ג אצל ג״ד כשעור ה״ג אצל ג״א.

[1] ואם ... ביחס] ⟨⟨*⟩⟩

إذا تَساوَتْ زاوِيَتانِ مِنْ مُثَلَّثَيْنِ، وَكانا مُتَساوِيَيْنِ، فَالأَضْلاعُ المُحِيطَةُ بالزّاوِيَتَيْنِ مُتَكافِئَةٌ. وَإِنْ كانَتِ الأَضْلاعُ المُحِيطَةُ بالزّاوِيَتَيْنِ مُتَكافِئَةً، فالمُثَلَّثانِ مُتَساوِيانِ.

כאשר השתוו שתי זויות משני משולשים, והיו שוים, הנה הצלעות המקיפות בשתי הזויות מספיקות[1]. ואם היו הצלעות המקיפות בשתי[2] הזויות מספיקות, הנה שני המשולשים שוים.

مِثالُهُ: أَنَّ زاوِيَتَيْ ج مِنْ مُثَلَّثَيْ ا ب ج و ج د ه مُتَساوِيَتانِ. وَالمُثَلَّثانِ مُتَساوِيانِ. فَأَقُولُ إِنَّ الأَضْلاعَ المُحِيطَةَ بِزاوِيَتَيْ ج مِنَ المُثَلَّثَيْنِ مُتَكافِئَةٌ—نِسْبَةُ ب ج إلى ج د كَنِسْبَةِ خَطِّ ه ج إلى ج ا.

המשל בו: כי שתי זויות ג׳ משני משולשי אב״ג ודה״ג[3] שוות. ושני המשולשים[4] שוים. הנה אומר כי הצלעות המקיפות בשתי זויות ג׳ משני המשולשים מספיקות—יחס ב״ג אל ג״ד כיחס ה״ג אל א״ג.

שרטוט—**ו**: נוסף ה״ב; **ופק**: $; **ס**: חסר א״ד ונוסף ה״ב [1]מספיקות] +~~הנה~~ **ב** ~~והיו שוים הנה הצלעות המקיפות~~+ **פ** [2]בשתי] בשני **ו** [3]ודה״ג] זד״ה‡ **ב** [4]המשולשים] משולשים **בס**

מופת זה: שנדביק א׳ בד׳. ומשלש אב״ג שוה למשלש גד״ה[1]. ויחס השוים אל דבר אחד הוא אחד. ואם כן שעור משלש אב״ג אצל משלש אג״ד כשעור משולש דה״ג אצל משלש אג״ד. ואמנם שעור משלש אב״ג אצל משלש אג״ד כשיעור ב״ג אל ג״ד. ושיעור משלש גד״ה אצל משלש גד״א כשיעור ה״ג אל ג״א. אם כן שיעור ב״ג אצל ג״ד כשיעור ה״ג אצל ג״א.

بُرْهانُهُ: أَنَّا نَصِلُ ا د بِخَطّ ا د. فَمُثَلَّثُ ا ب ج مِثْلُ مُثَلَّثِ د ه ج. وَنِسْبَةُ الأَشْياءِ المُتَساوِيَةِ إِلى الشَّيْءِ الواحِدِ—واحِدَةٌ. فَقَدْرُ ا ب ج عِنْدَ مُثَلَّثِ ا ج د كَقَدْرِ مُثَلَّثِ ج د ه عِنْدَ مُثَلَّثِ ج د ا. وَلٰكِنَّ قَدْرَ مُثَلَّثِ ا ب ج عِنْدَ مُثَلَّثِ ا ج د كَقَدْرِ ب ج عِنْدَ ج د. وَقَدْرُ مُثَلَّثِ ج د ه عِنْدَ مُثَلَّثِ ج د ا كَقَدْرِ ه ج عِنْدَ ج ا. فَقَدْرُ ب ج عِنْدَ ج د كَقَدْرِ ه ج عِنْدَ ج ا.

עוד נשים שיעור ב״ג אצל ג״ד כשעור ה״ג אצל ג״א. ועוד הצלעות המקיפות בשתי הזויות מתהפכות ביחס. ואומר שמשולשי (אב״ג)[2] וגד״ה שוים.

فَأَقولُ إِنَّ مُثَلَّثَيْ ا ب ج و ج د ه مُتَساوِيانِ.

המופת: כי שעור ב״ג אצל ג״ד כשעור ה״ג אצל ג״א. ושיעור ב״ג אצל ג״ד כשיעור משלש אב״ג אצל משלש אג״ד. ושיעור ה״ג אצל ג״א כשיעור משלש גד״ה אצל משלש אג״ד. ואם כן יהיה שיעור משלש אב״ג אצל משלש אג״ד כשיעור משלש גד״ה אל משלש אג״ד. אם כן המשולש אב״ג ומשלש גד״ה יחסם אל משלש אג״ד אחד. ואם כן הם שוים.

بُرْهانُهُ: أَنَّ قَدْرَ ب ج عِنْدَ ج د كَقَدْرِ ه ج عِنْدَ ج ا. وَقَدْرُ ب ج عِنْدَ ج د كَقَدْرِ مُثَلَّثِ ا ب ج عِنْدَ مُثَلَّثِ ا ج د. وَقَدْرُ ه ج عِنْدَ ج ا كَقَدْرِ مُثَلَّثِ ه ج د عِنْدَ مُثَلَّثِ ج ا د. فَقَدْرُ مُثَلَّثِ ا ب ج عِنْدَ ا ج د بِقَدْرِ مُثَلَّثِ ج د ه عِنْدَ مُثَلَّثِ ا ج د. فَمُثَلَّثا ا ب ج و ج د ه نِسْبَتُهُما إِلى مُثَلَّثِ ا ج د واحِدَةٌ، فَهُما مُتَساوِيانِ.

וזה מה שרצינו לבאר.

وَذٰلِكَ ما أَرَدْنا بَيانَهُ.

[1] גד״ה] ⟨⟨*⟩⟩ [2] אב״ג] אד״ג

بُرهانُه: أنَّا نُنْزِلُ ب ج مُتَّصِلاً[1] ب ج د على اسْتِقامَةٍ. فَنُصَيِّرُ ا ج مُتَّصِلاً ب ج ه على اسْتِقامَةٍ^. وَنُخْرِجُ ا د. فَمُثَلَّثا ا ب ج و د ه ج مُتَساوِيان. فَنِسْبَتُهُما إلى مُثَلَّثِ ا ج د واحِدَةٌ. لكِنَّ نِسْبَةَ مُثَلَّثِ ا ب ج إلى مُثَلَّثِ ا ج د كَنِسْبَةِ ب ج إلى ج د. وَنِسْبَةَ مُثَلَّث ج د ه إلى مُثَلَّث ج د ا كَنِسْبَةِ ه ج إلى ج ا. فَنِسْبَةُ ب ج إلى ج د كَنِسْبَةِ ه ج إلى ج ا.

המופת: אנחנו נניח ג״ב מדובק בג״ד על יושר. ונוציא א״ד. הנה שני משולשי אב״ג ודה״ג[1] שוים. ויחסם[2] אל משולש אג״ד אחד. אבל יחס משולש אב״ג[3] אל משולש אג״ד כיחס ב״ג אל ג״ד. ויחס משולש גד״ה אל משולש גד״א כיחס ה״ג[4] אל א״ג. אם כן יחס ב״ג אל ג״ד כיחס ה״ג אל א״ג.

ثُمَّ نَجْعَلُ نِسْبَةَ ب ج إلى ج د كَنِسْبَةِ ه ج إلى ج ا[2]. فَأَقولُ إنَّ مُثَلَّثَيْ ا ب ج و ج د ه مُتَساوِيان.

עוד נשים יחס[5] ב״ג אל ג״ד כיחס ה״ג אל א״ג. הנה אומר כי שני משולשי אב״ג וגד״ה שוים.

بُرهانُه: إنَّ التَّدبيرَ واحِدٌ. فَنِسْبَةُ ب ج إلى ج د كَنِسْبَةِ ه ج إلى ج ا. وَنِسْبَةُ ب ج إلى ج د كَنِسْبَةِ مُثَلَّث ا ب ج إلى مُثَلَّث ا ج د. فَنِسْبَةُ ه ج إلى ج ا كَنِسْبَةِ مُثَلَّث ا ب ج إلى مُثَلَّث ا ج د. وَنِسْبَةُ^ ه ج إلى ج ا[3] كَنِسْبَةِ مُثَلَّث ج د ه إلى مُثَلَّث ا ج د. فَمُثَلَّثا ا ب ج و ج د ه نِسْبَتُهُما إلى مُثَلَّثِ ا ج د واحِدَةٌ. فَهُما مُتَساوِيان[4].

מופתו: כי ההנהגה אחת. הנה[6] יחס ב״ג אל ג״ד כיחס ה״ג אל א״ג. ויחס ב״ג אל ג״ד כיחס משולש אב״ג אל משולש אג״ד. אם כן[7] יחס ה״ג אל א״ג כיחס משולש אב״ג אל משולש אג״ד. אבל יחס ה״ג אל א״ג כיחס משולש[8] הג״ד אל משולש אג״ד. אם כן שני משולשי אב״ג וגד״ה יחס[9] שניהם אל משולש אג״ד אחד[10]. הנה שניהם אם כן[11] שוים.

وذَلِكَ ما أَرَدْنا أَنْ نُبيِّنَ.

וזה מה שרצינו לבאר[12].

---

[1] نزل … متصلا] **M** [2] ثم … ج ا] **MPR2T2** [3] ونسبة … ج ا] **R2** [4] نسبتهما … متساويان] **MPR2**

[1] ודה״ג] ודה״ז **ב** [2] ויחסם] ויחס **ב** [3] אב״ג] ~~אג״ד~~+ **פ** [4] ה״ג] ד״ג **ק** [5] יחס] *+ **ס** [6] הנה] +* **ב** [7] אם כן] אבל **ס** [8] אב״ג … משולש] ליתא **בס** [9] יחס] יחד **פ** [10] אחד] ⟨*⟩ **ז** [11] אם כן] ליתא **ט** [12] וזה … לבאר] ונשלם ביאורו **בופק** ליתא **ס**

[Proposition VI.16]

[B15]

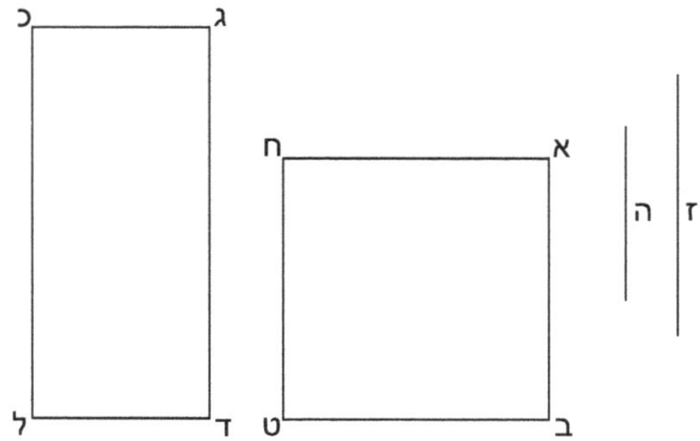

כל ארבעה קוים מתיחסים, יהיה העולה ממערכת הראשון על האחרון שוה למערכת אחד מן הנשארים באחר. ואם היה מערכת הראשון על האחרון שוה למערכת אחד הנשארים על האחר, יהיו הקוים מתיחסים.

إِذا كانَتْ أَرْبَعَةُ خُطوطٍ مُتَناسِبَةً، فَإِنَّ الَّذي يَكونُ مِنْ ضَرْبِ الخَطِّ الأَوَّلِ في الخَطِّ الآخَرِ مِثْلَ الَّذي يَكونُ مِنْ ضَرْبِ أَحَدِ الخَطَّيْنِ الباقِيَيْنِ في الآخَرِ. وَإِنْ كانَ الَّذي يَكونُ مِنْ ضَرْبِ الأَوَّلِ في الآخَرِ مِثْلَ الَّذي يَكونُ مِنْ ضَرْبِ أَحَدِ الخَطَّيْنِ الباقِيَيْنِ في الآخَرِ، فَإِنَّ الخُطوطَ الأَرْبَعَةَ مُتَناسِبَةٌ.

המשל בזה: כי הקוים[1] הארבעה המתיחסים עליהם א״ב וג״ד וה׳ וז׳, שיעור א״ב אל ג״ד כשעור ה׳ אל ז׳. ואומר[2] כי העולה ממערכת א״ב בז׳ שוה לעולה ממערכת ג״ד בה׳.

مِثالُهُ: أَنَّ خُطوطَ اب وجد وه وز الأَرْبَعَةَ مُتَناسِبَةٌ، قَدْرُ با عِنْدَ جد كَقَدْرِ ه عِنْدَ ز. فَأَقولُ إِنَّ الَّذي يَكونُ مِنْ ضَرْبِ اب في ز مِثْلَ الَّذي يَكونُ مِنْ ضَرْبِ جد في ه.

---

[1]כי הקוים] ש⟨⟨כי⟩⟩הקוים [2]ואומר] ⟨⟨*⟩⟩

كُلُّ أَرْبَعَةِ خُطوطٍ مُتَنَاسِبَةٍ، فضَرْبُ الأَوَّلِ في الأَخيرِ مِثْلَ ضَرْبٍ أَحَدِ الباقِيَيْنِ في الآخِرِ^. وَإِنْ كانَ ضَرْبُ الأَوَّلِ في الآخِرِ^[1] مِثْلَ ضَرْبِ أَحَدِ الباقِيَيْنِ في الآخِرِ، فالخُطوط مُتَنَاسِبَةٌ.

כל ארבעה קוים מתיחסים, הנה הכאת[1] הראשון באחרון כמו הכאת אחד מן הנשארים באחרון. ואם היה הכאת הראשון באחרון כמו הכאת אחד מן הנשארים באחר, הנה הקוים מתיחסים.

مِثَالُهُ: أَنَّ خُطوطَ اَبَ جَ دَ هَ زَ الأَرْبَعَةَ مُتَنَاسِبَةٌ—نِسْبَةُ اَبَ إلى جَ دَ كَنِسْبَةِ هَ إلى زَ. فَأَقولُ إِنَّ ضَرْبَ اَبَ الأَوَّلِ في زَ الآخر مثل ضَرْبِ جَ دَ الباقي^ في هَ الثَّالِثِ.

המשל בו: כי קוי א״ב וג״ד וה׳ ז׳ הארבעה מתיחסים—יחס א״ב אל ג״ד כיחס ה׳ אל ז׳. הנה אומר כי הכאת א״ב הראשון בז׳ הרביעי כהכאת ג״ד השני בה׳ השלישי.

[1]الاخر ... الاخر] נראה שיש כאן טעות והמשפט צריך להתחיל ולהסתיים במילה الاخير

שרטוט—ו: ה׳ משמאל לז׳; **בוספק**: הקווים משמאל למלבנים [1]הכאת] ליתא **ט**

מופת זה: שנוציא מנקדת א׳ וג׳ מצדי א״ב וג״ד שני קוים על זויות נצבות, והם א״ח וג״כ. ונשים קו ג״כ שוה לקו ה׳, וא״ח שוה לז׳. ונשלים שטח א״ט ושטח ג״ל. וכבר היה שיעור א״ב אצל ג״ד כשעור ה׳ אצל ז׳. וה׳ שוה לג״כ. וז׳ שוה לא״ח. יהיה שיעור א״ב אל ג״ד כשיעור ג״כ אל א״ח. אם כן צלעות א״ט וג״ל המקיפות בזויות השוות—מתהפכות ביחס. ובהכרח[1] א״ט שוה לג״ל. וא״ט הוא העולה ממערכת א״ב בז׳, לפי שא״ח שוה לז׳. וג״ל הוא העולה ממערכת ג״ד בה׳, לפי שג״כ הוא שוה לה׳. והתבאר כי העולה ממערכת א״ב בז׳ שוה לעולה ממערכת ג״ד בה׳.

بُرْهانُهُ: أَنَّا نُخْرِجُ مِنْ نُقْطَتَيْ ا وَج بَيْنَ خَطَّيْ ا ب[1] وَج د خَطَّيْ ا ح وَج ك عَلَى زاوِيَتَيْنِ قائِمَتَيْنِ. وَنَجْعَلُ ج ك مِثْلَ ه، وَا ح مِثْلَ ز. وَيُتَمِّمُ سَطْحَ [!] ا ط ج ل. فَقَدْرُ ا ب عِنْدَ ج د كَقَدْرِ ه عِنْدَ ز. ه مِثْلُ ج ك. (و ز)[2] مِثْلُ ا ح. فَقَدْرُ ا ب عِنْدَ ج د كَقَدْرِ ج ك عِنْدَ ا ح. فَأَضْلاعُ سَطْحَيْ ا ط وَج ل الَّتي تُحيطُ بِالزّاوِيَتَيْنِ المُتَساوِيَتَيْنِ مُتَكافِئَةٌ. [فَيَكونُ][3] ا ط مِثْلَ ج ل. وَا ط هُوَ الَّذي كانَ مِنْ ضَرْبِ ا ب في ز، لِأَنَّ ا ح مِثْلُ ز. وَج ل هُوَ الَّذي كانَ مِنْ ضَرْبِ ج د في ه، لِأَنَّ ج ك مِثْلُ ه. فَالَّذي يَكونُ مِنْ ضَرْبِ ا ب في ز هُوَ مِثْلُ الَّذي يَكونُ مِنْ ضَرْبِ ج د في ه.

עוד נשים העולה ממערכת א״ב בז׳ שוה למערכת ג״ד בה׳. ואומר ששעור א״ב אל ג״ד כשיעור ה׳ אל ז׳.

وَنَجْعَلُ أَيضًا الَّذي يَكونُ مِنْ ضَرْبِ ا ب في ز مِثْلَ الَّذي يَكونُ مِنْ ضَرْبِ ج د في ه. فَأَقولُ إِنَّ قَدْرَ ا ب عِنْدَ ج د كَقَدْرِ ه عِنْدَ ز.

[1] ובהכרח] +〈〈~~אם כן~~〉〉

[1] اب] ه+ [2] ون] دز [3] فيكون] 〈*〉

بُرهانُهُ: أنَّا نُخْرِجُ مِنْ نُقْطَتَيْ اج خَطَّيْ اح ج ك مثل زاوِيَتَيْنِ قائِمَتَيْنِ. ونَجْعَلُ ج ك مثل ه، و اح مثل ز. وَنُتَمِّمُ سَطْحَيْ اط ج ل. فَنِسْبَةُ اب إلى ج د كَنِسْبَةِ ه إلى ز. و ه مثل ج ك. و ز مثل اح. فَنِسْبَةُ اب إلى ج د كَنِسْبَةِ ج ك إلى اح^. فأَضلاع سَطْحَيْ اط ج ل المُحِيطَةُ بالزَّاوِيَتَيْنِ المُتَساوِيَتَيْنِ—مُتَكافِئَةٌ. ف اط مثل ج ل. و اط هو ضَرْبُ اب في ز، و ج ل هو ضَرْبِ ج د في ه. فضَرْب[1] اب في ز مثل ضَرْب ج د في ه.

מופתו: אנחנו נוציא משתי נקודות א׳ ג׳ שני קוי א״ח וג״כ על שתי זויות נצבות. ונשים ג״כ כמו ה׳, וא״ח כמו ז׳. ונשלים שני שטחי א״ט וג״ל. הנה יחס א״ב אל ג״ד כיחס ה׳[1] אל ז׳. וה׳ כמו ג״כ. וז׳ כמו א״ח. אם כן צלעות שטח א״ט וג״ל[2] המקיפות בשתי הזויות[3] השוות—מספיקות. אם כן א״ט כמו ג״ל. וא״ט הוא הכאת א״ב בז׳. וג״ל הוא הכאת ג״ד בה׳. אם כן הכאת א״ב בז׳ כמו הכאת[4] ג״ד בה׳[5].

ثُمَّ لِيَكُنْ ضَرْبُ اب في ز مثل ضَرْبِ ج د في ه. فَأَقولُ إنَّ نِسْبَةَ اب إلى ج د كَنِسْبَةِ ه إلى ز.

עוד יהיה הכאת א״ב בז׳ כמו הכאת ג״ד בה׳. הנה אומר כי יחס א״ב אל ג״ד כיחס ה׳ אל ז׳.

[1] فضرب] T2

[1] ה׳] ליתא **ב** [2] הנה ... וג״ל] ליתא **ז** [3] הזויות] זויות **בוסף** ⟨ה⟩זויות **ק** [4] א״ב ... הכאת] ליתא **ב** [5] בה׳] +~~הנה אומר כי יחס א״ב אל ג״ד כיחס ה׳ אל ז׳~~ **ט**

כי הנהגתם אחת. והעולה ממערכת א״ב בז׳ שוה למערכת ג״ד בה׳. וז׳ שוה לא״ח, וה׳ שוה לג״כ. ויהיה העולה ממערכת א״ב בא״ח שוה למערכת ג״ד כג״כ. והעולה ממערכת א״ב בא״ח הוא שטח א״ט. והעולה ממערכת ג״ד בג״כ הוא שטח ג״ל. ואם כן א״ט הוא שוה לג״ל. ואם כן[1] צלעות א״ט וג״ל המקיפות בזויות השוות מהם—מתהפכות ביחס. ויהיה שיעור א״ב אצל ג״ד כשיעור ג״כ אל א״ח. וג״כ שוה לה׳ וא״ח[2] שוה לז׳. התבאר אם כן כי שיעור א״ב אל ג״ד כשיעור ה׳ אל ז׳.

וזה מה שרצינו לבאר.

بُرْهانُهُ: أَنَّ تَدْبيرَهُما واحِدٌ. فالَّذي يَكونُ مِنْ ضَرْبِ اب في زَ مِثْلِ الَّذي يَكونُ مِنْ ضَرْبِ دج في ه. و زَ مِثْلُ اح. و ه مِثْلُ جك. فَالَّذي يَكونُ مِنْ ضَرْبِ ب ا في اح هو سَطْحُ اط. وَالَّذي يَكونُ مِنْ ضَرْبِ ج د في جك هُوَ سَطْحُ جل. فَأَضْلاعُ (اط)[1] وجل الَّتي تُحيطُ بالزّاوِيَتَيْنِ المُتَساوِيَتَيْنِ مُتَكافِئَةٌ. فَقَدْرُ اب عِنْدَ ج د كَقَدْرِ جك عِنْدَ اح. وجك مِثْل ه. واح مِثْل زَ. فَقَدْرُ اب عِنْدَ ج د كَقَدْرِ ه عِنْدَ زَ.

وَذَلِكَ ما أَرَدْنا بَيانَهُ.

---

[1] ואם כן] ו⟨⟨אם כן⟩⟩ [2] וא״ח] ~~וא״ב~~ ⟨⟨*⟩⟩

[1] اط] اك

والتَّدبير واحِدٌ. فضَرْب اب في زَ مثل ضَرْب ج د في ه. و ز مثل اح. و ه مثل ج ك. فضَرْب اب في اح مثل ضَرْب ج د في ج ك. وضَرْب اب في اح هو اط. وضَرْب ج د في ج ك هو ج ل. ف اط مثل ج ل. فأَضلاع اط ج ل المُحِيطَةُ بِالزَّاوِيَتَيْنِ المُتَساوِيَتَيْنِ مُتَكافِئَةٌ— نِسْبَةَ اب إلى ج د كَنِسْبَةِ ج ك إلى اح. وج ك مثل ه. و اح مثل ز. فنِسْبَةَ اب إلى ج د كَنِسْبَة ه إلى ز.

וההנהגה אחת. הנה הכאת א״ב בז׳ כמו הכאת ג״ד בה׳. וז׳ כמו א״ח. וה׳ כמו ג״כ. אם כן הכאת א״ב בא״ח כהכאת ג״ד בג״כ. והכאת א״ב בא״ח הוא א״ט. והכאת ג״ד בג״כ הוא ג״ל. אם כן[1] א״ט כמו ג״ל. אם כן צלעות א״ט וג״ל המקיפות בשתי הזויות השוות מספיקות—יחס א״ב אל ג״ד כיחס ג״כ אל א״ח. וג״כ כמו ה׳. וא״ח כמו ז׳. אם כן יחס א״ב אל ג״ד כיחס ה׳ אל ז׳.

وذَلِكَ ما أَرَدْنا أَنْ نُبَيِّنَ.

וזה מה שרצינו לבאר[2].

[1]כן] ליתא **ב** [2]לבאר] ליתא **ס**

[Proposition VI.17]

[B16]

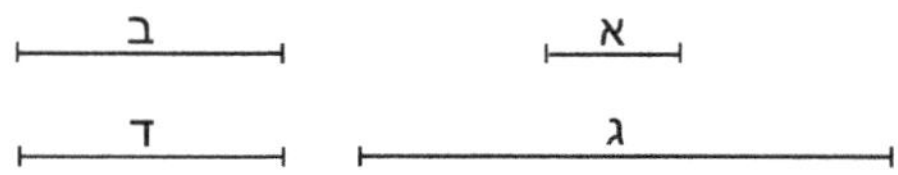

כשיהיו שלשה קוים מתיחסים, יהיה העולה ממערכת הראשון על האחרון שוה לעולה ממערכת האמצעי בעצמו. ואם היה מערכת הראשון על האחרון כמערכת האמצעי בעצמו, יהיו השלשה קוים מתיחסים.

إِذا كانَتْ (ثَلاثَةُ)[1] خُطوط مُناسِبَةً، فَإِنَّ الَّذي يَكونُ مِنْ ضَرْبِ الخَطِّ الأَوَّلِ في الخَطِّ الآخَرِ مِثْل الَّذي يَكونُ مِنْ ضَرْبِ الباقي في نَفْسِه. وَإِنْ كانَ الَّذي يَكونُ مِنْ ضَرْبِ الخَطِّ الأَوَّلِ في الخَطِّ الآخَرِ مِثْل الَّذي يَكونُ مِنْ ضَرْبِ الباقي في نَفْسِهِ، فَإِنَّ الخُطوطَ مُتَناسِبَةٌ.

המשל בזה: שקו א׳ וקו ב׳ וקו ג׳ מתיחסים—שיעור א׳ אצל ב׳ כשיעור ב׳ אצל ג׳. ואמר כי העולה ממערכת א׳ על [!] ג׳ כעולה ממערכת ב׳ בעצמו.

مِثالُهُ: أَنَّ خُطوطَ آ و بَ و جَ مُتَناسِبَةٌ، قَدْرُ آ عِنْدَ بَ كَقَدْرِ بَ عِنْدَ جَ. فَأَقولُ إِنَّ الَّذي مِنْ ضَرْبِ آ في جَ مِثْل الَّذي يَكونُ مِنْ ضَرْبِ بَ في نَفْسِهِ.

---

[1] ثلاثة] ثلثة

إذا كَانَتْ ثَلاثَةُ خُطوطٍ مُتَنَاسِبَةً، فضَرْبُ الأَوَّلِ في الأخير مثل ضَرْبِ الأَوْسَط في مِثْلِهِ. وإنْ كانَ ضَرْبُ الأَوَّلِ في الأخير مثل ضَرْبِ الأَوْسَطِ في مِثْلِهِ^، فالخُطوط مُتَنَاسِبَة.

مِثَالُهُ: أَنَّ خُطوطَ آ بَ جَ الثَّلاثَةَ مُتَناسِبَةٌ^— نِسْبَةُ آ إلى بَ كنِسْبَةِ بَ إلى جَ. فَأَقولُ إنَّ ضَرْبَ آ في جَ مِثْلِ ضَرْبِ بَ في نَفْسِهِ[1].

[1]نفسه] CT1T3 مثله CopT2

כאשר היו שלשה קוים מתיחסים, הנה הכאת הראשון באחרון כמו הכאת האמצעי בדומה לו[1]. ואם היתה[2] הכאת[3] הראשון באחרון כמו הכאת האמצעי בנפשו, הנה הקוים מתיחסים.

המשל בו[4]: יחס א׳ אל ב׳ כיחס ב׳[5] אל ג׳. הנה אומר כי הכאת א׳ בג׳ כמו הכאת ב׳ בעצמו.

שרטוט—**בוספק**: א׳, ב׳, ד׳, ג׳ מונחים זה מעל זה; **ט**: ג׳ מעל ד׳ [1]בדומה לו] בנפשו **ז** [2]היתה] היה **ט** [3]הכאת] ליתא **ס** [4]בו] ליתא **וספ** כי **ט** [5]כיחס ב׳] ליתא **ב**

מופת זה: שנשים ד׳ שוה לב׳. ויהיה שיעור א׳ אצל ב׳ כשיעור ד׳ אצל ג׳. אם כן[1] העולה מא׳ על ג׳ שוה לעולה מב׳ על ד׳, והוא כמו ב׳ בעצמו.

بُرْهانُهُ: أَنَّا نَجْعَلُ د مِثْلَ ب. فَقَدْرُ ا عِنْدَ ب كَقَدْرِ ب عِنْدَ ج. و ب مِثْلُ د. فَقَدْرُ ا عِنْدَ ب كَقَدْرِ د عِنْدَ ج. فَالَّذي يَكونُ مِنْ ضَرْبِ ا في ج مِثْلُ الَّذي يَكونُ مِنْ ضَرْبِ ب في د. وَالَّذي يَكونُ مِنْ ضَرْبِ ب في د هُوَ مِثْلُ ب في نَفْسِه، لِأَنَّ ب مِثْلُ د. فَالَّذي يَكونُ مِنْ ضَرْبِ ا في ج مِثْلُ الَّذي يَكونُ مِنْ ضَرْبِ ب في نَفْسِه.

עוד נשים העולה מא׳ על ג׳ שוה לעולה מב׳ בעצמו. ואומר כי שעור א׳ אצל ב׳ כשעור ב׳ אצל ג׳, לפי שהנהגתם אחת.

وَنَجْعَلُ الَّذي يَكونُ مِنْ ضَرْبِ ا في ج مِثْلَ الَّذي يَكونُ مِنْ ضَرْبِ ب في نَفْسِه. فَأَقولُ إِنَّ القَدْرَ الَّذي ل ا عِنْدَ ب كَقَدْرِ ب عِنْدَ ج.

מופת זה: כי העולה מא׳ על ג׳ שוה לעולה מב׳ בעצמו. והעולה מב׳ בעצמו שוה לעולה מב׳ בד׳, לפי שב׳ שוה לד׳. והעולה מא׳ על ג׳ שוה לעולה מב׳ על ד׳. אם כן שיעור א׳ אצל ב׳ כשעור ד׳ אצל ג׳. וד׳ שוה לב׳. אם כן התבאר כי שיעור א׳ אצל ב׳ כשיעור ב׳ אצל ג׳.

بُرْهانُهُ: أَنَّ تَدْبيرَهُما واحِدٌ. فَالَّذي يَكونُ مِنْ ضَرْبِ ا في ج مِثْلَ الَّذي يَكونُ مِنْ ضَرْبِ ب في نَفْسِه. والَّذي يَكونُ مِنْ ضَرْبِ ب في نَفْسِه مِثْل الَّذي يَكونُ مِنْ ضَرْبِ ب في د، لِأَنَّ ب مِثْلُ د. وَالَّذي يَكونُ مِنْ ضَرْبِ ا في ج مِثْلَ الَّذي يَكونُ مِنْ ضَرْبِ ب في د. فَقَدْرُ ا عِنْدَ ب كَقَدْرِ د عِنْدَ ج. و د مِثْلُ ب فَقَدْرُ ا عِنْدَ ب كَقَدْرِ ب عِنْدَ ج.

וזה מה שרצינו לבאר.

وَذٰلِكَ ما أَرَدْنا بَيانَهُ.

---

[1] אם כן] ⟨⟨*⟩⟩

بُرهانُهُ: أَنَّا نَجْعَلُ د مثل ب. فنِسْبَةَ ا إلى ب كنِسبَةِ د إلى ج^. فضَرْب ا في ج مثل ضَرْب ب في د. و د مثل ب. فضَرْب ا في ج مثل ضَرْب ب في نَفْسِهِ.

המופת: אנחנו נשים ד׳ כמו ב׳. הנה[1] יחס א׳ אל ב׳ כיחס ב׳ אל ג׳. וה״ב[2] כמו ה״ד[3]. אם כן יחס א׳ אל ב׳ כיחס ד׳ אל ג׳. אם כן הכאת א׳ בג׳ כהכאת ב׳ בד׳, וד׳ כמו ב׳. אם כן הכאת א׳ בג׳ כמו הכאת[4] ב׳[5] בעצמו.

ثُمَّ نَجْعَلُ ضَرْبَ ا في ج مثل^ ب في نَفْسِهِ. فَأَقولُ إنَّ نِسْبَةَ ا إلى ب كنِسْبَةِ ب إلى ج.

עוד נשים הכאת[6] א׳ בג׳ כהכאת ב׳[7] בעצמו. הנה אומר כי יחס א׳ אל ב׳ כיחס ב׳ אל ג׳.

والتَّدبيرُ واحِدٌ. فضَرْبُ ا في ج مثل ضَرْبِ ب في مِثْلِهِ. و ب مثل د. فضَرْب ا في ج مثل ضَرْبِ ب^ في د. فنِسْبَةُ ا إلى ب كنِسْبَةِ د إلى ج. و د مثل ب. فنِسْبَةُ ا إلى ب كنِسْبَةِ ب إلى ج.

וההנהגה[8] אחת. אם כן הכאת א׳ בג׳ כמו הכאת ב׳ בד׳. אם כן יחס א׳ אל ב׳ כיחס ד׳ אל ג׳. וד׳ כמו ב׳. אם כן יחס א׳ אל ב׳[9] כיחס ב׳ אל ג׳.

وذَلِكَ ما أَرَدْنا أَنْ نُبَيِّنَ.

וזה מה שרצינו לבאר.

---

[1]הנה] +אם כן **ק** [2]וה״ב] וב׳ **טק** [3]ה״ד] ד׳ **טק** [4]הכאת] +א׳ בג׳ הכאת **ב** [5]בד׳ ... ב׳] ליתא **ז** [6]הכאת] ליתא **ס** [7]א׳ ... ב׳] ליתא **ב** [8]וההנהגה] וההנהגת **ב** [9]כיחס ... ב׳] ⟨*⟩ **ק**

[Proposition VI.18]

[H19, B17]

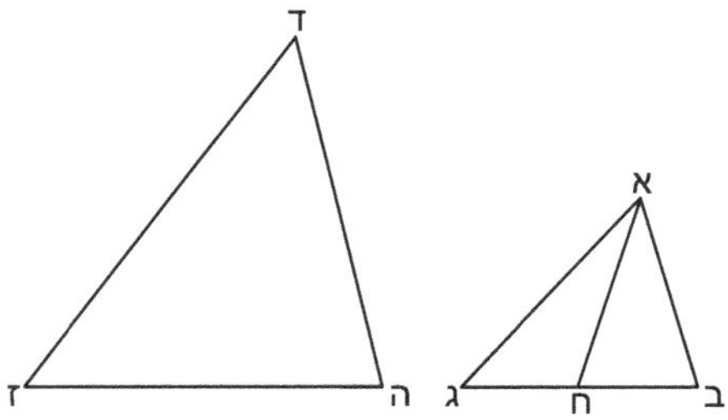

כל שני משולשים מתדמים, יהיה יחס אחד מהם אל האחר כיחס צלעו אל צלעו הדומה לו, נשנה בהשנות.

كُلُّ مُثَلَّثَيْنِ يَكونانِ مُتَشابِهَيْنِ فَإِنَّ نِسْبَةَ أَحَدِهِما إِلى الآخَرِ كَنِسْبَةِ أَحَدِ الأَضْلاعِ المَنْسوبِ إِلى نَظيرِهِ مِنْ أَضْلاعِ المَنْسوبِ إِلَيْهِ، مُثَنّاةً بِالتَّكْريرِ.

והנה התבאר לך הנה כי כל שלשה קוים מתיחסים, יהיה יחס הראשון אל השלישי כיחס השטח שיהיה על הראשון אל השטח שיהיה על השני כשידמהו, ויהיה על מצבו. והוא בסוף הצורה.

המשל בזה: שנצייר שני משלשים מתדמים, עליהם אב״ג ודה״ז. ואומר כי יחס משלש אב״ג אל משלש הד״ז כיחס צלע ב״ג אל צלע ה״ז הדומה לו, נשנה בהשנות.

مِثالُهُ: مُثَلَّثا ا ب ج و د ه ز المُتَشابِهانِ. فَأَقولُ إِنَّ نِسْبَةَ مُثَلَّثِ ا ب ج إِلى مُثَلَّثِ د ه ز كَنِسْبَةِ ضِلْعِ ب ج إِلى نَظيرِهِ، وَهُوَ ضِلْعُ ه ز مُثَنّاةً.

שרטוט—ד״ה ⊥

كُلُّ مُثَلَّثَيْنِ مُتَشَابِهَيْنِ، فَإِنَّ نِسْبَةَ أَحَدِهِمَا إِلى الآخَرِ هِيَ نِسْبَةُ ضِلْعِهِ إِلى ضِلْعِهِ الَّذِي هُوَ نَظِيرُهُ، مُثَنَّاةً.

כל שני משולשים דומים, הנה יחס אחד משניהם אל האחר הוא[1] יחס צלעו אל צלעו אשר הוא נכחי לו, שנוי.

مِثَالُهُ: أَنَّ مُثَلَّثَيْ ا ب ج د ه ز مُتَشَابِهَانِ. فَأَقُولُ إِنَّ نِسْبَةَ مُثَلَّثِ ا ب ج إِلى مُثَلَّثِ د ه ز كَنِسْبَةِ ضِلْعِ ب ج إِلى ضِلْعِ ه ز، مُثَنَّاةً.

המשל בו: כי שני משולשי אב״ג ודה״ז דומים. הנה אומר כי יחס משולש אב״ג אל משולש דה״ז כיחס[2] צלע ב״ג אל צלע ה״ז, שנוי.

שרטוט—**וס**: א״ח ⊥; **ס**: משולש דה״ז מימין למשולש אב״ג   [1]הוא] +⟨⟨כפל⟩⟩ **ב**   [2]כיחס] ⟨⟨כפל⟩⟩ כיחס **ב**

מופת זה: שנוציא לצלע ב״ג וה״ז צלע שלישי שיתיחס אליהם, והוא ב״ח. ונדביק א׳ בח׳. ויהיה שעור ב״ג אל ה״ז כשעור ה״ז אל ב״ח. ושיעור ב״ג אל ה״ז הוא כשעור א״ב אל ד״ה. אם כן שעור א״ב אל ד״ה כשעור ה״ז אל ב״ח. ויהיה משלש אב״ח ודה״ז צלעותם המקיפות בזויותיהם השוות מתהפכות ביחס. אם כן משלש אב״ח שוה למשולש דה״ז. ושעור ב״ג אל ה״ז כשיעור ה״ז אל ב״ח. ויהיו [!] יחס ב״ג אל ב״ח שוה ליחס ב״ג אל ה״ז, נשנה בהשנות. ואמנם שיעור ב״ג אל ב״ח כשיעור משולש אב״ג אל משלש אב״ח. אם כן[1] יהיה יחס משלש אב״ג אל משלש אב״ח שוה ליחס ב״ג אל ה״ז, נשנה בהשנות. ומשלש אב״ח שוה למשלש דה״ז. אם כן יחס משלש אב״ג אל משלש דה״ז כיחס ב״ג אל ה״ז, נשנה בהשנות.

بُرْهانُهُ: أَنَّا نَسْتَخْرِجُ لِضِلْعَيْ ب ج و ه ز ضِلْعًا ثالِثًا مُناسِبًا لَهُما، وَهُوَ ب ح. وَنَصِلُ ا ح بِخَطِّ ا ح. فَقَدْرُ ب ج عِنْدَ ه ز كَقَدْرِ ه ز عِنْدَ ب ج. وَقَدْرُ ب ج عِنْدَ ه ز كَقَدْرِ ا ب عِنْدَ د ه. [..] كَقَدْرِ ه ز عِنْدَ ب ح. فَمُثَلَّثا ا ب ح و د ه ز أَضْلاعُهُما الَّتي تُحِيطُ بِالزّاوِيَتَيْنِ المُتَساوِيَتَيْنِ مُتَكافِئَةٌ. فَمُثَلَّثُ ا ب ج مُساوٍ لِمُثَلَّثِ د ه ز. فَقَدْرُ ب ج عِنْدَ ه ز كَقَدْرِ ه ز عِنْدَ ب ح. وَأَضْعافُ نِسْبَةِ ب ج إِلى ب ح مِثْلُ أَضْعافِ نِسْبَتِهِ إِلى ه ز مُثَنّاةً. وَلَكِنَّ قَدْرَ ب ج عِنْدَ ب ح كَقَدْرِ مُثَلَّثِ ا ب ج عِنْدَ مُثَلَّثِ ا ب ح. وَنِسْبَةُ مُثَلَّثِ ا ب ج إِلى مُثَلَّثِ ا ب ح مِثْلُ نِسْبَةِ ب ج إِلى ه ز، مُثَنّاةً بِالتَّكْرِيرِ. مُثَلَّثُ ا ب ح مِثْلُ مُثَلَّثِ د ه ز. فَنِسْبَةُ مُثَلَّثِ ا ب ج إِلى مُثَلَّثِ د ه ز كَنِسْبَةِ ب ج إِلى ه ز، مُثَنّاةً.

[1] אם כן] ⟨⟨*⟩⟩

بُرهانُهُ: أَنَّا نَسْتَخْرِجُ لِضِلْعَي ب ج و ه ز ضِلْعًا ثَالِثًا مُنَاسِبًا لَهُما، وهو ب ح. وَنُخْرِجُ ا ح. فَنِسْبَة ب ج إلى ه ز كَنِسْبَةِ ه ز إلى ب ح. وَنِسْبَة ب ج إلى ه ز كَنِسْبَةِ ا ب إلى د ه، لِأَنَّ المُثَلَّثَيْنِ مُتَشابِهانِ[1]. فَنِسْبَةَ ا ب إلى د ه كَنِسْبَةِ ه ز إلى ب ح. فَأَضْلاعُ مُثَلَّثَيْ ا ب ج و د ه ز[2] المُحِيطَةُ بِزاوِيَتَيْنِ ال ب وال ه مِنْهُما[3] المُتَساوِيَتَيْنِ—مُتَكافِئَةٌ. فَهُما مُتَساوِيانِ. وَنِسْبَةَ ب ج إلى ه ز كَنِسْبَةِ ه ز إلى ب ح. فَنِسْبَةُ ب ج إلى ب ح هي ضِعْفُ[4] نِسْبَةِ ب ج إلى ه ز. أَعْنِي أَنَّ[5] نِسْبَةَ ب ج إلى ب ح هي نِسْبَةُ خطّ ب ج إلى خطّ ه ز، مثناة، أي مُكَرَّرَة. لكن نِسْبَةَ خطّ ب ج إلى خطّ ب ح[6] كَنِسْبَةِ مُثَلَّثِ ا ب ج إلى مُثَلَّثِ ا ب ح، لِأَنَّ ارتفاعَهُما بِقَدْرٍ واحد. فَنِسْبَةُ مُثَلَّثِ ا ب ج إلى مُثَلَّثِ ا ب ح هي نِسْبَةُ ب ج إلى ه ز، مُثَنَّاة. ومُثَلَّث ا ب ح مِثْل مُثَلَّث د ه ز^. فَنِسْبَةَ مُثَلَّث ا ب ح إلى مُثَلَّث د ه ز كَنِسْبَةِ ب ج إلى ه ز، مُثَنَّاة.

وَذَلِكَ ما أَرَدْنا أَنْ نُبَيِّنَ.

---

[1]لأن ... متشابهان] **MPR2** [2]فاضلاع ... و د ه ز] **MPR2** فمثلثا ا ب ج و د ه ز أضلاعهما [3]ال ب ... منهما] **MPR2** [4]ضعف] **MPR2** [5]أعني ان] **MPR2** [6]هي, ..ب ح] **MPR2**

המופת: אנחנו[1] נוציא לשתי צלעות ב״ג וה״ז צלע שלישי מתיחס לשניהם, והוא ב״ח. ונוציא א״ח. הנה יחס ב״ג[2] אל ה״ז כיחס ה״ז אל ב״ח. ויחס ב״ג אל ה״ז כיחס א״ב אל ד״ה, מפני כי שני המשולשים דומים. ויחס[3] א״ב אל ד״ה כיחס ה״ז אל ב״ח. אם כן צלעות שני משולשי אב״ח ודה״ז המקיפות בשתי זויות ה׳ ב׳ [!] משניהם השוות—מספיקות. הנה שניהם[4] שוים. ויחס ב״ג אל ה״ז כיחס ה״ז אל ב״ח. אם כן יחס ב״ג אל ב״ח כפל יחס[5] ב״ג אל ה״ז. אם כן יחס ב״ג[6] אל ב״ח הוא יחס ב״ג אל ה״ז, שנוי, כלומר כפול[7]. אבל יחס ב״ג אל ב״ח כיחס משולש אב״ג אל משולש[8] אב״ח, מפני כי גבהותם בשעור אחד. אם כן יחס משולש[9] אב״ג אל משולש אב״ח הוא יחס ב״ג אל ה״ז, כפול. אם[10] כן יחס משולש אב״ג אל משולש דה״ז כיחס ב״ג אל ה״ז, כפול.

וזה מה שרצינו לבאר.

---

[1]אנחנו] +~~נשים~~ **ט** [2]ב״ג] ⟨⟨*⟩⟩ **ב** [3]ויחס] אם כן יחס **ס** [4]שניהם] שתיהם **בו** ⟨⟨אם כן⟩⟩+ **ק** [5]יחס] +~~ב״ח~~ **ק** [6]ב״ג] ⟨*⟩ **ט** [7]כלומר כפול] ליתא **ו** [8]משולש] משולשי **ב** [9]משולש] ⟨*⟩ **ט** [10]אם] וכבר היה משולש אב״ח שוה למשולש דה״ז+ **ס** ⟨⟨ומשולש אב״ח כמו משולש דה״ז אם כן⟩⟩ **ק**

وَهُناكَ اسْتَبانَ أَنَّ كُلَّ (ثَلاثَة)[1] خُطوط مُتَناسِبَةٍ، فَقَدْرُ الأَوَّلِ عِنْدَ الثَّالِثِ كَقَدْرِ السَّطْحِ الَّذي يَكونُ عَلى الأَوَّل إِلى السَّطْحِ الَّذي يَكونُ عَلى الثَّاني، إِذا كانَ يُشْبِهُهُ عَلى خِلْقَتِهِ.

וזה מה שרצינו לבאר.

وَذَلِكَ ما أَرَدْنا بَيانَهُ.

והנה התבאר וכו׳.

[1] ثلاثة] ثلثة

وَهُنالِكَ اسْتَبَانَ أَنَّ كُلَّ ثَلاثَةِ خُطُوطٍ مُتَنَاسِبَةٍ، فَإِنَّ نِسْبَةَ الأَوَّلِ إِلى الثَّالِثِ كَنِسْبَةِ السَّطْحِ الَّذي يَكُونُ عَلى الأَوَّلِ إِلى السَّطْحِ الَّذي يَكُونُ على الثَّاني، إذا كانَ يُشْبِهُهُ، وكانَ على وَضْعِهِ.

وذَلِكَ ما أَرَدْنا أَنْ نُبَيِّنَ.[1]

ובכאן התבאר כי כל[1] שלשה קוים מתיחסים, הנה יחס[2] הראשון אל השלישי כיחס השטח אשר יהיה על הראשון אל השטח אשר יהיה על השני, כאשר יהיה ידמה אותו על מצבו.

ונשלם ביאורו.

[1]وذلك ... نبين] P

[1]כל] ליתא ז [2]יחס] ליתא בו

[Proposition VI.19]

[H20, B18]

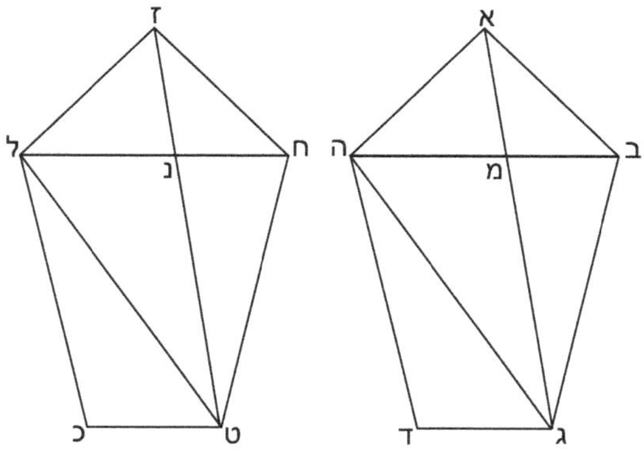

השטחים רבי הזויות המתדמים[1] יחלקו במשולש מתדמים יחסיהם כיחסיהם על מספרם. ויחס רב הזויות אל רב הזויות כיחס צלעו אל צלעו הדומה לו, נשנה בהשנות.

السُّطوحُ الكَثيرَةُ الزَّوايا المُتَشابِهَةُ فَإِنَّها مُنْقَسِمَةٌ بِمُثَلَّثاتٍ مُتَشابِهاتٍ، وَعِدَّةُ ما تَنْقَسِمُ مِنْ كُلِّ واحِدٍ مِثْلُ عِدَّةِ ما يَنْقَسِمُ مِنْ كُلِّ واحِدٍ. وَنِسْبَتُها مِثْلُ نِسْبَتِها. وَنِسْبَةُ السَّطْحِ الكَثيرِ الزَّوايا إِلى السَّطْحِ الكَثيرِ الزَّوايا كَنِسْبَةِ أَحَدِ أَضْلاعِ المَنْسوبِ إِلى نَظيرِهِ مِنْ أَضْلاعِ المَنْسوبِ إِلَيْهِ مُثَنَّاةً.

המשל בזה: שנצייר שני שטחים רבי הזויות מתדמים: על אחד מהם אבגד״ה, ועל האחר זחטכ״ל. ואומר כי הם יחלקו במשולשים מתדמים, ומספר מה שיחלק מאחד מהם כמספר מה שיחלק מן האחר, ויחסם שוה ליחסם, ויחס שטח אבגד״ה רב הזויות אל שטח זחטכ״ל רב הזויות, כיחס צלע א״ב אל צלע ז״ח שהוא דומה לו, נשנה בהשנות.

مِثالُهُ: سَطْحا ا ب ج د ه زح ط ك ل الكَثيرَةُ الزَّوايا المُتَشابِهانِ. فَأَقولُ: إِنَّهُما يَنْقَسِمانِ بِمُثَلَّثاتٍ مُتَشابِهاتٍ. وَعِدَّةُ ما يَنْقَسِمُ مِنْ أَحَدِهِما مِثْلُ عِدَّةِ ما يَنْقَسِمُ مِنَ الآخَرِ. وَنِسْبَتُها مِثْلُ نِسْبَتِها. وَسَطْحُ ا ب ج د ه الكَثيرُ الزَّوايا إِلى سَطْحِ (زح ط ك ل)[1] الكَثيرِ الزَّوايا كَنِسْبَةِ ضِلْعِ ا ب إِلى ضِلْعِ زح الَّذي هُوَ نَظيرُهُ مُثَنَّاةً.

[1] שרטוט—מרובעים בגד״ה וחטכ״ל הם ריבועים המתדמים] ⟨⟨*⟩⟩

[1] زح ط ك ل] زح ح ط ك ل

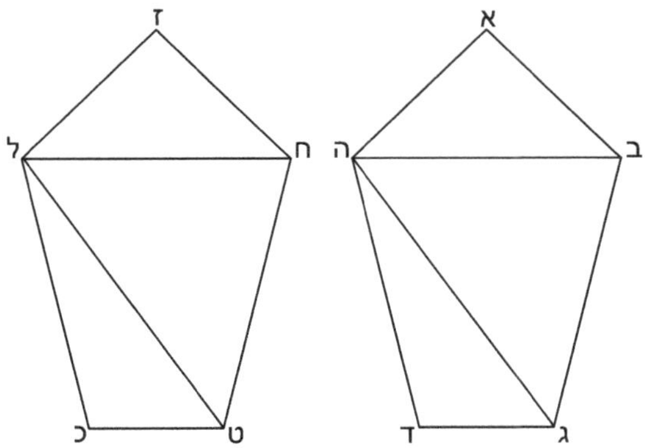

السُّطُوحُ الْكَثِيرَةُ الزَّوَايَا الْمُتَشَابِهَةُ تَنْقَسِمُ بِمُثَلَّثَاتٍ مُتَشَابِهَاتٍ، نِسْبَتُهَا كَنِسْبَتِهَا عَلَى عِدَّةٍ وَاحِدَةٍ. وَنِسْبَةُ الْكَثِيرِ الزَّوَايَا إِلَى الْكَثِيرِ الزَّوَايَا كَنِسْبَةِ ضِلْعِهِ إِلَى ضِلْعِهِ الَّذِي هُوَ نَظِيرُهُ فِي النِّسْبَةِ مُثَنَّاةً.

השטחים רבי[1] הזויות המתדמים יחלקו במשולשים מתדמים, יחסם כיחסם על מספר אחד. ויחס[2] רבי[3] הזויות אל[4] רבי[5] הזויות כיחס צלעו אל צלעו אשר הוא נכחי לו ביחס כפול.

مِثَالُهُ: أَنَّ سَطْحَيْ ا ب ج د ه و ز ح ط ك ل كَثِيرَيِ الزَّوَايَا مُتَشَابِهَانِ. فَأَقُولُ إِنَّهُمَا يَنْقَسِمَانِ بِمُثَلَّثَاتٍ مُتَشَابِهَاتٍ عَلَى عِدَّةٍ وَاحِدَةٍ، نِسْبَتُهَا كَنِسْبَتِهَا، وَنِسْبَةَ سَطْحِ ا ب ج د ه إِلَى سَطْحِ ز ح ط ك ل كَنِسْبَةِ ضِلْعِ ا ب إِلَى ضِلْعِ ز ح الَّذِي هُوَ نَظِيرُهُ فِي النِّسْبَةِ مُثَنَّاةً.

המשל בו: כי שני שטחי אבגד״ה וזחטכ״ל רבי[6] הזויות מתדמים[7]. הנה אומר כי שניהם יחלקו במשולשים מתדמים על מספר אחד, יחסם כיחסם, ויחס שטח אבגד״ה אל שטח זחטכ״ל כיחס צלע א״ב אל צלע ז״ח אשר הוא נכחי לו, ביחס שנוי.

---

[1]רבי] רובי **בוספ** [2]ויחס] ויהיה שטח **ס** +⟨⟨שטח⟩⟩ **ק** [3]רבי] רובי **בוספ** [4]אל] ליתא **ב** +שטח **ס** [5]רבי] רובי **וטספ** [6]רבי] רב **בוטספ** [7]מתדמים] +~~יחסם כיחס~~ **ט**

מופת זה: שנדביק ב׳ בה׳, וה׳ בג׳, וח׳ בל׳, ול׳ בט׳. ושטח אבגד״ה יתדמה לשטח זחטכ״ל[1]. ותהיה זוית בא״ה שוה לזוית חז״ל. ושעור א״ב אל ז״ח כשיעור א״ה אל ז״ל. ואם כן זוית משלש אב״ה שוות לזוית משלש זח״ל. ואחר שזוית אב״ה שוה לזוית זח״ל, וזוית אב״ג שוה לזוית זח״ט, נשאר זוית הב״ג שוה לזוית לח״ט. ויהיה שיעור א״ב אצל ז״ח כשיעור ב״ג אצל ח״ט, ושיעור א״ב אצל ז״ח כשיעור ב״ה אצל ח״ל. ואחר ששיעור ב״ג אצל ח״ט כשיעור ב״ה אצל ח״ל, וזוית גב״ה שוה לזוית טח״ל, ותהיינה זויות משלש בג״ה שוות לזויות משלש חט״ל, אחר שזוית בג״ה שווה לזוית חט״ל. וזוית בג״ד כולה שוה לזוית חט״כ כולה. ונשארה זוית הג״ד שוה לזוית לט״כ. וזוית ד׳ שוה לזוית כ׳. ותשאר זוית גה״ד שוה לזוית טל״כ. ותהיינה זויות משלש גד״ה שוות לזויות משלש טל״כ. וכבר חולקו משטח אבגד״ה ומשטח זחטכ״ל משלשים מתדמים, כל משלש מאחד השטחים ידמה למשלש הדומה לו מן השטח האחר, ומספרם כמספרם, ויחסם כיחסם.

[1]זחטכ״ל] ~~א״ח~~ ⟨⟨ז״ח⟩⟩טכ״ל

بُرْهانُهُ: أَنَّا نَصِلُ (ه ب ب)[1] و ج و ل ب ح و ط بِخُطوطِ ه ب و ه ج و ل ح و ل ط. فَسَطْحُ (ا ب ج د ه)[2] يُشْبِهُ سَطْحَ ز ح ط ك ل. فَزاوِيَةُ ب ا ه مِثْلُ زاوِيَةِ ح ز ل. وَقَدْرُ ا ب عِنْدَ ز ح كَقَدْرِ ا ه عِنْدَ ز ل[3]. فَزَوايا مُثَلَّثِ ا ب ه مُساوِيَةٌ لِزَوايا مُثَلَّثِ ح ز ل. فَزاوِيَةُ ا ب ه مُساوِيَةٌ لِزاوِيَةِ ز ح ل. وَزاوِيَةُ ا ب ج مِثْلُ زاوِيَةِ ز ح ط. وَبَقِيَتْ زاوِيَةُ ه ب ج مِثْلَ زاوِيَةِ ل ح ط. فَقَدْرُ ا ب عِنْدَ ز ح كَقَدْرِ ب ج عِنْدَ ح ط. وَقَدْرُ ا ب عِنْدَ ز ح كَقَدْرِ ب ه عِنْدَ ح ل. فَقَدْرُ ب ج عِنْدَ ط ح كَقَدْرِ (ب ه)[4] عِنْدَ ح ل. وَزاوِيَةُ ج ب ه مِثْلُ زاوِيَةِ ط ح ل. فَزَوايا مُثَلَّثُ ج ب ه مُساوِيَةٌ لِزَوايا مُثَلَّثِ ط ح ل. وَزاوِيَةُ (ب ج ه)[5] مِثْلُ زاوِيَةِ (ح ط ل)[6]. وَجَميعُ زاوِيَةِ ب ج د مِثْلُ جَميعِ زاوِيَةِ ح ط ك. وَبَقِيَتْ زاوِيَةُ ه ج د مِثْلَ زاوِيَةِ ك ط ل. وَزاوِيَةُ د مِثْلُ زاوِيَةِ ك. فَبَقِيَتْ زاوِيَةُ (ج ه د)[7] مِثْلَ زاوِيَةِ ط ل ك. فَزَوايا مُثَلَّثِ ج د ه مُساوِيَةٌ لِزَوايا مُثَلَّثِ ط ل ك. فَقَدْ قُسِمَ مِنْ سَطْحِ ا ب ج د ه وَمِنْ سَطْحِ ز ح ط ك ل مُثَلَّثانِ مُتَشابِهانِ، كُلُّ مُثَلَّثٍ مِنْ أَحَدِ السَّطْحَيْنِ يُشْبِهُ نَظيرَهُ مِنَ السَّطْحِ الآخَرِ، وَعِدَّتُها كَعِدَّتِها، وَنِسْبَتُها مِثْلُ نِسْبَتِها.

[1]ه ب ب] ه ب [2]ا ب ج د ه] ا ب ج د [3]وقدر ز ل] +* [4]ب ه] ب ح [5]ب ج ه] ج ب ه
[6]ح ط ل] ط ح ل [7]ج ه د] ج د ه

بُرهانُهُ: أَنّا نُخْرِجُ خُطوطَ ب ه ه ج و ح ل و ل ط. فَسَطْحا ا ب ج د ه و ز ح ط ك ل مُتَشابِهانِ. فزاويةُ ا مثل زاوية ز. ونِسْبَةُ ا ب إلى ز ح كَنِسْبَةِ ا ه إلى ز ل. فزوايا مُثَلَّث ا ب ه مُساوِيَةٌ لِزَوايا مُثَلَّث ز ح ل. وأَضلاعهُما مُتَناسِبَةٌ. فَهُما مُتَشابِهانِ. فزاوية ا ب ه مثل زاوية ز ح ل. لكِنَّ جميعَ زاوية ب مثل جميع زاويةِ ح. فزاوية ه ب ج الباقية مثل زاوية ل ح ط الباقية. ونِسْبَةُ ا ب إلى ز ح كَنِسْبَةِ ب ج إلى ح ط. ونِسْبَةُ ا ب إلى ز ح كَنِسْبَةِ ب ه إلى ح ل. فنِسْبَةُ ب ج إلى ح ط كَنِسْبَةِ ب ه إلى ح ل. وزاوية ج ب ه مثل زاويةِ ط ح ل. فمُثَلَّثا ب ج ه و ط ح ل. مُتَساوِيا الزَّوايا مُتَناسِبَا الأَضْلاعِ. فَهُما مُتَشابِهانِ. وزاوية د. مثل زاوية كَ. ونِسْبَةُ ج د. إلى ط ك كَنِسْبَةِ د ه إلى ك ل. فمُثَلَّثا ج ب ه و ط ك ل مُتَساوِيا الزَّوايا مُتَناسِبا الأَضْلاعِ. فَهُما مُتَشابِهانِ. فقَدْ قُسِمَ سَطْحا ا ب ج د ه و ز ح ط ك ل الكَثيرا الزَّوايا بمُثَلَّثاتٍ مُتَشابِهاتٍ على عِدَّةٍ واحِدَةٍ.

מופתו: אנחנו נוציא קוים ב״ה וה״ג ול״ח ול״ט. הנה[1] שני שטחי אבגד״ה וזחטכ״ל דומים. אם כן זוית א׳ כמו זוית ז׳. ויחס א״ב אל ז״ח כיחס א״ה אל ז״ל. אם כן זויות[2] משולש אב״ה שוות לזויות משולש זח״ל. וצלעות שניהם מתיחסות. אם כן שניהם מתדמים. אם כן זוית אב״ה כמו זוית זח״ל. אבל כל זוית ב׳ כמו כל זוית ח׳. אם כן זוית הב״ג הנשארת כמו זוית לח״ט הנשארת. ויחס א״ב אל ז״ח כיחס ב״ג אל ח״ט. ויחס א״ב אל ז״ח כיחס ב״ה אל ח״ל. אם כן יחס ב״ג אל ח״ט כיחס ב״ה אל ח״ל. וזוית גב״ה כמו זוית טח״ל. אם כן שני משולשי בג״ה וטח״ל שוים [!] הזויות מתיחסי הצלעות. אם כן שניהם מתדמים. וזוית ד׳ כמו[3] זוית כ׳. ויחס ג״ד אל ט״כ כיחס ד״ה אל כ״ל. אם כן שני משולשי גד״ה וטכ״ל שוי הזויות מתיחסי הצלעות. אם כן שניהם מתדמים[4]. הנה כבר חולקו שני שטחי אבגד״ה וזחטכ״ל רבי[5] הזויות[6] למשולשים מתדמים על מספר אחד.

[1] הנה] ליתא **בו** [2] זויות] זוית **ספ** [3] כמו] +* **ט** [4] וזוית ... מתדמים] +* **פ** [5] רבי] רובי **ס** [6] הזויות] +*̲ **ט**

ויחס רב הזויות אל האחר כיחס צלע המיוחס אל דומה לו מהצלעות המיוחס אליו, נשנה בהשנות.

מופת זה: שנדביק א׳ בג׳, וז׳ בט׳. ושטח אבג(ד)״ה[1] ידמה לשטח זחטכ״ל. ותהיה זוית אב״ג שוה לזוית זח״ט. ושיעור א״ב אל ז״ח כשעור ב״ג אל ח״ט. ויהיו זויות משלש אב״ג שוות לזויות זח״ט. וזוית בא״מ שוה לזוית חז״נ. וזוית אב״מ שוה לזוית זח״נ. ונשאר זוית אמ״ב שוה לזוית זנ״ח. ותהיינה זויות משלש אב״מ[2] שוות לזויות משלש זח״נ. ויהיה שיעור ב״מ אצל ח״נ כשיעור א״מ אצל ז״נ. ולפי זה יהיה שיעור ב״מ אצל ח״נ כשיעור מ״ג אצל נ״ט. ואם כן שיעור א״מ אצל ז״נ כשיעור מ״ג אצל נ״ט. עוד נחליף ויהיה שיעור א״מ אצל מ״ג כשעור ז״נ אצל נ״ט. ושיעור א״מ אצל מ״ג כשיעור משלש אב״מ אל משלש במ״ג, וכשיעור משלש אה״מ אל משלש מה״ג, וכשיעור אב״ה כולו אל משלש בג״ה כלו.

وَنِسْبَةُ السَّطْحِ الكَثيرِ الزَّوايا عِنْدَ الآخَرِ الكَثيرِ الزَّوايا كَنِسْبَةِ ضِلْعِ المَنْسوبِ إلى نَظيرِهِ مِنْ أَضْلاعِ المَنْسوبِ إِلَيْهِ مُثَنّاةً.

بُرْهانُهُ: أَنّا نَصِلُ ا ج و ز ط بِخَطَّيْ ا ج و ز ط. فَسَطْحُ ا ب ج د ه يُشْبِهُ (ز ح ط ك ل)[1]. فَزاوِيَةُ ا ب ج مِثْلُ زاوِيَةِ ز ح ط. فَقَدْرُ ا ب عِنْدَ ز ح كَقَدْرِ ب ج عِنْدَ ح ط. فَزَوايا مُثَلَّثِ ا ب ج مُساوِيَةٌ لِزَوايا مُثَلَّثِ ز ح ط. فَزاوِيَةُ (ب ا م)[2] مِثْلُ زاوِيَةِ (ح ز ن)[3]. ا ب م مِثْلُ زاوِيَةِ ز ح ن. وَبَقِيَتْ زاوِيَةُ ا م ب مِثْلُ زاوِيَةِ ز ن ح. فَزَوايا مُثَلَّثِ ا ب م مُساوِيَةٌ لِزَوايا مُثَلَّثِ ز ح ن. فَقَدْرُ ب م عِنْدَ ح ن كَقَدْرِ ا م عِنْدَ ز ن. وَمِنْ أَجْلِ ذَلِكَ قَدْرُ ب م عِنْدَ ح ن كَقَدْرِ م ج عِنْدَ ن ط. فَقَدْرُ ا م عِنْدَ ز ن كَقَدْرِ م ج عِنْدَ ن ط. ثُمَّ نُبَدِّلُ بِقَدْرِ ا م عِنْدَ م ج كَقَدْرِ ز ن عِنْدَ ن ط. وَقَدْرُ ا م عِنْدَ م ج كَقَدْرِ مُثَلَّثِ (ا ب م)[4] عِنْدَ مُثَلَّثِ ب م ج. وَقَدْرُ ز ن عِنْدَ ن ط كَقَدْرِ مُثَلَّثِ ز ح ن عِنْدَ مُثَلَّثِ ح ن ط. فَقَدْرُ ا ب م عِنْدَ ب م ج كَقَدْرِ ز ح ن عِنْدَ ح ز ط. وَقَدْرُ مُثَلَّثِ ا ب م عِنْدَ مُثَلَّثِ ب م ج كَقَدْرِ ا م عِنْدَ م ج. فَقَدْرُ ا ب م عِنْدَ (ب م ج)[5] كَقَدْرِ مُثَلَّثِ ا ه م عِنْدَ مُثَلَّثِ م ه ج. وَقَدْرُ واحِدٍ مِنَ المُقَدَّماتِ

[1]ز ح ط ك ل] ز ط ك ل [2]ب ا م] ب ا ج [3]ح ز ن] ح ز ل [4]ا ب م] ا ب ج [5]ب م ج] ب م م

[1]אבגד״ה] אבגב״ה [2]אב״מ] אב״ס ⟨⟨*⟩⟩

وأَقولُ إنَّ المُثَلَّثاتِ مُتَناسِبَةٌ، وأَنَّ نِسبَةَ السَّطْحِ إلى السَّطْحِ كنِسبَةِ ضِلْعِهِ إلى ضِلْعِهِ النَّظيرِ لَهُ في النِّسْبَةِ مُثَنّاةً.

ואומר כי המשולשים מתיחסים, ושיחס השטח אל השטח[1] כיחס צלעו אל צלעו[2] הנכחי לו ביחס שנוי.

بُرهانُهُ: أنَّ مُثَلَّثَيْ ا ب ه و ز ح ل مُتَشابِهانِ. فنِسْبَةُ مُثَلَّثِ ا ب ه إلى مُثَلَّثِ ز ح ل كنِسبَةِ ب ه إلى ح ل، مُثَنّاةً. وأيضًا فإنَّ مُثَلَّثَيْ ب ج ه و ح ط ل مُتَشابِهانِ. فنِسْبَةُ مُثَلَّثِ ب ج ه إلى مُثَلَّثِ ح ط ل كنِسْبَةِ ب ه إلى ح ل، مُثَنّاةً. فنِسْبَةُ مُثَلَّثِ ا ب ه إذَنْ إلى مُثَلَّثِ ز ح ل كنِسْبَةِ مُثَلَّثِ ب ج ه إلى مُثَلَّثِ ح ط ل. وكذَلكَ تَبَيَّنَ أنَّ نِسْبَةَ مُثَلَّثِ ب ج ه إلى مُثَلَّثِ ح ط ل كنِسْبَةِ مُثَلَّثِ ج د ه إلى مُثَلَّثِ ط ك ل. ونِسْبَةُ واحِدٍ مِنَ المُقَدَّماتِ إلى قرينِهِ مِنَ التَّوالي كنِسْبَةِ كُلِّ المُقَدَّماتِ إلى كُلِّ التَّوالي. فنِسْبَةُ مُثَلَّثِ ا ب ه إلى مُثَلَّثِ ز ح ل[1] كنِسْبَةِ ضِلْعِ ا ب إلى ضِلْعِ ز ح، مُثَنّاةً. فنِسْبَةُ سَطْحِ ا ب ج د ه إلى سَطْحِ ز ح ط ك ل كنِسْبَةِ ا ب إلى ز ح، مُثَنّاةً.

מופתו: כי שני משולשי אב״ה וזח״ל מתדמים[3]. אם כן יחס משולש אב״ה אל משולש זח״ל כיחס ב״ה אל ח״ל, שנוי. וגם כן הנה שני[4] משולשי בג״ה וחט״ל מתדמים. אם כן יחס משולש בג״ה[5] אל משולש חט״ל כיחס ב״ה אל ח״ל, שנוי. אם כן יחס משולש אב״ה אל משולש[6] [7] זח״ל כיחס משולש בג״ה אל משולש חט״ל. וכן התבאר כי יחס משולש בג״ה אל משולש חט״ל[8] כיחס משולש גד״ה אל משולש טכ״ל. ויחס אחד מן הקודמים אל קרובו מן הנמשכים כיחס כל הקודמים אל כל הנמשכים. אם כן יחס משולש אב״ה אל משולש זח״ל כיחס צלע א״ב אל צלע ז״ח[9], שנוי. אם כן שטח[10] אבגד״ה אל שטח זחטכ״ל כיחס[11] א״ב אל ז״ח, שנוי.

وذَلِكَ ما أَرَدْنا أَنْ نُبَيِّنَ.

וזה מה שרצינו לבאר.

[1] زح ل] + كنسبة سطح ا ب ج د ه إلى سطح زح ط ك ل ونسبة مثلث ا ب ه الى مثلث زح ل **CopPR2T2**

[1] אל השטח] ליתא **ב** [2] צלעו] צלעי **ב** [3] מתדמים] דומים **ט** [4] הנה שני] הנה **בו** שני **פ** [5] משולש בג״ה] דבג״ה **ט** [6] חט״ל ... משולש] ליתא **ו** [7] משולש] ⟨*⟩ **ט** [8] וכן ... חט״ל] +* **ס** [9] ז״ח] ~~ב״ז~~+ **פ** [10] שטח] ⟨יחס⟩+ **ק** [11] כיחס] שוה ~~אל~~ ⟨כ⟩יחס **ק**

عِنْدَ واحِدٍ مِنَ التَّوالي كَقَدْرِ جَميعِ المُقَدَّماتِ عِنْدَ جَميعِ التَّوالي. فَقَدْرُ مُثَلَّثِ ا ب م عِنْدَ مُثَلَّثِ ب م ج كَقَدْرِ جَميعِ مُثَلَّثِ ا ب ه عِنْدَ مُثَلَّثِ ب ج ه.

وَقَدْرُ ز ن عِنْدَ ن ط كَقَدْرِ مُثَلَّثِ ز ح ن عِنْدَ مُثَلَّثِ ح ن ط. وَقَدْرُ مُثَلَّثِ ز ل ن عِنْدَ مُثَلَّثِ ل ط ن كَقَدْرِ ز ن عِنْدَ ن ط. فَقَدْرُ مُثَلَّثِ ز ل ن عِنْدَ مُثَلَّثِ ل ن ط كَقَدْرِ مُثَلَّثِ ز ح ن عِنْدَ مُثَلَّثِ ح ن ط. وَقَدْرُ واحِدٍ مِنَ المُقَدَّماتِ عِنْدَ واحِدٍ مِنَ التَّوالي كَقَدْرِ جَميعِ المُقَدَّماتِ عِنْدَ جَميعِ التَّوالي. فَقَدْرُ مُثَلَّثِ ز ح ن عِنْدَ مُثَلَّثِ ح ن ط كَقَدْرِ جَميعِ مُثَلَّثِ ز ح ل عِنْدَ جَميعِ مُثَلَّثِ ح ط ل. وَقَدْرُ مُثَلَّثِ ز ح ن عِنْدَ مُثَلَّثِ ح ن ط كَقَدْرِ ا ب م عِنْدَ ب م ج. فَقَدْرُ مُثَلَّثِ ا ب ه عِنْدَ مُثَلَّثِ ب ج ه كَقَدْرِ ز ح ل عِنْدَ مُثَلَّثِ ح ط ل.

ושיעור ז״נ אל נ״ט כשיעור משלש זח״נ אל משלש חנ״ט, וכשיעור משלש זל״נ אל משלש נל״ט, וכשיעור משלש זח״ל כלו אל משלש חל״ט כלו. ושיעור משלש אב״ה אל משלש בה״ג כשיעור משלש זח״ל אל משלש חט״ל.

וכאשר החלפנו, היה שיעור משלש בא״ה אל משלש חז״ל כשיעור משולש בג״ה אל משלש חט״ל. וכמו כן שיעור משלש בג״ה אל משלש חט״ל כשיעור משלש הג״ד אל משלש לט[״כ]. ושיעור משולש בג״ה אל משלש חט״ל כשיעור משולש אב״ה אל משלש זח״ל. ושעור משלש אב״ה אל משלש זח״ל כשיעור משלש בג״ה אל משלש חט״ל, וכשיעור משלש הג״ד אל משלש לט״כ[1]. ושיעור אחד מן הקודמים אל הנמשך אליו כשיעור כל הקודמים אל כל הנמשכים אליהם. ואם כן שיעור משלש אב״ה אל משלש זח״ל כשיעור שטח אבגד״ה כלו אל שטח זחטכ״ל[2] כלו. ויחס משלש אב״ה אל משלש זח״ל כיחס צלע א״ב אל צלע ז״ח, נשנה בהשנות. אם כן יחס שטח אבגד״ה אל שטח זחטכ״ל כיחס א״ב אל ז״ח, נשנה בהשנות.

וזה מה שרצינו לבאר.

وَاذا بَدَّلْنا كانَ قَدْرُ مُثَلَّثِ ب ا ه عِنْدَ مُثَلَّثِ زح ل كَقَدْرِ مُثَلَّثِ ب ج ه عِنْدَ مُثَلَّثِ ح ل ط. وَلِذَلِكَ قَدْرُ مُثَلَّثِ ب ج ه عِنْدَ مُثَلَّثِ ح ط ل كَقَدْرِ مُثَلَّثِ ه ج د عِنْدَ مُثَلَّثِ ل ط ك. وَقَدْرُ مُثَلَّثِ ب ج ه عِنْدَ مُثَلَّثِ ح ط ل كَقَدْرِ مُثَلَّثِ ا ب ه عِنْدَ مُثَلَّثِ زح ل. فَقَدْرُ مُثَلَّثِ ا ب ه عِنْدَ مُثَلَّثِ زح ل كَقَدْرِ مُثَلَّثِ ب ج ه عِنْدَ مُثَلَّثِ ح ط ل وَكَقَدْرِ مُثَلَّثِ ه ج د عِنْدَ مُثَلَّثِ ل ط ك. وَقَدْرُ واحِدٍ مِنَ المُقَدَّماتِ عِنْدَ الَّذي يَتْلوهُ كَقَدْرِ كُلِّ المُقَدَّماتِ عِنْدَ كُلِّ التَّوالي الَّتي يَتْلوها. فَقَدْرُ مُثَلَّثِ ا ب ه عِنْدَ مُثَلَّثِ زح ل كَقَدْرِ جَميعِ سَطْحِ ا ب ج د ه الكَثيرِ الزَّوايا عِنْدَ سَطْحِ زح ط ك ل الكَثيرِ الزَّوايا. وَنِسْبَةُ مُثَلَّثِ ا ب ه إلى مُثَلَّثِ زح ل كَنِسْبَةِ ضِلْعِ ا ب إلى ضِلْعِ زح مُثَنَّاةً. فَنِسْبَةُ سَطْحِ ا ب ج د ه إلى سَطْحِ (زح ط ك ل)[1] كَنِسْبَةِ ضِلْعِ ا ب إلى ضِلْعِ زح مُثَنَّاةً بِالتَّكْريرِ.

وَذَلِكَ ما أَرَدْنا بَيانَهُ.

[1]לט״כ] ~~לט״ב~~ ⟨⟨*⟩⟩ [2]זחטכ״ל] זח״ט⟨⟨כ״ל⟩⟩

[1]زح ط ك ل] زح ط ل

[Proposition VI.20]

[B19]

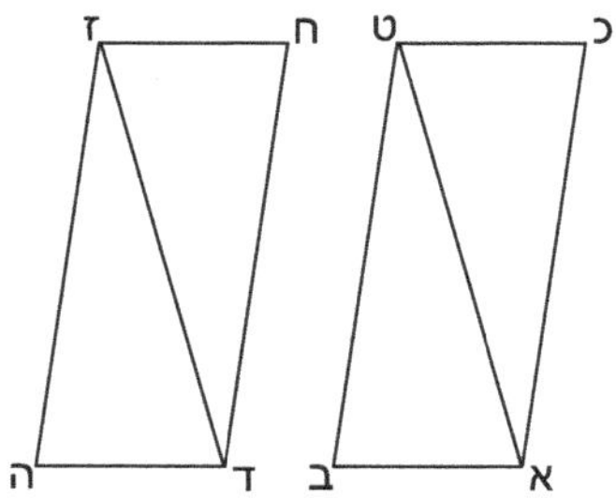

נרצה לעשות על קו ישר ידוע שטח ישר הקוים ידוע דומה לשטח ישר הקוים[1] ומוצב במצבו.

نُرِيدُ أَنْ نُبَيِّنَ كَيْفَ نَعْمَلُ عَلَى خَطٍّ مَعْلومٍ سَطْحًا شَبِيهًا بِسَطْحٍ مَعْلومٍ.

ונשים הקו הידוע קו א״ב. ושטח הידוע שטח ה״ח. וכאשר רצינו לעשות על קו א״ב הידוע שטח דומה לשטח ה״ח, נדביק ז׳ בד׳. ונעמיד על קו א״ב על נקודת ב׳ זוית שוה לזוית דה״ז, והיא זוית אב״ט. ונעמיד עוד על קו א״ב על נקודת א׳[2] זוית שוה לזוית הד״ז[3], והיא זוית בא״ט[4].

مِثالُهُ: خَطُّ ا ب المَعْلومُ، وَسَطْحُ ح ه المَعْلومُ. وَنُرِيدُ أَنْ نَعْمَلَ عَلَى ا ب المَعْلومِ سَطْحًا شَبِيهًا بِسَطْحِ ح ه المَعْلومِ. فَنَصِلُ ز د. وَنُقِيمُ عَلَى نُقْطَةِ ب مِنْ خَطِّ ا ب زاوِيَةَ ا ب ط مِثْلَ زاوِيَةِ ه. وَنُقِيمُ عَلَى نُقْطَةِ ا مِنْ خَطِّ ا ب زاوِيَةَ ب ا ط مِثْلَ زاوِيَةِ ز د ه.

---

[1] דומה ... הקוים] ⟨⟨*⟩⟩ [2] א׳] ~~ד~~ ⟨⟨*⟩⟩ [3] הד״ז] ~~הד״ז דה״ז~~ ⟨⟨*⟩⟩ [4] בא״ט] ~~מב״ט~~ ⟨⟨*⟩⟩ ~~ונעמיד עוד על קו א״ב על נקודת א׳ זוית שוה לזוית הד״ז והיא זוית בא״ט~~

نُرِيدُ أَنْ نَعْمَلَ على خَطٍّ مَعْلومٍ مُسْتَقِيمٍ شَكْلًا مُسْتَقِيمَ الخُطوطِ شَبِيهًا لِشَكْلٍ مُسْتَقِيمِ الخُطوط مَعْلوم، وَمَوْضُوعًا كَوَضْعِهِ.

נרצה לעשות על קו ידוע ישר תמונה[1] ישרת הקוים דומה בתמונה[2] ישרת הקוים ידועה[3] מונחת כמצבה.

وَلْيَكُنِ الشَّكْلُ المُسْتَقِيمُ الخُطوطِ المَعْلوم سَطْحَ ه ح. والخطّ المُسْتَقِيم المَعْلوم^ خطّ ا ب. ونُرِيدُ أَنْ نَعْمَلَ على خطِّ ا ب سَطْحًا شَبِيهًا بِسَطْحِ ه ح. فَنُخْرِجَ خَطَّ ز د. ونُقِيمَ على نُقْطَةِ ب زاويةَ ا ب ط مثل زاويةِ د ه ز، وعلى نُقْطَةِ ا زاويةَ ب ا ط مثل زاويةِ ز د ه. فَتَبْقَى زاوِيَةُ ا ط ب مثل زاويةِ د ز ه.

ותהיה התמונה ישרת הקוים הידועה[4] שטח ה״ח. והקו הישר קו א״ב. ונרצה שנעשה על קו[5] א״ב שטח דומה לשטח ה״ח. הנה נוציא קו ז״ד. ונעמיד על נקודת ב׳ זוית אב״ט כמו זוית דה״ז, ועל נקודת א׳ זוית בא״ט כמו זוית הד״ז. ותשאר זוית אט״ב כמו זוית דז״ה[6].

שרטוט—**בוטספק**: אבב״ט $ על הציר האופקי [1]ישר תמונה] תמונת **ז** [2]בתמונה] תמונת **ז** בתמונת **טפ** [3]דומה ... ידועה] ידוע **ס** [4]הידועה] דומה בתמונה ישרת הקוים ⟨ה⟩ידועה **ב** [5]קו] ליתא **ס** [6]ועל ... דז״ה] ליתא **ב**

ותשאר זוית אט״ב שוה לזוית דז״ה. ותהיינה זויות משלש אב״ט שוות לזויות משלש דה״ז. אם כן שיעור ט״א אל ד״ז כשיעור ט״ב אל ז״ה וכשיעור א״ב אל ד״ה. ונעמיד עוד על קו ט״א על נקדת ט׳ זוית שוה[1] לזוית דז״ח, והיא זוית אט״כ. ועוד נעמיד על קו א״ט, על נקדת א׳, זוית שוה לזוית זד״ח[2], והיא זוית טא״כ. ותשאר זוית טכ״א שוה לזוית זח״ד. ותהיינה זויות משלש טא״כ שוות לזויות דח״ז. אם כן שיעור ט״א אל ד״ז כשיעור כ״ט אל ז״ח וכשיעור א״כ אל ד״ח. ואמנם שיעור ט״א אל ד״ז כשיעור ט״ב אל ז״ח וכשיעור א״ב אל [ד״ח][3]. ויהיה שיעור א״כ אל ד״ה כשיעור א״ב אל ד״ח. וכשיעור ט״כ אל ז״ח וכשיעור ט״ב אל ז״ה. וזוית אט״כ שוה לזוית דז״ה. וזוית אט״כ שוה לזוית דז״ח. אם כן זוית בט״כ כולה שוה לזוית הז״ח כולה. וכמו כן זוית כא״ב שוה לזוית חד״ה. וזוית[4] ב׳ שוה לזוית ה׳. אם כן זויות שטח א״ט שוות לזויות שטח ה״ח. והצלעות המקיפות בזויותיהם השוות הן מתיחסות. ואם כן שטח א״ט דומה לשטח ה״ח. וכבר עשינו על קו א״ב הישר שטח דומה לשטח ה״ח הידוע, והוא שטח א״ט.

فَزَوايا مُثَلَّثِ ا ب ط مُساوِيَةٌ لِزَوايا مُثَلَّثِ د ه ز. فَقَدْرُ ط ا عِنْدَ د ز كَقَدْرِ ط ب عِنْدَ ز ه وكَقَدْرِ ا ب عِنْدَ ه د. ونُقيمُ أَيْضًا عَلى نُقْطَةِ ط مِنْ خَطِّ ط ا [..] زاوِيَةَ ط ا ك مِثْلَ زاوِيَةِ ز د ج. وبَقِيَتْ زاوِيَةُ ط ك ا مِثْلُ زاوِيَةِ زح د. فَزَوايا مُثَلَّثِ ط ا ك مُساوِيَةٌ لِزَوايا مُثَلَّثِ د ح ز. فَقَدْرُ ط ا عِنْدَ د ز كَقَدْرِ ك ط عِنْدَ ح ز وكَقَدْرِ ا ك عِنْدَ د ح. فَقَدْرُ ا ط عِنْدَ د ز كَقَدْرِ ا ك عِنْدَ د ح. وكَقَدْرِ ط ك عِنْدَ زح وَقَدْرِ ب ط عِنْدَ ه ز. وَكَقَدْرِ ا ب عِنْدَ د ه. وَزاوِيَةُ ب ا ط مِثْلُ زاوِيَةِ ه د ز. وَزاوِيَةُ ط ا ك مِثْلُ زاوِيَةِ ح د ز. وَكُلُّ زاوِيَةِ ب ا ك مِثْلُ كُلِّ زاوِيَةِ ه د ح. وَلِذَلِكَ زاوِيَةُ ك ط ب مِثْلُ زاوِيَةِ ه زح. وَزاوِيَةُ كَ مِثْلُ زاوِيَةِ [!]. وَزاوِيَةُ ب مِثْلُ[1] زاوية ه. فَزَوايا سَطْحِ ك ب مُساوِيَةٌ لِزَوايا سَطْحِ ه ح. وَالأَضْلاعُ الَّتي تُحيطُ بِزَواياهما المُتَساوِيَةِ—مُتَناسِبَةٌ. فَسَطْحُ ك ب نِسْبَةَ سَطْحِ ه ح. فَقَدْ عَمِلْنا عَلى خَطِّ ا ب المَعْلومِ سَطْحَ ك ب شَبيهًا بِسَطْحِ ه ح المَعْلومِ.

וזה מה שרצינו לבאר.

وَذَلِكَ ما أَرَدْنا بَيانَهُ.

[1]שוה] ש⟨ו⟩ה [2]זד״ח] ~~דז״ח~~ ⟨⟨*⟩⟩ [3]ד״ח] נראה שנכתב ד״ה ותוקן [4]וזוית] ~~וזוית ג׳ שוה לזוית ח׳~~+

[1]مثل] +*

فزوايا مُثَلَّث ا ب ط مُساوِيَة لزوايا مُثَلَّث د ه ز. فنِسْبَة ط ا إلى د ز كنِسْبَة ط ب إلى ز ه وكنِسْبَة ا ب إلى د ه. ونقيم أيضًا على نُقْطَة ط زاوية ا ط ك مثل زاوية د ز ح، وعلى نُقْطَة ا زاوية ط ا ك مثل زاوية ز د ح. فتبقَى زاوِيَةُ ط ك ا مثل زاوية ز ح د. فزوايا مُثَلَّث ط ا ك مُساوِيَةٌ لزوايا مُثَلَّث ز ح د. فنسْبة ط ا إلى د ز كنِسْبَة ك ط إلى ز ه، وكنِسْبَة ا ك إلى د ه. لكنَّ نِسْبَة ط ا إلى د ز كنِسْبَة ط ب إلى ز ح، وكنِسْبَة ا ب إلى د ح. فنِسْبَة ا ب إلى د ح كنِسْبَة ا ب إلى د ه وكنِسْبَة ط ك إلى ز ه وكنِسْبَة ط ب إلى ز ح. وزاوية ا ط ب مثل زاوية د ز ح. وزاوية ا ط ك مثل زاوية د ز ه. فكُلّ زاوية ب ط ك مثل كُلّ زاوية ه ز ح. وكذلِكَ زاوية ك ا ب مثل زاوية ه د ح. وزاوية ك مثل زاوية ه. وزاوية ب مثل زاوية ح. فزوايا سطح ا ط مُساوِيَة لزوايا سطح ه ح. وأضلاعهما المُحِيطَة بزواياهما المُتَساوِيَة مُناسبة. فَهُما مُتَشابِهانِ. فَقَدْ عَمِلْنا على ا ب سَطْحَ ا ط شَبِيهًا بِسَطْحِ ه ح ومَوْضُوعًا كَوَضْعِهِ.

وذلِكَ ما أَرَدْنا أَنْ نُبَيِّنَ.

אם כן זויות[1] משולש אב״ט שוות לזויות[2] משולש דה״ז. אם כן יחס א״ט אל ד״ז כיחס ט״ב אל ז״ה וכיחס א״ב אל ד״ה. ונעמיד גם כן על נקודת ט׳ זוית אט״כ כמו זוית דז״ח, ועל נקודת א׳ זוית[3] טא״כ כמו זוית זד״ח. ותשאר זוית טכ״א כמו זוית זח״ד. אם כן[4] זויות משולש טא״כ שוות לזויות משולש דח״ז. אם כן יחס א״ט אל ד״ז כיחס כ״ט אל ז״ה, וכיחס א״כ אל ד״ה. אבל יחס ט״א אל ד״ז כיחס ט״ב אל ז״ח וכיחס א״ב אל ד״ח. אם כן יחס א״ב אל ד״ח כיחס א״ב אל ד״ה[5] וכיחס ט״כ אל ז״ה וכיחס ט״ב אל ז״ח. וזוית אט״ב כמו זוית דז״ח. וזוית אט״כ[6] כמו זוית דז״ה. אם כן[7] כל זוית בט״כ כמו כל[8] זוית חז״ה. וכן זוית כא״ב כמו זוית חד״ה. וזוית[9] כ׳ כמו זוית ה׳. וזוית ב׳ כמו זוית ח׳. אם כן זויות[10] שטח א״ט שוות לזויות[11] שטח ה״ח. וצלעות שניהם המקיפות בזויותיהם השוות מתיחסות. ואם כן שניהם דומים. הנה כבר עשינו על קו א״ב שטח א״ט דומה לשטח ה״ח ומונח כמצבו.

וזה מה שרצינו לבאר[12].

[1]זויות] זוית **בס** [2]לזויות] לזוית **ב** [3]זוית] וזוית **ס** [4]כן] ליתא **ס** [5]ד״ה] ~~ד״ה~~+ **ט** [6]כמו ... אט״כ] ליתא **ז** [7]כן] ⟨⟨*⟩⟩ **ט** [8]כל] ליתא **ט** [9]חד״ה וזוית] ליתא **ב** [10]זויות] זוית **וספ** [11]לזויות] לזוית **ז** [12]וזה ... לבאר] ונשלם ביאורו **בספק**

[Proposition VI.21]

[B20]

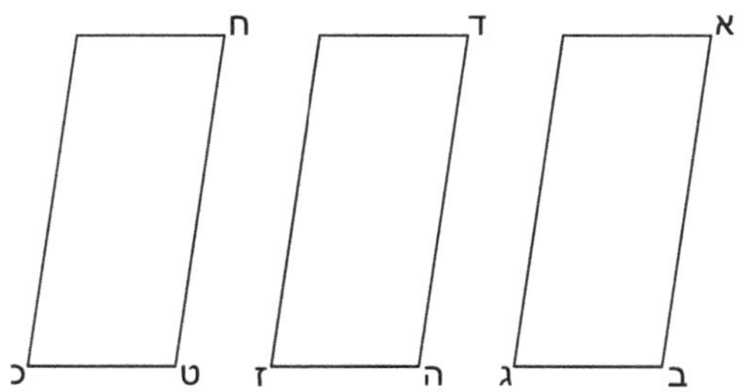

השטחים הדומים לשטח אחד הם מתדמים.

السُّطوحُ الَّتي تُشْبِهُ سَطْحًا واحِدًا فَإِنَّ بَعْضَها يُشْبِهُ بَعْضًا.

המשל בזה: שטחי א״ג וח״כ דומים לשטח ד״ז. ואומר כי שטחי א״ג וח״כ מתדמים.

مِثالُهُ: أَنَّ سَطْحَي اج وحك يُشْبِهانِ سَطْحَ د ز. فَأَقولُ إِنَّ سَطْحَيْ اج وحك مُتَشابِهانِ.

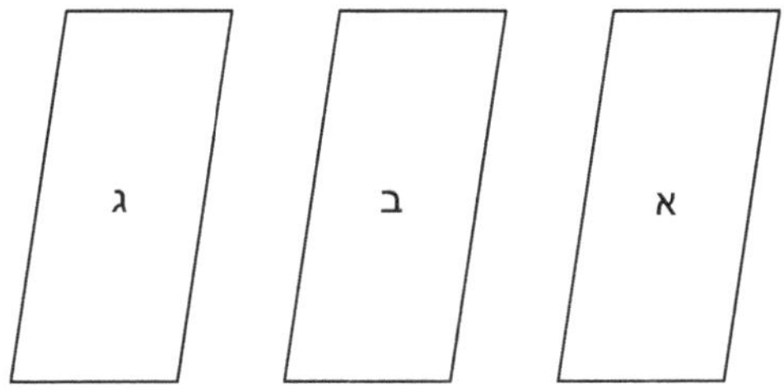

السُّطُوحُ الشَّبِيهَةُ بِسَطْحٍ واحِدٍ مُتَشابِهَةٌ.

השטחים הדומים לשטח אחד מתדמים.

مِثالُهُ: أَنَّ سَطْحَيْ ا جَ يُشْبِهانِ سَطْحَ بَ. فَأَقولُ إِنَّهُما مُتَشابِهانِ.

המשל בו: כי שני שטחי א׳ ג׳ ידמו שטח ב׳. הנה אומר כי שניהם מתדמים.

שרטוט—ז: אותיות על הקדקודים והן זהות לאותיות בתרגום הרב יעקב, בשני הבדלים: א ← ז, כ ← ד

بُرْهانُهُ: أَنَّ اج (يُشْبِهُ)[1] د ز. فَزاوِيَةُ بَ مِثْلُ زاوِيَةِ ه. وَقَدْرُ اب عِنْدَ د ه كَقَدْرِ ب ج عِنْدَ ه ز. وَأَيضًا ح ك يُشْبِهُ د ز. فَزاوِيَةُ طَ مِثْلُ زاوِيَةِ ه. وَقَدْرُ د ه عِنْدَ ح ط. كَقَدْرِ ه ز عِنْدَ (ط ك)[2]. وَزاوِيَةُ ه [..] مِثْلُ زاوِيَةِ ب. وَقَدْ كانَ قَدْرُ ا ب عِنْدَ د ه كَقَدْرِ ب ج عِنْدَ ه ز. وزاوِيَةُ بَ مِثْلُ زاوِيَةِ طَ. فَقَدْرُ ا ب عِنْدَ ح ط كَقَدْرِ ب ج عِنْدَ ط ك. فَسَطْحا ا ج ح ك مُتَشابِهانِ.

وَذَلِكَ ما أَرَدْنا بَيانَهُ.

[1] يشبه] نسبه [2] ط ك] ط ا ك

מופת [!]: לפי שא״ג דומה לד״ז, תהיה זוית ב׳ שוה לזוית ה׳. ושיעור א״ב אל ד״ה[1] כשיעור ב״ג אל ה״ז. וכמו כן ח״כ דומה לד״ז. אם כן זוית ט׳ שוה לזוית ה׳. ושיעור ד״ה אל ח״ט כשיעור ה״ז אל ט״כ. וזוית ה׳ שוה לזוית ט׳. ואם כן זוית ט׳ שוה לזוית ב׳. וכבר היה שיעור א״ב אל ד״ה כשיעור ב״ג אל ה״ז. וזוית ב׳ שוה לזוית ט׳. ויהיה שיעור א״ב אל ח״ט כשיעור ב״ג אל ט״כ. אם כן שטח א״ג וח״כ מתדמים.

וזה מה שרצינו לבאר.

[1] ד״ה] ב+

بُرهانُهُ: أَنَّ سَطْحَ آ شَبيهٌ بِسَطْحِ بَ. فَزواياهُما النَّظائِرُ مُتَساوِيَةٌ. وأَضلاعهما النَّظائِرُ مُتَناسِبَةٌ. وكَذلِكَ زوايا سَطْحَيْ بَ جَ النَّظائِرُ مُتَساوِيَةٌ. وأَضلاعهما النَّظائِرُ مُتَناسِبَةٌ. والمُساوِيَةُ لشيءٍ واحِدٍ بِعَيْنِه هي مُتَساوِيَةٌ[1]. فإِذَنْ[2] زوايا سَطْحَيْ آ جَ النَّظائِرُ مُتَساوِيَةٌ. وأَضلاعهما النَّظائِرُ مُتَناسِبَةٌ. فَهُما مُتَشابِهانِ^.

המופת: כי שטח א׳ דומה לשטח ב׳. אם כן זויות שניהם הנכחיות שוות. וצלעות שניהם הנכחיות מתיחסות[1]. וכן זויות שטחי ב׳ ג׳ הנכחיות שוות. וצלעותיהם הנכחיות מתיחסות. והשוים לדבר אחד בעינו הנה[2] הם[3] שוים. הנה אם כן זויות[4] שני שטחי א׳ ג׳ הנכחיים שוים. וצלעות שניהם הנכחיות מתיחסות. אם כן שניהם שוים[5].

وذَلِكَ ما أَرَدْنا أَنْ نُبَيِّنَ.

וזה מה שרצינו לבאר[6].

[1]والمساوية ... متساوية] MPR2 [2]فإذن] فإذا CPT1T3

[1]מתיחסות] +⟨נ״א המקיפות בזויות השוות מתיחסות⟩ ס [2]הנה] ליתא ק [3]הם] +שני פ [4]זויות] זוית ב [5]שוים] ⩲⟨⟨דומים⟩⟩ ק [6]וזה ... לבאר] ונשלם ביאו׳ ו

[Proposition VI.22]

[B21]

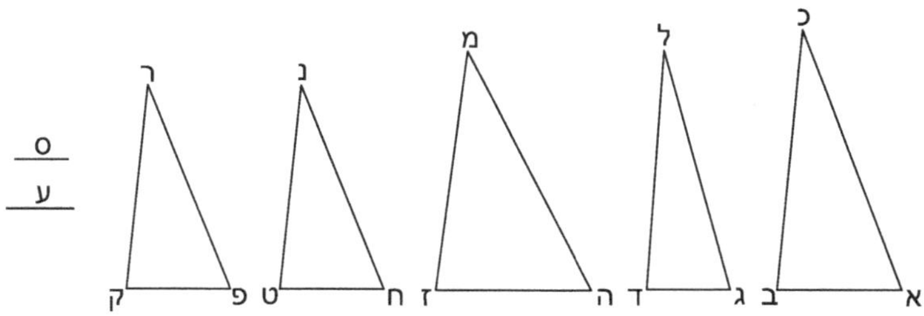

כשיהיו שטחים מתדמים על קוים מתיחסים עשוים עליהם מעשה אחד, יהיו השטחים מתיחסים. ואם היו שטחים מתדמים[1] מתיחסים ועל קוים[2] הם על מעשה אחד[3], יהיו הקוים מתיחסים.

إِذا كانَتْ خُطوطٌ مُتَناسِبَةً كَمْ كانَتْ، فَإِنَّ السُّطوحَ الَّتي تَكونُ عَلَيْها إِذا كانَتْ مُتَشابِهَةً، فَإِنَّها مُتَناسِبَةٌ، وَإِنْ كانَتْ سُطوحٌ مُتَشابِهَةً مُتَناسِبَةً عَلى خُطوطٍ، فَإِنَّ تِلْكَ الخُطوطِ أَيْضًا مُتَناسِبَةٌ.

שרטוט—B: המצולעים ح ن ط ف ر ق ه م ز הם מרובעים [1]מתדמים] ⟨⟨*⟩⟩ [2]ועל קוים] ו⟨⟨על קוים⟩⟩ [3]אחד] אפשר שמילה זו מחוקה

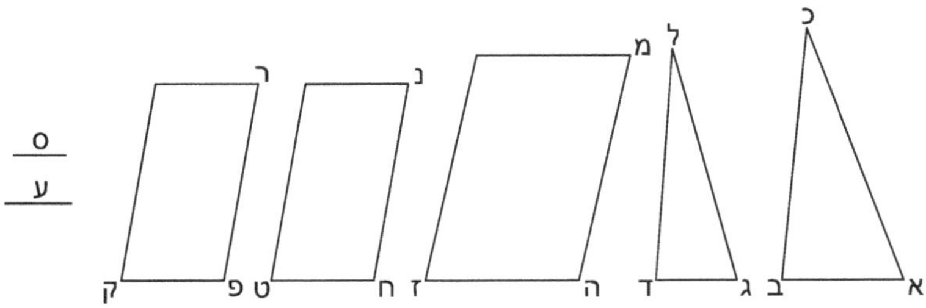

إذا كَانَتْ سُطُوحٌ متشابِهَةٌ على خُطوطٍ مُتَنَاسِبَةٍ وكَانَتْ مَعْمُولات عَلَيْها عَمَلًا واحِدًا، فَإِنَّ السُّطُوحَ مُتَنَاسِبَةٌ. وَإِنْ كانَتْ سُطُوحٌ متشابِهَةٌ مُتَنَاسِبَةً على خُطوط، وكان عَمَلُها عَمَلًا واحِدًا، فَإِنَّ الخُطوطَ مُتَنَاسِبَةٌ.

כאשר היו שטחים[1] דומים על קוים מתיחסים עשויים[2] עליהם מעשה אחד, הנה השטחים מתיחסים. ואם היו שטחים דומים[3] מתיחסים על קוים, והיה מעשיהם מעשה אחד, הנה הקוים[4] מתיחסים.

שרטוט ראשון: **ב**: $; בשני השרטוטים כל המשולשים שווי שוקיים בששת כתבי היד העבריים; בכתבי היד הערביים ממשפחה A מופיע מרובע שעליו האותיות ف, ق ו-ص; האות ر מונחת במקום האות ض בכתב יד T4 המשתייך למשפחה B: בכתבי היד הערביים ממשפחה A-1 אין מובא שרטוט לנסחה האחרת [1]שטחים] שני+ **ו** [2]עשויים] עשוי **ס** [3]שטחים דומים] שטחים **ו** השטחים מתיחסים **ס** [4]מתיחסים ואם ... הקוים] ~~מתיחסים על קוים והית והיה מעשיהם מעשה אחד הנה הקוים~~ **פ**

המשל בזה: שיהיו ארבעה קוים מתיחסים, והם א״ב וג״ד וה״ז וח״ט—שיעור א״ב אל ג״ד כשיעור ה״ז אל ח״ט. ונעשה על קוי א״ב וג״ד שני שטחים מתדמים, והם שטחי אכ״ב וגל״ד, ועל קוי ה״ז ח״ט שני שטחים אחרים מתדמים, והם שטחי המ״ז וחנ״ט. ואומר כי שיעור אכ״ב אל שטח גל״ד כשיעור שטח המ״ז אל חנ״ט.

مِثالُهُ: خُطوطُ اب وج د وه ز وح ط الأَرْبَعَةُ المُتَناسِبَةُ، قَدْرُ اب عِنْدَ ج د كَقَدْرِ ه ز عِنْدَ ح ط. ونَعْمَلُ على خَطَّيْ اب وج د سَطْحَ[يِ] اك ب و ل ج د المُتَشابِهَيْنِ. ونَعْمَلُ على خَطَّيْ ه ز وح ط سَطْحَيْ (ه م ز وح ن ط)[1] الآخَرَين المُتَشابِهَيْنِ. فَأَقولُ إِنَّ قَدْرَ سَطْحِ اك ب عِنْدَ سَطْحِ ج ل د كَقَدْرِ سَطْحِ ه م ز عِنْدَ سَطْحِ (ح ن ط)[2].

[1] ه م ز وح ن ط] ه ج ز وح ز ط [2] ح ن ط] ه ز ط

مِثَالُهُ: أَنَّ خُطوط ا ب و ج د و ه ز و ح ط مُتَنَاسِبَةٌ—نِسْبَةُ ا ب إلى ج د كنِسْبَة ه ز إلى ح ط. وعلى خَطَّيْ ا ب و ج د سَطْحا ا ك ب وج ل د، وَهُما مُتَشابِهانِ. وَقَدْ عَمِلا عَمَلًا واحدًا. وعلى خَطَّيْ ه ز وح ط سَطْحاه م ز وح ن ط، وَهُما مُتَساوِيانِ ومَعْمُولان عَمَلًا واحِدًا. فَأَقولُ إنَّ نِسْبَةَ سَطْحِ ا ك ب إلى سَطْحِ ج ل د كنِسْبَة سَطْحِ ه م ز إلى سَطْحِ ح ن ط.

המשל בו כי קוי א״ב וג״ד וה״ז וח״ט מתיחסים—יחס א״ב אל ג״ד כיחס ה״ז אל ח״ט. ועל שני קוי א״ב וג״ד[1] שני שטחי[2] אכ״ב וגל״ד, ושניהם דומים, וכבר[3] נעשו מעשה אחד. ועל שני קוי ה״ז וח״ט שני שטחי המ״ז וחנ״ט, ושניהם דומים, עשוים מעשה אחד. הנה אומר כי יחס שטח אכ״ב[4] אל שטח גל״ד[5] כיחס[6] שטח המ״ז אל שטח חנ״ט.

---

[1] א״ב וג״ד] ג״ד וא״ב **ט** [2] שני שטחי] ליתא **ז** [3] וכבר] +$\overset{*}{+}$ **ק** [4] אכ״ב] המ״ז **פ** [5] כי ... גל״ד] ⟨⟨*⟩⟩ **פ** [6] כיחס] כי יחס **פ**

המופת בזה: שנוציא לקוי א״ב וג״ד קו שלישי מתיחס להם, והוא ס׳. ולקוי ה״ז וח״ט קו שלישי, והוא קו ע׳. ואחר שקו ס׳ הוא מתיחס אל קוי א״ב וג״ד, יהיה שיעור א״ב אל ס׳ כשיעור שטח אכ״ב אל שטח גל״ד. וכמו כן שיעור ה״ז אל ע׳ כשיעור שטח המ״ז אל שטח חנ״ט. ושטח א״ב אל ג״ד כשיעור ה״ז אל ח״ט. ואמנם שיעור א״ב אל ג״ד כשיעור ג״ד אל ס׳. ושעור ה״ז אל ח״ט כשיעור ח״ט אל ע׳. ובשווי שיעור[1] א״ב אל ס׳ כשיעור ה״ז אל ע׳. ושיעור א״ב אל ס׳ כשיעור אכ״ב אל גל״ד. ושיעור ה״ז אל ע׳ כשיעור המ״ז אל חנ״ט[2]. אם כן שיעור אכ״ב אל גל״ד כשיעור המ״ז אל חנ״ט.

بُرهانُه: أنّا نَستخرِجُ لِخطَّي ا ب و ج د خطًّا ثالِثًا مُناسِبًا لَهُما، وهُوَ خطُّ س، ولِخطَّي ه ز و ح ط أيضًا خطًّا ثالِثًا مُناسِبًا لَهُما، وهُوَ خطُّ ع. فلِخطَّي ا ب و ج د خطٌّ ثالِثٌ مُناسِبٌ لَهُما، وهُوَ س. فقَدْرُ ا ب عِندَ س كقَدْرِ ا ك ب عِندَ ج ل د. وقَدْرُ ه ز عِندَ ع كقَدْرِ ه م ز عِندَ (ح ن ط)[1]. وقَدْرُ ا ب عِندَ ج د كقَدْرِ ه ز عِندَ ح ط. ولَكِنَّ قَدْرَ ا ب عِندَ ج د كقَدْرِ ج د عِندَ س. وقَدْرُ ه ز عِندَ ح ط كقَدْرِ ح ط عِندَ ع. فبِالمُساواةِ قَدْرُ ا ب عِندَ س كقَدْرِ ه ز عِندَ ع. وقَدْرُ ا ب عِندَ س كقَدْرِ ا ك ب عِندَ ج ل د. وقَدْرُ ه ز عِندَ ع كقَدْرِ (ه م ز)[2] عِندَ (ح ن ط)[3]. فقَدْرُ ا ك ب عِندَ ج ل د كقَدْرِ ه م ز عِندَ (ح ن ط)[4].

[1] שיעור] ושיעור תוקנה לנו״ן [2] חנ״ט] נכתב חכ״ט וכ״ף

[1] ح ن ط] ه ز ط [2] ه م ز] ه م ن [3] ح ن ط] ح ب ط [4] ح ن ط] ح ب ط

بُرهانُهُ: أَنَّا نَسْتَخْرِجُ خَطًّا^ ثَالِثًا مُنَاسِبًا لِخَطَّيْ ا ب وج د، وهو خطّ س^. وخَطًّا^ ثَالِثًا مُنَاسِبًا لِخَطَّيْ ه ز وح ط، وهو خطّ ع^. فنِسْبَةُ[1] خطّ ا ب إلى خطّ ج د كنِسْبَةِ خطّ ه ز إلى خطّ ح ط. ونِسْبَةَ خطّ ا ب إلى خطّ ج د كنِسْبَةِ خطّ ج د إلى خطّ س. ونِسْبَةَ خطّ ه ز إلى خطّ ح ط كنِسْبَةِ خطّ ح ط إلى خطّ ع. فَفِي نِسْبَةِ المُساواتِ تكونُ نِسْبَةُ خطِّ ا ب إلى خطِّ ال س كنِسْبَةِ خطّ ه ز إلى خطِّ ال ع. وَلِأَنَّ نِسْبَةَ خطِّ ا ب إلى خطِّ ج د كنِسْبَةِ خطّ ج د إلى خطّ ال س، تكونُ نِسْبَةُ خطِّ ا ب إلى خطِّ ال س كنِسْبَةِ خطّ ا ب إلى خطّ ج د، مُثَنَّاة. ونِسْبَة سَطْحِ ا ك ب أيضًا إلى سَطْحِ ج ل د كنِسْبَةِ خَطِّ ا ب إلى خطّ ج د. مُثَنَّاة. فنِسْبَة خطّ ا ب إلى خطّ ال س كنِسْبَةِ سَطْحِ ا ك ب إلى سَطْحِ ج ل د. وكَذَلِكَ تَبَيَّنَ أيضًا أَنَّ نِسْبَةَ خطِّ ه ز إلى خطِّ ال ع كنِسْبَةِ سَطْحِ ه م ز إلى سَطْحِ ح ن ط. وَقَدْ كُنَّا بَيَّنَّا أَنَّ نِسْبَةَ خَطِّ ا ب إلى خَطِّ ال س كنِسْبَةِ خَطِّ ه ز إلى خطِّ ال ع. فنِسْبَة سَطْحِ ا ك ب إلى سَطْحِ ج ل د كنِسْبَةِ سَطْحِ ه م ز إلى سَطْحِ ح ن ط.

[1] فنسبة] MPR2 + الأول

המופת: אנחנו נוציא קו ס׳[1] שלישי מתיחס לשני קוי א״ב וג״ד, וקו ע׳ שלישי מתיחס לשני קוי ה״ז וח״ט. הנה יחס א״ב אל ג״ד כיחס ה״ז אל ח״ט. ויחס א״ב אל ג״ד כיחס[2] ג״ד אל ס׳[3]. ויחס ה״ז אל ח״ט כיחס ח״ט אל ע׳. הנה ביחס השווי יהיה יחס[4] א״ב אל ס׳ כיחס ה״ז אל ע׳. ומפני כי יחס א״ב אל ג״ד כיחס ג״ד אל ס׳[5], יהיה יחס א״ב אל ס׳ כיחס א״ב אל ג״ד, שנוי. ויחס אכ״ב גם כן אל גל״ד כיחס א״ב[6] [7] אל ג״ד, שנוי[8]. אם כן יחס א״ב אל ס׳ כיחס אכ״ב אל גל״ד. וכן גם כן התבאר כי יחס ה״ז אל ע׳ כיחס המ״ז אל חנ״ט. וכבר ביארנו כי יחס א״ב אל ס׳ כיחס ה״ז אל ע׳. אם כן יחס אכ״ב אל גל״ד כיחס המ״ז אל חנ״ט.

[1] ס׳] ~~ה׳~~ ⟨⟨*⟩⟩ פ [2] כיחס ... כיחס] ⟨⟨*⟩⟩ ק [3] ס׳] ~~ה׳ נ׳~~+ פ [4] יחס] ליתא פ [5] ס׳] ~~ה׳~~+ פ [6] ויחס ... א״ב] אם כן יחס א״ב אל ס׳ כיחס אכ״ב ק [7] א״ב] א״ה ב [8] ויחס ... שנוי] ליתא ז

ונשים כמו כן שיעור אכ״ב אל גל״ד המתדמים כשיעור המ״ז אל חנ״ט המתדמים. ואומר שיהיה שיעור א״ב אל ג״ד כשיעור ה״ז אל ח״ט.

وَنَجْعَلُ أَيضًا قَدْرَ ا ك ب عِنْدَ ج ل د المُتَشابِهَيْنِ كَقَدْرِ ه م ز عِنْدَ (ح ن ط)[1] المُتَشابِهَيْنِ. فَأَقولُ إِنَّ قَدْرَ ا ب عِنْدَ ج د كَقَدْرِ ه ز عِنْدَ ح ط.

מופת זה: שנשים שיעור א״ב אל ג״ד כשיעור ה״ז אל פ״ק. ונצייר על קו פ״ק שטח דומה לשטח חנ״ט ועל תארו, והוא שטח פק״ר. ויהיה שיעור א״ב אל ג״ד כשיעור ה״ז אל פ״ק[1]. ויהיה שעור שטח[2] אכ״ב אל שטח גל״ד כשיעור המ״ז אל פק״ר[3]. ושיעור המ״ז אל חנ״ט כשעור המ״ז אל פק״ר. אם כן המ״ז יחסו אל חנ״ט ואל פק״ר אחד. ואחר שהוא כן, חנ״ט שוה לפק״ר. ויחס ה״ז אל פ״ק כיחסו אל ח״ט. אם כן פ״ק שוה לח״ט. ושעור א״ב אל ג״ד כשיעור ה״ז אל פ״ק. ואם כן שיעור א״ב אל ג״ד כשיעור ה״ז אל ח״ט.

بُرْهانُهُ: أَنَّا نَجْعَلُ قَدْرَ ا ب عِنْدَ ج د كَقَدْرِ ه ز عِنْدَ ف ق. وَنَخُطُّ عَلى خَطِّ ف ق سَطْحَ (ف ر ق)[2] شَبيهًا بِسَطْحِ (ح ن ط)[3] وَعَلى خِلْقَتِهِ. فَقَدْرُ ا ب عِنْدَ ج د كَقَدْرِ ه ز عِنْدَ ف ق. فَقَدْرُ ا ك ب عِنْدَ ج ل د كَقَدْرِ ه م ز عِنْدَ (ف ر ق)[4]. وَقَدْرُ ا ك ب عِنْدَ ج ل د كَقَدْرِ ه م ز عِنْدَ (ح ن ط)[5]. فَقَدْرُ ه م ز عِنْدَ (ح ن ط)[6] كَقَدْرِه م ز عِنْدَ (ف ر ق)[7]. ه م ز نِسْبَتُهُ إِلى (ح ن ط)[8] وَإِلى ف ق ر واحِدَةٌ. (ح ن ط)[9] مِثْلُ ف ق ر. وَنِسْبَةُ ه ز إِلى ف ق مِثْلُ نِسْبَتِهِ إِلى ح ط. ف ق مِثْلُ ح ط. وَقَدْرُ ا ب عِنْدَ ج د كَقَدْرِ ه ز عِنْدَ ف ق. فَقَدْرُ ا ب عِنْدَ ج د كَقَدْرِ ه ز عِنْدَ ح ط.

וזה מה שרצינו לבאר.

وَذَلِكَ ما أَرَدْنا بَيانَهُ.

[1] פ״ק] + ~~ונצייר על קו פ״ק שטח דומה לשטח חנ״ט~~
[2] ויהיה ... שטח] ז⟨⟨ויהיה שעור⟩⟩ שטח
[3] פק״ר] ~~חנ״ט~~ ⟨⟨*⟩⟩

[1] ح ن ط] ه ز ط [2] ف ر ق] ف ز ق [3] ح ن ط] ه ز ط [4] ف ر ق] ف ز ق [5] ح ن ط] ه ز ط [6] ح ن ط] ه ز ط [7] ف ر ق] ف ز ق [8] ح ن ط] ه ز ط [9] ح ن ط] ه ز ط

ثُمَّ نَجْعَلُ نِسْبَةَ سَطْحِ ا ك ب إلى سَطْحِ ج ل د المُتشابِهَيْنِ كنِسْبَةِ سَطْحِ ه م ز إلى سَطْحِ ح ن ط المُتَشابِهَيْنِ[1]. فأَقولُ إنَّ نِسْبَةَ خطِّ ا ب إلى خطِّ ج د كنِسْبَةِ خَطِّ ه ز إلى خطِّ ح ط.

עוד נשים יחס אכ״ב אל גל״ד המתדמים כיחס המ״ז אל חנ״ט המתדמים. הנה אומר כי יחס א״ב אל ג״ד כיחס ה״ז אל ח״ט.

بُرْهانُهُ: أنَّا نَجْعَلُ نِسْبَةَ خطِّ ا ب إلى ج د كَنِسْبَةِ ه ز إلى ف ق. ونَعْمَلُ^ على ف ق سَطْحَ ف ر ق شَبيهًا بِسَطْحِ ح ن ط ومَعْمُولًا كَعَمَلِهِ. فنِسْبَةُ خطِّ ا ب إلى ج د كَنِسْبَةِ ه ز إلى ف ق. ونِسْبَةُ سَطْحِ ا ك ب إلى سَطْحِ ج ل د كَنِسْبَةِ سَطْحِ ه م ز إلى سَطْحِ ف ر ق. وقَدْ كانَتْ نِسْبَةُ سَطْحِ ا ك ب إلى سَطْحِ ج ل د كَنِسْبَةِ سَطْحِ ه م ز إلى سَطْحِ ح ن ط. فنِسْبَةُ سَطْحِ ه م ز إلى ح ن ط ف ر ق واحِدَةٌ. فَهُما إذَنْ مُتَساوِيانِ وخَطُّ ف ق نَظيرٌ^ في النِّسْبَةِ لِخَطِّ ح ط. فَهُما مُتَساوِيانِ. ونِسْبَةُ ا ب إلى ج د كَنِسْبَةِ ه ز إلى ف ق. فنِسْبَةُ ا ب إلى ج د كَنِسْبَةِ ه ز إلى ح ط.

המופת: אנחנו נשים יחס א״ב אל ג״ד כיחס ה״ז אל פ״ק. ונשים על פ״ק שטח פק״ר, דומה לשטח חנ״ט ועשוי כמעשהו. אם כן יחס א״ב אל ג״ד כיחס ה״ז אל פ״ק. ויחס אכ״ב אל גל״ד כיחס המ״ז אל פק״ר. וכבר היה יחס אכ״ב אל גל״ד כיחס המ״ז אל חנ״ט[1]. אם כן יחס המ״ז אל חנ״ט ופק״ר אחד[2]. הנה[3] שניהם אם כן שוים וקו פ״ק נכחי לו ביחס ח״ט. אם כן שניהם שוים. ויחס א״ב אל ג״ד כיחס ה״ז אל פ״ק. אם כן יחס א״ב אל ג״ד כיחס ה״ז אל ח״ט.

وذَلِكَ ما أَرَدْنا أَنْ نُبيِّنَ.

וזה מה שרצינו לבאר.

[1]المتشابهين] MR2; במשפט זה יש הבדלים רבים בין הנוסח בכתבי היד הערביים ממשפחה A-1 לבין הנוסח בשאר כתבי היד ממשפחה A, והנוסח בתרגום אבן תיבון נמשך בעיקרו אחר הנוסח במשפחה A-1

[1]וכבר ... חנ״ט] ליתא ק [2]אחד] ליתא ו [3]הנה] ליתא ק

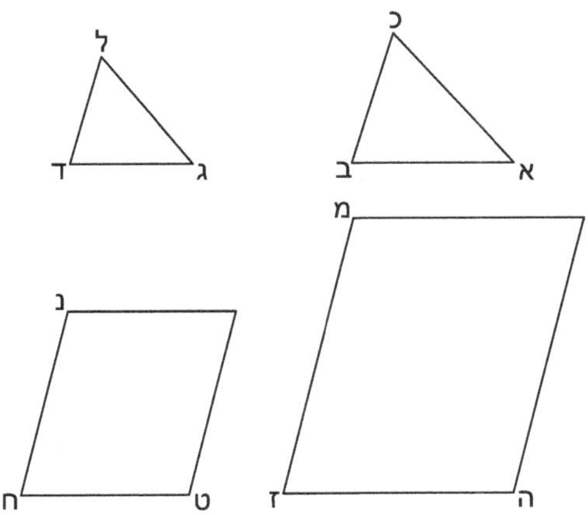

وَجَدْنا في نُسْخَةٍ أُخرى بِغَيْرِ هذا الطَّريقِ^:

وَهُوَ مِنْ أَجْلِ أَنَّ نِسْبَةَ سَطْحِ ا ك ب إلى ج ل د هي كَنِسْبَةِ ا ب إلى ج د، مُثَنَّاة. ونِسْبَةَ ه م ز إلى ح ن ط كَنِسْبَةِ ه ز إلى ح ط، مُثَنَّاة. ونِسْبَةَ ا ب إلى ج د كَنِسْبَةِ ه ز إلى ح ط. فتكونُ نِسْبَةُ ا ك ب إلى ج ل د كَنِسْبَةِ ه م ز إلى ح ن ط. وَأَيْضًا فَإِنَّا نَجْعَلُ نِسْبَةَ ا ك ب إلى ج ل د كَنِسْبَةِ ه م ز إلى ح ن ط. فَأَقولُ إِنَّ نِسْبَةَ ا ب إلى ج د كَنِسْبَةِ ه ز إلى ح ط.

מצאנו בנסחא אחרת בזולת זה המופת:

והוא מפני[1] כי יחס אכ״ב אל גל״ד כיחס א״ב אל[2] ג״ד[3], שנוי. ויחס המ״ז אל חנ״ט[4] כיחס ה״ז אל ח״ט, שנוי. ויחס א״ב אל ג״ד כיחס ה״ז אל ח״ט[5]. אם כן[6] יחס אכ״ב אל גל״ד כיחס המ״ז אל חנ״ט.

וגם כן הנה אנחנו[7] נשים יחס אכ״ב אל גל״ד כיחס המ״ז אל חנ״ט. הנה אומר כי יחס א״ב אל ג״ד[8] כיחס ה״ז אל ח״ט.

---

1 מפני] ⟨⟨*⟩⟩ **ק** 2 אל] ⟨⟨*⟩⟩ **פ** 3 א״ב ... ג״ד] אבג״ד **ב** 4 כיחס ... חנ״ט] ⟨⟨*⟩⟩ **ס** 5 ח״ט] +~~שנוי ויחס א״ב אל ג״ז~~ **ב** 6 כן] +* **ב** 7 אנחנו] ליתא **ו** 8 ג״ד] ~~גל״ד~~ ⟨*⟩ **ק**

فَلِأَنَّ^ نِسْبَةَ ا ك ب إلى ج ل د كَنِسْبَةِ ا ب إلى ج د، مُثَنَّاةً. وَأَنَّ سَطْحَ ه م ز نِسْبَتُهُ^ إلى سَطْحِ ح ن ط كَنِسْبَةِ خَطِّ ه ز إلى خَطِّ ح ط. مُثَنَّاةً. وَأَنَّ نِسْبَةَ سَطْحِ ا ك ب إلى سَطْحِ ج ل د فُرِضَتْ^ كَنِسْبَةِ سَطْحِ ه م ز إلى سَطْحِ ح ن ط، تكونُ نِسْبَةُ ا ب إلى ج د كَنِسْبَةِ ه ز إلى ح ط.

وذَلِكَ ما أَرَدْنا أَنْ نُبَيِّنَ.

הנה[1] יחס אכ״ב אל גל״ד כיחס א״ב אל ג״ד, שנוי. ויחס המ״ז אל חנ״ט כיחס ה״ז אל ח״ט, שנוי. ויחס אכ״ב אל גל״ד[2] כיחס המ״ז אל חנ״ט. הנה יהיה יחס א״ב אל ג״ד כיחס ה״ז אל ח״ט.

וזה מה שרצינו לבאר.

[1]הנה] כי מפני כי ז [2]גל״ד] +כיחס א״ב אל ג״ד שנוי ויחס המ״ז אל חנ״ט כיחס ה״ז אל ח״ט שנוי ויחס אכ״ב אל גל״ד כיחס א״ב אל ג״ד ויחס א״ב אל ג״ד ס

[Proposition VI.23]

[H23, B24, RJ25]

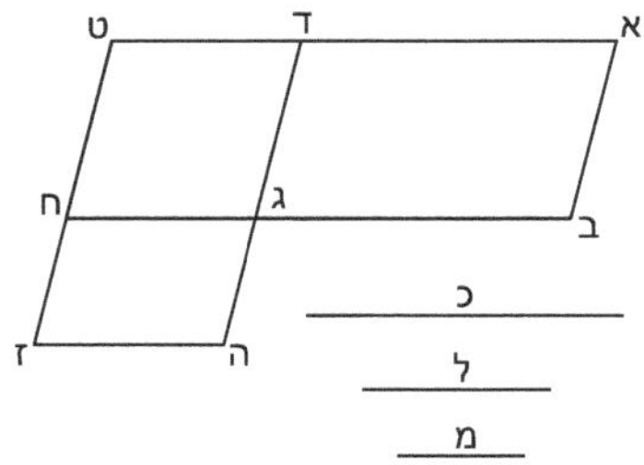

כשתהיין זויות שטחי נכוחי הצלעות שוות, יהיה יחס אחד מהם אל האחר מחובר מיחס צלעותם.

السُّطوحُ المُتَوازِيَةُ الأَضْلاعِ، إِذا كانَتْ زَوايا‌ها مُتَساوِيَةَ، فَإِنَّ نِسْبَةَ السَّطْحِ مِنْها [!] إِلى السَّطْحِ الآخَرِ كَنِسْبَةِ أَحَدِ ضِلْعَي المَنْسوبِ المُحيطَيْنِ بِنَظيرِ تِلْكَ الزَّاوِيَةِ مُثَنَّاةً، نِسْبَةُ الضِّلْعِ الآخَرِ [!] مِنَ المَنْسوبِ إِلى الضِّلْعِ الآخَرِ مِنَ المَنْسوبِ إِلَيْهِ.

המשל בזה: ששטח [!] א״ג וג״ז נכוחי הצלעות, וזויות האחד שוות לזויות האחר. ואומר שיחס שטח א״ג אל שטח ג״ז הוא אשר יתחבר[1] מיחס צלעותם.

مِثالُهُ: أَنَّ سَطْحَيْ اج وج ز مُتَوازِيَي الأَضْلاعِ، وَزَواياهما مُتَساوِيَةٌ. فَأَقولُ إِنَّ نِسْبَةَ سَطْحِ اج إِلى سَطْحِ ج ز هِيَ الَّتي كانَتْ مِنْ نِسْبَةِ أَضْلاعِهِا.

[1] יתחבר] ⟨⟨*⟩⟩

إذا تَساوَتْ زَوايا سَطْحَيْنِ مُتَوازِيَيْ الأَضْلاعِ، فَنِسْبَةُ أَحَدِهِما إلى الآخَرِ مُؤَلَّفَةٌ مِنْ نِسْبَةِ أَضْلاعِهِما.

כאשר תשתוינה זויות שני שטחים נכחיי הצלעות, הנה יחס אחד משניהם אל האחר מחובר מיחס צלעות שניהם.

مِثالُهُ: أَنَّ سَطْحَيْ اَجَ وجَ زَ مُتَوازيَا الأَضْلاعِ. وَزَوايا سَطْحِ اَجَ مُساوِيَةٌ لِزَوايا سَطْحِ جَ ز. فَأَقولُ إنَّ نِسْبَةَ سَطْحِ اَجَ إلى سَطْحِ جَ ز مُؤَلَّفَةٌ مِنْ نِسْبَةِ أَضْلاعِهِما أَعْنِي^ مِنْ نِسْبَةِ خَطِّ بَ جَ إلى خَطِّ جَ حَ وَمِنْ نِسْبَةِ خَطِّ دَجَ إلى خَطِّ جَ ه.

המשל בו: כי שני שטחי א״ג וג״ז[1] נכחיי הצלעות. וזויות שטח א״ג שוות לזויות שטח ג״ז. הנה אומר כי יחס א״ג[2] אל[3] ג״ז[4] מחובר מיחס צלעות שניהם—מיחס ב״ג אל ג״ח ומיחס ד״ג אל ג״ה.

כמו במשפט הקודם גם במשפט זה יש הבדלים בולטים בין הנוסח בכתבי היד ממשפחה A-1 לנוסח בכתבי היד האחרים במשפחה שממנה התפצל, ואולם כאן יש גם הבדלים בולטים בין הנוסח בתרגום אבן תיבון לבין הנוסח בכתבי היד ממשפחה A-1, בייחוד בחלק המרכזי של ההוכחה

שרטוט—**ב**: ה׳ וז׳ חסרות [1]וג״ז] +א״ט וג״ז **ס** [2]א״ג] ⟨⟨שטח⟩⟩+ **פ** [3]נכחיי ... אל] ליתא **ב** [4]ג״ז] ⟨⟨שטח⟩⟩+ **פ**

بُرْهانُهُ: أَنَّا نُتِمُّ سَطْحَ (ج ط)[1]. وَنَضَعُ (ثَلاثَةَ)[2] خُطوطٍ عَلَيْها ك و ل و م. وَنَجْعَلُ قَدْرَ ب ج عِنْدَ ج ح كَقَدْرِ ك عِنْدَ ل. وَقَدْرَ د ج عِنْدَ ج ه كَقَدْرِ ل عِنْدَ م. فَنِسْبَةُ ب ج إِلى ج ح مُثَنَّاةً بِنِسْبَةِ د ج إِلى ج ه، كَنِسْبَةِ ك إِلى ل، مُثَنَّاةً بِنِسْبَةِ ل إِلى م. وَلكِنَّ نِسْبَةَ ك إِلى ل، مُثَنَّاةً بِنِسْبَةِ ل إِلى م، هِيَ نِسْبَةُ ك إِلى م. فَنِسْبَةُ ك إِلى م هِيَ نِسْبَةُ ب ج إِلى ج ح، مُثَنَّاةً بِنِسْبَةِ د ج إِلى ج ه. وَقَدْرُ ب ج عِنْدَ ج ح كَقَدْرِ ا ج عِنْدَ ج ط. وَقَدْرُ ب ج عِنْدَ ج ح كَقَدْرِ ك عِنْدَ ل. فَقَدْرُ ك عِنْدَ ل كَقَدْرِ ا ج عِنْدَ ج ط. وَأَيْضًا قَدْرُ د ج عِنْدَ ج ه كَقَدْرِ ج ط عِنْدَ ج ز. وَقَدْرُ د ج عِنْدَ ج ه كَقَدْرِ ل عِنْدَ م. فَقَدْرُ ل عِنْدَ م كَقَدْرِ ج ط عِنْدَ ج ز. وَقَدْ كانَ قَدْرُ ك عِنْدَ ل كَقَدْرِ ا ج عِنْدَ ج ط. فَنِسْبَةُ ك إِلى م مِثْلُ نِسْبَةِ ا ج إِلى ج ز. فَنِسْبَةُ ك إِلى م هِيَ الَّتي قامَتْ مِنْ نِسْبَةِ الأَضْلاعِ. وَكَذٰلِكَ نِسْبَةُ ا ج إِلى ج ز هِيَ الَّتي قامَتْ مِنْ نِسْبَةِ أَضْلاعِهِما.

המופת בזה: שנשלים שטח ג״ט. ונניח שלשה קוים עליהם כ׳ ל׳ מ׳. ונשים שיעור ב״ג אל ג״ח כשיעור כ׳ אל ל׳. ושיעור ד״ג אל ג״ה כשיעור ל׳ אל מ׳. ויהיה יחס ב״ג אל ג״ח נשנה ביחס ד״ג אל ג״ה, כיחס כ׳ אל ל׳, נשנה ביחס ל׳ אל מ׳. ואמנם יחס כ׳ אל ל׳ נשנה ביחס ל׳ אל מ׳, הוא יחס כ׳ אל מ׳. ויהיה יחס כ׳ אל מ׳ כיחס ב״ג אל ג״ח, נשנה ביחס ד״ג אל ג״ה. ושיעור ב״ג אל ג״ח כשיעור א״ג אל ג״ט. ושיעור ב״ג אל ג״ח כשיעור כ׳ אל ל׳. ואם כן[1] יהיה שיעור כ׳ אל ל׳ כשיעור א״ג אל ג״ט. וכמו כן שיעור ד״ג אל ג״ה כשיעור ג״ט אל ג״ז[2]. ושיעור ג״ד אל ג״ה כשיעור ל׳ אל מ׳. ואם כן יהיה שיעור ל׳ אל מ׳ כשיעור ג״ט אל ג״ז. וכבר היה שיעור כ׳ אל ל׳ כשיעור א״ג אל ג״ט. אם כן יחס כ׳ אל מ׳ כיחס א״ג אל ג״ז. ויחס כ׳ אל מ׳ הוא אשר חובר מיחס הצלעות. וכמו כן יחס א״ג אל ג״ז הוא היחס אשר חובר מיחס צלעותם.

[1] ج ط] ا ج ط [2] ثلاثة] ثلثة

[1] ואם כן] ו⟨⟨אם כן⟩⟩ [2] ג״ז] ג״[·]+

بُرهانُهُ: أَنَّا نَجْعَلُ خَطَّيْ ب ج و ج ح مُتَّصِلَيْن على اسْتِقَامَةٍ. فيكونُ خَطُّ د ج على اسْتِقَامَةِ خَطِّ ج ه. ونُتِمُّ سَطْحَ ج ط. وَنَضَعُ خُطوطَ ال ك وال ل وال م. ونَجْعَلُ نِسْبَةَ خَطِّ ال ك إلى خَطِّ ال ل كَنِسْبَةِ خَطِّ ب ج إلى خَطِّ ج ح. ونَجْعَلُ نِسْبَةَ خَطِّ ال ل إلى خَطِّ ال م كَنِسْبَةِ خَطِّ د ج إلى خَطِّ ج ه. فالنِّسْبَةُ المُؤَلَّفةُ مِنْ خَطِّ ب ج إلى خَطِّ ج ح ومِنْ نِسْبَةِ خَطِّ د ج إلى خَطِّ ج ه كَنِسْبَةِ خَطِّ ال ك إلى خَطِّ ال م. ونِسبة خَطِّ ب ج إلى خطّ ج ح كَنِسْبَةِ سَطحِ ا ج إلى سَطْحِ ج ط وكَنِسْبَةِ خَطِّ ال ك إلى خَطِّ ال ل. ونِسْبَة خطّ د ج إلى خطّ ج ه كَنِسْبَةِ سَطحِ ج ط إلى سَطحِ ج ز وكَنِسْبَةِ خَطِّ ال ل إلى خَطِّ ال م. فَهِيَ نِسْبَةُ المُساواتِ—تكونُ نِسْبَةُ خَطِّ ال ك إلى خطّ ال م كَنِسْبَةِ سَطحِ ا ج إلى سَطْحِ ج ز^. وقَدْ كُنَّا بيَّنا أَنَّ نِسْبَةَ خَطِّ ال ك إلى خطِّ ال م مؤَلَّفةٌ مِنْ نِسْبَةِ خَطِّ ب ج إلى خَطِّ ج ح ومِنْ نِسْبَةِ خَطِّ د ج إلى خَطِّ ج ه. فنِسْبَة سَطْحِ ا ج إلى سَطحِ ج ز مؤَلَّفَةٌ مِنْ نِسْبَةِ خَطِّ ب ج إلى خَطِّ ج ح ومِنْ نِسْبَةِ خَطِّ د ج إلى خَطِّ ج ه.

המופת: אנחנו נשים ב״ג וג״ח מדובקים על יושר. ויהיה קו ד״ג על יושר קו ג״ה. ונשלים שטח ג״ט. ונניח קו כ׳. ונשים יחסו אל ל׳ כיחס ב״ג אל ג״ח. ויחס ד״ג אל ג״ה כיחס ל׳ אל מ׳. הנה יחס ב״ג אל ח״ג שינוי ביחס ד״ג אל ג״ה כיחס[1] הכ׳[2] אל הל׳ שנוי ביחס הל׳ אל המ׳[3]. אבל יחס הכ׳ אל הל׳ שנוי[4] ביחס הל׳ אל המ׳ הוא יחס הכ׳ אל המ׳. הנה יחס הכ׳ אל הל׳ שנוי ביחס[5] הל׳ אל המ׳ כיחס[6] [7] הכ׳ אל המ׳[8]. אם כן יחס הכ׳ אל המ׳ הוא מחובר מיחס ב״ג אל ג״ח ומיחס ד״ג אל ג״ה. ויחס ב״ג אל ג״ח כיחס א״ג אל ג״ט, וכיחס כ׳ אל ל׳. ויחס ד״ג[9] אל ג״ה כיחס ט״ג אל ג״ז וכיחס ל׳ אל מ׳. והיה יחס א״ג אל ג״ט כיחס כ׳ אל ל׳. ויחס ג״ט אל ג״ז כיחס ל׳[10] אל מ׳. הנה בשווי יהיה יחס א״ג אל ג״ז כיחס כ׳ אל מ׳. וכבר ביארנו כי יחס כ׳ אל מ׳[11] מחובר מיחס ב״ג אל ג״ח ומיחס ד״ג אל ג״ה. אם כן יחס שטח א״ג אל שטח ג״ז מחובר מיחס ב״ג אל ג״ח ומיחס ד״ג אל ג״ה.

---

[1]ביחס] הוא+ ס [2]הכ׳] כאן ובהמשך ה׳ הידיעה מושמטת לפני ציוני קווים כ׳ ל׳ ומ׳ בעד נוסח ז; השמטה דומה קיימת בעד נוסח ס אך היא בלתי עקיבה [3]אבל ... המ׳] שני המשפטים האלה נראים זהים מבחינת תכנם ואפשר שאחד מהם מיותר [4]שנוי] +⟨⟨ביחס ל׳ אל מ׳ אבל יחס כ׳ אל ל׳ שנוי⟩⟩ ק [5]הל׳ אל ... ביחס] ליתא ס [6]יחס ... כיחס] ליתא ב [7]כיחס] הוא יחס ס [8]וכבר ... מ׳] ליתא פ [9]ד״ג] ליתא ב [10]אל ... ל׳] אל א״ז כיחס א׳ ס [11]וכבר ... מ׳] ליתא פ

וזה מה שרצינו לבאר[1].

وَذَلِكَ ما أَرَدْنا بَيانَهُ.

[1]לבאר] +תמצא הצורה בדף השני

وذَلِكَ ما أَرَدْنا أَنْ نُبيِّنَ.

וזה מה שרצינו לבאר[1].

[1] וזה ... לבאר] ונשלם ביאורו **בספק**

[Proposition VI.24]

[H24, B22, RJ23]

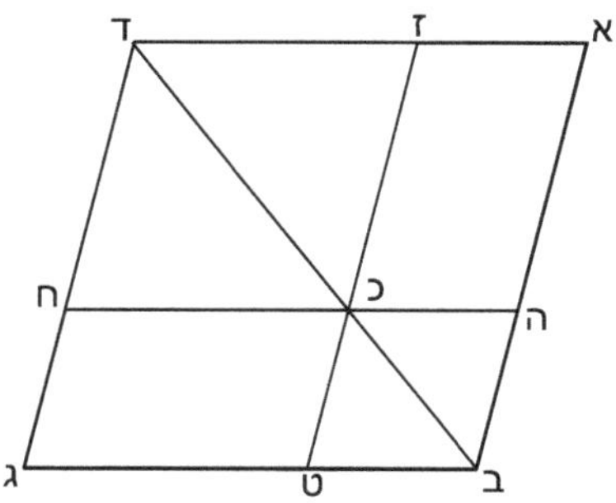

כל שטח נכוחי הצלעות אשר על אלכסונו שטחים נכוחי הצלעות, ידמה הוא אליהם ויהיו הם מתדמים.

السُّطوحُ المُتَوازِيَةُ الأَضْلاعِ الَّتي تَكونُ عَلى قُطْرِ سَطْحٍ مُتَوازي الأَضْلاعِ، فَإِنَّ كُلَّ واحِدٍ مِنْها نِسْبَتُهُ السَّطْحَ الأَعْظَمِ المُتَوازي الأَضْلاعِ وَنِسْبَةَ كُلِّ واحِدٍ مِنْهُما الآخَرَ.

המשל בו: ששטח אבג״ד נכוחי הצלעות. ואלכסונו ב״ד. ועל אלכסונו שני שטחים נכוחי הצלעות, והם ה״ט וח״ז. ואומר כי שטחי ה״ט וח״ז הם דומים לשטח א״ג. וכל אחד מהם ידמה לאחר.

مِثالُهُ: أَنَّ سَطْحَ ا ج مُتَوازي الأَضْلاعِ. وَقُطْرُهُ ب د. وَعَلى قُطْرِهِ سَطْحا ه ط وح ز المُتَوازي الأَضْلاعِ. فَأَقولُ إِنَّ سَطْحَيْ ه ط وح ز يُشْبِهانِ سَطْحَ ا ج، وَيُشْبِهُ كُلُّ واحِدٍ مِنْهُما الآخَرَ.

كُلُّ سَطْحٍ مُتَوازِي الأَضْلاعِ على قُطْرِهِ سُطُوحٌ مُتَوازِيَةُ الأَضلاعِ، فَهُوَ يُشْبِهُهُا وَهِيَ مُتَشابِهَةٌ.

כל שטח נכחי הצלעות על קוטרו שטחים נכחיי הצלעות, הנה הוא דומה להם והם מתדמים.

مِثَالُهُ: أَنَّ سَطْحَ ا ب ج د مُتَوازِي الأَضْلاعِ. وَعَلَى قُطْرِهِ، وَهُوَ ب د، سَطْحَا ه ط و زح وَهُمَا^ مُتَوازِيا الأَضْلاعِ. فَأَقولُ إنَّ سَطْحَيْ ه ط و زح يُشْبِهانِ سَطْحَ ا ج وَهُما^ مُتَشابِهانِ.

המשל בו: כי שטח אבג״ד[1] נכחי הצלעות. ועל קוטרו, והוא ב״ד, שטחי ה״ט וז״ח נכחי הצלעות. הנה אומר כי שני שטחי ה״ט וז״ח דומים לשטח א״ג ושניהם מתדמים.

שרטוט—$ על ציר א״ג   [1]אבג״ד] נכחיי הצלעות+ פ

מופת זה: כי משלש בג״ד כבר יצא מצלעו קו נכוחי לצלע ב״ג והוא כ״ח. אם כן שיעור ב״כ אל כ״ד כשיעור ג״ח אל ח״ד. וכמו כן משולש אב״ד כבר יצא מצלעו קו נכוחי לצלע א״ב, והוא קו ז״כ. אם כן שיעור ב״כ אל כ״ד כשיעור א״ז אל ז״ד. וכבר היה שיעור ב״כ אל כ״ד כשיעור ג״ח אל ח״ד. אם כן יהיה שיעור א״ז אל ז״ד כשיעור ג״ח אל ח״ד. וכאשר יוחסו על ההמשך ואחר כן הורכבו[1], היה שיעור א״ד אל ד״ז כשיעור ד״ג אל ד״ח. והצלעות המקיפות בזוית ד׳ מתיחסות. אם כן ז״ח ידמה לא״ג. ולזה גם כן ה״ט ידמה לא״ג. ואומר כי ז״ח וה״ט הם מתדמים. לפי שכל אחד מהם ידמה לא״ג. והדומים לשטח אחד כל אחד מהם ידמה לאחר. אם כן ז״ח ידמה לה״ט.

וזה מה שרצינו לבאר.

[1]הורכבו] ה‹ו›רכבו

بُرْهانُهُ: أَنَّ مُثَلَّثَ ب ج د قَدْ خَرَجَ مِنْ ضِلْعٍ لَهُ خَطُّ ح كَ مُوازِيًا لِضِلْعِ ب ج. فَقَدْرُ ب كَ عِنْدَ كَ د كَقَدْرِ ج ح عِنْدَ ح د. وَأَيْضًا مُثَلَّثُ ا ب د قَدْ خَرَجَ مِنْ ضِلْعٍ لَهُ خَطُّ ز كَ مُوازِيًا لِضِلْعِ ا ب. فَقَدْرُ ب كَ عِنْدَ كَ د كَقَدْرِ ا ز عِنْدَ ز د. وَقَدْ كانَ قَدْرُ ب كَ عِنْدَ كَ د كَقَدْرِ ج ح عِنْدَ ح د. فَقَدْرُ ا ز عِنْدَ (ز د)[1] كَقَدْرِ ج ح عِنْدَ ح د. وَإِذا جَعَلْنا بَعْضَها يَتْلو بَعْضًا، كانَ قَدْرُ ا د عِنْدَ د ز كَقَدْرِ د ج عِنْدَ د ح. وَالأَضْلاعُ الَّتي تُحيطُ بِزاوِيَةِ د مُتَناسِبَةٌ. ف ز ح (يُشْبِهُ)[2] ا ج. وَكَذَلِكَ ه ط يُشْبِهُ ا ج. فَأَقولُ إِنَّ ز ح و ه ط مُتَشابِهان. بُرْهانُهُ: أَنَّ كُلَّ واحِدٍ مِنْهُما (يُشْبِهُ)[3] ا ج. وَالسُّطوحُ كَسَطْحٍ واحِدٍ مُتَشابِهَةٌ. ز ح مِثْلُ نِسْبَةِ ه ط.

وَذَلِكَ ما أَرَدْنا بَيانَهُ.

[1]ز د] ز كَ [2]يشبه] نسبه [3]يشبه] نسبته

بُرْهانُهُ: أَنَّ كـ ح يُوازِي في مُثَلَّثِ د ب ج خَطَّ ب ج. فَنِسْبَةَ ب كـ إلى كـ د كَنِسْبَةِ ج ح إلى خطّ ح د. وَأَيْضًا فَإِنَّ خَطَّ ز كـ يُوازِي في مُثَلَّثِ ا ب د خَطَّ ا ب. فَنِسْبَةَ خَطِّ ب كـ إلى خَطِّ كـ د كَنِسْبَةِ ا ز إلى ز د. وَقَدْ كانَتْ نِسْبَةُ ب كـ إلى كـ د كَنِسْبَةِ ج ح إلى ح د. فَنِسْبَةَ ا ز إِذَنْ[1] إلى ز د كَنِسْبَةِ ج ح إلى ح د. وَإِذ رَكَّبْنَا كانَتْ نِسْبَةُ ا د إلى د ز كَنِسْبَةِ ج د إلى د ح. فَالأَضْلاعُ المُحِيطَةُ بِزاوِيَةِ ا د ج مُتَناسِبَةٌ. فَسَطْحُ ز ح يُشْبِهُ سَطْحَ ا ج. وكَذَلِكَ يَكونُ سَطْحُ ه ط أيضًا يُشْبِهُ سَطْحَ ا ج. فَسَطْحُ ز ح يُشْبِهُ سَطْحَ ه ط، لِأَنَّ كِلَيْهِما^ يُشْبِهانِ سَطْحَ ا ج.

וְذَلِكَ ما أَرَدْنا أَنْ نُبَيِّنَ.

המופת: כי כ״ח נכחי במשולש בג״ד לב״ג[1]. אם כן יחס ב״כ אל כ״ד כיחס ח״ג[2] אל ח״ד. וגם כן הנה ז״כ[3] נכחי לא״ב[4] במשולש אב״ד. אם כן יחס ב״כ אל כ״ד כיחס[5] א״ז אל ז״ד. וכבר היה יחס ב״כ אל כ״ד כיחס ג״ח אל ח״ד. הנה[6] יחס א״ז אם[7] כן אל ז״ד[8] [9] כיחס ג״ח אל ח״ד. וכאשר הרכבנו היה יחס א״ד אל ד״ז כיחס ג״ד[10] אל ד״ח. אם כן הצלעות המקיפות בזוית אד״ג מתיחסות. אם כן שטח ז״ח ידמה א״ג. וכן יהיה ה״ט גם כן[11] [12] ידמה א״ג. אם כן ז״ח ידמה ה״ט, מפני כי שניהם ידמו א״ג.

ונשלם ביאורו.

---

[1]إذن] P; במשפט זה כמו בקודמו רבים ההבדלים בין הנוסח ברוב כתבי היד הערביים לבין הנוסח בכתבי היד ממשפחה A-1, והנוסח בתרגום אבן תיבון נמשך אחר עדי הנוסח ממשפחה זו

---

[1]לב״ג] כמ״ג **ס** [2]ח״ג] חג״ב **ב** [3]ז״כ] קו+ **ס** ~~ז״ס~~+ **פ** [4]לא״ב] א״ב **וטפ** [5]כיחס] +~~ח״ג אל ח״ד~~ **ק** [6]הנה] אם כן+ **ק** [7]אם] גם **ק** [8]אם ... ז״ד] ליתא **ס** [9]ז״ד] ~~ח״ד~~ **ב** [10]ג״ד] ג״ב **פ** [11]וכן ... כן] וז״ח **ב** [12]גם כן] +ידמה א״ג וז״ח **ו** ליתא **פ**

[Proposition VI.25]

[H26, B23, RJ24]

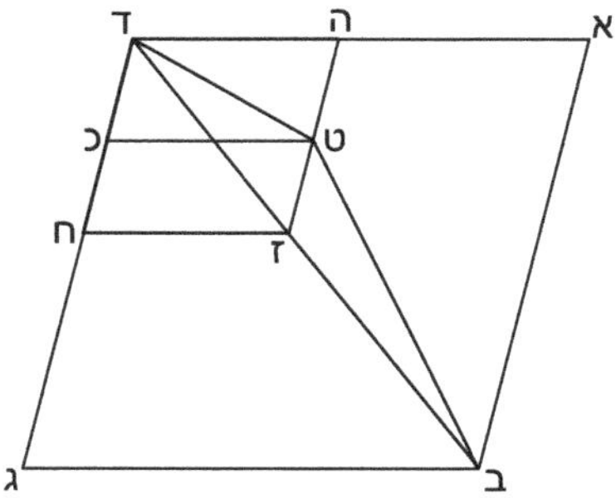

כאשר נחתך שטח משטח, והם נכוחי הצלעות מתדמים, הוא יהיה על אלכסונו. ע״א: כשנחתך שטח משטח, והוא דומה לו ועל תארו, ומוצב במצבו, וזוית משותפת להם, יהיה על אלכסונו.

إِذا (فُصِلَ)[1] مِنْ سَطْحٍ مُتَوازِي الأَضْلاعِ (سَطْحٌ)[2] مُتَوازِي الأَضْلاعِ (يُشْبِهُ)[3] السَّطْحَ الأَعْظَمَ الَّذي فُصِلَ مِنْهُ وَعَلى خِلْقَتِهِ، فَإِنَّهُ يَكونُ عَلى قُطْرِ السَّطْحِ الأَعْظَمِ.

המשל בזה: שנחתוך משטח א״ג נכוחי הצלעות שטח ה״ח נכוחי הצלעות[1] הדומה לא״ג, שהוא על תארו. ואומר כי שטח ה״ח הוא על אלכסון א״ג[2], והוא האלכסון אשר עליו ד״ב, אי אפשר אלא זה.

مِثالُهُ: أَنَّا نَفْصِلُ مِنْ اَجَ المُتَوازِي الأَضْلاعِ سَطْحَ ه حَ المُتَوازِي الأَضْلاعِ الشَّبيهِ اَجَ الَّذي عُمِلَ عَلى خِلْقَتِهِ. فَأَقولُ إِنَّ سَطْحَ ه حَ عَلى قُطْرِ سَطْحِ اَجَ، وَهُوَ القُطْرُ الَّذي عَلَيْهِ د ز ب.

[1] שטח ... הצלעות] ⟨⟨*⟩⟩ [2] א״ג] נראה שהמילה מחוקה

[1] فصل] فضل [2] سطح] سطحا [3] يشبه] نسبه

إذا فُصِلَ سَطْحٌ مِنْ سَطْحٍ، وَهُما مُتَوازِيا الأَضْلاعِ مُتَشابِهانِ، فَهُوَ على قُطْرِه[1].

ووَجَدْنا في نُسْخَةٍ أُخْرى مِنْ غَيْرِ إِصْلاحٍ ثابِتٍ: إذا فُصِلَ سَطْحٌ مِنْ سَطْحٍ، وَهُوَ شَبيهٌ بِهِ، وعلى خِلْقَتِه، ومَوْضوعٌ كَوَضْعِه، زاوِيَةٌ مُشْتَرَكَةٌ لَهُما^، فَهُوَ عَلَى قُطْرِه.

مِثالُهُ: أَنَّ سَطْحَ ه ح مَفصولٌ مِنْ سَطْحِ ا ج، وَهُما مُتَوازِيا الأَضْلاعِ مُتَشابِهانِ. فَأَقولُ إِنَّ ه ح على قُطْرِ سَطْحِ ا ج، والقُطْرُ د ز ب^ لا يُمْكِنُ غَيْرُه.

כאשר נבדל שטח משטח, ושניהם נכחיי הצלעות מתדמים, הנה[1] הוא על קוטרו[2] [3].

ומצאנו בנסחא אחרת מבלתי תקון תאבת: כאשר נבדל שטח משטח, והוא דומה לו, ועל שני צלעיו, ומונח כמצבו, וזוית משותפת[4], הנה הוא על קוטרו[5].

המשל בו: כי שטח[6] ה״ח נבדל משטח א״ג, ושניהם נכחי הצלעות מתדמים. הנה אומר כי ה״ח על קוטר שטח א״ג, והקוטר ד״ז [!] אי אפשר זולתו.

[1] إذا ... قطره] PR2

[1] הנה] ליתא ב [2] כאשר ... קוטרו] ליתא ז [3] קוטרו] קוטרי ט [4] משותפת] +לשניהם פ [5] ומצאנו ... קוטרו] ליתא ב [6] כי שטח] ליתא ב

بُرْهانُهُ: أَنَّهُ لا يُمْكِنُ إِلّا ذلِكَ. فَإِنْ كانَ يُمْكِنُ، فَلْيَكُنْ عَلى خَطِّ د ط ب. وَنُخْرِجُ مِنْ ط خَطَّ ك ط مُوازِيًا لِخَطِّ ز ح. فَسَطْحُ ه ك عَلى قُطْرِ سَطْحِ ا ج. فَقَدْرُ ا د عِنْدَ د ه كَقَدْرِ ج د عِنْدَ د ك. وَقَدْرُ ا د عِنْدَ د ه كَقَدْرِ ج د عِنْدَ د ح، لِأَنَّ سَطْحَ ا ج يُشْبِهُ سَطْحَ ه ح. فَقَدْرُ ج د عِنْدَ د ح كَقَدْرِ ج د عِنْدَ د ك. ج د نِسْبَتُهُ إِلى د ح وَإِلى د ك إِذًا واحِدَةٌ—هَذا خُلْفٌ لا يُمْكِنُ. فَلَيْسَ د ط ب قُطْرَ السَّطْحِ ا ج. وَكَذَلِكَ نُبَيِّنُ أَنَّهُ وَلا خَطٌّ آخَرَ يُمْكِنُ أَنْ يَكونَ عَلَيْهِ سَطْحُ ه ح إِلّا عَلى قُطْرِ ا ج الَّذي عَلَيْهِ د ز ب. فَسَطْحُ ه ح عَلى قُطْرِ سَطْحِ ا ج.

وَذَلِكَ ما أَرَدْنا بَيانَهُ.

ואם היה אפשר, יהיה על קו דט״ב. ונוציא מט׳ קו נכוחי לקו ב״ג, והוא ט״כ. ויהיה שטח ה״כ על אלכסון א״ג, והוא האלכסון אשר עליו דט״ב[1]. ויהיה שיעור א״ד אל ד״ה כשיעור ג״ד אל ד״כ. ושיעור א״ד אל ד״ה כשיעור ג״ד אל ד״ח, לפי ששטח א״ג הוא דומה לה״ח. ויהיה שיעור ג״ד אל ד״ח כשיעור ג״ד אל ד״כ. אם כן יהיה ג״ד יחסו אל ד״ח ואל ד״כ אחד—זה חלוף ודבר בטל. אם כן אין דט״ב אלכסון לשטח א״ג. וכמו כן יתבאר שאין קו אחר שיתכן שיהיה שטח ה״ח עליו, כי אם על אלכסון א״ג אשר עליו דז״ב. ואם כן שטח ה״ח הוא על אלכסון שטח א״ג.

וזה מה שרצינו לבאר.

[1] דט״ב] ⟨⟨*⟩⟩

فَإِنْ أَمْكَنَ، فَلْيَكُنِ^ القُطْرُ د ط ب. ونُخْرِج ط ك يوازي ب ج. فسطح ه ك على قُطْرِ سطح ا ج. فنِسْبَةَ ا د إلى د ه إِذَنْ[1] كنِسْبَةِ ج د إلى د ك. لكِنَّ نِسْبَةَ ا د إلى د ه كنِسْبَةِ ج د إلى د ح، لِأَنَّ سَطْحَيْ ا ج و ه ح مُتَشابِهانِ. فإِذَنْ نِسْبَةَ ج د إلى د ح و د ك واحدة. فَهُما مُتَساوِيان—هذا خُلْفٌ. فَلَيْسَ د ط ب بِقُطْرِ سَطْحِ ا ج، ولا غَيْرُهُ مِنَ الخُطوط غير د ز ب، وهو قُطْرُ سطح ا ج. فسطح ه ح على قُطْرِ سطح ا ج.

وذَلِكَ ما أَرَدْنا أَنْ نُبَيِّنَ.

[1] إذن] Cop

שאם היה אפשר, נאמר שיהיה הקוטר דט״ב. ונוציא ט״כ[1] נכחי לב״ג. הנה שטח ה״כ על קוטר שטח א״ג. הנה יחס א״ד אל ד״ה אם כן[2] כיחס ג״ד אל ד״כ. אבל יחס א״ד אל ד״ה כיחס ג״ד אל ד״ח, מפני כי שני[3] שטחי א״ג וה״ח מתדמים. הנה אם כן יחס ג״ד אל ד״ח וד״כ אחד. אם כן שניהם שוים—זה[4] שקר[5]. אם כן אין דט״ב בקוטר שטח א״ג, ולא זולתו מן הקוים זולת דז״ב, והוא קוטר שטח א״ג. אם כן שטח ה״ח[6] על קוטר שטח א״ג.

וזה מה שרצינו לבאר[7] [8].

[1] ט״כ] דט״כ ס [2] אם כן] ליתא ס [3] שני] ליתא בז [4] זה] ⟨*⟩ ט [5] שקר] +~~הוא~~ ט [6] ה״ח] ~~ה״ג~~+ ק [7] וזה ... לבאר] ונשלם ביאורו ס [8] לבאר] לבארו ק

[Proposition VI.26]

[H25, B25]

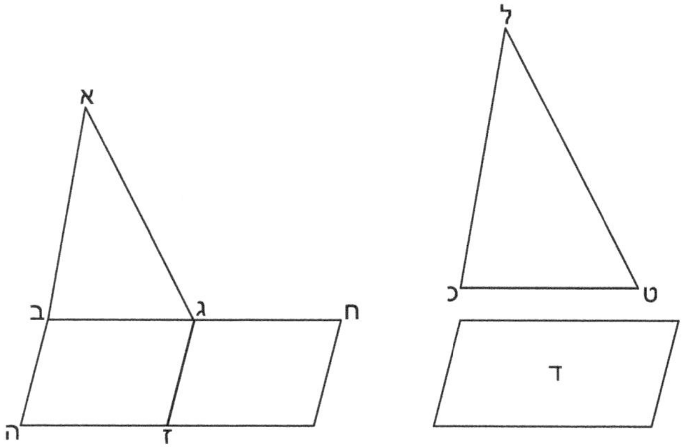

نُرِيدُ أَنْ نُبَيِّنَ كَيْفَ نَعْمَلُ سَطْحًا شَبِيهًا بِسَطْحٍ مَعْلُومٍ مُساوِيًا لِسَطْحٍ مَعْلُومٍ آخَرَ.

مِثالُهُ: أَنَّا نَجْعَلُ السَّطْحَ المَعْلُومَ الَّذِي يَكُونُ شَبِيهًا بِالسَّطْحِ المَعْمُولِ سَطْحَ ا ب ج. وَالسَّطْحُ الَّذِي يَكُونُ مُساوِيًا لَهُ سَطْحَ د. وَنُرِيدُ أَنْ نَعْمَلَ سَطْحًا شَبِيهًا بِسَطْحِ ا ب ج مُساوِيًا لِسَطْحِ د. فَنُضِيفُ إِلَى خَطِّ ب ج سَطْحًا مُساوِيًا لِسَطْحِ ا ب ج، وَهُوَ ه ج. وَنُضِيفُ إِلَى خَطِّ ج ز سَطْحَ ز ح مُساوِيًا لِسَطْحِ د. وَنَسْتَخْرِجُ خَطَّ ط ك مُناسِبًا لِخَطَّيْ (ب ج ج ح)[1] فِيما بَيْنَهُما. وَنَعْمَلُ عَلَيْهِ سَطْحَ ط ك ل شَبِيهًا بِسَطْحِ ا ب ج مُساوِيًا لِسَطْحِ د.

נרצה לעשות שטח שידמה לשטח ידוע ויהיה שוה לשטח אחר ידוע.

ונשים השטח שנרצה לעשות שטח בדמותו שטח אב״ג. והשטח שיהיה שוה לו שטח ד׳. נחבר [!] אל קו ב״ג שטח שוה לשטח אב״ג, והוא ה״ג. ונחבר אל קו ג״ז שטח שוה לשטח ד׳ והוא ז״ח. ונוציא קו בין קוי ב״ג וג״ח מתיחס להם, והוא קו ט״כ. ונעשה על קו ט״כ שטח דומה לשטח אב״ג, והוא טכ״ל. ואומר כי כבר עשינו השטח הדומה לשטח אב״ג השוה לשטח ד׳, והוא שטח טכ״ל.

[1] ب ج ج ح] ب ج ح

نُرِيدُ أَنْ نَعْمَلَ سَطْحًا يُشْبِهُ سَطْحًا مَعْلُومًا، وَيُساوِي سَطْحًا آخَرَ مَعْلُومًا.

فَلْيَكُنِ الشَّبِيهُ سَطْحَ ا ب ج، وَالمُساوِي سَطْحُ د. ونُرِيدُ أَنْ نَعْمَلَ سَطْحًا يُشْبِهُ سَطْحَ ا ب ج، وَيُساوِي سَطْحَ د. فَنُضِيفُ إلى خطّ ب ج سَطْحًا مُتَوازِيَ الأَضْلاعِ يُساوِي ا ب ج، وَهُوَ ه ج. وَنُضِيفُ إلى خطّ ج ز سَطْحَ ز ح مُتَوازِيَ الأَضْلاعِ يُساوِي سَطْحَ د. وَنَجْعَلَ زاوِيَةَ ز ج ح مِثْلَ زاوِيَةِ ج ب ه، لِيَكُونَ سَطْحًا ه ج و ز ح بين^ خَطَّيْنِ مُتَوازِيَيْنِ. وَنَسْتَخْرِجَ مِنْ[1] خَطَّيِ ب ج و ج ح خَطًّا مُناسِبًا لَهُما، وهو ط ك. وَنَعْمَلَ عَلَيْهِ سَطْحًا يُشْبِهُ ا ب ج، وهو ط ك ل.

[1] من] **T2**

נרצה שנעשה שטח ידמה שטח ידוע, ויהיה שוה לשטח אחר ידוע.

ויהיה הדומה שטח אב״ג, והשוה שטח ד׳. ונרצה לעשות שטח ידמה שטח אב״ג, ויהיה שוה לשטח ד׳. הנה נחבר אל קו ב״ג שטח נכחי הצלעות יהיה שוה אל[1] אב״ג, והוא ה״ג. ונחבר אל קו ג״ז שטח ז״ח נכחי הצלעות שוה[2] לשטח ד׳. ונשים זוית[3] זג״ח כמו זוית גב״ה, כדי שיהיה שטח ה״ג וז״ח מן שני קוים נכחיים. ונוציא משני קוי ב״ג וג״ח קו מתיחס לשניהם, והוא ט״כ. ונעשה עליו שטח ידמה אב״ג, והוא טכ״ל.

שרטוט—**ב**: ח׳ חסרה; **בטס**: שטח אב״ג אינו מחובר לשטח ג״ה; **בטספק**: מעל צלע ב״ג נכתב ״ראשון״, מעל צלע כ״ט נכתב ״שני״; **בטפ**: מעל צלע ג״ח נכתב ״שלישי״; **ופ**: שטח ד׳ מחובר לצלע כ״ט; **פ**: האות ח׳ מופיעה בתוך השטח ח״ז ולא לצד הקדקוד [1] אל] +* **ט** [2] שוה] נכחי ⟨*⟩ **ב** [3] זוית] ליתא **פ**

بُرْهانُهُ: أَنَّ خَطَّ ط ك بَيْنَ خَطَّيْ ب ج ج ح مُناسِباً لَهُما. فَقَدْرُ ب ج عِنْدَ ج ح كَقَدْرِ ا ب ج عِنْدَ ط ك ل. وَقَدْرُ ب ج عِنْدَ ج ح كَقَدْرِ ه ج عِنْدَ ز ح. فَقَدْرُ ا ب ج عِنْدَ ط ك ل كَقَدْرِ ه ج عِنْدَ ز ح. وَتُبَدِّلُ. فَقَدْرُ ا ب ج عِنْدَ ه ج كَقَدْرِ ط ك ل عِنْدَ ز ح. و ا ب ج مِثْلُ ه ج. ط ك ل مِثْلُ ز ح. ز ح مِثْلُ د. و ط ك ل مِثْلُ د. وَهُوَ (يُشْبِهُ)[1] ا ب ج.

وَذَلِكَ ما أَرَدْنا بَيانَهُ.

המופת בזה: כי בין קוי ב״ג וג״ח קו מתיחס להם, והוא קו ט״כ. ויהיה ב״ג אל ג״ח כשיעור אב״ג אל טכ״ל. ושיעור ב״ג אל ג״ח כשעור ה״ג אל ז״ח. ואם כן שיעור אב״ג אל טכ״ל כשיעור ה״ג אל ז״ח. וכאשר נחליף יהיה אב״ג אל ה״ג כשיעור טכ״ל אל ז״ח. ואב״ג שוה לה״ג. אם כן טכ״ל שוה לז״ח. ואמנם ז״ח שוה לד׳. אם כן טכ״ל שוה לד׳, והוא דומה לאב״ג.

וזה מה שרצינו לבאר.

[1] يشبه] نسبه

فخطّ ط ك بَيْنَ خَطَّيْ ب ج و ج ح، مُنَاسِبًا لَهُما. فنِسْبَة خَطّ ب ج إلى ج ح كنِسْبَة ا ب ج إلى ط ك ل. ونِسْبَة ب ج إلى ج ح كنِسْبَة ه ج إلى ز ح. فنِسبَة ا ب ج إلى ط ك ل كنِسبة ه ج إلى ز ح. وسطح ا ب ج مثل سطح ه ج. ف ط ك ل مثل ز ح^. لكن^ ز ح مثل د. فسطح ط ك ل مثل د وهو يُشْبِهُ ا ب ج.

הנה קו ט״כ בין שני קוי ב״ג וג״ח, מתייחס לשניהם. אם כן יחס ב״ג אל ג״ח כיחס אב״ג אל טכ״ל[1]. ויחס[2] ב״ג אל ג״ח כיחס[3] ה״ג אל ז״ח. אם כן יחס אב״ג אל טכ״ל כיחס ה״ג אל ז״ח. אבל ה״ג כמו אב״ג. אם כן יחס אב״ג אל[4] טכ״ל[5] וז״ח אחד. אם כן שניהם שוים[6]. וז״ח[7] וטכ״ל אחד. אם כן שניהם שוים. וז״ח כמו ד׳. אם כן טכ״ל כמו ד׳, והוא דומה אב״ג [!].

وذَلِكَ ما أَرَدْنا أَنْ نُبَيِّنَ.

וזה מה שרצינו לבאר[8].

[1] כיחס ... טכ״ל] ⟨*⟩ ט +~~ויהיה~~ פ [2] ויחס] ⟨⟨*⟩⟩ פ [3] כיחס] +שטח ס [4] טכ״ל ... אל] ⟨*⟩ ק [5] כיחס ... טכ״ל] ⟨⟨*⟩⟩ פ [6] שוים] +ע״כ ס [7] וז״ח] +כמו ד׳ אם כן טכ״ל כמו ד׳ ס [8] וזה ... לבאר] ומש״ל ב

[Proposition VI.27]

[B26]

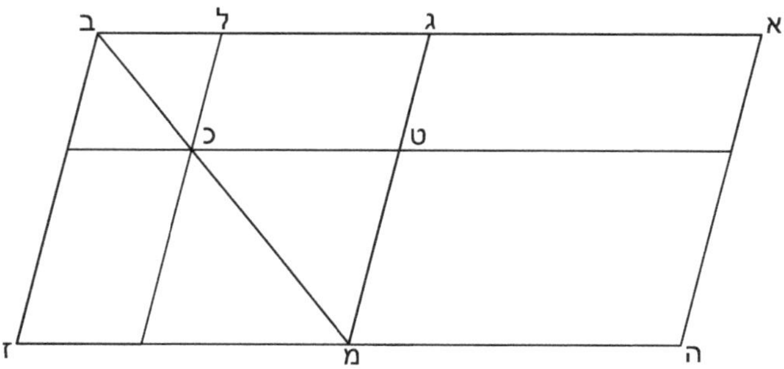

הגדול שבשטחים נכוחי הצלעות אשר יתחברו אל קו ישר, ויחסרו מן שלמותו שטחי׳ דומים לשטח נכוחי הצלעות העשוי על חצי הקו, ומצבם כמצבו, הוא השטח העשוי על חצי הקו ממה שימשך לשטח הדומה בשטחים שהם החסרונות.

السُّطوحُ المُتَوازِيَةُ الأَضْلاعِ الَّتي تُضافُ إلى خَطٍّ مَعْلومٍ، وَيَنْقُصُ مِنْ تَمامِ الخَطِّ سَطْحًا يُشْبِهُ السَّطْحَ المُضافَ إلى نِصْفِ الخَطِّ المَعْلومِ[1]، وَيَكونُ السَّطْحُ النّاقِصُ عَلى قُطْرِ السَّطْحِ المُضافِ إلى نِصْفِ الخَطِّ[2]، فَإِنَّ أَعْظَمَ تِلْكَ السُّطوحِ كُلِّها هُوَ السَّطْحُ المُضافُ إلى نِصْفِ الخَطِّ.

המשל בזה: שנשים קו הידוע קו א״ב. ונחלק אותו לחציים על נקודת ג׳. ונעשה על קו ב״ג שטח נכוחי הצלעות עליו ג״ז. ונשלים שטח ג״ה. ואומר כי השטח נכוחי הצלעות המחובר אל קו א״ב, והוא אשר יחסר משלמות הקו שטח ידמה ג״ז, והוא על תארו, יהיה הגדול שבהם שטח א״מ.

مِثالُهُ: خَطُّ ا ب المَعْلومُ. وَنَقْسِمُهُ بِنِصْفَيْنِ عَلى نُقْطَةِ ج. فَنَعْمَلُ عَلى خَطِّ ب ج سَطْحًا مُتَوازي الأَضْلاعِ عَلَيْهِ ج ز. وَنُتَمِّمُ سَطْحَ ج ه. فَأَقولُ إِنَّ السُّطوحَ المُتَوازِيَةَ الأَضْلاعِ المُضافَةَ إلى خَطِّ ا ب، وَهِيَ تَنْقُصُ مِنْ تَمامِ السَّطْحِ سَطْحًا يُشْبِهُ سَطْحَ ج ز عَلى خِلْقَتِهِ فَإِنَّ أَعْظَمَها سَطْحُ ا م.

שרטוט—B: ه בכתיבה אחרת

[1]المعلوم] +المتوازي الإضلاع [2]ويكون ... الخط]
אפשר שמשפט זה מיותר

أَعْظَمُ السُّطُوحِ الْمُتَوازِيَةِ الأَضْلاعِ الَّتِي تُضافُ إلى خَطٍّ مُسْتَقِيمٍ، وتَنْقُصُ عن تَمامِهِ سُطُوحٌ شَبِيهَةٌ بِالسَّطْحِ الْمُتَوازِي الأَضْلاعِ الْمَعْمُولِ على نِصْفِ الخَطِّ، ووضعها كَوَضْعِهِ، هو السَّطح الْمَعْمُول على نِصْفِ الخَطِّ الشَّبِيهِ بِالسُّطُوحِ الَّتِي هِيَ النُّقْصَانَات.

היותר גדול[1] שבשטחים נכחי הצלעות אשר יהיו סמוכים אל קו ישר, ויחסרו משלמותו שטחים דומים בשטח נכחי הצלעות עשוי על חצי הקו, ומצבם כמצבו, הוא[2] השטח העשוי על חצי הקו ממה שילוה השטח הדומה בשטחים אשר הם החסרונות[3].

مِثالُهُ: أَنَّ سَطْحَ ج ز مُتَوازِي الأَضْلاعِ مُضافٌ إلى خطّ ب ج، وَهُوَ نِصْف خطّ ا ب. وَقَدْ تَمَّ سَطْحُ ج ه. وأُضِيفَ إلى خطّ ا ب سَطْحٌ آخر^ مُتَوازِي الأَضْلاعِ، وهو سطح ا ك. ينقص عن تَمامِ الخطّ سطح ب ك الَّذي يُشْبِهُ سَطْحَ ج ز، وهو مَوْضُوعٌ كَوَضْعِهِ. فَأَقُولُ إنَّ سَطْحَ ا م أَعْظَمُ مِنْ سَطْحِ ا ك.

המשל בו: כי שטח ג״ז נכחי הצלעות סמוך אל קו ב״ג, והוא חצי קו א״ב. וכבר נשלם שטח ג״ה. ונסמך אל קו א״ב שטח נכחי הצלעות, והוא שטח א״כ. וחוסר משלמות הקו שטח ב״כ[4] אשר ידמה שטח ג״ז, והוא מונח כמצבו. הנה אומר כי א״מ יותר גדול מא״כ[5].

שרטוט—בכל כתבי היד הערביים שאינם ממשפחה A-1 הנחת האותיות על התרשים שונה מהנחתן בתרגום אבן תיבון [1]גדול] ⟨*⟩ **ט** [2]הוא] הנה+ **ס** [3]בשטחים ... החסרונות] בשטח הנחסר **ט** [4]ב״כ] ⟨⟨*⟩⟩ **פ** [5]מא״כ] בין א״כ **ט**

מופת זה: שנחבר אל קו א״ב שטח אחר נכוחי הצלעות, והוא א״כ, והוא שיחסר משלימות הקו שטח שידמה ג״ז, והוא כ״ב, והוא על תאר ג״ז[1]. ואומר כי א״מ גדול מא״כ, לפי שכ״ב ידמה ג״ז, הוא על אלכסונו. והאלכסון בכ״מ. ונשלים קוי הצורה. ויהיה קו א״ג שוה לקו ג״ב וא״ג שוה לה״מ. וג״ב שוה למ״ז. ואם כן ה״מ שוה למ״ז. ושטח ה״ט שוה לשטח ט״ז. אם כן ה״ט גדול מכ״ז. וכ״ז שוה לט״ל—יהיה ה״ט גדול מט״ל. וא״ט משותף. אם כן א״מ כלו גדול מא״כ כלו.

וזה מה שרצינו לבאר.

بُرْهانُهُ: أَنَّا نُضيفُ إِلى خَطِّ ا ب سَطْحَ ا ك مُتوازي الأَضْلاعِ، وَهُوَ يَنْقُصُ مِنْ تَمامِ الخَطِّ سَطْحًا يُشْبِهُ ج ز وَعَلى خِلْقَتِه. فَأَقولُ إِنَّ ا م أَعْظَمُ مِنْ ا ك. و ك ب يُشْبِهُ ج ز، وَهُوَ عَلى قُطْرِه. وَالقُطْرُ ب ك م. وَنُتِمُّ خُطوطَ الشَّكْلِ. فَخَطُّ ا ج مِثْلُ خَطِّ ه م. و ج ب مِثْلُ م ز. (و ه م)[1] مِثْلُ (م ز)[2]. وَسَطْحُ ه ط مِثْلُ سَطْحِ ط ز. ه ط أَعْظَمُ مِنْ ك ز. و ك ز مِثْلُ ط ل. و ه ط أَعْظَمُ مِنْ ط ل. و ا ط مُشْتَرَكٌ. فَكُلُّ ا م أَعْظَمُ مِنْ كُلِّ ا ك.

وَذٰلِكَ ما أَرَدْنا بَيانَهُ.

[1] ג״ז] ⟨⟨*⟩⟩

[1] و ه م] و ه ج [2] م ز] ج ز

بُرهانُهُ: أَنَّ سَطْحَ كـ ب يُشْبِهُ سَطْحَ ج ز. فَهُوَ على قُطْرِهِ. والقُطْرُ خَطُّ ب كـ م. ويُتِّمُّ[1] خُطوطَ الشَّكلِ. وخطّ ا ج مثل خطّ ب ج ومثل ہ م. و ب ج مثل م ز. ف ہ م مثل م ز. فسطح ہ ط مثل سطح ط ز. ف ہ ط أعظم مِنْ كـ ز. و كـ ز مثل ط ل. ف ہ ط أعظم مِنْ ط ل. و ا ط مُشْتَرَكٌ. فكُلّ سَطْحِ ا م أعظم مِنْ كُلِّ سَطْحِ ا كـ.

המופת: כי שטח ב״כ ידמה שטח ג״ז. אם כן הוא על קוטרו. והקוטר בכ״מ. ויושלמו קוי התמונה. וקו א״ג כמו קו ב״ג וכמו ה״מ. וב״ג כמו מ״ז. אם כן ה״מ כמו מ״ז. ושטח ה״ט כמו שטח ט״ז[1]. אם כן ה״ט יותר גדול מן כ״ז. וכ״ז כמו ט״ל. אם כן ה״ט יותר גדול מן ט״ל. וא״ט משותף. אם כן כל שטח[2] א״מ יותר גדול מכל שטח א״כ.

وذَلِكَ ما أَرَدْنا أَنْ نُبيِّنَ.

וזה מה שרצינו לבאר.

---

[1] ويتمم] T2

---

[1] ט״ז] +אבל ט״ז גדול מן כ״ז **ב** +אבל ט״ז יותר גדול מן כ״ז **ו**   [2] כל שטח] שטח **בו** שטח כל **ס**

[Proposition VI.28]

[B27]

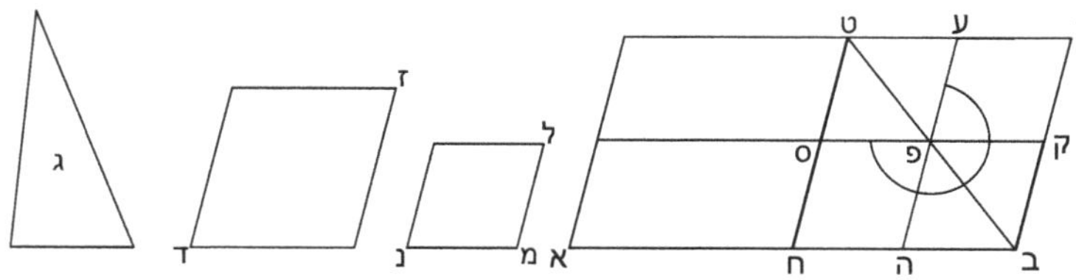

נרצה שנחבר אל קו ישר ידוע שטח נכוחי הצלעות יחסר אצל שלמותו שטח דומה בשטח נכוחי הצלעות ידוע, ויהיה המחובר שוה לצורה ישרת הקוים ידועה. ויצטרך שזאת הצורה ישרת הקוים לא תהיה גדולה מנכוחי הצלעות המחובר אל חצי הקו הידוע הדומה בשטח החסר.

نُرِيدُ أَنْ نُبَيِّنَ كَيْفَ نُضِيفُ إِلَى خَطٍّ مَعْلُومٍ سَطْحًا مُتَوَازِي الأَضْلاعِ مُسَاوِيًا لِسَطْحٍ مَعْلُومٍ يُحِيطُ بِهِ (ثَلاثَةُ)[1] خُطُوطٍ، وَنَنْقُصُ مِنْ تَمَامِ الخَطِّ سَطْحًا شَبِيهًا بِسَطْحٍ مَعْلُومٍ آخَرَ مُتَوَازِي الأَضْلاعِ. وَلا يَكُونُ السَّطْحُ المُسَاوِي المُضَافُ إِلَى الخَطِّ بِأَعْظَمَ مِنَ السَّطْحِ الَّذِي يَكُونُ عَلَى نِصْفِ الخَطِّ.

[1] ثلاثة] ثلثة

نُرِيدُ أَنْ نُضِيفَ إلى خَطٍّ مُسْتَقِيمٍ مَعْلومٍ سَطْحًا مُتَوازِيَ الأَضْلاعِ يَنْقُصُ عَنْ تَمامِهِ سَطْحًا شَبِيهًا بِسَطْحٍ مُتَوازِي الأَضْلاعِ مَعْلومٍ، ويَكونُ المُضافُ مُساوِيًا لِشَكْلٍ[1] مُسْتَقِيمِ الخُطُوطِ مَعْلومٍ. وَيَحْتاجُ إلى أَنْ يَكُونَ ذَلِكَ الشَّكلُ المُسْتَقِيمُ الخُطوطِ لَيسَ بِأَعْظَمَ مِنَ المُتوازِي الأَضْلاعِ المُضافِ إلى نِصْفِ الخَطِّ المَعْلومِ الشَّبِيهِ بِالسَّطْحِ النَّاقِصِ.

נרצה שנחבר אל קו ישר ידוע שטח נכחי הצלעות יחוסר משלמותו[1] שטח דומה לשטח נכחי הצלעות[2] ידוע. ויהיה המחובר שוה לתמונת[3] ישרת הקוים ידועה. ונצטרך שתהיה[4] התמונה ההיא[5] ישרת הקוים[6] בלתי יותר גדולה מן הנכחי[7] הצלעות המחובר אל חצי הקו הידוע הדומה בשטח החסר.

---

[1] لشكل] CPT1T3 +ما

שרטוט—**בופק**: מלבן א״כ מימין לשלוש הצורות האחרות; **פ**: חסרה ז׳, ונראה שלא שורטט מצולע למ״נ; הנחת הצורות זו לעומת זו אינה אחידה בששת כתבי היד העבריים

[1] משלמותו] משלמותיו **ב** [2] הצלעות] ⟨⟨*⟩⟩ **ט** [3] לתמונת] לתמונה **טק** +מה **ס** [4] שתהיה] ליתא **פ** [5] ההיא] ליתא **ט** [6] הקוים] ליתא **ס** [7] הנכחי] ⟨⟨השטח⟩⟩+ **ק**

ונשים הקו הישר הידוע קו א״ב, והשטח השוה למחובר אל הקו שטח ג׳. ואיננו גדול מנכוחי הצלעות אשר על חצי הקו. ונשים השטח הידוע האחר נכוחי הצלעות הדומה בשטח[1] החסר שטח ד״ז. וכאשר רצינו שנחבר אל קו א״ב שטח נכוחי הצלעות שוה לשטח ג׳, ויחסר אצל שלמות הקו שטח דומה לשטח ד״ז.

مِثالُهُ: خَطُّ ا بَ المَعْلومُ. وَالسَّطْحُ المُساوي المُضافُ إِلى الخَطِّ سَطْحُ جَ. وَلَيْسَ هُوَ بِأَعْظَمَ مِنَ المُتَوازِيَ الأَضْلاعِ الَّذي عَلى نِصْفِ الخَطِّ. وَالسَّطْحُ المَعْلُومُ الآخَرُ المُتَوازِيَ الأَضْلاعِ الشَّبيهُ بِالسَّطْحِ النَّاقِصِ سَطْحُ د زَ. وَنُريدُ أَنْ نُضيفَ إِلى خَطِّ ا بَ سَطْحًا مُتَوازِيَ الأَضْلاعِ مُساوِيًا لِسَطْحِ جَ. وَنَنْقِصُ مِنْ تَمامِ الخَطِّ سَطْحًا شَبيهًا بِسَطْحِ د زَ.

[1] בשטח] ⟨⟨*⟩⟩

فَنَجْعَلَ الخَطَّ المَعْلومَ خطَّ ا ب، وَالشَّكلَ المُسْتَقيمَ الخُطوطِ المُساوي لِلْمُضافِ شَكْلَ ج. وَلَيْسَ هذا الشَّكل بِأَعْظَمَ مِنَ السَّطحِ المُتوازي الأَضْلاعِ المُضاف إلى نِصْفِ الخَطِّ الشَّبيهِ بِالسَّطحِ الَّذي نُريدُ أَنْ يَكُونَ السَّطحُ النَّاقِصُ شَبيهًا بِهِ. ونَجْعَلَ السَّطحَ المُتَوازي الأَضْلاعِ الشَّبيهِ بالنَّاقِصِ سَطْحَ د ز. ونُريدُ أَنْ نُضيفَ إلى خطِّ ا ب سَطْحًا مُتَوازي الأَضْلاعِ مُساوِيًا لِشَكْلِ ج، يَنْقُصُ عن تَمامِ الخَطِّ سَطْحًا شَبيهًا بِسَطْحِ د ز.

הנה נשים הקו הידוע קו[1] א״ב, והתמונה ישרת הקוים השוה[2] למחובר תמונת ג׳. ואין התמונה הזאת יותר גדולה מהשטח[3] הנכחי הצלעות המחובר אל חצי הקו הדומה בשטח אשר נרצה שיהיה השטח החסר דומה לו[4]. ונשים השטח הנכחי הצלעות הדומה בחסרון שטח[5] ד״ז. ונרצה שנחבר אל קו א״ב שטח[6] נכחי הצלעות שוה לתמונת[7] ג׳, יחסר משלמות[8] הקו שטח דומה לשטח ד״ז.

---

[1] קו] ק⟨ו⟩ **ב** [2] השוה] +שתהיה התמונה ההיא ישרת הקוים כי לא יוכל להעשות זולת זה ממה שנתבאר בתמונה אשר לפניה השוה **ס** [3] מהשטח] מ⟨ה⟩שטח **ב** משטח **ס** [4] לו] ליתא **ב** [5] שטח] השטח **בוס** [6] שטח] ושטח **ב** ~~שוה~~ ⟨*⟩ **ק** [7] לתמונת] לתמונה **ט** [8] משלמות] משלמותו **ס**

אז נחלק קו א״ב לחציים על נקודת ח׳. ונצייר על ח״ב שטח[1] דומה לשטח ד״ז והוא ח״כ. ונשלים שטח א״ט. ואם היה א״ט שוה לשטח ג׳ כבר יצא מה שרצינו, כי כבר נעשה על קו א״ב שטח א״ט נכוחי הצלעות שוה לשטח ג׳, והוא יחסר משלמות הקו שטח כ״ח הדומה לד״ז. ואם היה א״ט גדול מג׳ יהיה מה שיוסיף על ג׳ מה שידמה[2] ד״ז. ויהיה מה שיוסיף עליו שטח נ״ל. ויהיה נ״ל בדמות ח״כ ועל תארו.[3] ודמות ט״ח כדמות נ״מ. ודמות ט״כ כדמות מ״ל. וט״ח[4] ארוך מנ״מ. ואם כן[5] ט״כ ארוך ממ״ל. ונשים מ״נ כמו ט״ס. ומ״ל כמו ט״ע. ונשלים שטח ס״ע. וס״ע על קוטר ח״כ. והקוטר ט״ב[6].

[1] שטח] ⟨⟨*⟩⟩ [2] מה שידמה] ~~שמחמת~~ ⟨⟨*⟩⟩ [3] תארו] +~~לפי שהוא דומה לב״א ⟨⟨לד״ז⟩⟩ וד״ז דומה לח״ב~~ [4] וט״ח] ~~ואם כן~~ ⟨⟨ו⟩⟩ט״ח [5] ואם כן] ~~ויהיה~~ ⟨⟨*⟩⟩ [6] ונשלים ... ט״ב] ⟨⟨*⟩⟩

فَنَقْسِمُ خطَّا ا ب بِنِصْفَيْنِ على ح. وَنَخُطُّ على ح ب سطحًا يُشْبِهُ سَطْحَ د ز، وَهُوَ ح ك. ونتِمُّ سَطْحَ ا ط. فَإِنْ كانَ ا ط مِثْل سَطْح ج. فَقَدْ كانَ ما أَرَدْنا، لِأَنَّهُ قَدْ عُمِلَ عَلى خَطِّ ا ب سَطْحُ ا ط المُتَوازِي الأَضْلاعِ المُساوِي لِسَطْحِ ج، وَهُوَ يَنْقُصُ مِنْ تَمامِ الخطِّ سَطْح ك ح الشَّبيه ب د ز. وَإِنْ كانَ ا ط أَعْظَم من ج، فَإِنَّ ما يَزِيدُ على ج سَطْحٌ يُشْبِهُ د ز. فَلْيَكُنْ ما يَزِيدُ عَلَيْهِ سَطْح ن ل م. فَلْيَكُن (ن ل)[1] شَبيهَ ح ك، وَهُوَ على خِلْقَتِهِ. وَشَبيهُ ح ط إلى ن م كَنِسْبَتِهِ إلى ط ك إلى م ل. و ح ط أَطْوَلُ مِنْ ن م. و ط ك أَطْوَلُ مِنْ (م ل)[2]. فَنَجْعَلُ م ن مِثْل ط س. و م ل مِثْل ط ع. ونتِمُّ سَطْحَ س ع. ف س ع هُوَ على قُطْرِ ح ك. وَالقُطْرُ ط ب. وَنُتِمُّ تَخْطِيطَ الشَّكْلِ. فَأَقولُ إِنَّا قد أَضَفْنا إلى خَطِّ ا ب سَطْحَ ا ف كما أَرَدْنا.

[1] ن ل] ز ن [2] م ل] م ز

فنقسم خطَّ ا ب بنصفين على ح. ونعمل على ب ح سطحًا يشبه د ز، وهو ك ح. ونتمّم سطح ا ط. فإن كان ا ط مثل ج، فقد كان ما أردنا، لأنّه قد عُمل على خطّ ا ب سطح ا ط وهو متوازي الأضلاع يساوي ج، وينقص عن تمام خطّ ا ب سطح ك ح الذي يشبه د ز. وإن كان غير مساوٍ لشكل ج، فهو أعظم منه. فنجعل فضل ا ط على ج شبيهًا ب د ز. وليكن سطح ن ل و د ز شبيه ح ك. و ن ل شبيه ب د ز. ف ن ل شبيه ب ح ك. فلتكن زاويته المساوية لزاوية ح ط ك زاوية م. وليكن ن م نظيرًا في النسبة لضلع ح ط، و م ل نظير ط ك. فلأنّ سطح ا ط أعظم من سطح ن ل، و ح ك مثل ا ط، يكون ح ك أعظم من ن ل، وهو شبيه به. فكلّ ضلع من أضلاع ح ك أطول من نظيره من أضلاع ن ل. ف ح ط ادن[1] أطول من ن م. و ط ك أطول من م ل. ولنفصل من ط ح مثل ن م، وهو ط س، ومن ط ك مثل م ل، وهو ط ع.

ונחלק א״ב בשני חציים על ח׳. ונעשה על ב״ח שטח ידמה לשטח ד״ז, והוא כ״ח. ונשלים שטח א״ט. הנה אם היה א״ט כמו ג׳, הנה כבר היה מה שרצינו, כי הוא כבר נעשה על קו א״ב שטח א״ט נכחי הצלעות[1] שוה אל ג׳, ויחסר משלמות קו א״ב שטח כ״ח אשר ידמה ד״ז. ואם היה בלתי[2] שוה לתמונת ג׳, הנה הוא יותר גדול ממנו. ונשים יתרון א״ט על ג׳[3] דומה אל ד״ז. ויהיה שטח נ״ל וד״ז דומה לח״כ[4]. ונ״ל דומה אל ד״ז. אם כן נ״ל דומה אל ח״כ. ותהיה זויתו[5] השוה לזוית חט״כ זוית מ׳[6]. ותהיה נ״מ גילו ביחס לצלע ח״ט, ומ״ל גילו לצלע כ״ט. ומפני כי שטח א״ט יותר גדול משטח נ״ל, וח״כ כמו א״ט, יהיה ח״כ יותר גדול מן נ״ל, והוא דומה לו. אם כן כל צלע מצלעות ח״כ יותר ארוכה מגילה[7] מצלעות נ״ל. הנה ח״ט אם כן יותר ארוך מן נ״מ. וט״כ יותר ארוך מן מ״ל. ונבדיל מן ט״ח[8] כמו נ״מ, והוא ט״ס, ומן ט״כ כמו מ״ל, והוא ט״ע.

[1] إدن] **CopMPT1** إذا

[1] הצלעות] ⟨⟨*⟩⟩ **ט** [2] בלתי] הבלתי **ס** [3] ג׳] ד+ **ב** [4] לח״כ] לז״כ ⟨מ׳⟩ **ב** [5] זויתו] זוויתו **ב** [6] מ׳] ⟨*⟩ **ב** ליתא **ס** [7] מגילה] מגולה **ב** [8] ט״ח] שטח **ב**

ונשלים קוי הצורה[1]. וח״כ כמו א״ט. וא״ט כמו ג׳ ונ״ל. ויהיה[2] אם כן ח״כ כמו ג׳ ונ״ל יחד. ונ״ל כמו ס״ע. ואם כן ח״כ כמו ג׳ וס״ע. ויחסר ס״ע המשותף. וישאר ג׳ כמו המסומן אשר עליו סחבכ״ע. וכ״פ כמו פ״ח. ויהיה פ״ב משותף. ויהיה יחס ה״ק כמו כ״ה יחד. ואמנם ח״ק כמו א״ס, לפי שא״ח שוה לח״ב. ויהיה אם כן א״ס כמו ה״כ. ונשים ח״ף משותף. ויהיה א״ף יחד מן המסומן אשר עליו סחבכ״ע. ואמנם המסומן היה כמו ג׳. ויהיה[3] א״פ כמו ג׳. וכבר[4] חברנו אל קו א״ב שטח נכוחי הצלעות, והוא א״פ שוה לשטח ג׳. ויחסר משלמות הקו שטח ה״ק הדומה לשטח ד״ז.

بُرْهانُهُ: أَنَّ ح ك مِثْلُ ا ط. و ا ط مِثْلُ ج و ن ل. وح ك مِثْلُ ج. و ن ل مِثْلُ س ع. وح ك مِثْلُ ج و س ع. فَنَنْقِصُ س ع المُشْتَرَكَ. فَيَبْقَى ج مِثْلُ العَلَمِ الَّذي عَلَيْهِ س ح ب ك ع. (و ف ح)[1] مِثْلُ ف ك. و ف ب مُشْتَرَكٌ. فَجَمِيعُ (ه ق)[2] مِثْلُ جَمِيعِ ك ه. وَلَكِنَّ ح ق مِثْلُ ا س. لِأَنَّ ا ح مِثْلُ ح ب، وَهُما بَيْنَ خَطَّيْنِ مُتَوازِيَيْنِ. و ا س مِثْلُ ه ك. وَنَجْعَلُ ح ف مُشْتَرَكـ[ـاً]. فَجَمِيعُ ا ف مِثْلُ العَلَمِ الَّذي عَلَيْهِ س ح ب ك ع. وَلَكِنَّ العَلَمَ مِثْلُ ج. و ا ف مِثْلُ ج. فَقَدْ أَضَفْنا إِلى خَطِّ ا ب سَطْحَ ا ب مُتَوازِيَ الأَضْلاعِ مُساوِيًا لِسَطْحِ ج. نَنْقُصُ مِنْ تَمامِ الخَطِّ سَطْحَ ه ق (يُشْبِهُ)[3] سَطْحَ د ز.

וזה מה שרצינו לבאר.

وَذَلِكَ ما أَرَدْنا بَيانَهُ.

[1]הצורה] ~~ח׳~~+ [2]ויהיה] ~~הנה~~ ⟨⟨*⟩⟩ [3]ויהיה] ⟨⟨ו⟩⟩יהיה [4]וכבר] ~~וכמו~~+

[1]و ف ح] وح [2]ه ق] ح ف [3]يشبه] نسبة

ونُتمّم سَطْحَ ط ف المُتوازي الأَضْلاعِ. ف ط ف مُساوٍ ل ن ل وشَبيه بِهِ. لكِنَّ ن ل شَبيهٌ بِح ك. ف ط ف شَبيهٌ ب ح ك. فَهُوَ على قُطْرِه. وَلْيَكُنْ قُطْرُهُ ط ف ب. ونُتمّم تَخْطيطَ الشَّكلِ. ف ا ط مثل ج و ن ل جَميعًا. و ا ط مثل ح ك. ف ح ك مثل ج و ن ل جَميعًا. لكِنَّ ن ل مثل ط ف. فعَلَمُ س ف ع الباقي مثل ج^. وَأَيْضًا فإِنَّ ك ف مثل ف ح. و ف ب مُشْتَرَكٌ. ف ك ه مثل ح ق. وح ق مثل ا س، لِأَنَّ ا ح مثل ح ب. ف ا س مثل ه ك. ونَجْعَلُ ح ف مُشْتَرَكًا. ف ا ف مِثْل العَلَمِ. وَقَدْ تَبيَّنَ أَنَّ العَلَمَ مِثْلُ شَكْلِ ج. ف ا ف مثل شكل ج. وح ك شَبيه ب د ز. و ف ب شبيه ب ح ك. ف ف ب شَبيه ب د ز. فَقَدْ أُضيفَ إلى خَطِّ ا ب سَطْحُ ا ف مساوٍ لِسَطْحِ ج، يَنْقُصُ عن تَمامِ ا ب سطح ف ب الشَّبيه ب د ز.

وذَلِكَ ما أَرَدْنا أَنْ نُبيِّنَ.

ונשלים שטח ט״פ[1] הנכחי הצלעות. אם כן ט״פ[2] שוה אל נ״ל ודומה לו. אבל נ״ל דומה אל ח״כ. אם כן ט״פ[3] דומה אל ח״כ. הנה הוא על קוטרו. ויהיה קוטרו טפ״ב. ונשלים לקוות התמונה. הנה א״ט כמו ג׳ ונ״ל יחד. וא״ט כמו ח״כ. אם כן ח״כ כמו ג׳ ונ״ל יחד[4]. אבל נ״ל כמו ט״פ. אם כן ח״כ שוה אל ג׳ וט״פ. ונחסר ט״פ המשותף. וישאר שטח ג׳ הידוע כמו הרושם. וגם כן הנה כ״פ[5] כמו פ״ח. ופ״ב משותף. אם כן כ״ה כמו ח״ק. וח״ק כמו א״ס, מפני כי א״ח כמו ח״ב. אם כן א״ס כמו ה״כ. ונשים ח״פ משותף. אם כן א״פ כמו הרושם. וכבר התבאר כי הרושם כמו תמונת ג׳. אם כן א״פ כמו ג׳. וח״כ דומה אל ד״ז. וב״פ דומה לח״כ[6]. אם כן פ״ב דומה אל ד״ז. הנה כבר חובר אל קו א״ב שטח א״פ שוה לשטח ג׳, יחסר משלימות א״ב שטח ב״פ הדומה אל ד״ז.

וזה מה שרצינו לבאר.

1ט״פ] ט״כ **ב** 2ט״פ] ט״כ **ב** 3ט״פ] ט״כ **ב** צ״פ **ס** 4יחד] ~~וא״ט כמו ח״כ~~ **ק** 5כ״פ] ~~ס״פ~~ **ס** 6לח״כ] אל ~~ד״ז~~ ח״כ **פ**

[Proposition VI.29]

[B28]

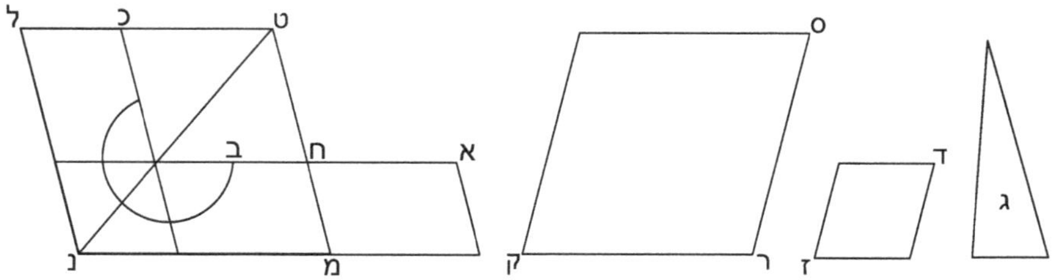

נרצה לחבר אל קו ישר ידוע שטח נכוחי הצלעות, ויעדף על שלמותו שטח דומה לשטח נכוחי הצלעות ידוע, ויהיה המחובר שוה לשטח ישר הקוים ידוע.

نُرِيدُ أَنْ نُبَيِّنَ كَيْفَ نُضِيفُ إِلى خَطٍّ مَعْلومٍ سَطْحًا مُتَوازِيَ الأَضْلاعِ مُساوِيًا لِسَطْحٍ مَعْلومٍ. وَنَزِيدُ عَلَى تَمامِ الخَطِّ سَطْحًا مُتَوازِيَ الأَضْلاعِ مَعْلومـ[ـًا] آخَرَ.

ונשים הקו הישר הידוע קו א״ב, והשטח השוה למחובר לקו שטח ג׳. והשטח הידוע האחר נכוחי הצלעות הדומה לשטח העודף שטח ד״ז. וכאשר רצינו לחבר אל קו א״ב שטח נכוחי הצלעות שוה לשטח ג׳, ויעדף על שלמות הקו שטח[1] דומה לד״ז,

مِثالُهُ: خَطُّ اب المَعْلومُ. وَالسَّطْحُ المُساوي لِلْمُضافِ إِلى الخَطِّ سَطْحُ ج. وَالسَّطْحُ المَعْلومُ الآخَرُ المُتَوازِي الأَضْلاعِ الشَّبِيهُ بِالسَّطْحِ الزَّائِدِ سَطْحُ د ز. وَنُرِيدُ أَنْ نُضِيفَ إِلى خَطِّ اب سَطْحًا مُتَوازِيَ الأَضْلاعِ مُساوِيًا لِسَطْحِ ج. وَنَزِيدُ عَلَى تَمامِ الخَطِّ سَطْحًا شَبِيهًا بِسَطْحِ د ز.

שרטוט—B: $, ز ← ه, ر ← ز והאות ه מסומנת על הקדקוד הרביעי של המקבילית שבקדקודיה האחרים س, ز ו-ق   [1]שטח] ⟨⟨*⟩⟩

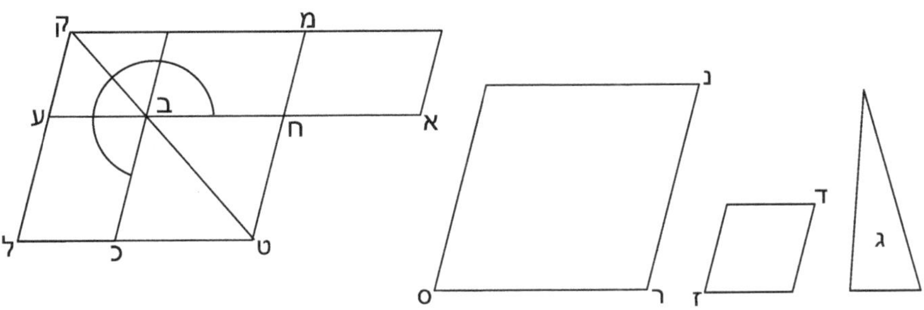

نُرِيدُ أَنْ نُضِيفَ إلى خَطٍّ مُسْتَقِيمٍ مَعْلومٍ سَطْحًا مُتوازِي الأَضْلاعِ، يَزِيدُ على تَمامِهِ سَطْحًا شَبيهًا بِسَطْحٍ مُتَوازِي الأَضْلاعِ مَعْلومٍ، ويكونُ المُضافُ مُساوِيًا لِشَكْلٍ ما^ مُسْتَقِيمِ الأَضْلاعِ مَعْلومٍ.

נרצה שנחבר אל קו ידוע ישר[1] שטח נכחי הצלעות[2], ויוסיף על שלמותו שטח דומה לשטח נכחי הצלעות ידוע, ויהיה המחובר שוה לתמונה ישרת הצלעות ידועה.

فَنَجْعَلَ الخَطَّ المَعْلومَ خطَّ اب، وَالسَّطحَ الشَّبيهَ بالزّائِدِ سَطْحَ دز، والشَّكْلَ المُساوي لِلْمُضافِ شَكْلَ ج. ونُرِيدُ أَنْ نُضيفَ إلى خَطِّ اب سَطْحًا مُتوازِي الأَضْلاعِ يَزِيدُ على تَمامِهِ سَطْحًا شَبيهًا بِسَطْحِ دز، وَيكونُ المُضافُ مُساوِيًا لِشَكْلِ ج.

הנה נשים הקו[3] הידוע קו א״ב, והשטח הדומה אל הנוסף ד״ז. והתמונה השוה המחוברת תמונת ג׳. ונרצה[4] שנחבר אל קו א״ב שטח נכחי הצלעות יוסיף על[5] שלמותו שטח דומה אל שטח ד״ז, ויהיה המחובר שוה לתמונת ג׳.

שרטוט—**ב**: הישר שקטע מ״ק עליו חתוך בשול; **ס**: ע׳ חסרה; **פ**: מ׳ חסרה; בכתבי היד הערביים הנחת הצורות בשרטוט זו לעומת זו נעשית באופנים מגוונים, אין אחידות גם בהנחת האותיות בקדקודי המצולע המסומן בתרגום אבן תיבון באותיות נר״ס, ואין כתב יד אחד שבו סממנים אלה של השרטוט זהים לסממניו בתרגום אבן תיבון

[1] ישר] ליתא **פ**
[2] הצלעות] +~~ידוע ויהיה המחובר שוה~~ **ק**
[3] הקו] אל+ **ס**
[4] ונרצה] ~~ונשים~~+ **פ**
[5] על] ליתא **ב**

אז נחלק קו א״ב לחציים על נקודת ח׳. ונעשה על קו ח״ב דמות שטח ד״ז, והוא שטח ח״כ. ונשים שטח ק״ס שוה לשטחי ח״כ וג׳ יחד, ודומה לשטח ד״ז. ויחס[1] ק״ר כמו יחס ח״ט. ויחס ר״ס כמו יחס ט״כ. וק״ר ארוך מח״ט. ור״ס ארוך מט״כ. ונשים ט״מ שוה לר״ק. וט״ל שוה לר״ס. ונשלים שטח ל״מ. ויהיה ח״כ על קוטר מ״ל. והקוטר ט״נ. ונשלים קוי הצורה. וק״ס שוה לח״כ וג׳ יחד. ואמנם ק״ס שוה למ״ל. ומ״ל שוה לח״כ ולג׳ יחד. ויחסר ח״כ המשותף. וישאר שטח ג׳ שוה לחמנלכ״ב המסומן. וא״ח שוה לח״ב. ויהיה א״מ שוה למ״ב, ומ״ב שוה לב״ל. וב״נ הוא משותף. ויהיה א״נ כלו שוה לחמנלכ״ב המסומן. ואמנם המסומן שוה לג׳. ואם כן יהיה א״נ שוה לג׳. וכבר חברנו לקו א״ב שטח נכוחי הצלעות שוה לשטח ג׳, ויעדף על שלמות הקו שטח ב״נ הדומה לד״ז, והוא שטח א״נ.

[1] ויחס] וח״כ+; אפשר שהמילה מחוקה

فنقسم خطّ ا ب بنصفين على ح. ونعمل على خطّ ح ب سطحًا يشبه سطح د ز، وهو ح ك. ونجعل سطح ق س مثل سطحي ح ك و ج شبيهًا بسطح د ز[1]. ونسبة ق ر إلى ح ط مثل نسبة ر س إلى ط ك. و ق ر أطول من ط ح. و ر س أطول من ط ك. فنجعل ط م مثل ر ق. (و ط ل)[2] مثل ر س. ونتمّم سطح م ل. ح ك هو على قطر ل م. والقطر ط ب ن. ونتمّم تخطيط الشكل. فأقول إنّا قد أضفنا إلى خطّ ا ب سطح ا ن كما أردنا. برهانه: ان ق س مثل ح ك و ج. ولكنّ ق س مثل (م ل)[3]. و م ل مثل ح ك و ج. فننقص ح ك المشترك. فيبقى سطح ج مثل ح م ن ل ك ب العلم. و ا ح مثل ح ب. و ا م مثل م ب. (م ب)[4] مثل ب ل. ا م مثل ب ل. ق س مشترك. فجميع ا ن مثل ح م ن ل ك العلم. ولكنّ العلم مثل ج. فقد أضفنا إلى خطّ ا ب سطحًا متوازي الأضلاع مساويًا لسطح ج، نزيد على تمام الخطّ سطح ب ن الشبيه ب د ز، وهو ا ن.

[1] د ز] ح ك ⟨*⟩ [2] و ط ل] و ط [3] م ل] ل [4] م ب] ליתא

فنقسِم خطَّ اب بنصفَيْنِ على ح. ونعمل على ب ح سطحَ ك ح يشْبِهُ د ز. ونجْعل سطح ن س مِثْلَ سطْحَيْ ك ح و ج جميعًا، وشَبيهًا بِسطح د ز. ونجْعَل ن ر نظيرًا في النِّسْبة^ لِضلْع ط ح، و ر س لِضلْع ط ك. ف ن س أعظم مِنْ ك ح. وهو شَبيهٌ بِهِ، لأنّهما جميعًا يُشبِهانِ د ز. فكُلّ ضِلْعٍ مِنْ أضْلاعِ ن س أطْوَلُ مِنْ نَظيرِهِ في النِّسْبَةِ مِنْ أضْلاعِ سَطْحِ[1]^ ك ح. ف ن ر أطول مِنْ ح ط. و ر س أطول مِنْ ط ك. فنَجْعَل ط م مثل ن ر. و ط ل مثل ر س. ونُتِمّ سَطْحَ م ل. ف م ل مُساوٍ ل ن س وشبيه بِهِ، و ن س شبيه ب ك ح. ف ك ح شبيه ب م ل. فهُوَ على قُطْرِهِ. والقُطْر ط ق. ونُتِمّ^ تخْطيطَ الشَّكْلِ. ف ن س مثل ك ح و ج جميعًا. لكنَّ ن س مثل ل م. ف م ل مثل ك ح و ج جميعًا. فلْيُلْقَ ك ح المُشْتَرَك. فيَبْقى ج مثل علَمِ ح ف ك. وخطّ ا ح مثل خطّ ب ح. فسطح ا م مثل سَطْحِ ب م. و ب م مثل ب ل. ف ا م مثل ب ل. و م ع مُشْتَرَكٌ. فجميع ا ق مثل العلَم. لكِنَّ العلَمَ مِثْلُ ج. ف ا ق مثل ج. وسطح ب ق شَبيهٌ بِسَطْحِ ط ق، لأنّهُ على قُطْرِهِ. وسطح ط ق شَبيهٌ بِسَطْحِ د ز. فَسَطْح ب ق شَبيهٌ بِسَطْحِ د ز[2]. فقَدْ أضَفْنا إلى خَطِّ ا ب سَطْحَ ا ق، ويَزيدُ على تَمامِ الخطّ سَطْحَ ب ق الشَّبيه بِسَطْحِ د ز، ويُساوي شَكْلَ ج المُسْتَقيمَ الأضْلاعِ.

[1] سطح] MPR2 [2] فسطح ... د ز] CopT3

ונחלק קו א״ב בשני חציים על ח׳[1]. ונעשה על ב״ח שטח כ״ח ידמה ד״ז. ונשים שטח נ״ס כמו שני שטחי כ״ח[2] וג׳ יחד, ודומה לשטח ד״ז. ונשים נ״ר גילו בדמיון[3] לצלע ח״ט, ור״ס לצלע ט״כ. הנה נ״ס[4] יותר גדול מן כ״ח. והוא דומה לו, מפני כי שניהם יחד ידמו ד״ז. אם כן כל צלע מצלעות נ״ס יותר ארוך מגילו ביחס מן שטח כ״ח. אם כן נ״ר יותר ארוך מן ח״ט. ור״ס יותר ארוך מן ט״כ. ונשים ט״מ כמו נ״ר, וט״ל כמו ס״ר. וישלים[5] שטח מ״ל. אם כן מ״ל שוה לנ״ס ודומה לו, ונ״ס לכ״ח[6]. אם כן כ״ח[7] דומה למ״ל. אם כן הוא על קוטרו. והקוטר[8] ט״ק. ותשלים הקוות[9] התמונה. הנה[10] נ״ס כמו כ״ח וג׳ יחד[11]. אבל נ״ס כמו ל״מ. אם כן מ״ל כמו כ״ח וג׳ יחד. והפיל[12] כ״ח המשותף. וישאר ג׳ כמו רושם ב׳. וקו א״ח כמו קו ב״ח[13]. אם כן שטח אח״מ כמו שטח[14] ב״מ. וב״מ כמו ב״ל. אם כן א״מ[15] כמו ב״ל. ומ״ע משותף. אם כן כל א״ק כמו הרושם. אבל הרושם[16] כמו ג׳. אם כן א״ק כמו ג׳. ושטח ב״ק דומה לשטח ט״ק, מפני שהוא על קוטרו. ושטח ט״ק דומה לשטח ד״ז. אם[17] כן שטח ב״ק דומה לשטח ד״ז[18]. הנה כבר חברנו אל קו[19] א״ב שטח א״ק, ויוסיף על שלמות הקו שטח ב״ק הדומה לשטח ד״ז, והוא שוה לתמונת ג׳ ישרת הצלעות.

[1] על ח׳] ליתא **ז** [2] כ״ח] ליתא **ב** [3] בדמיון] בדימיונו **ז** +נ״ד **ס** [4] ט״כ ... נ״ס] ~~ד״[..]~~ ט״כ הנה **ס** [5] וישלים] ונשלים **ב** [6] לכ״ח] דומה+ **ס** [7] כ״ח] +גילו **ס** [8] והקוטר] וקוטרו **ס** והקו⟨טר⟩ **פ** [9] הקוות] ⟨*⟩ **ב** [10] הנה] +אם כן **ק** [11] יחד] ליתא **ב** [12] והפיל] ותפיל **ב** [13] כמו ... ב״ח] +* **ב** [14] שטח] ליתא **ב** [15] כמו ... א״מ] ⟨⟨*⟩⟩ **ק** [16] אבל הרושם] ליתא **ז** [17] אם] הנה+ **ט** [18] אם ... ד״ז] ליתא **פ** [19] קו] ליתא **ט**

וזה מה שרצינו לבאר.	وَذَلِكَ ما أَرَدْنا بَيانَهُ.

وذلك ما اردنا ان نبين

וזה מה שרצינו לבאר[1].

[1] וזה ... לבאר] ונשלם ביאורו ז

[Proposition VI.30]

[B29]

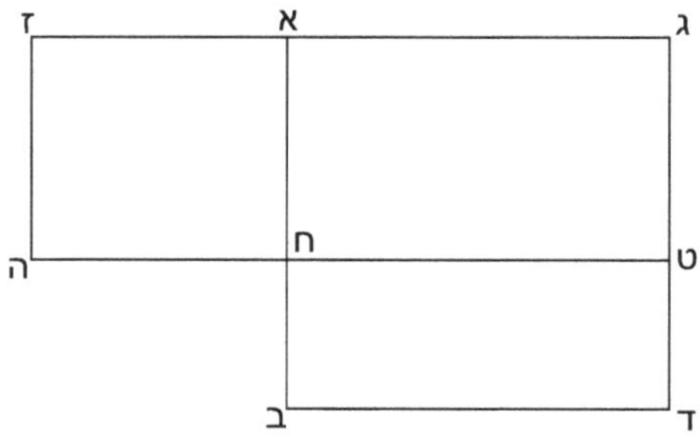

נרצה לחלק קו ידוע על יחס עצם ואמצע ושתי קצוות.

نُرِيدُ أَنْ نُبَيِّنَ كَيْفَ نَقْسِمُ خَطًّا مَعْلومًا عَلَى نِسْبَةٍ ذاتِ وَسَطٍ وَطَرَفَيْنِ.

ונשים הקו הידוע קו א״ב. וכאשר רצינו לחלק אתו [!] על יחס עצם אמצע ושתי קצוות,

مِثالُهُ: خَطُّ اب المَعْلومُ. وَنُرِيدُ قِسْمَتَهُ عَلَى نِسْبَةٍ ذاتِ وَسَطٍ وَطَرَفَيْنِ.

שרטוט—B: ه ← ح, ز ← ه, ح ← ز (ن?)

نُرِيدُ أَنْ نَقْسِمَ خَطًّا مَعْلُومًا على نِسْبَةٍ ذات وَسَطٍ وطَرَفَيْنِ.

נרצה שנחלק[1] קו ידוע על יחס בעל אמצע ושתי קצוות.

فَنَجْعَلَ الخَطَّ المَعْلومَ خَطَّ اب. ونُرِيدُ أَنْ نَقْسِمَهُ على نِسْبَةٍ ذات وَسَطٍ وطَرَفَيْنِ.

הנה נשים הקו הידוע קו א״ב. ונרצה לחלק אותו על יחס[2] בעל אמצע ושתי קצוות.

שרטוט—**בוטק**: $

[1] שנחלק] +על **ט**

[2] יחס] ליתא **בפ**

נעשה על קו א״ב שטח מרובע, והוא א״ד. ונחבר אל קו א״ג[1] שטח נכוחי הצלעות שוה לשטח א״ד[2], יעדיף על שלמות הקו שטח ז״ח, והוא שטח ז״ט. ונשים ז״ח דומה לא״ד. ויהיה ז״ט שוה לא״ד. ויחוסר[3] א״ט המשותף. וישאר ז״ח שוה לח״ד, והשטחים נכוחי הצלעות השוים אשר זויות אחד מהם שוות לזויות האחר, צלעותם המקיפות בזויות השוות הן מתהפכות. ויהיה שעור ט״ח אל ה״ח כשעור א״ח אל ח״ב. וט״ח שוה לא״ב. וח״ה[4] שוה לא״ח. אם כן שיעור א״ב אל א״ח כשיעור א״ח אל ח״ב. וכבר חלקנו קו א״ב על יחס עצם אמצע ושתי קצוות על נקודת ח׳, והחלק הארוך הוא א״ח.

וזה מה שרצינו לבאר.

فَنَعْمَلُ عَلَى خَطِّ اب سَطْحَ اد المُرَبَّعَ. وَنُضيفُ إِلى خَطِّ اج سَطْحَ ه ط مُتَوازِيَ الأَضْلاعِ مِثْلَ سَطْحِ اد، يَزيدُ عَلى تَمامِ الخَطِّ سَطْحَ ز ه. وَنَجْعَلُ ز ه (يُشْبِهُ)[1] ا د. فَأَقولُ إِنَّا قَدْ قَسَمْنا خَطَّ ا ب المَعْلومَ عَلى نِسْبَةِ ذاتِ وَسَطٍ وَطَرَفَيْنِ عَلى نُقْطَةِ ه. وَنِسْبَةُ ا ب إِلى ا ز كَنِسْبَةِ ا ز إِلى ز ب.

بُرْهانُهُ: أَنَّ ه ط مِثْلُ ا د. وَنَنْقُصُ ا ط المُشْتَرَكَ. فَيَبْقَى ز ه مِثْلَ ط ب. وَالسُّطوحُ المُتَوازِيَةُ الأَضْلاعِ المُتَساوِيَةَ الَّتي زَواياها مُتَساوِيَةٌ فَأَضْلاعُهُما الَّتي تُحيطُ بِالزَّوايا المُتَساوِيَةِ — مُتَكافِئَةٌ. فَقَدْرُ ط ز عِنْدَ ز ح كَقَدْرِ ا ز عِنْدَ ز ب. و ط ز مِثْلُ ا ب. و ز ح مِثْلُ ا ز. فَقَدْرُ ا ب عِنْدَ ا ز كَقَدْرِ ا ز عِنْدَ ز ب. فَقَدْ قَسَمْنا خَطَّ ا ب المَعْلومَ[2] عَلى نِسْبَةِ ذاتِ وَسَطٍ وَطَرَفَيْنِ عَلى نُقْطَةِ ز، وَالقِسْمُ الأَطْوَلُ ا ز.

وَذَلِكَ ما أَرَدْنا بَيانَهُ.

---

[1] א״ג] +~~שוה לשטח א״ד~~ [2] שוה ... א״ד] ⟨⟨*⟩⟩
[3] ויחוסר] ויח⟨ו⟩סר [4] וח״ה] ⟨⟨*⟩⟩

[1] يشبه] نسبة [2] المعلوم] ~~المعلوم~~+

فنَعْمَلَ عَلَيْهِ مُرَبَّعَ ا د. ونُضِيفَ إلى خطّ ا ح سَطْحًا مُتَوازِي الأَضْلاعِ مثل ا د، وهو ز ط، يَزِيدُ على تَمامِ الخَطِّ سَطْحًا يُشْبِهُ مُرَبَّعَ ا د، وَهُوَ ز ح. ف ز ط مثل ا د. فيَنْقُصُ ا ط المُشْتَرَكُ. فيَبْقَى ز ح مثل د ح. والسُّطُوحُ المُتوازِيَةُ الأَضْلاعِ المُتَساوِيَةِ أَضلاعهُما المُحِيطَةُ بِالزَّوايا المُتَساوِيَةِ مُتَكافِئَةٌ. فَنِسْبَةَ ط ح إلى ح ه كَنِسْبَةِ ا ح إلى ح ب. و ط ح مثل ا ب. و ح ه مثل ا ح. فَنِسْبَةَ ا ب إلى ا ح كَنِسْبَةِ ا ح إلى ح ب. فَقَدْ قَسَمْنَا خطّ ا ب على نِسْبَةٍ ذات وَسَطٍ وطَرَفَيْنِ، وَقِسْمة الأَطْوَلِ ا ح.

הנה נעשה עליו מרובע א״ד. ונחבר אל קו א״ג שטח נכחי הצלעות כמו א״ד[1], והוא ז״ט, יוסיף על שלמות הקו שטח ידמה מרובע[2] א״ד, והוא ז״ח. אם כן ז״ט כמו[3] א״ד. ויחוסר א״ט המשותף. וישאר ז״ח כמו ד״ח. והשטחים הנכחי הצלעות השוים הנה צלעותם[4] המקיפות בזויות השוות מספיקות. אם כן יחס ט״ח אל ח״ה כיחס א״ח אל ח״ב. וט״ח כמו א״ב. וח״ה כמו א״ח. אם כן יחס א״ב אל א״ח כיחס א״ח אל ח״ב. הנה כבר חלקנו קו א״ב על יחס בעל אמצע ושתי קצוות, וחלוקת היותר ארוך א״ח.

وَذَلِكَ ما أَرَدْنا أَنْ نُبَيِّنَ.

וזה מה שרצינו לבאר.

---

1 א״ד] א״ט **ס**   2 מרובע] ⟨*⟩ **ט**   3 כמו] ליתא **ב**   4 צלעותם] צלעותיהם **ק**

[Proposition VI.31]

[H32, B30]

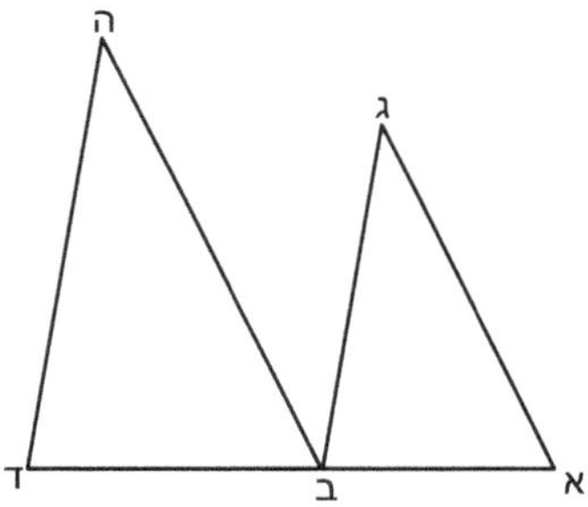

כשהרכבו [!] שני משולשים על זוית אחת, ושתי צלעות המקיפות בה הן נכוחיות לשתי צלעות אחרות מהם, ויחס הנכוחיות כל אחת מהן אל הנכוחיות לה—אחד, תהיין שתי הצלעות הנשארות נדבקות על יושר.

كُلُّ مُثَلَّثَيْنِ يكونانِ عَلى زاوِيَةٍ واحِدَةٍ، وَيَكونُ الضِّلْعانِ المُحيطانِ بِتِلْكَ الزّاوِيَةِ مُوازِيَيْنِ لِضِلْعَيْنِ آخَرَيْنِ مِنْ أَضْلاعِ[1] المُثَلَّثَيْنِ، فَتَكونُ نِسْبَةُ الضِّلْعِ مِنْ أَحَدِ المُثَلَّثَيْنِ إِلى المُوازي لَهُ مِنَ المُثَلَّثِ الآخَرِ مِثْلَ نِسْبَةِ الآخَرِ إِلى الَّذي يُوازيهِ، فَإِنَّ المُثَلَّثَيْنِ يَكونانِ عَلى خَطٍّ واحِدٍ مُسْتَقيمٍ.

המשל בזה: כי משולשי אב״ג (בד״ה)[1] על זוית אחת, והיא זוית גב״ה. וצלע א״ג נכוחית לצלע ב״ה. וב״ג נכוחי לד״ה. ויחס א״ג אל ב״ה כיחס ב״ג אל ד״ה. ואומר כי שני המשולשים הם על קו אחד ישר והוא קו א״ד.

مِثالُهُ: أَنَّ مُثَلَّثَيْ ا ب ج و ب د ه عَلى زاوِيَةٍ واحِدَةٍ، وَهِيَ زاوِيَةُ ج ب ه. وَضِلْعُ ا ج مُوازٍ لِضِلْعِ ب ه. و ب ج مُوازٍ لِضِلْعِ ه د. وَنِسْبَةُ ا ج إِلى ب ه كَنِسْبَةِ ب ج إِلى د ه. فَأَقولُ إِنَّ المُثَلَّثَيْنِ عَلى خَطٍّ واحِدٍ مُسْتَقيمٍ، وَهُوَ خَطُّ ا ب د.

[1] בד״ה] וגד״ה

[1] أضلاع] ⟨*⟩

إذا رُكِّبَ مُثَلَّثانِ على زاوِيَةٍ واحِدَةٍ يُحِيطُ بِها ضِلْعانِ مِنْهُما، يُوازِيانِ ضِلْعَيْنِ آخَرَيْنِ مِنْهُما، وَنِسْبَةُ^ المُتَوازِيَيْنِ كُلٍّ^ واحِدٍ مِنْهُما إلى المُوازي لَهُ—واحِدَةٌ، فَإِنَّ الضِّلْعَيْنِ الباقِيَيْنِ مِنَ المُثَلَّثَيْنِ مُتَّصِلانِ على اسْتِقامَةٍ.

כאשר הורכבו[1] שני משולשים על זוית אחת יקיפו בהן[2] שתי צלעות מהם[3], יהיו נכחיות לשתי צלעות אחרות מהם, יחס[4] הנכחית מכל אחת מהם אל הנכחית לו—אחד, הנה השתי צלעות[5] הנשארות משני המשולשים[6] דבקות על יושר.

مِثالُهُ: أَنَّ مُثَلَّثَيْ ا ب ج د ب ه مُرَكَّبانِ^ على زاوِيَةِ ج ب ه. وضلع ا ج يُوازي ب ه. و ب ج يُوازي د ه. ونِسْبَةَ ا ج إلى ب ه المُوازي لَهُ كَنِسْبَةِ ب ج إلى د ه المُوازي لَهُ. فَأَقولُ إِنَّ ضِلْعَيْ ب د ب ا الباقِيَيْنِ مُتَّصِلانِ على اسْتِقامَةٍ.

המשל בו: כי שני משולשי אב״ג דב״ה על[7] זוית גב״ה. וצלע א״ג נכחי אל ב״ה. וב״ג נכחי[8] ד״ה. ויחס א״ג אל ב״ה הנכחי לו כיחס ב״ג אל ד״ה הנכחי לו. הנה אומר כי שתי צלעות ד״ב א״ב הנשארות דבקות על יושר.

שרטוט—**ס**: שני משולשים נפרדים זה מזה, האחד בג״ד ועל האחר מסומנים קדקודים ד׳ וה׳ בלבד   1 הורכבו] ⟨*⟩ **ט**   2 בהן] נראה שצ״ל ״בה״   3 מהם] מהן **ק**   4 יחס] ויחס **ס**   5 צלעות] הצלעות **בספ**   6 המשולשים] משולשים **פ**   7 על] הורכב+ **ס**   8 נכחי] +אל **ק**

מופת זה: לפי שקו א״ג נכוחי לקו ב״ה, ונפל עליהן ב״ג, תהיין שתי הזויות המתחלפות, והן זוית ג׳ וזוית גב״ה, שוות. וכמו כן לפי שקו ב״ג נכוחי לקו ד״ה, ונפל עליהן קו ב״ה, תהיין שתי הזויות המתחלפות, שהן זוית גב״ה וזוית בה״ד, שוות. ואמנם זוית גב״ה שוה לזוית ג׳. ואם כן זוית ג׳ שוה לזוית ה׳. ושיעור א״ג על ב״ה כשיעור ב״ג על ד״ה. והצלעות המקיפות בזויות ג׳ וה׳ השוות הן מתיחסות. ואם כן משולש אב״ג ובד״ה זויותיהם שוות—זוית בא״ג שוה לזוית דב״ה, וזוית גב״א שוה לזוית הד״ב. וזוית ג׳ שוה לזוית ה׳. וזוית ג׳ שוה לזוית גב״ה. וזוית א׳ שוה לזוית הב״ד. וזוית אב״ג משותפת. ותהיין הזויות השלש, והן זוית א׳ וזוית ג׳ וזוית גב״א[1], שוות לזויות השלש, והן זוית גב״א וזוית גב״ה וזוית הב״ד. ואמנם זוית א׳ וג׳ וזוית גב״א יחד שוות לשתי זויות נצבות. ואם כן זויות גב״א וגב״ה והב״ד שוות לשתי זויות נצבות. וכבר יצא מקו ב״ג מנקודת ב׳ שני קוים לשני צדדים מתחלפים, והם ב״א וב״ד. ושתי הזויות אשר אצל צדי קו ב״ג, שהן זוית אב״ג וזוית גב״ד, שוות לשתי זויות נצבות. ואם כן קו א״ב וב״ד הם נדבקים על יושר. וכבר היו קו אחד ישר.

וזה מה שרצינו שנבאר.

[1] גב״א] ~~גב״ה~~ ⟨⟨*⟩⟩

بُرْهانُه: أَنَّ خَطَّ ا ج مُوازٍ لِخَطِّ ب ه. وَوَقَعَ عَلَيْهِما خَطُّ ب ج. فَصارَتْ زاوِيَتا (ا ج ب وج ب ه)[1] المُتبادَلَتانِ مُتَساوِيَتَيْنِ. وَأَيضًا خَطُّ ب ج مُوازٍ لِخَطِّ د ه. وَوَقَعَ عَلَيْهِما خَطُّ ب ه. فَزاوِيَتا ج ب ه و ب ه د المُتبادَلَتانِ مُتَساوِيَتَيْنِ. وَلٰكِنَّ زاوِيَةَ ج ب ه مِثْلُ زاوِيَةِ ه. فَقَدْ[رُ] ا ج عِنْدَ ب ه كَقَدْرِ ب ج عِنْدَ ه د. فَالأَضْلاعُ الَّتِي تُحيطُ بِزاوِيَتَيْ ج و ه المُتَساوِيَتَيْنِ هِيَ مُتَناسِبَةٌ. فَزاوِيَتا مُثَلَّثَيْ ا ب ج و ب د ه مُتَساوِيَةٌ. زاوِيَةُ ب ا ج مِثْلُ زاوِيَةِ د ب ه. وَزاوِيَةُ ج ب ا مِثْلُ زاوِيَةِ ه د ب. وَزاوِيَةُ ج مِثْلُ زاوِيَةِ ه. فَزاوِيَةُ ج مِثْلُ زاوِيَةِ ج ب ه. وَزاوِيَةُ ا مِثْلُ زاوِيَةِ ه ب د. وَزاوِيَةُ ا ب ج مُشْتَرَكَةٌ. فَزَوايا ا وج وج ب ه (الثَّلاثُ)[2] مِثْلُ زَوايا ج ب ا وج ب ه و ه ب د مِثْلُ زاوِيَتَيْنِ قائِمَتَيْنِ الثّلث. وَلٰكِنَّ زَوايا ا و ج و ج ب ا هُنَّ جَميعًا مِثْلُ زاوِيَتَيْنِ. فَزَوايا ج ب ا وج ب ه و ه ب د مِثْلُ زاوِيَتَيْنِ قائِمَتَيْنِ. فَقَدْ خَرَجَ مِنْ نُقْطَةِ بَ مِنْ خَطِّ ب ج خَطّا ب ا و ب د في جِهَتَيْنِ مُخْتَلِفَتَيْنِ. وَزاوِيَتا ا ب ج وج ب د اللَّتانِ عَنْ جَنْبَتَيْ خَطِّ ب ج هُما مُتَساوِيَتانِ لِزاوِيَتَيْنِ. فَخَطّا ا ب و ب د مُتَّصِلانِ عَلى اسْتِقامَةٍ.

وَذٰلِكَ ما أَرَدْنا بَيانَه.

[1] ا ج ب وج ب ه] ~~ج وه د ج~~ ⟨⟨*⟩⟩ [2] الثلاث] ألثلث

بُرهانُهُ: أَنَّ ا ج يُوازِي ب ه. وَقَدْ وَقَعَ عَلَيْهِما ب ج. فزاويةُ ب ج ا مثل زاويةِ ج ب ه المُبادِلَةِ لها. و ب ج يُوازِي د ه. وَقَدْ وَقَعَ عَلَيْهِما ب ه. فَزاوِيَتا ج ب ه ب ه د المُتَبادِلَتانِ مُتَساوِيَتانِ. لٰكِنَّ زاوِيَةَ ج ب ه مثل زاويةِ ج. فزاويةُ ج مثل زاوية ه. ونِسْبَةُ ا ج إلى ب ه كَنِسْبَةِ ب ج إلى د ه. والأَضْلاعُ المُحِيطَةُ بِزاوِيَتَيْ ج ه المُتَساوِيَتَيْنِ مُتَناسِبَةٌ. فَزَوايا مُثَلَّثَيْ ا ب ج و د ه ب مُتَساوِيَةٌ—زاويةُ ا مثل زاويةِ د ب ه، وزاويةُ ج مثل زاويةِ ج ب ه^. فَجَميعُ زاويةِ ج ب د مثل زاوِيَتَيْ ا ج. وزاويةُ ا ب ج مُشْتَرَكَةٌ. فَزَوايا مُثَلَّثِ ا ب ج الثَّلاث مثل زاوِيَتَيْ ج ب ا ج ب د. لٰكِنَّ زوايا المُثَلَّثِ[1] مثل زاوِيَتَيْنِ قائِمَتَيْنِ. فَزَوايا ج ب ا ج ب د مُعادِلَتانِ لِزاوِيَتَيْنِ قائِمَتَيْنِ. فَقَدْ خَرَجَ مِنْ ب ج مِنْ نُقْطَةِ ب خَطّانِ في جِهَتَيْنِ مُخْتَلِفَتَيْنِ، والخَطّانِ ب ا ب د. وعَنْ جَنْبَتَي ب ج زاويتا ا ب ج د ب ج. وَهُما مُساوِيَتانِ لِزاوِيَتَيْنِ^ قائِمَتَيْنِ. فَخَطّا ا ب ب د مُتَّصِلانِ على اسْتِقامَةٍ.

وَذٰلِكَ ما أَرَدْنا أَنْ نُبَيِّنَ.

המופת: כי א״ג נכחי[1] ב״ה. וכבר נפל על שניהם ב״ג. אם כן זוית בג״א כמו זוית גב״ה המומרת[2] לה. וב״ג נכחי[3] ד״ה. וכבר נפל על שניהם ב״ה[4]. אם כן שתי זויות גב״ה בה״ד המומרות[5] שוות. אבל זוית גב״ה כמו זוית ג׳. אם כן זוית ג׳ כמו זוית ה׳. ויחס א״ג אל ב״ה כיחס ב״ג אל ד״ה[6]. אם כן הצלעות המקיפות בשתי[7] זויות ג׳ ה׳ השוות מתיחסות. אם כן שתי זויות שני משולשי אב״ג בה״ד שוות—זוית א׳ כמו זוית דב״ה, וזוית ד׳ כמו זוית אב״ג. הנה כל זוית גב״ד כמו שתי[8] זויות א׳ ג׳[9]. וזוית אב״ג משותפת. אם כן זויות משולש אב״ג השלש כמו שתי זויות גב״א גב״ד[10]. אבל זויות המשולש כמו שתי[11] זויות נצבות. אם כן[12] זויות גב״א גב״ד שוות לשתי זויות נצבות[13]. הנה כבר יצא מקו ב״ג מנקודת ב׳ ממנו שני קוים בשני צדדים מתחלפים, והשני קוים א״ב[14] ב״ד. ומשני צדדי ב״ג שתי זויות אב״ג דב״ג. ושתיהן שוות[15] לשתי נצבות. אם כן קוי א״ב ב״ד דבקים על יושר.

ונשלם ביאורו.

---

[1]المثلث] CT_1T_3 +الثلاث

---

[1]נכחי] +נכחית **ט** +אל **ק** [2]המומרת] המומרות **ס** [3]נכחי] +אל **ק** [4]ב״ה] קו+ **ז** [5]המומרות] המומרת **ב** [6]ד״ה] +וזויות ג״ה שוות **ס** [7]בשתי] שתי **ז** [8]שתי] כל+ **ז** [9]ג׳] +⟨יחד⟩ **ק** [10]אבל ... גב״ד] ⟨⟨*⟩⟩ **ק** [11]שתי] כל+ **ז** [12]כן] ליתא **ס** [13]אם ... נצבות] ליתא **ב** [14]א״ב] *+ **ס** [15]ושתיהן שוות] ושתי השות **ב** ושתיהם שוות **וספ**

وَلَهُ وَجْهٌ آخَرُ، وهو هذا[1]:

إذا رُكِّبَ مُثَلَّثانِ مُتَشابِهانِ على زاوِيَةٍ واحِدَةٍ، فَأَحاطَ بِها ضِلْعانِ مِنْهُما، وكانا مُوازِيَيْنِ لِضِلْعَيْنِ آخَرَيْنِ مِنْهُما، وكانَتِ الأَضلاعُ المُتَوازِيَةُ نَظائِرَ في النِّسْبَةِ، فَإِنَّ الضِّلْعَيْنِ الباقِيَيْنِ مُتَّصِلانِ على اسْتِقامَةٍ.

مِثالُهُ: أَنَّ مُثَلَّثَيْ ا ب ج د ب ه مُتَشابِهانِ مُرَكَّبانِ على زاوِيَةِ ج ب ه. وَضِلع ا ج يُوازي ب ه. و ب ج يُوازي د ه. ونِسْبَةَ ا ج إلى ب ه كَنِسْبَةِ ب ج إلى د ه. فَأَقولُ إنَّ خَطَّيْ ا ب و ب د مُتَّصِلانِ على اسْتِقامَةٍ.

ולו פנים אחרים, והוא זה[1]:

כאשר הורכבו שני משולשים מתדמים על זוית אחת, והקיפו בה שתי צלעות משניהם, והיו נכחיות לשתי צלעות אחרות משניהם, והיו הצלעות הנכחיות גילות[2] ביחס, הנה השתי צלעות[3] הנשארות דבקות על יושר.

המשל בו: כי שני משולשי אב״ג דב״ה מתדמים מורכבים על זוית גב״ה. וצלע א״ג נכחי אל ב״ה. וב״ג נכחי[4] אל ד״ה. ויחס א״ג אל ב״ה כיחס ב״ג אל ד״ה. הנה אומר כי שני קוי א״ב ב״ד דבקים על יושר.

---

[1] وهو هذا] CopT2

[1] זה] ליתא **ב** [2] גילות] דומות **ס** [3] צלעות] הצלעות **וטס** [4] נכחי] +אל **ק**

بُرْهانُه: أَنَّ ا ج يُوازي ب ه. وَقَدْ وَقَعَ عَلَيْهِما ب ج. فزاوية ا ج ب مثل زاويةِ ج ب ه. وَمُثَلَّثا ا ج ب ب ه د مُتَشابِهانِ. فزاوية ج ا ب مثل زاوية ه ب د. وزاوية ا ب ج مُشْتَرَكَة. فزوايا ا ج ب ج ب ا ب ا ج مثل زوايا ج ب ا ج ب ه ه ب د. لَكِنَّ^ زوايا ا ج ب ج ب ا ب ا ج مثل زاوِيَتَيْنِ قائِمَتَيْنِ. فزوايا ج ب ا ج ب ه ه ب د مثل قائِمَتَيْنِ[1]. فَقَدْ خَرَجَ مِنْ ب ج مِنْ نُقْطَةِ ب مِنْهُ خَطّانِ في جِهَتَيْنِ مُخْتَلِفَتَيْنِ، وَالخَطّانِ ب ا ب د. وَعَنْ جَنْبَتَيْ ب ج زاوِيَتا ا ب ج ج ب د مُساوِيَتَيْنِ لِقائِمَتَيْنِ. فَخَطّا ا ب ب د مُتَّصِلانِ على اسْتِقامَةٍ.

وَذَلِكَ ما أَرَدْنا أَنْ نُبَيِّنَ^.

[1]زوايا ... قائمتين] ComMPR2T2

המופת: כי א״ג נכחי[1] אל ב״ה. וכבר נפל על שניהם ב״ג. אם כן זוית אג״ב כמו זוית גב״ה. ושני משולשי אג״ב בה״ד מתדמים. אם כן זוית גא״ב כמו זוית הב״ד. וזוית אב״ג משותפת. אם כן זויות[2] אג״ב[3] גב״א בא״ג כמו זויות[4] גב״א גב״ה הב״ד[5]. וזויות[6] אג״ב גב״א בא״ג כמו שתי זויות נצבות. אם כן זויות[7] גב״א גב״ה הב״ד כמו שתי נצבות. הנה כבר יצא מקו[8] ב״ג מנקודת ב׳ ממנו שני קוים בשני צדדים מתחלפים, ושני הקוים ב״א ב״ד. ומשני צדדי[9] ב״ג שתי[10] זויות אב״ג גב״ד שוות לשתי נצבות. אם כן שני קוי[11] א״ב ב״ד דבקים על יושר.

ונשלם ביאורו.

[1]נכחי] ליתא **ס** [2]זויות] זוית **בוק** [3]אג״ב] ~~אב״ג~~+ **ק** [4]זויות] זוית **בס**⟨זוית⟩ **ו** [5]הב״ד] ~~גב״ד~~+ **ב** [6]וזויות] וזוית **בוספ** [7]זויות] זוי⟨ו⟩ת **ב** זוית **ס** [8]מקו] ~~מנקדת~~ ⟨⟨*⟩⟩ **ק** [9]צדדי] צדדיו **ס** [10]שתי] כמו+ **ס** [11]שני קוי] ליתא **ס**

[Proposition VI.32]

[H31, B31]

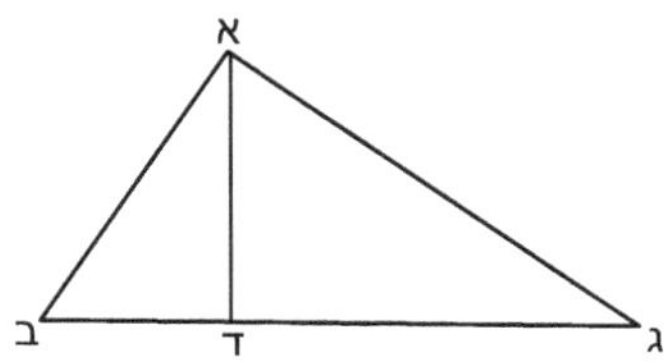

כל משלש נצב[1] הזוית, יהיה השטח ישר הקוים המחובר אל מיתר הזוית הנצבה שוה לשני שטחים ישרי הקוים המחוברים אל שתי צלעותיו הנשארות יחד כשיהיו דומים לו והם על תארו.

كُلُّ مُثَلَّثٍ قائِمِ الزَّاوِيَةِ فَإِنَّ السَّطْحَ الَّذي يَكونُ عَلى ضِلْعِهِ الَّذي يوتِرُ الزَّاوِيَةَ القائِمَةَ، هُوَ مِثْلُ السَّطْحَيْنِ جَميعًا اللَّذَيْنِ يَكونانِ عَلى الضِّلْعَيْنِ المُحيطَيْنِ بِالزَّاوِيَةِ القائِمَةِ إِذا كانَا يُشْبِهانِهِ.

המשל בזה: כי משולש אב״ג נצב הזוית, והוא זוית א׳. ואומר כי השטח שיהיה על צלע ב״ג שוה לשני השטחים יחד אשר יהיו על צלעות ב״א וא״ג אשר ידמו לו והם על תארו.

مِثالُهُ: أَنَّ زاوِيَةَ ب ا ج مِنْ مُثَلَّثِ ا ب ج قائِمَةٌ. فَأَقولُ إِنَّ السَّطْحَ الَّذي يَكونُ عَلى ضِلْعِ ب ج مِثْلُ السَّطْحَيْنِ جَميعًا اللَّذَيْنِ يَكونانِ عَلى ضِلْعَيْ ب ا و ا ج اللَّذَيْنِ يُشْبِهانِهِ وَعَلى خِلْقَتِهِ (بِخَطَّيْنِ)[1].

[1] נצב] תושב ⟨*⟩

[1] بخطين] بخطان

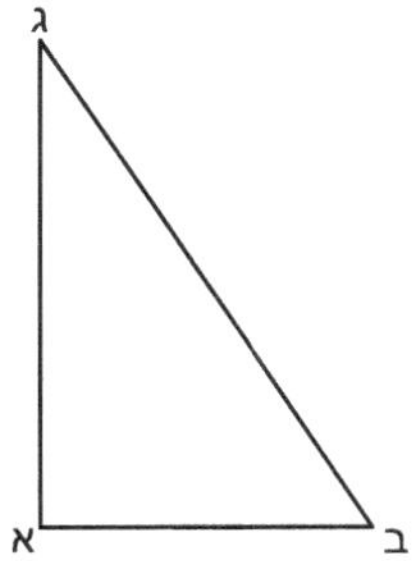

كُلُّ مُثَلَّثٍ قائِمِ الزَّاوِيَةِ فَإِنَّ الشَّكلَ المُسْتَقِيمَ الأَضْلاعِ المُضافُ إلى وَتَرِ الزَّاوِيَةِ القائِمَةِ مِنْهُ مِثلَ الشَّكْلَيْنِ المُسْتَقيمَيْ الأَضْلاعِ المُضافَيْنِ إلى ضِلْعَيْهِ الباقِيَيْنِ جَمِيعًا إِذا كانا يُشْبِهانِه وَكانا على وَضْعِهِ.

כל משולש נצב הזוית[1], הנה[2] התמונה ישרת הקוים המחוברת אל מיתר הזוית הנצבת[3] ממנו כמו שתי התמונות ישרות הצלעות המחוברות אל שתי צלעיו הנשארות יחד כאשר היו דומות אליו והיו על מצבו.

مِثالُهُ أَنَّ زاوِيَةَ اَ مِنْ مُثَلَّثِ اب ج قائِمَةٌ. فَأَقولُ إنَّ الشَّكلَ المُسْتَقِيمَ الأَضْلاعِ المُضاف إلى وَتَرِ زاوِيَةِ آ، وهُوَ ضِلْعُ ب ج، مثل الشَّكْلَيْنِ المُسْتَقيمَيْ الأَضْلاعِ المُضافَيْنِ إلى ضِلْعَيْ ب ا ا ج جَمِيعًا إذا كانا يُشْبِهانِه وَكانا على وَضْعِهِ.

המשל בו: כי זוית א׳[4] ממשולש אב״ג נצבת. הנה אומר כי התמונה ישרת הצלעות הסמוכה אל מיתר זוית א׳, והוא צלע ב״ג, כמו השתי תמונות ישרות הצלעות הסמוכות אל שתי צלעות א״ב א״ג[5] יחד כאשר היו דומות[6] לה ועל מצבה[7].

שרטוט—**ס**: התרשים הראשון זהה לתרשים השני חוץ מזווית א׳ שאינה ישרה בו [1]הזוית] הזויות **וספ** [2]הנה] ליתא **ב** [3]הנצבת] ~~הנצבת~~+ **ב** [4]א׳] ח׳ **ב** [5]א״ג] ~~ג א״ד~~+ **ב** [6]דומות] דומים **בוספ** [7]מצבה] מצבו **בוספ**

מופת זה: שנוציא עמוד א״ד. ויהיה משלש אב״ג דומה למשולש אב״ד. ואם כן שיעור ב״ג אל ב״א כשיעור א״ב אל ב״ד. ושיעור ב״ג אל ב״ד כשיעור הדומה אשר יהיה על ב״ג, כשיעור הדומה אשר יהיה אל א״ב הדומה לו בתאר ובקוים. ולפיכך גם כן שיעור ב״ג אל ג״ד כשיעור הדומה אשר יהיה על ב״ג אל הדומה שיהיה על א״ג הדומה לו בתאר ובקוים. ויהיה שעור ב״ג אל ב״ד ואל ג״ד יחד כשיעור הדומה שיהיה על קו ב״ג אל הדומים יחד שיהיו על ב״א ועל א״ג הדומים לו בתאר ובקוים. וב״ג שוה לב״ד וד״ג יחד. והדומה שיהיה על ב״ג שוה לדומים שיהיו על ב״א וא״ג יחד הדומים לו בקוים ובתאר.

بُرْهانُهُ: أَنَّا نُخْرِجُ عَمودَ ا د. فَمُثَلَّثُ ا ب ج يُشْبِهُ مُثَلَّثَ ا ب د. فَقَدْرُ ب ج عِنْدَ ا ب كَقَدْرِ ب ا عِنْدَ ب د. وَقَدْرُ ب ج عِنْدَ ب د كَقَدْرِ الشَّبيهِ الَّذي يكونُ عَلى ج ب عِنْدَ الشَّبيهِ الَّذي يكونُ عَلى ا ب الشَّبيهِ بِهِ وَعَلى خِلْقَتِه. وَمِنْ أَجْلِ ذَلِكَ قَدْرُ ب ج عِنْدَ ج د كَقَدْرِ الشَّبيهِ الَّذي يَكونُ عَلى ب ج عِنْدَ الشَّبيهِ الَّذي يَكونُ عَلى ا ج الشَّبيهِ بِهِ في الخِلْقَةِ وَالتَّخْطيطِ. فَقَدْرُ ب ج عِنْدَ ب د وَعِنْدَ ج د كَقَدْرِ الشَّبيهِ الَّذي يَكونُ عَلى ب ج عِنْدَ الشَّبيهَيْنِ اللَّذَيْنِ يكونانِ عَلى ب ا و ا ج الشَّبيهَيْنِ بِهِ في الخِلْقَةِ وَالتَّخْطيطِ. و ب ج هُوَ مِثْلُ د ج و ب د. وَالشَّبيهُ الَّذي يَكونُ عَلى ب ج مِثْلَ الشَّبيهَيْنِ اللَّذَيْنِ يكونانِ عَلى ب ا و ا ج جَميعًا الشَّبيهَيْنِ بِهِ في التَّخْطيطِ وَالخِلْقَةِ.

וזה מה שרצינו באורו.

وَذَلِكَ ما أَرَدْنا بَيانَهُ.

بُرهانُه: أَنَّ نِسْبَةَ مُرَبَّعِ ب ج إلى مُرَبَّعِ ب ا كِنِسْبَةِ ب ج إلى ب ا، مُثَنَّاةً. ونِسْبَةَ الشَّكْلِ المُسْتَقِيمِ الأَضلاعِ المُضافِ إلى ب ج إلى الشَّكْلِ المُسْتَقِيمِ الأَضلاعِ المُضافِ إلى ب ا الشَّبيه بِهِ والمَوْضوعِ كَوَضْعِهِ، كِنِسْبَةِ ب ج إلى ا ب، مُثَنَّاةً. فنِسبَةُ مُرَبَّعِ ب ج إلى مُرَبَّعِ ب ا كِنِسْبَةِ الشَّكْلِ المُضافِ إلى ب ج إلى الشَّكْلِ المُضافِ إلى ب ا. وكَذَلِكَ تكونُ نِسْبَةُ مُرَبَّعِ ب ج إلى مُرَبَّعِ ج ا كِنِسْبَةِ الشَّكْلِ المُضافِ إلى ب ج إلى الشَّكْلِ المُضافِ إلى ج ا. فإِذَنْ[1] نِسْبَةُ مُرَبَّعِ ب ج إلى مُرَبَّعَيْ ب ا ج ا كِنِسْبَةِ الشَّكْلِ المُضافِ إلى ب ج إلى الشَّكْلَيْنِ المُضافَيْنِ إلى ب ا ج ا. لكِنَّ مُرَبَّعَ ب ج مثل مُرَبَّعَيْ ب ا ج ا. فالشَّكل المُسْتَقيم الأَضْلاعِ المُضاف إلى ب ج مثل الشَّكْلَيْنِ المُسْتَقيمَيْ الأَضْلاعِ الشَّبيهَيْنِ بِهِ والمَوضُوعَيْنِ كَوَضْعِهِ المُضافَيْنِ إلى ب ا ج ا.

وَذَلِكَ ما أَرَدْنا أَنْ نُبَيِّنَ.

[1] فإذن] **Cop**

המופת: כי יחס מרובע ב״ג אל מרובע א״ב כיחס ב״ג אל א״ב[1], שנוי. ויחס התמונה[2] ישרת הצלעות הסמוכה אל ב״ג אל התמונה[3] ישרת הצלעות הסמוכה אל[4] א״ב הדומה אליו והמונח כמצבו, כיחס ב״ג אל א״ב, שנוי. אם כן יחס מרובע ב״ג אל מרובע א״ב כיחס התמונה הסמוכה אל ב״ג אל התמונה הסמוכה אל א״ב. וכן יחס מרובע ב״ג אל מרובע א״ג כיחס התמונה הסמוכה אל[5] ב״ג אל התמונה הסמוכה אל א״ג. הנה אם כן יחס מרובע ב״ג אל שני מרובעי[6] א״ב א״ג כיחס התמונה הסמוכה אל ב״ג אל שתי התמונות הסמוכות אל א״ב א״ג[7]. אבל מרובע ב״ג כמו שני מרובעי א״ב א״ג. אם כן התמונה ישרת הצלעות[8] הסמוכה אל ב״ג כמו שתי התמונות[9] ישרות הצלעות הדומות אליה והמונחות כמצבה הסמוכות אל א״ב[10] א״ג.

וזה מה שרצינו לבאר[11].

[1] אל א״ב] ⟨*⟩ **ק** [2] התמונה] התמונת **ב** [3] התמונה] התמונת **ב** [4] אל] ~~אליו~~+ **ק** [5] א״ב ... אל] ליתא **ב** [6] מרובעי] מרובעו **ב** [7] א״ב א״ג] א״ג+ **ב** א״ב ~~א״ג~~ ב״ג **ס** ~~א״ג~~ א״ג א״ב **ק** [8] הצלעות] הקוים **ס** [9] שתי התמונות] השתי תמונות **בו** [10] א״ב] ליתא **ב** [11] לבאר] ליתא **ס**

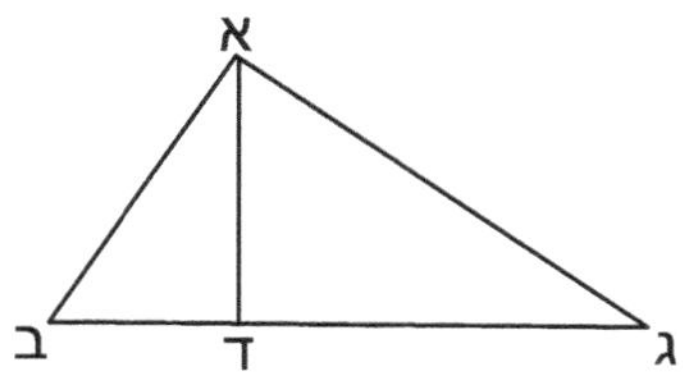

وَلَهُ[1] وَجْهٌ آخَر:

وهو أَنْ يُخْرَجَ عَمودُ ا د. فَمُثَلَّثا ا ب ج ا ب د مُتَشابِهانِ. فَنِسْبَةَ ب ج إلى ا ب كَنِسْبَةِ ا ب إلى ب د. فَنِسْبَةَ ج ب إلى ب د كَنِسْبَةِ السَّطْحِ المُضافِ إلى ج ب إلى السَّطحِ المُضافِ إلى ا ب الشَّبِيهِ بِهِ. وَكَذَلِكَ نِسْبَةُ ب ج إلى ج د كَنِسْبَةِ السَّطْحِ المُضافِ إلى ب ج إلى السَّطحِ المُضافِ إلى ج ا الشَّبِيهِ بِهِ. فَنِسْبَةَ ب ج إلى ب د د ج جَميعًا كَنِسْبَةَ السَّطحِ المُضافِ إلى ب ج إلى السَّطْحَيْنِ الشَّبِيهَيْنِ بِهِ المُضافَيْنِ إلى ب ا ا ج جَميعًا. و ب ج مثل ب د د ج جَميعًا. فالسَّطح المُضاف إِلَيْهِ مثل السَّطْحَيْنِ المُضافَيْنِ إِلَيْهِما جَميعًا.

وَذَلِكَ ما أَرَدْنا أَنْ نُبَيِّنَ.

ולו פנים אחרים:

והוא שיוצא[1] עמוד א״ד. הנה שני[2] משולשי אב״ג אב״ד מתדמים. אם כן יחס ב״ג אל א״ב כיחס א״ב אל ב״ד. ויחס ג״ב אל ב״ד כיחס השטח הסמוך אל ג״ב אל השטח הסמוך אל א״ב הדומה אליו. וכן יחס ב״ג אל ג״ד כיחס השטח הסמוך אל ב״ג אל השטח הסמוך אל א״ג[3] הדומה אליו. אם כן יחס ב״ג אל ב״ד ד״ג יחד[4] כיחס השטח הסמוך אל ב״ג[5] אל שני השטחים הדומים אליו הסמוכים אל א״ב א״ג יחד. וב״ג[6] כמו ב״ד ד״ג יחד. אם כן השטח הסמוך אליו כמו שני השטחים הסמוכים אל שניהם יחד.

וזה מה שרצינו לבאר.

---

1 وله] **MPR2**

1 שיוצא] יושר **ב** 2 שני] השני **ספ** 3 א״ג] א״ב+ **ק** 4 יחד] יחס **בט** 5 ב״ג] ~~א״ג~~+ **ב** 6 וב״ג] ~~וב״~~[..] ⟨*⟩ **ס**

[Proposition VI.33]

[B32]

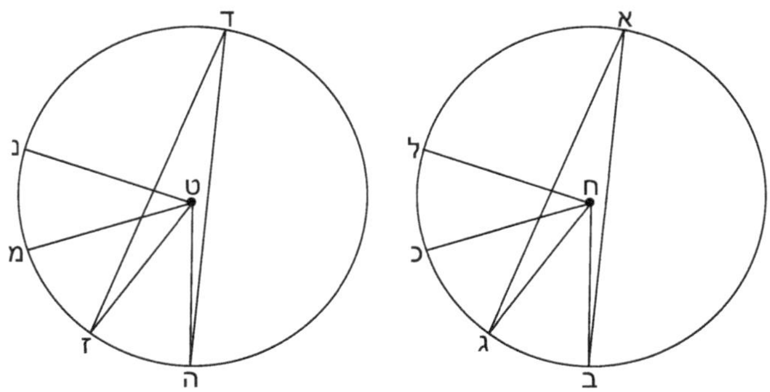

כשהיו בשני עיגולים שוים שתי זויות על המרכז או על קו המקיף, יהיה יחס הזוית לזוית כיחס הקשתות אשר הם עליהן.

الزَّوايا الَّتي تكونُ في الدَّوائِرِ المُتَساوِيَةِ، فَإِنَّ نِسْبَةَ الزَّاوِيَةِ إِلى الزَّاوِيَةِ كَنِسْبَةِ إِحْدى القَوْسَيْنِ اللَّتَيْنِ هُما[1] عَلَيْهِما إِلى الأُخْرى، إِذا كانَتِ الزَّوايا مُرَكَّبَةً عَلى المَراكِزِ أَو عَلى القُسِيِّ.

המשל בזה: ששני עיגולי אב״ג ודה״ז שוים. ועל מרכזיהם שתי זויות, והן זוית בח״ג וזוית הט״ז. ועל הקוים המקיפים גם כן שתי זויות, זוית א׳ וזוית ד׳. ואומר שיחס קשת ב״ג אל קשת ה״ז כיחס זוית בח״ג אל זוית הט״ז וכיחס זוית א׳ אל זוית ד׳.

مِثالُهُ: أَنَّ دائِرَتَيْ ا ب ج و د ه ز، مُتَساوِيَتانِ عَلى مَرْكَزَيْهِما زاوِيَتا (ب ح ج)[2] و ه ط ز. وَعَلى القُسِيِّ أَيْضًا زاوِيَتا ا و د. فَأَقولُ إِنَّ نِسْبَةَ قَوْسِ ب ج إِلى قَوْسِ ه ز كَنِسْبَةِ زاوِيَةِ (ب ح ج)[3] إِلى زاوِيَةِ ه ط ز المُرَكَّبَتَيْنِ عَلى مَرْكَزَيْ ح و ط، وَكَنِسْبَةِ زاوِيَةِ ا إِلى زاوِيَةِ د المُرَكَّبَتَيْنِ عَلى قَوْسِ ز د ه وج ا ب.

[1] هما] +* [2] ب ح ج] ب ج ح [3] ب ح ج] ب ج ح

إِذا كانَ في دائِرَتَيْنِ مُتَساوِيَتَيْنِ زاوِيَتانِ عَلى الْمَرْكَزِ أو عَلى الخَطِّ المُحِيطِ[1]، فَإِنَّ نِسْبَةَ الزَّاوِيَةِ إلى الزَّاوِيَةِ كَنِسْبَةِ إِحْدَى[2] القَوْسَيْنِ اللَّتَيْنِ هُما عَلَيْهِما إلى الأُخْرَى[3].

مِثالُهُ: أَنَّ^ في دائِرَتَيْ ا ب ج د ه ز المُتَساوِيَتَيْنِ عَلى مَرْكَزَيْهِما زاوِيَتا ب ح ج ه ط ز، وعَلى الخَطَّيْنِ المُحِيطَيْنِ زاوِيَتا ج ا ب و ه د ز^. فَأَقُولُ إِنَّ نِسْبَةَ قَوْسِ ب ج إلى قَوْسِ ه ز كَنِسْبَةِ زاوِيَةِ ب ح ج إلى زاوِيَةِ ه ط ز وكَنِسْبَةِ زاوِيَةِ ج ا ب إلى زاوِيَةِ ه د ز.

כאשר היו בשתי[1] עגולות שוות שתי זויות על המרכז[2] או על הקו המקיף, הנה יחס הזוית[3] אל הזוית[4] כיחס שתי הקשתות אשר הן[5] עליהם אחת אל אחת.

המשל בו בשתי עגולות אב״ג דה״ז השוות על מרכזיהם שתי זויות גח״ב הט״ז[6]. הנה אומר כי יחס קשת ב״ג אל קשת ה״ז כיחס זוית בח״ג אל זוית הט״ז וכיחס זוית גא״ב אל זוית[7] הד״ז.

شرטוט—כל אחד מהמעגלים $, הנקודה ח׳ אינה נמצאת בין מיתרים א״ב וא״ג, והנקודה ט׳ אינה נמצאת בין מיתרים ד״ה וד״ז   [1]בשתי] שתי **ב**   [2]המרכז] ⟨ה⟩מרכז **ק**   [3]הזוית] זוית **ספ**   [4]אל הזוית] ליתא **ב**   [5]הן] הם **בופק** ליתא **ס**   [6]הט״ז] +על שני הקוים המקיפים שתי זויות גא״ב הד״ז **ס** +⟨⟨ועל שני הקוים המקיפים שתי זויות גא״ב הד״ז הנה⟩⟩ **ק**   [7]זוית] +*̲ **ק**

---

[1]الخط المحيط] المحيطة **CopQT2** **CT1T3**

[2]إحدى] **MPR2**   [3]إلى الأخرى] **MPR2**

ומפת [!] זה: שנחתוך מעגול אב״ג כשיעור קשת ב״ג כמו שנרצה, והם קשת ג״כ וקשת כ״ל; ומעיגול דה״ז כמו שנרצה בשיעור קשת ה״ז, והן קשת ז״מ וקשת מ״נ. ונדביק ח׳ בכ׳ ובל׳, וט׳ במ׳ ובנ׳. וקשתות ב״ג וג״כ וכ״ל שוות. ואם כן זויות בח״ג וגח״כ וכח״ל כמו כן שוות. ונמצא שעור כפלי קשת ב״ל לקשת ב״ג[1] כשיעור כפלי זוית[2] בח״ל לזוית בח״ג. ולפי עניין זה יהיה שיעור כפלי קשת ה״נ לקשת ה״ז כשיעור כפלי זוית הט״נ לזוית הט״ז. ושיעור מה שיוסיף קשת ב״ל על קשת ה״נ כשיעור מה שיוסיף זוית בח״ל על זוית הט״נ. ואם היתה שוה לה גם היא שוה לה, ואם היתה חסרה ממנה גם היא חסרה ממנה בזה החסרון. והשיעורים ארבעה—קשת ב״ג וקשת ה״ז וזוית בח״ג וזוית הט״ז. וכבר בארנו שכפלי קשת ב״ג וכפלי זוית בח״ג הם קשת ב״ל וזוית בח״ל. וכמו כן התבאר שכפלי קשת ה״ז וכפלי זוית הט״ז הם קשת ה״נ וזוית הט״נ. וכבר בארנו שאם היו מוסיפים—האחרים מוסיפים. ואם היו שוים—האחרים שוים. ואם היו חסרים—האחרים חסרים. אם כן שיעור קשת ב״ג אל שיעור קשת ה״ז כשיעור זוית בח״ג אל זוית הט״ז. וחצי זוית בח״ג היא זוית א׳. וחצי זוית הט״ז היא זוית ד׳. ואם כן התבאר ששיעור קשת ב״ג אל קשת ה״ז כשיעור זוית בח״ג אל זוית הט״ז וכשיעור זוית א׳ אל זוית ד׳.

[1] ב״ג] +~~כשיעור כפלי קשת כ״ל לקשת ב״ג~~ [2] זוית] ~~זויות~~ ⟨⟨*⟩⟩

بُرْهانُهُ: أَنْ نَفْصِلَ مِنْ دائِرَةِ ا ب ج مِثْلَ قَوْسِ ب ج، كَمْ شِئْنا، وَهِيَ قَوْسا ج ك و ك ل؛ وَمِنْ دائِرَةِ ه د ز مِثْلَ قَوْسِ ه ز، كَمْ شِئْنا، وَهِيَ قَوْسا ز م و م ن. وَنَصِلُ ح و ك و ل (و ط)[1] و م و ن بِخُطوطِ ح ك و ح ل و ط م و ط ن. فَقُسِيُّ ب ج و ج ك و ك ل مُتَساوِيَةٌ. وَزَوايا ب ح ج و ج ح ك و ك ح ل مُتَساوِيَةٌ. فَقَدْرُ أَضْعافِ قَوْسِ ب ل لِقَوْسِ ب ج كَقَدْرِ أَضْعافِ زاوِيَةِ ب ح ل لِزاوِيَةِ ب ح ج. وَمِنْ أَجْلِ ذَلِكَ قَدْرُ أَضْعافِ[2] قَوْسِ ه ن لِقَوْسِ ه ز كَقَدْرِ أَضْعافِ زاوِيَةِ ه ط ن لِزاوِيَةِ ه ط ز. وَقَدْرُ ما تَزيدُ قَوْسُ ب ل عَلَى قَوْسِ ه ن كَقَدْرِ ما تَزيدُ زاوِيَةُ ب ح ل عَلَى زاوِيَةِ ه ط ن. وَإِنْ كانَتْ مُساوِيَةً لَها، فَهِيَ مُساوِيَةٌ لَها. وَإِنْ كانَتْ ناقِصَةً عَنْها، فَهِيَ ناقِصَةٌ عَنْها بِذَلِكَ القَدْرِ. بِالأَقْدارِ الأَرْبَعَةِ قَوْسُ ب ج وَقَوْسُ ه ز وَزاوِيَةُ ب ح ج وَزاوِيَةُ ه ط ز. وَقَدْ تَبَيَّنَ أَنَّ أَضْعافَ قَوْسِ ب ج وَأَضْعافَ زاوِيَةِ ب ح ج هِيَ قَوْسُ ب ل وَزاوِيَةُ ب ح ل. وَقَدْ تَبَيَّنَ أَيْضًا أَنَّ أَضْعافَ قَوْسِ ه ز وَزاوِيَةَ ه ط ز هِيَ قَوْسُ ه ن وَزاوِيَةُ ه ط ن. وَقَدْ بَيَّنّا أَنَّها إِنْ كانَتْ تَزيدُ—فَالأُخْرى تَزيدُ. وَإِنْ كانَتْ مُساوِيَةً لَها—فَالأُخْرى مُساوِيَةٌ لَها. وَإِنْ كانَتْ أَنْقَصَ مِنْها—فَالأُخْرى أَنْقَصُ مِنْها. فَقَدْرُ قَوْسِ

[1] و ط] و ز [2] أضعاف] ⟨*⟩

بُرْهانُهُ: أَنَّا نَفْصِلُ مِنْ دائِرَةِ ا ب ج مِثْلَ قَوْسِ ب ج كَمْ شِئْنا، فَلْنَفْصِلْ^ ج ك ك ل. وَمِنْ دائِرَةِ د ه ز أَيْضًا مثل قَوْسِ ه ز كَمْ شِئْنا، فَلْنَفْصِلْ^ ز م م ن. وَنُخْرِجُ خَطَّيْ ك ح ح ل، وَخَطَّيْ ط م ط ن. فَقِسِيُّ ب ج ج ك ك ل مُتَساوِيَةٌ. فَزَوايا ب ح ج و ج ح ك ك ح ل مُتَساوِيَةٌ. فَأَضْعافُ قَوْسِ ب ل لِقَوْسِ ب ج كَأَضْعافِ زاوِيَةِ ب ح ل لِزاوِيَةِ ب ح ج. وَكَذَلِكَ أَضْعافُ قَوْسِ ه ن لِقَوْسِ ه ز كَأَضْعافِ زاوِيَةِ ه ط ن لِزاوِيَةِ ه ط ز. فَإِنْ كانَتْ ب ل تَزِيدُ عَلَى قَوْسِ ه ن، فَإِنَّ زاوِيَةَ ب ح ل تَزِيدُ عَلَى زاوِيَةِ ه ط ن. وَإِنْ كانَتْ تُساوِيها. فَهِيَ تُساوِيها. وَإِنْ كانَتْ تَنْقُصُ عَنْها، فَهِيَ تَنْقُصُ عَنْها. فالأَقْدارُ أَرْبَعَةٌ—قَوْس ب ج وقَوْس ه ز وزاوية ب ح ج وزاوية ه ط ز. وأَضْعاف قَوْس ب ج. وزاوية ب ح ج المُتَساوِيَةَ المَرّات هي قَوْس ب ل. وزاوية ب ح ل وأَضْعاف قَوْس ه ز وزاوية ه ط ز هي قَوْس ه ن وزاوية ه ط ن. وَقَدِ اسْتَبانَ أَنَّ قَوْسَ ب ل، إِنْ كانَتْ زائِدَةً على قَوْسِ ه ن، فَإِنَّ زاوِيَةَ ب ح ل زائِدَةٌ على زاوِيَةِ ه ط ن. وإِنْ كانَتْ مُساوِيَةً لَها، فَإِنَّها مُساوِيَةٌ لَها. وَإِنْ كانَتْ تَنْقُصُ عَنْها، فَهِيَ تَنْقُصُ عَنْها. فَنِسْبَةُ قَوْسِ ب ج إلى قَوْسِ ه ز كَنِسْبَةِ زاوِيَةِ ب ح ج إلى زاوِيَةِ ه ط ز. وزاوية ا الَّتِي على

המופת: אנחנו נבדיל מעגולת אב״ג כמו קשת ב״ג כמה שרצינו, נאמר שנבדילי[1] ג״כ כ״ל. ומעגולת דה״ז גם כן כמו קשת ה״ז כמה שנרצה, נאמר שנבדיל ז״מ מ״נ. ונוציא שני קוי כ״ח ח״ל, ושני קוי[2] ט״מ ט״נ[3]. הנה קשתות ב״ג ג״כ כ״ל שוות. אם כן זוויות[4] בח״ג גח״כ כח״ל שוות. אם כן כפלי קשת ב״ל לקשת ב״ג ככפלי זוית[5] בח״ל לזוית בח״ג. וכן כפלי[6] קשת ה״נ לקשת ה״ז ככפלי זוית הט״נ לזוית הט״ז[7]. ואם היה ב״ל[8] יוסיף על קשת ה״נ, הנה זוית בח״ל תוסיף על זוית הט״נ. ואם יהיה[9] שיהיו שוים אליה, הנה היא[10] תשוה אליה. ואם היו[11] שיחסרו ממנה, הנה היא תחסר ממנה. אם כן השעורים ארבעה—קשת ב״ג וקשת ה״ז וזוית[12] בח״ג וזוית הט״ז[13]. וכפלי קשת ב״ג וזוית בח״ג השוי הפעמים[14] היא קשת ב״ל וזוית בח״ל. וכפלי קשת ה״ז וזוית הט״ז[15] היא קשת ה״נ וזוית הט״נ. וכבר התבאר כי קשת ב״ל, אם היתה נוספת על קשת ה״נ, הנה זוית בח״ל נוספת על זוית הט״נ[16]. ואם היתה שוה לה, הנה היא שוה לה[17]. ואם היתה חסרה ממנה, הנה היא חסרה ממנה. אם כן יחס קשת ב״ג אל קשת ה״ז כיחס זוית בח״ג אל[18] זוית הט״ז. וזוית א׳ אשר על המקיף הוא חצי זוית בח״ג אשר על המרכז. וזוית הד״ז היא חצי זוית הט״ז. אם כן יחס קשת ב״ג אל קשת ה״ז גם כן[19] כיחס זוית[20] א׳ אל זוית ד׳.

---

[1] שנבדיל] +~~גם כן~~ **ט** [2] ושני קוי] +גם כן **ו** [3] ט״נ] +אם כן **ק** [4] זוויות] זוית **ס** [5] זוית] זווית **ס** [6] כפלי] ככפלי **ב** [7] לזוית הט״ז] ⟨⟨*⟩⟩ **פ** [8] ב״ל] ⟨⟨קשת⟩⟩+ **פ** [9] יהיה] היה **ק** [10] היא] ליתא **ק** [11] היו] היה **ק** [12] וזוית] וקשת⟨זוית⟩ **ו** [13] הט״ז] +מתיחסים **ס** [14] השוי הפעמים] השוה הפעמים **בטפ** השוה בערכים **ס** השות הפעמים **ק** [15] הט״ז] +השוה פעמים **ס** [16] על ... הט״נ] ליתא **ו** [17] לה] ~~ממנה~~ **ב** [18] אל] אשר על+ **ב** [19] גם כן] ליתא **ס** [20] זוית] ליתא **פ**

بَ جَ عِنْدَ قَوْسِ هـ زَ كَقَدْرِ زاوِيَةِ ب ح ج عِنْدَ زاوِيَةِ ه ط ز. وَنِصْفُ زاوِيَةِ ب ح ج هِيَ زاوِيَةُ اَ. وَنِصْفُ زاوِيَةِ ه ط ز هِيَ زاوِيَةُ دَ. فَقَدْرُ قَوْسِ بَ جَ عِنْدَ قَوْسِ ه زَ كَقَدْرِ زاوِيَةِ ب ح ج عِنْدَ زاوِيَةِ ه ط ز وَكَقَدْرِ زاوِيَةِ اَ عِنْدَ زاوِيَةِ دَ.

وَذَلِكَ ما أَرَدْنا بَيانَهُ.

וזה מה שרצינו לבאר.

נשלם המאמר הששי בכח השם ועזרו.

المُحِيط هي نِصْفُ زاوِيَةِ ب ح ج الّتي على المَرْكَز. وَزاوِيَةُ د هي نِصْفُ زاوِيَةِ ه ط ز. فنِسْبَةُ قَوْسِ ب ج إلى قَوْسِ ه ز أَيْضًا كَنِسْبَةِ زاوِيَةِ ا إلى زاوِيَةِ د.

وذلك ما اردنا ان نبين.

وזה מה שרצינו לבאר[1] [2].

تمت المقالة السادسة من كتاب اقليدس[1]. وهي ثلثة وثلثون شكلا[2].

נשלם[3] המאמר הששי מספר אקלידס[4]. ומספר תמונותיו ל״ג[5].

---

[1] من ... اقليدس] CopM [2] وهي ... شكلا] CT1T3

[1] וזה ... לבאר] ונשלם ביאורו ז [2] שרצינו לבאר] +* ב [3] נשלם] ונשלם ט [4] מספר אקלידס] +החכם וס ליתא טק [5] ומספר ... ל״ג] ומספר תמונתו ל״ג תמונות ב תל״ח ז ⟨*⟩ ט +ת״ל ס +תמונות ק

CORRIGENDUM

The Approximation of the Arabic Source of MIT's Definition II.2

The approximate reconstruction of the Arabic source of **MIT** Book II, presented in the first volume of this edition, mistakenly includes Definition II.1 instead of Definition II.2. Here is the correct approximation of Definition II.2.

[2] وكُلّ سَطْحٍ مُتوازي الأَضْلاع، فَلْيُسَمَّ أَحَدَ السَّطْحَيْنِ المُتوازِيَيِ الأَضْلاعِ اللَّذَيْنِ على قُطْرِهِ—أَيُّهُما كانَ مع كِلَيِ^ السَّطْحَيْنِ المُتَمِّمَيْنِ—العلمَ.

 | DOI:10.1163/9789004707801_011

Glossary

Arabic terms translated in the definitions and enunciations in **RJ** are provided on the basis of **I/T**; those translated in its proofs are taken from **B**.

א

אורך

באורכו

lit. "in its length," in a straight extension of another straight line (when a straight line is produced), فِي طُولِهِ

RJ: III: prop. 35

See also: דבק–מדובק, דבקות, יושר

אות

label

RJ: IV: prop. 12

אחרון

last, أَخِيرٌ

MIT: V: props. 20–21, 23, 25; VI: props. 16–17

RJ: V: props. 20–21, 25; VI: props. 16–17

See also: ראשון ונמשך

אי זה ... שיהיה

lit. "whichever ... may be," arbitrarily, أَيّ ... كَانَ

MIT: V: def. 7, props. 3–4, 7; VI: def. 3

See also: כמה שיהיה

אי זה ... שרצינו

lit. "whichever ... we wish," arbitrarily, أَيّ ... شِئْنَا/أَرَدْنَا

MIT: VI: prop. 12

See also: כמה—כמה שנרצה, כמו—כמו שנרצה

איחור—See: קדימה ואיחור

איך שנפל, איך מה שנפל

arbitrarily placed, كَيْفَمَا وَقَعَ

MIT: III: def. 8, props. 1–2, 5–6, 9, 30–31; IV: prop. 3
See also: כמו שנפל

אלכסון
(a) diameter (of a circle), قُطْرٌ
RJ: III: def. 1, props. 3, 14
See also: קוטר
(b) diameter (or: diagonal; of a parallelogram), قُطْرٌ
RJ: VI: props. 24–25, 27
See also: קוטר

אמצע (אמצעי)
middle (term), وَسَط (أَوْسَط)
MIT: V: def. 19 (no Arabic source); VI: def. 4, props. 8, 17
RJ: VI: def. 4, props. 8, 17
See also: יחס—יחס אמצעי ביחס, יחס—יחס בעל אמצע

ארוך (וגם: הארוך, יותר ארוך)
long (longer, longest, أَطْوَلُ (أَطْوَلُ مِن، ٱلأَطْوَلُ)
MIT: III: props. 2, 7–9, 11–12, 14–15, 17; VI: props. 28–30
RJ: III: 2, 5, 7–8, 11–12, 14, 17, 24, 27, 34; VI: props. 28–30
See also: אורך

ב

בא בתוך
enter (a circle), دَخَلَ
RJ: III: prop. 8
See also: נכנס

באר (v.; בארנו, נבאר)
establish, explain, بانَ (نبيِّنُ، بيَّنّا; only in **ARAB**)
MIT: IV: props. 12, 15; V: prop. 6; VI: props. 22, 23
RJ: III: props. 7–8

התבאר (כבר התבאר, יתבאר)
has been established, has been explained, تبيَّنَ (قَدِ ٱسْتَبَانَ، قَدْ كانَ تبيَّنَ، يتبيَّنُ)
MIT: III: props. 1–2, 4–12, 15–16, 24–25, 28, 34; IV: props. 4–6, 8, 12–15; V: props. 7–8, 10, 14, 17–18, 20–23; VI: props. 1, 5, 7–8, 13, 18, 22, 28, 33

RJ: III: props. 1–8, 10–12, 14–16, 18, 21–23, 26, 34, 36; IV: props. 3–6, 8–9, 12–13, 15; V: props. 6–8, 10, 12, 17–18, 20–23, 25; VI: props. 1, 5, 7–9, 12–13, 16–19, 25, 33

(ו)זה מה שרצינו ביאורו
Q.E.D., وَذَلِكَ ما أَرَدْنا بَيانَهُ (ARAB), وَذَلِكَ ما أَرَدْنَا أَنْ نُبَيِّنَ (B)
MIT: III: props. 1, 5–6, 10, 12
RJ: III: props. 13, 22–23; IV: prop. 5; V: props. 4, 14, 32

(ו)זה מה שרצינו לבאר
Q.E.D., وَذَلِكَ ما أَرَدْنا بَيانَهُ (ARAB), وَذَلِكَ ما أَرَدْنَا أَنْ نُبَيِّنَ (B)
MIT: III: props. 2, 4, 7, 9, 11, 13, 15–24, 25–26, 28–34, 36: IV: props. 1–16; V: all props.; VI: props. 1–8, 10–12, 15–22, 25, 26–30, 32–33
RJ: III: props. 1–6, 8, 10, 12, 14–15, 20–21, 26–33, 35; IV: prop. 4: V: props. 1–3, 5, 8–13, 15–18, 20, 23–25; VI: props. 1–7, 10–13, 15–31, 33

(ו)זה מה שרצינו שנבאר
Q.E.D., this is what we wished to demonstrate, وَذَلِكَ ما أَرَدْنا بَيانَهُ
RJ: V: props. 6–7, 19, 21–22; VI: prop. 31

נרצה לבאר
we wish to demonstrate (or: explain), نُرِيدُ أَنْ نُبَيِّنَ
MIT: III: prop. 1
RJ: III: prop. 1

(ו)נשלם ביאורו
Q.E.D., this is what we wished to demonstrate, وَذَلِكَ ما أَرَدْنَا أَنْ نُبَيِّنَ
MIT: III: props. 3, 14, 27, 35; VI: props. 9, 13–14, 24, 23, 31

בדל (v.)
הבדיל (נבדל)
set out (or: cut; let us cut), فَصِلَ (نَفْصِلُ)
MIT: III: props. 14, 27–28, 33; V: props. 8, 17–19, 25; VI: props. 1, 11–12, 25, 28, 33
See also: חִלק, חתך, לקח, שם

בלבל (v.)
התבלבל
perturbed (ratio), اَضْطَرَبَت
MIT: V: props. 21, 23
See also: יחס—יחס מבולבל, יחס—יחס מתחלף, יחס—יחס נפתל במערכת

ג

גבהות
height, إِرْتِفاع
MIT: V: props. 1–2, 18
See also: גובה

גבול
limit, boundary, حَدّ
MIT: V: def. 10
RJ: V: def. 10
See also: תכלית

(יותר) גדול
bigger, أَعْظَمُ (اَلْعُظْمى, היותר גדול), أَكْبَرُ (اَلْكُبْرى, היותר גדול); larger, أَطْوَلُ
MIT: III: def. 5, props. 1–2, 4–5, 7–8, 14–15, 17, 22–23, 26–27, 30; IV: prop. 1; V: defs. 1–3, 9, props. 6, 8–10, 12, 14, 18, 20–21, 25; VI: defs. 3–4, props. 7–8, 11, 27–29
RJ: III: def. 5, props. 1–2, 4, 6–8, 14–15, 17, 22–24, 26–27, 30, 32; IV: prop. 1; V: defs. 1–3, 9, props. 6, 8–10, 12, 14, 18, 20–21, 25; VI: def. 4, props. 7–8, 27–28
See also: רב

גובה
height, اَرْتِفاع
MIT: VI: def. 3, prop. 1
RJ: VI: def. 3, prop. 1
See also: גבהות

גיל
similar (to), نَظِير
MIT: VI: props. 28–29, 31
See also: דומה, כל אחד ל

גרע
be less than, نَقَصَ
MIT: V: def. 7, prop. 4
RJ: V: def. 7
See also: הפיל, השליך, חִסר, סר, פחות

ד

דבק (v.)

הדביק (נדביק)

(a) connect, join (two points, thus producing a line), وَصَلَ (نَصِلُ)

RJ: III: props. 1–7, 9–13, 15–17, 19–25, 27–29, 30–31, 34–35; IV: props. 1–2, 5–6, 9–16; VI: props. 1–3, 9–11, 12–13, 15, 18–19, 20, 33

See also: חִבר

(b) connect, join (two straight lines, on a point), وَصَلَ (نَصِلُ)

MIT: IV: props. 4 (ARAB: نُخْرِجَ), 5

See also: חִבר

דבק

(a straight line) continuing (or: extending; another straight line), مُتَّصِلٌ

MIT: VI: props. 11, 31

מדבק

(a straight line) connecting, joining (two points or lines)

RJ: III: prop. 11

מדובק

meeting (two straight lines), مُتَّصِلٌ

MIT: VI: props. 4, 14–15, 23

See also: יושר ,דבקות ,בארכו–אורך

נדבק

(a) tangent to a point on a circle, وَصَلَ (?)

RJ: III: def. 8

(b) (a straight line) continuing (or: extending; another straight line), مُتَّصِلٌ

RJ: VI: prop. 31

דומה

(a) similar (to), شَبِيهٌ

MIT: VI: props. 20–21, 25, 26, 27–29, 32

RJ: VI: props. 8, 20–22, 24–25 (24: also a translation of كَ), 26–30, 32

(b) similar, مُتَشَابِهٌ

MIT: III: props. 22–23; VI: props. 8, 18–19, 22

See also: כל אחד לדומה לו—כל אחד ל ,הדמות היחסים—יחס ,דמיון ,דמות

(b) parallel (lit. similar), نَظِيرٌ

RJ: VI: props. 7, 18–19
See also: גיל, כל אחד ל

מתדמה
similar, مُتَشَابِهٌ
MIT: III: def. 11, prop. 22; VI: def. 1, props. 8, 19, 21–22, 24–25, 31–32
RJ: III: def. 11, props. 22–23, 25; VI: def. 1, props. 8, 18–19, 21–22, 24–25

דמה (v.)
be similar, تُشْبِهُ
MIT: III: props. 22, 25; VI: props. 8, 18, 21, 24, 26–30, 32
RJ: III: prop. 22; VI: props. 8, 18, 24 (also a translation of مُتَشَابِهٌ and of مِثْلُ نِسْبَةِ), 26–30, 32

התדמה (מתדמה)
be similar
RJ: III: prop. 25

כדמות ... דמות, בדמות
similar. شَبِيهٌ
RJ: VI: props. 26, 28–29
See also: דומה, דמיון, התדמה, יחס—הדמות היחסים, כל אחד ל—כל אחד לדומה לו

דמיון
בדמיון
similar (to), فِي ٱلنِّسْبَةِ (?)
MIT: VI: prop. 29
See also: דומה, דמות, יחס—הדמות היחסים, כל אחד ל—כל אחד לדומה לו

דמיונים
multiples, أَمْثَال
RJ: V: props. 3, 6
See also: כפל

ה

הִדָּמוּת
similarity, تَشَابُهٌ
MIT: V: def. 5
RJ: V: def. 5

הווה (v.)
being, existing (e.g. a square on a given line), كائِنٌ
MIT: III: props. 13, 34–36

הכאה
multiplication (of one given straight line by another, thus producing a rectangle), ضَرْبٌ
MIT: VI: props. 16–17
RJ: III: props. 13, 34, 36

הלך (v.)
pass (lit. went), مَرَّ (**RJ** also جازَ)
MIT: III: props. 7–9, 24; IV: props. 4–8, 14
RJ: III: props. 11, 13 (جازَ)
See also: עבר

המיר (v.)
alternate (terms in a ratio), بَدَّلَ
MIT: V: props. 16, 19, 25
See also: התחלף

הנהגה
order (or: course, of construction or proof), تَدْبِيرٌ
MIT: V: props. 7–8; VI: props. 2, 7, 14–17
RJ: V: props. 7–8; VI: props. 2–3, 7, 16–17

הפך—See: יחס—הפך היחס, יחס—התהפכות היחס

הִקוות
delineation, تَخْطِيطٌ
MIT: IV: prop. 29
See also: קוה

ז

זווית
angle, زَاوِيَة
MIT: III: defs. 8–11, props. 2–4, 7–9, 13, 15–17, 19–22, 24–33, 36; IV: defs. 1–2, props. 2–4, 6–15; VI: def. 1, props. 3–8, 10–16, 18–21, 23–26, 28, 31, 33
RJ: III: defs. 7–11, props. 1–4, 7–9, 13, 15–22, 24–36; IV: defs. 1–2, props. 2–15; VI: def. 1, props. 3–8, 10, 12, 14–16, 18–21, 23–26, 31, 33

See also: מרובע שוה הצלעות נצב הזוויות—מרובע, מחומש שוה הצלעות והזוויות—מחומש, משולש רחב הזוויות—משולש, משולש נצב הזווית—משולש, משולש חד הזוויות—משולש, שטח רב הזוויות—שטח, שטח נצב הזוויות—שטח, משולש שוה הצלעות והזוויות—משולש

זוויות זו לעומת זו
opposite angles, زَاوِيَةٌ تُقَابِلُ الزَّاوِيَةَ، زَاوِيَتَانِ مُتَقَابِلَتَانِ
RJ: IV: props. 3, 5
See also: זוית מתנגדות, זוית מקבילות

זוויות מומרות
alternate angles, زَاوِيَتَانِ مُتَبَادَلَتَانِ
MIT: VI: props. 3, 31
See also: זוויות מתחלפות

זוויות אשר יקבילו, זוויות מקבילות
opposite angles, زَاوِيَتَانِ مُتَقَابِلَتَانِ
MIT: III: def. 11, props. 30–31
RJ: III: props. 21, 30–31
See also: זוויות מתנגדות, זוויות זו לעומת זו

זוויות מתחלפות
alternate angles, زَاوِيَتَانِ مُتَبَادَلَتَانِ
RJ: VI: props. 3, 31
See also: זוויות מומרות

זוויות מתנגדות
opposite angles, زَاوِيَتَانِ مُتَقَابِلَتَانِ
MIT: III: prop. 21; IV: props. 3, 7
See also: זוויות מקבילות, זוויות זו לעומת זו

זווית חדה
acute angle, زَاوِيَةٌ حَادَّةٌ
MIT: III: props. 15, 30
RJ: III: props. 15, 30 (in 17 زَاوِيَةٌ حَادَّةٌ is translated into זוית קטנה מנצבה); IV: prop. 5
See also: משולש חד הזוויות

זווית חוץ ממשולש
angle exterior to a triangle, زَاوِيَةٌ خَارِجَةٌ مِنْ مُثَلَّثٍ
MIT: III: prop. 2
See also: זווית יוצאת מן משולש, זווית חיצונה

זווית חיצונה
angle exterior to a triangle, زَاوِيَةٌ خَارِجَةٌ
MIT: III: props. 15, 22, 30; IV: prop. 10; VI: prop. 3
RJ: III: props. 2, 15, 22, 30; IV: props. 3, 5, 10; VI: prop. 3
See also: זווית חוץ ממשולש, זווית יוצאת מן משולש

זווית חצי נצבת ,זווית חצי נצבה
half a right angle, زَاوِيَةٌ نِصْفُ قائِمَةٍ
MIT: IV: prop. 9
RJ: IV: prop. 9

זווית יוצאת מן משולש
angle exterior to a triangle, زَاوِيَةٌ خَارِجَةٌ
MIT: III: prop. 22
See also: זווית חיצונה ,זווית חוץ ממשולש

זווית ישרת הצלעות
rectilinear angle, زَاوِيَةٌ مُسْتَقِيمَةُ ٱلْخَطَّيْنِ
MIT: III: prop. 22
See also: זווית ישרת הקוים

זווית ישרת הקוים
rectilinear angle, زَاوِيَةٌ مُسْتَقِيمَةُ ٱلْخَطَّيْنِ
MIT: III: props. 15, 30, 32–33
RJ: III: props. 15, 30, 32–33
See also: זווית ישרת הצלעות

זווית נכנסת
interior angle, زَاوِيَةٌ داخِلَةٌ
MIT: III: prop. 15
See also: זווית פנימית

זווית נצבת ,זווית נצבה
right angle, زَاوِيَةٌ قائِمَةٌ
MIT: III: props. 1, 3–4, 8–10, 13, 15–18, 21, 24, 29–32, 35–36; IV: props. 3, 8–9, 13, 15; VI: props. 7–9, 16, 31–32
RJ: III: props. 1, 3–4, 9–10, 13, 15–18, 21, 24, 29–32, 35–36; IV: props. 2–3, 5–9, 12–13, 15; VI: props. 7–9, 16, 31–32
See also: שטח נצב הזוויות—שטח

זווית נרווחת
obtuse angle, زَاوِيَةٌ مُنْفَرِجَةٌ
RJ: III: prop. 30; IV: prop. 5
See also: זווית נרחבת, משולש—משולש רחב הזוויות

זווית נרחבת
obtuse angle, زَاوِيَةٌ مُنْفَرِجَةٌ
MIT: III: prop. 30
See also: זווית נרווחת, משולש—משולש רחב הזוויות

זווית פנימית
interior angle, زَاوِيَةٌ دَاخِلَةٌ
MIT: III: props. 15, 22; IV: prop. 7; VI: prop. 3
RJ: III: props. 15, 22; IV: props. 5, 7; VI: prop. 3
See also: זווית נכנסת

ח

חבר (v.)
(a) applied to (area to a straight line), أَضَافَ
MIT: VI: props. 26, 28–30
RJ: VI: props. 26–30 (in VI. 27 משך is also employed)
See also: מחובר, סמך
(b) combine, add, جَمَعَ
RJ: V: props. 3, 17
See also: מחובר
(c) combine (ratio, of ratios), قامَ
RJ: VI: prop. 23
See also: מחובר
(d) connect, join (two straight lines), وَصَلَ (نَصِلُ)
RJ: VI: prop. 9
See also: הדביק

התחבר
combine (ratio, of ratios), كان
RJ: VI: prop. 23
See also: מחובר

חָבֵר
associated with (area, to a straight line), قَرِينٌ، مُقارِنٌ
RJ: V: props. 1 (מחובר), 13
See also: קרוב ,סמוך ,מחובר

חוץ
exterior (to a circle or a triangle), خارِج
MIT: III: props. 2, 11, 15, 35–36; IV: prop. 10
RJ: III: props. 2, 8, 11, 15, 23, 35–36; VI: prop. 3
See also: חיצון ,זווית חוץ ממשולש—זווית

חזק (v.)
  החזיק
  subtend, قَبِلَ ,حازَ (def. III.11)
  RJ: III: defs. 9–11
  See also: סבב ,(היה) מיתר ,הקיף

חיצון
exterior, خارِج
RJ: III: props. 8, 35
See also: חוץ ,זווית חיצונה—זווית

חִלוק
point of intersection, قَطْعٌ
RJ: III: prop. 9
See also: חתך

חלוקה
division, قِسْمَةٌ
MIT: IV: prop. 10; VI: def. 5, prop. 13
See also: חלוק היחס—יחס ,חתך

חלף (v.)
  החליף (v.)
  alternate, بَدَّلَ
  RJ: V: props. 19; VI: 19

  התחלף (v.)
  alternate, بَدَّلَ
  RJ: V: props. 16, 21
See also: מתחלף ,המיר

חִלק (v.)
(a) divide, قَسَمَ
MIT: IV: props. 10, 16; V: props. 1, 3, 15; VI: defs. 4–5, props. 13, 19–20, 30
RJ: IV: props. 10, 16; V: props. 1, 3, 15; VI: props. 13, 19–20, 30
(b) cut or bisect, قَطَعَ ، تقَاطَعَ
RJ: III: props. 4–6, 9–10, 28–29; IV: prop. 6 (تقاطَع)
See also: לקח ,חִלק ,הבדיל
(c) set out (or: cut) (let us cut), فَصْلَ (نَفْصِلُ)
RJ: III: props. 14, 27–28, 33; V: props. 17–18; VI: prop. 1
See also: שם ,הבדיל

חלק בחלקים
bisect (lit. cut into parts), قَسَمَ بِنِصْفَيْنِ
RJ: III: prop. 34
See also: חתך לחציים

חלק לחצאים
bisect, قَطَعَ بِنِصْفَيْنِ ,قَسَمَ بِنِصْفَيْنِ
MIT: III: props. 1, 9–10, 13, 24, 32, 35; IV: props. 4–5, 8, 11–14, 16; VI: props. 3, 28–29
RJ: III: props. 1, 4, 9–10, 13, 24, 29, 32, 34; IV: props. 4–5, 8, 11, 13–14, 16; VI: props. 3, 27–28
See also: חתך לחציים

חלק
(a) part (or: portion, of a whole), جُزْءٌ
MIT: IV: prop. 16; V: def. 1, props. 15–16, 23; VI: props. 1, 12
RJ: IV: prop. 16; V: def. 1, props. 1, 5, 15–16, 23; VI: props. 12–13
(b) part (of a straight line, created by its division), قِسْمٌ
MIT: III: prop. 34; V: props. 1, 3, 15; VI: def. 4, props. 3, 8–9, 12–13, 30
RJ: III: prop. 34; V: props. 3, 15; VI: def. 4, props. 3, 8–9, 13, 30
See also: חִלק ,חלוקה
(c) half (lit. part), نِصْفُ
RJ: III: props. 34–35

חִסר (v.)
remove, subtract, نَقَّصَ
MIT: III: prop. 34: V: props. 7, 16–17, 19: VI: props. 1, 27–28, 30, 33
RJ: V: props. 17, 19: VI: props. 27–30
See also: השליך ,הפיל ,הסיר ,הוציא ,גרע

חסר
be less than, ناقِص
MIT: V: def. 7, props. 7, 11–13, 16–17, 22–23: VI: props. 1, 28, 33
RJ: V: props. 7, 11–13, 16–17, 23: VI: props. 1, 28, 33
See also: מחוסר, פחות

חסרון
defect, نَقْص، ناقِص
MIT: VI: props. 27–28
RJ: V: props. 11, 16; VI: props. 27, 33
See also: מחוסר

חצי
half, نِصْف
MIT: III: props. 1, 3–4, 9–10, 13, 15, 24, 26, 29–31, 34–35; IV: props. 4–6, 8–9, 11–16; VI: props. 3, 9, 27–29, 33
RJ: III: props. 1, 3–4, 9–10, 13, 15, 24, 26, 29–32, 34–35; IV: props. 4–6, 8–9, 11, 13–16; VI: props. 3, 9, 27–29, 33
See also: זווית—זווית חצי נצבה, חלק—חלק בחצאים, עגולה—חצי העגולה

חתיכת העגולה, חתיכת העגול
segment of circle, a figure contained by a straight line and the circumference cut off by it, قِطْعَة الدَّائِرَة
MIT: III. defs. 6–8, 11, props. 20–25, 28, 30–33, 36; IV: props. 2, 10
RJ: III. defs. 6–8, 11, props. 20–25, 30–33; IV: props. 2, 10

חתך (v.)
(a) cut, قَطَعَ (فَصَلَ)
MIT: III: props. 1, 3–4, 9–10, 13, 29, 35–36: VI: prop. 3
RJ: III: props. 1, 3–4, 9–10, 33–35; IV: prop. 11; VI: props. 1, 3, 12, 25, 33
See also: הבדיל, חִלק, לקח, שם
(b) intersect, قَطَعَ
MIT: III: defs. 2–3, props. 3–5, 8, 10, 13, 31–32, 34–36; IV: props. 2, 6–7, 10; VI: prop. 2
RJ: III: defs. 2–3, props. 3–5, 8–10, 27, 31–36; IV: props. 2, 7, 10; VI: props. 1–2

חתך לחציים
bisect, قَسَمَ بِنِصْفَيْنِ, قَطَعَ بِنِصْفَيْنِ
RJ: III: props. 1, 3–4, 10, 13, 29, 35; VI: prop. 3
RJ: III: props. 1, 3–4, 34–35; VI: prop. 11
See also: הבדיל, חִלק, לקח, שם

חֶתֶך
intersection point, قَطْعٌ (فَصْلٌ)
MIT: III: prop. 9
See also: חלוק ,חלוקה ,חתיכת העגולה

י

יַדוּע
given (lit. known), مَعْلُومٌ
MIT: III: props. 16, 24, 29, 32–33; IV: props. 1–9, 11–16; VI: props. 9–13, 20, 26, 28–30
RJ: III: props. 1, 16, 24, 29, 32–33: IV: 1–9, 11–16; VI: props. 9–13, 20, 26–30
See also: מונח

יושר
straight extension, اَسْتِقَامَةٌ
MIT: III: prop. 35; IV: prop. 15; VI: def. 3, props. 4, 9, 11, 14–15, 23, 31
RJ: III: prop. 11; IV: prop. 15: VI: props. 9, 14, 31
See also: דבקות ,מדובק–דבק ,בארכו–אורך

יחד
taken together, مَعًا، جَمِيعًا (found mainly in Book V)
MIT: III: props. 5, 16, 19, 21, 30–31; IV: props: 10, 13; V: def. 7, props. 1, 4, 6–8, 11–13, 16–17, 22–23; VI: props. 1, 28–29, 32
RJ: III: props. 1, 7–8, 11, 16, 21, 30–31, 34–35; IV: props. 10, 13; V: def. 7, props. 1–3, 8, 13, 23–24; VI: props. 28–29, 31–32
See also: מקובץ ,מחובר ,כל

יחס (v.)

התיחס
proportional (to), نَاسَبَ
MIT: VI: props. 6–7

(שיעור) יוחס
brought into ratio, نُسِبَ
RJ: V: prop. 8; VI: props. 19, 24

נתיחס
brought into ratio, نُسِبَ
MIT: V: prop. 8

יחס
ratio, نِسْبَةٌ
MIT: v: defs. 4–9, 11–21, props. 4, 7–25; vi: defs. 4–5, props. 1–8, 10–20, 22–26, 28–33; vi: props. 6–7, 10–20, 22–26, 28–33
RJ: v: defs. 4–8, 11–17, 19–21, props. vi: 4, 7–15, 19–24; vi: defs. 4–5, props. 1–3, 6–7, 9–10, 13–16, 18–19, 22, 23, 30–31, 33

אמצעי ביחס
mean proportional, وَسَطٌ فِي ٱلنِّسْبَةِ
MIT: vi: prop. 8
RJ: vi: prop. 8 (porism attached to the enunciation)

הבדל היחס
separation of a ratio, تَفْصِيلُ ٱلنِّسْبَةِ
MIT: v: def. 17
See also: חלוק היחס

הפך היחס
inverse ratio, عَكْسُ ٱلنِّسْبَةِ
MIT: v: def. 15
RJ: v: def. 14
See also: תמורת היחס

הרכבת היחס
composition of a ratio, تَرْكِيبُ ٱلنِّسْبَةِ
MIT: v: def. 16, prop. 24; vi: prop. 24
RJ: v: def. 16, prop. 24; vi: prop. 24
See also: שיעור—שיעורים מורכבים

התהפכות היחס
conversion of a ratio, قَلْبُ ٱلنِّسْبَةِ
MIT: v: def. 18
RJ: v: def. 18, props. 14–16, 18
See also: מתהפכות ביחס

התיחס
proportion, تَنَاسُبٌ
MIT: v: defs. 5, 10
RJ: v: defs. 5, 10

חִלוף היחס
alternate ratio, تَبْدِيلُ ٱلنِّسْبَةِ
RJ: v: def. 15
See also: תמורת היחס

חִלוק היחס
separation of a ratio, تَفْصِيلُ ٱلنِّسْبَةِ
RJ: v: def. 17
See also: הבדל היחס

יחס בעל אמצע וקצוות
نِسْبَةٌ ذاتُ وَسَطٍ وطَرَفَيْنِ
MIT: vi: def. 4, prop. 30; See also: v: def. 19
RJ: vi: def. 4, prop. 30 (יחס עצם ואמצע ושתי קצוות)

יחס הולך על סדר
ordered proportion, مُناسَبَةٌ مُنْتَظِمَةٌ
MIT: v: def. 20
RJ: v: def. 20

יחס השווי
ratio ex aequali, نِسبَةُ ٱلْمُساواةِ
MIT: v: def. 19, props. 20–24; vi: props. 22–23
RJ: v: def. 19, props. 20–24; vi: prop. 22 (שווי)

יחס מבולבל
perturbed proportion, نِسْبَةٌ اختِلُفَ في الأَقْدارِ ,نِسْبَةٌ مُخْتَلِفَةٌ في الأَقْدارِ
MIT: v: props. 21, 23
See also: יחס נפתל במערכת ,יחס מתחלף

יחס מתחלף (בשיעור)
perturbed proportion, نسبة مختلفة في الأَقْدارِ
RJ: v: props. 21, 23
See also: יחס נפתל במערכת ,יחס מבולבל

יחס נפתל במערכת
perturbed proportion, مُناسَبَةٌ مُضْطَرِبَةٌ
MIT: v: def. 21
RJ: v: def. 21
See also: יחס מתחלף ,יחס מבולבל

מיוחס
proportional to, مَنْسُوبٌ
RJ: VI: prop. 19

מתיחס
proportional, proportional to, مُناسِبٌ، مُتَنَاسِبٌ
MIT: V: defs. 8, 11–12, props. 14, 16–18, 25; VI: defs. 1–2, props. 4–11, 16–24, 26, 31
RJ: V: defs. 8, 11–12, props. 14, 16–18, 23, 25; VI: defs. 1–2, props. 4–7, 9–11, 16–18, 20, 22, 24, 26, 31
See also: שיעור—שיעורים מתייחסים

תמורת היחס
alternate ratio, تَبْدِيلُ ٱلنِّسْبَةِ
MIT: V: def. 14, prop. 16
See also: חילוף היחס

יסף (v.)
הוסיף
add, extend زَادَ
MIT: V: prop. 16; VI: props. 1, 12, 29–30, 33
RJ: V: props. 8, 17; VI: props. 1, 28, 33

הוסף
added, extended, زَادَ
RJ: III: prop: 35; VI. prop. 28
See also: יצא, משך, עבר, עדף

יצא (v.)
extend (a line), مَرَّ، خَرَجَ
MIT: III: def. 1, props. 2–4, 7–8, 12, 15–18, 21, 24, 31–36 (prop. 34 also مَرَّ); IV: props. 2, 10, 12, 15–16; VI: props. 2–4, 8–9, 11, 31
RJ: III: def. 1, props. 3–4, 7–9, 12–13, 15–18, 24–25, 27–28, 31–36; IV: props. 2–3, 7, 10, 12; V: prop. 10; VI: def. 3, props. 2–4, 8–13, 24, 28, 31 (also as translation of أَخْرَجَ in: III: prop. 35; IV: prop. 10; VI: def. 3, prop. 2)
See also: הוסף, משך, עבר

הוציא (נוציא)
(a) produce (a straight line), أَخْرَجَ
MIT: III: props. 1–17, 19–22, 24–25, 27–32, 34–36; IV: props. 1–4, 6–11, 13–16; VI: props. 1–4, 9–13, 15–16, 18–20, 23, 25, 26, 33

RJ: III: props. 1–3, 5–10, 12–17, 19, 22, 24–25, 29–30, 32, 34–36; IV: props. 1, 3–8, 10, 12–13, 15–16; VI: props. 1–4, 9–13, 16, 25, 32
(b) produce a straight line (proportional to another line), اِسْتَخْرَجَ
MIT: VI: props. 18, 22, 26
RJ: VI: props. 18, 22, 26
(c) produce multiples
RJ: V: prop. 12
See also: הגיע, הדביק, משך, נדבק, קוה, רחק

הוצא
(a straight line) produced, أُخْرِجَ
MIT: III: defs. 2, 4, props. 2, 8–9, 13, 15, 31, 35–36; IV: props. 3, 7–8; VI: def. 3, props. 2–3, 8, 32
RJ: III: def. 2, props. 2, 15, 35; IV: prop. 5
See also: הגיע, הדביק, משך

יציאה
extension, خُروج
MIT: III: prop. 11
See also: מוצא

ישר—See: קו ישר–קו

יתרון
excess, addition, فَضْل
MIT: VI: prop. 28
See also: תוספת, עודף, מותר

כ

כל
(a) entire, جَميع، كُلّ
MIT: III: props. 19, 21, 30–31, 35–36; IV: props. 6, 10–11, 13, 15–16; V: props. 1–2, 5, 13, 15; VI: props. 19, 29, 31
RJ: III: props. 11, 19, 21, 35–36; IV: props. 8, 15, 19; VI: props. 12, 19, 20, 27, 29
(b) whole, كُلّ
MIT: V: props. 1, 5, 13, 19
RJ: III: props. 18; V: props. 1, 5, 13, 19
See also: יחד, מחובר, מקובץ

כל אחד ל ...
respectively, كُلُّ واحِدٍ لِنَظيرِهِ
See also: דומה ,גיל

כُلّ واحِدٍ في نَفْسِهِ ,כל אחד בעצמו
RJ: III: props. 13, 34–35

كُلُّ واحِدٍ مِثْلُ الآخَرِ ,כל אחד לאחר
RJ: III: props. 4, 14

כל אחד לדומה לו ,كُلّ واحِدٍ لِنَظيرِهِ , كل واحدة مثل نظيرِهِ (only **RJ**)
MIT: IV: prop. 12
RJ: III: props. 3, 7–9, 13, 16, 24–25; IV: props. 4–5, 12–14; VI: props. 5–6

כל אחד לנכחי לו ,כל אחד כמו הנכחי לו ,כל אחד אצל הנכחי לו ,كُلّ واحِدٍ لِنَظيرِهِ , كُلُّ واحِدٍ مثل
كُلّ واحِدٍ مثل نظيره ,نَظيرِهِ
MIT: III: props. 13, 24, 36; IV: props. 4–5, 13–15; VI: props. 5–7, 31

כלה (אל) (v.)
terminate, أنْتَهى (الى) (**RJ** also وَقَعَ عَلى)
MIT: III: props. 8, 11, 36; IV: prop. 10; V: prop. 8
RJ: III: props. 1, 8 (وَقَعَ عَلى), 36; IV: prop. 10

כלל
entire, أَسْر
MIT: VI: def. 4
See also: כל

כלל (יכלול; v.)
take up, إِسْتَغْرَقَ
MIT: V: def. 2
RJ: V: def. 2

כמה

כמה שהיה
any whatever, or: any number of, كَيْفَما كانَت، أيُّ كانَتْ، ما كانَتْ، كَمْ كانَتْ
MIT: V: def. 19, props. 13, 23
RJ: V: defs. 7, 19, props. 3–4, 13, 22–23
See also: אי זה ... שיהיה

כמה שנרצה
any number we wish, كَمْ شِئْنا
MIT: VI: props. 1, 33
See also: כמו שנרצה—כמו ,אי זה ... שרצינו—אי זה ... שיהיה
See also: כמו

כמו
(a) (also: כ), same as (or: like; many times in the sense of "equal"), مِثْلُ
MIT: III: props. 1–9, 11, 13–17, 19–22, 25–35; IV: all props.; V: props. 1–10, 12, 14–15, 25; VI: props. 1–8, 10–14, 16–17, 19–20, 26–33
RJ: III: prop. 11; V: props. 1–3, 5–10, 15; VI: props. 13, 17, 28–29
See also: כמה, שוה
(b) e.g., مِثْلُ
MIT: III: props. 2, 11–12, 15, 23
RJ: III: props. 2, 11, 15, 23

כמו שהוא
arbitrarily (placed, numbered), كَيْفَ ما كانَ ,كَيْفَ ما وَقَعَ
RJ: III: prop. 2 (كَيْفَ ما وَقَعَ); V: props. 4, 17 (كَيْفَ ما كانَ)

כמו שנזדמן
arbitrarily, كَيْفَما اتَّفَقَتْ
RJ: III: def. 8
See also: איך שנפל

כמו שנפל
arbitrarily placed, كَيْفَ ما وَقَعَ
RJ: III: props. 1, 5–6, 22, 30–31; IV: prop. 3
See also: איך שנפל

כמו שנרצה
any number we wish, كَمْ شِئْنا
RJ: VI: props. 1, 33
See also: כמה שנרצה—כמה ,אי זה ... שרצינו—אי זה ... שיהיה

כנס (v.)
נכנס
enter (into), دَخَلَ
MIT: III: props. 8, 15
See also: זווית נכנסת—זווית ,בא בתוך

כפול
multiplied (by itself), مُكَرَّرَة, مُثَنَّاةٌ
MIT: VI: props. 18–19
See also: שנוי, נשנה בהשנות

כפל (v.)
  נכפל
  (a) doubled, أُضْعِفَ
  MIT: III: prop. 13
  RJ: III: prop. 13
  See also: לקח כפלים
  (b) multiplied, ضَاعَفَ
  MIT: V: def. 6, prop. 8; VI: def. 5
  RJ: V: def. 6, prop. 8; VI: def. 5

  הוכפל
  multiplied, ضَاعَفَ
  RJ: V: prop. 8
See also: לקח כפלים

כפל
(a) multiple (usually in the sense of double), مِثْلٌ
MIT: III: props. 13, 19–20, 25, 30; IV: props. 10–12, 15; VI: prop. 12
RJ: III: props. 13, 19–20, 25, 30; IV: props. 8, 10–14; V: prop. 3; VI: prop. 12
(b) multiple, ضِعْفٌ
MIT: IV: prop. 12; V: defs. 3, 7, 9, 11–12, props. 1–8, 11–13, 15–17, 22–23; VI: props. 1, 12, 18, 33
RJ: V: defs. 3, 7, 9, 11–12, props. 1–8, 11–13, 15–17, 22–23; VI: props. 1, 33
See also: מערכת, הכאה, דמיונים

  כפלים שוי הפעמים
  equimultiples, أَضْعافٌ مُتَسَاوِيَةُ ٱلْمَرَّاتِ
  MIT: V: defs. 7, 9, props. 3–4; VI: prop. 33
  RJ: V: defs. 7, 9, props. 3–4

ל

לוה (v.)

נלוה

face towards, وَلِيَ

MIT: VI: prop. 8

See also: נמשך

לקח (v.)

cut off, أَخَذَ

RJ: V: props. 20–23

See also: הבדיל, חִלק, חתך, שם

לקח כפלים

multiply, أَخَذَ أَضْعافَ

MIT: V: props. 3–4, 11, 13, 16–17, 22–23; VI: prop. 1

RJ: V: props. 3–4, 7, 11–13, 16–17, 22–23; VI: prop. 1

See also: כפל

מ

מומר

alternate, مُتَبَادَلٌ

MIT: III: props. 31–33; IV: props. 2, 10

See also: זווית—זוויות מומרות, מתחלף

מונח

(a) given, مَفْرُوضٌ

MIT: III: props. 1, 32; IV: prop. 8

RJ: III: props. 1, 22

(b) situated, مَوْضُوعٌ

MIT: VI: props. 20, 25, 27, 32

See also: ידוע, מוצב

מוסכמות

postulates

RJ: III: prop. 18

מופת
demonstration, بُرْهان
MIT: all props. except III.1–2, 4–6, 12, 15, 17–18, 22, 24, 26–27, 29, 32–33, 36; IV.1–3, 5–16, V.9–10, VI.7, 9–13, 20, 24, 25–26, 28–30
RJ: all props. except III.1–2, 4–6, 10, 12, 15, 17–18, 22, 26–28, IV.8–10, 12–14, 16, VI.7, 9–13, 20, 24, 28–30

(ונשלם המופת) ונשלם מופתו
Q.E.D. (lit. its demonstration has been completed)
RJ: III: props. 16, 24

מוצא
extension, خُرُوج
RJ: III: prop. 11
See also: יציאה

מוצב
situated, مَوْضُوعٌ
RJ: VI: props. 20, 25
See also: מצב, מונח

מור (v.)
המיר
alternate, بَدَّلَ
MIT: V: props. 19, 25
See also: החליף

מורכב
(a) added together, مَجْمُوعٌ
RJ: V: prop. 24
See also: שיעורים מורכבים—שיעור ,מקובץ ,מחובר ,הרכבת היחס—יחס
(a) (an angle) in (an arc; lit. composed), مُرَكَّبٌ
MIT: III: prop. 30
RJ: III: prop. 30

מורשם
inscribed, مَرْسُومٌ
MIT: IV: def. 1
RJ: IV: prop. 16
See also: מצויר

מותר
excess, addition, فَضْلٌ
MIT: V: defs. 17–18
RJ: V: defs. 17–18
See also: תוספת, עודף, יתרון

מחובר (מחוברים)
(a) combined, added, مجموعٌ (مجموعانِ), مُرَكَّبٌ (RJ)
MIT: III: props. 7, 11, 14; V: prop. 25
RJ: V: prop. 25 (enunciation: مجموعٌ; proof: مُرَكَّبٌ)
See also: מקובץ, מורכב, כל, יחד
(b) associated with, مُقارِنٌ
RJ: V: prop. 1
See also: קרוב, סמוך, חָבר
(c) combined (ratio, of other ratios), مُؤَلَّفةٌ
MIT: VI: def. 5, prop. 23
RJ: VI: def. 5
See also: חִבר
(d) connected, joined (two straight lines), مُتَّصِلًا
RJ: VI: prop. 14
See also: נגיע, חבר
(e) applied (polygon, to a straight line), مُضافٌ
MIT: VI: props. 28–29
RJ: VI: props. 27–29
See also: סמוך, מחובר

מחומש
pentagon, مُخَمَّسٌ
MIT: IV: prop. 15
RJ: IV: prop. 15

מחומש שוה הצלעות והזוויות
equilateral pentagon, مُخَمَّسٌ مُتَساوي ٱلأَضْلاعِ وَٱلزَّوايا
MIT: IV: props. 11–14, 16
RJ: IV: props. 11–14, 16

מחוסר
remainder, مَنْقُوصٌ
MIT: V: props. 5, 19
RJ: V: props. 5, 19
See also: חסרון

מטה
below, تَحْتَ
RJ: III.35
See also: מעלה

מישוש
contact, تَمَاسٌّ
MIT: III: props. 11–12, 17–18, 31–33; IV: props. 2–3, 7, 10, 12
RJ: III: props. 11–12, 17–18, 31–33, 35–36; IV: props. 2–3, 7, 10

מקום המישוש
point of contact, مَوْضِعُ الَّذي يَمَاسُّ، مَوْضِعُ ٱلتَّمَاسِّ
MIT: III: props. 11–12, 17–18, 31–33; IV: props. 2, 7, 12
RJ: III: props. 11–12, 17–18, 31–33, 35–36; IV: props. 2, 7

היה מיתר (v.)
subtend (or: to be a chord), وَتَرَ
RJ: III: props. 2–3, 16–17, 28; IV: props. 4, 13; VI: props. 5–6
MIT: III: props. 2, 16, 36; IV: props. 4, 12–13, 15; VI: props. 5–6, 8

השלים מיתר
be a chord
RJ: IV: prop. 15 (porism)
See also: הקיף ,החזיק

עשה מיתר
be a chord, وَتَرَ
MIT: VI: prop. 5
See also: הקיף ,החזיק

מיתר
chord, وَتَرٌ
MIT: III: def. 6, props. 1, 3–4, 9, 13–15, 27–30, 34; IV: props. 1, 10, 16; VI: props. 4, 6, 8, 32
RJ: III: def. 6, props. 3–4, 9, 13–15, 27–28, 34; IV: props. 1, 11, 15–16; VI: props. 4, 6, 32

מסומן
gnomon, عَلَمٌ
RJ: VI: props. 28–29
See also: רושם

מספיק
reciprocally related, مُتَكافّ
MIT: VI: def. 2, props. 14–16, 18, 30
See also: שטח מספיק הצלעות—שטח ,מתהפך

מספר
multitude, number, عَدَدٌ ,عِدّةٌ
MIT: V: def. 19, props. 1–5 (prop. 3: عَدَدٌ), 13, 15, 20–23; VI: prop. 19
RJ: V: def. 19, props. 1–4, 13, 15, 20–23; VI: prop. 19

מעל ,(ל)מעלה
above, فَوْقَ
RJ: III: prop. 35
See also: מטה

מערכת
multiplication, ضَرْبٌ
RJ: III: props. 34–35; VI: props. 16–17
see also: הכאה

מעשה
construction, عَمَلٌ
MIT: V: prop. 6; VI: prop. 22
RJ: VI: prop. 22

מצב
כמצבו (or: על מצבו)
similarly situated, كَوَضْعِهِ
MIT: VI: props. 18, 20, 25, 27
RJ: VI: props. 18, 20, 25, 27
See also: מוצב ,עשה—עשוי כמעשהו ,תואר

מצויר
inscribed, مَرْسُومٌ
RJ: IV: defs. 1–2
See also: מורשם

(מקובצים) מקובץ
added together, جَمِيعًا، مجموعِينَ

MIT: III: props. 7–8, 11; V: props. 1, 13, 17, 24–25
See also: מחובר ,מורכב ,כל ,יחד

מקום
place, position, مَوْضِعٌ
MIT: III: props. 10–12, 17
RJ: III: props. 1, 7, 10–12, 36
See also: מקום המישוש—מישוש

מקיף
perimeter, مُحِيطٌ
MIT: VI: def. 3

מקיף העגול, מקיף העגולה
circumference, مُحِيطُ ٱلدَّائِرَةِ
MIT: III: props. 5, 36
RJ: III: prop. 8
See also: קו מקיף—קו

מרובע
square, مُرَبَّعٌ
MIT: III: props. 13, 34–36; IV: props. 6–9; VI: props. 30, 32
RJ: III: props. 35–36; IV: props. 6–9; VI: prop. 30
See also: שטח מרובע—שטח

מרובע שוה הצלעות
square (lit. equilateral quadrilateral), مُرَبَّعٌ مُتَسَاوِي الأَضْلَعِ
MIT: IV: props. 6–7

מרובע שוה הצלעות נצב הזוויות
square (lit. equilateral and rectangular quadrilateral), مُرَبَّعٌ مُتَسَاوِي الأَضْلَعِ قائِمُ الزَّوايا
RJ: IV: props. 6–7

מרחק
distance (between a straight line and point), بُعْدٌ
MIT: III: def. 5, props. 13–14, 16, 24, 32; IV: props. 1, 4–5, 8–10, 13–15
RJ: III. defs. 4–5, props. 13–14, 24, 32; IV: props. 1, 4–5, 8–10, 13
See also: שיעור מרחק—שיעור

מרכז
center (of a circle), مَرْكَزٌ
MIT: III: defs. 1, 4–5, 10, props. 1–20, 24–28, 30–36; IV: props. 1, 3–5, 7–10, 12–15; VI: prop. 33
RJ: III: defs. 1, 4–5, 10, props. 1–20, 24–28, 30–32, 34–36; IV: props. 1, 3–5, 7–10, 12–15; VI: prop. 33

משולש
triangle, مُثَلَّثٌ
MIT: III: props. 2–3. 7–8, 13, 15–17, 24, 30–31, 36; IV: props. 2–5, 10–15; VI: props. 1–8, 11, 13, 15, 18–20, 24, 31–32
RJ: III: props. 2–3, 7–8, 13, 15–16; IV: props. 2–5, 10–15; VI: props. 1–13, 15, 18–20, 24, 31–32

משולש חד הזוויות
acute-angled triangle, مُثَلَّثٌ حادُّ ٱلزَّوايا
RJ: IV: prop. 5

משולש נצב הזווית
right-angled triangle, مُثَلَّثٌ قائِمُ ٱلزَّاوِيَةِ
MIT: VI: prop. 32
RJ: IV: prop. 5; VI: props. 8, 32

משולש רחב הזוויות
obtuse-angled triangle, مُثَلَّثٌ مُنْفَرِجُ ٱلزَّاوِيَةِ
RJ: IV: prop. 5
See also: זווית רחבה

משולש שוה הצלעות והזוויות
equilateral and (or) equiangular triangle, مُثَلَّثٌ مُتَسَاوِي ٱلْأَضْلاعِ وَٱلزَّوايا
MIT: IV: props. 15–16
RJ: IV: props. 11, 15–16

משולש שוה השוקיים
isosceles triangle, مُثَلَّثٌ مُتَسَاوِي ٱلسَّاقَيْنِ
MIT: IV: props. 10–11
RJ: IV: props. 10–11

משושה
hexagon, مُسَدَّسٌ
MIT: IV: prop. 15 (also משושה שוה הצלעות והזוויות)
RJ: IV: prop. 15 (also משושה שוה הצלעות והזוויות)

משותף
common, مُشْتَرَكٌ
MIT: III: props. 1, 3, 7–10, 21, 24, 29, 31, 34–36; IV: props. 4–6, 10–15; V: props. 5–6, 17, 25; VI: props. 6, 8, 25, 27–31
RJ: III: props. 1, 3, 7–9, 21, 24, 29, 31, 34–36; IV: props. 5–6, 10–15; V: props. 5–6, 17, 25; VI: props. 5–6, 8, 25, 27–31

משך (v.; נמשוך)
(a) extend (a straight line; let us extend), خَرَجَ
RJ: III: prop. 31; VI: prop. 27
See also: עבר ,יצא ,הוסף
(b) face towards, وَلِيَ
RJ: VI: prop. 8.
See also: נלוה
(c) successive (multiples) be repeated, تَلَا
RJ: V: prop. 8
See also: קודמים ונמשכים

משש (v.)
touch, مَاسَّ
MIT: III: defs. 2–3, props. 11–12, 16–18, 31–33, 35–36; IV: defs. 1–2, props. 2–8, 10, 12–13, 15
RJ: III: defs. 2–3, props. 6, 11–12, 15–18, 31–33, 35–36; IV: defs. 1–2, props. 2–4, 7, 10, 12–13
See also: ממשש לעגולה

התמשש
touch (mutually), تَمَاسَّ
MIT: III: props. 6, 11

מתהפך
reciprocally related, مُتَكَافِئَةٌ
RJ: VI: props. 14–16, 18, 30
See also: שטח מתהפך הצלעות—שטח ,התהפכות היחס—יחס ,מספיק

מתחלף (מתחלפים, מתחלפות)
(a) alternate, مُتَبَادَلٌ
RJ: III: props. 31–33; IV: props. 2–10
See also: זווית—זוויות מתחלפות ,יחס—חלוף היחס ,יחס מתחלף—מומר
(b) unequal, مُخْتَلِفٌ
MIT: V: prop. 8; VI: prop. 31
RJ: V: prop. 8; VI: prop. 31

נ

נגד (v).

התנגד

be opposite, مُتَقَابِلٌ

MIT: IV: props. 3, 7; VI: prop. 14

See also: זוויות מתנגדות—זווית ,הקביל

(כ)נגד

opposite (an angle exterior to a triangle, to an interior angle), مُقَابِلٌ، قَابَلَ

RJ: IV: prop. 5; VI: prop. 3

See also: זוויות מתנגדות—זווית

נגע (v.)

(נגיע) הגיע

(a) produce a line (by connecting two points; lit. made something touch another thing), وَصَلَ

MIT: III: props. 9–11, 24, 29, 34–36; IV: props. 10, 12–13, 15

See also: משך ,יצא ,חִבר ,הדביק

(b) extended such that it touches (a straight line, a point or another line), وَصَلَ

MIT: III: def. 8, prop. 5; VI: prop. 9

See also: חִבר ,הדביק

נוסף

greater (lit. added, extended), زَائِدٌ

MIT: V: props. 7–8, 12–13, 16–17; VI: props. 1, 28–30, 33

RJ: V: props. 8, 17; VI: props. 1, 28, 33

נח (v.)

(מונח) הונח

given, مَفْرُوضٌ

MIT: III: props. 1, 32; IV: prop. 8

RJ: III: props. 1, 22

See also: ידוע

(נניח) הניח

situate, وَضَعَ ,نَزَّلَ

MIT: VI: props. 4, 14, 15, 23 (وَضَعَ)

See also: מונח

נכוחי ,נכחי
parallel; مُوازٍ or مُوَازِي (parallel to), مُتَوازٍ or مُتَوَازِي (two or more lines, to each other)
MIT: III: props. 13–14, 24, 36; IV: props. 4–5, 7–8, 13–15; V: def. 13; VI: props. 1–7, 10–14, 18–19, 21–31
RJ: III: prop. 14; IV: props. 5, 7–8; VI: props. 1–4, 10–14, 23–25, 27–31
See also: כל אחד ל—כל אחד לנכחי לו—שטח נכחי הצלעות, שטח ,שיעור—שיעורים נכחיים

נמשך—See: ראשונים ונמשכים

נפל
(a) fall (in a certain place), وَقَعَ
MIT: III: props. 2–3, 11–15, 18, 21, 28, 31–34, 36; IV: props. 2, 10, 16
RJ: III: props. 2–3, 9, 11–15, 18, 28, 30–33, 35–36 (in 12 and 36 also a translation of كَانَ); IV: props. 1, 2, 10, 16
(b) intersect, وَقَعَ
MIT: III: def. 5, prop. 2; IV: prop. 7; VI: props. 3, 31
RJ: III: def. 5, props. 2, 34; IV: props. 5, 7; VI: props. 3, 31
See also: חתך, נפגש
(c) congruent to (or: superposed on), وَقَعَ، اَنْطَبَقَ
MIT: III: prop. 23
RJ: III: prop. 23
See also: נדבק, רכב
(d) subtract (a common element from each of two equal or unequal expressions), أَسْقَطَ
MIT: III: prop. 7
See also: גרע, הפיל, השליך, חִסר

(נפיל) הפיל
subtract (a common element from each of two equal or unequal expressions), أَسْقَطَ
MIT: V: def. 19; VI: prop. 29
See also: גרע, הפיל, השליך, חִסר

See also: איך שנפל—כמו ,כמו שנפל—שיעור ,יפול עליו שיעור

נקודה
point, نُقْطَةٌ
MIT: III: def. 8, props. 1–3, 5–18, 22, 24–26, 29–36; IV: props. 2–8, 10, 12–16; VI: def. 3, props. 3–7, 9, 11, 13, 16, 20, 31
RJ: III: def. 8, props. 1–2, 4–12, 14–18, 20, 22–25, 26, 29–36; IV: props. 2–4, 6–7, 10, 12–13, 15; VI: def. 3, props. 3–7, 9, 11–13, 16, 20, 27–31

נקף (v.)

הקיף

(a) contain (a figure), أَحَاطَ

MIT: III: defs. 6–10, props. 15–16, 30, 34–36; IV: props. 3–16; VI: def. 1, props. 6–7, 11–12, 14–16, 18, 20, 24, 30–31, 33

RJ: III: defs. 6–10, props. 15–16, 25, 34–36; IV: props. 3–9, 11–16; VI: def. 1, props. 6–7, 10, 11–16, 18, 20, 24, 30–31

See also: החזיק, מיתר (היה)

(b) draw a circle, خطط (**RJ**: also عَمِلَ)

MIT: III: prop. 32; IV: props. 4, 9, 15

RJ: III: props. 16, 24, 32; IV: props. 1, 4 (عَمِلَ), 5, 8–10, 13–15

See also: קוה

נשאר

remaining, بَاقٍ

MIT: III: props. 7–8, 13–14, 16, 25, 31; IV: props. 4, 10–14; V: props. 5–6, 19, 25; VI: props. 2–3, 7, 16, 19, 31–32

RJ: III: props. 7–8, 13–16, 27, 31; IV: props. 4, 10–14; V: props. 5–6, 19, 25; VI: props. 2–3, 6–7, 31–32

ס

סבב (v.)

(a) draw a circle, أَدَارَ, خَطَّ

MIT: III: prop. 24 (خَطَّ); IV: props. 5, 8, 13

See also: הקיף

(b) contain (an arc on two radii), حازَ

MIT: III: def. 9

See also: החזיק, הקיף, מיתר (היה)

סביב

in (lit. around, or: about; inscribe), عَلَى، بِ، حَوْلَ

MIT: IV: def. 2 (حَوْلَ)

RJ: IV: def. 2 (حَوْلَ); IV: props. 4 (بِ), 7 (عَلَى)

סדר

לקח על סדר, הוקש בסדר

taken (successively) in (corresponding) order, قِيسَ/أُخِذَ على ٱلْوَلَاءِ

MIT: V: def. 7, prop. 8

RJ: v: def. 7
See also: יחס הולך על סדר—יחס

סמוך
applied to (area, to a straight line), مُضَافٌ
MIT: v: props. 27, 32
See also: קרוב ,מחובר ,חָבר

סמך (v.)
נסמך
applied (parallelogram, to a straight line), أَضَافَ
MIT: vi: prop. 27
See also: סמוך ,חִבר

סר (v.)
הסיר
remove, subtract, نَزَعَ
RJ: v: prop. 17
See also: חִסר ,השליך ,הפיל ,גרע

ע

עבר (v.)
(a) pass (through something, usually on a point), مَرَّ، جازَ
MIT: iii: props. 1, 4, 7–8, 11–12, 31; iv: props. 2–3, 12–13
RJ: iii: props. 1, 3–4, 7–8, 11, 31–32, 34–35; iv: props. 2–3, 5, 13
See also: הלך
(b) extend, بعد
MIT: iii: prop. 31

העביר (נעביר)
(a) produce (a straight line such that it passes on a given point), أَجازَ (**RJ**: also خطط)
MIT: iii: props. 33; iv: props. 2–3, 12
RJ: iii: props. 27 (خطط), 33; iv: props. 2–3
See also: יצא ,הגיע
(b) extend, نَفَذَ
MIT: iii: props. 9–10
See also: משך ,יצא ,הוסף

עגול, עגולה
circle, دائِرَةٌ
MIT: III: defs. 1–4, 6–8, 10–11, all props.; IV: all props. save for 29; VI: props. 9, 33
RJ: III: defs. 1–4, 6–8, 10–11, all props.; IV: all props.; VI: props. 9, 33

חותך העגול ,חותך העגולה
sector of a circle, قِطاعُ ٱلدَّائِرَة
MIT: III: def. 10
RJ: III: def. 10

חצי עגול ,חצי עגולה
semicircle, نِصْفُ دائِرَةٍ
MIT: III: props. 15, 30; IV: prop. 6; VI: prop. 9
RJ: III: props. 15, 24, 30; IV: prop. 6; VI: prop. 9
See also: מקיף העגולה—מקיף ,חתיכת העגולה

עדף (v.)
exceed, فَضَلَ (def. V.6), زَادَ
MIT: V: def. 6
RJ: V: def. 6, props. 12–13; VI: props. 29–30
See also: הוסף

עודף
exceeds, زَائِدٌ
RJ: VI: prop. 29
See also: תוספת ,מותר ,יתרון

עמד (v.)
stand, set up, قَامَ
MIT: III: props. 22–23
RJ: III: props. 1, 3, 7–9 (in 8 also a translation of خَرَجَ), 22–23
See also: שם ,קם ,צייר ,עשה

העמיד
construct, erect, build, أَقَامَ
MIT: III: props. 7–8, 24, 26, 32–33; IV: props. 2–3; VI: props. 5–7; VI: prop. 20
RJ: III: props. 7–8, 22, 24–26, 32–33; IV: props. 2–3; VI: props. 5–7, 20

עמוד
perpendicular, عَمُودٌ

MIT: III: defs. 4–5, III: props. 13–15, 17, 24, 31, 35; IV: props. 3–4. 7, 13; VI: def. 3; VI: props. 8–9, 32
RJ: III: defs. 4–5, props. 13–15, 17, 31, 34–35; IV: props. 3–4, 7, 13; VI: def. 3, props. 8–9, 32

עשה (v.)
construct (lit. make), عَمِلَ
MIT: IV: props. 2–3, 7, 11, 15–16; VI: props. 22, 26–30
RJ: III: prop. 4; IV: props. 7, 11–12, 15–16; VI: props. 22, 26–30
See also: שם ,קם ,צייר ,העמיד—עמד

נרצה שנעשה ,נרצה לעשות (only MIT)
we wish to construct, نُرِيدُ أَنْ نَعْمَلَ
MIT: III: prop. 32; IV: props. 2–16; VI: props. 20, 26
RJ: III: prop. 32; IV: props. 2–16; VI: props. 20, 26

עשוי כמעשהו, similarly constructed
MIT: VI: prop. 22

עשוי
is constructed, مَعْمُولٌ
MIT: IV: props. 2–3, 7, 15; VI: props. 22, 27
RJ: VI: props. 22, 27

פ

פגש (v.)
נפגש
meet, intersect, اَلْتَقَى
MIT: III: defs. 2–3, props. 8, 24, 32; IV: prop. 14; VI: props. 3–4
RJ: III: defs. 2–3, prop. 24; IV: props. 4–5, 13–14; VI: props. 3–4
See also: נפל ,חתך

פחות
less (then half a circle), أَقَلُّ (MIT), أَصْغَرُ (RJ)
MIT: III: props. 24, 32
RJ: III: props. 24, 30
See also: חסר ,גרע

פְּנִים
within (a circle), دَاخِلٌ
MIT: III: props. 11–12, 18
RJ: III: props. 11–12, 23
See also: תוך

פעמים
times, أمْثالٌ
RJ: VI: prop. 12
See also: כפל—כפלים שוי הפעמים

צ

צד
direction, side, جِهَةٌ, جَنَبَةٌ, جَنْبٌ (MIT only)
MIT: III: def. 2, props. 7–9, 22–23, 31–33; IV: props. 2–3, 10; VI: props. 1, 8, 31
RJ: III: def. 2, props. 1, 3, 7–9, 22–23, 31–33; IV: props. 3, 10, 12; VI: props. 1, 8, 16, 31

צורה
(a) geometrical figure, شَكْلٌ
RJ: VI: props. 27–29
See also: תמונה
(b) diagram
RJ: III; prop. 36; VI: prop. 23
See also: תמונה
(c) proposition
RJ: VI: prop. 18
See also: תמונה
(d) fashion, way (of demonstration), صِفَةٌ
MIT: III: 21

צייר (v.; נצייר)
draw, construct (let us draw), خُطَّ
RJ: III: props. 15, 25–28, 30; IV: prop. 1; VI: props. 1–2, 4–5, 9, 18–19, 22, 28
See also: העמיד, סבב, עשה, קם, שם

צלע
side, ضِلْعٌ
MIT: III: props. 2–3, 7–8, 13, 15–17, 21, 24, 30–32, 36; IV: defs. 1–2, props. 3–16; VI: defs. 1–2, props. 1–8, 11, 13–16, 18–21, 23–32

RJ: III: props. 2–3, 7–8, 13, 15–17, 21, 24, 30–32; IV: defs. 1–2, props. 3–16; VI: defs. 1–2, props. 1–8, 10–16, 18–20, 23–25, 27–32

See also: זווית—מרובע, מחומש שוה הצלעות והזוויות—מחומש, זווית ישרת הצלעות שטח בעל—שטח, שוק, משולש שוה הצלעות והזוויות—משולש, שוה הצלעות נצב הזוויות שטח נכחי—שטח, שטח מתהפך הצלעות—שטח, שטח מספיק הצלעות—שטח, ארבע צלעות תמונה בעלת—תמונה, תמונה בעלת ארבע צלעות—תמונה, שטח שוה הצלעות—שטח, הצלעות תמונה ישרת הצלעות—תמונה, חמש עשרה זווית שוות הצלעות והזוויות

ק

קבל (v.)

קִبِلَ, (תקביל) הקביל

(a) be opposite (segment of a circle, to an angle in or on the circle)

MIT: III: def. 11, props. 32–33

RJ: III: props. 32–33

(b) be opposite (sides or angles in parallelograms), قابَلَ

MIT: IV: prop. 8

See also: זווית מקבילות—זווית, התנגד

הקדמה ואיחור, קדימה ואיחור

prior and posterior, تقديم وتأخير

RJ: V: props. 21, 23; VI: def. 2

קו

line, خَطٌّ

MIT: III: defs. 1, 5, 10, props. 1–19, 22–36; IV: all props.; VI: defs. 2–4, props. 1–7, 9–13, 16–20, 22–23, 25–33

RJ: III: defs. 1, 5, 10, props. 1–19, 22–36; IV: all props.; VI: def. 3–4, props. 1–7, 9–14, 16–18, 20, 22–33

See also: צלע

קו ישר

straight line, خَطٌّ مُسْتَقِيمٌ

MIT: III: defs. 2, 4, 6–10, props. 2, 11, 15, 22–23, 31, 34–36; IV: prop. 1; VI: def. 4, props. 11, 20, 27–29

RJ: III: defs. 2, 4, 6–10, props. 1–2, 11, 15, 22–23, 31–32; IV: prop. 1; VI: def. 4, props. 20, 27–29, 31

See also: זווית—תמונה ישרת הקוים—תמונה, קוה, זווית ישרת הקוים

קו מקיף
circumference, اَلْخَطُّ المُحِيطُ
MIT: III: defs. 1, 6–7, props. 1–2, 7, 9, 15, 19, 25–26, 35–36; IV: props. 3, 12, 15–16; VI: prop. 33
RJ: III: defs. 1, 6–7, props. 1–2, 7–9, 12, 15, 19–20, 25–26, 27–28; IV: props. 3, 16; VI: prop. 33
See also: מקיף העגולה—מקיף

קו קשת
arc, قَوْسٌ، خُطَّ قَوْسٌ
RJ: III: props. 15, 23
See also: קשת

קודם ונמשך
antecedent and consequent, مُقَدَّمٌ وَتَالِي
MIT: VI: def. 2, prop. 19
RJ: VI: prop. 19
See also: ראשון ונמשך, אחרון

קוה (v.; נקוה)
draw (let us draw a geometrical figure), خُطَّطُ
MIT: III: props. 1, 14, 16, 24; IV: props. 1, 5, 8, 10, 12–13, 15–16; VI: props. 9, 11 (نجد؟), 28 (تَخْطِيط, לקוות)

הקוה
circumscribe, أَحَاطَ
MIT: VI: prop. 14
See also: רשם, צייר, עשה, סבב, הקיף, הקוות

קטר/קוטר
(a) diameter of a circle, قُطْرٌ
MIT: III: def. 1, props. 3, 7–8, 14–16, 30, 36; IV: props. 1, 6–8, 15
RJ: III: props. 7–8, 15–16, 30; IV: props. 1, 6–7, 15
(b) diameter of a parallelogram, قُطْرٌ
MIT: VI: props. 24–25, 27–29
RJ: VI: props. 28–29
See also: אלכסון

קטן (יותר)
(a) smaller, أَقْصَرُ، أَصْغَرُ، صُغْرَى

MIT: III: props. 1, 4–5, 7–8, 11, 15, 24, 26–27, 30; V: defs. 1, 3, props. 6, 8–10, 14, 18, 20–22, 25; VI: def. 4, prop. 7
RJ: III: props. 1, 6, 8, 15, 17, 24, 26–27, 30, 32; IV: prop. 1; V: defs. 1, 3, props. 6, 8–10, 14, 18, 20–21, 25; VI: def. 4, prop. 7
(b) less, أَقَلّ
RJ: V: prop. 4
(c) defect, ناقِص
RJ: V: prop. 22
See also: קצר

קם (v.)
construct, قَامَ
MIT: III: prop. 22
See also: שם ,צייר ,עשה ,העמיד

קצה
extremity, طَرَفٌ
MIT: III: def. 8, props. 7–8, 15–16, 36; IV: props. 7–8
RJ: III: def. 8, props. 1, 7–8, 15–16; IV: prop. 16
See also; יחס בעל אמצע וקצוות—יחס

(יותר) קצר
shorter, أَقْصَرُ
MIT: III: props. 2, 7–9; IV: prop. 1
RJ: III: props. 2, 7–8, 11, 24; IV: prop. 1
See also: קטן

קצת
some, بَعْض
RJ: III: props. 27; V: def. 2
See also: חלק

קצתם ... (אל) קצת/קצתם
one (to/opposite/etc.) another, بَعْضُها ... (إلى) بَعْضٍ
MIT: III: defs. 1, 3; IV: prop. 15; V: defs. 6–7, 15–16, 23; VI: def. 5, props. 1, 6
RJ: III: defs. 1, 3, prop. 18; V: defs. 6–7, 15; VI: prop. 1

קצת שווה לכול
some equals the whole, الكُبْرى مِثْلُ الصُّغْرى
RJ: III: prop. 18

קרב (v.)
be near, قُرْب
MIT: III: props. 7–8
RJ: III: props. 7–8

(יותר) קרוב
(a) near, أَقْرَب
MIT: III: prop. 14
RJ: III: props. 7–8, 14
(b) applied to (area, to a straight line), مُقارِن
MIT: V: prop. 1
See also: חָבר, מחובר, סמוך

קָרוב
counterpart, or: associated with, قَرِين
MIT: V: props. 1, 13, 15; VI: prop. 19
See also: מחובר, קרוב

קשת
arc, قَوْس
MIT: III: defs. 6–10, props. 2, 5, 12, 15, 19, 22–23, 25–31; IV: props. 11–12, 15–16; VI: prop. 33
RJ: III: defs. 6–10, props. 2, 5, 12, 15, 19–20, 22–31; IV: props. 11–12, 15–16; VI: prop. 33
See also: קו—קו קשת

ר

ראש
vertex (point, of a perpendicular), رَأْس
MIT: VI: def. 3
RJ: VI: def. 3

ראשון ונמשך
antecedent and consequent, مُقَدَّم وَتَالي
MIT: V: defs. 13–18, 20–21; VI: prop. 19
RJ: V: defs. 13–18, 20–21, prop. 15
See also: אחרון, קודם ונמשך

רב
greater (ratio), أَكْبَرُ, أَعْظَمُ

RJ: V: props. 8–10, 12, 14 (also أَعْظَمُ), 18 (أَعْظَمُ), 20–21
See also: שטח רב הזוויות—שטח ,גדול

רוחק
distance (radius), بُعْدٌ
MIT: III. def. 4
RJ: IV: prop. 14
See also: שיעור מרחק—שיעור ,מרחק

רושם
gnomon, عَلَمٌ
MIT: VI: props. 28–29
See also: מסומן

(יותר) רחוק
distant, أَبْعَدُ
MIT: III: prop. 14
RJ: III: def. 5 (أَعْظَمُ ?), props. 7–8, 14; IV: prop. 14

רחק (v.)
moved away, بَعُدَ
MIT: III: props. 7–8, 24
RJ: III: props. 7–8, 24

רכב (v.)
    (הרכבנו) הרכיב
    (a) superpose, apply to, رَكَّبَ
    MIT: III: prop. 23
    RJ: III: prop. 23
    See also: שם ,נרכב ,נפל ,נדבק ,הושם
    (b) placed, رَكَّبَ
    MIT: VI: prop. 31
    RJ: VI: prop. 31

רשם (v.; נרשום)
(a) mark, inscribe (a point), عَلَّمَ, (**RJ**: also وَضَعَ ,تَعَلَّمَ)
MIT: III: props. 2, 10, 22, 30–31, 35
RJ: III: def. 8 (تَعَلَّمَ), props. 2, 7, 8–9 (وَضَعَ), 22, 25, 30–31, 35; IV: props. 12, 16
See also: צייר

(b) draw (a geometrical figure). رَسَمَ, خُطَطَّ
RJ: IV: props. 10, 16
See also: קוה, צייר, עשה, סבב, מורשם

נרשם
(a) marked, inscribed (a point), تَعَلَّمَ
MIT: III: def. 8, props. 2, 35
RJ: III; props. 2, 35
(b) drawn (a geometrical figure), مَرْسُومٌ
MIT: IV: defs. 1–2
See also: מצויר, מורשם

ש

שאר (v.)
נשאר
remain, بَقِيَ
MIT: III: props. 1, 8, 13, 30, 34–35; IV: props. 2–3, 7, 15–16; V: props. 5–6; VI: props. 5–8, 20, 28–30
RJ: III: props. 1, 8, 13, 30, 34–35; IV: props. 2–3, 7, 15–16; V: props. 5–6, 19; VI: props. 5–8, 19–20, 28–30

שוה (v.)
be equal, تَسَاوَى
RJ: V: def. 7

השתוה
be equal, تَسَاوَى
MIT: VI: props. 4, 6, 7, 14–15, 23
RJ: V: prop. 21

שוה
equal (to), مُسَاوٍ, سَاوَى, تَسَاوَى; equal, مُتَسَاوٍ, سَوَاءٌ
MIT: III: defs. 1, 4, 11, props. 7–10, 13–14, 16, 18–21, 23–27, 28, 30, 34–36; IV: props. 1–5, 7–9, 11–16; V. def. 7, props. 1–4, 6–9, 11–17, 20–22; VI: def. 1, props. 1–8, 14–16, 18–23, 25–26, 28–31, 33
RJ: III: defs. 1, 4, 11, props. 1, 7–9, 13–14, 16, 20–21, 23, 25–30, 34–36; IV: props. 2–5, 7–9, 11–15; V: props. 4, 6–9, 11–17, 20–23; VI: defs. 1, props. 1–8, 12, 14–16, 18–20, 23, 26, 28–31, 33

equal (to), مِثْلُ
MIT: III: prop. 5: IV: prop. 10; VI: props. 3, 28
RJ: III: props. 1–9, 1–17, 21–35; IV: props. 1–16; V: props. 10, 14; VI: props: 2, 4–8, 10, 13–23, 26–32
See also: מחומש—מחומש, כפלים שוי הפעמים—כפל, כמו, יחס השווי—יחס, השתוה משולש שוה הצלעות—משולש, מרובע שוה הצלעות נצב הזוויות—מרובע, שוה הצלעות והזוויות תמונה—תמונה, שטח שוה הצלעות—שטח, משושה, משולש שוה השוקיים—משולש, והזוויות בעלת חמש עשרה זוויות שוות הצלעות והזוויות

שׂוּם (v.)
(a) let … be (lit. put), جَعَلَ
MIT: III: props. 7–9, 14, 16, 20, 24, 28–36; IV: props. 1–15; V: props. 3, 9–10, 28–30; VI: props. 9–10, 12–13, 28–30
RJ: III: props. 1, 16, 24, 26, 29, 32–34; IV: props. 1–4, 9–15; V: props. 3, 7; VI: props. 7, 9–10, 12–13, 20, 26–30
See also: לקח, חִלֵק, חתך, הבדיל
(b) make (something be in a certain relation [e.g. equal] to something), جَعَلَ (**RJ**: also كَانَ)
MIT: III: props. 1, 3, 7–11, 13, 21, 34–35; IV: props. 11, 13; V: props. 5–6, 8, 20–21, 25; VI: props. 14–17, 22–23, 26, 28–29
RJ: III: props. 21, 23, 34–35; IV: props. 10–13; V: props. 5–6, 7 (كَانَ), 8, 10, 12, 20–21, 25; VI: props. 2, 4–5, 14–17, 22–23, 28–30
(c) construct, عَمِلَ, جَعَلَ
MIT: IV: prop. 16; VI: props. 5, 22
RJ: IV: prop. 16
(d) superpose, apply to, وَضَعَ
RJ: III: prop. 23
See also: קם, צייר, עשה, העמיד

שוק—See: משולש—משולש שוה השוקים

שטח
area, surface, سَطْحٌ
MIT: IV: props. 7–8; VI: def. 1, props. 1, 13–14, 16, 18–30, 32
RJ: IV: props. 3, 6–7; VI: def. 1, props. 13–14, 16, 18–30, 32
See also: תמונה

שטח בעל ארבע צלעות
Quadrilateral, سَطْحٌ ذو أَرْبَعَةِ أَضْلاعٍ
RJ: III: props. 30–31; IV: prop. 3
See also: תמונה—תמונה בעלת ארבע צלעות

שטח ישר הקוים
polygon, شَكْلٌ مُستَقِيمُ الخُطُوطِ
RJ: VI: props. 20, 29, 32
See also: תמונה—תמונה ישרת הצלעות, תמונה—תמונה ישרת הקוים

שטח מרובע
square (lit. square area), سَطْحٌ مُرَبَّعٌ
MIT: IV: prop. 6
RJ: IV: prop. 6; VI: prop. 30
See also: מרובע

שטח נכחי הצלעות
parallelogram, سَطْحٌ مُتَوَازِي ٱلْأَضْلَاعِ
MIT: IV: props. 7–8; VI: props. 1, 4, 13, 23–24, 26–30
RJ: IV: props. 7–8; VI: props. 1, 14, 23–24, 27–30

שטח נצב הזוויות
rectangle (lit. right-angled area), سَطْحٌ ٱلقائِمُ ٱلزَّوايا
MIT: III: props. 34–36
RJ: III: props. 34–36; IV: props. 6–7

שטח רב הזוויות
polygon (lit. multi-angled area), سَطْحٌ كثيرُ ٱلزَّوَايا
MIT: VI: prop. 19
RJ: VI: prop. 19

שטח שוה הצלעות, سَطْحٌ مُتَساوِي الأَضْلاعِ
MIT: IV: prop. 7
RJ: IV: props. 6–7

שטחים מספיקי הצלעות
reciprocally related areas, سُطُوحُ المُتَكافِئَةُ الأَضْلاعِ
MIT: VI: def. 2
See also: שטחים מתהפכי הצלעות

שטחים מתהפכי הצלעות
reciprocally related areas, سُطُوحُ المُتَكافِئَةُ الأَضْلاعِ
RJ: VI: def. 2
See also: שטחים מספיקי הצלעות

See also: מרובע שוה הצלעות נצב הזוויות—מרובע, מחומש שוה הצלעות והזוויות—מחומש, תמונה בעלת—תמונה, משולש שוה השוקיים—משולש, משולש שוה הצלעות והזוויות—משולש חמש עשרה זוויות שוות הצלעות והזוויות

שיעור
magnitude, قَدْرٌ
MIT: v: defs. 1, 3–4, 6–8, 11–13, 19, props. 1–23, 25; vi: def. 5, props. 1, 12, 18, 33
RJ: v: defs. 1, 3–4, 6–8, 11–13, 19, props. 1–3, 5–11, 12, 13–23, 25; vi: def. 5, props. 1, 14, 19, 33

יפול שיעור על
measure, وَقَعَ ٱلتَّقْدِيرُ عَلَى
MIT: v: def. 3
RJ: v: def. 3

שיעור מרחק
magnitude of a distance, بُعْدٌ, قَدْرُ بُعْدٍ, أَقْرَبُ
RJ: iii: props. 7 (أَقْرَبُ), 16, 24, 32; iv: prop. 15
See also: מרחק

שיעורים מורכבים
composite magnitudes, أَقْدَارٌ مُرَكَّبَةٌ
MIT: v: props. 17–18
RJ: v: props. 17–18, 24

שיעורים מסודרים ביחס
properly ordered magnitudes, مَقَادِيرُ مُتَّسِقَةٌ فِي ٱلنِّسْبَةِ
MIT: v: def. 13
RJ: v: def. 13

שיעורים מתיחסים
proportional magnitudes, مَقَادِيرُ مُتَنَاسِبَةٌ
RJ: v: defs. 8, 11–12, props. 8, 14, 16–18, 23, 25
MIT: v: defs. 8, 11–12, props. 8, 14, 16–18, 25

שיעורים נכחיים
corresponding magnitudes, مَقَادِيرُ نَظِيرَةٌ
MIT: v: def. 13

שלך (v.)

(נשליך) השליך

subtract (or: remove; a common element from each of two equal or unequal expressions), أَلْقى

MIT: V: props. 5–6

RJ: III: props. 7, 34; V: props. 5–6

See also: סר ,חִסר ,הפיל ,גרע

שלם (v.)

(נשלים) השלים

complete (lines so as to construct a complementary area), تَمَّمَ

MIT: III: prop. 24; VI: props. 14, 16, 23, 27–29

RJ: III: prop. 24; VI: props. 14, 16, 23, 27–29

See also: השלים מיתר—מיתר

נשלם

complete (an area), تَمَّ

MIT: VI: prop. 27

שלמות

whole, تَمَامٌ

MIT: III: prop. 7; VI: props. 27–30

RJ: III: prop. 7; VI: props. 27–30

שנה (v.)

נשנה בהשנות

multiplied (by itself), مُثَنَّاةٌ

MIT: V: def. 11

RJ: V: def. 11 (**B**: مُثَنَّاة بِالتَّكرِير); VI: props. 18–19, 23 (נשנה)

See also: שנוי ,כפול

שִנוי

multiplied (by itself), مُثَنَّاة

MIT: VI: props. 18–19, 22–23, 32

See also: נשנה בהשנות—שנה ,כפול

שער (v.)

measure, كانَ بِقَدْرِ

MIT: V: defs. 1–2

RJ: V: defs. 1–2

ת

(על) תואר
(a) similarly shaped, or: similarly described, عَلَى خِلْقَتِهِ
RJ: VI: props. 22, 25, 27–28, 32
(b) similarly situated, على وَضْع
MIT: VI: prop. 32
See also: מצבו, עשה—עשוי כמעשהו

תוך (usually בתוך)
within, (فِي) داخِلِ
MIT: III: props. 2, 8–9, 12
RJ: III: def. 8, props. 2, 8–9, 12, 15, 18, 30, 36; IV: def. 1, props. 2, 4, 6, 8, 11–13, 15–16
See also: פנים

תוספת
excess, addition, زِيادَةٌ
RJ: V: props. 7, 11–12, 16–17, 22–23; VI: prop. 1
See also: יתרון, מותר, עודף

תושבת
base, قاعِدَةٌ
MIT: III: def. 8, props. 1, 3, 7–9, 13–14, 16, 19, 23–25, 27–29, 36; IV: props. 5–6, 10–15; VI: def. 3, props. 1–3, 5–6, 8–9, 11
RJ: III: def. 8, props. 1, 3, 7–9, 13–14, 16, 19–20, 23–25, 27–29; IV: props. 5–6, 10–14; VI: def. 3, props. 1–3, 5–6, 8–9

תחת
under, تَحْتَ
RJ: III: prop. 35

תכלית
(a) limit, boundary, نِهايَةٌ (?)
RJ: III: prop. 36
See also: גבול
(b) extremity (terminates in)
MIT: III: prop. 5
See also: כלה

לא תכלית
unlimited, infinite, غَيْرُ نِهايَةٍ
MIT: III: prop. 24
RJ: III: prop. 24
See also: כלה

תמונה
(a) geometrical figure, شَكْلٌ
MIT: III: def. 1, props. 21; IV: defs. 1–2; VI: def. 3, props. 27–29, 32
RJ: III: def. 1, prop. 21; IV: defs. 1–2; VI: def. 3
See also: צורה

תמונה בעלת ארבע צלעות
Quadrilateral, شَكْلٌ ذو أَرْبَعَةِ أَضْلاعٍ
MIT: III: props. 21, 30–31; IV: prop. 3
RJ: III: prop. 21
See also: שטח בעל ארבע צלעות—שטח

תמונה אשר לה חמשה עשר זויות שוות, תמונה בעלת חמש עשרה זויות שוות הצלעות והזויות
شَكْلًا ذا خَمْسَ عَشْرَةَ زاوِيَةً مُتساوي الأَضْلاعِ وَالزَّوايا, הצלעות והזויות
MIT: IV: prop. 16
RJ: IV: prop. 16

תמונה ישרת הצלעות
polygon, شَكْلٌ مُسْتَقيمُ الأَضْلاعِ
MIT: VI: prop. 29

תמונה ישרת הקוים
polygon, شَكْلٌ مُسْتَقيمُ الخُطوطِ
MIT: VI: props. 20, 28, 32

(b) proposition, شَكْلٌ (**I/T**)
MIT: III: props. 9–10, 30; VI: colophon
RJ: III: props. 25, 34
See also: צורה
(c) diagram, شَكْلٌ
RJ: III: prop. 36
See also: צורה

Bibliography

Brentjes, Sonja. "Observations on Hermann of Carinthia's Version of the *Elements* and its Relation to the Arabic Transmission." *Science in Context* 14 (2001): 39–84.

Brentjes, Sonja. "Who Translated Euclid's *Elements* into Arabic?" In *Translation and Transmission*, edited by Jaakko Hämeen-Anttila and Ilkka Lindstedt, 21–54. Münster: Ugarit, 2018.

De Young, Gregg. "The Arithmetic Books of Euclid's *Elements* in the Arabic Tradition" PhD diss., Harvard University, 1981.

De Young, Gregg. "The Latin Translation of Euclid's *Elements* attributed to Gerard of Cremona in relation to the Arabic Transmission." *Suhayl* 4 (2004): 311–383.

De Young, Gregg. "New Traces of the Lost al-Ḥajjāj Arabic Translations of Euclid's *Elements*." *Physis* 28 (1991): 647–666.

De Young, Gregg. "Two Hitherto Unknown Arabic Euclid Manuscripts." *Historia Mathematica* 42 (2015): 133–154.

De Young, Gregg. "MS Paris, BULAC, Ara 606 and the Ḥajjāj Tradition." *Historia Mathematica*. Forthcoming in 2024.

Elior, Ofer. "The Hebrew Translations of Euclid's *Elements* by Moses Ibn Tibbon and by Jacob Ben Makhir—A Study of Differences in Book I." *Journal of Semitic Studies* 64, no. 2 (2019): 481–506.

Elior, Ofer. *Euclid's Elements, Books I–II, in Hebrew Garbs: Critical Editions of the Translation by Moses Ibn Tibbon and the Translation Ascribed to Rabbi Jacob with an Introduction and Glossary*. Leiden: Brill, 2021.

Elior, Ofer. "The Hebrew Translation of Euclid's *Elements* ascribed to Rabbi Jacob: A New Analysis following the "Discovery" of the Arabic Version in MS Paris, BULAC ARA. 606." *Historia Mathematica*. Forthcoming in 2024.

Engroff, John W. "The Arabic Tradition of Euclid's *Elements*: Book V." PhD diss., Harvard University, 1980.

Lévy, Tony. "Les *Éléménts* d'Euclide en hébreu (XIIIe–XVIe siècles)." In *Perspectives arabes et médiévales sur la tradition scientifique et philosophique grecque*, edited by Mārūn ʿAwwād, Ahmad Hasnawi, and Abdelali Elamrani-Jamal, 79–94. Leuven: Peeters, 1997.

Margoliouth, George. *Catalogue of the Hebrew and Samaritan Manuscripts in the British Museum*. 3 Vols., London: Trustees of the British Museum, 1965.

Rashed, Rushdi, ed. *Lexique historique de la langue scientifique arabe*. Hildesheim: Georg Olms, 2017.

www.ingramcontent.com/pod-product-compliance
Lightning Source LLC
LaVergne TN
LVHW010636300926
849161LV00005B/963
* 9 7 8 9 0 0 4 7 0 7 7 9 5 *